Series Editor

R. Magjarevic

For further volumes:
http://www.springer.com/series/7403

The International Federation for Medical and Biological Engineering, IFMBE, is a federation of national and transnational organizations representing internationally the interests of medical and biological engineering and sciences. The IFMBE is a non-profit organization fostering the creation, dissemination and application of medical and biological engineering knowledge and the management of technology for improved health and quality of life. Its activities include participation in the formulation of public policy and the dissemination of information through publications and forums. Within the field of medical, clinical, and biological engineering, IFMBE's aims are to encourage research and the application of knowledge, and to disseminate information and promote collaboration. The objectives of the IFMBE are scientific, technological, literary, and educational.

The IFMBE is a WHO accredited NGO covering the full range of biomedical and clinical engineering, healthcare, healthcare technology and management. It is representing through its 60 member societies some 120.000 professionals involved in the various issues of improved health and health care delivery.

IFMBE Proceedings Vol. 40
Vo Van Toi, Nguyen Bao Toan, Truong Quang Dang Khoa,
and Tran Ha Lien Phuong (Eds.)

4th International Conference on Biomedical Engineering in Vietnam

Springer

Editors
Prof. Vo Van Toi
International University Vietnam
Ho Chi Minh City
Vietnam

Nguyen Bao Toan
International University Vietnam
Ho Chi Minh City
Vietnam

Truong Quang Dang Khoa
International University Vietnam
Ho Chi Minh City
Vietnam

Tran Ha Lien Phuong
International University Vietnam
Ho Chi Minh City
Vietnam

ISSN 1680-0737
ISBN 978-3-642-32182-5
DOI 10.1007/ 978-3-642-32183-2
Springer Heidelberg New York Dordrecht London

e-ISSN 1433-9277
e-ISBN 978-3-642-32183-2

Library of Congress Control Number: 2012948630

The IFMBE Proceedings is an Official Publication of the International Federation for Medical and Biological Engineering (IFMBE)

Printed on acid-free paper

Springer is part of Springer Science+Business Media (www.springer.com)

Preface

Welcome to the Fourth International Conference on the Development of Biomedical Engineering in Vietnam.

Vietnam is a rapidly developing, socially dynamic country, where interest in biomedical engineering activities has grown considerably in recent years. The leadership of the Vietnamese government, and of research and educational institutions, are well aware of the importance of this field for the development of the country and have instituted policies to promote its development. The political, economic and social environment within the country offers unique opportunities for the international community and this conference was intended to provide a vehicle for the sharing of experiences; development of support and collaboration networks for research; and exchange of ideas on how to improve the educational and entrepreneurial environment to better address the urgent needs of Vietnam.

In January 2004, under the sponsorship of the U.S. National Science Foundation, Professor Vo Van Toi of the Biomedical Engineering Department of Tufts University, Medford Massachusetts USA, led a U.S. delegation that consisted of Biomedical Engineering professors from different universities in the United States, and visited several universities and research institutions in Vietnam to assess the state of development of this field. This delegation proposed a five year plan that was enthusiastically embraced by the international scientific communities to actively develop collaborations with Vietnam.

Within this framework, in July 2005, the First International Conference on the Development of Biomedical Engineering in Vietnam was held in Ho Chi Minh City. From that conference a Consortium of Vietnam- International Universities was created to advise and assist the development of Biomedical Engineering in Vietnamese universities.

In July 2007, the Second International Conference on the Development of Biomedical Engineering in Vietnam was held in Hanoi. During this event the Vietnamese Association of Biomedical Science and Engineering was endorsed by the Asia-Pacific International Molecular Biology Network (AIMBN), Biomedical Engineering Society Singapore (BESS), International Federation for Medical and Biological Engineering (IFMBE), Société Française de Génie Biologique et Médical (SFGBM) and IFMBE Asia-Pacific Working Group.

In March 2009, International University (IU), established its Biomedical Engineering (BME) Department and the first accredited Engineer in BME degree in Vietnam (code: 52.42.02.04). IU is a member of Vietnam National Universities – Ho Chi Minh City (VNU-HCM), one of the two elite university networks in Vietnam, and is the first public university in Vietnam that teaches all courses in English. It was created as a new model to modernize the higher education in Vietnam. The BME Department at IU has since coordinated with its national and international counterparts to promote the activities of this field in Vietnam.

In January 2010, the Third International Conference on the Development of Biomedical Engineering in Vietnam was organized by IU in Ho Chi Minh City. It reflected the steady growth of the activities in this field in Vietnam, and featured the contributions of researchers of 21 countries, including: Australia, Belgium, Canada, Denmark, France, India, Japan, Korea, Malaysia, New Zealand, Philippines, Poland, Russia, Singapore, Spain, Switzerland, Taiwan, Thailand, United Kingdom, the United States and Vietnam. The Conference was endorsed by the International Federation for Medical and Biological Engineering (IFMBE). It also hosted the Clinical Engineering Workshop of the IFMBE Asia Pacific Working Group. The contributed papers were published in the IFMBE Proceedings Series by Springer (ISBN 978-3-642-12019-0).

This year, the Fourth International Conference on the Development of Biomedical Engineering in Vietnam is organized in Ho Chi Minh City as a ***Mega-conference***. It is kicked off by the **Regenerative Medicine Conference** (Jan 8–10, 2012) with the theme "BUILDING A FACE" USING A REGENERATIVE MEDICINE APPROACH", endorsed mainly by the Tissue Engineering and Regenerative Medicine International Society (TERMIS) and co-organized by Professor Stephen E. Feinberg, University of Michigan Health System, USA, Professor Anh Le, University of Southern California, USA and Professor Vo Van Toi, International University-VNU HCM, Vietnam. It is followed by the **Computational Medicine Conference**, endorsed mainly by the Computational Surgery International Network (COSINE) and the Computational Molecular Medicine of German National Funding Agency; and the **General Biomedical Engineering Conference**, endorsed mainly by the

International Federation for Medical and Biological Engineering (IFMBE) (Jan 10–12) and co-organized by Professor Paolo Carloni, German Research School for Simulation Sciences GmbH, Germany, Professor Marc Garbey, University of Houston, USA and Professor Vo Van Toi, International University-VNU HCM, Vietnam. It featured the contributions of 435 scientists from 30 countries, including: Australia, Austria, Belgium, Canada, China, Finland, France, Germany, Hungary, India, Iran, Italy, Japan, Jordan, Korea, Malaysia, Netherlands, Pakistan, Poland, Russian Federation, Singapore, Spain, Switzerland, Taiwan, Turkey, Ukraine, United Kingdom, United States, Uruguay and Viet Nam.

The editors would like to thank the leadership and staff of VNU HCM and IU, local and international sponsors, and the staff and students of Biomedical Engineering Department for their valuable support for the conference and their assistance in the editing and publication of this volume.

Editors
Vo Van Toi
Nguyen Bao Toan
Truong Quang Dang Khoa
Tran Ha Lien Phuong

and Editing Team
Nguyen Van Hoa
Luu Gia Loc
Nguyen Huynh Minh Tam

Biomedical Engineering Department
International University
Vietnam National Universites Ho Chi Minh City
Viet Nam

Organizing Institutions

Vietnam National Universities
HCM City (VNU-HCM)

International University
of VNU-HCM

Honorary Chairs

Professor Nguyen Thi Kim Tien, Minister of Vietnam Ministry of Health

Professor Phan Thanh Binh, President of Vietnam National Universities - HCM City

Professor Ho Thanh Phong, Rector of International University VNU-HCM

Conference General Chair

Vo Van Toi, PhD
Professor and Chair of Biomedical Engineering Department
International University - Vietnam National Universities
Quarter 6, Linh Trung, Thu Duc District
Ho Chi Minh City, Vietnam
Tel: (84-8) - 37 24 42 70 Ext. 3237
Fax: (84-8) - 37 24 42 7
Website: www.hcmiu.edu.vn/bme

Organizing Committee

Paolo Carloni, PhD
Professor, Head of Laboratory Computational Biophysics
German Research School for Simulation Sciences GmbH
Jülich, Germany

Hoang Dung, PhD
Professor, Head of Office of Science and Technology Development
Vietnam National Universities
Ho Chi Minh City, Vietnam

Stephen E. Feinberg, DDS, MS, PhD
Professor & Associate Chair of Research
Department of Oral & Maxillofacial Surgery
University of Michigan Health System
Ann Arbor, MI, USA

Marc Garbey, PhD
Professor, Department of Computer Science
University of Houston
Houston, TX, USA

Anh Le DDS, PhD
Professor, Center for Craniofacial Molecular Biology
School of Dentistry
University of Southern California
Los Angeles, CA, USA

Vo Van Toi, PhD
Professor and Chair of Biomedical Engineering Department
International University
Vietnam National Universities Ho Chi Minh City, Vietnam

Organizing Committee

Scientific Committee

Anh Le	University of Southern California, USA
Benjamin S. Kelley	Baylor University, USA
Beom-Jin Lee	Kangwon National University, Korea
Catherine Le Visage	Inserm, Paris, France
Dang Mau Chien	Laboratory for Nanotechnology, VNU-HCM, Vietnam
Dang Thanh Tin	University of Technology, VNU-HCM, Vietnam
Dao Tien Tuan	Compiegne University, France
Dietmar Hutmacher	Queensland University of Technology, Australia
Duong Tan Nhut	Da Lat Institute of Biology, Vietnam
Evelyn Yim	National University of Singapore, Singapore
Frederic Lesage	Ecole Polytechnique de Montreal, Canada
Hanry Yu	National University of Singapore, Singapore
Hoang Dung	Vietnam National University in Hochiminh City, Vietnam
Huynh Luong Nghia	Le Qui Don Technical University, Vietnam
Huynh Quang Linh	University of Technology, VNU-HCM, Vietnam
Lawrence Le	University of Alberta, Canada
Le Thi Ly	International University, VNU-HCM, Vietnam
Le Thi Tuyet Lan	HCM University of Medicine and Pharmacy, Vietnam
Marc Garbey	Houston University, USA
Marie-Christine Ho Ba Tho	Compiègne University, France
Michael Raghunath	National University of Singapore, Singapore
Nguyen Bao Toan	International University, VNU-HCM, Vietnam
Nguyen Linh Trung	University of Engineering and Technology, VNU-Hanoi, Vietnam
Nguyen Thanh Hai	International University, VNU-HCM, Vietnam
Nguyen Thi Hue	International University, VNU-HCM, Vietnam
Nguyen Van Thuan	Hanoi University of Science and Technology, Vietnam
Pam Yellick	Tufts University, USA
Paolo Carloni	German Research School for Simulation Sciences, Germany
Paul Krebsbach	University of Michigan, USA
Phan Kim Ngoc	University of Natural Science, VNU-HCM, Vietnam
Stephen E. Feinberg	University of Michigan Health System, USA
Tong Duy Hien	Laboratory for Nanotechnology, VNU-HCM, Vietnam
Tran Cong Toai	Pham Ngoc Thach Medical University, Vietnam
Tran Diep Tuan	HCM University of Medicine and Pharmacy, Vietnam
Tran Ha Lien Phuong	International University, VNU-HCM, Vietnam
Tran Le Bao Ha	University of Science, VNU-HCM, Vietnam
Truong Quang Dang Khoa	International University, VNU-HCM, Vietnam
Vo Van Toi	International University-VNU-HCM, Vietnam

Local Organizing Committee (LOC) & Secretaria

Acknowledgements

We would like to express our deepest appreciation to the following sponsors:

Air Force Office of Scientific Research – United States Air Force

Asian Office of Aerospace R&D – Bridging Science & Technology

U.S. Army RDECOM

Office of Naval Research Global – Science & Technology

termis – Tissue Engineering International & Regenerative Medicine Society

NAFOSTED – National Foundation for Science & Technology Development

International Association of Oral and Maxillofacial Surgeons

Cosine

American Association of Oral and Maxillofacial Surgeons

Quintessence Publishing

Shimadzu – Solutions for Science since 1875

Trung Sơn T.S.S.E

Korea United Pharm.Inc – For the Best Global Healthcare Company

Springer

LifeCell – A KCI Company

IFMBE

Table of Contents

Biomedical Engineering

Computational Medicine

Erratum

Investigating Physiology of Untruth in Cerebral Cortex by Functional Near-Infrared Spectroscopy (fNIRS)

Huynh Luong Nghia[1], Truong Quang Dang Khoa[2], Tran Xuan Tan[2], Duong Duc Thien[2], Nguyen Ngoc Phuong Trinh[2], and Vo Van Toi[2]

[1] Biomedical Electronic Department, Le Quy Don Technical University, Hanoi, Vietnam
[2] Biomedical Engineering Department, International University-HCMC, Vietnam

Abtract — **Physiological measurement has been deeply investigated in microenvironment. The physiology on molecular or biochemical levels is examined to diagnose disease in early state. With new brain imaging techniques such as multi - channel near – infrared spectroscopy (NIRS), the functions of the cognitive brain activities can be determined. In this study, the relationship between specific physiology of brain cortex and psychology of untruth is assessed through measurements of the changes in concentration deoxygenated hemoglobin (HHb) and oxygenated hemoglobin (HbO2)by multi - channel NIRS. Moreover, we found the complex combination at prefrontal cortex (Cz) is related to physiology of truth and untruth.**

Keywords — **physiology of truth, physiology of untruth, prefrontal cortex, central zone, NIRS, FUSION.**

I. Introduction

With brain imaging techniques such as PET, fMRI…, the functions of cognitive brain activities can be determined. These techniques require strictly motion restriction, so most of participants feel stressful during experiments. However, NIRS is a non-invasive measurement.

Due to important role of untruth physiology of human being, the United States and Russia produced many devices such as Polygraph (Lie detector), детектор лжи, etc.… These techniques were used to measure and analyze data based on blood pressure, heart rate, breathe rate and electrical skin impedance…. These parameters are determined to be linked to the physiology of untruth in different levels.

In this study, multi -channel NIRS technique was used as an equipment for investigating of particular relationship between physiology of untruth and the change concentration of HbO2 and HHb in brain cortex.

II. Methodology

A. Participant

Nine healthy adults (2 female, age 19 - 22 years) have been included in our experiment. Each subject was tested three times in 180 seconds. Between tests the rest from 30 to 60 second is taken for preparing next tests. During NIRS measurements, participants had to communicate face to face with researcher and answer to his questions related to psycho-physiology of untruth.

B. Material

We acquired data of the cerebral tissue oxygenation from all participants by Shimadzu FOIRE-3000 NIRS machine (Fig. 1). This new imaging technique is designed with three different wave lengths of 780, 805 and 830 nm to monitor changes in the oxyhemoglobin (HbO), deoxyhemoglobin (DeHbO) and total hemoglobin (TotalHb).

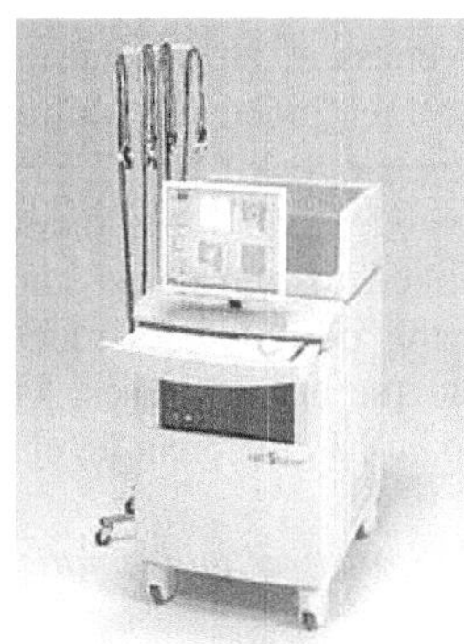

Fig. 1 Shimadzu FOIRE-3000 NIRS machine

V. Van Toi et al. (Eds.): 4th International Conference on Biomedical Engineering in Vietnam, IFMBE Proceedings 40, pp. 1–3, 2013.
www.springerlink.com

III. Data analysis

There are two approach to be applied for data analyses in our study:

- Analysis based on FUSION software.
- Application by statistical mathematic.

1. *Approach 1(FUSION software)*

- *Method*

We used MRI-FUSION software as a tool for analyzing data in real time with NIRS measurement.

- *ROI (Region of interest)*

Right hemisphere was checked as significant cortical region where the intensive changes of HHb and HbO2 in channel 3 (Fig. 2) were observed.

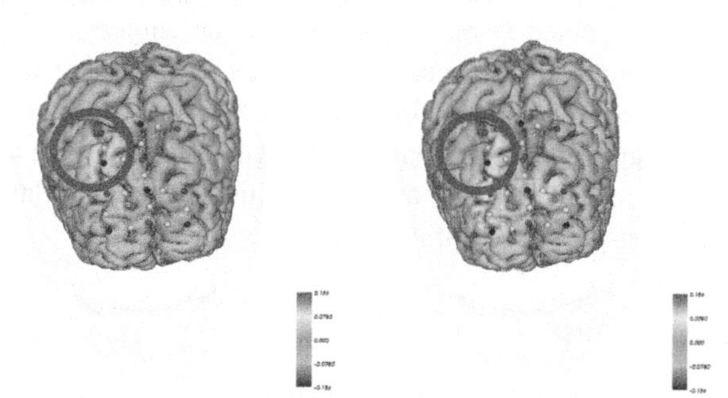

Fig. 2 The changes of HHb and HbO2 in channel 3

- *Results*

Three periods in accordance with 5th, 25th and 45th seconds of the tests are important to be under carefully examination on the changes of HHb and HbO2 concentration. Through the different colors expressing different intensiveness of HbO2 consumption at Channel 3 in three periods mentioned the untruth is checked. The period where the changes of HbO2 concentration increased higher and repeated much more than in other periods is considered as linked to untruth psycho-physiology. It was recognized that the period that much more changing in color indicated to the untruth cases. Moreover, we noticed that each participant had different level of changed colors.

2. *Approach 2 (Statistical mathematic)*

- *Method*

Data of each experiment was separated into three stages, in accordance with three periods (5th, 25th and 45th seconds). For each stage the statistical characteristic such as variance, mean, etc were determined. Then these characteristics were compared to that of other stages. Their maximum or minimum values (in dependence on gender) are considered as linked to untruth psycho-physiology (untruth cases).

- *Result*

Applying this method for Channel 3 given results presented in Table 1.

Table 1 The result of true test analyzed by statistical mathematic method

Date	Name	Sex	Test1	Test2	Test3
	N.M.N	Male	T	T	T
26/9	L.H.D	Male	T	1/2T	1/2T
	T.C	Female	T	2/3T	1/2T
	V.N.T	Male	T	T	T
27/9	B.N.H	Female	T	T	F
	N.M.D	Male	F	T	T
	T.N.N.M	Male	T	T	F
28/9	T.Q.V	Male	T	F	T
	D.N.H	Male		F	F
	Overall		**87%**	**70%**	**56%**

- *Notices*

- The method correctness is about 70% of tests with taking into account the participants' gender and individual differences in level adaptation and in psycho-physiology.

- The decreasing precision of method according to tests order shows the different adaptation levels of each participant in experiment

IV. Conclusion

Near – infrared spectroscopy (NIRS) is sensitive to physiology of truth. There is complex combination of cerebral cortex regions, related to physiology of untruth. The untruth states can be detected by FUSION and statistical mathematic methods. The more subtle algorithms are needed to improve results of data analysis.

V. Acknowledgement

This study was supported in partnership between Biomedical Electronic Department, Le Quy Don Technical University, Hanoi, Vietnam and Biomedical Engineering Department, International University-HCMC.We would like to thank Shimadzu Cooperation, Japan for technical support. Finally, we highly appreciate the friends who helped us in collecting information.

References

Hemodynamic Activities of Motor Cortex Related to Jaw and Arm Muscles Determined by Near Infrared Spectroscopy L.M. Hoa, D.N. Huan, N.V. Hoa, D.D. Thien, T.Q.D. Khoa, and V.V. Toi IFMBE Proceedings 35, p. 229 ff.

Changes in Cortical Blood Oxygenation Responding to Arithmetical Tasks and Measured by Near-Infrared Spectroscopy N.N.P. Trinh, N.H. Binh, D.D. Thien, T.Q.D. Khoa, and V.V. Toi IFMBE Proceedings 35, p. 279 ff.

Scott C. Bunce, Ajit Devaraj, Meltem Izzetoglu, Banu Onaral and Kambiz Pourrezaei, "Detecting deception in the brain: a functional near-infrared spectroscopy study of neural correlates of intentional deception", Proc. SPIE 5769, 24 (2005); doi:10.1117/12.600601.

Truong Quang Dang Khoa, Masahiro Nakagawa, "Functional Near Infrared Spectroscope for Cognition Brain Tasks by Wavelets Analysis and Neural Networks," International Journal of Biological and Life Sciences 4:1 2008).

High Order Moment Features for NIRS-Based Classification Problems

Tuan Hoang[1], Dat Tran[1], Khoa Truong[2], Phuoc Nguyen[1], Toi Vo Van[2], Xu Huang[1], and Dhamendra Sharma[1]

[1] Faculty of Information Sciences and Engineering, University of Canberra, Canberra, Australia
[2] Department of Biomedical Engineering, HCMC International University, Ho Chi Minh City, Viet Nam

Abstract — **This paper aims to experiment high order moment features in two well-known problems which are motor imagery and person authentication in Brain Computer Interface (BCI) systems using Near Infrared Spectroscopy (NIRS) technique. To improve performance of the systems, we propose a new feature by combining 2nd order and 4th order moments of signal together. Our results show that such the feature not only achieves very high recall and precision ratios but also is practical for online NIRS-based BCI systems. Our systems can achieve recall and precision ratio at 99.2% for the left-hand and right-hand imagery problem, and up to 100% for the person authentication problem.**

Keywords — **NIRS-based BCI, high order moment, Brain Computer Interface, motor imagery, person authentication.**

I. Introduction

Brain-Computer Interface (BCI) as defined by Wolpaw et al in [1] is "a system for controlling a device by human intentions, which does not depend on the brain's normal output pathways of peripheral nerves and muscles". A typical BCI system includes the following stages: data acquisition, data pre-processing, feature extraction, classification, device controller and feedback [2]. Many noninvasive methods have been developed to monitor brain activity. Such typical methods include Electro-encephalography (EEG), Magnetoencephalography (MEG), functional Magnetic Resonance Imaging (fMRI), Positron Emission Topography (PET), and Near-infrared spectroscopy (NIRS). This paper focuses on BCI systems based on NIRS technology.

Near-infrared spectroscopy (NIRS) is a new non-invasive optical method to acquire brain signals. This method provides researchers a range of advantages including flexibility of use, metabolic specificity, good temporal resolution, localized information, and high sensitivity in detecting small substance concentrations [3]. It also has some disadvantages such as slow operation due to the inherent latency of the hemodynamic response and weaken signal due to hair on the head. Nevertheless, NIRS's ability to record localized brain activity with a spatial resolution in the order of centimeter (depending on the probe geometry) provides us with a good mean to measure a variety of motor and cognitive activities in BCI systems [1,2].

This paper aims to experiment high order moment features in two well-known problems which are motor imagery and person authentication in Brain Computer Interface (BCI) systems using Near Infrared Spectroscopy (NIRS) technique. For NIRS-based motor imagery problem, Sitaham et al. [4] reported that they could achieve average accuracy at 73% and 89% on 2-class problem, depending on classifiers they used. Niide et al. [5] conducted similar experiments also with the 2-class problem and achieved accuracy up to 94% in general experiment and 97% in subject-specific experiment. Recently, Gottemukula et al. [6] used classification-guided feature selection in the same problem to improve the system's accuracy. For NIRS-based person authentication problem, Rowe et al [7] used discriminant analysis and Mahalanobis distance for verification task of 288 different people. The Near-infrared tissue spectra are obtained by projecting near-infrared radiation into skin on the underside of human forearms and capturing the light reflected back and out through the tissue. The instrumentation utilized was a near infrared Fourier transfer spectrophotometer manufactured by Perkin Elmer 2000. Result can attain false positive rate of 0.17% and false negative error rate of 0.0%. All of those investigations used raw or simple filtered signal as input to classifiers.

We propose a new feature by combining 2nd order and 4th order moments of NIRS signal together. RBF SVMs were used as classifiers. Our experimental results on four healthy male subjects who voluntarily participated in the experiments show that such the feature not only achieve very high recall and precision ratios but also are practical for online NIRS-based BCI systems. Our systems can achieve recall and precision ratio at 99.2% for the left-hand and right-hand imagery problem, and up to 100% for the person authentication problem.

II. Materials and Methods

A. Subjects

Four healthy subjects who are all male and have ages range from 20 to 30 voluntarily participated in the study.

V. Van Toi et al. (Eds.): 4th International Conference on Biomedical Engineering in Vietnam, IFMBE Proceedings 40, pp. 4–7, 2013.
www.springerlink.com

None of the recruited subjects had neurological or psychiatric history or was on medication. Each of them gave written informed consent for the experiment. All experiments were performed at Laboratory of Biomedical Engineering Department of International University. During the experiment, subjects were sitting on a comfortable arm-chair within a dark and sound–isolated room.

B. Experimental Procedure

All of subjects were asked to sit on an arm-chair in front of a screen which displayed the stimuli in a dark and sound-isolated room. We used three types of protocols. The first protocol (Fig. 1a) consisted of five loops including interchange of two left-hand and two right-hand imagery loops, and a resting loop at final which totally last 130 seconds. The second (Fig. 1b) (the third protocol – Fig. 1c) consisted of four loops including three left-hand imagery loop (right-hand correspondingly) and a resting loop at final which totally last 100 seconds. A normal loop (Fig. 1d) consists of 10 seconds of resting and then 20 seconds of (left or right) imagery task. In imagery time, the subjects would clutch a ball by left hand or right hand corresponding to stimuli on screen. Each subject was asked to finish all three protocols. In total, for each subject, there were 5 trials for each type of imagery.

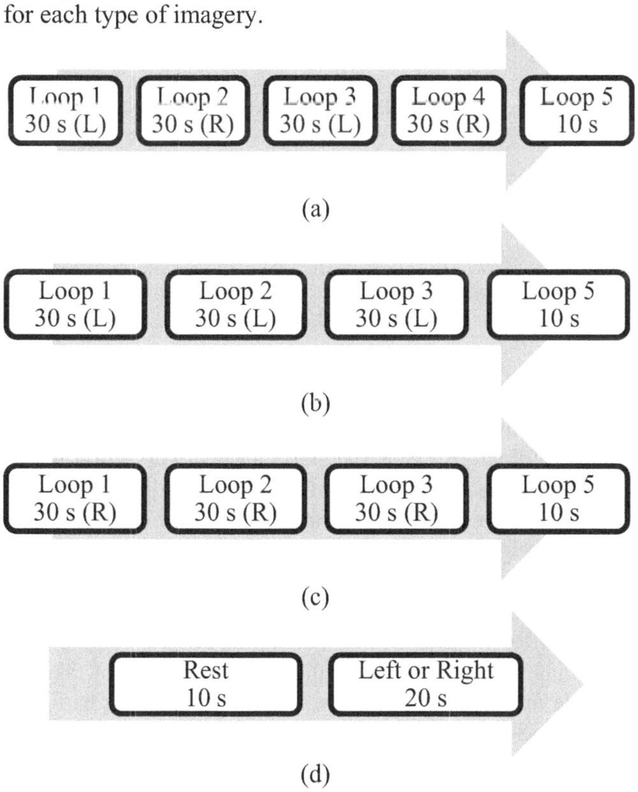

Fig. 1 (a) First protocol, (b) Second protocol, (c) Third protocol, (d) Performed tasks in a loop.

C. Data Acquisition

A 14-channel NIRS instrument (Shimadzu FOIRE3000) was used for acquiring oxygenated and deoxygenated hemoglobin concentration changes during our experiments. It operated at three different wavelengths of 780 nm, 805 nm and 830 nm, emitting an average power of 3 $mWmm^{-2}$. The detector optodes were fix placed at a distance of 3 cm from the illuminator optodes. The optodes were arranged on the left and right hemisphere of the subject's head, above the motor cortex, around C3 (left hemisphere) and C4 (right hemisphere) areas (International 10–20 System) as illustrated in Fig. 2. For each of hemisphere, we use three illuminators and three detectors. A pair of illuminator and detector optodes formed one channel. So in total, we have seven channels on each hemisphere. The photomultiplier cycles through all the illuminator–detector pairings to acquire data at every sampling period. The sampling rate is approximately 18.2 Hz. Finally, the acquired signal is digitized by the 16–bit analog to digital converter.

Fig. 2 Three illuminators and three detectors in arrangement result in 7 channels on each hemisphere where red optode illustrates illuminator and blue optode illustrates detector.

D. Data Pre-processing

Our fNIRS imaging system adopts near–infrared lasers of three wavelengths 780 nm, 805 nm, and 830 nm. For each channel j, optical densities $A_{j,\lambda}$ in wavelengths λ are measured. Then, from the changes in the optical density $\Delta A_{j,\lambda}$, relative concentration changes of oxy-genated hemoglobin Δoxy_j and of deoxy-genated hemoglobin $\Delta deoxy_j$ are estimated by following formula

$$\begin{pmatrix} \Delta oxy_j \\ \Delta deoxy_j \end{pmatrix} = \begin{pmatrix} -1.4887 & 0.5970 & 1.4847 \\ 1.8545 & -0.2394 & -1.0947 \end{pmatrix} \begin{pmatrix} \Delta A_{j,780} \\ \Delta A_{j,805} \\ \Delta A_{j,830} \end{pmatrix} \quad (1)$$

which is derived from the modified Beer–Lambert law and the extinction coefficients of the tissue reported by Matcher et al 2004.

Each trial was segmented into 38-data point chunks which last nearly 2 seconds. We chose 2-second chunk because it is short enough for real practical BCI systems and long enough for further processing. Two consecutive segments are overlapped by 27 data points, meaning one and a half second overlapping. Each label was labeled corresponding to class of action resting, left imagery, right imagery. We experimented responses of oxy-generated hemoglobin and deoxy-generated hemoglobin separately.

E. *High Order Moment Feature Extraction*

Let *y(t)* be the single channel signal we would like to extract feature. We used combination of 2nd order moment, naming mobility, and 4th order moment, naming complexity to form feature vector v of NIRS signal *y(t)*. In signal processing, mobility and complexity are very good at describing properties of signal.

$$Activity(y(t)) = VAR(y(t)) \quad (2)$$

$$Mobility(y(t)) = \sqrt{\frac{Activity(dy(t))}{Activity(y(t))}} \quad (3)$$

$$Complexity(y(t)) = \frac{Mobility(dy(t))}{Mobility(y(t))} \quad (4)$$

Because each segment has 14 channels, we extract mobility and complexity for each channel j. So in total, we have 28-dimension feature vectors as following $v = (Mobility_j, Complexity_j)$ with j ranges from channel 1 to channel 14.

F. *Classification*

Feature vectors then were put into classifier for training and prediction. We used only feature vector associating with left or right imagery, skipping resting segments. Two classifiers are built and trained to classify imagery tasks, left or right, and to authenticate subject. The former problem is 2-class classification and the later one is 4-class classification.

SVM has been proven as a powerful classier for solving BCI classification problems. The popular RBF kernel function $K(x,x') = e^{-\gamma\|x-x'\|^2}$ was used in our experiments. We applied grid search to get the optimal classifiers. The parameter γ was searched in $\{2^k: k = -15, -13, -11, \ldots, 1, 3\}$. The trade-off parameter C was searched over the grid $\{2^k: k = -5, -3, -1, \ldots, 13, 15\}$. We used 3-fold cross validation test on the training set. To eliminate effects of randomness, we used 3 repetitions in training and classification.

III. Results

A. *Motor Imagery Classification Problem*

Table 1 shows recall and precision for left/right motor imagery classification. It shows that our system can achieve at 99.2% for both recall and precision ratios with responses of oxy-generated hemoglobin. With responses of deoxy-generated hemoglobin, recall and precision ratios slightly increase to 99.5%.

We also tried experiments in which responses of oxy-generated and deoxy-generated hemoglobin are added together naming total hemoglobin. Results show that with total hemoglobin, our systems slightly decrease to 98.7% for both recall and precision ratios.

Table 1 Left/Right motor imagery classification results

	Oxy	Deoxy	Total
Recall	99.2	99.5	98.7
Precision	99.2	99.5	98.7

B. *Person Authentication Problem*

Table 2 shows recall and precision for person authentic-cation problem. It is seen that our systems could achieve very high recall and precision ratios which are nearly 100%.

Table 2 Person authentication results

	Oxy	Deoxy	Total
Recall	99.9	100	99.9
Precision	99.9	100	99.9

IV. Discussion

A. *General*

The above results show that the combination of mobility and complexity of NIRS signals can be a very good feature for NIRS-based BCI systems. As seen from Equation 3, mobility is calculated by taking first order differentiation of original signal. By doing so, mobility is naturally filtered without any explicitly specified filters. Similarly, as seen from Equation 4, complexity is calculated by taking second order differentiation of original signal. Hence, complexity is also naturally filtered. It is the reason we did not take any simple or complicated filters as other researchers had done. To our best knowledge, our research is the first one has used combination of mobility and complexity of signal to form feature in NIRS-based BCI systems.

Our proposed feature is not only practical in accuracy perspective but also in practical perspective. Each segment lasts approximately 2 seconds. It means that our systems need slightly more than 2 seconds (plus processing time) to make classification result based on imagery action from subjects. Moreover, by setting segment step 9 data point, resulting in one and a half second overlapping region, from the second segment, we can make decision in every half of second. This response time is very practical in almost BCI systems.

B. Motor Imagery Classification

As mentioned in above sections, our systems could achieve very high accuracy, up to 99.5%. Comparing with work of Gottemukula et al., our systems could produce comparable accuracy but much simpler than their system. We use no filters and no feature selection methods. Hence, our system could be more practical than their systems. Comparing with work of Niide et al. our systems could achieve better accuracy with an offset ranging from 2% to 5%. However, our experiments have less number of subjects than theirs. It could be better if we can have a standard dataset for evaluation. We also aim to conduct future experiments with more subjects and trials.

C. Person Authentication

For person authentication problem, our systems could achieve nearly 100% of accuracy. These results are similar with previous work of Rowe [7]. However, our experiments have much less number of participating subjects than theirs. Another important difference is that the method we used to approximate change of hemoglobin. Comparison between our results and theirs therefore is very difficult. We proved that we could apply NIRS signal in person authentication problem in small scale. In the future, we will extend the experiments size including more participants and collecting data time points.

V. CONCLUSION

We have proposed a new feature based on high order moment for NIRS-based BCI systems. To our best knowledge, there is no previous work proposing similar features for NIRS-bases BCI systems. By combining the 2nd order and 4th order moments together and using RBF SVMs as classifiers, our systems could achieve very high recall and precision ratios which are nearly 99.2% and 100% for both the left-hand and right-hand imagery problem and person authentication problem. These systems are not only high performance but also practical when running online. We believe that these features are also good at other NIRS-based problems. Our future work is to extend number of subjects participating in experiments and to collect data at some different time points.

ACKNOWLEDGMENT

The authors would like to express grateful thanks to Laboratory of Biomedical Engineering Department of International University, Viet Nam for allowing and supporting us conducting experiments. We would like to say thanks to lecturers and students of International University, Hochiminh City, Viet Nam who participated in our experiments. This research is supported by Faculty of Information Science and Engineering, University of Canberra, ACT, Australia.

REFERENCES

1. Wolpaw J R, Birbaumer N, McFarland D J, Pfurtscheller G, Vaughan T M (2007) Brain-Computer Interfaces for Communication and Control. J. Clinical Neurophysiology, Volume 113, pp. 767 – 791.

Krepki R, Curio G, Blankertz B, Muller K R. (2007) Berlin Brain-Computer Interface - The HCI communication channel for discovery. International Journal of Human-Computer Studies, Volume 65, Issue 5, Pages 460-477.

Villringer A, Obrig H. (1996) Near-infrared Spectroscopy and Imaging in A. W. Toga and J. C. Mazziotta, editors. Brain Mapping: The Methods. Academic Press, Inc., San Diego, 1996.

Sitaram R, Zhang H, Guan C, Thulasidas M, Hoshi Y, Ishikawa A, Shimizu K, Birbaumer N. (2007) Temporal classification of multichannel near-infrared spectroscopy signals of motor imagery for developing a brain–computer interface. J. NeuroImage, Volume 34, Issue 4, 15 February 2007, Pages 1416-1427.

Niide W, Tsubone T, Wada Y. (2009) Identification of moving limb using near infrared spectroscopic signals for brain activation. IJCNN 2009. International Joint Conference on Neural Networks, pp. 2264-2271.

Gottemukkula V, Derakhshani R. (2011) Classification-guided feature selection for NIRS-based BCI. Proceedings of the 5th International IEEE EMBS Conference on Neural Engineering, Cancun, Mexico.

Rowe R K, Miller W A, Ge N, Robinson R M. (2007). *U.S. Patent No. 7203345*. Washington, DC: U.S.

The address of the corresponding author:

Author: Dat Tran
Institute: Faculty of Information Science and Engineering, University of Canberra
City: ACT 2601
Country: Australia
Email: Dat.Tran@canberra.edu.au

Experiments on Synchronous Nonlinear Features for 2-Class NIRS-Based Motor Imagery Problem

Tuan Hoang[1], Dat Tran[1], Khoa Truong[2], Phuoc Nguyen[1], Toi Vo Van[2], Xu Huang[1], and Dhamendra Sharma[1]

[1] Faculty of Information Sciences and Engineering, University of Canberra, Canberra, Australia
[2] Department of Biomedical Engineering, HCMC International University, Ho Chi Minh City, Viet Nam

Abstract — **This paper aims to experiment several synchronous nonlinear features in the well-known 2-class motor imagery problem in Brain Computer Interface (BCI) systems using Near Infrared Spectroscopy (NIRS) technique. Those features including phase synchronizations and nonlinear interdependences are well known and widely applied on several neural-related problems such as epilepsy prediction. However, only a few publications are related to NIRS-based BCI systems. We conducted several experiments using NIRS technique to analyze how useful those synchronous nonlinear features can be applied on NIRS-based BCI systems. Results show that while the nonlinear interdependences can produce quite good recall and precision ratios, the phase synchronizations are not good for classification because the accuracy is as low as that in random guessing.**

Keywords — **NIRS-based BCI, nonlinear feature, synchronous feature, Brain Computer Interface, motor imagery.**

I. Introduction

Near-infrared spectroscopy (NIRS) is a new non-invasive optical method to acquire brain signals. This method provides researchers a range of advantages including flexibility of use, metabolic specificity, good temporal resolution, localized information, and high sensitivity in detecting small substance concentrations [3]. It also has some disadvantages such as slow operation due to the inherent latency of the hemodynamic response and weaken signal due to hair on the head. Nevertheless, NIRS's ability to record localized brain activity with a spatial resolution in the order of centimeter (depending on the probe geometry) provides us a good mean to measure a variety of motor and cognitive activities in Brain Computer Interface (BCI) systems [1, 2].

This paper aims to experiment several synchronous nonlinear features in the well-known 2-class motor imagery problem in BCI systems using Near Infrared Spectroscopy (NIRS) technique. Nonlinear synchronous features are widely analysed and used in several neural-related problems such as epilepsy prediction. Le et al. [8] analysed phase synchrony between neuronal signals to find large scale interactions in the brain. Their results showed that there are minor differences between the two methods and that phase synchronization measure can be used for study of neuroelectrical signals. Mirowski et al [7] conducted an important research in seizure prediction from intracranial EEG. They extracted bivariate features of several EEG synchronization measures then put them in classifiers including support vector machine (SVM), logistic regression and convolutional neural networks. Their best method outperformed previous seizure detection methods on the Freiburg dataset with 71% sensitivity and 0% false positives. Quian et al [9] experiment several synchronization measures on EEG signals. They concluded that these synchronization measures could be used in the study of synchronization in EEG data and could be of clinical relevance. For motor imagery problem based on NIRS technology, Sitaham et al. [4] reported that they could achieve average accuracy at 73% and 89% on 2-class problem, depending on classifiers they used. Niide et al. [5] conducted similar experiments also with the 2-class problem and achieved accuracy up to 94% in general experiment and 97% in subject-specific experiment. Recently, Gottemukula et al. [6] used classification-guided feature selection in the same problem to improve the accuracy.

Our research is inspired from the work of Quian et al [9]. We experiment several synchronization measures including phase synchronizations and nonlinear interdependences in NIRS-based BCI systems to see how useful those synchronization measures can provide. Results show that while the nonlinear interdependences can produce quite good recall and precision ratios, the phase synchronizations are not good for classification because the accuracy is as low as that in random guessing.

II. Materials and methods

A. Subjects

Four healthy subjects who are all male and have ages range from 20 to 30 voluntarily participated in the study.

V. Van Toi et al. (Eds.): 4th International Conference on Biomedical Engineering in Vietnam, IFMBE Proceedings 40, pp. 8–12, 2013.
www.springerlink.com

None of the recruited subjects had neurological or psychiatric history or was on medication. Each of them gave written informed consent for the experiment. All experiments were performed at Laboratory of Biomedical Engineering Department of International University in Vietnam. During the experiment, subjects were sitting on a comfortable arm-chair within a dark and sound–isolated room.

B. *Experimental Procedure*

All of the subjects were asked to sit on an arm-chair in front of a screen which displayed the stimuli in a dark and sound-isolated room. We used three types of protocols. The first protocol (Fig. 1a) consisted of five loops including interchange of two left-hand and two right-hand imagery loops, and a resting loop at final which totally last 130 seconds. The second (Fig. 1b) (the third protocol – Fig. 1c) consisted of four loops including three left-hand imagery loop (right-hand correspondingly) and a resting loop at final which totally last 100 seconds. A normal loop (Fig. 1d) consists of 10 seconds of resting and then 20 seconds of (left or right) imagery task. In imagery time, the subjects would clutch a ball by left hand or right hand corresponding to stimuli on screen. Each subject was asked to finish all three protocols. In total, for each subject, there were 5 trials for each type of imagery.

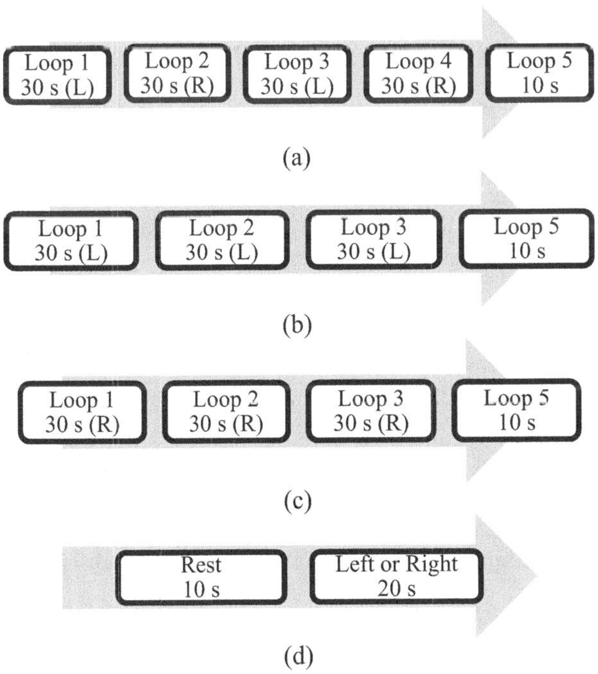

Fig. 1 (a) First protocol, (b) Second protocol, (c) Third protocol, (d) Performed tasks in a loop.

C. *Data Acquisition*

A 14-channel NIRS instrument (Shimadzu FOIRE3000) was used for acquiring oxygenated and deoxygenated hemoglobin concentration changes during our experiments. It operated at three different wavelengths of 780 nm, 805 nm and 830 nm, emitting an average power of 3 $mWmm^{-2}$. The detector optodes were fixed at a distance of 3 cm from the illuminator optodes. The optodes were arranged on the left and right hemisphere of the subject's head, above the motor cortex, around C3 (left hemisphere) and C4 (right hemisphere) areas (International 10–20 System) as illustrated in Fig. 2. For each of hemisphere, we use three illuminators and three detectors. A pair of illuminator and detector optodes formed one channel. So in total, we have seven channels on each hemisphere. The photomultiplier cycles through all the illuminator–detector pairings to acquire data at every sampling period. The sampling rate is approximately 18.2 Hz. Finally, the acquired signal is digitized by the 16–bit analog to digital converter.

Fig. 2. Three illuminators and 3 detectors in arrangement result in 7 channels on each hemisphere where red optode illustrates illuminator and blue optode illustrates detector.

D. *Data Pre-processing*

Our fNIRS imaging system adopts near–infrared lasers of three wavelengths 780 nm, 805 nm, and 830 nm. For each channel j, optical densities $A_{j,\lambda}$ in wavelengths λ are measured. Then, from the changes in the optical density $\Delta A_{j,\lambda}$, relative concentration changes of oxy-genated hemoglobin Δoxy_j and of deoxy-genated hemoglobin $\Delta deoxy_j$ are estimated by the following formula

$$\begin{pmatrix} \Delta oxy_j \\ \Delta deoxy_j \end{pmatrix} = \begin{pmatrix} -1.4887 & 0.5970 & 1.4847 \\ 1.8545 & -0.2394 & -1.0947 \end{pmatrix} \begin{pmatrix} \Delta A_{j,780} \\ \Delta A_{j,805} \\ \Delta A_{j,830} \end{pmatrix} \quad (1)$$

which is derived from the modified Beer–Lambert law and the extinction coefficients of the tissue.

Each trial was segmented into 38-data point chunks which last nearly 2 seconds. We chose 2-second chunk because it is short enough for real practical BCI systems and long enough for further processing. Two consecutive segments are overlapped by 27 data points, meaning one and a half second overlapping. Each label was labeled corresponding to class of action resting, left imagery, right imagery.

E. Nonlinear Interdependences

Let *x(t)* and *y(t)* be two discrete time series length N. State space of a signal is reconstructed and defined by embedding dimension m and time lag τ. Corresponding delay vectors are $\boldsymbol{x_n} = (x_n, x_{n-\tau}, \ldots, x_{n-(m-1)\tau})$ and $\boldsymbol{y_n} = (y_n, y_{n-\tau}, \ldots, y_{n-(m-1)\tau})$. Let $r_{n,j}$ and $s_{n,j}$ be the time indices of the k nearest neighbors of $\boldsymbol{x_n}$ and $\boldsymbol{y_n}$, respectively.

We define three distances naming mean squared Euclidean distance, **Y**-conditioned mean squared Euclidean distance and mean squared distance to random points as followings in order:

$$R_n^{(k)}(\boldsymbol{X}) = \frac{1}{k}\sum_{j=1}^{k}(\boldsymbol{x_n} - \boldsymbol{x_{r_{n,j}}})^2 \quad (2)$$

$$R_n^{(k)}(\boldsymbol{X}|\boldsymbol{Y}) = \frac{1}{k}\sum_{j=1}^{k}(\boldsymbol{x_n} - \boldsymbol{x_{s_{n,j}}})^2 \quad (3)$$

$$R_n\,(\boldsymbol{X}) = \frac{1}{N-1}\sum_{j\neq n}(\boldsymbol{x_n} - \boldsymbol{x_j})^2 \quad (4)$$

Naturally by construction in Equation (2) and (3), we always get $R_n^{(k)}(\boldsymbol{X}|\boldsymbol{Y}) \geq R_n^{(k)}(\boldsymbol{X})$. It is easy to see that the smaller the greater difference is the stronger two signals *x(t)* and *y(t)* are correlated. It is the main idea for us to define following nonlinear interdependences.

$$S^{(k)}(\boldsymbol{X}|\boldsymbol{Y}) = \frac{1}{N}\sum_{n=1}^{N}\frac{R_n^{(k)}(\boldsymbol{X})}{R_n^{(k)}(\boldsymbol{X}|\boldsymbol{Y})} \quad (5)$$

$$H^{(k)}(\boldsymbol{X}|\boldsymbol{Y}) = \frac{1}{N}\sum_{n=1}^{N} ln\frac{R_n\,(\boldsymbol{X})}{R_n^{(k)}(\boldsymbol{X}|\boldsymbol{Y})} \quad (6)$$

According to Equation (3), $S^{(k)}(\boldsymbol{X}|\boldsymbol{Y})$ has values ranging from 0 to 1. The lower value of $S^{(k)}(\boldsymbol{X}|\boldsymbol{Y})$ is, the more independent the two signals are. The higher value of $S^{(k)}(\boldsymbol{X}|\boldsymbol{Y})$ is, the more synchronized two signals are. $H^{(k)}(\boldsymbol{X}|\boldsymbol{Y})$, on the other hand, considers relations Y-condition distance with distance with random points. If $H^{(k)}(\boldsymbol{X}|\boldsymbol{Y}) = 0$, the two signals are completely independent. If it is positive, nearness in Y could lead to nearness in X for equal time partners. Otherwise, nearness in Y could lead to farness in X.

The last nonlinear interdependence is an normalized version of $H^{(k)}(\boldsymbol{X}|\boldsymbol{Y})$, defined as following.

$$N^{(k)}(\boldsymbol{X}|\boldsymbol{Y}) = \frac{1}{N}\sum_{n=1}^{N}\frac{R_n\,(\boldsymbol{X}) - R_n^{(k)}(\boldsymbol{X}|\boldsymbol{Y})}{R_n\,(\boldsymbol{X})} \quad (7)$$

We can see that $S^{(k)}(\boldsymbol{X}|\boldsymbol{Y})$, $H^{(k)}(\boldsymbol{X}|\boldsymbol{Y})$ and $N^{(k)}(\boldsymbol{X}|\boldsymbol{Y})$ are not symmetric. So in general $S^{(k)}(\boldsymbol{X}|\boldsymbol{Y})$, $H^{(k)}(\boldsymbol{X}|\boldsymbol{Y})$ and $N^{(k)}(\boldsymbol{X}|\boldsymbol{Y})$ are different from $S^{(k)}(\boldsymbol{Y}|\boldsymbol{X})$, $H^{(k)}(\boldsymbol{Y}|\boldsymbol{X})$ and $N^{(k)}(\boldsymbol{Y}|\boldsymbol{X})$, correspondingly.

F. Phase Synchronization

Analytic signals of signal *x(t)* are defined by Hilbert transform method as followings.

$$\tilde{x}(t) = (Hx)(t) = \frac{1}{\pi}P\int_{-\infty}^{\infty}\frac{x(t')}{t-t'}dt' \quad (8)$$

$$Z_x(t) = x(t) + i\tilde{x}(t) = A_x^H(t)e^{i\phi_x^H(t)} \quad (9)$$

Similarly, we can derive A_y^H and ϕ_y^H of signal *y(t)*. We then define *(a,b)* phase difference of *x(t)* and *y(t)*.

$$\phi_{xy}^H(t) = a\phi_x^H(t) - b\phi_y^H(t) \quad (10)$$

We say that *x(t)* and *y(t)* are a.b synchronized if their phase difference $\phi_{xy}^H(t)$ remains bounded for all *t*. If so, we define phase synchronization index as following.

$$\gamma_H = |\langle e^{i\phi_{xy}^H(t)}\rangle_t| \quad (11)$$

where brackets denote average over time. γ_H will be zero if the phases are completely not synchronized and will be 1 if they are synchronized.

G. Feature Vector

We considered four types of features: phase synchronizations, combinations of $S^k(X|Y)$ and $S^k(Y|X)$, $H^k(X|Y)$ and $H^k(Y|X)$, and $N^k(X|Y)$ and $N^k(Y|X)$. As mentioned in above section, there are 14 channels in our experiments. So in total we have 91 pairs of channel. Correspondingly, feature vectors of phase synchronizations have 91 elements while the feature vectors of the others have 182 elements.

H. Classification

Feature vectors then were put into classifier for training and prediction. We used only feature vector associating with left or right imagery, skipping resting segments. Classifiers are built and trained to classify imagery tasks, left or right motor imagery. It is a 2-class classification problem.

SVM has been proven as a powerful classifier for solving BCI classification problems. The popular RBF kernel function $K(x, x') = e^{-\gamma\|x-x'\|^2}$ was used. We applied grid search to get the optimal classifiers. The parameter γ was searched in $\{2^k: k = -15, -13, -11, \ldots, 1, 3\}$. The trade-off parameter C was searched over the grid $\{2^k: k =$

$-5, -3, -1, \ldots, 13, 15\}$. We used 3-fold cross validation test on the training set. To eliminate effects of randomness, we used 3 repetitions in training and classification.

III. Results

There are four parameters we need to set when running experiments: time lag, embedding dimension, Theiler correction window size, and number of nearest neighbors. In our experiments, we fixed Theiler correction window size as 5 data points, and number of nearest neighbors. For the two others, time lag varies from 1 to 3 and embedding dimension varies from 5 to 10. We also consider oxy-, deoxy- and total-hemoglobin in separate experiments. Results show that phase synchronization index is not a good feature for the 2-class NIRS-based motor imagery BCI systems. Its accuracies are about 50% which equal to random guessing for 2-class problems. In the mean while, combinations of $S^k(X|Y)$ and $S^k(Y|X)$, $H^k(X|Y)$ and $H^k(Y|X)$, and $N^k(X|Y)$ and $N^k(Y|X)$ are quite good at our systems. Table 1 shows an example of results with predefined parameters.

Table 1 Performance in accuracy (%) of systems using nonlinear interdependences as features. These experiments use following parameters: time lag=1, embedding dimension=5, Theiler correction window=5, Number of neighbors=10.

	$S^k(X\|Y)$ & $S^k(Y\|X)$	$H^k(X\|Y)$ & $H^k(Y\|X)$	$N^k(X\|Y)$ & $N^k(Y\|X)$
Oxy	78.2	74.9	75.5
Deoxy	77.6	76.6	77.1
Total	80.7	76.1	77.3

IV. Discussion

One of the challenges in nonlinear feature extraction step is parameter selection. It is also true in our experiments, especially with NIRS-based systems. The disadvantage of low sampling frequency of NIRS-based systems leads to the small number of data points in single window which is 38 data point window in our systems. We cannot choose large embedding dimension and time lag because the larger embedding dimension is, the smaller size of vector in state space and the larger time lag is, the less practical information the systems are. That is also the reason our time lag varies from 1 to 3 data points, corresponding from 1/8 to 1/6 seconds, and embedding dimension varies from 5 to 10 data points, corresponding from a quarter to half of second.

Synchronization measures seem not to be a good feature for NIRS-based systems. This bad result might come from low sampling frequency of NIRS systems. Lacking of enough data points could make phase estimation and then phase synchronization index estimation turn to very bad situation. Possibly, 38 data point windows are not appropriate selection for extracting synchronization measures. However, as we discussed above, increasing window size could lead to less practical NIRS-based BCI systems.

Results also show that there are minor differences among responses of oxy-, deoxy-, and total-hemoglobin. This means that we can use any of responses without considering much about accuracy decreasing.

V. Conclusion

We have presented experiments using NIRS technique with different types of synchronous nonlinear features for 2-class motor imagery problem. It seems that the phase synchronization index does not work well with the problem, while the nonlinear interdependences could achieve good results with recall and precision ratios up to 80%. Future research will focus on improving performance of these systems and analyze statistical significance of the results.

Acknowledgment

The authors would like to express grateful thanks to Laboratory of Biomedical Engineering Department of International University, Viet Nam for allowing and supporting us conducting experiments. We would like to say thanks to lecturers and students of International University, who participated in our experiments. This research is supported by Faculty of Information Science and Engineering, University of Canberra, ACT, Australia.

References

1. Wolpaw J R, Birbaumer N, McFarland D J, Pfurtscheller G, Vaughan T M (2007) Brain-Computer Interfaces for Communication and Control. J. Clinical Neurophysiology, Volume 113, pp. 767 – 791.

Krepki R, Curio G, Blankertz B, Muller K R. (2007) Berlin Brain-Computer Interface - The HCI communication channel for discovery. International Journal of Human-Computer Studies, Volume 65, Issue 5, pp. 460-477.

Villringer A, Obrig H. (1996) Near-infrared Spectroscopy and Imaging in A. W. Toga and J. C. Mazziotta, editors. Brain Mapping: The Methods. Academic Press, Inc., San Diego, 1996.

Sitaram R, Zhang H, Guan C, Thulasidas M, Hoshi Y, Ishikawa A, Shimizu K, Birbaumer N. (2007) Temporal classification of multichannel near-infrared spectroscopy signals of motor imagery for developing a brain–computer interface. J. NeuroImage, Volume 34, Issue 4, 15 February 2007, pp. 1416-1427.

Niide W, Tsubone T, Wada Y. (2009) Identification of moving limb using near infrared spectroscopic signals for brain activation. IJCNN 2009, pp. 2264-2271.

Gottemukkula V, Derakhshani R. (2011) Classification-guided feature selection for NIRS-based BCI. Proceedings of the 5th International IEEE EMBS Conference on Neural Engineering, Cancun, Mexico.
Mirowski P, et al. (2009) Classification of patterns of EEG synchronization for seizure prediction. Journal of Clinical Neurophysiology, vol. 120, 2009, pp 1927-1940.
Le Van Quyen M, Foucher J, Lachaux J, Rodriguez E, Lutz A, Martinerie J, Varela FJ. (2001) Comparison of Hilbert transform and wavelet methods for the analysis of neuronal synchrony. J Neurosci Methods. 2001 Oct 30; 111(2):83-98.
Quian Q R, Kraskov A, Kreuz T, Grassberger P. (2002) Performance of different synchronization measures in real data: a case study on electroencephalographic signals. Phys. Rev. E, vol. 65, no. 4, pp. 041903.

The address of the corresponding author:

Author: Dat Tran
Institute: Faculty of Information Science and Engineering, University of Canberra
City: ACT 2601
Country: Australia
Email: Dat.Tran@canberra.edu.au

Application of Transillumination Imaging to Injection Assist System

K. Shimizu, N. Tobisawa, T.T. Nghia, T. Namita, and Y. Kato

Graduate School of Information Science and Technology, Hokkaido University, Sapporo, Japan

Abstract — **A technique of near-infrared (NIR) trans-illumination imaging was applied to develop an injection assist system. We have developed a simultaneous observation system for a body surface image and a transillumination image of blood vessels. The system consists of a high-intensity and low-leak light source, near-infrared CMOS camera and a small and light one-eye head mounted display. In transillumination imaging, we could visualize the veins in the adult forearm with the thickness of as much as 67 mm. A model phantom of the veins in the forearm was fabricated with the blood-contained tubes buried in pork thigh meat. The blood tube was hardly observable with a naked eye. Using the developed system, we could identify the blood tube in the NIR transillumination image, and could extract the blood successfully with the assist of the NIR image.**

Keywords — **transillumination, near-infrared light, injection assist, head mounted display, see-through.**

I. Introduction

In clinical practice, it is sometimes difficult to locate an appropriate vein for venous injection or blood sampling. This becomes a problem particularly for the patients with thin or deep veins, such as for a child or an obese person. Finding an appropriate vein is also significant for the patient on long-term hemodialysis.

Near-infrared (NIR) light with the wavelength of 700-1200 nm has relatively high transmission through our body. By transillumination imaging using the NIR light, we can visualize the blood vessels inside the body. The blood vessel images have been used for biometrics and intraoperative support to check the vasculature. However, the transillumination image becomes severely blurred because of the strong scattering in the body tissue, as the depth of the vessels increase. Therefore, their applications have been limited only to the surface structure and to the thin object such as a finger or a palm.

We have developed some techniques to suppress the effect of scattering in transillumination imaging of deep located blood vessel [1-4]. If we can obtain a clearer transillumination image of deep-seated blood vessels, it can be used to assist the intravenous injection, the varicose-vein treatment, arterial catheterization, intraoperative observation of blood vessels and lymph ducts, and so on. This paper reports the development of a simultaneous observation system for body surface image and transillumination image of blood vessels to assist the intravenous injection. The feasibility of the injection assist was examined in an experiment.

II. Method

Figure 1 schematically illustrates the principle of the developed system. This system is designed to assist the intravenous injection at a human forearm. For trans-illumination imaging, an NIR light source is placed under the forearm. The transmitted light is captured by the CMOS camera through an NIR filter. The image obtained with the camera is processed in a personal computer and presented on a head mounted display (HMD). The HMD is hanged in front of the single eye from a head belt. It is small and does not interrupt the observation with both eyes in visible light. Thus, the user can observe the NIR transillumination image in the HMD by slightly looking up and manipulate the injection with both eyes by slightly looking down. In this way, the user can search the blood vessel, locate an appropriate injection spot in NIR transillumination image and perform the injection using both views.

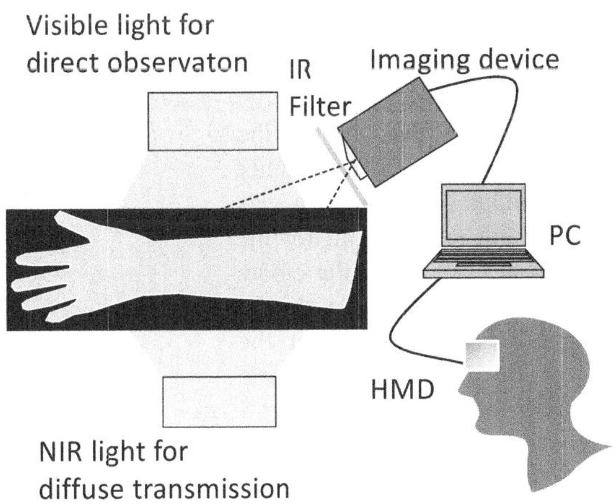

Fig. 1 Principle of simultaneous observation system.

V. Van Toi et al. (Eds.): 4th International Conference on Biomedical Engineering in Vietnam, IFMBE Proceedings 40, pp. 13–16, 2013.
www.springerlink.com

III. SIMULTANEOUS OBSERVATION SYSTEM

Wide-area and high-intensity illumination is required to make the transillumination of a thick body part possible. We arranged 400 LED's (850 nm wavelength, SFH4350; Osram Co., Ltd) in a 25 x 16 array on a flexible circuit board of 150 x 100 mm^2. Figure 2 shows the appearance of the light source.

The maximum optical power density was 689 W/m^2. Since the circuit board was flexible, it can be fitted along the body surface to minimize the light leak. The diameter of each LED was as small as 3 mm and the high density array arrangement of the LED's made homogeneous illumination possible.

An optical band-pass filter with the passband of 800 - 1100 nm was attached in front of a CMOS camera with a night-vision function (DC-NCR30U; Digital Cowboy Co., Ltd.). The HMD (Data Glass 3/A; Shimadzu Co. Ltd.) weighs 120 g and can display color image with 800 x 600 pixels resolution. The user can see the equivalent image to the 10 inch wide monitor screen at 60 cm in front of him/her. Figure 3 shows the appearance of the HMD.

Fig. 2 Appearance of LED array on flexible printed circuit board.

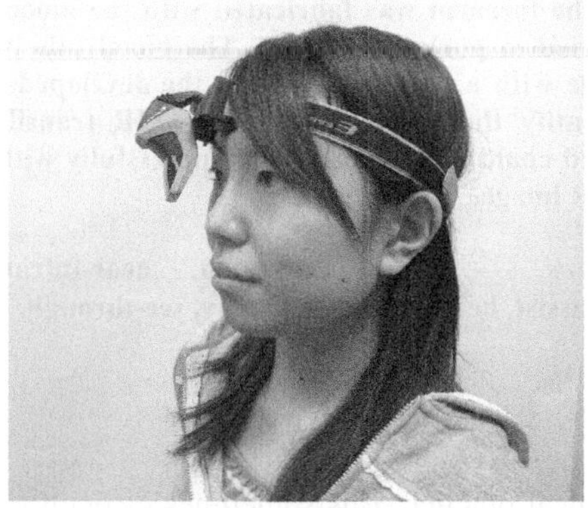

Fig. 3 Appearance of head mounted display.

IV. TRANSILLUMINATION IMAGES OF FOREARM

Figure 4 shows the results of transillumination imaging of an adult forearm. The subject was a healthy male volunteer of 24 years old. Figure 4(a) is a visible image of the forearm of the subject observed under common fluorescent lamp lighting. The surface veins were hardly recognized. Figure 4(b) is the NIR transillumination image of the wrist area observed from its inner side. Figures 4(c) and 4(d) are the NIR transillumination images of the central part of the forearm observed from inner and outer side of the arm, respectively. Figure 4(e) is the NIR transillumination image around the elbow joint observed from its inner side. The thickness of the arm was 47 mm, 53 mm and 67 mm for the wrist, the central part and the elbow area, respectively. Even with the thick object as much as 67 mm, the blood vessels were made visible with a high signal to noise ratio.

In some subjects, thick surface veins are often visible with the naked eye. However, the comparison between the NIR transillumination image and the

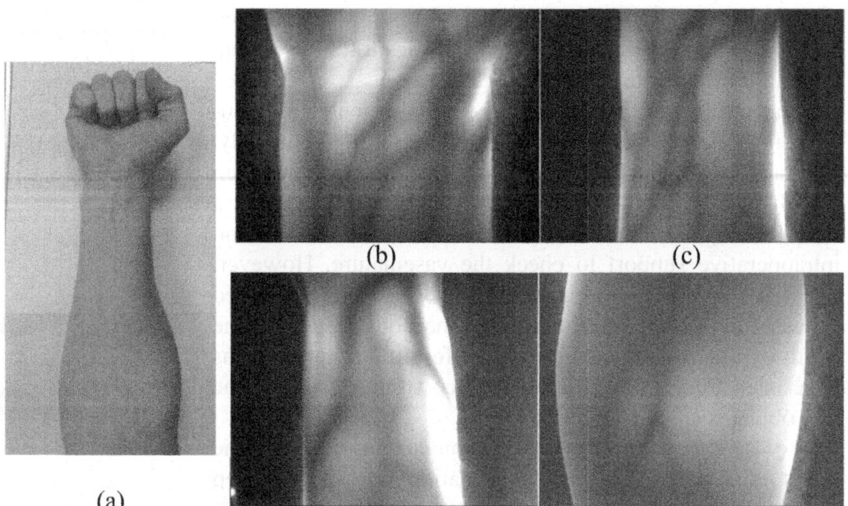

Fig. 4 Transillumination images: (a) visible image of the forearm, (b) wrist, (c) central forearm of inner side, (d) central forearm of outside, (e) elbow.

surface visible image showed that many blood vessels hardly seen in the visible image could be visualized in the NIR transillumination image. This result suggested the useful applicability of the developed system to assist the intravenous injection in clinical practice.

V. Experimental simulation of injection assist

The feasibility of the injection assist was examined in an experiment. A model phantom to simulate a human forearm was fabricated. Figure 5 shows the schematic diagram of the phantom. Pork thigh meat was packed in a transparent container of Plexiglas (100x60x65 mm^3). Preserved horse blood was filled in silicone tubes with the inner diameter of 1.5 mm. Two blood tubes were buried at a depth of 2.75 mm from the meat surface.

Figure 6 shows a frame of the video image for visible and NIR illumination. They correspond to the images observed with the naked eye and in the HMD, respectively. The hardly visible blood tubes were visualized in the transillumination image with NIR light. Using the developed system, we could locate the blood tube, insert an injection needle, and extract the blood easily. This system allowed the user to confirm the position of the blood vessel by quickly comparing the visible and NIR images as often as he/she wanted. It did not hinder the manual operation of the syringe with both eyes and both hands. Through this experiment, the feasibility of the injection assist using the developed system was verified.

VI. Conclusions

To assist the intravenous injection, we have developed a simultaneous observation system for a body surface image and a transillumination image of blood vessels. The system basically consists of a light source, a CMOS camera, a personal computer and a head mounted display. As a light source, many LED's were arranged on a flexible circuit board. It made the high-intensity and small-leak illumination possible. The small one-eye HMD presented a transillumination image with sufficient view angle and image resolution. It does not hinder the user from manipulating tools with both eyes and both hands.

Using the developed system, the blood vessels in all area of the adult forearm were visualized. We could obtain the transillumination images of the body part with as much as 67 mm thickness. In an experiment, the feasibility of the injection assist using the developed system was verified.

These results suggest that this system can assist the injection with the patient whose blood vessels are hardly recognized with the naked eye.

Acknowledgment

This research was supported in part by a Grant-in-Aid for Scientific Research from Japan Society for the Promotion of Science.

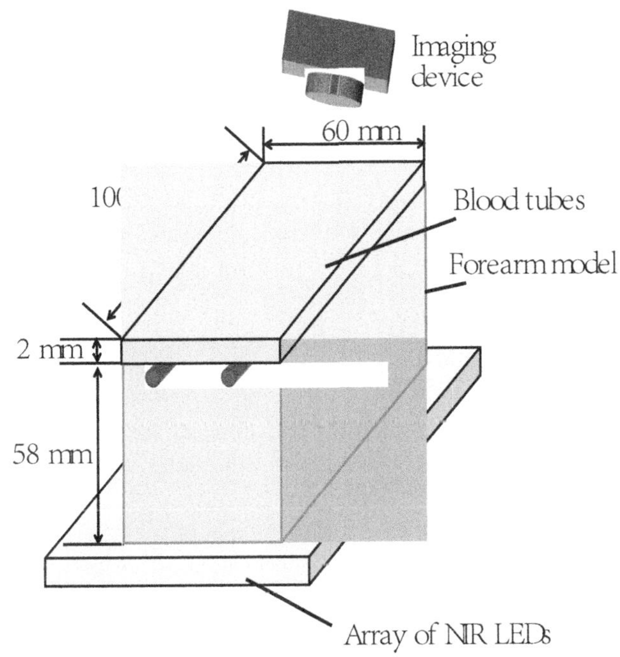

Fig. 5 Phantom model for veins in forearm.

Fig. 6 Surface and transillumination images of phantom model.

REFERENCES

1. K. Shimizu, and M. Kitama, "Fundamental study on near-axis scattered light and its application to optical computed tomography," Opt. Rev. vol. 7, pp. 383-388 (2000).
2. K. Shimizu, K. Tochi and Y. Kato, "Improvement of transcutaneous fluorescent images with a depth-dependent point-spread function", Applied Optics, Vol.44, No.11, pp.2154－2161 (2005)..
3. K. Takagi, Y. Kato and K. Shimizu, "Extraction of near-axis light for transillumination imaging", Applied Optics, Vol.48, No.10, pp.36－44 (2009).
4. K. Takagi, H. Kakinuma, Y. Kato and K. Shimizu, "CW trans- illumination imaging by subtracting weakly scattered light from strongly diffused light", Optics Express, Vol.17, No.10, pp.8332－8342 (2009).

Author: Koichi Shimizu
Institute: Graduate School of Information Science and Technology, Hokkaido University
Street: N14 W9, 060-0814
City: Sapporo,
Country: Japan
Email: shimizu@bme.ist.hokudai.ac.jp

Design Current Stabilization Circuit for Transcranial Impulse Electro Stimulation

Viet Dung Nguyen[1], Quang Huan Dao[1], Huu Phuong Trung Lai[1], Ngoc Tien Pham[1], Duc Thuan Nguyen[1], A.V. Kipensky[2], and A.P. Lastovka[2]

[1] Department of Biomedical Engineering, School of Electronics and Telecommunications, H anoi University of Science and Technology, Hanoi, Vietnam

[2] Laboratory of Biomedical Electronics, National Technical University "Kharkov Polytechnic Institute" NTU "KhPI", Kharkov, Ukraine

Abstract — **For more than 30 years, researchers and doctors from Russian, the U.S, France, Israel and other countries have studied methods of transcrebral or transcranial electrotherapy stimulation (TES). Transcranial impulse electrotherapy is a specific type of transcranial electrotherapy that uses low power impulse current to stimulate the antinoceptive system of the encephalon to provide procedures of electrodream and transcranial electroanagelsia. The stimulation current must be kept constant during treatment session. Traditionally, it can be done by employing close-loop control that continuously supervises the stimulation current. The fact that the impulse current for TES is low power can cause difficulties in measuring and controlling the current. Therefore, in this paper, we will introduce detailed design of current stabilization circuit that produces constant impulse current but doesn't require closed loop control.**

Keywords — **transcranial electrotherapy stimulation, impulse current stabilization.**

I. INTRODUCTION

For more than 30 years, researchers and doctors from Russian, the U.S, France, Israel and other countries have studied methods of transcrebral or transcranial electrotherapy stimulation (TES) [1]. Transcranial impulse electrotherapy is a specific type of transcranial electrotherapy that uses low power impulse current to stimulate the antinoceptive system of the encephalon to provide procedures of electrodream and transcranial electroanagelsia. It is conducted by applying impulse current to electrodes placed at various allocations such as oculomammiform, oculoocipital, frontomammiform or frontojugular.

Earliest prototypes of transcranial electrostimulation devices originated in Russia. These original designs, although successfully employed for several different treatment modalities, had a severe drawback with regard to the comfort of the wearer or patient. In some cases, these earlier cranial electrostimulation devices even subjected the wearer to pain. It has been discovered that the reason for the discomfort of these earlier designs was a result of the use of direct current as part of the overall operation of the devices. The direct current was used to break down or lower skin resistance to allow the treatment alternating current signals to penetrate the brain and nervous systems to cause the desired effect established by the placement of the electrodes on the head of the patient.

In these earlier types of machines, the wearer received a combination of direct current and alternating current electrical waveform packages through a series of electrodes affixed to the head with straps. With this DC current based design, the wearer was required to place a thick pad between any electrode and the skin. Typically, the pad was comprised of several layers of unbleached and uncolored cotton flannel, or an equivalent product. For best results, the fabric pads were soaked with water to provide a conductive path between the electrodes and the skin of the wearer. Without the presence of the pads (which were only required because of the presence of the DC current), such devices could either burn the skin of the wearer, or cause relatively intense pain before a usable level of the treatment modality of the currents at the AC frequency could be reached.

An additional and potentially harmful drawback of the DC based designs was that of iontophoresis. A characteristic of a DC circuit application of this type is that molecular sized parts of metal, toxins and other undesirable impurities can be caused to migrate in the direction of current flow through the skin and into the bloodstream of the wearer of such DC based CES devices. Consequently, care had to be taken to ensure that no substance was present other than water used to create good electrical contact with the pad to the skin of the wearer. Since practically all CES treatment modalities require repeated treatments, the potential for iontophoresis being a harmful factor was escalated.

The stimulation current must be kept constant during treatment session. Traditionally, it can be done by employing close-loop control that continuously supervises the stimulation current. Low power impulse current used for TES can lead to difficulties in measuring and controlling the current. Therefore, in this paper, we will introduce detailed design of current stabilization circuit that produces constant impulse current but doesn't require closed loop control.

V. Van Toi et al. (Eds.): 4th International Conference on Biomedical Engineering in Vietnam, IFMBE Proceedings 40, pp. 17–21, 2013.
www.springerlink.com

In Fig. 1 is a typical block diagram of transcranial impulse electro stimulation device [2]. Applied current to electrodes is converted from output voltage of waveform generator. During treatment session, patient's impedance or impedance along electrode to electrode pathway changes. This impedance change causes variations in stimulation current while it must be constant. To keep the stimulation current constant, provided impulse current must be supervised continuously. Hence, a closed loop control is employed.

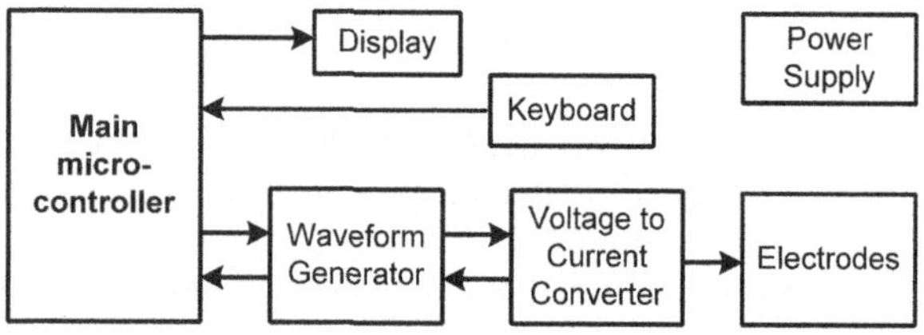

Fig. 1 Typical block diagram of transcranial impulse electro stimulation device.

The fact that the impulse current for transcranial impulse electro stimulation is low power [3] (micro amp to milliamp) can cause difficulties in measuring and controlling the current. Therefore, a different block diagram has been proposed in our previous publication [4]. It is shown in Fig. 2. It is obvious that the impulse current is stabilized by current stabilizer module. Consequently, the closed loop control is no more needed.

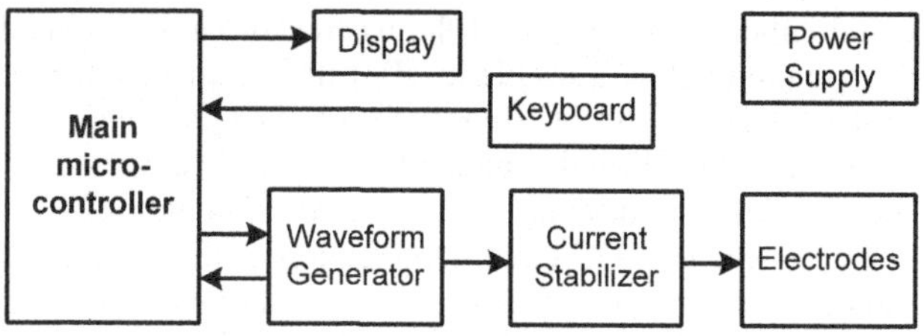

Fig. 2 Block diagram of transcranial impulse electro stimulation device that does not require closed loop control.

It was found out that constant voltage impulses are effective in combination with DC. Contrary constant current impulses of the equal amplitude had the same efficacy without additional DC. This result gave an opportunity to reduce significantly the level of current applied. Further improvements of electrical regimen were directed to diminution of local irritating action on the skin under electrodes. To realize that the bipolar impulse with zero net charge was used [5]. Zero net charge means positive part or phase is equal to negative part. Fig. 3 illustrates the waveform of this impulse current.

There are various types of TES modality that have been being used mainly for pain relief treatment [6, 7]. However, all of them provide bipolar zero net charge rectangular impulse current at one or some fixed frequency and width. Our aim is to provide impulse current having the same waveform but with changeable impulse amplitude up to 5mA (in 0.02mA steps), changeable impulse frequency up to 2000Hz (in 1Hz steps) and changeable impulse width up to 90% (in 10% steps). It is another reason why closed loop control is hardly employed in this circumstance. Our proposed current stabilization circuit satisfied all above specifications. In next part, we will present detailed design of our proposed impulse current stabilization circuit.

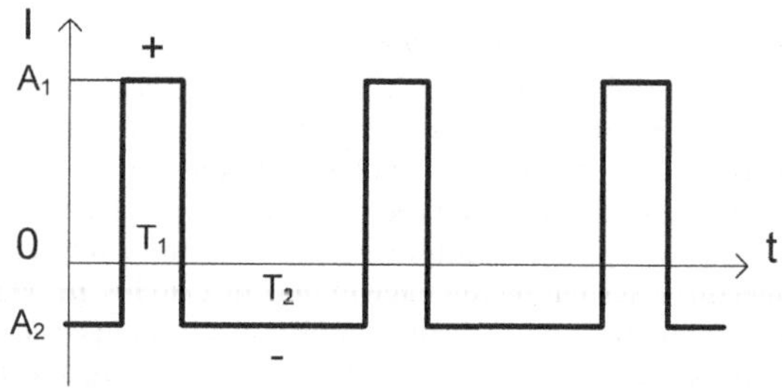

Fig. 3 Waveform of bipolar impulse current with zero net charge

II. DESIGN CURRENT STABILIZATION CIRCUIT

The block diagram of the proposed current stabilization circuit is shown in Fig. 4 and has already been given in our previous publication [4]. It composes of 4 blocks: Buffer Amplifier, Opto Coupler, Power Amplifier and Current Stabilizer block.

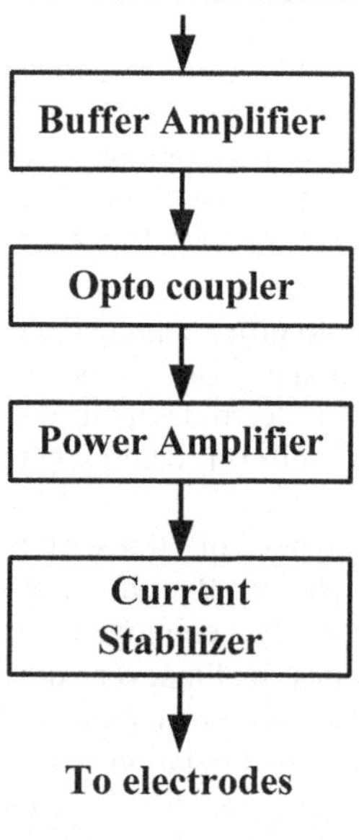

Fig. 4 Block diagram of current stabilization circuit.

To couple with input signal from waveform generator, a high impedance operational amplifier is used as buffer amplifier. In our design, operational amplifier LF4124 having high impedance JFET input is used. The schematic circuit of buffer amplifier block is shown in Fig. 5.

The maximal peak-to-peak impulse voltage from waveform generator is 5V while input pulse width is from 10% to 90%. So that amplification coefficient is chosen as 0.5. R1 and R2 are selected as 100kΩ and 150kΩ to ensure high input impedance. VR1 is tuned so as to comply with power amplifier stage.

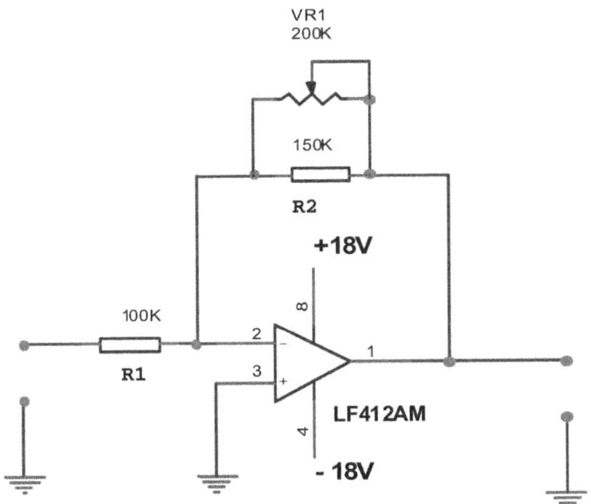

Fig. 5 Schematic circuit of buffer amplifier block.

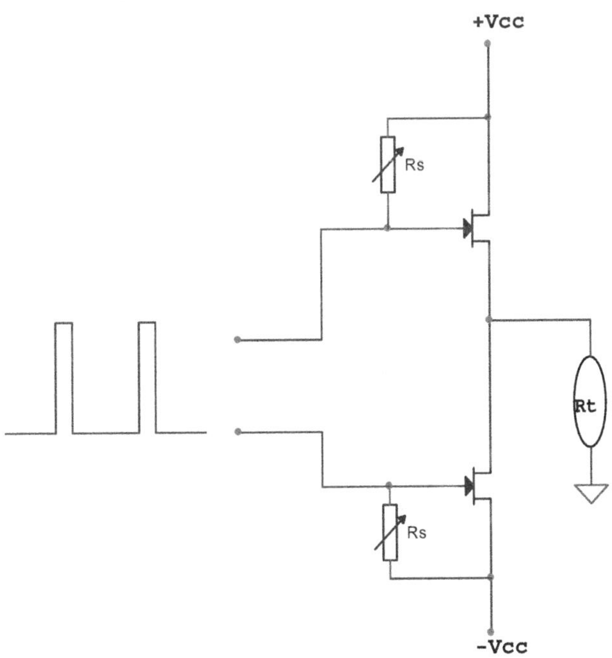

Fig. 6 Using high voltage JFETs to amplify power and stabilize output currtent.

To amplify power and to stabilize output current, JFETs are normally used. We would like to produce constant impulse current up to 5mA while patient's impedance changes up to 10 kΩ so high viltage JFETs are required. In this design 2 JFETs TN5277 from Sanyo are chosen to form symmetrical current source (fig. 6).

Practical characteristic of TN5277 (fig. 7) indicates that controlling output current is nonlinear. It is desired to have linear characteristic as in fig. 8. Therefore, Rs must be selected to have characteristic opposite to that of TN5277 (fig. 9).

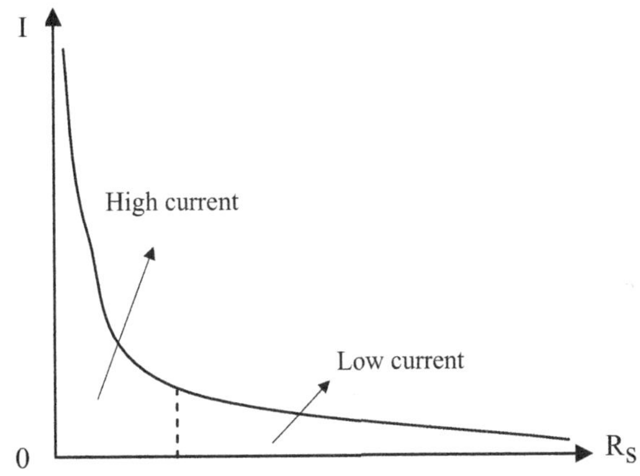

Fig. 7 Practical characteristic curve of TN5277.

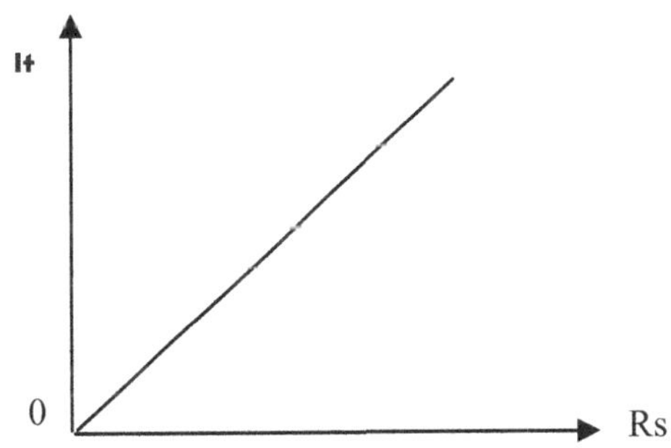

Fig. 8 Ideal characteristic curve.

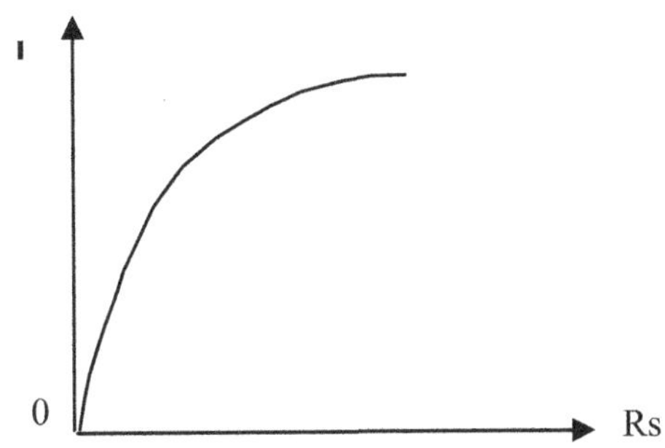

Fig. 9 Desired characteristic curve.

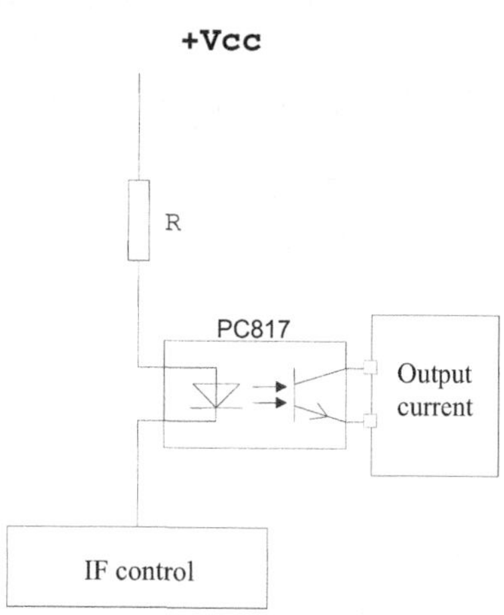

Fig. 10 Opto coupler schematic circuit.

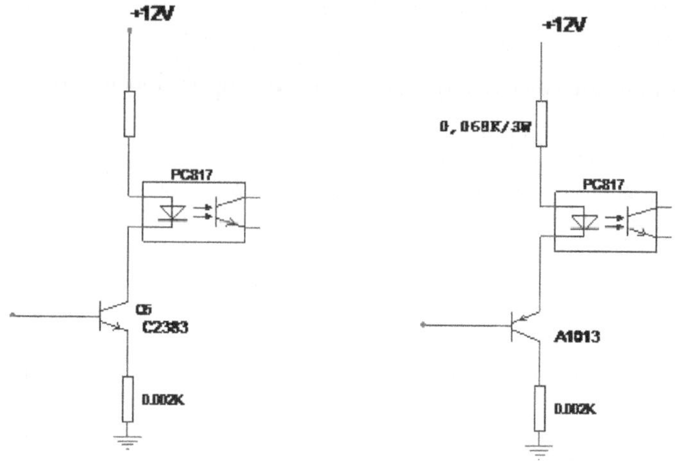

Fig. 11 IF control and opto coupler circuit.

Output current to patient must be isolated from input. It can be done by using transformer or opto coupler. However, isolation by transformer cannot provide good response in wide frequency range (as up to 2000Hz in this design). Hence, opto coupler is used. Opto coupler PC817 used in our design meets this requirement. The schematic circuit is given in fig. 10. Output impulse current is zero net charge so that to control IF we must choose 2 elements working in different parts of output impulse current. In our design, 2 transistors C2383 and A1013 are chosen. Schematic circuits of IF control and opto coupler are given in fig. 10 to 11.

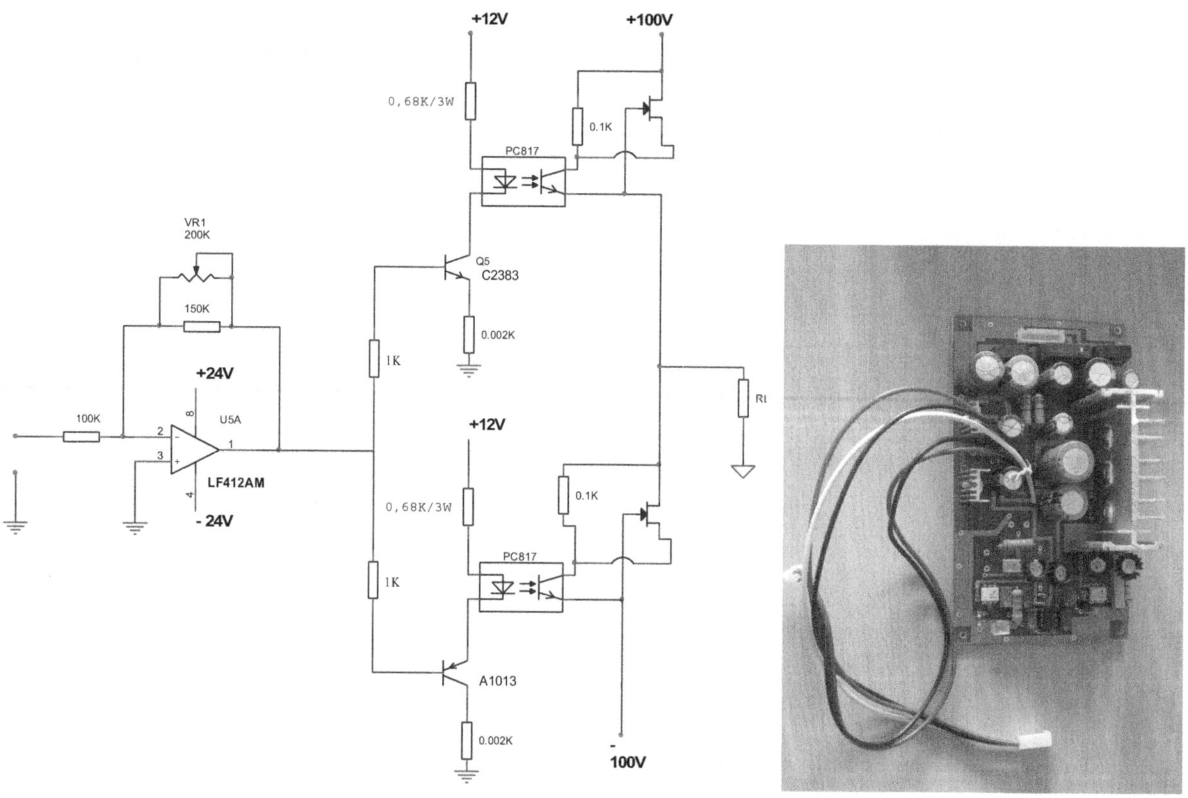

Fig. 12 Impulse current stabilization schematic and manufactured circuits.

III. CONCLUSIONS

In figure 12 is schematic diagram and manufactured circuit of impulse current stabilization module. In our laboratory, we performed testing on the ability of producing accurate constant impulse current. We changed impulse frequency, impulse width and used different sham loads. Output impulse current was monitored using digital oscilloscope.

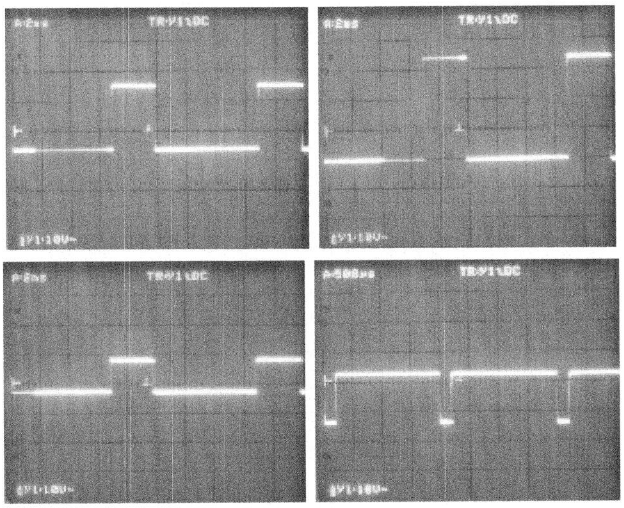

Fig. 13 Test results.

The results as given in fig. 13 shows that our proposed circuit can produce constant impulse current up to 5mA in peak-to-peak amplitude while impulse frequency changes up to 2000Hz, impulse width changes up to 90% and load changes up to 10kΩ.

ACKNOWLEDGMENT

This research is a part of joint Vietnam-Ukraine project No. 05/2351/2009/HD-NDT supported by Vietnamese Ministry of Science and Technology.

REFERENCES

1. A. Limoge, C. Robert and T. H. Stanley (1999) Transcutaneous cranial electrical stimulation (TCES). Neurosci and Biobehavioral Reviews, vol. 23, no. 4 pp. 529-538

Y. S. Katsnelson (2005) Transcranial electrostimulation apparatus and method. US Patent No US6904322B2

V. P. Lebedev, A. B. Savchenko, A. B. Fan and S. Yu. Zhilyayev (1988) Transcranial Electroanalgesia in Rats: Optimal Electrical Parameters. Fiziologischesky zhurn SSSR imeni Sechnova, vol. 74, no. 8, pp. 1094-1101

V. D. Nguyen, Q. H. Dao, P. T. Lai Huu, N. T. Pham (2010) Design of Microcontroller-based Transcranial Impulse Electrotherapy Stimulation Device – Premilinary Results. Proceeding of the 3rd International Conference on Communications and Electronics, Hanoi, Vietnam

V. P. Lebedev, A. V. Malygin, A. V. Kovalevski, S. V. Rychkova, V. N. Sisoev, S. P. Kropotov, E. M. Krupitski, L. I. Gerasimova, D. V. Glukhov and G. P. Kozlowski (2001) Device for non-invasive transcranial electrostimulation of the brain endorphinergic system: Application for improvement of human psycho-physiological status. Proceedings in the 7th Vienna International Workshop on Functional Electrical Stimulation, Vienna, Austria

http://www.tes.spb.ru/

http://www.fisherwallace.com/

Author: Viet Dung Nguyen
Institute: Hanoi University of Science and Technology
Street: 1 Dai Co Viet
City: Hanoi
Country: Vietnam
Email: nvdung-bme@mail.hut.edu.vn

Feasibility Study for 3D Reconstruction of Internal Structure of Animal Body Using Near-Infrared Light

Tran Trung Nghia, Takeshi Namita, Yuji Kato, and Koichi Shimizu

Graduate School of Information Science and Technology, Hokkaido University, Sapporo, Japan

Abstract — **Using an array of near-infrared (NIR) LED's, we can obtain transillumination images of an animal body and visualize the veins and arteries in the adult forearm as thick as 6 cm. To examine the feasibility of three dimensional (3D) reconstruction using these transillumination images, experiments were conducted to evaluate the image of an absorber in a tissue-equivalent phantom. An experimental system was designed to obtain a transillumination image of a model of blood vessels in the adult forearm. Imaging was accomplished with a high-resolution cooled digital CCD camera, placed at the opposite side of the phantom to the light source. The transillumination images were obtained with different depths of the absorber from the container wall.**

Keywords — **3D reconstruction; turbid medium; transillumination; optical CT3.**

I. INTRODUCTION

The importance of optical computed tomography (optical CT) in clinical applications has been recognized. However, it has not been widely used in practical applications. The intensity of transmitted light decreases rapidly as the object thickness increases. The image of the light-absorbing structure inside the body becomes severely blurred as its depth from the surface increases. To overcome these problems, we have devised some techniques to supply incident light to inside the body efficiently and to obtain transillumination image sensitively through a human body. Using these techniques, the veins and arteries were visualized in the adult forearm as thick as 6 cm [1-5].

With these 2D NIR transillumination images, we can expect to realize 3D optical CT of a human body. As the first step for the 3D imaging, we have conducted a basic study to examine the feasibility of this methodology. In this report, we present some preliminary results of the analysis on the detectable depth of an absorber in tissue-equivalent phantom.

I. MATERIALS AND METHODS

Measurement system: Figure 1 shows a schematic of the experimental system to obtain a transillumination image of the absorber in a cylindrical scattering medium. This phantom is a model of blood vessels in human body.

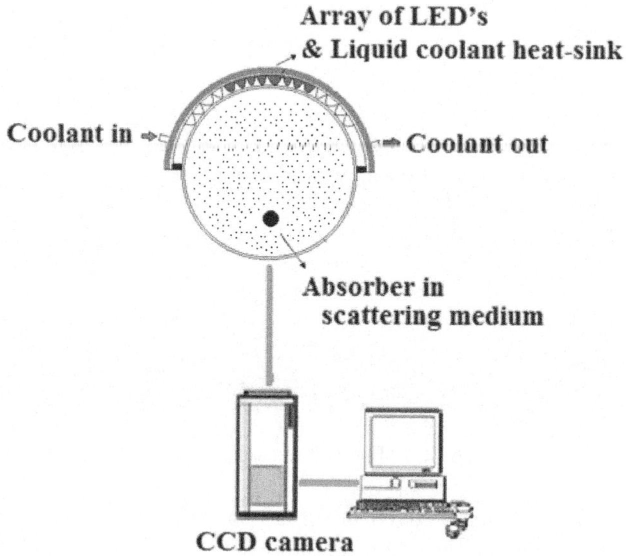

Fig. 1 Schematic of experimental system.

Light source and imaging system: The light source consisted of an array of LED's on a flexible printed circuit board and a liquid-coolant system to cool the LED's from behind. 400 LED's (850 nm peak wavelength) were arranged in a 25 x 16 array on a flexible PCB of 150 x100 mm^2. The maximum optical power density was 689 W/m^2. Since the circuit board was flexible, it could fit along the forearm or phantom surface in close contact to minimize light leakage. The diameter of each LED was as small as 3 mm and the high-density array arrangement of the LED's ensured homogeneous illumination.

Imaging was accomplished with a high-resolution cooled digital CCD camera (ORCA-R2 C10600-10B Hamamatsu) interfaced to a desktop computer. The CCD camera was placed at the opposite side of the phantom to the light source. The automatic exposure control of camera was turned off and the exposure was adjusted manually to provide a visually acceptable projection image. The transillumination

V. Van Toi et al. (Eds.): 4th International Conference on Biomedical Engineering in Vietnam, IFMBE Proceedings 40, pp. 22–24, 2013.
www.springerlink.com

images were obtained with different depths of the absorber from the container wall.

Phantom Model: In this experiment, the scattering media with four kinds of absorption coefficient (μa) were prepared. The scattering coefficient (μs') was made same in all the four phantoms, or μs'=1.0 /mm. Intralipid suspension and black ink were mixed in distilled water to make tissue-equivalent medium (μs'=1.0 /mm, μa=0.01–0.10 /mm). It was contained in a cylindrical container (with 50 mm outer diameter) made of transparent polycarbonate. The wall thickness and the height of the container were 0.5 mm and 100 mm, respectively.

As an absorbing target structure, a cylindrical post with 5 mm diameter made of black-painted metal was used. This simulated a blood vessel in an adult arm. It was placed inside the medium at the far side of the container from the light source as shown in Fig. 1. Using a mechanical rotation system, the transillumination images were obtained from different directions around the curved surface of the cylinder.

3D reconstruction from 2D images: Prior to carrying out tomographic reconstruction, all images obtained were reviewed using an image processing program and corrections were made if necessary. For tomographic reconstruction and 3D reconstruction from 2D transillumination images, the techniques based on the principles of the filtered back-projection, the multi-view 3D reconstruction and the cone beam CT 3D reconstruction are used.

II. Results

Figure 2 shows the example of transillumination images of phantoms with different optical properties and the same absorber depth (1 mm).

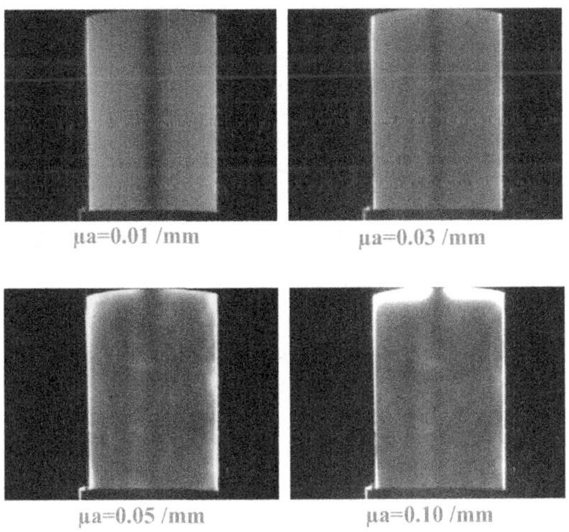

Fig. 2 Transillumination images of phantoms with different optical properties and the same absorber depth (1 mm).

We calculated the contrast and sharpness at the center of transillumination images. Figure 3 shows the contrast (Michelson contrast) of the absorber image for different absorber depths and different μa's. The contrast decreases rapidly with the depth of the object absorber. However, the rate of the decrease is largely dependent on the μa of the scattering medium. The larger the μa, the contrast decreases more rapidly.

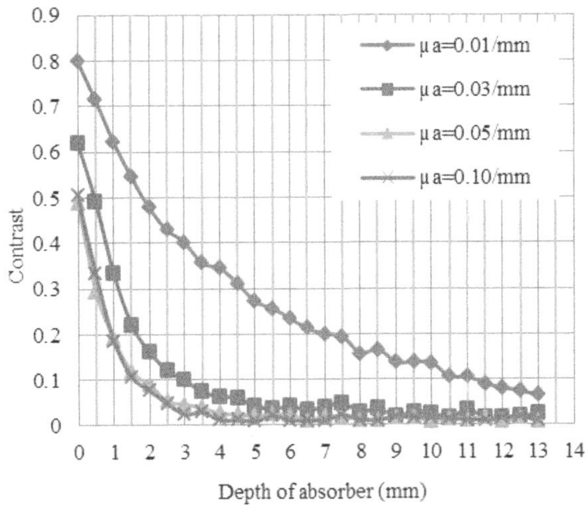

Fig. 3 Dependence of image contrast on μa of medium.

Figure 4 shows the measured sharpness of the absorber image at 1 mm absorber depth. As the μa increases, the scattering effect is suppressed, and the image sharpness increases.

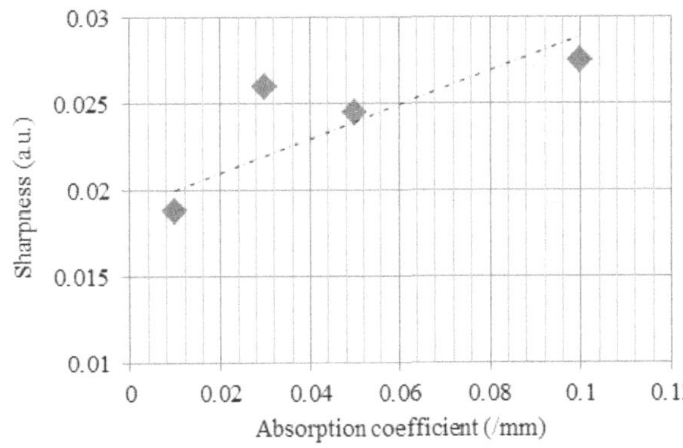

Fig. 4 Dependence of sharpness on μa of medium.

These results suggest that we should design the imaging system to optimize its performance for the specific μs' and μa of the scattering medium around the target object.

III. Conclusions

With a view toward the 3D reconstruction of internal structure of an animal body, a feasibility study was conducted. The images of an absorber in tissue-equivalent phantom were analyzed for different optical properties of scattering medium.

The feasibility was verified and the major factors to obtain clear transillumination image were identified. They are useful to provide appropriate projection images for the reconstruction of the 3D image of an animal body.

Acknowledgement

This research was partially supported by a grant for Hokkaido University Global COE program, "Center for Next-Generation Information Technology based on Knowledge Discovery and Knowledge Federation", from the Ministry of Education, Culture, Sports, Science and Technology of Japan.

References

1. N. Tobisawa, T. Namita, Y. Kato, and K. Shimizu, "Injection assist system with surface and transillumination images," Proc. Bioinfo. and Biomed. Eng., (iCBBE) 2011 5th Int'l. Conf. on, no.31863, Wuhan, China, May 2011.
2. M. Shimano, T. Namita, Y. Kato, and K. Shimizu, "Reconstruction of three-dimensional structure in turbid medium by scattering suppression," Proc. Photonics and Optoelectro. (SOPO), 2011 Sympo. On, no.21040, Wuhan, China, May 2011.
3. K. Takagi, H. Kakinuma, Y. Kato, and K. Shimizu, "CW transillumination imaging by subtracting weakly scattered light from strongly diffused light," Opt. Express, vol.17, no.10, pp.8332-8342, May 2009.
4. K. Takagi, Y. Kato, and K. Shimizu, "Extraction of near-axis light for transillumination imaging," Appl. Opt., vol.48, no.10, pp.D36-44, Apr. 2009.
5. K. Shimizu, and M. Kitama, "Fundamental study on near-axis scattered light and its application to optical computed tomography," Opt. Rev., vol.7, no.5, pp. 383-388, Sept./Oct. 2000.

Author: Tran Trung Nghia
Institute: Graduate School of Information Science and Technology, Hokkaido University
Street: Kita 14, Nishi 9, Kita-ku
City: Sapporo, 060 - 0814
Country: Japan
Email: tran@bme.ist.hokkudai.ac.jp; ttnpfiev@gmail.com

Fabrication of Nano-piezomaterials for Powering Bioelectronics and Interfaced Cellular Biomechanics

Thanh D. Nguyen and Michael C. McAlpine

Department of Mechanical and Aerospace Engineering, Princeton University, Princeton 08544 NJ, USA

Abstract — **Nanostructures of piezoelectric materials are attractive building blocks for useful bioimplant devices such as self-powering nanogenerators and nano-biosensors. Yet, one of the most efficient piezo-nanomaterials - $PbZr_xTi_{1-x}O_3$ (PZT) nanowires is difficult to synthesize without suffering from polycrystallinity or poor stoichiometric control. Current large-scale nanofabrication methods have been restricted by their complexity and cost, while bottom-up synthesis of nanowires has been limited in methods to assemble these structures at precisely-defined locations. Here, we report a novel fabrication method which requires only low resolution photolithography to generate ultra smooth, high performance PZT nanowires over wafer scales. These fabricated PZT nanomaterials also shows a biocompatible interface with living cells to offer a great potential for its applications in biomedicine such as self-powering implant devices and cellular biomechanical sensors.**

Keywords — **nanomanufacturing, energy conversion, piezoelectric nanowires, piezo-cell interface.**

Nanowires (NWs) have emerged as powerful building blocks for Biomedical applications such as nano-biosensors [1], nano-implant generators [2] and nanoelectronics [3]. Key to achieving high device performance is the generation of NWs whose characteristics – including geometry, composition, and defect density – are well controlled at the point of their growth or fabrication. Further, many applications require nanowires which are hierarchically patterned over large areas and at precisely defined locations. Thus, to generate NW arrays for practical applications, a fabrication method which is high throughput, not exotic, applicable to a broad range of functional materials, and capable of generating highly ordered NWs on a large-scale area is requisite. For example, bottom-up synthesis of NWs by vapor-liquid-solid (VLS) [4] or electrochemical [5] methods has successfully produced NWs with broad compositions [6] which can be assembled via flow alignment [7], Langmuir-Blodgett assembly[8], dry transfer[9], or dielectrophoresis[10] over macroscopic scales.

Recently, top-down methods, including nanoimprint lithography (NIL)[11,12], superlattice nanowire pattern transfer (SNAP)[1,13] and lithographically patterned nanowire electrodeposition (LPNE)[14] have regained attractiveness in their ability to pattern NWs that can be deterministically positioned. For example, LPNE has shown the preparation of NWs with diameters down to sub-50 nm of materials that can be electrodeposited (such as noble-metals and lead telluride)[14,15] while SNAP is applicable to single-crystalline materials by etching metal templates into host substrates[16]. Here, we report a new top-down method – inspired by LPNE and free of complex processing – for generating functionaxl NW arrays with accurate placements over wafer scales. Our technique can generate NWs of complex compositions and functionalities. Indeed, we demonstrate its power by realizing large area fabrication of high performance piezoelectric NWs, which may have applications in nanodevice power generation.

Our approach, which we term Photolithography & Etching for Nanoscale Lithography (PENCiL), produces a metal nanomask template over wafer scales. Figure 1a outlines the PENCiL process. First, a 100 nm Ni thin film is deposited onto any substrate. Next, arrays of 1-2 μm wide resist lines are patterned using standard photolithography. Third, and key to PENCiL, the Ni layer is electrochemically etched in concentrated phosphoric acid by "under-cutting" the resist windows in order to define the NW structures. Variation in the applied voltage, photo-resist patterns/ heights, and etching times can be utilized to directly determine NW features of various shapes and sizes[17]. Most significantly, Figure 1b shows that NWs patterned using the PENCiL technique can reproducibly achieve sub-40 nm diameters, as well as curved and ring-like shaped NWs. Because PENCiL relies only on photolithography and etching, large-area scalability of this process was achieved by patterning a 3 inch wafer with Ni NW arrays, as shown by the diffraction pattern in Figure 1c. indeed, ordered, uniform NW arrays are realized on a large scale (Fig. 1d), and monolithically patterned to larger contact pads to

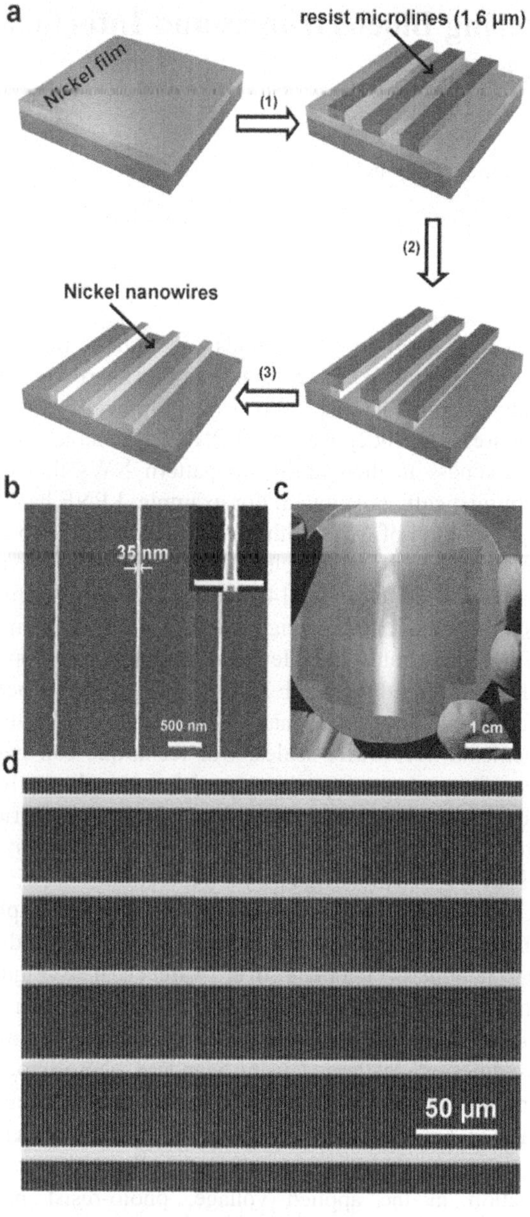

Fig. 1 PENCiL, fabrication of metal nanowires. a. Schematic of fabrication. b. SEM image of NWs. c. Diffraction image of whole wafer NWs. d. Sem of large scale NWs.

fabricate nanoscale devices from the simple two-step PENCiL process (Fig. 1d). PENCiL can be performed on a wide variety of substrates, including materials with inherent functionalities. One particularly interesting class is piezoelectric crystals, which represent promising materials for electromechanical energy conversion technologies [2,18]. Electromechanically derived energy harvesting via low-dimensional piezoelectric nanostructures has garnered substantial research attention, due to the synthesis of various one dimensional piezoelectric nanostructures including ZnO[19], $BaTiO_3$[20], GaN[20], CdS[21] and PVDF NWs[22] and the possibility of harvesting energy from the environment for self-powered nanosystems and biomedical devices[2,18]. Lead zirconate titanate ($PbZr_xTi_{1-x}O_3$, PZT) is a particularly interesting piezoelectric material for high performance applications due to its exceptionally large piezoelectric constant (~250 pm/V)[23]. However, this potential is greatly restricted by the lack of an efficient method for its synthesis at the nanoscale. This is a result of the complex stoichiometric composition of PZT, as most techniques (e.g., VLS) for producing high quality NWs cannot be easily applied due to issues such as phase separation and the lack of suitable catalysts[24]. Synthesis of PZT NWs to date has thus been limited to sol-gel templates and hydrothermal methods[25-28], resulting in polycrystalline wires and/or wires with limited stoichiometric control. By contrast, the controlled synthesis of stochiometric, epitaxial thin films of PZT with precise atomic compositional control has been well studied[29,30]. Accordingly, we performed the PENCiL process on an underlying layer of crystalline PZT. A PZT thin film (300–500 nm thickness, 52/48 Zr/Ti) was first deposited epitaxially on Pt/MgO or on a Si substrate via RF-sputtering with a 52/48 PZT target to ensure a composition close to the morphotropic phase boundary (MPB)[31]. The film was post-annealed at 700 °C for 30 minutes to induce a phase transition to a perovskite structure[18,32]. After performing PENCiL, a subsequent dry reactive ion etching (RIE) step was performed with a Micro RIE (150 mTorr SF_6, 150 W), selectively removing the exposed PZT film and resulting in PZT NW arrays under the PENCiL NW mask (Fig.2a). Finally, the Ni nanomask was removed using 63% aq. nitric acid at 60 °C for 10 min to complete the process and reveal the defined PZT NWs.

Figure 2b shows that even over large areas, the PENCiL nanomask can be directly translated into highly ordered PZT NW arrays (diameter ~ 75 nm). Energy Dispersive Spectroscopy (EDS) was locally performed at multiple points along the axial direction of individual PZT NWs (Fig.2c). Importantly, the EDS spectrum shows that the stoichiometry closes matches that of a bulk standard PZT sample, and confirms that the PENCiL translation process does not leave residual Ni from the nanomask. Next, the functional properties of the defined PZT NWs were investigated using piezo-force microscopy (PFM)[33]. A 80 nm Pt film was deposited prior to PZT film deposition for use as the bottom conducting electrode in the PFM measurement. An AFM (Asylum Research, Santa Barbara, CA) equipped with a doped diamond conducting tip (Olympus AC240TM Ti/Pt coating, k = 2.1 N/m, f = 77 kHz) was operated in contact mode. A lock-in amplifier (Stanford Instruments SR830) was used, and an applied AC signal frequency of 283 kHz was chosen to avoid unnecessary topographic crosstalk near the cantilever resonance. An intermediate force (2000 nN) was applied to ensure that the tip deflection was electromechanically-response dominated[33].

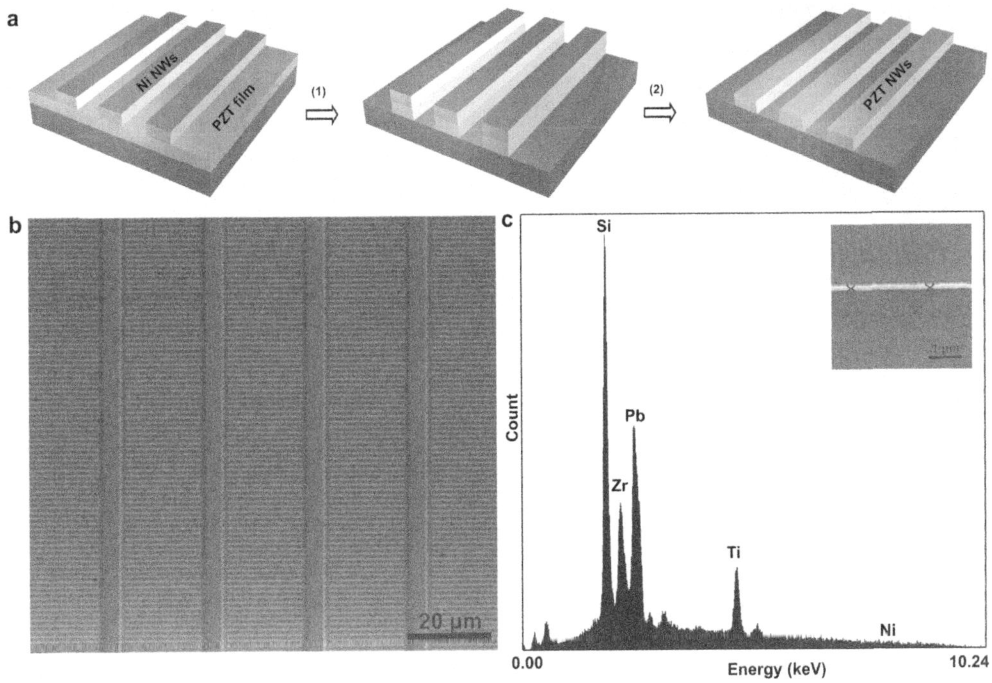

Fig. 2 Fabrication of high performance PZT nanowires .a. Schematic of fabrication. b. Large scale PZT nanowires. c. EDS for composition investigation of PZT nanowires.

Figure 3a shows representative local height (left), piezoelectric amplitude (middle), and phase (right) scans of the PZT NW array collected via the conductive cantilever tip. The observed contrast in the amplitude and phase traces of Figure 3a indicates that the PZT NW array is of uniform response (standard deviation = 7.8%), and shows distinct differences from the collected character of the substrate base layer. To illustrate this more clearly, Figure 3b shows a three-dimensional overlay of the induced piezoelectric amplitude imposed on the topographic features of the PZT NW array over a 20 × 20 μm^2 area. Again, the induced piezoelectric response is found to be quite uniform along the length of each NW.

An effective piezoelectric coefficient, d_{eff}, was derived by calculating the average deformation (~300 pm without amplification) per applied AC voltage amplitude (V_{AC} = 3 V), yielding a value of ca. 100 pm/V for the unpoled PZT NWs. Finally, ramping mode PFM, in which the cantilever tip is held in direct contact with the PZT NW surface while the applied AC bias to the tip is continuously increased (V_{AC} = 0 – 2.5 V), was performed to quantitatively determine the effective piezoelectric coefficient of individual PZT NWs as defined by PENCiL. Figure 3c shows the collected linear piezoelectric response of an individual, unpoled PZT NW using ramping mode PFM, with the resulting slope indicating an effective piezoelectric coefficient of 114 pm/V. The observed value increased to d_{eff} = 145 pm/V following 30 min poling under an applied DC voltage of +10 V, scanned over a selected area of the PZT NW.[32] To our knowledge, this value represents the highest piezoelectric coefficient reported to date from any piezoelectric nanostructure[28,34]. We attribute this result to the ability of the PENCiL technique to define smooth nanowire structures from high quality thin films. The achievement of high performance PZT nanomaterials offers great potentials for using the materials as biomechanical sensors or self-powering devices for human implants.

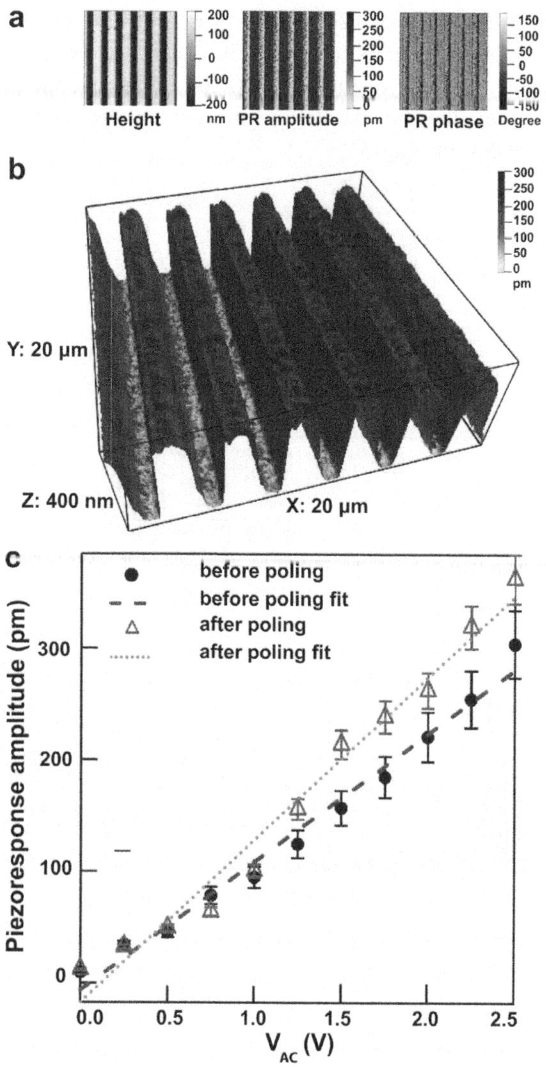

Fig. 3 Piezo-Force Microscopy measurement. a. PFM measurement for piezoelectric signals of height, phase and amplitude. b. Piezoelectric signal lays out on morphology of PZT nanowires. c. PFM measurement of piezoelectric signal on ramping mode.

I. Conclusion

In summary, a combined photolithographic and electrochemical etching method – which we term PENCiL – was able to pattern highly ordered PZT NW arrays over large areas, with the highest reported piezoelectric coefficient to date. It should be emphasized that this process is universally applicable to a host of metallic nanomask systems and functional substrates, including semiconducting, magnetic, and optoelectronic materials, so long as selective etch reagents are available. We anticipate that the results demonstrated here will open exciting avenues in fundamental research on piezoelectric size scaling in nanomaterials, as well as applications in biomedical engineering fields such as self-powering implant devices or biomechanical sensors.

References

1. McAlpine, M.C., Ahmad, H., Wang, D.W. & Heath, J.R. Highly ordered nanowire arrays on plastic substrates for ultrasensitive flexible chemical sensors. **6,** 379-384 (2007).
2. Xu, S. *et al.* Self-powered nanowire devices. *Nat. Nanotechnol.* **5,** 366-373 (2010).
3. Nam, S., Jiang, X.C., Xiong, Q.H., Ham, D. & Lieber, C.M. Vertically integrated, three-dimensional nanowire complementary metal-oxide-semiconductor circuits. *Proc. Natl. Acad. Sci. USA* **106,** 21035-21038 (2009).
4. Morales, A.M. & Lieber, C.M. A laser ablation method for the synthesis of crystalline semiconductor nanowires. *Science* **279,** 208-211 (1998).
5. Singh, M. *et al.* Synthesis and Superconductivity of Electrochemically Grown Single-Crystal Aluminum Nanowires. *Chem. Mater.* **21,** 5557-5559 (2009).
6. Tian, B., Kempa, T.J. & Lieber, C.M. Single nanowire photovoltaics. *Chem. Soc. Rev.* **38,** 16-24 (2009).
7. Huang, Y., Duan, X., Wei, Q. & Lieber, C.M. Directed Assembly of One-Dimensional Nanostructures into Functional Networks. *Science* **291,** 630-633 (2001).
8. Whang, D., Jin, S. & Lieber, C.M. Nanolithography Using Hierarchically Assembled Nanowire Masks. *Nano Lett.* **3,** 951-954 (2003).
9. Fan, Z.Y., Ho, J.C., Jacobson, Z.A., Razavi, H. & Javey, A. Large-scale, heterogeneous integration of nanowire arrays for image sensor circuitry. *Proc. Natl. Acad. Sci. USA* **105,** 11066-11070 (2008).
10. Freer, E.M., Grachev, O., Duan, X., Martin, S. & Stumbo, D.P. High-yield self-limiting single-nanowire assembly with dielectrophoresis. *Nat. Nanotechnol.* **5,** 525-530 (2010).
11. Ahn, S.H. & Guo, L.J. Large-Area Roll-to-Roll and Roll-to-Plate Nanoimprint Lithography: A Step toward High-Throughput Application of Continuous Nanoimprinting. *ACS Nano* **3,** 2304-2310 (2009).
12. Chou, S.Y., Krauss, P.R. & Renstrom, P.J. Imprint lithography with 25-nanometer resolution. *Science* **272,** 85-87 (1996).
13. Melosh, N.A. *et al.* Ultrahigh-density nanowire lattices and circuits. *Science* **300,** 112-115 (2003).
14. Menke, E.J., Thompson, M.A., Xiang, C., Yang, L.C. & Penner, R.M. Lithographically patterned nanowire electrodeposition. *Nat. Mater.* **5,** 914-919 (2006).
15. Yang, Y. *et al.* Wafer-Scale Patterning of Lead Telluride Nanowires: Structure, Characterization, and Electrical Properties. *ACS Nano* **3,** 4144-4154 (2009).
16. Boukai, A.I. *et al.* Silicon nanowires as efficient thermoelectric materials. *Nature* **451,** 168-171 (2008).
17. Shenoy, R.V., Datta, M. & Romankiw, L.T. Investigation of island formation during through-mask electrochemical micromachining. *J. Electrochem. Soc.* **143,** 2305-2309 (1996).
18. Qi, Y. *et al.* Piezoelectric ribbons printed onto rubber for flexible energy conversion. *Nano Lett.* **10,** 524-528 (2010).
19. Wang, Z.L. & Song, J.H. Piezoelectric nanogenerators based on zinc oxide nanowire arrays. *Science* **312,** 242-246 (2006).
20. Duan, X.F. & Lieber, C.M. Laser-assisted catalytic growth of single crystal GaN nanowires. *J. Am. Chem. Soc.* **122,** 188-189 (2000).
21. Routkevitch, D., Bigioni, T., Moskovits, M. & Xu, J.M. Electrochemical fabrication of CdS nanowire arrays in porous anodic aluminum oxide templates. *J. Phys. Chem.* **100,** 14037-14047 (1996).

22. Chang, C.E., Tran, V.H., Wang, J.B., Fuh, Y.K. & Lin, L.W. Direct-write piezoelectric polymeric nanogenerator with high energy conversion efficiency. *Nano Lett.* **10,** 726-731 (2010).
23. Li, J.F., Dai, X.H., Chow, A. & Viehland, D. Polarization Switching Mechanisms and Electromechanical Properties of La-Modified Lead-Zirconate-Titanate Ceramics. *J. Mater. Res.* **10,** 926-938 (1995).
24. Han, S. *et al.* Transition metal oxide core-shell nanowires: Generic synthesis and transport studies. *Nano Lett.* **4,** 1241-1246 (2004).
25. Lin, Y.R., Liu, Y.T. & Sodano, H.A. Hydrothermal synthesis of vertically aligned lead zirconate titanate nanowire arrays. *Appl. Phys. Lett.* **95,** 122901 (2009).
26. Zhang, X.Y. *et al.* Synthesis and piezoresponse of highly ordered $Pb(Zr_{0.53}Ti_{0.47})O_3$ nanowire arrays. *Appl. Phys. Lett.* **85,** 4190-4192 (2004).
27. Limmer, S.J. *et al.* Electrophoretic growth of lead zirconate titanate nanorods. *Adv. Mater.* **13,** 1269-1272 (2001).
28. Luo, Y. *et al.* Nanoshell tubes of ferroelectric lead zirconate titanate and barium titanate. *Appl. Phys. Lett.* **83,** 440-442 (2003).
29. Blanco, O., Heiras, J., Siqueiros, J.M., Martinez, E. & Castellanos-Guzman, A.G. PZT films grown by RF sputtering at high oxygen pressure. *J. Mater. Sci. Lett.* **22,** 449-453 (2003).
30. Takayama, R. & Tomita, Y. Preparation of epitaxial $Pb(Zr_xTi_{1-x})O_3$ thin-films and their crystallographic, pyroelectric, and ferroelectric properties. *J. Appl. Phys.* **65,** 1666-1670 (1989).
31. Kim, Y.K., Kim, S.S., Lee, B., Shin, H. & Baik, S. Piezoelectric effect in epitaxial $PbZr_{1-x}Ti_xO_3$ thin films near morphotropic phase boundary region. *J. Mater. Res.* **20,** 787-790 (2005).
32. Granzow, T., Kounga, A.B., Aulbach, E. & Rodel, J. Electromechanical poling of piezoelectrics. *Appl. Phys. Lett.* **88,** 252907 (2006).
33. Kalinin, S.V. & Bonnell, D.A. Imaging mechanism of piezoresponse force microscopy of ferroelectric surfaces. *Phys. Rev. B* **65,** 125408-125418 (2002).
34. Wang, J., Sandu, C.S. & Setter, N. Large-scale fabrication of titanium-rich perovskite PZT submicro/nano wires and their electromechanical properties. *IEEE T. Ultrason. Ferr.* **56,** 1813-1819 (2009).

CORRESPONDING AUTHOR
Thanh D. Nguyen
Princeton University
01 olden street –school of engineering
Princeton, United States
Email: tdnguyen@princeton.edu

The Image Reconstruction Principle in PET/CT Scanner

Huynh Luong Nghia and Nguyen Minh Hong

Biomedical Electronic Department, Le Quy Don Technical University

Abstract — **Thanking to the ability of providing images of slices of organs with very high accuracy currently the PET / CT scanner is widely used for early diagnosis of many cancers and some other diseases. This paper examines the reconstruction algorithms in PET/CT scanner and builds mathematical model describing the operation of this device.**

Keywords — **PET / CT scanner, reconstruction algorithms, radiopharmaceuticals, attenuation correction factor.**

I. INTRODUCTION [1]

In radioactive emission computed tomography the heterogeneous structure, the hydrodynamics of biological subject and the radionuclide redistribution are determined by concentration distribution of the radiopharmaceuticals (FDG) injected onto patient's body or it's parts. Besides, the intensity of recorded γ- rays is under influence of tissue absorption that is characterized by the linear attenuation coefficient $\mu = \mu(\vec{r}) = \mu(x, y, z)$. Then, complete setting up the fundamental problem of radioactive emission computed tomography will include estimating both the distribution of FDG uptake $f(\vec{r}) = f(x, y, z)$ and the attenuation coefficient $\mu(\vec{r})$ based on intensity of γ- rays recorded from outside body.

In practice, the fundamental problem mentioned above is simplified by only calculating the distribution of FDG uptake uptake $f(\vec{r}) = f(x, y, z)$ with suggestion of negligible absorption, or estimating the attenuation coefficient $\mu(\vec{r})$ by another method such as CT -scanner . Just the latest approach is used nowadays in PET/CT equipment.

II. THE IMAGE RECONSTRUCTION ALGORITHM OF PET/CT SCANNER [2,3]

A. *Operation Principles of PET/CT Scanner*

Using the geometry depicted in Fig 1 the collected emission projection data can be represented mathematically by:

$$p(x',\phi)=\left[\exp\left(-\int_{-\infty}^{\infty}\mu(x,y)dy'\right)\right]\int_{-\infty}^{\infty}f(x,y)dy' \qquad (1)$$

where the exponential term (in square brackets) represents the attenuation along the line of response (LOR) at detector position x' and projection angle ϕ, and where f(x,y) represents the distribution of FDG or other positron tracer in the patient

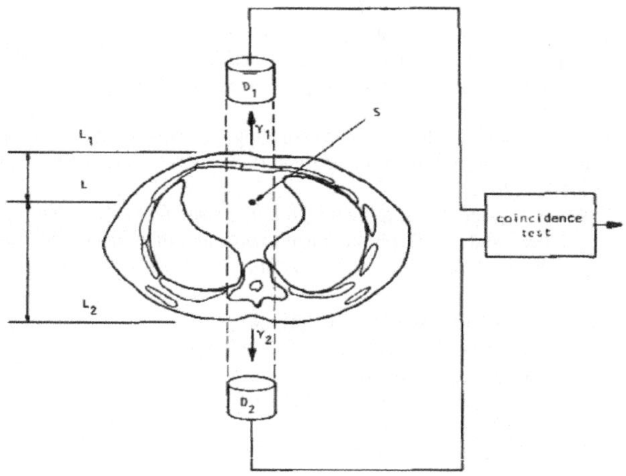

Fig. 1 A photon emitted at S and traveling toward the D1 detector is attenuated over a distance of $L_1 - L$, while a photon traveling toward the D2 detector undergoes an attenuation proportional to L – L2.

The form of Eq. 1 can be simplified by expressing the inverse of the exponential term as an attenuation correction factor

$$a(x',\phi)=\exp\left(\int_{-\infty}^{\infty}\mu(x,y)dy'\right) \qquad (2)$$

which represents inverse of the dual-photon attenuation along the LOR for detector position x' and projection angle ϕ. A multiplicative correction for photon attenuation is then given by

V. Van Toi et al. (Eds.): 4th International Conference on Biomedical Engineering in Vietnam, IFMBE Proceedings 40, pp. 30–32, 2013.
www.springerlink.com

$$p_{AC}(x',\phi)=a(x',\phi)p(x',\phi)=\int_{-\infty}^{\infty} f(x,y)dy' \qquad (3)$$

where $p_{AC}(x', \phi)$ are the emission data (sinograms) corrected for attenuation. The corrected projection values $p_{AC}(x', \phi)$ can be used in a tomographic image reconstruction algorithm, such as FBP, to estimate the two-dimensional distribution of radionuclide concentration represented by the function f(x,y). Alternatively, the acquired projection data $p(x, \phi)$ can be reconstructed directly with an iterative method, where the attenuation correction factors (ACFs) are used to provide proper statistical weighting to the data. An example of this approach is attenuation-weighted OSEM (AWOSEM), which has been implemented on most commercial PET scanners.

The Attenuation Compensation for positron emission tomography is easier in comparison with single photon emission tomography. This ease is explained as follows.

Let's say that the detectors D1 and D2 in Fig. 1 are being used to measure one ray in a projection and let's also assume that there is a source of positron emitters located at the point S. Suppose for a particular positron annihilation, the two annihilation gamma-ray photons labeled y1 and y2 in the figure are released toward D1 and D2, respectively. The probability of y1 reaching detector D, is given by

$$\exp(-\int_{L}^{L1} \mu(x)dx) \qquad (4)$$

where $\mu(x)$ is the attenuation coefficient at 511 keV as a function of distance along the line joining the two detectors. Similarly, the probability of the photon $\gamma 2$ reaching the detector D2 is given by

$$\exp(-\int_{L2}^{L} \mu(x)dx) \qquad (5)$$

Then the probability that this particular annihilation will be recorded by the detectors is given by the product of the above two probabilities

$$\exp(-\int_{L}^{L1} \mu(x)dx) . \exp(-\int_{L2}^{L} \mu(x)dx) \qquad (6)$$

which is equal to

$$\exp(-\int_{L1}^{L2} \mu(x)dx) \qquad (7)$$

This is a most remarkable result because, first, this attenuation factor is the same no matter where positron annihilation occurs on the line joining D1 and D2, and, second, the factor above is exactly the attenuation that a beam of monoenergetic photons at 511 keV would undergo in propagating from L1 at one side to L2 at the other. Therefore, one can readily compensate for attenuation by first doing a transmission study .

B. *Determining Attenuation Correction Factors [4]*

As shown in Eqs. 2 and 3, attenuation correction factors $a(x', \phi)$ must be derived from transmission data to correct the PET data for photon attenuation. With PET/CT scanners a 511 keV attenuation map can be generated from the CT image to correct the PET emission data for photon attenuation: Because the CT images contained attenuation information for effective photon energies of about 70 keV, given in HU, they had to be transformed to PET photon energies of 511 keV (μ-maps). For this purpose, different approaches (such as Scaling, Segmentation, Hybrid and Hybrid Dual-Energy methods) have been developed and implemented in commercially available tomographs. For example, the following established bilinear algorithm can be used:

$$\mu^{PET}(CT \le 0HU) = \mu^{PET}_{H_2O}(CT + 1{,}000)/1{,}000 \qquad (8)$$

$$\mu^{PET}(CT > 0HU) = \mu^{PET}_{H_2O} + CT\frac{\mu^{PET}_{H_2O}(\mu^{PET}_{Bone} - \mu^{PET}_{H_2O})}{1{,}000(\mu^{CT}_{H_2O} - \mu^{CT}_{H_2O})} \qquad (9)$$

where CT denotes the CT image value in HU and the different μ values represent the linear absorption coefficients of bones and water at CT and PET energies, respectively. The following values were used:

$$\begin{aligned} \mu^{PET}_{H_2O} &= 0.096cm^{-1} \\ \mu^{PET}_{Bone} &= 0.172cm^{-1} \\ \mu^{CT}_{H_2O} &= 0.184cm^{-1} \\ \mu^{CT}_{Bone} &= 0.428cm^{-1} \end{aligned} \qquad (10)$$

Thus, an analysis done above enables to define The Image Reconstruction Algorithm of PET/CT scanner as showed in Fig 2.

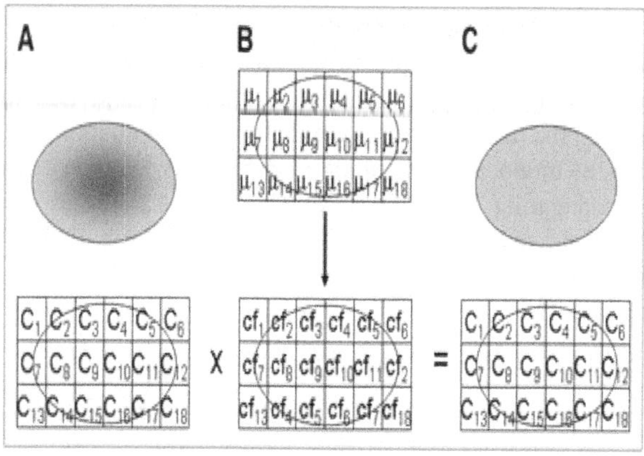

Fig. 2 PET images are corrected by Attenuation maps.

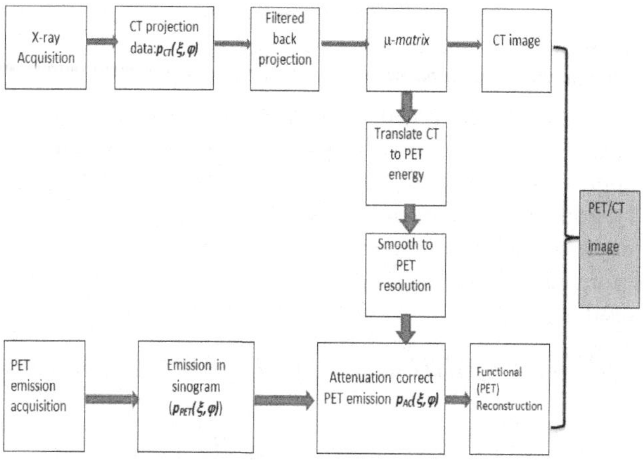

Fig. 3 Image Reconstruction algorithm in PET/CT scanner.

III. Mathematic Modeling and Simulation of PET/CT Scanner' Operation

To appreciate the correctness of Reconstruction algorithm and to illustrate the operation of PET/CT scanner it's mathematic simulation had been realized by software MATLAB.

Mathematic model is built from Reconstruction algorithm presented in Figure 3. Physical processes implemented in model of PET/CT scanner include next stages:

- CT and PET acquisions
- Forming CT projection data and PET sinogram
- Filtered back projection for receiving CT-image and µ-matrice
- Translating CT µ-matrice to PET energy
- Correcting Attenuation on PET emission:
- PET Functional Reconstruction
- Building PET/CT image

All these stages are imitated in software " PET/CT Simulator " with user's interface as follows:

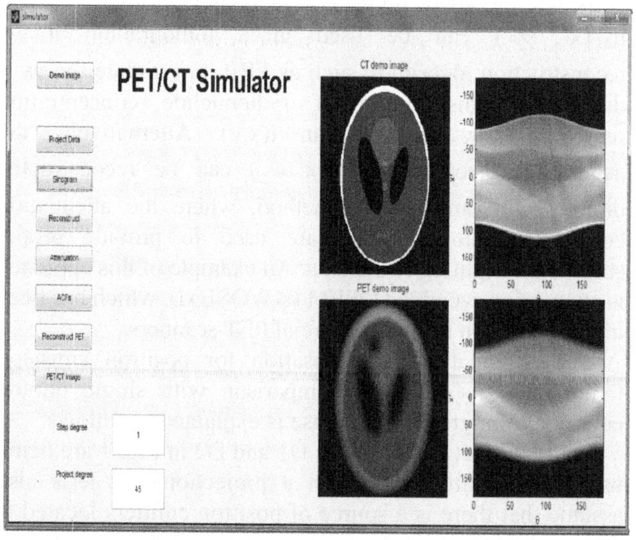

Fig. 4 The user's interface of software" PET/CT Simulator"

IV. Conclusions

- The algorithm and built upon it mathematic model describes properly the Image Reconstruction Principle in PET/CT scanner.
- The software " PET/CT Simulator " is useful for learning and reseaching Reconstruction principles of PET/CT scanner.

References

1. А.С.КРАВЧУК, (1996) *Вычислительная Томография*, Москва, МГАПИ.
2. NGHIA HUYNH LUONG, *"The Algorithm for image reconstruction in CT- Scanne "* , in Proceedings 2002, REV8, Hanoi, pp 369-371.
3. Gopal B. Saha, PhD, (2009), ***Basics of PET Imaging***, Springer New York
4. James A. Patton1 and Timothy G. Turkington2, (March 2008), *"SPECT/CT Physical Principles and Attenuation"*, JOURNAL OF NUCLEAR MEDICINE TECHNOLOGY • Vol. 36 • No. 1, North Carolina.

X-Ray I_O Monitor Device for Primary Intensity Measurement in Computed Tomography (CT) Scanner

Rafay Mehmood Siddiqui[1], Inam Ul Ahad[1], Samreen Amir[2], Bassim Aklan[3], and Tahir Uddin[4]

[1] Sir Syed University of Engineering and Technology, Department of Biomedical Engineering, Karachi, Pakistan
[2] Mehran University of Engineering and Technology, Institute of Information and Communication, Jamshoro, Pakistan
[3] University of Erlangen-Nuremberg, Institute of Medical Physics, Nuremberg, Germany
[4] University of Debrecen, Debrecen, Hungary

Abstract — **For a Computed Tomography (CT) Scanner used as a Non Destructive Testing (NDT) machine, 3D volume is reconstructed mathematically using un-attenuated (primary I_O) and attenuated (secondary I) intensities. Primary intensity (I_O) values are acquired from area of the imaging detector that is not covered by the object being tested. This procedure is prone to errors due to detector artifacts, nonlinear detector behavior, and scattered radiations detection as primary intensities, etc. This study is an attempt to make this procedure efficient and unimpeachable in which an X-ray I_O monitor device is designed and validated that permit precise detection of the primary intensity to obtain better normalization and consequently higher quality image. Using TSL235 photodiode from Texas Instrument the X-ray intensity measured in current is directly converted into frequency which provides high resolution and precise X-ray detection. The device is designed to be easily integrate with an existing CT machine and could be interfaced and read out with a standard Personal Computer (PC) without any need of additional hardware. Furthermore, this device could also be used for other applications, like direct measurement of scattered radiations to apply correction to data set of scan obtained.**

Keywords — **X-Ray I_O Monitor, Non Destructive Testing, Un-attenuated, Intensity Detection.**

I. Introduction

The intensity of light is usually measured using photodiode sensors, which provide proportional current or voltage to input intensity of incident beam. However frequency is a very informative sensor parameter with various advantages over other parameters like current and voltage. Within a wide range of wavelength of incident light, the output frequency signal will be noiseless, thus eliminating need of filters and resulting hardware structure will be simplified. With integrated current to frequency converter many photodiodes (TSL 235 from Texas Advanced Optoelectronic Solutions [TAOS] being particular) provide efficient solution to detect input intensity in the form of frequency output signal [1]. For a Computed Tomography machine, the 3D volume image reconstruction is dependent primarily on un-attenuated and attenuated x-ray intensities. The conservative method for primary intensity detection is utilization of uncovered area of scintillation detector, but naturally the un-covered area of detector is exposed to scattered radiations. In this study designing and implementation of an x-ray I_O monitor device is described which provide efficient method for detection of primary intensity. The most prominent distinction of this design with respect to other x-ray intensity measurement systems is utilization of TSL235 photodiode which provide frequency output signal in proportional to incoming x-ray intensity. The photodiode is coupled with scintillation material which converts x-ray beam into visible light and light is converted into proportional frequency signal by photodiode.

The study is organized as follows. In second section, the x-ray sensing element is discussed with detailed description of conversion of x-ray into frequency signal. In the third section, signal processing of captured frequency signal is discussed. The microcontroller programming for conversion of frequency signal into readable signal is also described in this section. Last but not least, the Graphical User Interface (GUI) which provide user friendly environment to visualize the measured intensities is discussed in section IV. Section V concludes the whole work.

II. Sensing Element

Cesium Iodide (CsI) is efficient scintillation material used widely due to its high light yield. TSL235 photodiode is primarily a light sensor with frequency output. In order to detect x-ray, CsI is coupled with TSL235 to measure x-ray intensity. Use of frequency output sensor in x-ray IOdevice significantly abridges the design process in various ways. These types of sensors are primarily a photodiode with integrated current to frequency converter available as single integrated circuit. The outputs of these sensors are in two formations. In one case a pulse train is generated in direct proportion to input intensity or a square wave in otherwise. As the output signal is frequency, the dynamic rage is not

V. Van Toi et al. (Eds.): 4th International Conference on Biomedical Engineering in Vietnam, IFMBE Proceedings 40, pp. 33–36, 2013.
www.springerlink.com

restricted by sensor supply voltage and prominently noise is unable to make any change in output signal. TSL235 gives high resolution conversion of light to frequency signal without requirement of external components; communicate directly with microcontroller and single power supply operation. An aluminum foil is wrapped over the photodiode to provide the sensor dark environment, so that no visible light incident on it except those resulting from illumination of CsI scintillating material. With low tube voltage and current, lower intensities are critical to measure. In order to increase the sensitivity of the designed monitor, a wide detector area is indispensable. For TSL235, the photodiode window area is 1.36 *mm²*. A prominent advantage of frequency output signal sensor is immunity with noise. Virtually, noise cannot be combined with frequency signal and transformation of measured signal from remote sensors to other components of system is simple. The designed x-ray I_O monitor is multichannel system and concurrently measured data from two sensors can be displayed simultaneously on GUI of designed system. Due to the pulse data format of sensor output signal, more than one sensor can be easily multiplexed into a single microcontroller.

III. MATERIALS AND METHODS

Fig. 1 represents the block diagram of a virtual instrument for x-ray intensity measurements. As discussed, the sensor utilized has a frequency output. The aim is to convert frequency signal into count number, which is based on a Program-oriented Conversion Method (PCM). A PCM is the processor algorithm of measurement, incarnated in the functional-logic structure of a computer or a microcontroller through the software [2]. In this design all circuitry

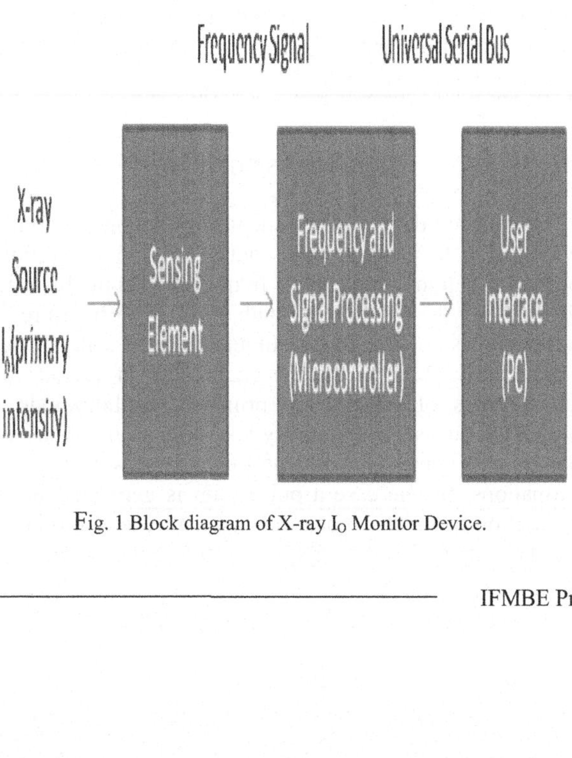

Fig. 1 Block diagram of X-ray I_O Monitor Device.

is embedded. Data is transferred via USB cable into the computer.

PICDEM FS USB Board

To build the complete X-ray I0 monitor, a PICDEM FS USB board has been used because this board provides the compact electronic hardware, to read the sensor output data and a USB interface. The heart of the PICDEM FS USB board is the PIC18F4550, a versatile microcontroller made by Microchip Technology. This board operates at 5V. Power can be drawn directly from the USB cable.

A. Frequency to Count Number Conversion

The TSL235 sensor has the frequency output in the range of 200 kHz to 300 kHz, where the minimum frequency is 200 kHz and the maximum frequency is 300 kHz. As described by fig. 2, the frequency signal proportional to input x-ray intensity is delivered to timer of the microcontroller PIC18F4550, which counts the pulses in administered frequency signal. Concurrently another timer is measuring time period, and hence the counting is performed. Since the counter is 16 bit, maximum 65,536 events can be counted by the device. The frequency count is delivered to HyperTerminal window via USB port.

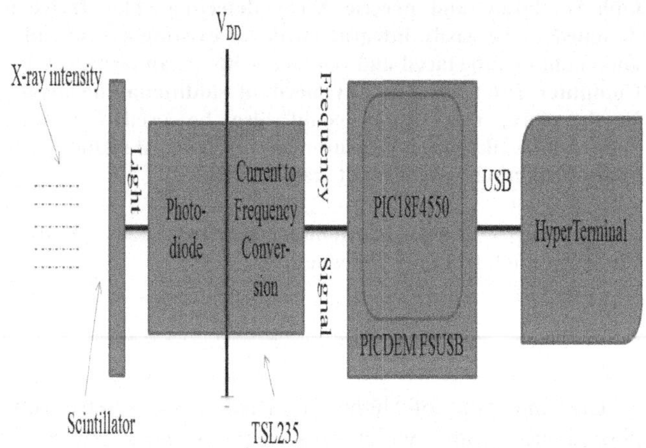

Fig. 2 Flow of x-ray intensity signal with various conversions.

B. Measurement Display

For display of measured x-ray intensity, HyperTerminal window is utilized. The sensor starts taking continuous measurements as soon connected to supply. Once the device is connected to HyperTerminal window, user may press 1 on keyboard, and the measured frequency count of current intensity will be displayed on screen.

IV. Results and Discussion

Before the measurement of x-rays, it is important to see whether the x-ray sensor is producing any output signal in the absence of x-ray beam. Since the photodiode used is very sensitive, even low incident beam of visible light from environment can produce frequency signal.

In fig. 3 it can be observed that with direct exposure of light on designed sensor with no scintillation covering, HyperTerminal window shows random frequency count. Here random readings basically represent the placement of sensor under various light intensity conditions (like direct exposure to lamp, under the table, etc.).

In order to calibrate the sensor for isolation from environment visible light, sensor is completely covered by an aluminum foil along with scintillation material (CsI), so that no visible light ray from environment can enter the detection window of sensor. In order to validate this setup, the output of the device is measured with no exposure of x-ray beam. As observed by fig. 4, the data read by device for sensor TSL235 without x-ray exposure is zero over many incidences.

Once the device is successfully calibrated, the x-ray intensities are exposed on the sensor and measured values are recorded. The tube voltage for this experiment is kept constant at 180kB, distance between the sensor and x-ray tube is 3.5cm, while tube current is varied from 30μA to 38 μA. As seen in fig. 5, the HyperTerminal window depicts frequency count from 2848 to 2856.

The x-ray I_O device is successfully able to continuously read intensities from x-ray sensors, convert it into digital data and transmission of digital data to a PC. The characterization of the x-ray sensor with x-ray I_O monitor device was performed with CT Machine. The result from the above experiments validates the device to be qualified by sensing primary intensities and processing the signal in microcontroller board as the results have been displayed on HyperTerminal window. The frequency count represents primary intensities detected by x-ray sensor. By implementing x-ray I_O device with frequency output, the device facilitates the user with wide range of work operation point and maximum counting, which is 65536. The final results acquired from X-ray I_O device so far are very promising. There is a need for multi channel sensors device which could provide a complete monitoring of primary and secondary X-ray intensities.

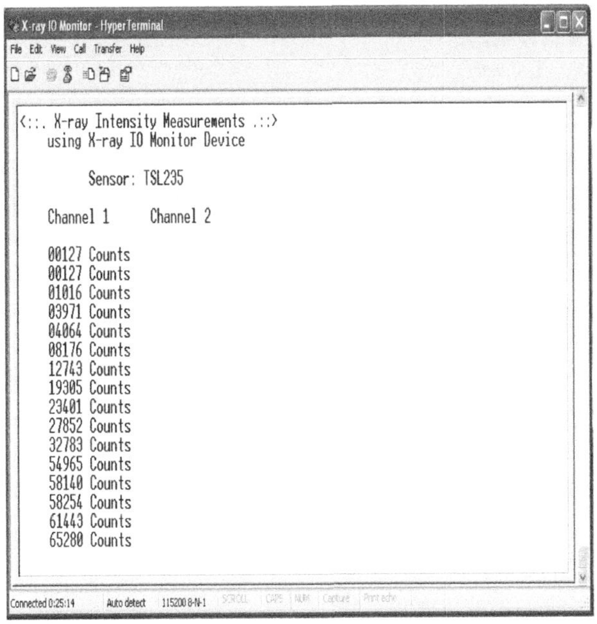

Fig. 3 HyperTerminal Window showing visible light intensity measurements.

Fig. 4 HyperTerminal Window showing measurements from sensor coupled with scintillation material and aluminum foil, without x-ray exposure.

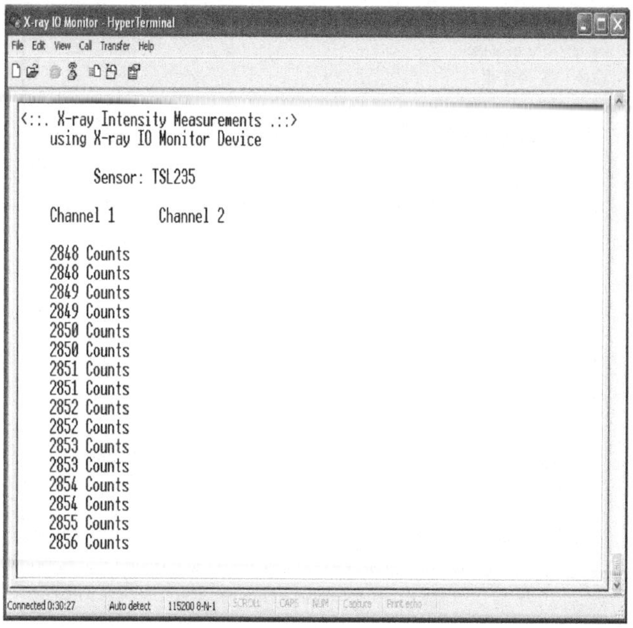

Fig. 5 HyperTerminal Window showing measured x-ray intensities at various tube currents.

V. Conclusions

The described design approach for X-ray I_O monitor and x-ray sensor has a frequency output. The device is successfully calibrated in presence and absence of visible light. The design process could also be simplified by making some improvements like integration of the I_O monitor into the CT machine application software. Thus for each CT-projection acquired by the CT flat panel detector, the primary intensity is measured by the I_O monitor. Furthermore the comparison of the two normalization methods (imaging detector and I_O monitor) could be used to observe the effect of the I_O monitor method on the reconstructed volume.

References

1. Sergey, Y. Yurish (2005) Intelligent Opto Sensors' Interfacing Based on Universal, Frequency-to-Digital Converter, Sensor and Transducer Magazine, Vol 56, Issue 6, 326–334

Nikolay V. Kirianaki, Sergey Y. Yurish, Nestor O. Shpak, Vadim P. Deynega (2002) Data Acquisition and Signal Processing for Smart Sensors. Wiley, England

Author: Rafay Mehmood Siddiqui
Institute: Sir Syed University of Engineering and Technology
Street: University Road
City: Karachi
Country: Pakistan
Email: bmengr@gmail.com

Characterization of X-Ray Sensors and Io Monitor Device Testing for Primary and Secondary Intensities Measurement

Rafay Mehmood Siddiqui[1], Inam Ul Ahad[1], Syedah Sadaf Zehra[1], and Anurag[2]

[1] Sir Syed University of Engineering and Technology, Department of Biomedical Engineering, Karachi, Pakistan
[2] Furtwangen University of Applied Sciences, Mechanical and Process Engineering, Villingen Schwenningen, Germany

Abstract — **This study is an attempt to characterize the specially designed x-ray sensors for the predesigned x-ray intensity monitoring device by evaluating their linearity and sensitivity. Two different sensors are designed using S8559 and S8193 photodiodes from Hamamatsu Photonics, Japan. These photodiodes coupled with Cesium Iodide (CsI) scintillator are used to convert x-ray intensities into variable voltage signals. The linearity and sensitivity of the two sensors are evaluated by finding the operating range of sensors under the same conditions and parameters. This has been accomplished by varying x-ray tube voltage and current and by varying position of sensor from fixed source (x-ray tube). The term sensors characterization indicates investigation of x-ray sensors behavior. For this purpose some experiments are developed in which x-ray sensor and x-ray Io device are examined. Computed Tomography (CT 500) machine which is used in Non Destructive Testing of objects generates the primary and secondary x-ray intensities.**

Keywords — **X-ray Io device, Characterization of X-ray Io device, Sensor Models S8193/S8559.**

I. Introduction

Computed Tomography (CT) is a well known technique employed to test various objects non-destructively. The x-ray intensities produced by CT machine are detected by placing scintillating detector behind object to be tested. The un-attenuated and attenuated x-ray intensities are measured to form 3D volume image. MacroScience Technology GmbH (MST) provides an efficient tool MSX-CT500 for non-destructive testing of lifeless objects [1]. Already designed x-ray Io monitor is a tool developed to measure primary and secondary intensities for such a CT scan to improve the quality of image, which is formed on the basis of such measurements [2].

Ideal characteristics of systems defined during their planning phase in engineering domain are critical, particularly when physical measurements are involved. The characteristics of sensor and the measuring device as a whole are altered by various working field effects leading to create disturbances in operating ranges of sensors and consequently of the measuring device. For an x-ray Io device, designed to measure un-attenuated (primary) and attenuated (secondary) x-ray intensities, the characterization of sensors and the measuring system as whole is crucial to identify the operating range, linearity and sensitivity. Two photodiodes from Hamamatsu Japan S8559 and S8193 are used to develop x-ray intensity sensors [3].

Since the operating range is particularly dependent on working of sensors beside some issues regarding delay involvements by microcontroller and other discrete components, in this paper, the two sensors composed of photodiodes prominently are utilized to characterize not only their own working properties but also for work operation point of the designed device.

This study is organized in a way that section I presents the introduction. Section II describes the materials and methods utilized to characterize the two sensors. Section III depicts the results of the characterization. Conclusion and further possibilities are discussed in last section.

II. Materials and Methods

A couple of experiments have been designed to characterize the sensors and testing of x-ray Io monitor. As explained earlier, the testing of x-ray Io device is dependent upon sensor behavior, it is important to note that the x-ray intensities fall on sensor depend upon the x-ray tube voltage and current. The radiations delivered from x-ray tube (source) to sensor (detector) also effected by the distance between the source and detector. CT500 provides an object inspection manipulator with a table turn for object rotation as shown in fig. 1. A safety switch on each side avoids collision of manipulator with the x-ray tube assembly or the x-ray detector assembly. On top of the x-ray tube assembly are two manually adjustable x-ray cabinet survey cameras. These cameras are used to survey the x-ray cabinet during the inspection of an object. This setup provides the opportunity to vary the distance between the x-ray tube and sensor. Experimental setups designed on the key point that x-ray intensities are directly proportional to x-ray tube current.

V. Van Toi et al. (Eds.): 4th International Conference on Biomedical Engineering in Vietnam, IFMBE Proceedings 40, pp. 37–40, 2013.
www.springerlink.com

These setups comprise of different x-ray intensities measurement conditions. The sensors are placed individually in the inspection manipulator to observe their behaviors. In order to carry out the robust results the sensors were tested under the same conditions and parameters, starting from a reference tube voltage, tube current and distance.

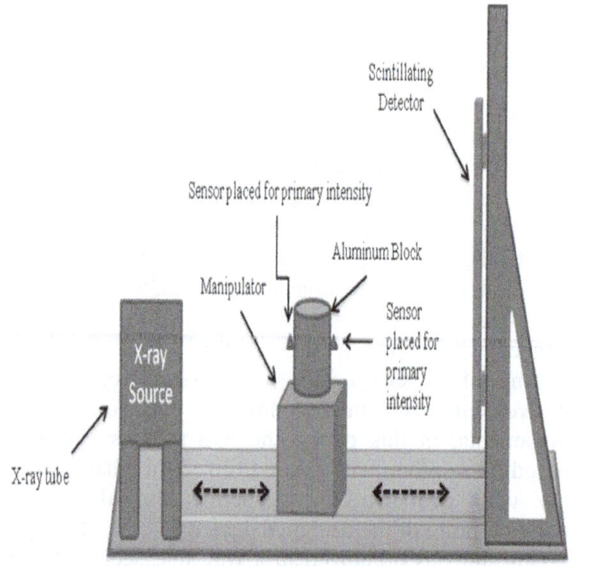

Fig. 1 Experimental setup for characterization of x-ray sensor and designed x-ray Io monitor device.

By keeping two out of three quantities (tube voltage, tube current and distance between tube and sensor) constant, the third quantity is varied. Hence the impacts of all these three factors are observed. In order to get robust results, the two sensors were tested first under same conditions and parameters, and afterwards with different conditions and parameters with reference to tube voltage, tube current and distance.

X-ray Io monitor device provides two type of display for the measured values of x-ray intensities, oscilloscope and HyperTerminal. Since the oscilloscope shows voltage measured proportional to incoming x-ray intensity and HyperTerminal shows decimal count of these voltages, normalization technique is utilized to get consistency in the observations.

III. Results and discussion

In this study six experiments along with results are discussed to characterize the working of sensors (S8559 and S8193).

A. Experiment I

Measurement Conditions: In the first experiment, the tube voltage and distance between the two sensors (placed one by one) kept constant while the tube current varied from 0 μA (starting current) to 400 μA (ending current). Graphical User Interface (GUI) is utilized to measure the incoming intensities.

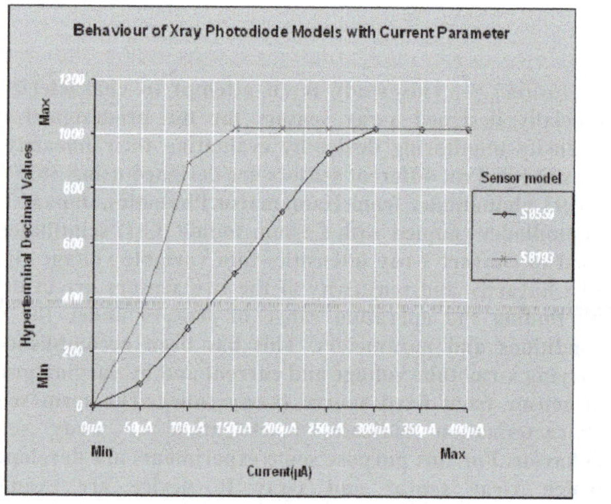

Fig. 2 Sensors behavior with varied current parameters (HyperTerminal Measurements).

Results and Discussion: The results from Experiment I as shown in fig. 2 depict information about maximum saturation working point of two sensor models. S8559 sensor model shows that maximum working point is around 150 μA while the S8193 sensor model demonstrates that maximum working point is around 300 μA. Conclusively, before the saturation region, the two sensors show the directly proportional relationship between the tube current and measured HyperTerminal values which are dependent upon measured signal by sensor hence proved that x-ray intensity is directly proportional to x-ray tube current.

B. Experiment II

Measurement Conditions: The tube voltage and distance between the two sensors (placed one by one) kept constant while the tube current varied from 0 μA (starting current) to 400 μA (ending current). Oscilloscope is used to measure the incoming intensities.

Results and Discussion: Fig. 3 shows information about maximum saturation working point of two sensor models. S8559 sensor model shows that maximum working point is around 150 μA. S8193 sensor model tells that maximum working point is around 300 μA. In conclusion, from above two experiments, it is validated that the behavior of sensors according to maximum operating range on HyperTerminal is exactly similar to that of oscilloscope.

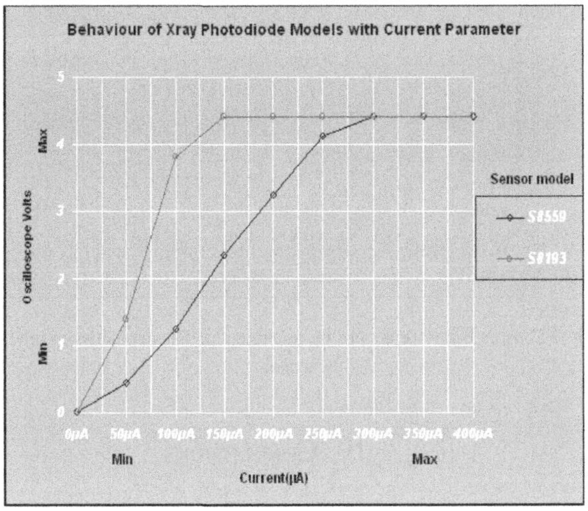

Fig. 3 Sensors behavior with varied current parameters (Oscilloscope Measurements).

C. *Experiment III*

Measurement Conditions: In the third experiment, the measuring conditions are exactly the same as in previous two, however this time; the intensity of x-ray is measured on HyperTerminal and Oscilloscope at same time. The measured values are normalized to have combined view.

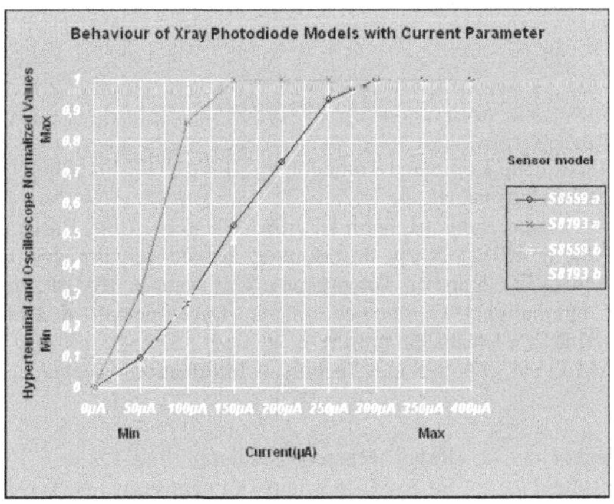

Fig. 4 Sensors behavior with normalized Oscilloscope and HyperTerminal values.

Results and Discussion: Fig. 4 shows information about maximum saturation working point of two sensor models. S8559 sensor model shows that maximum working point is around 150 µA. S8193 sensor model depicts maximum working point around 300 µA. In conclusion, these two sensor models show expected proportional output signal with variation at input intensity.

D. *Experiment IV*

Measurement Conditions: The x-ray Io monitor provides the facility of measuring the un-attenuated (primary) and attenuated (secondary) intensities simultaneously. In this experiment, both intensities are measured simultaneously by placing S8559 directly in front of x-ray tube at distance of 3 cm for primary and by placing S8193 behind the aluminum block which is located in front of x-ray tube. The total distance between x-ray tube and sensor S8193 is 13.5 cm. The tube voltage is 180kV and the intensities measured over current from 0µA to 750 µA. The measurements are taken through HyperTerminal window.

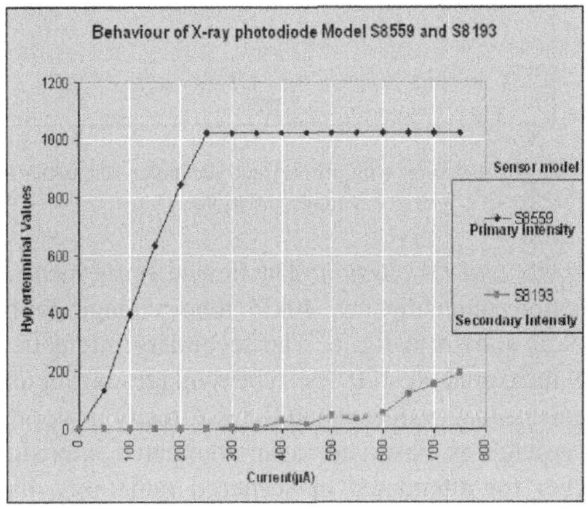

Fig. 5 Sensors behavior with primary and secondary intensities (with tube current variation).

Results and Discussion: It can be observed from fig. 5 that un-attenuated intensities are measured readily by sensor S8559 with increase in current. The saturation region started at 250 µA. The operating range of S8559 is also dependent upon tube voltage, as in previous experiments, the saturation region for S8559 started at 300µA with 120kV tube voltage. Determinedly more the tube voltage less the operating region of the sensor. Moreover the behavior of sensor S8193 shows scattered intensities measurements growing steadily at about 300 µA from 4 decimal counts and reached 200 decimal counts at 750 µA. The low scattered intensity measurements depict need of replacing LT1006 amplifiers with Field Effect Transistor (FET) Amplifiers.

E. *Experiment V*

Measurement Conditions: Now the x-ray tube current is kept constant at 200 µA and voltage varies from 10kV to

225kV. Sensor S8559 placed directly in front of x-ray tube at a distance of 3 cm for measuring un-attenuated intensities and Sensor S8193 is placed behind the aluminum block, which is placed in front of x-ray tube. The total distance of S8193 sensor and x-ray tube is 13.5 cm. the monitoring device used in this experiment is HyperTerminal.

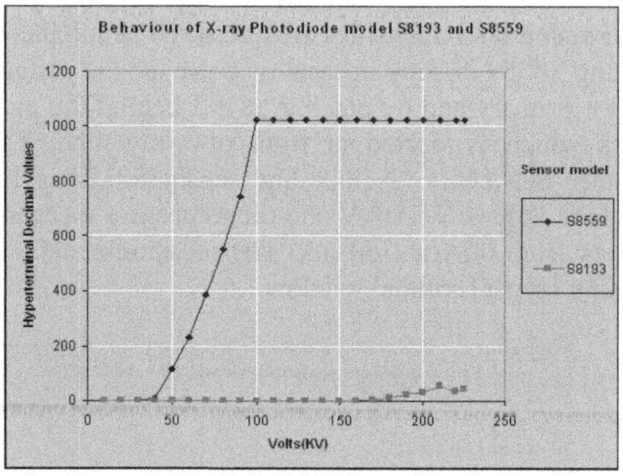

Fig. 6 Sensors behavior with primary and secondary intensities (with tube voltage variation).

Results and Discussion: The device starts measuring un-attenuated intensities at 40kV tube voltage with sensor S8559 as shown in fig. 6. The secondary intensities started to be measured at 170 kV tube voltages. Principally, the HyperTerminal values should be directly proportional to tube voltage as observed in un-attenuated intensities case, however for attenuated or scattered radiations, the sensor depicts very slight non-linear behavior.

F. Experiment VI

Measurement Conditions: Sensor characterization primarily dependent upon distance between sensor and x-ray tube. With increase in distance between source and sensor, the intensity should be decreased. In order to validate the device and sensor over this argument, the tube current and voltage kept constant at 200μA and 120kV respectively and the sensors placed in x-ray moving cabinet. The distance of sensors from x-ray tube varied from 3.5 cm to 33.5 cm and intensities are measured by sensor model S8559.

Results and Discussion: From fig. 07, it is evident that by increasing distance between sensor and x-ray tube, measured intensity depicts inverse relationship with distance.

Moreover, normalized values have been calculated for oscilloscope and HyperTerminal measurements and HyperTerminal shows increased operating region as compare to oscilloscope.

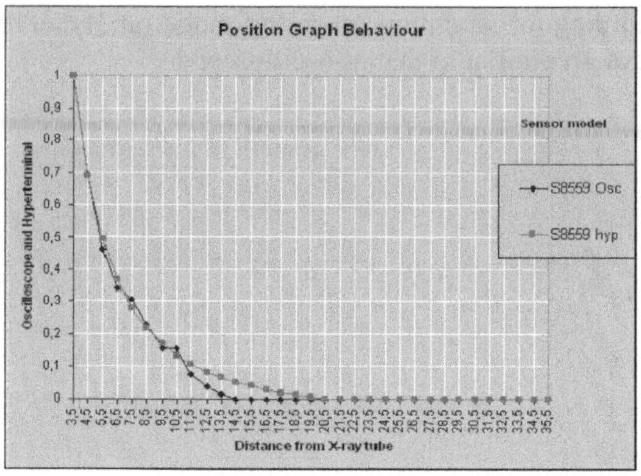

Fig. 7 Sensors behavior according to change in distance between tube and sensor.

IV. Conclusions

Two x-ray sensors were tested under the similar and different measurement parameters and are characterized on the basis of linearity and operating range successfully. Since analog detector circuit is used to detect x-ray intensities, inherent noise and offset disturb the signal accuracy. In the strictest view these drawbacks are unavoidable. Therefore selecting the right operational amplifier for the detector could minimize the noise and offset levels thus increasing signal to noise ratio (SNR). The sensor can be designed to give frequency response as an output instead of a variable voltage then this feature will allow performing a better response in linearity.

References

1. MSXCT500 Computed Tomography.: 1 V01.1, Feb.1, 2008 Operator and service manual.

Rafay Mehmood Siddiqui, Inam Ul Ahad, Samreen Amir, Bassim Aklan, Tahir Uddin (2012) X-ray Io Monitor Device for Primary Intensity measurement in Computed Tomography (CT) scanner, IFMBE Proc., The Fourth International Conference on the Development of Biomedical Engineering in Vietnam, (in press)

HAMAMATSU. Photodiode Technical Information. Technical report, 2002.

Author: Rafay Mehmood Siddiqui
Institute: Sir Syed University of Engineering and Technology
Street: University Road
City: Karachi Country: Pakistan
Email: bmengr@gmail.com

Scheelite Coupled Photodiode X-Ray Sensor Designing and Characterization

Inam Ul Ahad[1], Rafay Mehmood Siddiqui[1], Bassim Aklan[2], and Syedah Sadaf Zehra[1]

[1] Sir Syed University of Engineering and Technology, Department of Biomedical Engineering, Karachi, Pakistan
[2] University of Erlangen-Nuremberg, Institute of Medical Physics - Section for MR Imaging, Nuremberg, Germany

Abstract — **Efficient detection of un-attenuated and attenuated x-rays intensities is crucial to reconstruct 3D volume in a Computed Tomography (CT) machine employed for Non-Destructive Testing. In this study, an x-ray sensor is developed and characterized to be used with an existing X-ray Io device by evaluating its linearity and operating range. The sensor is developed by using Photodiode BPX61 from Siemens coupled with Sheelite (Calcium Tungstate [CaWO4]) scintillator. Sheelite illuminate by exposure of x-rays and BPX61 converts illuminated light into relative current signal. The current signal is amplified and for further processing transmitted to x-ray Io device virtual instrument which consist of a PIC microcontroller and a Personal Computer (PC). As the x-ray intensity depend upon x-ray tube voltage and current, the linearity and operating range of developed sensor is evaluated by performing special experiments in which the two quantities (tube voltage and current) are accreting one at a time while keeping the other constant. Using normalized values obtained from Oscilloscope and HyperTerminal measurements, the conversion time error is observed and reduced thus increasing linear and operating regions.**

Keywords — **X-ray sensor development and characterization, X-ray Io device.**

I. Introduction

For a Computed Tomography (CT) scan system, the data acquisition system is developed on the basis of formation of image. Image formation is fundamentally based on reconstruction algorithm which requires the un-attenuated (I_o) and attenuated/transmitted (I_t) intensities to be calculated. Many manufacturers measure un-attenuated intensity which also called primary intensity, through system calibration while others measure them through uncovered edges of scintillating detector [1]. Both methods are prone to errors and inefficient as slight variation in x-ray tube current/voltage, scattering, attenuation (even in air), and detector artifacts are few reasons from many to name. In order to provide efficient tool for detection of primary (and even secondary) intensity, an x-ray Io monitor device is designed. Since the device primarily dependent upon the working of x-ray sensor, various sensors are developed and characterize to identify the operating range and linearity for the most efficient production. The sensors are primarily composed of a scintillating material, which illuminates in response to x-ray exposure and a photodiode which absorb the illuminated visible light from scintillator and reemit proportional energy in the form of current. In this study, Sheelite scintillation material and BPX61 photodiode from SIEMENS are selected to design the sensor. Various properties of scintillation material and photodiode are discussed along with characterization of sensor through various experiments to identify the operating region and linearity.

II. Previous Work

X-ray Io monitor device is developed to measure un-attenuated and attenuated intensity efficiently [2]. The device consists of x-ray sensor(s), signal conditioning circuit, and display of measured intensities on Personal Computer (PC) through microcontroller and Digital Oscilloscope. X-ray sensor is basically composed of photodiode coupled with scintillation material. Signal conditioning circuit (SCC) is dependent upon the type of photodiode and its driving mode to make the measured signal readable by microcontroller. Microcontroller is used for hardware interface with PC. In this study, this monitoring device is used for characterization of Sheelite coupled x-ray sensor. Since the SCC is altered according to designed photodiode, its construction is also discussed. The signal is processed according to voltage levels produced by SCC, the communication between PC and SCC through microcontroller is also discussed.

III. Sensing Material

As discussed, the efficiency of x-ray Io monitor device is primarily depends upon the competence of sensor used. The sensor is essentially composed of a photodiode coupled with a scintillator as shown in fig. 1. The scintillation material upon exposure to ionizing radiation scintillates reemit energy in visible light form. The light is converted by photodiode into proportional current or voltage signal depending upon mode selected. Efficient detection of incident x-ray intensity is reliant on selection of right scintillator. For semiconductor scintillators, very high purity is a technological requirement as compare to metallic scintillators. In metallic scintillators, organic scintillators produce very fast response as compare to inorganic scintillators, however in this project, inorganic scintillator are more advantageous as they have high efficiency for x-ray absorption.

Correspondingly the photodiode selection is important to efficiently detect illuminated light from scintillator and to produce proportional signal. BPX61 photodiode from SIEMENS is selected which is a silicon PIN photodiode, with spectral sensitivity range 400 to 1100 nanometer (nm), possess large radiant sensitive area i.e. 7 mm^2 and its maximum output signal is 30nA.

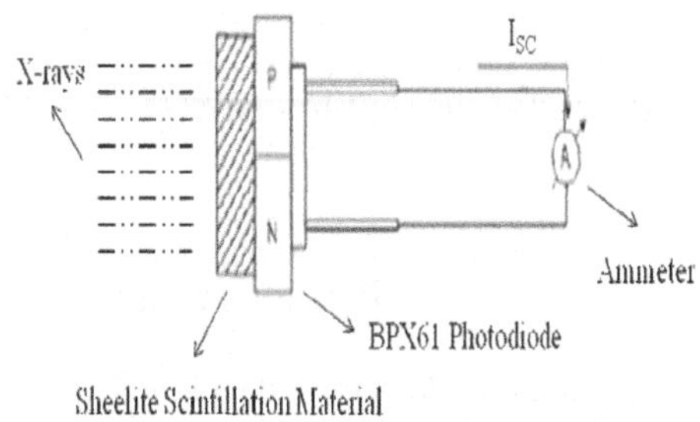

Fig. 2 Construction of an x-ray sensor.

IV. Signal conditioning circuit

The output of scintillator coupled photodiode sensor is a current signal commonly known as photocurrent (I_{sc}). Signal conditioning circuit (SCC) transforms the range of photocurrent at its input into proportional voltage levels. SCC is rather simple in composition.

The x-ray intensity is converted into proportional current signal through scintillator coupled BPX61 photodiode. The current output of sensor is delivered to SCC to be converted into voltage signal as the input of microcontroller requires voltage levels. The input range of microcontroller is 0 to 5 V, thus signal conditioning circuit is constructed in a way to produce 0 volts at minimum sensor output current and 5 volts at maximum sensor output current. In order to accomplish this task, a fundamental transfer function is developed according to the model of SCC. This transfer function limit output of SCC from 0 V to 5 V by utilizing following relationship.

$$V_O = I_{sc} R_f \tag{1}$$

Where

V_O	=	Voltage output
Rf	=	Feedback resistance
Isc	=	Photocurrent

V. Signal processing and communcation

Once the voltage signal presenting x-ray intensity values is scaled to 0 to 5 V, the voltage levels are delivered to digital oscilloscope and microcontroller's Analog to Digital Conversion (ADC) port. The system is designed to be multichannel, so that two sensors could be attached to system at same time. The microcontroller is programmed to convert these voltage levels (which are finally in digital form), into proportional decimal code. The decimal code is also scaled from 0000 to 1023 units. The decimal values proportional to incoming measured x-ray intensity are displayed on HyperTerminal Window.

VI. Results and disucssion

Linearity and sensitivity of the sensor is evaluated by performing specially designed experiments in which tube current and voltages are escalated individually by keeping other parameters constant. Thus the operating regions, saturating fields and linearity are observed depicting behavior of sensor under different environment (x-ray tube voltage/current values). Furthermore the normalized values for measured signals are calculated by employing oscilloscope and HyperTerminal simultaneously. These values are used to calculate conversion error depict by PIC controller while consuming time in analog to digital conversion.

A. Operating Range and Linearity (Varied Current)

In the first experiment the operating range, linearity and saturation regions have been identified on the basis of varied current. The tube voltage is kept constant at 180kV and distance between tube and sensor is kept 3.5 cm. The tube current is varied from 0μA to 1000μA. Intensity measurements are taken on oscilloscope and HyperTerminal simultaneously and are normalized from 0 to 1.

From the results of first experiment as shown in fig. 2, it is obvious that Sheelite coupled x-ray sensor provide a wide range of operation from 0μA to 1000μA over Oscilloscope and 100 μA to 1000μA over HyperTerminal window and no saturation region is found over the whole range of tested tube current. For the current below 100μA, the HyperTerminal shows zero intensity signals. This limits the operating range of sensor over HyperTerminal, and this limitation is present due to conversion error. For characterizing the linearity of the sensor operation, the correlation coefficient (R-squared) is calculated for HyperTerminal values, yielding 0.976, which pretty much depicts the linear behavior of designed sensor.

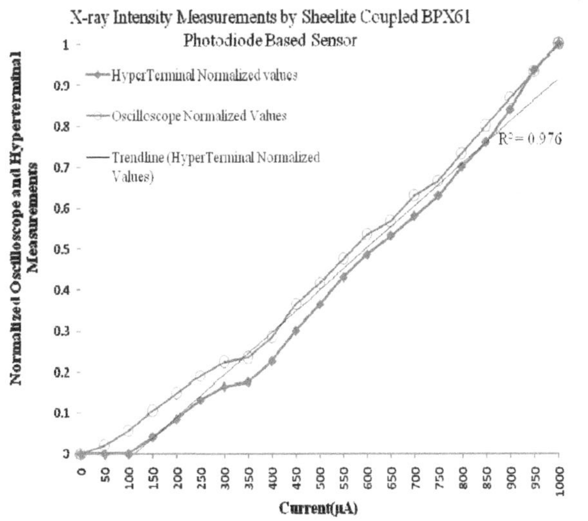

Fig. 2 Normalized Oscilloscope and HyperTerminal Window values for designed sensor (with conversion error).

B. Operating Range and Linearity (without Conversion Error)

In the second experiment, the operating range for HyperTerminal window is increased by the removal of conversion error. Remaining configuration for the experimental setup remains same as that of previous experiment, however the tube current is only varied till 850μA.

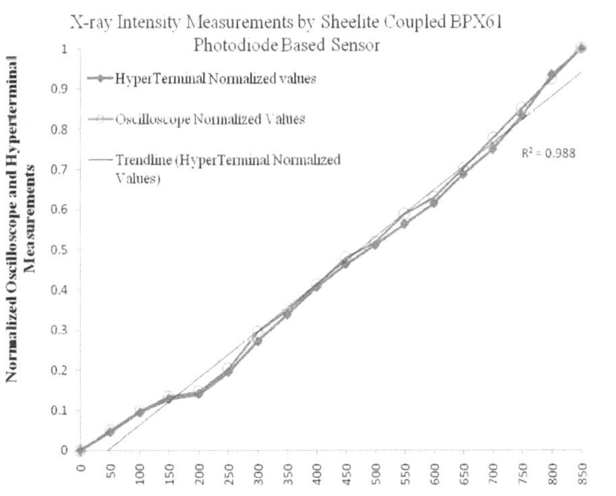

Fig. 3 Normalized Oscilloscope and HyperTerminal Window values for designed sensor (without conversion error).

Results for second experiment as shown in fig. 3, depicts increased operating region. For HyperTerminal window, now the intensity measurements start right from 50μA. moreover the linearity of sensor operating region is also increased as correlation coefficient value is now 0.988. The values from oscilloscope remain same. It is worthy to note that the sensor shows no saturation behavior throughout the current range from 0μA to 850μA.

C. Oscilloscope and HyperTerminal Window Comparison

From above two experiments, it is obvious that even the normalized values of oscilloscope and HyperTerminal values are different to a very little extent. The values recorded originally by oscilloscope were in voltage units and HyperTerminal window values were in decimal code. Since both measurements corresponds to incoming x-ray intensity, their trend on graph is approximately same. In order to validate linearity between them, the measurements from both displays are normalized and plotted on graph to calculate the correlation coefficient.

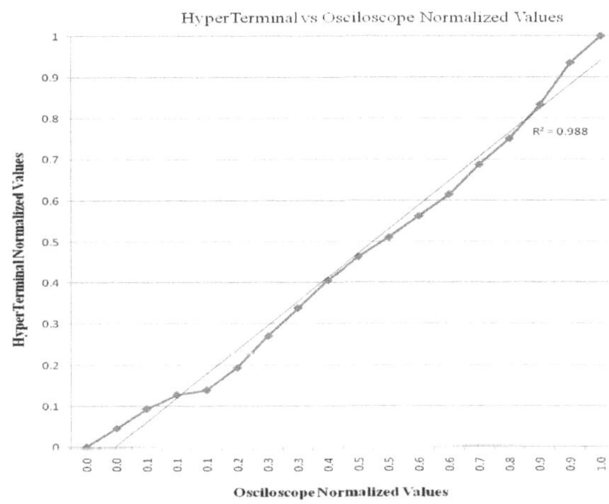

Fig. 4 Normalized Oscilloscope and HyperTerminal Window measurements comparison.

As shown in fig. 4, the calculated correlation coefficient for oscilloscope HyperTerminal measurements is 0.988. Although oscilloscope measurements would be more accurate as compare to HyperTerminal decimal code, still the transmitted intensity also measured through microcontroller, so the net effect is assumed to be zero.

D. Operating Range and Linearity (Varied Voltage)

The characteristics of sensor with respect to operating range, linearity and saturation region also dependent upon x-ray tube voltage. In order to identify sensor behavior on basis of varied tube voltage, the tube current is kept constant at 200μA and distance between x-ray tube and sensor is kept 3.5cm. Now the tube voltage is varied from 0 kV to 210 kV. Oscilloscope and HyperTerminal are used to record the measured values.

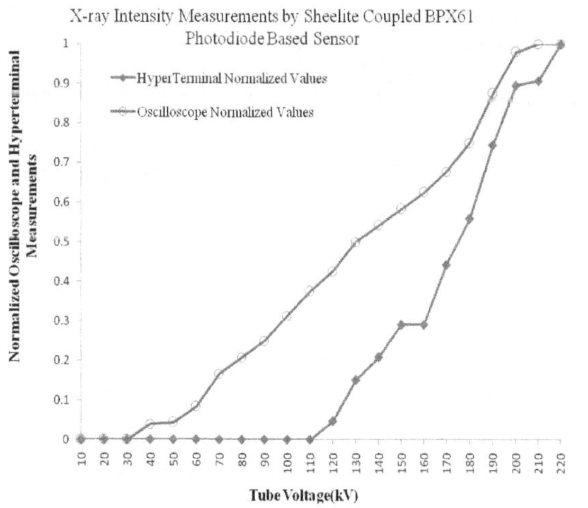

Fig. 5 Designed sensor measurements over oscilloscope and HyperTerminal with varied voltage.

Fig. 5 shows the results of varied tube voltage experiment. The operating region for sensor on HyperTerminal window is much limited. The operating region for oscilloscope is somewhat limited in this experiment though depicts very good linear region between 60kV to 200kV. After 200kV, it seems to notice existence of saturation region. Conclusively the current variation experiment shows more operating and linear regions as compare to behavior of sensor over voltage variation.

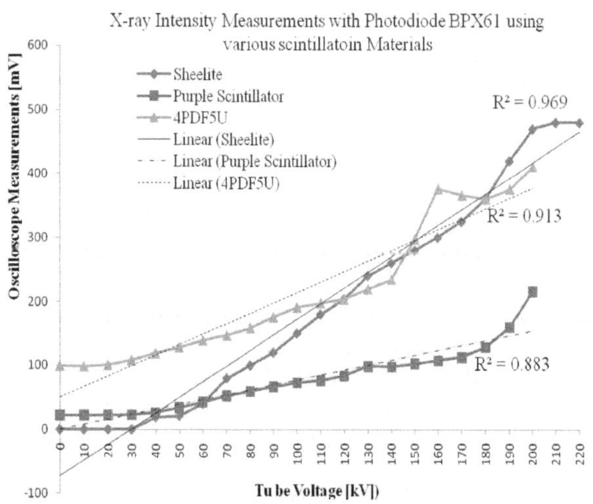

Fig. 6 Characterization of various sensors with respect to operation.

E. *Comparison of Sheelite with Other Scintillation Materials*

Sheelite scintillator is compared with other scintillators to characterize its performance with respect to others. The 4PDF5U and Purple scintillators are used to be coupled with photodiode BPX61 in this experiment. The tube voltage is varied from 10kV to 210 kV while tube current is kept constant. Fig. 6 presents measurements of all three sensor models. It can be concluded that although 4PDF5U and Purple scintillators depicts more operating ranges as compare to Sheelite, however the operating region of Sheelite scintilator (with correlation coefficient = 0.977) is more linear as compare to operating regions of 4PDF5U and Purple (with correlation coefficients 0.913 and 0.883). The correlation coefficient were calculated with respect to oscilloscope measurements and tube voltages.

VII. Conclusions

Considering normalized values of HyperTerminal window, it can be observed that by changing the tube current while keeping constant tube voltage and constant distance between sensor and source, significant linear region exists within operating ranges. Similar observations are obtained by changing voltage while keeping current and distance constant. However the behavior of sensor is more non-linear at different tube voltages as compare to varying tube current. Furthermore the normalize values for measured signals are calculated by employing Oscilloscope and HyperTerminal simultaneously and concluded that analog to digital conversion of signal consume remarkable time. This time error provides delayed starting point for sensor operation. The conversion time error is reduced by changing amplifier resistance values and thus better results are obtained as discussed in second experiment. The results depicts that the sensor designed using Photodiode BPX61 could be efficiently used within certain linear operating range with a CT machine to detect primary (un-attenuated) intensity.

References

1. Bronzino, Joseph D. (2000) The Biomedical Engineering Handbook, CRC Press, United States of America
2. Rafay Mehmood Siddiqui, Inam Ul Ahad, Samreen Amir, Bassim Aklan, Tahir Uddin (2012) X-ray Io Monitor Device for Primary Intensity measurement in Computed Tomography (CT) scanner, IFMBE Proc., The Fourth International Conference on the Development of Biomedical Engineering in Vietnam, (in press)

Author: Inam Ul Ahad
Institute: Sir Syed University of Engineering and Technology
Street: University Road 75300
City: Karachi
Country: Pakistan
Email: engrinam@gmail.com

Bending Stiffness of IM Nails for Orthopedic Applicatins: A Comparative Study between 12 Specimen Group from 9 Established Manufacturers

A. Ahmady, S.H. Safavai, and N.A. Abu Othman

Department of Biomedical Engineering, Faculty of Engineering, University of Malaya, Kuala Lumpur, Malaysia

Abstract — **Bending stiffness of 12 specimen groups of Stainless steel and Titanium IM nails from 9 established manufacturers were measured and compared. With this respect, a rapid literature review is conducted. Further, the difference of stainless steel and titanium bending stiffness is discussed. The results are provided as a reference for manufacturers and regulatory authorities as an indication for effectiveness and safety of IM nails.**

Keywords — **Biomedical, Biomaterials, Intramedullary nails, IM Nails, Bending stiffness.**

I. Introduction

Intramedullary nails (IM Nails) are used as internal supports to fix major bone fractures. However, this application requires the addition of long screws. Many designs are accessible, from solid to cylindrical, with shapes like diamond, cloverleaf, and slotted cylinders [1]. In recent years, these devices were modified to their present and more rotationally stable clover-leaf shape. In terms of material, there are mostly two types of these devices available in the market comprising stainless steel and titanium. Most recently majority of orthopedic prosthesis manufacturers are introducing stainless steel types into the market. Although studies show many advantages for stainless steel, this tendency may be due to the cheaper cost of this material in comparison with titanium, taking care to this fact that 90% of road traffic deaths occur in developing countries [2]. This study compares bending stiffness of different existing products on the market concentrating on titanium and stainless steel IM nails. This is to give an industrial standard value for bending stiffness according to their material type and diameter; apart from their anatomical position.

II. Methods

In this study bending stiffness of 12 specimen group from 9 established manufacturers are investigated. Each specimen group comprised 5 samples meaning that totally 60 samples were tested. The reason to choose bending stiffness to study, relies on this fact that bending stiffness is one of the most important indication of mechanical characteristic of IM nails and also its importance compared with other mechanical specifications in cases which loading on IM nails is significant [3,4]. According to the test result, we will discuss on different aspects which may result this difference and its impact on bending stiffness as a matter of safety.

ASTM F1264 standard was used to measure the bending stiffness of IM nails. Figure 1 shows the system setting for measurement of bending stiffness according to mentioned standard.

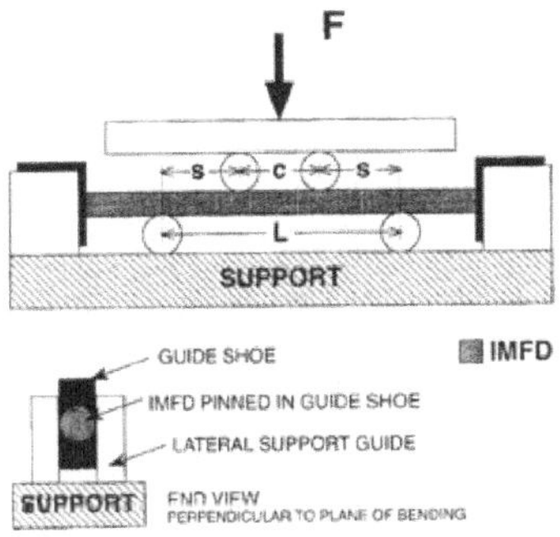

Fig. 1 Point Bent Test with Guide Shoes [3].

The bending stiffness could be computed by following expression:

$$EI_e = \frac{s^2(L+2c)\left(\frac{F}{y}\right)}{12}$$

or

$$EI_e = \frac{s^2(3L-4s)\left(\frac{F}{y}\right)}{12}$$

where F/y is the slope of the elastic portion of the load displacement curve, s is the span from a load point to the nearest support, c is the center span, and L is the total span (c + 2s).

III. Results

Table 1 shows the average bending stiffness of 12 different specimen groups of stainless steel IM nails distinguished by company number, anatomic location, and diameter. Also, this value for a titanium group specimen is put forward in mentioned table. Furthermore, figure 2 shows the graph of bending stiffness when diameter increases.

V. Van Toi et al. (Eds.): 4th International Conference on Biomedical Engineering in Vietnam, IFMBE Proceedings 40, pp. 45–46, 2013.
www.springerlink.com

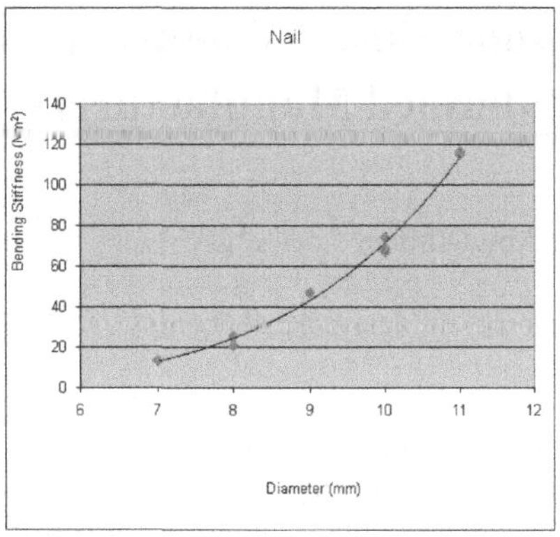

Fig. 2 Bending Stiffness of Stainless Steel IM nails.

Results show bending stiffness of different IM nails consisting Tibia, Femor and Hemorus manufactured by several producers has same values with respect to the bending stiffness. Taking into account that bending stiffness depends on the material and cross-sectional shape of product, presents this fact that all manufactures follow a slightly unique path way for design of their products.

Table 1 Average of Bending Stiffness of Experienced IM nails.

Row	*Manufacturer*	*Anatomic Location*	*Diameter (mm)*	*Bending Stiffness (Nm2)*
1	Company No.1	Humerus	8	20.57±0.24
2	Company No.1	Femure	11	114.67±0.61
3	Company No.1	Tibia	10	74.11±0.21
4	Company No.2	Tibia	9	46.88±0.61
5	Company No.3	Tibia	10	66.95±0.51
6	Company No.4	Humerus	7	13.32±0.64
7	Company No.5	Femure	10	68.31±0.66
8	Company No.6	Femure	11	116.19±2.42
9	Company No.6	Femure	8	25.45±1.56
10	Company No.7	Trauson	11	115.74±0.69
11	Company No.8	Humerus	7	13.16±0.46
12	Company No.9	Tibia	9	26±0.35

Test of the group of 9mm titanium tibia IM nails showed 26±0.35N.m^2 bending stiffness which is almost 50% less than that of stainless steel one. As it was presented before, bending stiffness of different nail brands rise with respect to the diameter increase on a curve. When it came to the test of a titanium tibia nail, however, the result showed less bending stiffness.

IV. Disscousion

Although this study suffers from lack of study on other sample diameters rather than 9mm, its result could raise concern on the safety issue of this type of IM nail. It should not be ignored that apart from different materials which is used to produce implant, logically, same performance and safety is expected. In contrast, the same performance may be achieved by considering different indication for use, meaning that titanium nail implants might be useful for special group of patients. Although some studies demonstrate that titanium nails increase biomechanical stability compared to stainless steel nails [5,6], need for comparative clinical study to follow performance of titanium IM nails in compare with stainless steel one is evident.

V. Conclusion

Bending stiffness of 12 specimen groups of Stainless steel and Titanium IM nails from 9 established manufacturers were measured and compared. The results provide a reference for manufacturers and regulatory authorities as an indication for effectiveness and safety for IM nails. However, more clinical investigation for mechanical safety evaluation of titanium in comparison with stainless steel is suggested.

References

1. Timothy Witzigreuter, "The Origin, History and Use of the Intramedullary Nail," National Association of Orthopaedic Technologists. http://www.naot.org/onlineceusdec2005.html, 2008.
2. Margie Peden, "World report on road traffic injury prevention: summary," 2004.
3. Standard Specification and Test Methods for Intramedullary Fixation Devices.: ASTM F1264, 2000.
4. AK Bhat, SK Rao, and K Bhaskaranand, "Mechanical failure in intramedullary interlocking nails," vol. 4(2), pp. 1:138-41,
5. Lee SS, Lalonde FD, Impelluso T, Newton PO Mahar AT, Biomechanical comparison of stainless steel and titanium nails for fixation of simulated femoral fractures.: J Pediatr Orthop, 2004.
6. Mahar AT, Lee SS, Newton PO. Fricka KB, Biomechanical analysis of antegrade and retrograde flexible intramedullary nail fixation of pediatric femoral fractures using a synthetic

Author: Amir Ahmady
Institute: Department of Biomedical Engineering, Faculty of Engineering, University of Malaya
City: Kuala Lumpur
Country: Malaysia
Email: amir.ahmady@hotmail.my

Retracted: A Pipelined Analog – To – Digital Converter (ADC) Using Umc 0.25μm Technology for Pacemaker Application

Phan Vo Kim Anh and Hoang Trang

University of Technology, VietNam National University at HoChiMinh City, HoChiMinh City, VietNam
pvkanh@hcmut.edu.vn, hoangtrang@hcmut.edu.vn

Abstract — **Electronics used in medical applications attract lots of researches on it, consequently more patients will be treated and have a healthy life. The analog-to-digital converter (ADC) is the inter-connection between the analog and digital part of a device, plays an important role transferring signals in real life to signals in digital domain. One requirement for this ADC is how to reduce power dissipation that can afford devices using battery source. A MDAC (Multiplying Digital to Analog Converter) using charge-pump capacitor technique is introduced for pipelined ADC topology. Compared with the switch-cap technique, MDAC using charge-pump technique consumes less power, hence reducing the total power used in the device. This MDAC is proposed for 1.5-bit stage pipelined ADC, using UMC 0.25μm technology, 2.5V supply and 0.5Mhz sampling rate.**

Keywords — **Analog-to-digital converter (ADC), pacemaker, low-power, MDAC, charge-pump.**

I. INTRODUCTION

First invented in 1950s, pacemaker is now used widely for treating arrhythmias. Arrhythmias are problem with the rate or rhythm of the heartbeat. During arrhythmias, the heartbeat may be too slow, too fast, or with abnormal rhythm. Pacemaker device helps people who have irregular heart rhythms remain an active lifestyle.

A pacemaker can be divided into four major function blocks: sensing system (amplifiers, filter and ADC); output (high voltage multiplier and high voltage pulse generator); battery management system and algorithm therapy controls. In this research, we concentrate on building an ADC block that appropriates to pacemaker usage.

The primary goal for pacemaker design is low-power dissipation. The battery should work more than ten years, and all techniques used have to be considered carefully to meet this requirement that is also the aim of designing ADC block in this work.

There are varieties of ADC architectures, such as Flash ADC, SAR ADC, Delta-Sigma ADC… In this paper, the pipelined ADC architecture is used due to its high sampling-rate, high resolution suitable for applications that require exactness. Moreover, the speed of pipelined ADC doesn't rely on the number of stages, and it only depends on one stage delay. Since those reasons, pipelined ADC is usually used in medical devices.

This paper is organized as follows. The overview of pipelined ADC is given in section II. After that, we concentrate on MDAC block in section III which is one of the main blocks in one pipelined stage. A charge-pump structure is proposed in section IV. Next section shows simulation results. Conclusion is given in the last section.

II. PIPELINED ANALOG-TO-DIGITAL CONVERTER

A pipelined ADC has many stages connected in series, the resolution of each stage depends on the whole ADC resolution and the number of stages which is decided by the designer. In this research, a 1.5-bit stage is used for low-power reason [1][2].

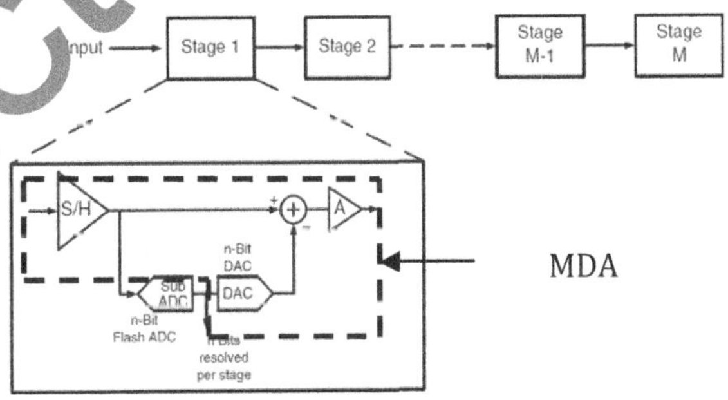

Fig. 1 M-stage Pipelined ADC

Fig. 1 shows an overview of one M-stage pipelined ADC and the structure of each stage. Each stage of a pipelined ADC consists of one sub-ADC and one MDAC – a circuit which can plays the role of a sample-and-hold, a DAC, a subtractor and an inter-stage amplifier.

The operation of a pipelined stage consists of four steps. First, the input is captured by sample and hold amplifier. Second, this signal is quantized by the sub-ADC, which produces a digital output. Third, this digital signal comes to the sub DAC which converts it to analog signal. This analog signal is subtracted from the original sampled signal, leaving a residual signal. Fourth, this residual signal is increased to full-scale by the inter-stage amplifier, and passed to the next stage.

One important requirement of an implanted medical device like pacemaker is how to reduce power dissipation so

V. Van Toi et al. (Eds.): 4th International Conference on Biomedical Engineering in Vietnam, IFMBE Proceedings 40, pp. 47–49, 2013.
www.springerlink.com

that the battery source can ensure the device work well as long as possible, hence, each part of the design has to be consider carefully to satisfy this requirement, and ADC is not an exception.

In each stage of a pipelined ADC, MDAC is the main part, which consumes large power, so it attracts many researchers working on it in order to find a perfect solution for power saving problem.

III. MULTIPLYING DIGITAL TO ANALOG CONVERTER (MDAC)

In an MDAC, the multiplying and subtracting circuit are concentrated to improve for low-power reason. Two main techniques used commonly are switch-cap and charge-pump.

Output voltage of an MDAC is:

$$v_o = \begin{cases} 2v_i & v_i < 3/8\, v_{ref} \\ 2v_i - 1/2\, v_{ref} & 3/8\, v_{ref} < v_i < 5/8\, v_{ref} \\ 2v_i - v_{ref} & v_i > 5/8 v_{ref} \end{cases} \quad (1)$$

With v_i is the input and v_o is the output of a stage.

Or we can summarize:

$$v_o = 2v_i - v_{dac} \quad (2)$$

With v_{dac} is the output of sub-DAC stage, and can have the value 0, $1/2v_{ref}$, v_{ref} relied on the output of sub-ADC. Fig.2 shows the equation (2) that is transfer function.

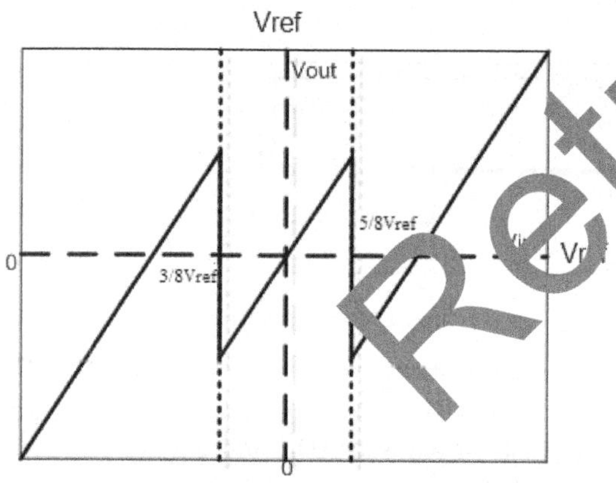

Fig. 2 Transfer function of 1.5-bit stage MDAC.

A. Switch-Cap Circuit

Fig.3 presents the Switch-cap circuit used in this work, with A_f is OP-AMP gain

$$v_o = \frac{1}{1+\frac{1}{A_f}}\left[\left(1+\frac{C_S}{C_F}\right)V_{in} - \left(\frac{C_S+C_F}{C_F}\right)V_{dac}\right] \quad (3)$$

When $A_f \rightarrow \infty$ and $C_S = C_F$, $v_o = 2V_{in} - V_{dac}$. However, C_S cannot exactly equal to C_F as well as A_f cannot be unlimited, it is hard to get the correct v_0. Besides, OP-AMP consumes large power due to on – off transitions that create thermal noise.

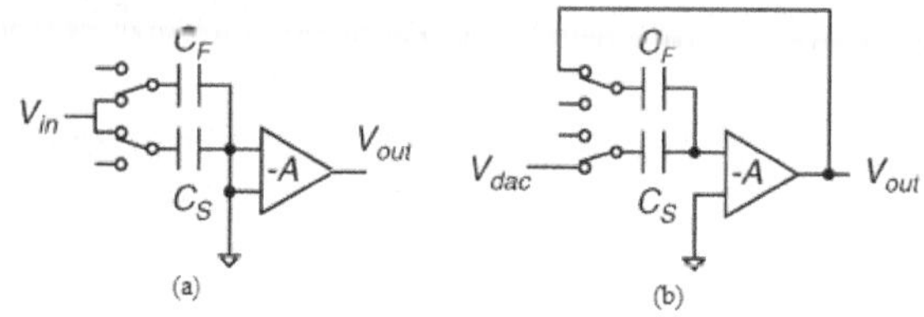

Fig. 3 Switch-cap circuit (a) Sample phase (b) Processing phase.

B. Charge-Pump Circuit

This topology consist only switches and capacitors, hence power usage is reduced a lot (as in Fig.3).

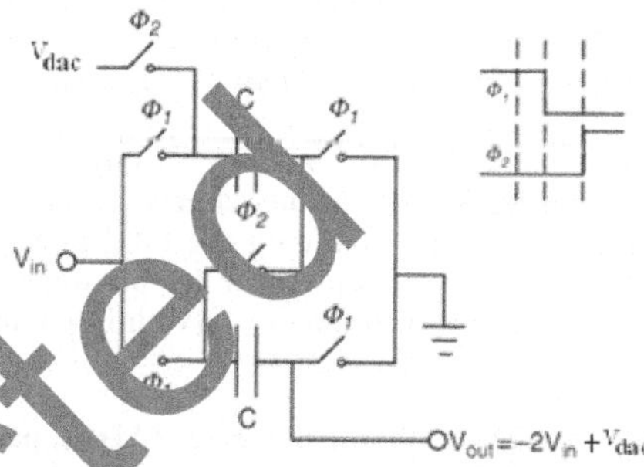

Fig. 4 Charge-pump circuit.

One disadvantage remain here is the parasitic capacitor effect which exists while switching at high frequency. In order to overcome this problem, a dummy switch is used and is presented in next section.

IV. CHARGE-PUMP TOPOLOGY

The parasitic capacitor existing in the circuit while operating at high frequency has two disadvantages. First of all, it consumes more power, and the second, it affects the exactness of the circuit. Eliminating this parasitic capacitor is one important requirement while designing an MDAC.

One solution for this problem is using a dummy capacitor which is equal to the parasitic capacitor.

When the switch is changing from *on* to *off* state, the charge from parasitic capacitor will be transferred to C_d instead of C (as in Fig.5). It helps to ensure that the output voltage is not affected by the parasitic capacitor.

This research introduces a structure that has charge injection self-cancellation ability. The operation principle bases on the symmetric configuration, one switch is the main switch in a cycle while it is dummy switch in the others. Fig.6 shows the charge-pump structure which is used in an MDAC, all switches here are NMOS transistors and have equal area.

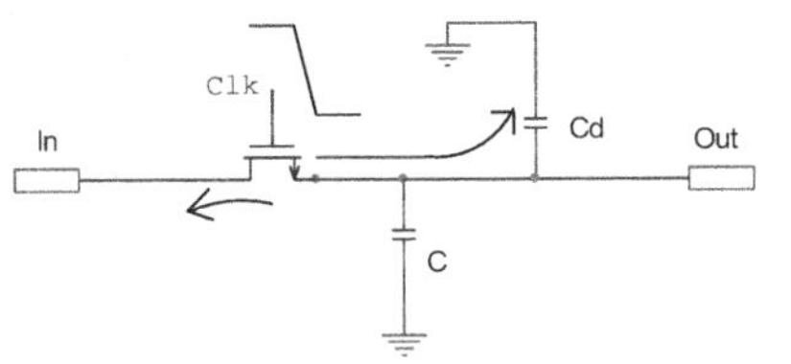

Fig. 5 Discharging through dummy capacitor.

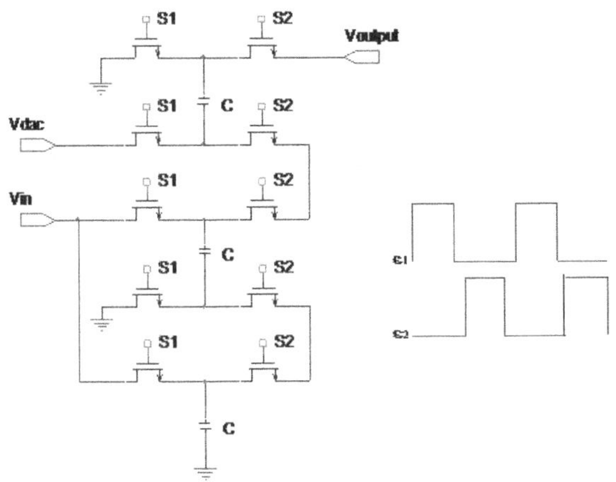

Fig. 6 Proposed symmetric charge-pump structure.

The circuit operation in Fig.6 consists of two phase. In the first cycle, S1 is on while S2 is off, V_{in} and V_{dac} charge three capacitors. In the second cycle, switch state is revised, and V_{output} is the total charge which is contained in three capacitors in previous cycle.

When S1 change from *on* to *off* state, it starts discharging while S2 starts charging due to changing from *off* to *on* state. The amount of charge is equal to the amount discharge because S1 and S2 are alike. The parasitic capacitor on S1 transfers to S2, supply charge for S2 in charging cycle and vice versa. Consequently, in first cycle, S1 is the main switch and S2 plays the role of a dummy switch, and in the second cycle, S1 in turn is the dummy switch for the main switch S2.

Through upper analysis, this proposed structure has eliminated the effect of parasitic capacitor, which is called "charge injection self-cancellation".

V. Simulation Results

The simulations are aimed to test the operation of the proposed MDAC. These simulation results are done by Cadence tools, with UMC 0.25um technology (as in Fig.7, Fig.8, and Fig.9). The input V_{in} has been sampled, V_{in} and V_{dac} amplitude varies so that three output regions in Equ. (2) are all tested (V_{ref} are chosen to be 1V).

The simulation shows that the proposed structure works as desired, the results are right with Equ.(2).

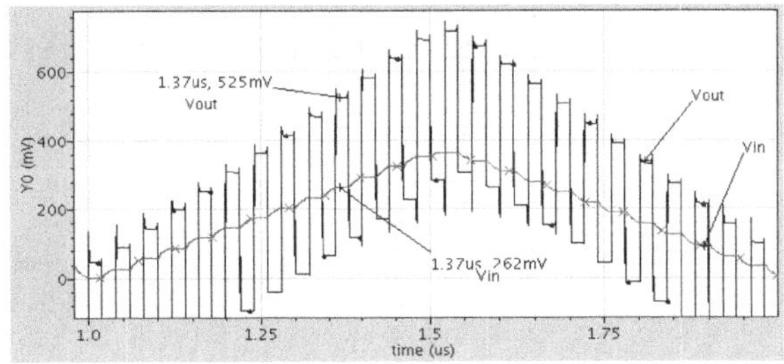

Fig. 7 $V_{in} = 0 – 375$ mV; $V_{dac} = 0V$.

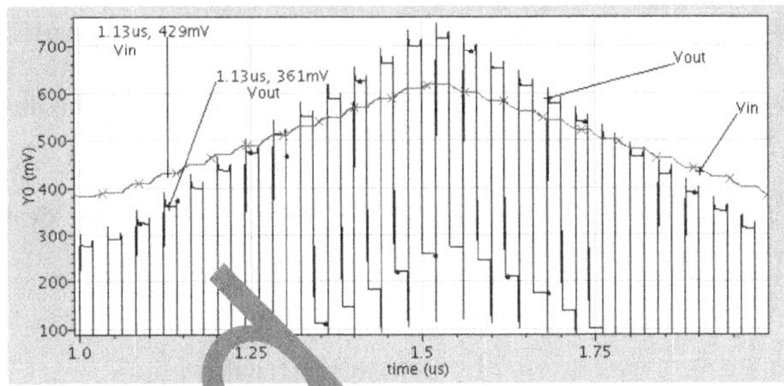

Fig. 8 $V_{in} = 375 – 625$ mV; $V_{dac} = 0.5V$.

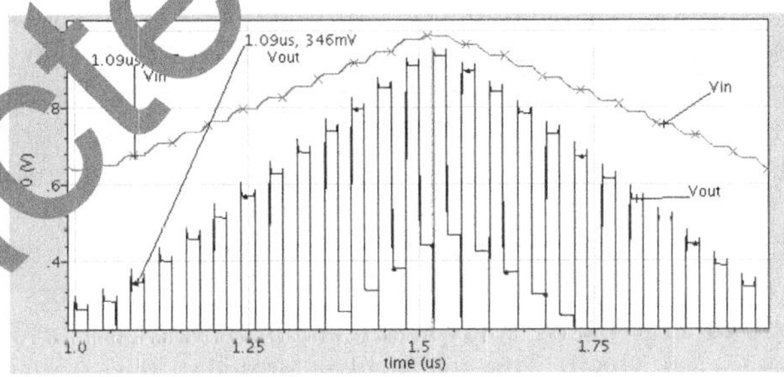

Fig. 9 $V_{in} = 625mV – 1V$; $V_{dac} = 1V$.

VI. Conclusions

A charge-pump MDAC topology has been introduced. It has symmetric configuration that can eliminate the effect of parasitic capacitor, so the power dissipation is reduced and remain the circuit's exactness. Moreover, this topology is simple, only contains capacitor and NMOS transistor act as a switch, hence the total power used is much smaller than the common switch capacitor. Since these reasons, this structure is suitable for applying in implanted medical devices especially pacemaker.

References

1. S. H. Lewis, and P. Gray, "A pipeline 5-Msample/s 9-bit analog-to-digital converter," IEEE J. Solid-State Circuits, vol. SC-22, NO. 6, pp. 954-961, Dec. 1987.
2. T. Cho, Low-power Low-voltage Analog-to-Digital Conversion Technique using Pipelined Architectures, Ph.D thesis, University of California, Berkeley, 1995.

Design an Optimized CPU Architecture for Pacemaker Applications

Le Trung Khoa and Hoang Trang

University of Technology, VietNam National University at HoChiMinh City, HoChiMinh City, VietNam
k hoa.le,hoangtrang@hcmut.edu.vn, hoangtrang@hcmut.edu.vn

Abstract — **The design technique of using gray code addressing to reduce power dissipation in CPU of pacemaker is presented in this paper. The experimental results of reducing power up to 20% would be a promising result. This work is implemented by using Altera Quartus II 9.0, the device Cylone II EP2C20F484C7 is used.**

Keywords — **Pacemaker CPU design, low-power design, Gray code addressing.**

I. INTRODUCTION

In the 1950s, the first pacemaker that has only a few transistors was introduced. Nowadays, an implantable pacemaker system has millions of transistor based on the greatly advancing of technology. In the United States of America, heart disease is one of the main causes of death, its ratio is equal to deaths associated with the combination of accident, cancer and diabetes. One of common heart diseases is Bradycardia and it has been successfully treated by using pacemaker devices. The pacemaker monitors the rate of heart and the rhythm. If the heart does not beat or beats too slowly, the pacemaker provides the electrical stimulation thought the electrode. Therefore, it is important to improve these devices for better performance and expand their operational ranges. Depend on the number of heart chamber monitored and stimulated, the pacemakers are classified in single chamber and dual-chamber, as in Fig.1. Dual-chamber pacemaker offers more advantages than the single one in term of patient life quality, such as reduction in atria fibrillation, pacemaker syndrome and heart failure [1]. However, a problem of dual-chamber pacemaker is that it consumes much more energy than the single chamber. Because pacemaker is installed in the body of patient, frequently changing its battery is not easy and safe. This problem limits the use of many patients and actually, many of them choose the single pacemaker instead of the dual one. The improvement, which keeps the advantages of dual-pacemaker but offers the power consumption competitive with the single pacemaker, is needed.

In a pacemaker, the CPU is the controlling central that controls all peripherals, therefore it is considered as the heart of the device. This is the reason why the CPU actions can affect the power consumption. The good design of the CPU not only reduces its own power consumption but also minimize the power dissipation of all peripherals, therefore, this design can reduce the total power consumption much more. This work presents an optimized CPU architecture for dual-chamber pacemakers. It contains a way to reduce the power consumption in reading the program memory that is Gray code addressing.

The design is implemented in the Altera's Quartus II 9.0 web edition software and the Powerplay power Analyzer is used to estimate the power consumption. The device used is Cylone II EP2C20F484C7.

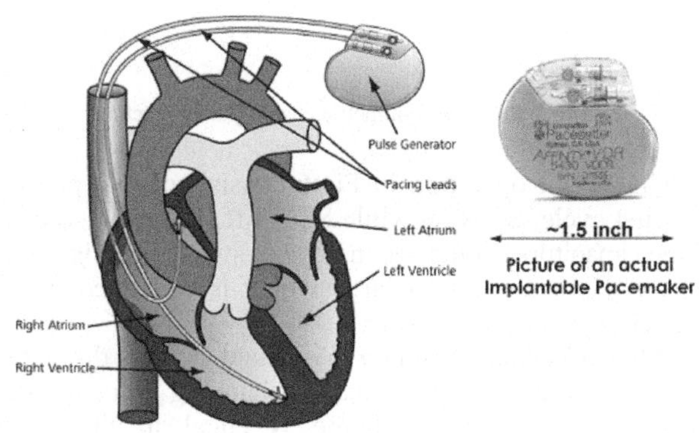

Fig. 1 A cardiac pacemaker system and a picture of an actual implantable pacemaker. [1]

The remaining of this paper is structured as follows. Pacemaker CPU is presented in section II. Section III shows our proposed gray code addressing in pacemaker CPU and experimental results. The conclusion is given in the last section.

II. PACEMAKER CPU

Pacemakers are composed of four main parts: Input stage, output stage, housekeeping and control unit as in Fig.2. The input stage has the task of detecting the heart activity and of notifying the control unit that an event has occurred. Commonly, it is composed of an electronic amplifier that amplifies the cardiac signal, a band pass filter to clean the signal and a comparator with a programmable sensitivity threshold generator to compare the input signal

V. Van Toi et al. (Eds.): 4th International Conference on Biomedical Engineering in Vietnam, IFMBE Proceedings 40, pp. 50–53, 2013.
www.springerlink.com

with a fixed reference. The output stage has a function of delivering a pacing pulse to the heart tissue through the pacing lead. All of these characteristics are controlled by the control unit. The housekeeping is the part where the good functioning of pacemaker device and any irregular change in the battery are monitored. If a voltage below a threshold is detected, this block will reset the control unit in order to ensure that the data is correct. It also has a battery charge detector, where the control unit can verify how much charge remains. Finally, the control unit is composed by the needed logic to keep time, pace parameters, develop the pacing algorithm, save data, communicate with doctor by telemetry...It is the part that controls the functionalities of the pacemaker. In this part, digital timers are used to keep time and pulse width while registers or memory store the pace parameters. Custom logic or a microprocessor is needed to develop the pacing algorithm. This CPU is the researching object of this paper.

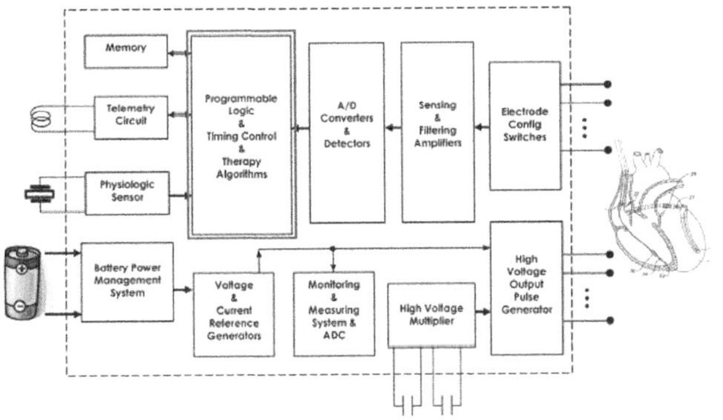

Fig. 2 Implantable pacemaker system outline.

A CPU is composed of an arithmetic logic unit (ALU), temporal storage devices called registers, a data path and control unit that together can run the software. The software is stored in an external memory. The operation of a CPU includes fetching, decoding and executing the instructions. Fetching is reading the instructions from the program memory, it include putting the address on the address bus and read the data from the data bus, decoding is the action to know what the CPU will do and executing is do this giving job. Each of action (Fetching, decoding, and executing) has its own power dissipation characteristics, In Fetching, if the number of bit switching on the address bus is decreased, the power consumption will be reduced, and one of solutions is using Gray code addressing that will be presented in section III. In executing, each instruction is represented by a sequence of bits called opcode, if the number of bit "1" in the opcode is large, the power consumption will be increased [4], [5]. The recommended solution in this paper is that the most frequently used instructions gets the op-code with less 1's while less used instructions have op-codes with more 1's, this solution will be presented in section IV.

The CPU for the pacemaker applications in this paper is Little Computer 3 (LC3) presented in Fig.3 with a Reduced Instruction Set Computer (RISC) architecture. This architecture offers many advantages for low power operation. A 16-bit address bus provides sufficient memory space to save users' histograms, electrograms, battery status and other important data. The total of 16 instructions is enough for pacemaker applications [5].

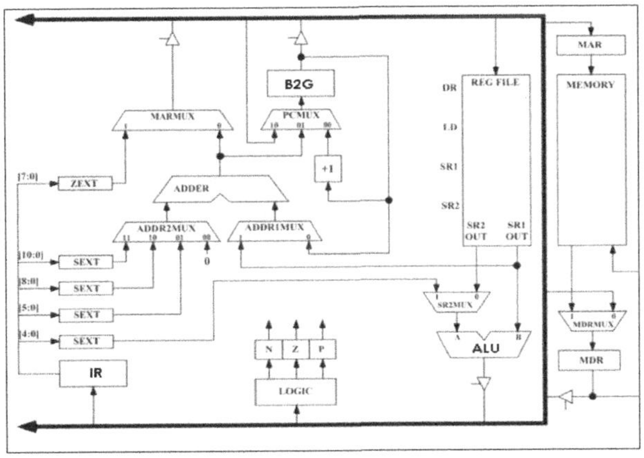

Fig. 3 The block diagram of Little Computer 3 (LC3).

III. GRAY CODE ADDRESSING

In CMOS, there are three components of power consumption: dynamic power consumption, short circuit power consumption and static power consumption. The total power could be calculated by the dynamic power consumption. The two later components can be reduced and are negligible compared to the dynamic power consumption [2]. Total power consumption can be:

$$P_d = \alpha C_L V_{dd}^2$$

Where: α: Switching activity.

C_L: The average capacitive load of the circuit.

V_{dd}: Supply voltage.

From the equation, if C_L and V_{dd} are omitted, the fewer switching activity occurs, the fewer power be consumed.

In the processor operation, many actions such as fetching instruction, decoding instruction and executing instruction are operated. Each of them has its own power consumption characteristic, the way to reduce power consumption in fetching is considered in this section.

Fetching includes putting the address value on the address bus and reading the data (instructions) back. The driving load of address line has many high capacitive loads like Program Counter (PC), Memory Address Register

(MAR) and Buses, so it seems to be effective in reducing power dissipation if the switching activity on the address line is minimized. There are some works done in [2], [3], [4] to reduce the switching activity using Gray code addressing. The Gray code differ from the Binary is that Gray code changes only one bit from one value to the next value. Moreover, the program instructions are read in sequence, therefore a significant number of bit switches can be eliminated using Gray code addressing [2].

In the designed CPU, one block called Binary to Gray converter (B2G) is inserted (as in Fig.5). In order to evaluating the efficiency in reducing power consumption, the models of experiment are designed (Fig 4 and Fig 5). The address value is increased every rising edge of the clock signal, the bit-width of address bus is 16 so the value of address will be 0 to $2^{16} - 1$. The binary code block output is the same of the output of the PC register in CPU. The Binary to Gray Converter Block in the Model of Gray code addressing (as in Fig.5) converses the 16'b binary to 16'b Gray code, the 16'b output of the two models is the input of the PC, MAR and buses, which are high capacitance loads. In this experiment, this high capacitive load is represented by the capacitor C. The structure without inserting the B2G block is called binary addressing (as in Fig.4). The structure with B2G block is called gray code addressing (as in Fig.5).

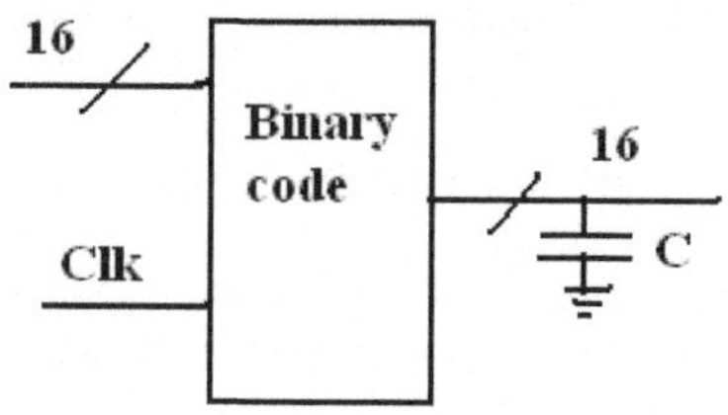

Fig. 4 The Binary code addressing.

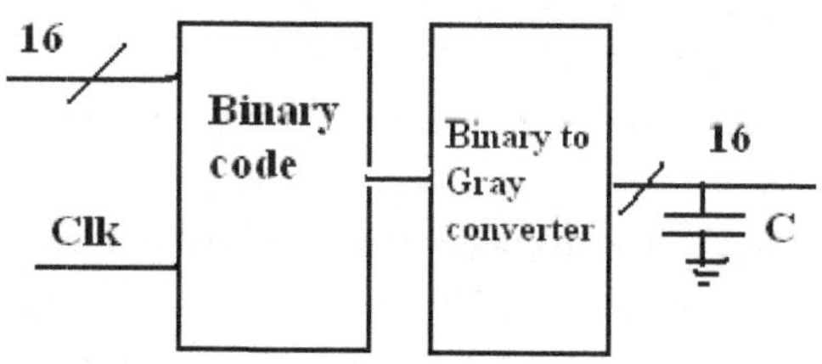

Fig. 5 The Gray code addressing .

Table 1 presents the static power consumption results of Binary addressing and Gray code addressing

Table 1 Static power consumption.

Binary addressing	Gray code addressing
68,60 mW	68,62 mW

Due to the gray code addressing model has one block more, it consumes more energy than the Binary addressing in static power. The binary to gray block is a small combinational block so it consumes a small amount of energy.

The total power of two addressing modes are experimented and presented in Fig.6 and Fig.7. The Binary code addressing power consumption is more than the Gray code addressing at all values of capacitor. Moreover, the difference is increased at the larger value of capacitor. At 5pF and 50pF output load, the average reduction is 5,2% and 15,73%. Figure 7 shows that the percentage of reduced power is increased when the value of capacitor increase. It means using Gray code addressing is more effective in power saving than the normal addressing.

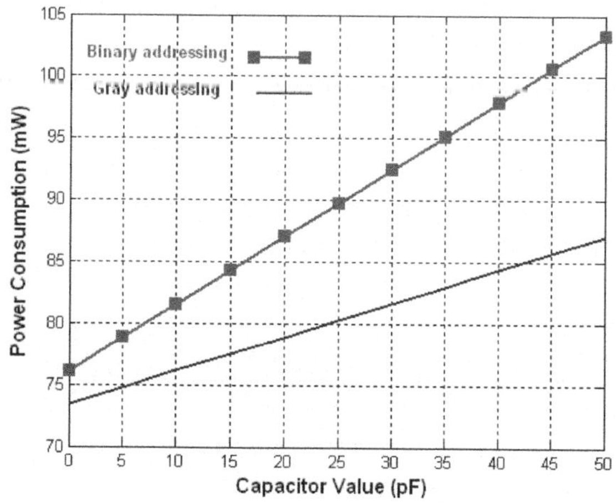

Fig. 6 Power consumption when Capacitor value is changed.

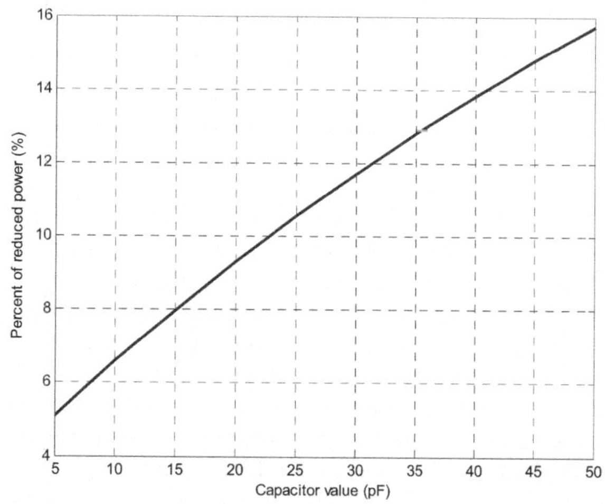

Fig. 7 Percent of power reduced when Gray code addressing is used over Binary addressing.

The branching range is about 16 bit. The percent of branch instructions is changed. When an instruction is executed, the new address is the addition result of at this time address and the 16 bits random value (0 to 65535). In theory, the branch instructions make the next executed instructions address is not the next value, so the number of switching is increased, and it increase the total power consumption. Because of the random added value, the power consumption is a stochastic value. In order to have a general view, three experiments are implemented

a) When branch instructions are executed, the added value is the value that invert all of 16 output bits. This experiment is the largest power consumption case because the number of changed bits is largest. When the percent of branch instructions is changed, it makes the upper bound of power consumption.

b) When branch instructions are executed, the added value is 0. The output value is not changed. This experiment is the lowest power consumption case because the number of changed bits is lowest. When the percentage of branch instructions is changed, it makes the lower bound of power consumption.

c) When branch instructions are executed, the added value is a random value. The output value is unpredictable. The power consumption is within the lower and upper bound.

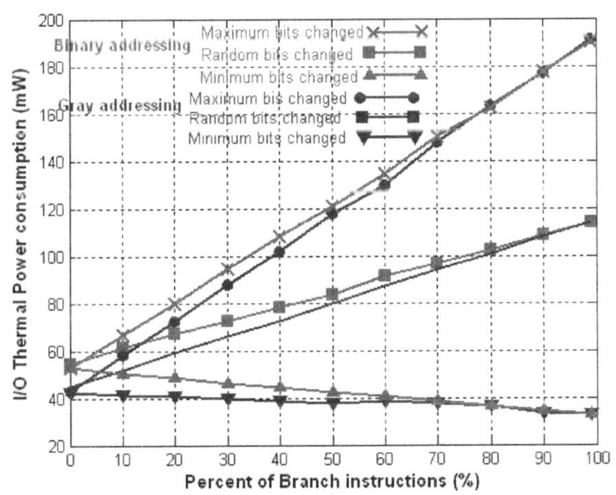

Fig. 8 The I/O Thermal power consumption of Binary and gray code addressing.

The power consumption of gray code addressing is always less than or equal to binary addressing. It is more close to the binary code addressing if the percentage of branch instructions is increased. In a random case, the power consumption change between the maximum and minimum bound. In the random case of Fig.8, the maximum percentage of reduced power is about 20% at 0% of branch instructions and the minimum is 0% at 99% of branch instructions.

One important thing should be considered in the results of our experiments is that all of the above value only make the relative comparison, the exact value in a particular experiment depend on the fabrication technology, mask design...

IV. Conclusions

The maximum saving power of the CPU is about 20% of applying gray code addressing method that is proposed in this paper There are other methods that could reduce more the consumption power: design the interrupt mode, design the idle mode of the CPU... These works should be the future work of this paper.

References

1. Wong L.S.Y., Hossain S., Ta A., Edvinsson J., Rivas D.H., Naas H., "A very low-power CMOS mixed-signal IC for implantable pacemaker applications", IEEE Journal of Solid-State Circuits, pp. 2446 – 2456, 2004.

Mehta H; Owens R.M; Irwin, M.J, "some issues in gray code addressing", VLSI, 1996. Proceedings, Page(s) 178 – 181.

Ching-Long Su; Chi-Ying Tsui; Despain, A.M. "Low power architecture design and compilation techniques for high-performance processors" Compcon Spring '94, Digest of Papers, IEEE Conferences, page(s): 489 – 498.

Luca Benini, Giovanni De Micheli, Enrico Macii, "Designing Low-Power Circuits: Practical Recipes", Circuits and Systems Magazine, IEEE Journals, pp. 6 25, 2001.

Edgar Marti Arbona , "Design of a custom cpu architecture for a pacemaker application", Master'thesis, University of Puerto Rico Mayaguez campus, May 2009.

T. Jamil. "RISC versus CISC". Potentials, IEEE, 14(3): 13-16, Aug/Sep 1995.

Simple Fabrication of Exfoliated Graphene/Nafion Hybrid as Glucose Bio-sensor Electrodes

Sun Uk Lee, Bong Gill Choi, and Won Hi Hong*

Department of Chemical & Biomolecular Engingeering (BK21 program),
KAIST, 335 Gwahangno, Yuseong-gu, Daejeon 305-701, Korea
Fax: +82-(42)-350-3959
whhong@kaist.ac.kr

Abstract — **New paradigm of biosensor electrode material, carbon-based composite materials (carbon nanotube (CNT), graphite, and graphene), have received great attentions because of high electronic conductivity and suitable mechanical properties. Among them, graphene modified electrodes show remarkable performance in biosensor applications with its high surface area, electrical and chemical properties. However, the graphene lose their intrinsic electrical and chemical properties during fabrication process by aggregation and oxidation. Here we suggest graphene and Nafion polymer composite not only enhancing processability but also maintaining properties. The easily fabricated glucose biosensor shows good performances with fast electron transfer rate, high selectivity, and low concentration range of detection with linear response.**

Keywords — **glucose sensor, graphene, graphene/Nafion hybrid, bio sensor.**

I. INTRODUCTION

Glucose biosensor is a historical homecare application to monitor diabetes with medical purpose. Developing more sensitive and accurate sensing system has been studied to increase patient's life quality. Here we report simple but highly sensitive glucose sensor based on graphene/Nafion hybrid system.

Graphene which is consisted of a sheet of carbon atoms in honeycomb lattice has been get great attentions with its unique nanostructure and properties such as high thermal conductivity, current density, electron mobility, zero band gap, and strong mechanical strength [1]. Ever since the discovery of graphene by Geim et al,.[2] with its remarkable properties it consider as a next generation material. The properties of graphene provide much potential in applications such as high sensitive sensor, super capacitor, battery, field effect transistors, and fabrications of electrical devices. However the problem is irreversible aggregation of graphene, becoming graphite, due to strong π-π stacking and Van der Waals interactions between graphene sheets. And the aggregation problem make hard to scale-up of process for mass production. Moreover the unique properties are valid only with individual sheet, so prevention of aggregation and exfoliation as a sheet of graphene from graphite is of particular importance. To solve problems there are various attempts have been conducted. Chemical vapor deposition (CVD) method was used to fabricate graphene film. Also chemically growing on metal catalyst was also proceeded as a bottom up process. In recently, method of chemically reduction of graphene oxide into graphene has been reported as cheap and up-scalable process for various applications [3].

To fabricate a sheet of graphene and developing various applications is motivated by CNTs processing techniques such as pH control, small molecule coating, chemical functionalization, and physical mixing with polymers. Among them, chemical functionalization method with various polymers such as functional polymers, perfluorosulfonated ionomer, and Nafion is applicable for graphene because these polymers are consisted with hydrophobic backbone and hydrophilic side chains which is suitable structure to directly dispersion onto graphene. Nafion can efficiently combine with graphene not only as an interfacial binder but also as physical supporting agents with better physical and chemical properties than other polymer additives. Moreover Nafion can effectively immobilize enzymes without loss of enzyme activity.

In this study, we used Nafion as a reduction agent of graphene oxide into graphene and also as functional groups to increase electron transfer rate. The glucose sensing system was fabricated with surface modified gold electrode with glucose oxidase (GOx) and graphene/Nafion hybrid (GN) was investigated.

II. EXPERIMENTAL [4]

A. *Materials*

Graphite powder (< 20 μM), hydrazine solution (35 wt% in water), and Nafion (perfluorinated resin solution, 5 wt % in lower aliphatic alcohol and water mixture) were purchased from Aldrich. 1-propanol was purchased from Junsei.

* Corresponding author.

V. Van Toi et al. (Eds.): 4th International Conference on Biomedical Engineering in Vietnam, IFMBE Proceedings 40, pp. 54–56, 2013.
www.springerlink.com

B. Synthesis of Graphite Oxide (GO)

GO was prepared following Hummers method. Priory graphite powder (2g) was immersed in a solution which was consisted with H_2SO_4 (10ml), $K_2S_2O_8$(2g), and P_2O_5 (2g) at 85 °C and then stirring for 5 hr. The dried mixture react with H_2SO_4 (100 ml) and gradually added $KMNO_4$ (10g) for 1hr under ice bathing condition. The mixture was washing with DI and H_2O_2 (30%) and then filtered. After drying, subsequently it was washed with 10 wt% of HCl and DI to remove remaining metal ions and acid. Dialysis were conducted for 1 week to purify GO.

C. Synthesis of Graphene/Nafion Hybrid (GN)

The prepared GO were dispersed in solution of DI and 1-propane (1:1 v %). 0.1 – 50 wt % of Nafion solution were mixed with 0.5 mg/ml of GO in 20 ml of DI and 1-propanol solution. This solution was sonicated for 1hr and then added 100 μM of hydrazine solution at 85 °C for 24 hr. The mixture washed several times with water and ethanol. To remove impurities dialysis also conducted for 1 week.

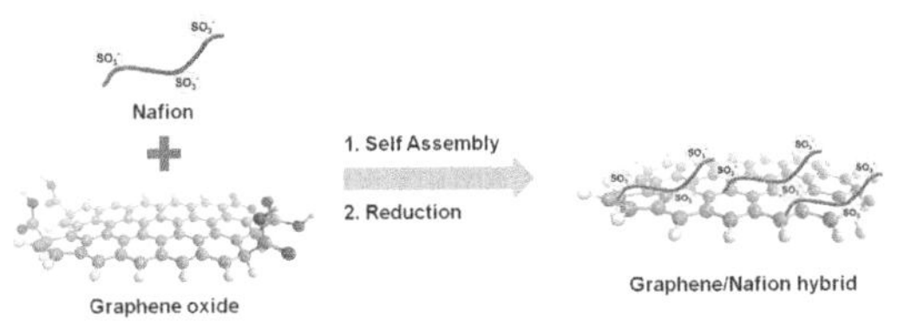

Fig. 1 Graphene/Nafion hybrid synthesis scheme by self assembly and reduction step.

D. Preparation of GOx/GN Modified Electrode

First the gold electrodes were polished with 0.05 μM of alumina powders and cleaning with DI and ultrasonication. To fabricate surface modified electrode, 5 μL of GN solution were dropping onto gold electrode surface and dried it overnight. After that, the gold/GN electrode was dipping in 5 mg/ml of GOx solution in phosphate-buffered-saline (PBS, pH 7.4). Dipped electrode was washed with PBS to remove loosely bounded GOx on electrode surface.

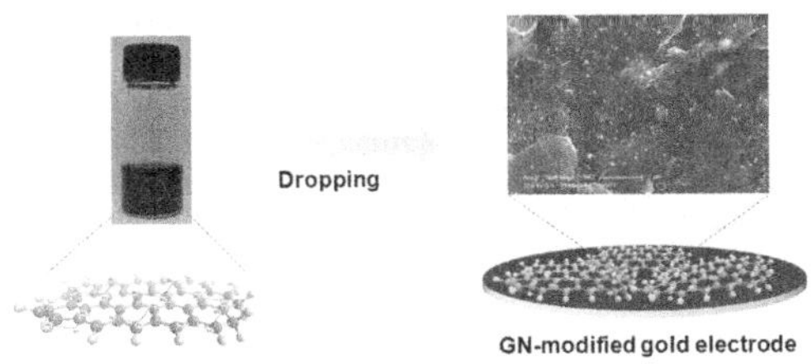

Fig. 2 Scheme of fabrication procedure of GN modified gold electrode as glucose sensor.

E. Analytical Methods

Transmission electron microscope (TEM) analysis was proceeded with JE-2100F HR-TEM at 200 kV and scanning electron microscope (SEM) images also obtained. Atomic force microscope (AFM, Park systems) utilized with non-contact mode with silicon probes at 1 Hz of scanning rate. CV was analyzed using CHI 660C electrochemical workstation (CH instrument) with traditional three electrode system, gold as a working electrode, KCl-Ag/AgCl as reference electrode, and Pt wire as a counter electrode. All electrochemical analysis was performed at room temperature and all results were averaged value with 5 times of experiments.

III. RESULTS AND DISCUSSION

A. Characteristics of GN Composite Sheet

To maintain electrical and mechanical properties and immobilize enzyme onto electrode, we have to accomplish the complete exfoliation of graphite and stabilization of exfoliated graphene in suspension. The TEM image shows that morphologies of GN hybrid successfully exfoliated a sheet of graphene and fabricated Nafion hybrid even after reduction process of graphene oxide into graphene. The thickness of GN hybrid sheet was confirmed about 1 nm by AFM image. And also with SEM analysis we can figure out that the morphologies of surface modified gold electrode with GOx/GN and the globular GOx was uniformly dispersed and absorbed onto GN sheet.

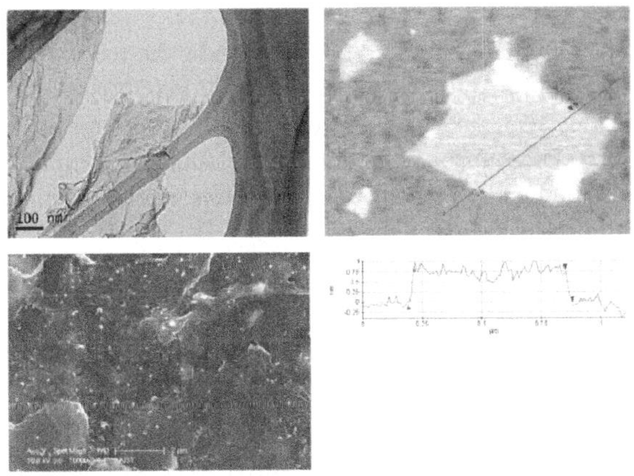

Fig. 3 (A) TEM image of the GN sheet (B) SEM image of surface of GN-modified gold electrode (C) Tapping mode-AFM image of the graphene oxide sheet deposited on the mica surface (D) AFM height profile of graphene oxide sheet

B. Electrochemical Analysis of GN Modified Gold Elecdtrode

To verify electrochemical properites of modified electrode, the CVs were used. Comparing of the bare gold electrode and GN modified electrode, the peak current drastically increased with decreasing potential separation (ΔE_p) from 73 mV for bare electrode to 68 mV for GN modified electrode. The high peak current and small peak separation (ΔE_p) mentioned that electron transfer of GN modified electrode was promoted.

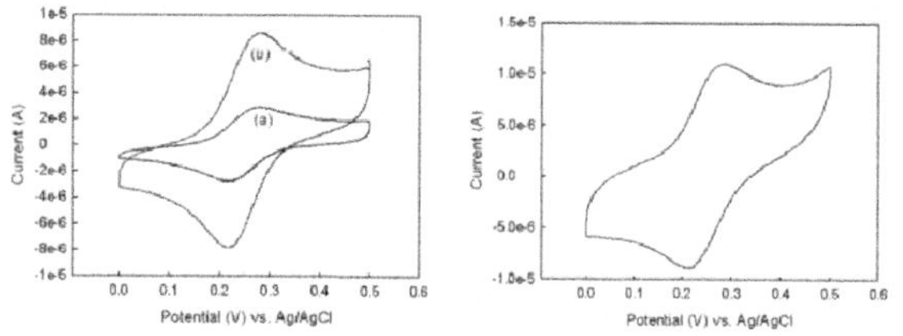

Fig. 4 (A) Cyclic voltammograms in solution containing 5 mM of $K_3Fe(CN)_6$ in 0.1 M KCl at scan rate of 50 mV/s (a) Bare gold electrode and (b) GN modified gold elecdtrode. (B) Cyclic voltammograms of GOx/GN modified gold electrode in same condition (Right).

C. *Direct Electrochemisty of GOx/GN Modified Gold Electrode*

The reaction mechanism of direct electrochemistry of GOx based on redox center, flavine adenine dinucleotide (FAD), was proposed previous study[5] and was confirmed by CV examination under potential range from -0.8V to 0V. In case of oxygen depiction, the mechanism is

$$GOx(FAD) + 2H+ 2e^- \leftrightarrow GOx(FADH_2) \quad (1)$$

The GOx/GN modified gold electrode, the redox reaction of FAD and $FADH_2$ was appeared in $K_3Fe(CN)_6$ solution at a various scan rate from 25 to 300 mV/s showing in Fig 5. And also the quasi reversible series showing symmetrical anodic and cathodic peaks indicated linear relationship with scan rate and current density.

Moreover we also verified direct bioelectrocatalytic activity of GOx/GN gold electrode for mechanism of oxidation of glucose by CV analysis under various concentration of H_2O_2 in solution. The mechanism of oxidation of glucose proposed following reaction [6].

$$GOx(FADH_2) + O_2 \rightarrow GOx(FAD) + H_2O_2 \quad (2)$$

$$GOx(FAD) + Glucose$$

$$\rightarrow GOx(FADH_2) + Gluconolactone \quad (3)$$

The typical current time response showing in Fig. 5 represents sharply decreasing of current with increasing concentration of H_2O_2 from 2 to 18 mM. The relation of linear detection shows as inset in Fig. 5 is suitable concentration range for practical application of sensing glucose in blood.

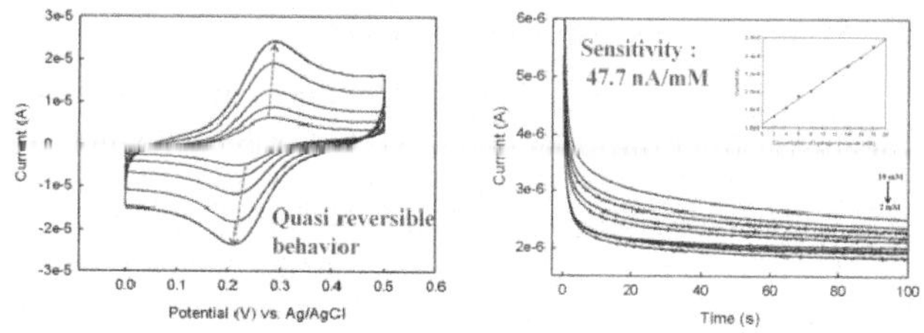

Fig. 5 (A) Cyclic volammograms of GOx/GN modified gold electrode in solution containing 5mM of $K_3Fe(CN)_6$ in 0.1 M KCl at different scan rate (25, 50, 100, 200 and 300 mV/s) (B) Typical current-density response of GOx/GN modified gold electrode to successive 2 mM increments in the level of hydrogen peroxide. Inset is plot of the current vs. concentration of hydrogen peroxide.

IV. CONCLUSION

We fabricated simple GN hybrids composite to improve processability of graphene by stabilization in suspension and apply for glucose biosensor. The easy fabrication of glucose bio sensor showed good electrochemical performances for detection of low concentration of H_2O_2 based on direct electron transfer. The large surface area of exfoliated graphene sheet and Nafion hybrid system with well dispersed globular GOx can increase the electron transfer rate and sensing ability. The high performance of GOx/GN-composite as glucose sensor guaranteed a new way to apply for electrocatalyst, advanced electrochemical electrode, and various electrochemical biosensors.

REFERENCES

1. Ming Zhou, Yueming Zhai, and Shaojun Dong, (2009) Electrochemical Sensing and Biosensing Platform Based on Chemically Reduced Graphene Oxide, Anal. Chem, 81, 5603-5613
2. Novoselov, K. S., Geim, A. K., Morozov, S. V., Jiang, D., Zhang, Y., et al. (2004) Electric Field Effect in Atomically Thin Carbon Film, Science, 2004, 306, 666-669
3. (A) Buch, J. S., Zande, A. M., Verbridge, S. S., Frank, I. W., et al, 2007,. Detection of Individual Gas Molecules Adsorbed on Graphene. Nat. Mater, 6, 652-655, (B) Wang, X., Zhi, L., Mullen, K., 2008, Transparent, Conductive Graphene Electrodes for Dye-Sensitized Solar Cells, Nano Lett. 8, 323-327, (C) Yoo, E., Kim, J., Hosono, E., Zhou, H. S., Kudo T., Honma, I., 2008, Large Reversible Li Storage of Graphene Nanosheet Families for Use in Rechargeable Lithium Ion Batteries, Nano Lett. 8, 2277-2282
4. B. G. Choi, H. S Park, W. H. Hong, S. Y. Lee, et al, 2010, Solution Chemistry of Self-Assembled Graphene Nanohybrids for High-Performance Flexible Biosensors, ACS NANO, 4, 2910-2918
5. G. Liu, M. N. Paddon-Row, J. Gooding, 2007, A molecular wire modified glassy carbon elecdtrode for achieving direct electron transfer to native glucose oxidase, ElectroChem, commun, 9, 2218-2223
6. (A) P. D. Hale, T. Inagaki, H. I. Karan, Y. Okamoto, T. A. Skotheim, 1989, A new class of amperometric biosensor incoprating a polymeric electron transfer mediator, JACS, 111, 3482-3484, (B) Avinash J. P, Jemma L. V, Thomas B. S, Stephen M, 2009, Aqueous Stabilization and Self-Assembly of Graphene Sheets into Layered Bio-Nanocomposites using DNA, Adv. Mater, 21, 3159-3164

Pharmacodynamics and Pharmacokinetics of Exendin-4 Orally Delivered via an Enteric Coated Capsule Containing Nanoparticle in Diabetic Rats

Ho-Ngoc Nguyen[1] and Ha-Giang Nguyen[2]

[1] Nanomega Medical Corp., CA 92630, United States
[2] Department of Chemical Engineering, Cantho University, Cantho, Vietnam

Abstract — **An orally available exendin-4 formulation, using an enteric-coated capsule containing pH-sensitive NPs, was studied. In the *in vivo* dissolution study, it was found that the enteric-coated capsule remained intact while in the stomach; the capsule was completely dissolved in the proximal segment of the small intestine and the loaded contents were then released. Oral administration of the capsule containing exendin-4 loaded NPs showed a maximum plasma concentration at 5 h after treatment; the bioavailability, relative to its subcutaneous counterpart, was found to be 14.0 ± 1.8 %. The absorbed exendin-4 could then stimulate the insulin secretion and provide a prolonged glucose-lowering effect. The aforementioned results suggest that the orally available exendin-4 formulation developed warrants further exploration as a potential therapy for diabetic patients.**

Keywords — **Oral delivery, exendin-4, pharmacodynamics, pharmacokinetics.**

I. INTRODUCTION

Recently, we prepared an exendin-4 loaded NPs using a simple ionic-gelation method self-assembled by the positively-charged CS and the negatively-charged γ-PGA in an aqueous environment [1,2]. The suspended NPs were then freeze-dried in the matrix of trehalose [3] and filled in hard gelatin capsules coated with an enteric-coating polymer Eudragit® L100-55 [4]. Characteristics of both suspended NPs and lyophilized NPs were examined using a Zetasizer. The dissolution rates of the enteric-coated capsules were evaluated *in vitro* at distinct pH environments and *in vivo* in a rat model; their pharmacodynamic (PD), pharmacokinetic (PK), and insulin secretion profiles were studied in diabetic rats.

II. MATERIALS AND METHODS

A. pH Stability of Test NPs

The exendin-4 loaded NP prepared in the presence of Fe^{+3} was reported in our previous study; the drug LE and LC were 60.9 ± 2.0% and 14.7 ± 1.9% (n = 5, $P < 0.05$), respectively. Test NPs, before and after freeze-drying, were characterized. The particle size and zeta potential value of test NPs at distinct pH values, simulating the pH environments in the small intestine, were measured using a Zetasizer (3000HS, Malvern Instruments Ltd., Worcestershire, UK).

B. In Vitro and in Vivo Dissolution Rates of the Enteric-Coated Capsules

Hard gelatin capsules (size 1, Torpac Inc., Fairfield, NJ, USA) were manually filled with trypan blue/$CaCO_3$ as a dye indicator as per the manufacturer's instructions. The capsules were then immersed in a methanol solution containing Eudragit® L100-55 (10% or 15% w/v, Evonik Industries, Parsippany, NJ, USA) [4], followed by drying at room temperature using an air-blower. The *in vitro* dissolution rates of the coated capsules were performed in an aqueous solution at pH 1.2 (adjusted by HCl) and in a phosphate buffer solution at pH 6.0 (adjusted by HCl) at 37°C under agitation (100 rpm).

In the *in vivo* dissolution study, test capsules (size 9, Torpac Inc.) were filled with barium sulfate and then coated with Eudragit® L100-55 as described above; barium sulfate is a commonly used radio-contrast agent for the gastrointestinal contrast radiography [5]. The above-mentioned SPECT/CT scanner was used for the animal imaging. The X-ray scans of rats (n = 3) were acquired at predetermined time intervals after oral administration of test capsules. To visualize anatomical landmarks, a reference CT scan was performed subsequent to the X-ray imaging

C. In Vivo PD/PK and Insulin Secretion

Diabetes was induced in Wistar rats (180–200 g body weight, male) via an intraperitoneal injection of streptozotocin (STZ, dissolved in a 10*mM* citrate buffer at pH 4.0) at a single dose of 50 mg/kg body weight [6]. Rats were considered diabetic when their (fasting or random) blood glucose levels were higher than 300 mg/dL, two weeks after the STZ treatment [6]. Diabetic rats were fasted overnight prior to and remained fasted during the experiment, but were allowed water *ad libitum*. The formulations following were individually administered to the diabetic rats: the

V. Van Toi et al. (Eds.): 4th International Conference on Biomedical Engineering in Vietnam, IFMBE Proceedings 40, pp. 57–60, 2013.
www.springerlink.com

enteric-coated capsule containing exendin-4 loaded NPs (300 μg/kg) and that filled with empty NPs blended with free-form exendin-4 (300 μg/kg) via oral administration (using a dosing syringe); and the free-form exendin-4 solution (50 μg/kg) via SC injection (n = 5 for each study group). Blood samples were collected from the tail veins of rats prior to drug administration and at different time intervals after dosing. The blood glucose levels were then determined by a glucose meter (LifeScan Inc., Milpitas, CA, USA).

For the analysis of plasma exendin-4 levels, blood samples were centrifuged (1600 x g, 4°C for 15 min) and then quantified using the exendin-4 EIA kit (EK-070-94, Phoenix Pharmaceuticals Inc., Burlingame, CA, US). The relative bioavailability (BA_R) of exendin-4 loaded NPs after oral administration was calculated using the following formula [2].

$$BA_R\ (\%) = \frac{AUC_{(oral)} \times Dose_{(SC)}}{AUC_{(SC)} \times Dose_{(oral)}} \times 100\%$$

where AUC is the total area under the plasma exendin-4 concentration vs. time curve. For the analysis of insulin secretion, centrifuged blood samples were tested by the rat insulin ELISA kit (10-1250-01, Mercodia AB, Uppsala, Sweden).

D. *Statistical Analysis*

Comparison between two groups was analyzed by the one-tailed Student's *t*-test using statistical software (SPSS, Chicago, Ill, USA). Data are presented as mean ± SD. A difference of $P < 0.05$ was considered statistically significant.

III. Results and discussion

A. *pH-Sensitivity of Test NPs*

In the study, trehalose was used as a cryoprotectant during the process of freeze-drying NPs. Aqueous trehalose is known to become glassy at its glass transition temperature [3]; the immobilization of test NPs within a glassy matrix of cryoprotectants can prevent their aggregation and protect them against the mechanical stress of ice crystals [3]. After freeze-drying, the NPs in the absence of trehalose resulted in the formation of NP aggregates, while in the company of trehalose, NPs were found to be embedded in a matrix and could be readily turned into powder form and resuspended freely in DI water for further characterization.

Table 1 Mean particle sizes (nm) and zeta potential values (mV) of exendin-4 loaded nanoparticles (NPs) before and after freeze-drying at distinct pH environments (n = 5).

pH value	NPs before Freeze-drying		NPs after Freeze-drying	
	Particle Size (nm)	Zeta Potential (mV)	Particle Size (nm)	Zeta Potential (mV)
6.0	260.6 ± 26.4	32.5 ± 0.1	261.0 ± 22.4	33.7 ± 1.3
6.6	438.8 ± 33.1	13.3 ± 0.3	450.2 ± 36.2	13.9 ± 0.2
7.0	1025.1 ± 56.3	3.6 ± 0.8	1049.2 ± 48.3	4.1 ± 0.3
7.4	N/A	1.2 ± 1.0	N/A	1.5 ± 0.9

N/A: data not available because NPs were disintegrated and precipitated.

The pH-sensitivity of test NPs before and after freeze-drying was evaluated at distinct pH environments, representing the small intestine conditions. As shown in Table 1, test NPs before and after freeze-drying had similar pH-stability profiles. With increasing pH, the size of test NPs increased significantly, while their zeta potential values decreased remarkably ($P < 0.05$). It was noted that test NPs became unstable and disintegrated at pH 7.0–7.4. These results indicate that the freeze-drying process did not significantly affect the pH sensitivity of test NPs.

B. *In Vitro and in Vivo Dissolution of the Enteric-Coated Capsules*

Eudragit® are copolymers derived from esters of acrylic and methacrylic acid, which enables a subject to achieve the desired drug release profile by releasing the drug at the right place and at the right time [4]. Eudragit® L100-55 is dissoluble above pH 5.5 which is suitable for protecting active drugs from the gastric fluid and improving the drug intestinal absorption [4].

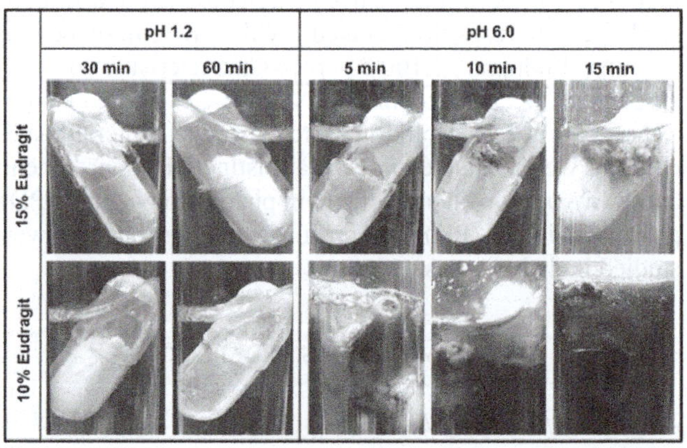

Fig. 1 Results of the *in vitro* dissolution of test capsules (coated in 10% or 15% Eudragit® L100-55) at pH 1.2 (mimicking the acidic environment in the stomach) or pH 6.0 (simulating the pH environment in the proximal end of the small intestine).

Results of the *in vitro* dissolution of test capsules (coated in 10% or 15% Eudragit® L100-55) are presented in Fig. 1. As shown, both test capsules remained intact at pH 1.2 (mimicking the acidic environment in the stomach), indicating the dissolution-resistance feature of the enteric-coating polymer used in the acidic environment. In contrast, the dissolution rate of the capsule coated in 10% Eudragit® L100-55 was significantly faster than its counterpart coated in 15% Eudragit® L100-55 at pH 6.0 (simulating the pH environment in the proximal end of the small intestine, the duodenum). Due to the prompt release of its loaded content in the pH environment of the duodenum, the capsule coated in 10% Eudragit® L100-55 was selected for the following *in vivo* studies.

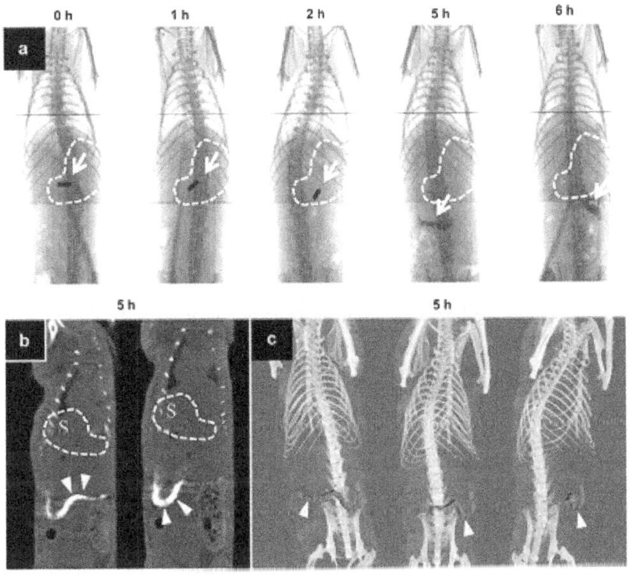

Fig. 2 (a) Sequential plain X-ray images; (b) CT-scan images; and (c) 3D-volume-rendering images of a rat orally treated with the barium sulfate-filled capsule (indicated by arrows). The released barium sulfate was located in the intestine (indicated by arrow heads). S: stomach.

Fig. 2a presents the X-ray images of a rat orally treated with the barium sulfate-filled capsule [5]. As shown, the capsule remained intact while in the stomach, confirming the gastroresistant feature of Eudragit® L100-55. Once in the small intestine, the capsule was completely dissolved and the loaded barium sulfate was clearly seen thereafter. The subsequent CT-scan and 3D-volume-rendering images (Figs. 2b and 2c) demonstrate that the released contrast agent propagated through the intestine, validating the total dissolution of test capsule.

C. *In Vivo PD, PK and Insulin Secretion*

The obtained PD profiles showed that the group treated with the free-form exendin-4 solution subcutaneously produced a sharp decrease in blood glucose level, which then returned to the basal level with time (Fig. 7a). No glucose-lowering effect was observed after oral administration of the capsule filled with empty NPs blended with free-form exendin-4, indicating a poor oral absorption of free-form exendin-4 in the small intestine. In contrast, oral administration of the capsule containing exendin-4 loaded NPs produced a slower but prolonged reduction in blood glucose level.

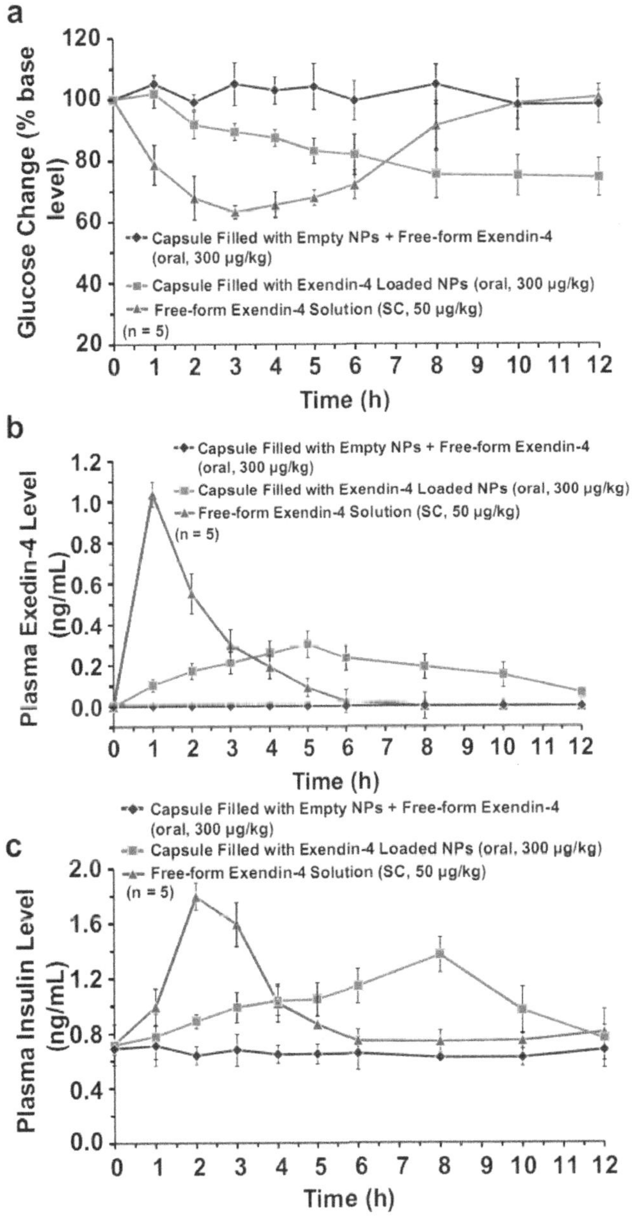

Fig. 3 (a) Blood glucose change vs. time profiles; (b) plasma exendin-4 level vs. time profiles of the diabetic rats following the administration of different exendin-4 formulations.

The plasma exendin-4 concentration vs. time profiles is shown in Fig. 3b. No detectable plasma exendin-4 was found in the group orally treated with the capsule filled with empty NPs blended with free-form exendin-4, whereas the subcutaneously-treated group resulted in a maximum plasma exendin-4 concentration at 1 h post injection. In contrast, oral administration of the capsule containing exendin-4 loaded NPs showed a maximum plasma concentration at 5 h after treatment. In this case, the AUC was 2140.6 ± 266.8 pg h/mL; its corresponding BA_R was 14.0 ± 1.8 %, which was significantly greater than those reported in the literature [1] as mentioned above.

As demonstrated in our *in vivo* dissolution study, upon oral administration, the capsule remained intact in the acidic environment of the stomach; however, it was dissolved promptly in the proximal segment of the intestine and then released its loaded content (Figs. 2a, 2b and 2c). As a result, the delivered capsule could prevent NPs from contacting the gastric medium, thus limiting the release of exendin-4 in the stomach. Additionally, most NPs loaded in the capsule could be delivered into the small intestine, consequently enhancing their possibility of adhering and infiltrating into the local mucus layer. The exendin-4 released from the disintegrated NPs was then able to permeate through the opened paracellular pathway to the systemic circulation, hence increasing its BA_R significantly.

Exendin-4 is a potent insulinotropic agent in nondiabetic and diabetic subjects [7]. The plasma insulin level vs. time curves are given in Fig. 3c. After SC injection of the free-form exedin-4 solution, the insulin level was significantly increased and reached its maximum value at about 2 h after injection and then returned to its basal level 3 h later; this observation was similar to that reported in the literature [8]. In contrast, oral administration of the capsule containing exendin-4 loaded NPs increased the insulin level in a slower but prolonged manner ($P < 0.05$), which was comparable with the PD/PK profiles (Figs. 3a and 3b). It is known that exendin-4 is able to stimulate ß-cells to secrete insulin in the case of hyperglycemia, whereas it does not stimulate the pancreatic insulin release when the glucose concentration is below the normal range [9]. This glucose dependence is thought to assure a potential safety advantage over agents (e.g., sulfonylureas and meglitinides) that increase insulin secretion via glucose-independent mechanisms.

IV. Conclusions

An enteric-coated capsule containing pH-responsive NPs for the oral delivery of exendin-4 was developed in this study. The capsule remained intact while in the stomach, whereas it was totally dissolved in the small intestine, thus releasing the loaded contents. The administered exendin-4 was then absorbed into the systemic circulation, stimulated the insulin secretion, and provided a prolonged glucose-lowering effect. These results suggest that the oral formulation might be considered as an alternative product for improving patient convenience and therapeutic efficacy.

References

1. Nielsen LL, Young AA, Parkes DG. (2004) Pharmacology of exenatide (synthetic exendin-4): A potential therapeutic for improved glycemic control of type 2 diabetes. Regul Pept 117(2):77–88.
2. Sonaje K, Lin YH, Juang JH, et al. (2009) *In vivo* evaluation of safety and efficacy of self-assembled nanoparticles for oral insulin delivery. Biomaterials 30(12):2329–2339.
3. Chang L, Shepherd D, Sun J et al. (2005) Mechanism of protein stabilization by sugars during freeze-drying and storage: Native structure preservation, specific interaction, and/or immobilization in a glassy matrix? J Pharm Sci 94(7):1427–1444.
4. Khan MZI, Prebeg Z, Kurjakovic N. (1999) A pH-dependent colon targeted oral drug delivery system using methacrylic acid copolymers: I. manipulation of drug release using Eudragit® L100-55 and Eudragit® S100 combinations. J Controlled Release 58(2):215–222.
5. O'Connor SD, Summers RM. (2007) Revisiting oral barium sulfate contrast agents. Acad Radiol 14(1):72–80.
6. Srinivasan K, Viswanad B, Asrat L et al. (2005) Combination of high-fat diet-fed and low-dose streptozotocin-treated rat: A model for type 2 diabetes and pharmacological screening. Pharm Res 52(4):313–320.
7. Egan JM, Clocquet AR, Elahi D. (2002) The insulinotropic effect of acute exendin-4 administered to humans: Comparison of nondiabetic state to type 2 diabetes. J Clin Endocrinol Metab 87(3):1282–1290.
8. Kolterman OG, Buse JB, Fineman MS et al. (2003) Synthetic exendin-4 (exenatide) significantly reduces postprandial and fasting plasma glucose in subjects with type 2 diabetes. J Clin Endocrinol Metab 88(7):3082–3089.
9. Xu G, Stoffers DA, Habener JF et al. (1999) Exendin-4 stimulates both beta-cell replication and neogenesis, resulting in increased beta-cell mass and improved glucose tolerance in diabetic rats. Diabetes 48(12):2270–2276.

Corresponding author:
Author: Ho-Ngoc Nguyen
Institute: Nanomega Medical Corp., Taiwan Branch
Street: Kuang-Fu Road
City: Hsinchu
Country: Taiwan R.O.C.
Email: ng.hngoc@gmail.com

Roles of Nanoparticles during Magnetic Resonance Navigation and Bacterial Propulsion for Enhanced Drug Delivery in Tumors

Sylvain Martel

NanoRobotics Laboratory, Department of Computer and Software Engineering, Institute of Biomedical Engineering, École Polytechnique de Montréal (EPM), Montréal, Canada

Abstract — **It is well known that the uses of nanoparticles (NPs) can enhance medical imaging, diagnostics, and drug delivery. But for these applications and for drug delivery in particular, the difficulty in targeting specific organs in the body limits the role of these NPs for medical interventions. In cancer therapy for instance, systemic injections of drug-loaded nanoparticles result into an increase of toxicity in the body coupled with a reduction of the therapeutic outcome due to a lack of efficient targeting. Therefore, although the small size of drug-loaded NPs allows them to preferentially accumulate at tumor sites because tumors lack an effective lymphatic drainage system, most of the NPs may not reach the tumor sites but would rather reach healthy organs through systemic circulations. Hence, the challenge would be to deliver the higher percentage of the drug close enough to the tumor sites while avoiding systemic circulations. This can be achieved by adding an additional capability known as Magnetic Resonance Navigation (MRN) to the NPs without compromising their potentials for medical imaging, diagnostics, and drug delivery. MRN used for targeted drug delivery relies on magnetic nanoparticles (MNPs) embedded in therapeutic magnetic microcarriers (TMMCs) where such MNPs being fully saturated in a high homogeneous magnetic field, allow the induction of a pulling force through the use of 3D directional gradients for vascular navigation along a pre-planned path. Although the effectiveness of MRN is independent of the depth at which it operates unlike the use of an external magnet, travel is limited to larger diameter blood vessels. As such, MRN is complemented by bacterial propulsion where drug-loaded MC-1 magnetotactic bacteria (MTB) relying on a chain of NPs known as magnetosomes for directional control and previously transported through MR-compatible microcarriers, are being considered as vehicles capable of reaching the tumor sites through the microvasculature.**

Keywords — **Magnetic nanoparticles, targeted drug delivery, cancer therapy, magnetic resonance imaging, magnetotactic bacteria.**

I. Introduction

The roles of nanoparticles for delivering drug in cancer therapy are very well known [1]. But such nanoparticles still typically travel through the systemic blood circulatory network, which often leads to an increase in toxicity with retention of many of these nanoparticles in unwanted regions. Magnetic nanoparticles (MNPs) have also been considered to enhance targeting in tumoral sites but the force acting on the MNP from an external magnet decreases exponentially when the region of interest in located deeper below the skin, thus limiting the effective usage for targets located only near below the skin. Recently, a new technique referred to as Magnetic Resonance Navigation (MRN) has been proposed, allowing for the navigation of therapeutic carriers relying on MNP to navigate in the vascular network with a targeting efficacy being independent of the depth. Although MRN was first demonstrated in the carotid artery of live swine using a standard clinical magnetic resonance imaging (MRI) system [2], travelling in smaller diameter vessels requires a substantial increase of the magnetic gradients generated by clinical MRI scanners. A recent example of the use of the MRN approach with an upgraded clinical MRI scanner is the successful targeting in specific lobes of rabbit livers after MR navigation in the hepatic artery of a drug loaded in special vehicles referred to as therapeutic magnetic microcarriers (TMMCs) [3]. Because MRN typically relies on an upgraded clinical MRI system referred to as a MRN system, the magnetic gradients that can be generated to induce a propelling (pulling) force on the MNPs embedded in the TMMCs although much greater than for standard MRI scanners, is still limited to TMMCs having a minimum diameters of a few tens of micrometers (μm) which remains too large for travelling in the narrow vessels such as capillaries prior to reach a tumor. As such, bacterial propulsion has been proposed [4] to complement MRN to enable the delivery of therapeutic agents deeper in tumors only accessible by passing through microvascular networks.

Bacterial propulsion relies on a special type of cells known as Magnetotactic Bacteria (MTB) [5]. Although the flagella are used to provide self-propulsion to the bacteria carrying the therapeutic loads, a chain of MNP known as magnetosomes embedded in the cells allows for the directional control of the swimming direction of the cells towards a target using a directional magnetic field with a magnitude much lower than what would be required to induce a pulling force during MRN. But unlike MRN, MTB proved to be much less effective in larger blood vessels. As such, efforts are presently done to complement MRN with

V. Van Toi et al. (Eds.): 4th International Conference on Biomedical Engineering in Vietnam, IFMBE Proceedings 40, pp. 61–64, 2013.
www.springerlink.com

MTB. But for such integration to be successful, the limits of the roles of the MNPs in both approaches need be identified and confirmed experimentally to evaluate if such integration is indeed possible within physiological and technological constraints.

II. THEORY, HYPOTHESES AND MAIN CHALLENGES

A first hypothesis suggests that bacterial propulsion could complement MRN within physiological and technological constraints. To do so, flagellated cells such as MC-1 magnetotactic bacteria could be encapsulated in special MRN-compatible microcarriers to be released near the microvasculature where the flagellated propulsion of MTB becomes more effective. But in order to investigate the possibility for such approach, it essential to evaluate if the limit in MRN would occur in regions in the vasculature where bacterial propulsion would still be sufficiently effective for targeting purpose. As such, the roles and the limits of MNPs responsible for how deep MRN can be performed must be investigated against bacterial propulsion.

A. Pulling Force during MRN

Although the blood flow provides a great propelling force for TMMCs or microcarriers capable of encapsulating drug-loaded MTBs, the magnitude of the pulling force induced on the MNPs embedded in such microcarriers would impact the steering capability of the microcarriers towards the branches leading to the final targeted area at vessel's bifurcations. For a given maximum magnetic gradients that have been estimated in the range of 400-500mT/m for a MRN platform capable of accommodating a human adult torso, the pulling force induced would then be mainly related to the effective volume of MNPs in the microcarriers and the magnetization saturation level of the MNPs. The first main challenge is then to synthesize MNPs having a sufficiently high magnetization saturation level while being compatible for interventions in humans. The second challenge would be to provide a sufficiently high homogenous magnetic field covering the entire MRN workspace that would fully saturate the MNPs. In other words, if ideally the high homogeneous field provided by a 1.5T and higher clinical MRI scanner would be sufficient to fully saturate MNPs having the highest magnetization saturation levels. Finally, to be able to embed sufficient MNPs in the microcarriers considering the limits in diameters of the microcarriers for navigating in thinner vessels while providing sufficient room for therapeutic cargos such as the drug-loaded MTBs.

B. Tracking during MRN

Unless performed in open loop, closed-loop navigation control would require tracking information about the position of the microcarriers. This tracking information gathered during MRN is required in order to apply at the right time the directional magnetic gradients responsible for inducing the pulling force on the MNPs. Indeed, applying such gradients at the wrong time or when the microcarriers are in the wrong location would most likely reduce MRN performance below an optimal level. This would not only result in a reduction of therapeutic efficacy but also in an increase of secondary toxicity and potential severe unwanted side-effects for the patients.

Again, the MNPs embedded in the microcarriers are responsible for the tracking information. Indeed, MNPs can be used as MRI contrast agents. When placed in a high homogeneous magnetic field such as the one provided by MRI scanners, ferromagnetic materials induce a distorted region in MR images due to the field inhomogeneity caused by their high magnetic susceptibility. The size of such distortions can be much larger than the overall size of the microcarriers. This is an important fact since MRN due to the need of a high homogeneous field and superposed magnetic gradients is only compatible with real-time tracking based on MRI techniques. The challenge is that MRI like other medical imaging modalities lack sufficient spatial resolutions to detect microcarriers that would be sufficient small for embolization in narrower vessels that would be close enough to allow the release of the drug-loaded MTBs where they would swim effectively for targeting purpose.

Indeed, the spatial resolution of clinical MRI scanners or the size of a voxel is in the order of a few hundred micrometers. This is substantially larger than the overall size (in the order of a few tens of micrometers) the microcarrier intended to launch drug-loaded MTBs. The challenge would then be to be able from the limited amount of magnetic material provided by the MNPs embedded in the microcarriers to create a distortion beyond the size of a single voxel in order to be detectable with MRI techniques.

C. Directional Control for Bacterial Propulsion

Bacterial propulsion provided by the MC-1 bacteria acting as therapeutic microcarriers in the microvasculature is only valid if adequate directional control can be performed during tumor targeting. Unlike for other types of bacteria, the swimming direction of MTBs can be directly controlled using magnetotaxis. For magnetotaxis control intended for drug delivery in cancer therapy, a chain of nanoparticles in the form of crystals and known as magnetosomes embedded in the MC-1 cell is the key element that would allow effective targeting. Knowing that the thrust propulsive force of the MC-1 bacterium of 4pN and higher, exceeds the MRN capability for spherical microcarriers having the highest magnetization saturation level of identical size (i.e. 2 micrometers in diameter) which appears to be ideal for navigating in the tiniest capillaries found in humans while providing maximum volume or surface to attach the therapeutic loads, the challenge is to determine if a sufficient low directional magnetic field can be used to guide these MTBs towards a tumor at any depth inside the body.

Indeed, the chain of single magnetic domain MNPs (approx. 70 nanometers in diameter) acts like a miniature magnetic compass needle embedded in the cell and therefore, its orientation will follow the lines of magnetic field which could lead towards a tumor. Because the magnetic field is only used to induce a directional torque (preventing its usage in a MRI scanner) while the propulsive force is provided by the flagella, the magnitude of the magnetic field required for adequate directional control is expected to be much less than the one required for generating a directional pulling force equivalent to the flagella propulsive force of the MC-1 cell. The challenge is then to evaluate such magnitude and to determine its applicability for drug delivery in cancer therapy.

III. Materials and methods

For MRN, a 1.5 T Siemens clinical scanner was upgraded with 400mT/m proprietary propulsion coils designed to fit small animals. As specified earlier, the limit of 400mT/m has been identified as the upper effective technological limit applicable to a human adult. Special software modules were developed allowing MR navigation by inducing a propelling force on the MNPs to maintain the carriers on a planned trajectory based on tracking information gathered from the artifacts generated by the same MNPs. Special fluidic platforms mimicking various vascular conditions such as flow rates and distances between successive bifurcations were developed to assess MRN capabilities using iron-oxide and higher magnetization saturation iron-cobalt MNPs.

For MR-tracking assessments, knowing that image distortion is apparent on both gradient echo (GE) and the spin echo (SE) sequences, GE sequences were used because the echo shifting effect only appears in GE images. Experiments were carried out in phantoms placed in the MRI scanner. Chrome-steel microspheres (measured saturation magnetization of $\sim 1.3\times10^{6}$ A/m) of various diameters were used for the experiments. Imaging was performed using a 1.5 T (Magnetom Siemens) MRI system with images acquired with a standard 8 channel head coil. The following GE parameters were used: FOV = 100 mm × 160 mm, TR = 500 ms, imaging matrix = 256, slice thickness = 3.5 mm, flip angle = 25°, pixel spacing = 0.5 mm, TE = 10 ms.

For bacterial propulsion, MC-1 flagellated MTBs were cultivated. Microfluidic platforms were developed to assess the capabilities of the bacteria recorded from images gathered through an optical microscope. A special platform referred to as a Magnetotaxis system was developed to allow the generation of low intensity 3D magnetic fields with various magnitudes and to allow the bacteria to aggregate and to control their swimming directions within a 3D volume.

IV. Results and discussion

For MRN, all MNPs were at magnetization saturation when placed in a 1.5T scanner, providing maximum induced propelling force when submitted to 3D gradients. For the same volume, iron-cobalt MNP coated with a thin layer of graphite to prevent oxidation provided close to 3 times propelling force compared to iron-oxide MNPs. Even with the highest magnetization saturation level achieved with iron-cobalt MNPs, the minimum diameter vessels in human that can be navigated by MRN is estimated from preliminary experimental results and in worst case conditions to approximately 50 to 100 micrometers.

Using a clinical MRI system, a single 15 μm microsphere was successfully detected in gradient-echo scans. These results suggest that MRN closed-loop (feedback) control could potentially be performed for the smallest MRN-compatible microcarriers. Nonetheless, the real-time performance would need further investigations while the fact that the exact positioning of the microcarriers within each voxel cannot be determined, would have to be taken into account during MRN.

MTB were very effective in capillaries, including vessels in the upper range of the MRN limits. MTB average swimming velocities were 200 micrometers/s with a loss in motility after 40 minutes in human physiological environments. Less than 15 Gauss were necessary to direct MTB towards a tumor, suggesting its applicability in cancer therapy.

V. Conclusion

The low magnitude of magnetic field required to induce a torque on the chain of MNPs in MTBs for directional control purpose suggests that bacterial carriers encapsulated in MRN-compatible carriers could potentially be used to target a larger range of tumors.

Acknowledgment

This work was supported in part by the Canada Research Chair (CRC), the Fond Québecois de la Recherche sur la Nature et les Technologies (FQRNT), and grants from the National Sciences and Engineering Research Council of Canada (NSERC).

References

1. Brannon-Peppas L, Blanchette J (2004) Nanoparticle and targeted systems for cancer therapy. Adv Drug Delivery Rev 56:1649-1659
2. Martel S, Mathieu J-B, Felfoul O et al (2007) Automatic navigation of an untethered device in the artery of a living animal using a conventional clinical imaging system. Appl Phys Lett 90:114105
3. Pouponneau P, Leroux J-C, Soulez G et al (2011) Co-encapsulation of magnetic nanoparticles and doxorobicin into biodegradable microcarriers for deep tissue targeting by vascular MRI navigation. Biomaterials 32(13):3481-3486
4. Martel S, Felfoul O, Mathieu J-B et al (2009) MRI-based nanorobotic platform for the control of magnetic nanoparticles and flagellated bacteria for target interventions in human capillaries. Int J of Robotics Research 28(9):1169-1182
5. Blakemore R (1975) Magnetotactic bacteria. Science 190(4212):377-379

Author: Sylvain Martel
Institute: École Polytechnique de Montréal
Street: 2900 boul. Édouard-Montpetit
City: Montréal
Country: Canada
Email: sylvain.martel@polymtl.c

Novel Nanoparticles for Oral Exendin-4 Delivery

Ho-Ngoc Nguyen[1], Thy-Anh Bui[2], and Ha-Giang Nguyen[2]

[1] Nanomega Medical Corp., CA 92630, United States
[2] Department of Chemical Engineering, Cantho University, Cantho, Vietnam

Abstract — **Exendin-4 is a potent insulinotropic agent in diabetes patients; however, its therapeutic utility is limited due to the frequent injections required. In this study, an orally available exendin-4 formulation, using a pH-responsive NP, was developed. Following oral administration of ^{123}I-labeled-exendin-4 loaded NPs in rats, the biodistribution of the administered drug was investigated using a dual isotope dynamic SPECT/CT scanner. The results showed that the radioactivity of ^{123}I-exendin-4 propagated from the esophagus, stomach, and small intestine and then was absorbed into the systemic circulation; with time progressing, ^{123}I-exendin-4 was metabolized and excreted into the urinary bladder.**

Keywords — **Oral delivery, biodistribution, exendin-4.**

I. INTRODUCTION

Exendin-4, a 39-amino-acid peptide, is an incretin mimetic found in the lizard saliva. It shares several glucoregulatory activities with the mammalian incretin hormone glucagon-like peptide-1 (GLP-1) such as glucose-dependent enhancement of insulin secretion, suppression of glucagon secretion, reduction of gastric mobility and food intake. Exenatide, a synthetic version of exendin-4, has been approved as adjunctive therapy for patients with type 2 diabetes failing to achieve glycemic control with oral antidiabetic agents [1]; one potential disadvantage in its clinical applications is the frequent subcutaneous (SC) injections required. SC injections could cause pain, side effects and possible infections at the sites of injection which could adversely affect patient compliance [2].

The oral route is considered to be the most convenient and comfortable means of drug administration for patients [2]. However, oral administration of protein drugs is encountered with many difficulties such as their proteolytic instabilities and limited abilities to traverse biological barriers. Consequently, the development of a delivery vehicle for oral administration of protein drugs, including that of exendin-4, presents an interesting challenge. Recently, we reported a pH-sensitive nanoparticle (NP) vehicle, composed of chitosan (CS) and poly(γ-glutamic acid) (γ-PGA), as a platform technology for oral delivery of protein drugs [2]. CS, a cationic polysaccharide, has special features of adhering to the mucus layer and transiently opening the tight junctions between epithelial cells, thus increasing the permeation of protein drugs via the paracellular pathway [2,3]. γ-PGA, an anionic peptide, is water-soluble, biodegradable, and nontoxic [2].

In this study, exendin-4 loaded NPs were prepared using a simple ionic-gelation method self-assembled by the positively-charged CS and the negatively-charged γ-PGA in an aqueous environment. Characteristics of the prepared NPs were examined using a scanning electron microscopy (SEM) and transmission electron microscopy (TEM). The biodistribution of exendin-4 loaded NPs after oral administration in rats was investigated by the single-photon emission computed tomography (SPECT)/computed tomography (CT).

II. MATERIALS AND METHODS

A. *Preparation of Exendin-4 Loaded NPs*

CS (MW 80 kDa) with a degree of deacetylation of approximately 85% was acquired from Koyo Chemical Co. Ltd. (Japan), while γ-PGA (MW 60 kDa) was purchased from Vedan Co. Ltd. (Taichung, Taiwan). In the preparation of NPs, an aqueous exendin-4 (2 mg/mL, 250 μL, pH 7.4, Sigma-Aldrich, St. Louis, MO, USA) was premixed with an aqueous CS (1.2 mg/mL, 2.5 mL, pH 6.0) in the presence of Fe^{+3} [($FeCl_3.6H_2O$) in a concentration of 0.009% w/v, to enhance the drug loading efficiency] under magnetic stirring at 4°C overnight. Subsequently, an aqueous γ-PGA (2 mg/mL, 250 μL, pH 7.4) was added, by flush mixing with a pipette tip, into this mixture to form NPs at room temperature. The obtained NPs were collected by centrifugation (15000rpm, 4 C for 50 min) and then redispersed in deionized (DI) water and stored at 4°C until used.

B. *MD Simulations*

MD simulations of the interaction between CS and exendin-4 in the absence/presence of Fe^{+3} were performed [4]; simulations were accomplished with the program NAMD [5] using parameters adapted from the CHARMM 27 force field [6]. The models were minimized to remove

V. Van Toi et al. (Eds.): 4th International Conference on Biomedical Engineering in Vietnam, IFMBE Proceedings 40, pp. 65–68, 2013.
www.springerlink.com

unfavorable contacts, brought to 310 K by velocity rescaling and equilibrated for 1 ns. Before any MD trajectory was run, 40 ps of energy minimization were performed to relax the conformational and structural tensions. This minimum structure was the starting point for the MD simulations. For this purpose, the molecule was embedded into a cubic simulation box of 80 Å. A cutoff distance of 12 Å was employed for the nonbonded and electrostatic interactions. The heating process was performed from 0 to 310 K through Langevin damping with a coefficient of 10 ps^{-1}. A time step of 2 fs was employed for rescaling the temperature. After 20 ps heating to 310 K, equilibration trajectories of 1 ns were recorded, which provided the data for the structural and thermodynamic evaluations. The equations of motion were integrated with the Shake algorithm with a time step of 1 fs. Figures displaying atomistic pictures of molecules were generated using UCSF Chimera.

C. *Characterization of Test NPs*

The morphology of test NPs were examined by SEM (JSE-6500F, JEOL, Tokyo, Japan) and TEM (JEOL 2010F, Tokyo, Japan). The drug loading efficiency (LE) and loading content (LC) of test NPs were determined by assaying the amount of free exendin-4 remaining in the supernatants using reverse-phase high-performance liquid chromatography (HPLC). The exendin-4 LE and LC in NPs were calculated using the equations listed below [2].

$$\text{LE (\%)} = \frac{\text{total exendin-4 added} - \text{free exendin-4}}{\text{total exendin-4 added}} \times 100\%$$

$$\text{LC (\%)} = \frac{\text{total exendin-4 added} - \text{free exendin-4}}{\text{weight of nanoparticles}} \times 100\%$$

D. *Transepithelial Electrical Resistance Measurements*

Evaluation of test NPs (0.2 mg/mL, 1 mL) in opening cellular tight junctions at pH 6.6, pH 7.0 and pH 7.4 (in the donor compartment; at pH 7.4 in the receiver compartment) was investigated in Caco-2 cell monolayers cultured in Costar Transwell 6 wells/plates [2,3]. After the addition of test NPs, the change of transepithelial electrical resistance (TEER) for the tightness of cell monolayers was measured with a Millicell®-Electrical Resistance System (Millipore Corp., Bedford, MA, USA).

E. *Biodistribution Study*

In the study, exendin-4 was radiolabeled with Iodine-123 (^{123}I) using the iodogen precoated tubes (Pierce Iodination Tubes, Thermo Fisher Scientific, Rockford, IL, USA) as per the manufacturer's instructions; the free ^{123}I was removed using a centrifugal dialysis device (MWCO: 3 kDa, Amicon Ultra 4, Millipore, Billerica, MA, USA). The radiochemical purity of ^{123}I-exendin-4 was determined by a reversed-phase HPLC system equipped with a radioactivity flow detector [7]. The ^{123}I-exendin-4 was then used to prepare NPs as described in *Section II.A.*

The biodistribution of ^{123}I-exendin-4 loaded NPs suspending in DI water, after oral administration in a rat model (n = 3), was studied using a high-resolution dual modality system (NanoSPECT/CT, Bioscan, Washington DC, USA). This system features the helical scanning for both SPECT and CT acquisitions using a translation stage with variable axial scanning ranged from 30 to 300 mm. Animal imaging was performed under the controlled temperature (37°C) and anesthesia (1.5% isoflurane in 100% oxygen).

The SPECT and CT fusion images were displayed in PMOD (PMOD Technologies Ltd, Zurich, Switzerland) and the volumes of interest (VOIs) are drawn manually over the whole body (Wb), stomach (S), intestine (Int), and urinary bladder (Ub). The average radioactivity in the VOIs was converted to percent of the ingested dose per cubic centimeter (%ID/cc) using a calibrated cylinder factor. The distribution of ^{123}I-exendin-4 in the peripheral tissue/plasma (PP) was calculated using the following formula:

$$\text{PP}(t) = \text{Wb}(t) - \text{S}(t) - \text{Int}(t) - \text{Ub}(t)$$

where t is the duration after the ingestion of ^{123}I-exendin-4 loaded NPs.

F. *Statistical Analysis*

Comparison between two groups was analyzed by the one-tailed Student's *t*-test using statistical software (SPSS, Chicago, Ill, USA). Data are presented as mean ± SD. A difference of $P < 0.05$ was considered statistically significant.

III. RESULTS AND DISCUSSION

A. *Characteristics of Exendin-4 Loaded NPs*

Metal ions have been used to stabilize peptides/proteins; additionally, the ability of CS to chelate metal ions in aqueous solution has been reported [8]. It was found that in the presence of Fe^{3+}, the drug LE in NPs was significantly increased to 60.9 ± 2.0% ($P < 0.05$); in this case, the LC of exendin-4 in NPs was 14.7 ± 1.9% (n = 5). The SEM and TEM micrographs of test NPs were shown in Fig. 1.

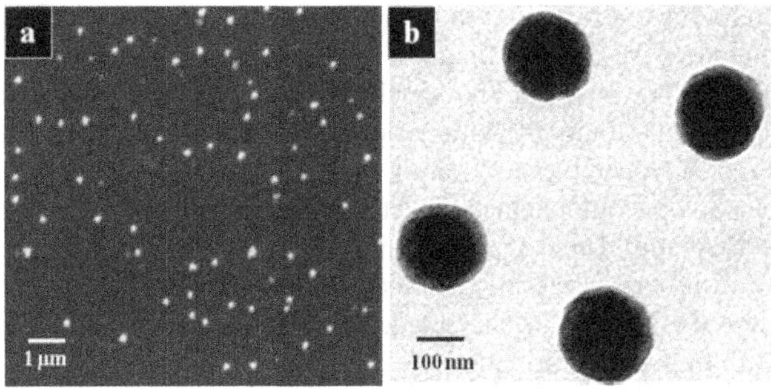

Fig. 1 (a) SEM and (b) TEM micrographs of exendin-4 loaded NPs.

B. MD Simulations

MD simulations were performed in a full-atom model to gain insight into the mechanism of the interaction between CS and exendin-4 in the absence/presence of Fe^{3+}; the results are shown in Fig. 2. In the study, 9 exendin-4 and 4 20-mer CS molecules were used in the MD simulation at pH 6.5, in which the ratio of protonated ($-NH_3^+$) and deprotonated ($-NH_2$) amine groups on CS was about 1:1. The pI of exendin-4, containing 5 glutamic acids (Glu) and 1 aspartic acid (Asp), is around 4.2 [1]; therefore, it is negatively charged at pH 6.5.

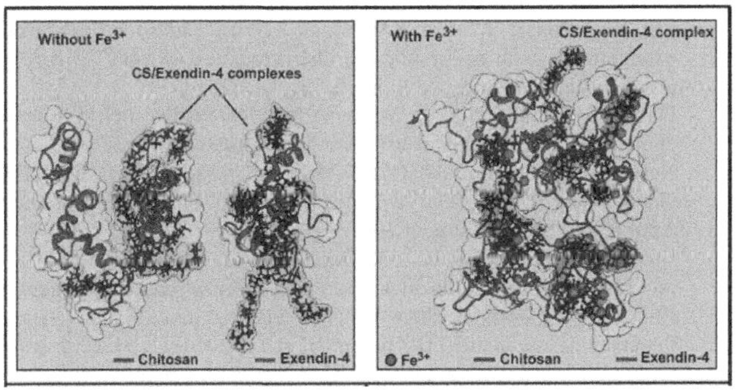

Fig. 2 Results obtained by the molecular dynamic simulations showing the interaction between 4 20-mer chitosan (CS) and 9 exendin-4 molecules in the absence/presence of Fe^{3+} at pH 6.5.

As shown in Fig. 2, in the absence of Fe^{3+}, 2 sets of complexes, each containing 2 CS and 3 exendin-4 molecules, were formed due to the electrostatic interaction, while 3 exendin-4 molecules were left alone. In contrast, in the presence of Fe^{3+}, a single complex consisting of 4 CS and 9 exendin-4 molecules was found. It is known that transition metal ions (such as Fe^{3+}) may form coordination in the vicinity of nitrogen as well as glycosidic, carboxylic, and hydroxyl oxygen atoms [8]. Hence, besides the electrostatic interaction, there were coordinate-covalent bonds formed between CS and exendin-4 in the presence of Fe^{3+}, thus significantly increasing the drug LE in the preparation of test NPs.

C. TEER Measurements

Measurements of TEER of cell monolayers can be used to predict their paracellular permeability [3]; as the tight junctions open, the TEER of Caco-2 cell monolayers is significantly reduced because of the passing of ions via their paracellular route.

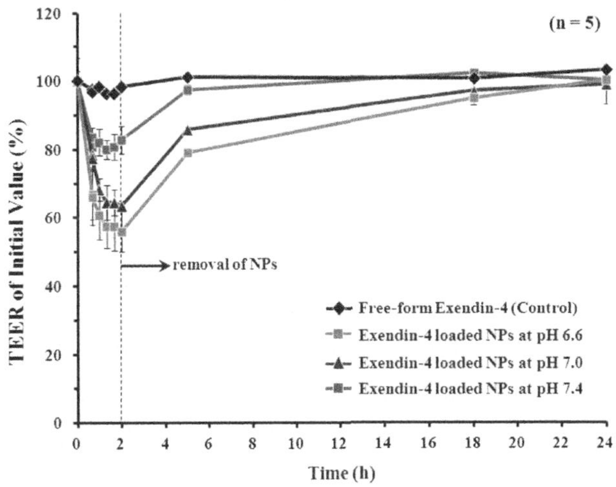

Fig. 3 Effects of the test NPs on TEER values of Caco-2 cell monolayers at different pH environments.

The incubation of freeze-dried exendin-4 loaded NPs on the apical side of Caco-2 cell monolayers at distinct pH values led to a significant reduction in TEER as compared to that treated with the free-form exendin-4 ($P < 0.05$, Fig. 3). These observations were pH-dependent; with decreasing pH, the reduction in TEER was significantly enhanced. After removal of the incubated NPs, a gradual increase in TEER was observed. Approximately complete recovery of the TEER was seen within 24 h, suggesting that changes in permeability of Caco-2 cell monolayers were not mediated through damaging of tight junctions. These observations indicate that the prepared NPs could transiently and reversibly open the tight junctions between Caco-2 cells.

D. Biodistribution Study

Fig. 4a shows the three-dimensional (3D) volume-rendering SPECT/CT images of ^{123}I-exendin-4 distributed at distinct time points after oral administration of test NPs. The radioactivity of ^{123}I-exendin-4 propagated from the esophagus, stomach, and intestine with time and then was absorbed into the systemic circulation (as shown in the kidney and the urinary bladder).

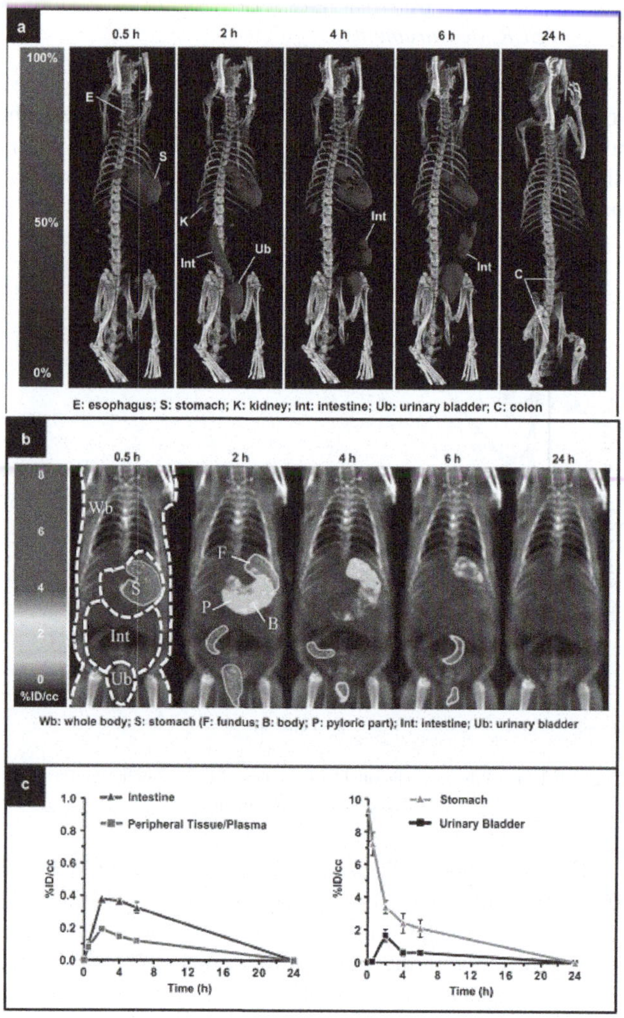

Fig. 4 (a) Three-dimensional (3D) volume-rendering SPECT/CT images of ^{123}I-exendin-4 distributed in rats at distinct time points after oral administration of test NPs; (b) maximum intensity projections of ^{123}I-exendin-4 in rainbow pseudo-color scale; and (c) radioactivity concentrations (%ID/cc) of ^{123}I-exendin-4 observed in different organs.

The maximum intensity projections of ^{123}I-exendin-4 in rainbow pseudo-color scale are illustrated in Fig. 4b; the radioactivity concentrations (%ID/cc) of ^{123}I-exendin-4 observed in different organs are presented in Fig. 4c. Shortly after oral intake, the radioactivity of ^{123}I-exendin-4 appeared quickly in the stomach. At 2 h post administration, approximately 55% of the ID of ^{123}I-exendin-4 entered into the small intestine; the clearance rate of ^{123}I-exendin-4 in the stomach became slower subsequently. A two-phase gastric emptying behavior of ^{123}I-exendin-4 was found in the stomach: a fast elimination rate in the pyloric/body regions and a relatively slow rate in the fundus portion (Fig 4b). The radioactivity of ^{123}I-exendin-4 reached to its peak value at 2 h post ingestion in the small intestine and PP and then eliminated gradually with a clearance half-life of 11.7 h and 12.3 h, respectively. A similar pattern in biodistribution in these two components indicates a significant absorption of ^{123}I-exendin-4 into the systemic circulation via the intestine. With time progressing, ^{123}I-exendin-4 was metabolized and excreted into the urinary bladder; its radioactivity fluctuated with time due to urination of the test animal.

IV. Conclusions

We developed a formulation, a pH-responsive NP, for the oral delivery of exendin-4. The *in vitro* TEER and *in vivo* biodistribution results suggest that the orally available formulation developed in the study might be employed as an alternative for mediating the glucoregulatory activity in diabetes subjects.

References

1. Gedulin BR, Smith PA, Jodka CM et al. (2008) Pharmacokinetics and pharmacodynamics of exenatide following alternate routes of administration. Int J Pharm 356(1-2):231–238.
2. Lin YH, Sonaje K, Lin KM et al. (2008) Multi-ion-crosslinked nanoparticles with pH-responsive characteristics for oral delivery of protein drugs. J Controlled Release 132(2):141–149.
3. Dodane V, Amin Khan M, Merwin JR. (1999) Effect of chitosan on epithelial permeability and structure. Int J Pharm 182(1):21–32.
4. Nelson MT, Humphrey W, Gursoy A et al. (1996) NAMD: A parallel, object-oriented molecular dynamics program. Int J Suppercomput Appl High Perform Comput 10(4):251–268.
5. Brooks BR, Bruccoleri RE, Olafson BD et al. (1983) CHARMM: A program for macromolecular energy, minimization, and dynamics calculations. J Comput Chem 4(2):187–217.
6. Pettersen EF, Goddard TD, Huang CC et al. (2004) UCSF Chimera-A visualization system for exploratory research and analysis. J Comput Chem 25(13):1605–1612.
7. Sodoyez JC, Sodoyez-Goffaux F, Guillaume M et al. (1983) [123I]Insulin metabolism in normal rats and humans: External detection by a scintillation camera. Science 219(4586):865–867.
8. Varma AJ, Deshpande SV, Kennedy JF. (2004) Metal complexation by chitosan and its derivatives: A review. Carbohydr Polym 55(1):77–93.

Corresponding author:

Author: Ho-Ngoc Nguyen
Institute: Nanomega Medical Corp., Taiwan Branch
Street: Kuang-Fu Road
City: Hsinchu
Country: Taiwan R.O.C.
Email: ng.hngoc@gmail.com

Identifying Ovine Transcranial Acoustic Windows

Tony de Souza-Daw[1], Philip M. Lewis[2], Robert Stewart[3], Paul Junor[4], Jerome Maller[5], Thang Manh Hoang[6], Tien Dzung Nguyen[6], Richard Manasseh[3,7,8]

[1] Centre of Technology, RMIT University, Vietnam
[2] Department of Neurosurgery, The Alfred Hospital, Australia
[3] Fluid Dynamics Group, CSIRO Materials Science and Engineering Division, Australia
[4] Department of Electronic Engineering, La Trobe University, Bundoora, Victoria, Australia
[5] Monash Alfred Psychiatry Research Centre, The Alfred & Monash University School of Psychology and Psychiatry, Australia
[6] School of Electronics and Telecommunications, Hanoi University of Science and Technology
[7] Faculty of Engineering and Industrial Sciences, Swinburne University of Technology, Hawthorn, Victoria, Australia
[8] Department of Mechanical Engineering, University of Melbourne, Melbourne, Victoria, Australia

Abstract — **Transcranial Doppler ultrasound (TCD) is used to measure cerebral blood flow velocity in patients with cerebrovascular pathology. The task of locating an adequate acoustic window, through which TCD may be performed, is frequently time consuming or impossible for patients with thick or dense temporal bones. We examined the possibility of utilising B-Mode ultrasound images to guide detection of acoustic windows prior to TCD. B-Mode images of the temporal bone were acquired on deceased sheep shortly after death. We found that on selected sheep, the inner table of the temporal bone was visible on B-Mode imaging. We suggest that visibility of the inner table of the temporal bone on B-Mode imaging is associated with a high probability of successful TCD at that location.**

Keywords — **Transcranial Doppler Ultrasound, Acoustic Windows, Temporal Bone.**

I. Introduction

Transcranial Doppler ultrasound (TCD) is used to measure cerebral blood flow velocity from the intracranial arteries of patients with intracranial disease. The procedure requires skull bone be suitably transparent to ultrasound, so that reflected waves are of sufficient strength to permit Doppler shift and therefore velocity measurement. Any point at which velocity signals may be obtained is generally referred to as an "acoustic window" [1]. Generally, the squamous temporal bone provides the best access to those vessels most implicated in cerebrovascular disease; typically these also follow a course permitting insonation angle minimization and thus more accurate velocity measurements.

Locating a suitable acoustic window requires manually scanning the temporal bone [1-3] until Doppler signals are obtained, at which point vessel identification can be attempted. Blind acoustic window detection can therefore be a time consuming process, particularly in those patients with poor or absent acoustic windows [4-9]. Hence success rates for the detection of intracranial arteries with TCD are typically around 80% when insonating through the temporal bone [3]. In patients with absent acoustic windows, procedural abandonment will only occur once a rigorous and exhaustive search for Doppler signals has been attempted. Therefore a need exists for an improved method of determining the location and/or absence of transcranial acoustic windows, particularly when the TCD practitioner is inexperienced.

The aim of this study was to investigate the utility of B-mode ultrasound imaging for quickly locating a suitable temporal acoustic window.

II. Method

Eight sheep heads were obtained from a local butcher, within two hours of slaughter. Five heads were used to prototype the experiments, using a GE Vingmed System V ultrasound scanner (GE Healthcare, Chalfton St. Giles, UK). Prior to scanning, the skin overlying the temporal bone was shaved and acoustic coupling gel applied. The probe was positioned on the temporal bone near the animal's ear. The probe was translated towards the eye, and then downwards and scanned back towards the ear, until the brain surface was visible in the B-Mode image or the entire area was covered. Final images were obtained from a Sonix RP scanner with L9-4 handheld probe (Ultrasonix Ultrasound Systems, Richmond, British Columbia, Canada), with the ultrasound frequency set at 10MHz. Images were reviewed and different media identified in a freehand fashion. The temporal bone was scanned whilst observing the B-Mode image for both brightness and details at the approximate depth of the cortical surface. It was hypothesize that if the far side of the temporal muscle was

V. Van Toi et al. (Eds.): 4th International Conference on Biomedical Engineering in Vietnam, IFMBE Proceedings 40, pp. 69–71, 2013.
www.springerlink.com

visible then an ultrasound could penetrate the temporal bone for TCD purposes. To demonstrate the differences in B-mode contrast across layers, the median and interquartile range of intensity values for each were computed. All image and statistical analysis was performed using MATLAB (Mathworks, Natick, Massachusetts, USA).

III. Results

A typical B-Mode image of the sheep temporal bone is presented in Figure 1, which demonstrates the scalp, jaw bone, temporal muscle and skull bone that form the acoustic window.

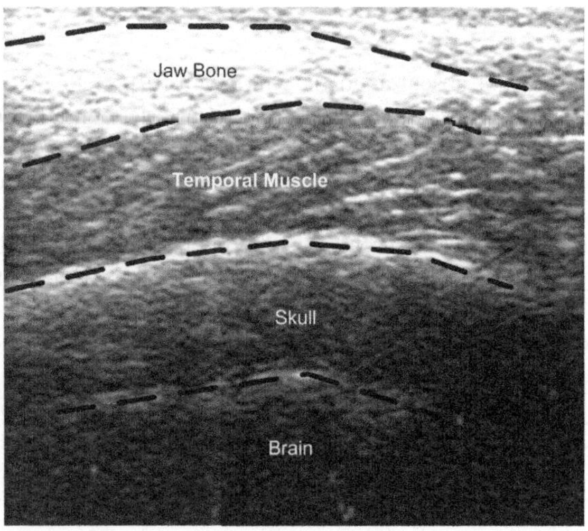

Fig. 1 A 10 MHz, B-Mode ultrasound image of the sheep temporal bone.

Median intensity values for the three acoustically-distinct layers seen are shown in Table 1. The temporal muscle values were included in the scalp. The results are presented as box-plots to demonstrate the variability of the media median intensity (Figure 2). The width of a notch is computed such that non-overlapping box plot notch widths have different medians at the 5% significance level.

IV. Discussion

Generally, we found the visibility of the sheep brain surface to be poor, because the ultrasound waves must travel through two layers of bone. The anatomy of the sheep skull is such that the temporal bone is partially obscured by the coronoid process of the mandible (Figure 3). Thus, scanning for acoustic windows in sheep is more difficult than would be expected in humans. We were not always able to isolate a part of the temporal bone not partially obscured by jaw, therefore the ultrasound waves were often penetrating both jaw and skull bone.

Despite this, the distribution of intensity values across layers seen in the B-Mode images contrast sufficiently with the transcranial media (scalp, temporal bone and brain) to enable manual detection, shown by the statistical differences of each layer in Table 1 and Figure 2. A large distribution is an indication of the signal strength and scattering; the scalp generally has a wider distribution because the signal strength is relatively high.

Ultrasound frequencies of 1-2MHz provide the best penetration of the skull to maximise the chance of procedural success. Therefore the 10MHz broadband frequency used in our experiments might not produce optimal results, given the greater attenuation by bone at this frequency. Therefore, one of the limitations of this study is that failure to detect an acoustic window at 10MHz, may not predict failure at 1-2MHz. However, in those images where the sheep brain was visible, we suggest that this indicates the presence of an acoustic window amenable to TCD at frequencies closer to 2MHz. Further work in humans will be required to validate these findings, as well as establishing whether a lack of visible brain in B-mode imaging at 10MHz can predict failure of TCD at 1 or 2MHz.

Table 1 Basic statistics for intensity values of different layers identified on B-mode imaging of the sheep skull.

	Scalp	Skull	Brain
Upper Percentile	235	74	51
75th Percentile	130	59	49
High-Low Notch	91-89	52-52	48-48
Median	90	52	48
25th Percentile	60	49	48
Lower Percentile	43	47	46

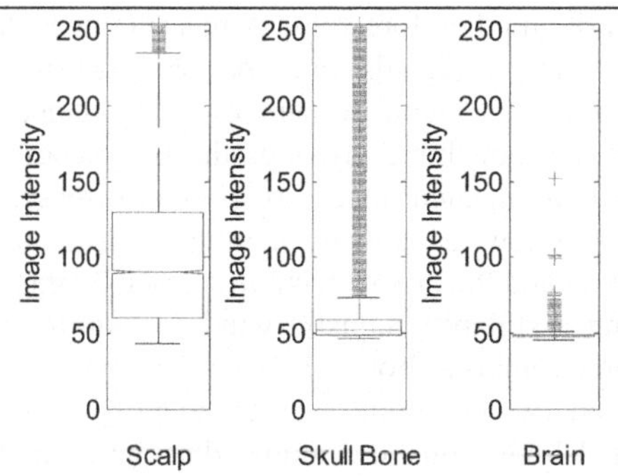

V. Conclusion

B-mode imaging may be used to identify scalp, skull and brain in the sheep head. The median intensities of all layers were found to be statistically different. We suggest that this finding may lead to a rapid, automated method of identifying acoustic windows in temporal bone, through which transcranial Doppler ultrasound may be performed successfully. We propose to test this hypothesis in human trials.

Acknowledgment

This work was supported by La Trobe University, Write-up award.

References

1. Newell DW, Aaslid R. (1992) Transcranial Doppler: Raven Press
2. Viken L. Babikian LRW. (1993) Transcranial doppler ultrasonography: Mosby-Year Book
3. Baumgarner RW. (2006) Handbook on Neurovascular Ultrasound,. Frontiers of Neurology and Neuroscience.
4. Kollar J, Schulte-Altedorneburg G, Sikula J, et al. (2004) Image quality of the temporal bone window examined by transcranial Doppler sonography and correlation with post-mortem computed tomography measurements. J Cerebrovasc Dis: 139
5. Grolimund P. Transmission of Ultrasound Through the Temporal Bone. Transcranial Doppler Sonography (1986). Berlin;. p 10.
6. Jarquin-Valdivia AA, McCartney J, Palestrant D, et al. (2004) The thickness of the temporal squama and its implication for transcranial sonography. J Neuroimaging., 139.
7. Fry FJ, Barger JE. (1978) Acoustical properties of the human skull. J Acoust Soc Am., 1576.
8. Pfaffenberger S, Devcic-Kuhar B, Kollmann C, et al. (2005) Can a Commercial Diagnostic Ultrasound Device Accelerate Thrombolysis?: An In Vitro Skull Model., Stroke, 124.
9. Behrens S, Daffertshofe M, D S, et al. (1999) Low frequency, low intensity ultrasound accelerates thrombolysis through the skull., J. Ultrasound Med Bio., 269.

Author: Tony de Souza-Daw
Institute: Centre of Technology, RMIT University
Street: 702 Nguyen Van Linh Blvd., District 7, HCMC
City: Ho Chi Minh City
Country: Vietnam
Email: Anthony.desouza-daw@rmit.edu.vn

Sleep Stages Recognition Based on Combined Artificial Neural Network and Fuzzy System Using Wavelet Transform Features

Chuang-Chien Chiu[1], Bui Huy Hai[2], and Shoou-Jeng Yeh[3]

[1] Department of Automatic Control Engineering Feng Chia University, Taichung, Taiwan, R.O.C.
[2] Electrical and Communications Engineering, Feng Chia University, Taichung, Taiwan, R.O.C.
[3] Section of Neurology and Neurophysiology, Cheng-Ching General Hospital, Taichung, Taiwan, R.O.C.

Abstract — **Improving the quality of sleep is an important issue for many researches. A number of biomedical signals, such as EEG, EMG, and EOG were used to classify sleep stages. Based on those signals, one can detect and diagnose the sleep related disorders. There were many researches focused on automatic sleep stages classification. In this research, a new classification method is presented by applying Elman neuron network combined with fuzzy rules and features are extracted by wavelets packets. Nine subjects were recruited from Cheng-Ching General Hospital, Taichung, Taiwan. The sampling frequency is 250Hz and the single channel (C_3-A_1) EEG signal was acquired for each subject. Combined network was used to recognize the sleep stages in each epoch (a 10 second segment data). The classification results relied on the strong points of neural network and fuzzy logic with average sensitivity is 88.48%, average specificity achieves 95.96%, and average accuracy is 93.79%. The data samples and the length of sleep intervals will be increased for experiment in the future to improve the accuracy.**

Keywords — **neural network, sleep stages, fuzzy system, wavelets.**

I. Introduction

Having a good sleep quality is very important to everybody. To this problem, the scientists aware that the sleep stages recognition for treatment of sleep related disorders are very essential tasks. There are many approaches to classify the sleep stages, such as polysomnography (PSG), which measures breathing efforts, blood oxygen levels, brain electrical activity (EEG), electrocardiogram (ECG), electrooculogram (EOG), and muscle activity (EMG), are now the main diagnostic tools in sleep medicine. EEG signal is the most important signal in sleep stage classification, while the EMG and EOG are the most important signals in differentiating the rapid eye movement (REM) sleep from the other stages. Because a careful inspection of EEG signal can reveal unusual patterns, these signals provide mended scoring rules for the recognition of sleep EEG patterns based on the manual of Rechtschaffen and Kales (R&K) 1968 [1]. The frequency ranges of the EEG signal are segmented into some sub-bands, which are the Delta band (0.3~4Hz), Theta band (4~8Hz), Alpha band (8~12Hz) and Beta band (>13Hz). Sleep can be divided into two entirely different behavioral states: REM (rapid eye movement) sleep, and non-REM sleep-NREM. After the sleeper falls asleep, he or she enters into what is known as the NREM sleep. It consists of 4 stages as shown in Fig. 1. Stage 1 sleep is the period that we drift off here being a transition period from wakefulness to the other 3 stages of NREM sleep. It is of short duration, usually lasting only from 1 to 7 minutes. Sleep stage 1 is characterized by low voltage with well-defined alpha and theta band, mixed frequency EEG, as well as some slow, rolling eye movements and some relatively higher EMG activity. Stage 2 makes up the bulk of an average person's sleep each night, around 40% to 45%. A sleep researcher can easily recognize sleep stage 2 because of the presence of sleep spindles and K-complexes in the EEG waves. A sleep spindle is a rapid wax and wane of the EEG waves in a 1 to 2 second interval (about 12-14Hz), while a K-complex is a large wax and wane of the wave that somewhat resembles a mountain. Sleep stage 3 and sleep stage 4 are similar and both fall into the category of slow wave sleeps. They are so named because of the high amplitude waves in the EEG. In order to qualify as sleep stage 3, these slow waves must be presented in no more than 50 percent of the epoch. The waves are also referred to as the delta waves.

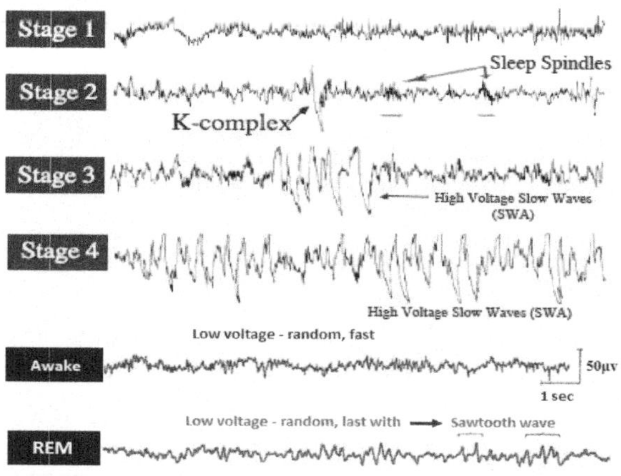

Fig. 1 10-second epochs of an EEG signal in different sleep stages.

V. Van Toi et al. (Eds.): 4th International Conference on Biomedical Engineering in Vietnam, IFMBE Proceedings 40, pp. 72–76, 2013.
www.springerlink.com

Sleep stage 4 is the other slow wave sleep period. In order to qualify as sleep stage 4, the high amplitude delta waves must be presented in more than 50 percent of the epoch, like in the EEG recording as shown in Fig. 2. Besides these four basic stages of sleep, a unique sleep stage exists, that is so called the REM. This stage gets its name from the darting eye movements that accompany it with a main component, which is beta wave. REM stage is nearly similar to stage 1. In order to distinguish this one to the other, we must apply more bio-signals, such as EMG, EOG. From the properties of six sleep stages as known in R&K scoring, our goal is to discriminate automatically the four stages, which are Awake + REM, Stage 1, Stage 2, and Slow Wave Sleep (stage 3 + stage 4 of in the NREM period).

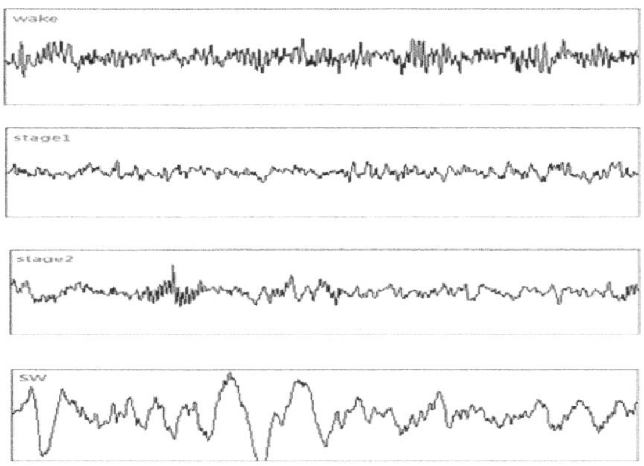

Fig. 2 Four discriminated sleep stages.

Since the EEG signal is non-stationary, the wavelet analysis can be a useful tool to analyze for non-stationary signals. Thus, wavelet transform is applied to extract features for recognition network. There were a lot of approaches that used neural network to recognize. In this research, we improve the quality of recognition depending on the combined artificial neural network for pattern recognition and fuzzy system for normalization process [4]. The scoring performed on two classification steps obtained the satisfactory results based on both strong points of neural network and fuzzy logic [9] [10].

II. Materials and Methods

***A) Data Acquisition**:* The EEG brain wave acquisition machine was the NEC|SYNAFIT1000 model. Data acquisition was performed using a general-purpose data acquisition unit (NI USB6210, National Instruments). The acquisition and processing software was developed by us. The acquisition system configuration is shown in Fig. 3. EEG signals were recorded from 9 subjects (5 males and 4 females with age ranged from 20 -23 years old). The sampling frequency was set to be 250 Hz. The subject information for each sleep stage classified manually by sleep experts is listed in Table1.

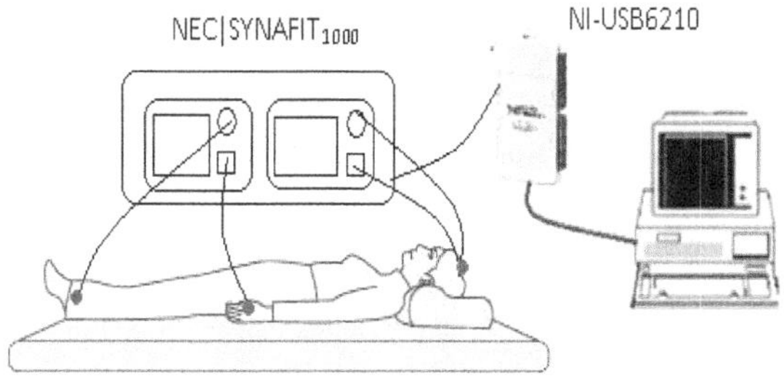

Fig.3. Data acquisition for a single channel EEG signal

The acquisition period lasted approximately one hour. Sleep stages were scored according to R&K rules based on 10-second epochs of FP1, FP2, F3, F4, C3, C4 channels. Artifacts were manually removed to work with clean data. Only the single (C_3-A_1) channel was used in our evaluation.

Table 1 Subject information.

	wake	Stage1	Stage2	SWS	total
01	98	--	1	--	99
02	24	2	13	136	175
03	61	5	8	--	74
04	9	2	62	36	109
05	13	25	18	--	56
06	--	29	48	--	77
07	11	108	--		119
08	--	--	29	43	72
09	10	9	43	17	79
total	226	180	202	232	860

B) Wavelet features: We have seen that the STFT (Short Time Fourier Transform) can be used to find time and frequency information at a set resolution. Also, it is known that the resolution must be set in order to extract certain time-frequency information from a signal. If we want to examine a signal across different time-frequency resolution scales, we must look into multi-resolution analysis (MRA). MRA is intended to allow for high time resolution with poor frequency resolution at high frequency levels, and poor time resolution with high frequency resolution at low frequency levels. To perform this task, a wavelet is used instead of a windowing function. The name wavelet means a small wave or a wave which is not of infinite length (as in a sinusoidal wave). As time-domain analysis of bio-signals such as EEG does not provide frequency details. EEG signals are dynamic, sometime are spikes/bursts, mostly non-stationary. Non-stationary signals are characterized by numerous transitory drifts trends and abrupt changes. For practical analysis, we not only need to know their frequency components but also the time at which they occur. So wavelet transform (WT) is such a tool. The wavelet transform can be viewed as transforming the signal from the time domain to the wavelet domain. This new domain contains more complicated basis functions called wavelets, mother wavelets, or analyzing wavelets.

If we have a signal $x(t)$, the wavelet decomposition is given as follows.

$$x(t) = \sum_{j=1}^{L} \sum_{k=-\infty}^{\infty} d(j,k)\psi(2^{-j}t - k) + \sum_{k=-\infty}^{\infty} a(L,K)\phi(2^{-L}t - k)$$

The function $\psi(t)$ is known as the mother wavelet, while $\varphi(t)$ is known as the scaling function. The numbers $a(L,k)$ are known as the approximation coefficients at scale L, while $d(j, k)$ are known as the detail coefficients at scale j.

Wavelet decompositions: The Discrete Wavelet Transform (DWT) coefficients can be computed by using Mallat's Fast Wavelet Transform algorithm. This algorithm is sometimes referred to as the two-channel sub-band coder and involves filtering the input signal based on the wavelet function used.

Wavelet decomposition analyzed EEG inputs to sub-bands at depth 6 (6 levels) is depicted in Fig. 4.

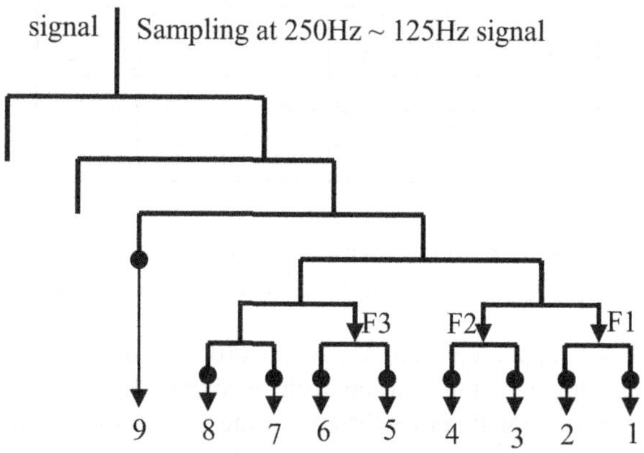

Fig. 4 Wavelet decomposition with 6 levels.

Features extraction: From wavelet decomposition, we received 9 frequency sub-bands which are consisted of the delta1, delta2, theta1, theta2, alpha1, alpha2, spindle, beta1, and beta2 as listed in Table 2. However, there are of total 13 input vectors for the input of the proposed neural network. E1, ... , E9 stand for the mean quadratic value of 9 frequency bands, which are {delta1, delta2, theta1, theta2, alpha1, alpha2, spindle, beta1, beta2}. E10 is the energy or mean quadratic of total input signal. E11 is calculated by F1/(F2+F3). E12 is calculated by F2/(F1+F3), and E13 is F3/(F2+F1), in which F1 is the combined power in delta1 and delta2, F2 is the combined power in theta1 and theta2, and F3 is the combined power in alpha1 and alpha2. Note that we can receive F1, F2, and F3 directly from wavelet decomposition as shown in Fig. 4.

Table 2 13 power bands used for features extraction.

No	Name	Frequencies range
1	Delta1 band	0~1.95Hz
2	Delta2 band	1.95~3.9Hz
3	Theta1 band	3.9~5.85 Hz
4	Theta2 band	5.85~7.8 Hz
5	Alpha1 band	7.8~9.65Hz
6	Alpha2 band	9.65~11.7Hz
7	Spindle band	11.7~13.65Hz
8	Beta1	13.65~15.6hz
9	Beta2	15.6~31.25Hz
10	S: Total signal	0 ~ 125Hz
11	F1	0 ~ 3.9 Hz
12	F2	3.9 ~ 7.8 Hz
13	F3	7.8 ~ 11.7 Hz

C) Classification: there were some researches that classify the sleep stages using neural networks, wavelet analysis, or fuzzy systems [3] [11] [12]. Each system has its own particular strong points. We intend to combine those strong points of these three systems above to improve the quality of recognition.

As one can see that the input vectors of our proposed neural network are the energy of sub-bands. However, the biomedical signals are drifted off and with complicated types. Therefore, the power of each sub-band of different sleep stages may be overlapped. In this research, we used a recognition network as classifier combined with the fuzzy system to standardize the inputs [2]. In this way, neural network and fuzzy system work independently [4] [10]. However, their inputs or outputs are interconnected in order to augment each other's capability.

Classifier: we used Elman back-propagation (EBP) network with 13 inputs, one hidden layer (with 8 neurons for best results), and one output layer (with 4 nodes) as the classifier. 13 input vectors are the power of delta1, delta2, theta1, theta2, alpha1, alpha2, spindle, beta1, beta2, full-band, E3/(E1+E2), E1/(E2+E3), and E2/(E1+E3) with the power of delta(E1), theta(E2), and alpha(E3). We can set the standard network to converge to local minimum, but there are some cases that the network was stabilized at local minima of error surface which would prevent to find the desired lowest error solution for the matrix of the weights. In our network, the private parameters-setting is set in training for best-case. There are 4 neurons in the output layer to discriminate between 4 sleep stages (Awake, Stage 1, Stage 2 and SWS). In the classification stage, the outputs were constrained by input patterns, usually represented by values, restricted in the range from 0 to 1.

Fuzzy system: A fuzzy system is implemented to standardize for input vectors of the neural network. To standardize the input vectors, we predefined the power ranges in each sub-band for every stage, based on experimental statistics and R&K [1] scoring results as listed in Table 3.

Table 3 R&K scoring results.

Sleep Stages	Signal Activity
Wake	Alpha & Beta waves
Stage 1	Theta > 50%, Alpha < 50%
Stage 2	Theta, K-complex, Sleep spindle
Stage 3	Theta, 20% <Delta<50%, EOG
Stage 4	Delta > 50%
REM	EOG, Alpha & Beta waves

There're ten basis frequency bands, but we have only been standardized for seven bands, because the difference of their power range are relatively obvious, the powers for three other bands have a similar range, we don't need to standardize them. Depend on the difference of power ranges in sleep stages, features are mapped into three or four classes (low, medium, and high) or (verylow, low, medium, and high) as shown in Fig. 5 below.

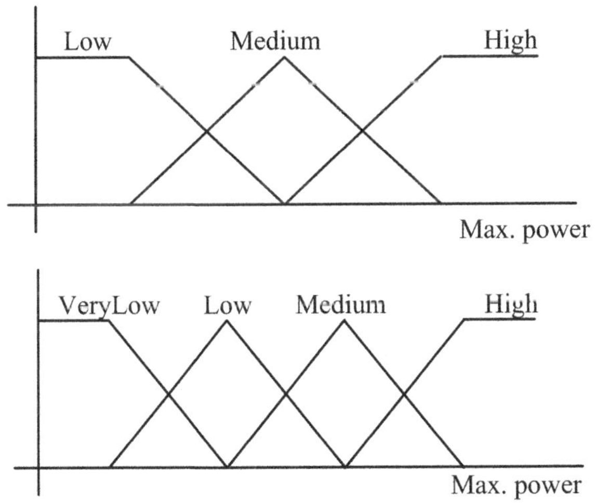

Fig. 5 (above) Features are mapped into 3 classes (below) Features are mapped into 4 classes

The standardization procedure is implemented on the basis of some rules. A correlating variable was used to impact the coefficient μ of fuzzy model. One of the two inputs in the inference fuzzy system was used to change the fuzzy weight of a fuzzy system. The process of changing a real scalar into a fuzzy rule is called the fuzzification [5]. IF-Than rules are the fundamental laws for this purpose. We used three correlating variables, Z_1, Z_2, and Z_3, represented for three cases to adjust μ value. They are the ratios between the powers of the sub-band frequencies. Z_1 is the power ratio of alpha1 band with the combined power in delta1 and delta2. Z_2 is the power ratio of alpha1 band with the combined power in delta1, delta2 and theta1. Z_3 is the ratio of total energy in spindle band with the combined power in delta1 and delta2. Each correlating variable is selected by the selector, and the selector activities are based on the thresholds to sure that each inference fuzzy system has only 2 inputs (one feature and one correlating variable). From the classes procedure, statistical experiments, and R&K rules, we have standardized our features as listed in Table 4.

Table 4 Standardized features.

Bands	Wake	Stage 1	Stage 2	SWS
Delta1	verylow	low	medium	high
Delta2	verylow	low	medium	high
Theta1	low	medium	high	high
Theta2	--	--	--	--
Alpha1	high	low	medium	medium
Alpha2	--	--	--	--
Beta1	--	--	--	--
Beta2	high	low	medium	low
Spindle	medium	low	high	medium
Fullband	low	verylow	medium	high

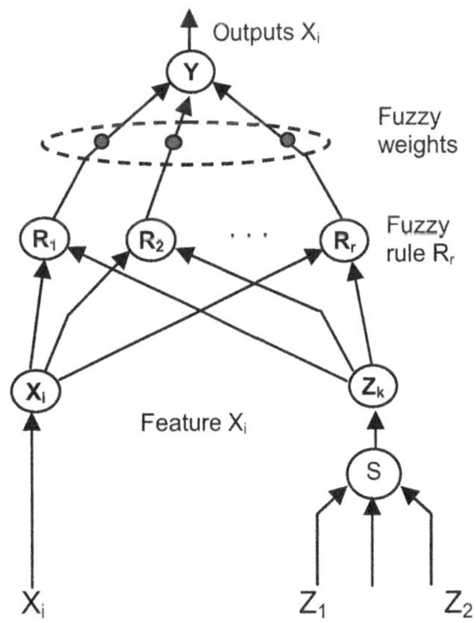

Fig. 6 Model of the fuzzy system (Nauck and Krause [6]). X_i is input features, Z_k is correlating variables, and S stands for the selector.

III. Results

The features were extracted by wavelets decomposition from 10-second segments of (C_3-A_1) channel EEG signals. There are totally 860 epochs being adopted for the experiments with clean data by removing artifact epochs manually (Table 1). Segment features were extracted in the single scales. Each feature was standardized by changing weights based on mean energy ratios of fuzzy rules. Following the features extraction and standardization, we used EBP network with momentum and adaptive learning rate to classify these features. Note that the total number of the resulting trained models for each stage might be different. Four sleep stages consisting of Wake, Stage 1, Stage 2, and Slow Wave Stage (SWS) were classified automatically by the proposed neuro-fuzzy network. The results are listed in Table 5. The average sensitivity is 88.48%, average specificity achieves 95.96%, and average accuracy is 93.79%. The results outperform that of using neural network only.

Table 5 Classification results using neuro-fuzzy network.

	Wake	Stage1	Stage 2	SWS	Average
Sensitivity	87.6	85.71	89.06	91.56	88.48
Specificity	94.85	94.80	95.63	98.57	95.96
Accuracy	92.99	91.91	93.68	96.59	93.79

IV. Discussions

We have proposed a systematic approach based on neuro-fuzzy network to perform sleep stage classification automatically. The key to any automated sleep monitoring system is the accurate detection of the sleep stages. The system is a hybrid system, this model included strong point of fuzzy logic, neural network, and wavelet transform with satisfactory scores. In the intrinsic nature, the neuro-fuzzy only can handle a limited number of inputs and can usually be identified from the empirical data. In these experiments, we used Mamdani form model. Mamdani method is widely accepted for capturing expert knowledge. It allows us to describe the expertise in more intuitive, more human-like manner. However, Mamdani-type fuzzy inference entails a substantial computational burden [7][8]. Most data on inter-scorer agreement are based on the study of normal subjects. The EEG waveforms recorded in this study for each subject lasted approximately only one hour which was relatively short period of time interval. Therefore, only four sleep stages, i.e., wake, Stage1 sleep, Stage2, and SW stages, were chosen to be classified.

V. Conclusion

The combined model with artificial neural network and fuzzy system using wavelet transform features is suitable for bio-mecical signal in sleep stage classification. The recognition result using neuro-fuzzy network is satisfactory. The data samples, rules, and the length of sleep intervals will be increased to our experiments in future work to improve the recognition rate.

Acknowledgement

The authors would like to thank both Feng Chia University and Cheng-Ching General Hospital, for supporting this research under the joint contract FCU/CCGH-99-01.

References

1. A. Rechtschaffen, A. Kales, ed., "A Manual of Standardized Terminology, Techniques and Scoring System for Sleep Stages of Human Subjects", Los Angeles: Brain Information Service/Brain Research Institute, 1968.
2. G. Sharada and O.B.V. Ramanaiah. "An Artificial Intelligence Based Neuro-Fuzzy System with Emotional Intelligence", International Journal of Computer Applications, Vol. 1, No. 13, pp. 74-79, 2010.
3. F. Ebrahimi, M. Mohammad, and E. Edson. "Automatic Sleep Stage Classification Based on EEG Signals by Using Neural Networks and Wavelet Packet Coefficients", 30th Annual International IEEE EMBS Conference, Vancouver, British Columbia, Canada, August 20-24, 2008.
4. S.A.S.Ab. Rahman, F. A.MD. Yuof, and M.Z.A. Bakar. "The Method Review of Neuro-Fuzzy Applications in Fault Detection and Diagnosis System", International Journal of Engineering & Technology IJET-IJENS Vol. 10, No. 03, pp. 65-68, 2010.
5. H.B. Verbruggen, "Fuzzy Logic Control". World Scientific Series in Robotics and Intelligent Systems, Vol. 23, 1999
6. V. Gomathi, K. Ramar, and A.S. Jeevakumar. "A Neuro Fuzzy Approach for Facial Expression Recognition Using LBP Histograms", International Journal of Computer Theory and Engineering, Vol. 2, No. 2, pp. 245-249, 2010.
7. K. Guney, "Comparison of Mamdani and Sugeno Fuzzy Inference System Model for Resonant Frequency Calculation of Rectangular Micro-strip Antennas", Progress In Electromagnetics Research B, Vol. 12, pp. 81-104, 2009.
8. K. Tamás and L.T. Kóczy, "Mamdani Type Inference in Fuzzy Signature Based Rule Bases", Magyar Kutatók, Nemzetközi Szimpóziuma 8th International Symposium of Hungarian Researchers on Computational Intelligence and Informatics, November 15-17, 2007.
9. J.E. Heiss, C.M. Held, P.A. Estévez, C.A. Perez, C.A. Holzmann, J.P. Pérez. "Classification Of Sleep Stages In Infants, A Neuro Fuzzy Approach". Proceedings of the 23rd Annual EMBS International Conference, Istanbul, Turkey 25-28 Oct. 2001.
10. A. Yousefpour and Z. Ahmadpour, "The Prediction of Air Pollution by Using Neuro-Fuzzy GMDH", The Journal of Mathematics and Computer Science. Vol. 2, No. 3, pp. 488-494, 2011.
11. A.L. Fraiwan, N.Y. Khaswaneh, and K.Y. Lweesy, "Automatic Sleep Stage Scoring with Wavelet Packets Based on Single EEG Recording", World Academy of Science, Engineering and Technology, 54, 2009.
12. L. Zoubek and F. Chapotot. "Automatic Classification of Sleep/Wake Stages Using Two-Step System", Proceedings of the 17th World Congress, The International Federation of Automatic Control, Seoul, Korea, July 6-11, 2008.

Lossless Image Compression Using Ideal Cross Point Regions with Wavelet Transform

Quan Tran Nhu, Canh X. Huynh, and Tin T. Dang

Faculty of Electrical and Electronics Engineering, Ho Chi Minh City University of Technology, Ho Chi Minh City, Vietnam

Abstract — **This paper presents the scheme WICRIC (Wavelet based Ideal Cross point Regions for lossless Image Compression) for lossless compression and decompression of digital images. The main idea is to use the Ideal Cross point Regions for lossless Image Compression (ICRIC) scheme together with wavelet transform. The base of this statement is the effect of Gray coding on cross point regions of the transformed matrix of the original image. Before Gray coding data sets of cross points are determined, they are called the ideal cross point regions (ICRs). The wavelet transform makes images more compact, thus increases the number of usable ICRs. After Gray coding these regions always contain only bits 1 or bits 0 depending on the number of each bit plane after the bit plane decomposition. The bit state in ideal cross point regions is very good at entropy coding because the probability of data bits 1/0 is 1 there. This optimization of probability like that has important effects on encoding and decoding processes of lossless image compression for data transmission. This scheme of WICRIC is developed from the traditional scheme ICRIC (Ideal Cross point Regions for lossless Image Compression) and the Wavelet transform, it can give high compression ratios for both medical and photometric images, and so it can be installed in medical imaging systems like digital radiology, CT, MRI,….**

Keywords — **cross points, integer wavelet, lifting scheme, bitplane decomposition, cross points regions.**

I. Introduction

This paper is a development of the papers [1],[2] that presented the definition of cross points and cross point regions, the propositions of bit states of cross points and the consequences about entropy of obtained data in cross point regions after Gray code transformation.

Cross points are neighbor points around the points of grey levels 2^n. The points of grey levels 2^n may or may not exist in data. The original data points whose values are less than 2^n have bit states being much different from those of the data points being greater than or equal to 2^n [3], [4]. The change of bit states for Gray code transformation is studied by many authors [5]-[8]; however, the number of bits and the distribution of these bits of Gray codes in the ideal cross point regions are not mentioned yet. This leads to a new scheme for data compression. The data is arbitrary, but in ideal cross point regions the change of bit states is systematically determined after Gray coding, so the probability of data bits in ideal cross point regions is maximal.

Wavelet is a popular kind of transform that has been used widely for image compression. Many types and kinds of wavelets and wavelet functions have been developed and studied. For the scheme of WICRIC, we use the lifting scheme [8], the main advantage of this transform is that it can be inverted. This can be viewed as the pre-processing of compression stage. It makes the values of cross points appear more frequently in a small area. As other wavelet-base compression algorithm, the choosing suitable wavelet functions and wavelet transform kinds is essential. In this lossless scheme, the integer wavelet transform is used.

This paper has five sections. After this introduction, the section 2 mentions the theory of cross points regions: ideal cross points regions with the definition, the proposition for the states of data bits used in the step "Modeling" of entropy coding [8]. The section 3 introduces WICRIC scheme (Wavelet based Ideal Cross point Regions for lossless Image Compression) that manipulates the data after wavelet transform in cross point regions in the process of entropy coding with stage "Coding" using C. B. Jones algorithm [9] with the improvement from [10]-[11]. Section 4 shows the experimental results from compressing images with the scheme WICRIC. Also, the results of original ICRIC scheme [12], and another compression programs such as Lossless JPEG 2000 [13], Winzip and Winrar [14] are shown for comparison. Finally, section 5 is the conclusion and the scope for future researches.

II. Ideal cross points regions and bit states

DEFINITION 1. Let the positive integer N be the bit length of data points. The region of cross points $A_o(n, p)$, with n from $(N-1)$ to 1, p from 1 to n, and $n \geq p$, is a set of data points whose grey values are from $(2^n - 2^{n-p})$ to $(2^n + 2^{n-p} - 1)$. The point of grey value 2^n (if it exists) is called the center point of the cross point region, and the grey value 2^n is called the central value. These regions are the ICRs.

V. Van Toi et al. (Eds.): 4th International Conference on Biomedical Engineering in Vietnam, IFMBE Proceedings 40, pp. 77–81, 2013.
www.springerlink.com

Data points in ICRs have grey values that satisfy the rule:

$$V_A(n, p) = \{2^n - 2^{n-p}, \ldots, 2^n + 2^{n-p} - 1\} \quad (1)$$

The value n is the exponent of central values 2^n. The value p gives the range of ICRs where the data points of grey values satisfying (1) are in. The number of bit plane for data bits compression is (n-p). In general, the set $V_A(n, p)$ can be divided into two groups: $V_{A\,l}(n, p)$ and $V_{A\,g}(n, p)$ with $V_{A\,l}(n, p) = \{2^n - 2^{n-p}, 2^n - 2^{n-p} + 1, \ldots, 2^n - 2, 2^n - 1\}$, and $V_{A\,g}(n, p) = \{2^n, 2^n+1, \ldots, 2^n + 2^{n-p} - 2, 2^n + 2^{n-p} - 1\}$.

For example, when $n = 3$, $p = 1$, $V_A(3, 1) = \{2^3 - 2^{3-1}, \ldots, 2^3 + 2^{3-1} - 1\} = \{4, 5, 6, 7, 8, 9, 10, 11\}$, at that time $V_{A\,l}(3, 1) = \{4, 5, 6, 7\}$, and $V_{A\,g}(3,1) = \{8, 9, 10, 11\}$.

PROPOSITION 1. Let n be the exponent of the central value 2^n in the ICR $A_o(n, p)$, where n is from $(N - 1)$ to 1, N is a positive integer representing bit length, p from 1 to n, and $n \geq p$. Bits of Gray codes in the ICR $A_o(n, p)$ on the bit plane $(n - p)$ are always bits 1 (if $p = 1$) or 0 (if otherwise).

Proof of Proposition 1.

According to (1), with a value of n in the interval from 1 to $(N - 1)$ and $p = 1$, grey values of data points in the region $A_o(n, 1)$ are from $(2^n - 2^{n-1})$ to $(2^n + 2^{n-1} - 1)$, so they may be expanded under the form of polynomials of radix 2 (2) (for data points in $V_{A\,l}$) and/or (3) (for data points in $V_{A\,g}$) as the followings:

$$0.\ 2^{N-1} + \ldots + 0.2^n + 1.\ 2^{n-1} + x.\ 2^{n-2} + \ldots + x.\ 2^0 \quad (2)$$

$$0.2^{N-1} + \ldots + 0.2^{n+1} + 1.2^n + 0.2^{n-1} + x.2^{n-2} + \ldots + x\ .2^0 \quad (3)$$

where x are bits 1 or 0.

After Gray code transformation, (2) and (3) become (4) and (5) respectively:

$$0.2^{N-1} + \ldots + 0.2^n + 1.2^{n-1} + x.2^{n-2} + \ldots + x.2^0 \quad (4)$$

$$0.2^{N-1} + \ldots + 0.2^{n+1} + 1.2^n + 1.2^{n-1} + x.2^{n-2} + \ldots + x.\ 2^0 \quad (5)$$

By combining (4) and (5), the region $A_o(n, 1)$ on the bit plane (n–1) always contains bits 1.

For example, when $n = 3$, the central value is 2^3, $A_o(3, 1) = \{4, 5, 6, 7, 8, 9, 10, 11\}$. After Gray coding, the Gray codes of those values are 6, 7, 5, 4, 12, 13, 15, 14 respectively, all of them have bits 1 on the bit plane 2 (= 3 – 1). This is very good for compressing real data because the probability of bit 1 in ICRs is always 1, and the probability of bit 0 in ICRs is always 0.

When $p = 2, 3, \ldots$ or n, grey values of data points in the region $A_o(n, p)$ are from $(2^n - 2^{n-p})$ to $(2^n + 2^{n-p} - 1)$, so they may be expanded under the form of polynomials of radix 2 (6) and/or (7):

$$0.2^{N-1} + \ldots + 0.2^n + 1.2^{n-1} + 1.2^{n-2} + \ldots + 1.2^{n-p} + x.2^{n-p-1} + \ldots + x.2^1 + x.2^0 \quad (6)$$

$$0.2^{N-1} + \ldots + 0.2^{n+1} + 1.2^n + 0.2^{n-1} + 0.2^{n-2} + \ldots + 0.2^{n-p} + x.2^{n-p-1} + \ldots + x.2^1 + x.2^0 \quad (7)$$

where x is bit 1 or 0.

After Gray code transformation, (6) and (7) become:

$$0.2^{N-1} + \ldots + 0.2^n + 1.2^{n-1} + 0.2^{n-2} + \ldots + 0.2^{n-p} + x.2^{n-p-1} + \ldots + x.2^1 + x.2^0 \quad (8)$$

$$0.2^{N-1} + \ldots + 0.2^{n+1} + 1.2^n + 1.2^{n-1} + 0.2^{n-2} + \ldots + 0.2^{n-p} + x.2^{n-p-1} + \ldots + x.2^1 + x.2^0 \quad (9)$$

Equation (8) shows that the data points satisfying $V_{A\,l}$ on the bit plane $(n - p)$ always contains bits 0. By combining (8) and (9), the region $A_o(n, p)$ on the bit plane $(n - p)$ always contains bits 0 too.

For example, when $n = 3$, $p = 2$, we have $A_o(3, 2) = \{6, 7, 8, 9\}$, Gray codes of these decimal values (5, 4, 12, 13) have bits 0 on the bit plane 1 (= 3 – 2).

This statement is convenient for the process of optimizing the probability of data bits in the scheme of lossless image compression, especially in the step of modeling of entropy coding.

PROPOSITION 2. In the ideal cross point regions A_o, after Gray code transformation the entropy of obtained data on a certain bit plane is minimum.

Proof of Proposition 2.

We can easily see that in the ideal cross point regions $A_o(n, p)$ on the bit plane (n–p), before Gray code transformation, the probabilities of bit 1 and bit 0 are random, therefore the entropy of the corresponding bit string is often $H = -P(1).\ \log_2 P(1) - P(0).\ \log_2 P(0) > 0$.

After Gray coding, these ideal cross point regions contain all same bits (bits 1 or bits 0), so the probability of bit 1 in the regions is 1, and the probability of bit 0 is 0 there, and inversely. This is the reason for that the entropy of data bit string is always $H = -P(1).\ \log_2 P(1) - P(0).\ \log_2 P(0) = 0$, that means the average information of this region is 0.

III. WICRIC SCHEME FOR LOSSLESS IMAGE COMPRESSION

Figure 1 presents the general WICRIC scheme for lossless image compression. The beginning of this scheme is the "Integer wavelet transform" box, and the ending is the "Save to output data" box. The arrows state the sequences of this scheme.

The first step, "Integer wavelet transform", is used to transform the image matrix into a new set of integer data. We need 9 to 10 bits to present each value in this data set, depending on the type of wavelet we used.

The "Data slicing" step aims to reduce the random rate of bits by processing the 8-higher bits and the remaining lower bits separately. For the higher bits, we compress them by ideal cross point algorithm. For the lower bits, other algorithm is suitable, such as Huffman coding or Run-length coding. In the experimental results of this paper, these lower bits are left unchanged to show the difference between WICRIC scheme and traditional ICRIC scheme.

The "Ideal cross point regions" step looks for the cross point regions where we can optimize the probability of data bits. Grey values in these regions must satisfy (1). With this scheme, we can find more cross point regions on the transformed data than on the image itself with the cost of processing more bit planes.

The "Cross point map" step will establish the map of ideal cross point regions. Notice that with each central value 2^n we have many different ideal cross point regions $A_0(n, p)$ on bit planes $(n-p)$, in those regions we encode data bits. This step evaluates and ignores small ideal cross point regions containing 1, 2 or 3 data points depending on the number of bit plane $(n-p)$, these areas don't affect much the compression ratio of processes of coding and decoding.

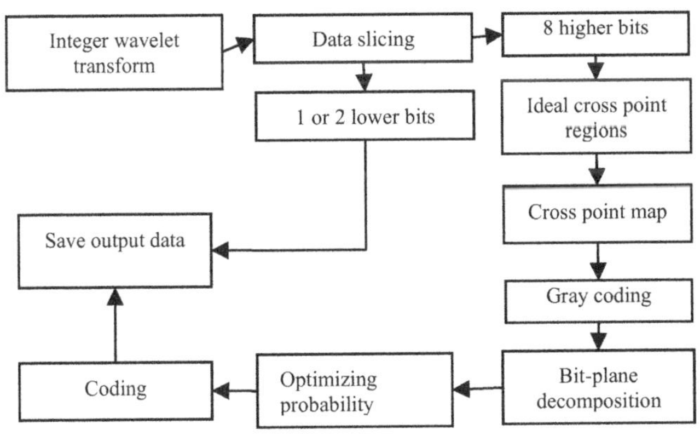

Fig. 1 WICRIC scheme.

The "Gray coding" step makes Gray code transformation to the data, according to [3].

The "Bit plane decomposition" decomposes data into 8 separate bit planes to do the bit-plane encoding.

The "Optimizing probability" step calculates the probability of data bits outside ideal cross point regions and optimizes probabilities of data bits in ideal cross point regions. This optimization is based on the cross point map beforehand.

The "Coding" step uses some algorithm in the process of entropy coding, like arithmetic coding. In this scheme, we use C.B Jones' algorithm [9]. The process of encoding data bits should be carried out from the most to the least significant bit planes because of the random of data on the less significant bit plane.

The last step saves the output data, the code word, of this process to files that have smaller size than the image file itself, thus achieves the compression.

IV. EXPERIMENTAL RESULTS

Figure 2 shows some (not all) images we have used to compress with the scheme WICRIC. These images are commonly found in many literatures. For medical images there are some areas where the grey levels don't have too much difference from one data point to another. The others are photometric images, and then some images that have many details. All these images are in .BMP format with the depth of 8 bits.

Table 1 shows the results when the test images are compressed by WICRIC scheme. It also shows the results of ICRIC [12] without using wavelet transform, and the results from another compression programs: Lossless JPEG 2000 [13] (with Integer wavelet transform), Winzip and

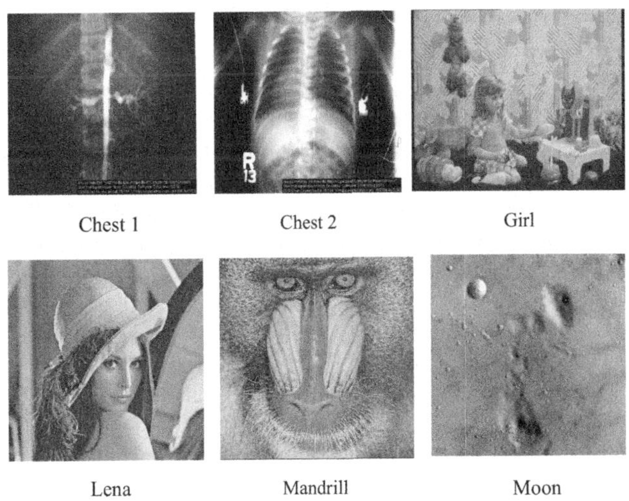

Fig. 2 Some test images.

Table 1 Experimental results for compression ratio.

Image	Size	WICRIC	ICRIC	JPEG 2000	WinRar 3.90	Winzip 14.5
Chest1	512x480	1.9132 : 1	1.9096 : 1	1.8775 : 1	1.6738 : 1	1.5034 : 1
Chest2	512x480	1.8523 : 1	1.8178 : 1	1.8749 : 1	1.4573 : 1	1.2473 : 1
Chest 3	512x480	2.2693 : 1	2.2600 : 1	2.2684 : 1	2.2015 : 1	1.5483 : 1
Chest 4	512x480	1.9325 : 1	1.9322 : 1	1.9226 : 1	1.7282 : 1	1.3417 : 1
Chest	256x256	1.7541 : 1	1.7035 : 1	1.2432 : 1	1.5739 : 1	1.5292 : 1
Joint	512x400	2.1849 : 1	2.0800 : 1	2.2253 : 1	1.9327 : 1	1.4565 : 1
Girl	720x576	1.8879 : 1	1.7026 : 1	1.8516 : 1	1.6031 : 1	1.2590 : 1
Lena	512x512	1.7392 : 1	1.6146 : 1	1.7522 : 1	1.5671 : 1	1.1735 : 1
Zelda	256x256	1.7324 : 1	1.7255 : 1	1.6031 : 1	1.5116 : 1	1.3830 : 1
France	672x496	8.5217 : 1	8.4476 : 1	3.0869 : 1	23.0391 : 1	19.4232 : 1
Mandrill	512x512	1.2691 : 1	1.2369 : 1	1.2910 : 1	1.2066 : 1	1.1030 : 1
Moon	256x256	1.7969 : 1	1.4009 : 1	1.4751 : 1	1.3830 : 1	1.2037 : 1

Winrar [14]. From these results we can see that with medical images (Chest 1, Chest 2, Chest 3, Chest 4, Chest, Joint) the two schemes WICRIC and JPEG2000 are better than another, and the results of WICRIC are better than the results from JPEG2000 because of the using ideal cross point regions. For the photometric images (Girl, Lena, Zelda, France), depending on images the compression ratio from WICRIC is better than another. For images with many details (Moon, Mandrill), the scheme WICRIC proves the power to get higher compression ratio because of getting range of data points in ideal cross points after wavelet transform.

V. Conclusion

This paper illustrated the scheme of WICRIC that uses the theory of cross point regions with optimizing probabilities of data bits in ideal cross point regions and Integer wavelet transform. The data received from this scheme is reversible, so this scheme can be strongly used in lossless images compression systems, so it can be installed in medical imaging systems like digital radiology, CT, MRI,…. This scheme can be also used in cryptographic systems with keys depending on n-order estimating, number of bit planes, number of ideal cross point regions, this problem will be mentioned in another papers.

The basic concepts have been introduced in [1], [2] which are now the base for the theory of optimizing probabilities of cross points with the new concept of ideal cross point regions of coefficients after the Integer wavelet transform. A meaningful improvement in compression ratio has been obtained, compared to the other methods. From this, we can see the scheme mentioned is good for losslessly compressing medical and many details images in which there are many areas that have approximate grey values, or grey values in the range satisfying the characteristic of ideal cross points regions. From these concrete bases, the problem of improving the compression ratio of image processing and transmission can be developed in the future.

References

Dang Thanh Tin, Vu Dinh Thanh, Nguyen Kim Sach, Seihaku Higuchi, "Optimizing probability of data bits based on the theory of cross point regions," MII journal, vol.22, No. 2, May 2005.

Dang Thanh Tin, Vu Dinh Thanh, Nguyen Kim Sach, Seihaku Higuchi, "Effect of Gray coding on the state and the distribution of data bits in cross point regions," in Proc. 3rd Int. Conf. Computational Electromagnetics and Applications, IEEE catalog No. 04EX875, Bejing, China, Nov 2004, pp. 553–556.

C. J. Lu, S. C. Tsai, "A note on iterating an α-ary Gray code," SIAM J. Discrete Math, vol. 14, No. 2, pp. 237-239, 2001.

Nguyen Kim Sach, "Digital images and video processing," Science and Technology Publisher, Vietnam, 1997, pp. 170-174.

M.C. Er, "On generating the N-ary reflected Gray codes," IEEE Trans. Computers, vol. 33, pp. 739-741, 1984.

W. Gilbert, "A cube- filling Hilbert curve," Math. Intell. , vol. 6, pp. 78-81, 1984.

Matthew B. Squire, "Gray codes and efficient generation of combinatorial structures," Doctor thesis, Dept. Computer science, North Carolina State University, 2001.

Rafael C. Gonzalez, Richard E. Woods, "Digital image processing," Prentice Hall, 2002, pp 438- 440, pp 506-510.

Jones C. B, "Efficient coding system for long source sequences," IEEE Trans. Inform. Theory, vol. 27, No. 3, pp. 280-291, 1981.

Dang Thanh Tin, Vu Dinh Thanh, Nguyen Kim Sach, "An improvement of Jones' method of lossless data compression," Vietnam posts and telecommunications journal, No. 7, pp. 65-69, May 2002.

Dang Thanh Tin, Vu Dinh Thanh, Nguyen Kim Sach, Seihaku Higuchi, "Lossless image compression with the theory of cross point regions," 4th IEEE Int. Conf. Computer Sciences, Research, Innovation and Vision for the Future, Collection Informatique ISSN: 1621-0865, Hors Serie ISSN: 1621-0065, Hochiminh city, Vietnam, Feb 2006, pp. 19-25.
Dang Thanh Tin, Vu Dinh Thanh, Nguyen Kim Sach, Seihaku Higuchi, "A lossless coding scheme for images with ideal cross point regions on multiple bit planes," WMWA, 2008.
Tinku Acharya, Ping-sing Tsai, "JPEG2000 Standard for image compression," Wiley, 2005.
Federal information processing standards publication 197, "Announcing the Advanced Encryption Standard," National Institute of Standards and Technology (NIST), Nov 2001.

Author: DANG Thanh Tin
Institute: Ho Chi Minh city University of Technology
Street: Ly Thuong Kiet
City: Ho Chi Minh
Country: Viet Nam
Email: dttin@hcmut.edu.vn / tindt@yahoo.com

Identification and Characterization of Novel Autoantibody Biomarkers for Rheumatoid Factor-Negative and Accp- Negative Rheumatoid Arthritis

Anh Nguyen-Hung[1,2,3], Binh Le-Thanh[2], Cindy Govart[3], Veerle Somers[3], and Son Nguyen-Hong[2]

[1] Biotechnology Center of Ho Chi Minh City, Ho Chi Minh City, Vietnam
[2] GBEC Bioengineering Company Limited, Ho Chi Minh City, Vietnam
[3] Biomedical Research Institute, Transnational University Limburg, Belgium

Abstract – **Rheumatoid arthritis (RA) is an inflammatory autoimmune disease in which 30% of patients are seronegative for the two serological RA markers, rheumatoid factor (RF) and anti-cyclic citrullinated peptide antibodies (ACCP). However, both markers also have lower sensitivities in early stages of the disease. The lack of effective biomarkers for this subpopulation of RA patients causes a delayed diagnosis. Therefore, discovery of novel circulating autoantibody biomarkers for these early RF-negative and ACCP-negative RA patients are advance planning of these subsets. 136 enriched antigenic markers were identified by high-throughput screening an RA library with autoantibodies in the sera of RA patients (positive selection) and healthy control (negative selection). Phage-ELISA-based re-screening determined 13 out of 136 enriched antigenic targets in which increased antibody levels rised up. Two of the 13 identified targets, namely special A-T rich DNA binding protein (SATB1) and EGF-containing fibulin-like extracellular matrix protein 1 (EFEMP1), which had highest increased antibody levels, were selected for further characterization. Reactivity of serum antibodies against these proteins was tested in a protein ELISA using the sera of 137 RA patients and 159 healthy controls. Autoantibodies against SATB1 were detected with 79% sensitivity and 82% specificity, while 87% sensitivity and 52% specificity were obatained for antibodies directed against EFEMP1. With regard to cytokine assay, interestingly, pro-inflammationary cytokines including TNF (tumour necrosis factor) increased over 1000 pg/ml or IL-4 (interleukin-4) or IL-13 similarly increased up to 90 pg/ml and 102 pg/ml, respectively for co-culture with anti-SATB1 monoclonal antibody. Under treatment with EFEMP1 mAb resulting IL-1, TNF and IL-4 triggered up to 1300 pg/ml, 1250 pg/ml and 180 pg/ml, respectively. In T cell phenotype characterization, two subsets of CD4^{+}T cell and CD8^{+} cell were stimulated and differentiated from peripheral blood mononuclear cells (PBMCs) at 29.7% and 1.2% for anti-SATB mAb; 31.3% and 2.8% for anti-EFEMP1 mAb.**

Keywords – **SATB1, EFEMP1, biomarkers, rheumatoid arthritis, phage display.**

I. Introduction (briefly)

The presence of autoantibodies is characteristic of rheumatoid arthritis (RA), in which they serve as biomarker for diagnostics and therapeutic development. These humoral biosignatures are usually taken as a sign that autoreactivity is driving the disease. Moreover, an imbalance of pro-inflammatory and anti-inflammatory cytokines, autoreative and inflammatory T helper 1 (Th1) cells, regulatory and suppressive T cells results in the loss immune tolerance, the breakdown of immune homeostasis and the subsequent appearance of exacerbated autoreactive B cells that form autoantibodies (**Gonzalez-Rey E et al., 2007).** Theories on the pathogenesis of RA have focused on autoantibodies and immune complexes. T-cell-mediated antigen-specific responses (**Atsuo Nakajima et al., 2006**) and T-cell independent cytokines networks (**Firesctein G et al., 1996**) have also been implicated. The SATB1 was selected for further characterization based on the previous publications on SATB1 in which its functions were described in **(1)** densely looped, transcriptionally active chromatin for coordinated expression of cytokine genes (**Cai S et al, 2006**), **(2)** the initiation of human Th1 or Th2 cell commitment (**Lund R et al., 2005**), **(3)** nuclear targeting determinant, as a genome organizer in the T cell lineage (**Nakayama Y et al., 2005**), **(4)** the development of CD8SP thymocytes and peripheral T lymphocytes (**Nie H et al., 2005**), **(6)** tissue-specific nuclear architecture and gene expression (**Cai S et al., 2003**), **(7)** regulation of gene expression in peripheral blood T cells from patients with atopic dermatitis (**Matsumoto Y et al., 2002**), **(8)** detachment from chromatin early in T-cell apoptosis (**Galande S et al., 2001**), **(9)** in B lymphocytes quiescence, tolerance and activation of global gene expression **(Glynne R et al., 2000**), **(10)** apoptotic mechanisms (**Gotzmann J et al., 2000**) and **(11)** orchestration temporal and spatial expression of multiple genes during T-cell development (**Alvarez J.D et al., 2000**). Moreover, to our knowledge this was the first report for the presence of SATB1 in RA patient sera. EFEMP1 plays an important role in **(1)** negative regulation of chondrocyte differentiation (**Wakabayashi T et, al., 2010**), **(2)** promotion of angiogenesis and acceleration of the growth of cervical cancer in vivo (**Song E.L et al., 2011**), **(3)** unique up-regulation in malignant gliomas and promotion of tumor cell mobility and invasion (**Hu B et al., 2009**), **(4)** binding of the epidermal growth

V. Van Toi et al. (Eds.): 4th International Conference on Biomedical Engineering in Vietnam, IFMBE Proceedings 40, pp. 82–87, 2013.
www.springerlink.com

factor (EGF) receptor and activation of mitogen-activated protein kinase (MAPK) and serine/threonine-protein kinase (Akt/PKB) pathway in pancreatic carcinoma cells **(5)** stimulation of DNA synthesis in fibroblasts (**Lecka-Czernik B et al, 2005**), **(6)** proliferation, differentiation of synovial fibroblast cells (**Masuda K et al, 2002**) and **(7)** formation of thickness of synovial tissue layers (**Nishimura et al., 2007**).

II. MATERIAL AND METHODS (PAGES NO ENOUGH)

III. RESULTS

A. Transcriptomic Profiles of Clones Amplified from RA cDNA Library

Screening for novel autoantibodies was performed on cDNA phage display library, as previously described (**Somers K et al, 2009**). This screening resulted in 13 in-frame and enriched clones, of which 5 clones were derived from selections using the early RA patients, and 8 clones from selections on the RF-negative and ACCP-negative RA patients (data not shown).

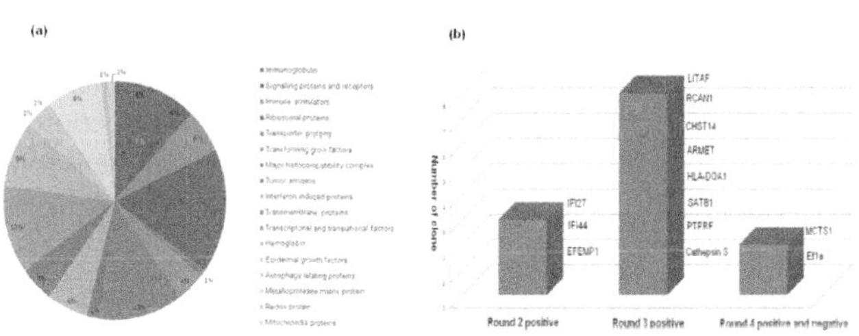

Fig. 1 Pie charts showing the detail distribution of protein profiles screened from RA cDNA library. Ribosomal proteins, immune stimulators and transcriptional and translational factor complexes that highly account for differential expressed protein is 17%, 12% and 12%, respectively.

Homology search analysis at the nucleotide and protein level showed 100% identity with DNA and amino acid sequences of known genes for these candidates on gene bank. Sequencing the cDNA insert from the clones and comparison with genome databases revealed a correspondence with peptides or proteins displayed on the phage surface. To confirm that enrichment of phage clones occurred based on specific interactions with RA patient autoantibodies and to further characterize the immunoreactivity of the identified enriched phage clones, two clones, SATB1 and EFEMP1 were expressed and purified (data not shown). The finding of SATB-1 and EFEMP-1 autoantibodies in the RA patients could be associated with overexpression of these proteins in the development of T-lymphocytes and infiltration of lymphocytes into synovial tissues.

B. Association between RA and the Presence of Circulating Autoantibodies against EFEMP1 and SATB1

To investigate the association between RA and the presence of autoantibodies directed against EFEMP1 and SATB1, immunoreactivity of 159 healthy controls and 137 RA patients against these proteins was investigated by means of ELISA.

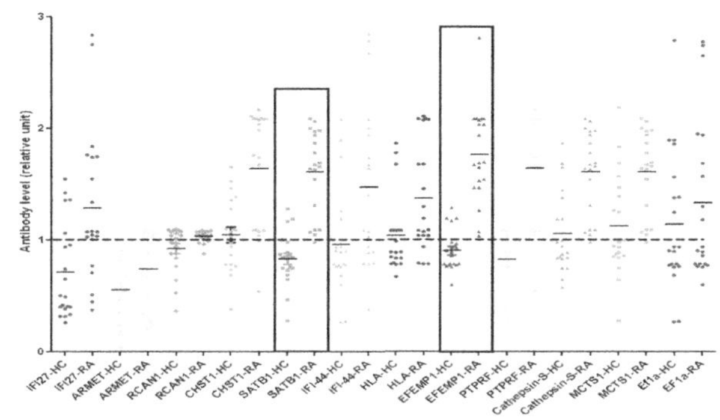

Fig. 2 Autoantibody levels were analyzed by means of protein ELISA in RA patients (n=20) and healthy controls (n=20). Background plasma reactivity was determined by measuring immunoreactivity against the V5 epitope fragment. Scatterplot dots represent the mean of OD (autoantigenic targets)/OD (V5 epitope) for each serum sample tested.

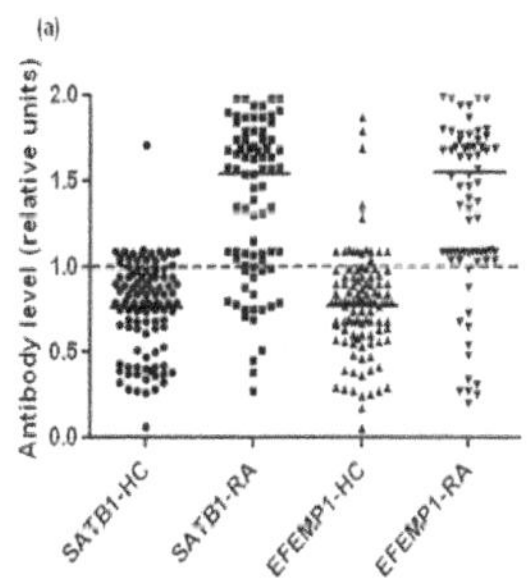

(b) (SATB1)

Cut-off	Sensitivity	Specificty
>0.9910	81.37	81.31
>0.9970	81.37	82.24
>1.003	80.39	82.24
>1.008	79.41	82.24
>1.018	79.41	83.18
>1.032	79.41	84.11
>1.037	79.41	85.05

(c) (EFEMP1)

Cut-off	Sensitivity	Specificity
>0.9931	87.25	84.11
>1.012	87.25	85.05
>1.021	86.27	85.05
>1.035	85.29	85.05
>1.036	84.31	85.05
>1.037	84.31	85.98
>1.037	83.33	85.98
>1.056	82.35	85.98
>1.074	82.35	86.92
>1.074	81.37	86.92

Fig. 3 Protein ELISA were performed on 137 RA patients and 159 healthy controls. Scatterplots show the level of antibody reactivity against EFEMP1 and SATB1. Background reactivity was determined by measuring the ELISA signal against the V5 epitope. The cut-off for positive samples as high as 1.5 times the standard deviation (SD) above the mean ratio OD (antigenic targets)/OD (V5 epitope) was arbitrarily set at 1 relative unit for EFEMP1 and SATB1.

Analysis of the area under the ROC curve (AUC) and p-value quantified the overall ability to test for antibody reactivity against each protein in order to differentiate between RA patients and healthy controls. The ROC results showed difference between RA patients and healthy controls by identification of autoantibody specificities against EFEMP1 and SATB1 (Fig. 4). The presence of autoantibodies against these autoantigens via analysis of ROC AUC and p-values might contribute to diagnostic differentiation between RA patients and healthy controls. The finding provides an evidence of presence of anti-EFEMP1 and anti-SATB1 autoantibody in RA patients.

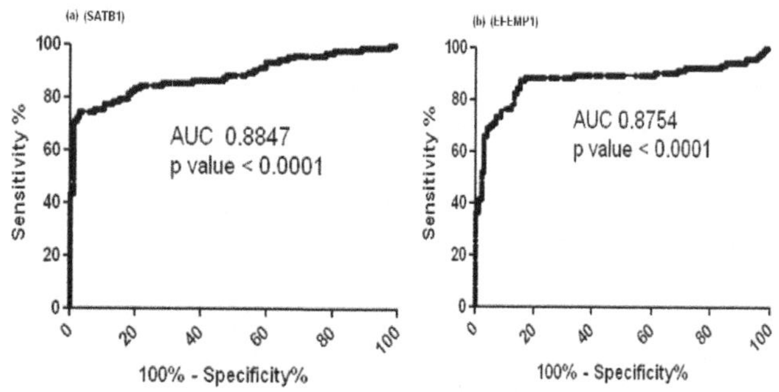

Fig. 4 ROC curves of autoantibody detection against EFEMP1 and SATB1 in RA patients and controls. The area under the ROC curve (AUC) and p-value of ROC analysis quantify the whole probability of the testing for antibodies against EFEMP1 and SATB1, to valuably distinguish between healthy controls and RA patients. The AUC value is approximately 0.5 for a useless test and 1.0 for perfectly discriminative test. The p-value of ROC was considered statistically significant when p<0.05.

C. Proinflammatory Cytokines Rising Up by Stimulation of Anti-SATB1/EFEMP1 mAb

We examined the differences in cytokine production between when cultured PBMCs after stimulation with anti-SATB1/EFEMP1 mAb.

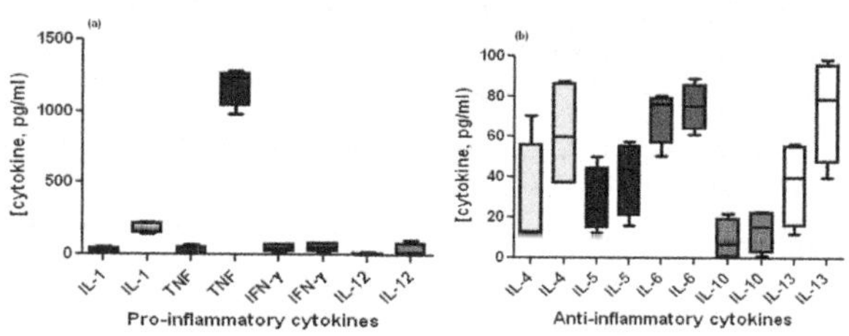

Fig. 5 Cytokine production was determined with supernatants from PBMCs culture with treatment of anti-SATB1 mAb. (a) TNF, pro-inflammatory cytokine, was stimulated up to 1000 pg/ml. (b) anti-inflammatory cytokine, significant difference from untreated cell culture assay with treated cell culture assay was measured. IL-4 and IL-13 similarly increased up to 90 pg/ml and 102 pg/ml, respectively.

Data showed a wide range of values for each of the generated cytokines. Anti-SATB1 mAb was much more effective in generating TNF synthesis than the other pro-inflammatory cytokines (**Figure 5a**). However, almost anti-inflammatory cytokines generated from treatment with anti-SATB1 mAb were variety of concentration and no significant difference from untreated and treated assay (**Figure 5b**).

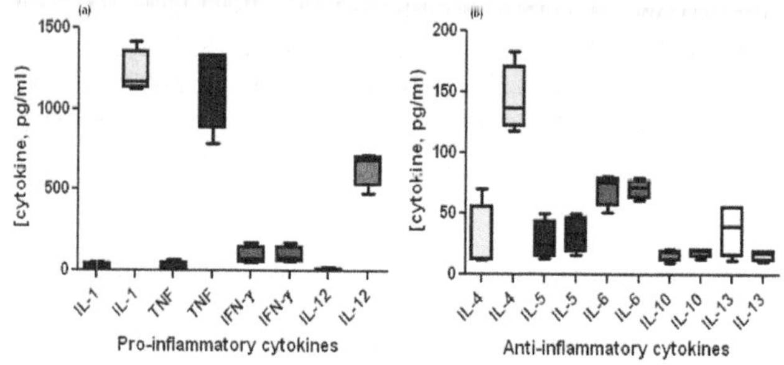

Fig. 6 Cytokine production was determined with supernatants from PBMCs culture with treatment of anti-EFEMP1 mAb. (a) TNF and IL-1, pro-inflammatory cytokine, was stimulated up to 1000 pg/ml, while IL-12 triggered up to 500 pg/ml (b) anti-inflammatory cytokine, significant difference from untreated cell culture assay with treated cell culture assay was measured. IL-4 increased up to 150 pg/ml.

The results of cytokine production following anti-EFEMP1 treatment are given in Figure 6 and again demonstrated a relatively wide range of value. In pro-inflammatory cytokine profile, TNF and IL-1 were highly synthesized **(Figure 6a)**. In anti-inflammatory cytokine profile, IL-4 production in anti-EFEMP1 mAb treated-cell culture showed wide range of values and significant difference from anti-EFEMP1 mAb treatment and un-treatment. Unexpected predominance of IL-4 secrection in PBMCs cell culture led us to hypothesize that a compensation between pro-inflammatory and anti-inflammatory cytokines.

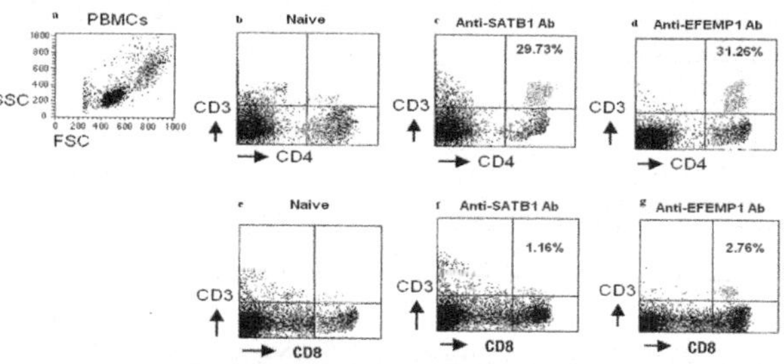

Fig. 7 Flow cytometry assay for anti-SATB1/EFEMP1 mAb-specific T cells.

(a) Lymphocytes were selected by forward and side scatter and data from at least 300,000 small lymphocytes were generally collected. (b,c,d) in most samples data from at least 30,000 to 40,000 $CD4^+$ cells were collected for analysis. (e,f,g) in most samples data from at least 30,000 to 40,000 $CD8^+$ cells were collected for analysis. Percentage of T-cell phenotypes were defined as the percentage of $CD3/CD4^+/CD8^+$ cells. A positive response was considered to be 1% $CD3/CD4^+$ or $CD3/CD8^+$ cells above background.

Examination of anti-EFEMP1 or anti-SATB mAb-stimulated $CD4^{+}T$ in PBMCs culture showed significantly more positive results in which the percent of stimulation of 29.7% for anti-SATB1 mAb and 31.3% for anti-EFEMP1 mAb. $CD8^{+}T$ cells from PBMCs had significantly less than $CD4^{+}$ that produced 1.16% for anti-SATB1 mAb and 2.76% for anti-EFEMP1 mAb. In addition to differences in the proportion of subjects, the total magnitude of cytokine-specific T cells (percentage of $CD4^{+}$ or $CD8^{+}$ T cells that had a cytokine response after anti-SATB1 or EFEMP1 mAb stimulation) was examined. In PBMCs, more $CD4^{+}$ highly produced TNF and IL-1 than IFN-γ and IL-12.There is no difference in the percentage of $CD8^{+}T$ cell followed by treatment of anti-SATB1/EFEMP1 mAb (Figure 7f,g).

IV. Discussion

In the light of autoantibody profiles in analysis of the humoral immune response in RA, finding of biomarkers for early and RF-negative and ACCP-negative RA and analysis of the autoantibody profiles in these patients can supply important insights into RA pathogenesis. The enriched and in-frame phage clones presented in the RA phage display library reflected the presence of the identical autoantibodies in serum, which suggests an essential involvement of these antibodies in the aetiology and disease process of RA. Moreover, these findings correlate with mechanisms of long-lived plasma and memory B cells, including autoreactive cells (**Dorner T et al, 2009**). Elevated immunoreactivity could only be found in the RA group (137 patients) compared to the control group (159 patients) (Fig. 3). Moreover, differences in immunoreactivity did allow discriminating between healthy controls and RA patients. EFEMP1 in early RA may be associated with proliferation and differentiation of synovial fibroblast cells (**Masuda K et al, 2002**). Interestingly, EFEMP1 stimulates DNA synthesis in fibroblasts (**Lecka-Czernik B et al, 2005**), whereas SATB1 is a genome organizer connecting multiple genomic loci and recruiting specific enzymes for chromatin-remodelling and gene expression (**Han H.J et al, 2008**). Therefore, the presence of autoantibodies against EFEMP1 and SATB1 in RA patients might be interrelated. In 137 RA patients, the discrimination of antibody levels against EFEMP1 and SATB1 were significant. Power of test on screening of the sample group including 137 RA patients and 159 healthy controls showed sensitivities (87%) and specificities (52%) for the EFEMP1 markers for early RA and RF(-) anti CCP (-) RA. Clinical parameters of fibrosis score and expression level associating with EFEMP1 expression (**Masuda K et al, 2002**) indicated that EFEMP1 could be involve in the pathology of ACCP-negative RA patients through regulation of MMPs.

SATB1 essentially regulates chromatin-loop architecture and transcription and organization of the major histocompatibility (MHC) class I locus into higher-order chromatin-loop structures (**Pavan K P, et al., 2007**). Silencing SATB1 by small interfering RNA (siRNA) or mutating SATB1 alters chromatin architecture and affects the expression profile of cytokine genes. Although the role of B cell subsets in RA is unclear, anti-SATB1 antibody found in serum of RF-negative and ACCP-negative RA patients might play a role in differentiation of pre-mature B cells into RF-negative and ACCP-negative RA associated B cell subsets.

Another study showed that SATB1 promotes lymphocyte regulation and production from primitive hematopoietic stem and progenitor cells (**Wen J.H et al., 2004**). A suggestion is whether the absence of RF and ACCP-antibody in the RA sub-population is regulated at the level of hematopoietic stem cells and progenitor cells. High-affinity antibodies could be relevant to pathogenesis and early RA diagnostics. Nonetheless, no association existed between antibody positivity towards these antigenic targets and available clinical parameters. A larger population testing with different disease subtypes could be possible to find antibody positivity with clinical parameters. Another hypothesis is that under SATB1 regulation at genome level in lymphocytes, a different B-cell subpopulation is differentiated from stem lymphocytes.

This study determined normal ranges for IL-1, IFN-γ, IL-12 and for IL-5, IL-6, IL-10 followed by anti-SATB1 mAb stimulation. Anti-EFEMP1 mAb showed normal ranges for IFN-γ and for IL-5, IL-6, IL-10. Increased levels of IL-1 and TNF noted to anti-SATB1/EFEMP1 mAb might have been due to dual stimulation of both T and B lymphocytes. The B cells might have produced an amplificative factor which then stimulated the T lymphocytes to increase IL-1, TNF production. The association between cytokine levels for each autoantibody and for all individuals was assessed. Stimulation with these autoantibody demonstrated direct correlations between pro-inflammatory could be stimulated and regulated by anti-SATB1/EFEMP1 autoantibody. Interleukin-1 (IL-1) and tumour necrosis factor (TNF) produce inflammatory and tissue destruction (**Charles A et al., 2000**). This indicates that anti-EFEMP1 mAb could play both functions of direct tissue destruction and IL-1 and TNF-mediated tissue destructinon. IL-4 is potent anti-inflammatory agents; it suppresses genes for pro-inflammatory cytokines such as IL-1 and TNF. This suggests that the larger magnitude reflected responses from population of T cells that regconized a large diversity of cytokine gene products rather than a single population of T cell that regconized a single gene product. The mechanism of how the autoantibodies generated the cytokine production was not known. IL-5 is involved with eosinophil proliferation. Therefore, in the healths would not expect the lymphocyte to be primed to produce increased IL-5 levels. Cytokine levels vary not only in cell differentiation in vitro but also in vivo of different clinical disease. The roles of cytokines within a complex regulatory network are related to specific immunological, chronic inflammation and tissue destruction. In synovial lavage of individuals with very early rheumatoid arthritis revealed elevated levels of IL-4 and IL-13 (**Raza et al., 2005**) and can stimulate T cell proliferation. In addition, IL-1 elicits of histamine from mast cell.TNF and IL-1 share several pro-inflammatory properties. Both cytokine can stimulate the production of collagenase by synovial cells (**Warren J.S et al., 1990**).

How autoantibodies such as anti-SATB1 mAb and anti-EFEMP1 mAb, known to stimulate the T-cell differentiation, could first stimulate IL-1, TNF and IL-4 synthesis and then involved into T cell differentiation. To elucidate the roles of SATB1 and EFEMP1 antibodies in RA, cytokine assay and T-cell phenotyping, functions of $CD4^+$ T cells are mediated by cytokines. The increasing of the IL-1, TNF and IL-4 in cytokine assays was associated with rising up of $CD4^+$T cell percentage. Anti-SATB1 mAb stimulation also demonstrated a positive relation of TNF and IL-13, and IL-4, and fit the elucidation of $CD4^+$/$CD8^+$ T cells (**Borish L et al., 1996).** The explain of why the each autoantibody stimulated the cells to produce different levels of cytokines was not clear, although the FACS determined the formation of the stimulated T cell subsets from PBMCs culture.

It is through that antibodies specific for nuclear autoantigens are frequently present in serum of RA patients, however their specificity and sensitity are often low and are used as an additional test when RF and anti-CCP antibodies showed an unclear results (**Lorenz H.M et al., 2005**). Understanding of their functions in RA will direct further research on the antigens that are associated with these autoantibodies in RA. It is useful for development of novel diagnostic approaches to identify the expression profiles of antigenic targets in RA associated cells, should be further investigated for their use as novel biomarkers or additional biomarkers in available validated biomarker panels.

Acknowledgement

Kindly thanks to Biotechnology center of Ho Chi Minh City, Biomedical Research Institute, GBEC Bioengineering Company Limited for suppy a state-of the art to complete the research.

The below research sections were performed at Biomedical Research Institute, Transnational University Limburg, Belgium.

1. Amplification, fingerprinting and sequencing of the clones retrieved from cDNA phage display library
2. Cloning, expression and immunoblotting of SATB1 and EFEMP1
3. Mass spectrometry (MS)

Flow cytometry were researched at GBEC Bioengineering Company limited and Protein ELISA was done at Biotechnology Center of Ho Chi Minh City

Correspodant author: Anh Nguyen-Hung
Department for Medical Biotechnology
Biotechnology Center of Ho Chi Minh City
Tel: (84.8)37157631
Fax: (84.8) 38222567
E.mail: hunganh.nguyen@hcmbiotech.com.vn

References

1. Lund R, Ahlfors H, Kainonen E, Lahesmaa AM, Dixon C, Lahesmaa R. Identification of genes involved in the initiation of human Th1 or Th2 cell commitment. Eur J Immunol. 2005; 35(11): 3307-3313
2. Alvarez, J.D, Yasui D.H, Niida H, Joh T, Loh D.Y, and Kohwi-Shigematsu T. The MAR-binding protein SATB1 orchestrates temporal and spatial expression of multiple genes during T-cell development. Genes Dev. 2000; 14: 521-526
3. Atsuo N. Application of cellular gene therapy for rheumatoid arthritis. Mod Rheumatol. 2006; 16(5): 269–275
4. Borish L and Rosenwasser L. Update on cytokines. J. Allergy Clin. Immunol. 1996. 97: 719-730.
5. Cai S, Han HJ, Kohwi-Shigematsu T.Tissue-specific nuclear architecture and gene expression regulated by SATB1. Nat Genet. 2003; 34(1) :42-51
6. Cai S, Han, H.J, Kohwi-Shigematsu, T. Tissue-s pecific nuclear architecture and gene expression regulated by SATB1. Nature Genet. 2003: 34; 42 – 51
7. Dörner T, Jacobi A.M and Lipsky P.E. B cells in autoimmunity. Arthritis Res. 2009; 11:247-252
8. Firestein, G. S., Alvaro-Gracia, J. M. & Maki, R. Quantitative analysis of cytokine gene expression in rheumatoid arthritis. J Immunol. 1990; 144: 3347–3353
9. Galande S, Dickinson LA, Mian IS, Sikorska M, Kohwi-Shigematsu T. SATB1 cleavage by caspase 6 disrupts PDZ domain-mediated dimerization, causing detachment from chromatin early in T-cell apoptosis. Mol Cell Biol. 2001; 21(16): 5591-5596
10. Glynne R, Ghandour G, Rayner J, Mack DH, Goodnow CC. B-lymphocyte quiescence, tolerance and activation as viewed by global gene expression profiling on microarrays. Immunol Rev. 2000; 176: 216-220
11. Gonzalez-Rey E, Chorny A, Delgado M. Regulation of immune tolerance by anti-inflammatory neuropeptides. Nature Rev Immunol. 2007; 7: 52-63
12. Gotzmann J, Meissner M, Gerner C. The fate of the nuclear matrix-associated-region-binding protein SATB1 during apoptosis. Cell Death Differ. 2000 ; 7(5): 425-438
13. Govart C, Somers K, Huppert R, Stinissen P, Somers V. Exploring cDNA phage display for autoantibody profiling in the serum of multiple sclerosis
14. Han J.H, Russo J, Kohwi Y, Kohwi-Shigematsu T. SATB1 reprogrammes gene expression to promote breast tumour growth and metastasis. Nature. 2008; 452-455
15. Hu B, Thirtamara-Rajamani K.K, Sim H, Viapiano M.S. Fibulin-3 is uniquely upregulated in malignant gliomas and promotes tumor cell motility and invasion. Mol Cancer Res. 2009 ; 7(11): 1756-1770
16. Lecka-Czernik, B., Lumpkin, C. K. J., and Goldstein, S., An overexpressed gene transcript in senescent and quiescent human fibroblasts encoding a novel protein in the epidermal growth factor-like repeat family stimulates DNA synthesis. Mol. Cell. Biol. 1995; 15 (1): 22-25

17. Lorenz H.M, Grunke M, Wendler J, Heinzel P.A, Kalden J.R. Safety of 15-deoxyspergualin in the treatment of glomerulonephritis associated with active systemic lupus erythematosu. Ann Rheum Dis. 2005; 64:1517-1519
18. Matsumoto Y, Oshida T, Obayashi I, Imai Y, Matsui K, Yoshida NL, Nagata N, Ogawa K, Obayashi M, Kashiwabara T, Gunji S Nagasu T, Sugita Y, Tanaka T, Tsujimoto G, Katsunuma T, Akasawa A, Saito H.Identification of highly expressed genes in peripheral blood T cells from patients with atopic dermatitis. Int Arch Allergy Immunol. 2002; 129(4): 327-340
19. Nakayama Y, Mian I.S, Kohwi-Shigematsu T, Ogawa T.A nuclear targeting determinant for SATB1, a genome organizer in the T cell lineage. Cell Cycle. 2005; 4(8): 1099-1106
20. Nie H, Maika SD, Tucker PW, Gottlieb PD.A role for SATB1, a nuclear matrix association region-binding protein, in the development of CD8SP thymocytes and peripheral T lymphocytes. J Immunol. 2005; 174(8): 4745-52
21. Nishimura K, Sugiyama D, Kogata Y, Tsuji G, Nakazawa T, Kawano S, Saigo K, Morinobu A, Koshiba M, Karen M, Kuntz, Kamae I, Kumagai S. Meta analysis: diagnostic accuracy of anti-cyclic citrullinated peptide antibody and rheumatoid factor for rheumatoid arthritis. Ann. Intern Med. 2007; 146: 797-808
22. Pavan Kumar P, Oliver Bischof, Prabhat Kumar Purbey, Dimple Notani, Henning Urlaub, Anne Dejean & Sanjeev Galande. Functional interaction between PML and SATB1 regulates chromatin-loop architecture and transcription of the MHC class I locus. Nature Cell Biol. 2007; 9: 45 – 56
23. Raza, K. et al. Early rheumatoid arthritis is characterized by a distinct and transient synovial fluid cytokine profile of T cell and stromal cell origin. Arthritis Res. Ther. 7, R784–R795 (2005).
24. Song E.L, Hou Y.P, Yu S.P, Chen S.G, Huang J.T, Luo T, Kong L.P, Xu J, Wang H.Q..EFEMP1 expression promotes angiogenesis and accelerates the growth of cervical cancer in vivo. Gynecol Oncol. 2011; 121(1): 174-1780
25. Wakabayashi T, Matsumine A, Nakazora S, Hasegawa M, Iino T, Ota H, Sonoda H, Sudo A, Uchida A. Fibulin-3 negatively regulates chondrocyte differentiation. Biochem Biophys Res Commun. 2010; 391(1): 1116-1121
26. Warren J.S : Interleukins and tumor necrosis factor. Crit Rev Clin Lab Sci. 1990: 28 (1) 37-59.
27. Wen J.H.S, Rogers H, Dickinson L.A, Kohwi-Shigematsu T, Noguchi C.T. SATB1 family protein expressed during early erythroid differentiation modifies globin gene expression. Blood. 2005; 105 (8): 3330-3339.

Detection of Biomarker in Biopsies Based on Hr-Mas 2D HSQC Spectroscopy Indexation

A. Belghith[1], C. Collet[1], and J.P. Armspach[2]

[1] University of Strasbourg, LSIIT - UMR CNRS 7005 - France
{belghith,c.collet}@unistra.fr
[2] University of Strasbourg, LINC- UMR CNRS 7237 - France
jparmspach@unistra.fr

Abstract — **Multiple Sclerosis MS is one of the principal causes of morbidity worldwide. One of the strategies employed by the emergent science of metabolomics is MS biomarker identification. In this context, the technique of High-Resolution Magic Angle Spinning (HR-MAS) Nuclear Magnetic Resonance (NMR) spectra is widely used in metabolomic analysis involving tissue studies. Indeed, the NMR offers the potential to study molecular structures and their associations and interactions. In this paper, we present a new content-based object indexing and retrieval scheme for biomarkers detection. The biomarker identification is addressed by comparing 2D NMR MS spectrum groups for metabolite change detection using kernel based support vector data description (SVDD). Preliminary results obtained on synthetic spectrum database are shortly presented.**

Keywords — **Biomarker, indexation, NMR, SVDD, Multiple Sclerosis.**

I. Introduction

The medical signal acquisition techniques are constantly evolving in recent years (Nuclear Magnetic Resonance NMR spectroscopy, Magnetic Resonance Imaging MRI) and providing an increasing amount of data which should be then analyzed. Indeed, the signal acquisition techniques are potentially carrying the disease signature (biomarkers) well beyond the punctual clinical examination. In this context, automatic signal processing methods are regularly proposed to assist the expert in the qualitative and quantitative analysis of these images and to facilitate their interpretation. These methods should take into account the physics of signal acquisition, the *a priori* we have on the signals and the amount of data to analyze for a more accurate and reliable interpretation.

Among the new signal acquisition techniques, the 2D $H^1 - C^{13}$ Heteronuclear Single Quantum Coherence (HSQC) NMR spectra is widely used in metabolic studies explored to identify new biomarkers (objects). These biomarkers could be used to help the diseases diagnosis or to monitor therapeutic effects. To this end, it is important to improve the analysis tools associated with these techniques. In this context, we propose a new content-based object indexing and retrieval scheme for biomarkers detection.

A classical indexing and information retrieval scheme operates as follows: a user submits a query and the system identifies the relevant information to the submitted query and then returned it to the user. The most ancient and widely used indexing and document (*e.g.*, medical images, respiratory sounds, etc) retrieval scheme is the indexing by keywords where the document is described by a set of keywords. The keywords can consist of a word, phrase, or alphanumerical term. However, this technique remains limited by the low expressive power of words, by the language constraints (the transition from one language to another, semantic ambiguity) and the subjective nature of the annotations (two physicians can differently annotate a medical image).

In order to overcome these drawbacks, the content-based information indexing and retrieval approach (henceforth called indexing approach) was proposed [1]. Indeed, content-based means that the search will analyze the actual contents of the information rather than the meta-data such as keywords, tags, and/or descriptions associated with the information. For example, the term 'content' might refer to colors, textures, or any other information that can be derived from an image itself. Therefore, the content-based information indexing and retrieval approach allows us not only to automatically index the documents and query a database directly from their information content without human intervention but also to objectively analyze the database content. For example, if we consider a cerebral tumor as a query, we can easily identify with an objective similarity measurement function the similar objects (e.g., a document in a first step) belonged to the requested database.

Basically, the indexing scheme requires:

1. Object alignment step,
2. Object codifying and similarity measurement steps. The first step aims at codifying different

V. Van Toi et al. (Eds.): 4th International Conference on Biomedical Engineering in Vietnam, IFMBE Proceedings 40, pp. 88–91, 2013.
www.springerlink.com

objects into a compact description whereas the second one consists in establishing the object similarity measurement procedure.

In order to accelerate the large database queries, the indexing scheme can be divided into two phases:

1. An off-line phase in which the content database alignment and coding is performed. Indeed, the off-line coding consists in extracting the signatures associated with the contents of the database. The latter are then stored in a reverse dictionary (file name and signature), which allows to quickly find the object (document) associated with a given signature.
2. An online phase in which the user queries the database using an object request. The online alignment and coding steps only concern the object request. A similarity measure between the request object signature and those calculated in the reverse dictionary is then performed. Finally, the objects belonging to the database are classified by order of similarity.

Although this indexing scheme was successfully applied on web databases document retrieval [1], it is not suited as well to the biomarker identification task, where the object is a tumor. Indeed, the later requires the classification of medical signal profiles (groups) for changes detection. For example, if we consider two profile classes: the healthy class and the pathologic class, the biomarker identification task returns to classify a group of medical signals into the healthy and the pathologic classes (*e.g.*, cancer or psychological diseases) and to detect then the differences (changes) between them. As a matter of fact, healthy class can be considered as the "unchanged class" whereas the pathologic class as "change" class. Consequently, adding a classification step to the classical indexing scheme would allow us to detect biomarkers from HSQC spectra.

This paper is divided as follows: in the next section, we address the problem of biomarker identification through the new content-based object indexing and retrieval scheme. Then, results obtained on synthetic spectrum database are detailed.

II. Biomarker identification scheme

Unlike the classical indexing scheme, the proposed indexing scheme contains two additional steps: an object detection step (detection of HSQC peaks) and an object classification (change detection) step. The new indexing scheme is then divided into two phases:

1. An off-line phase in which we perform on each spectrum: a peak detection and alignment step, a peak coding and similarity measurement step and finally a peak classification step.
2. An online phase in which the user queries the database using a request (new individual/ group of spectra). The same steps as in the off-line phase are applied on the spectra request. Finally, the later is assigned to a previously defined profile (in the off line step). Note that unlike the classical indexing scheme, the object similarity step aims here at grouping the similar objects belonging to a given medical signal group allowing then the assignment of this group to the appropriate profile. In other word, the assignment of a new group/individual task is addressed using the classification step and not the similarity measurement step as in the classical indexing scheme.

In the following, we develop the object detection and alignment, the object coding and similarity measurement and the object classification steps.

A. Object Detection and Alignment

The problem of object detection consists in dividing signal into meaningful groups (objects) based on the spatial arrangement and/or pixel intensity. The object alignment task is the process of overlaying two or more objects taken at different times, and/or from different viewpoints, and/or by different modalities. The object detection and alignment task is a crucial step in the indexing scheme since all other steps depend on it. Consequently, in order to lead to optimal object detection and alignment results, all *a priori* knowledge that we have on the data need to be properly integrated in the proposed object detection and alignment methods. To this end, we recently presented a new peak detection-alignment method based on the use of evidence theory [2]. Indeed, the evidence theory allows us to handle the uncertainty models and the imprecision that characterize the HSQC spectra. Therefore, we proposed the coupling between the fuzzy set theory and Bayesian theory to model and quantify the degree of spectrum imprecision which will be used to define the mass functions (*i.e.*, a function that models the belief degree on a given hypothesis). Once these peaks are detected and aligned, we proceed to the peak coding and similarity measurement.

B. Object Coding and Similarity Measurement

A spectrum is composed of several peaks. These peaks are the responses of metabolite presence. Therefore, peaks belonging to a given metabolite share common properties. In order to improve the peak coding and similarity measurement task, these properties should be modeled and integrated within the proposed scheme. Thus, it is better to manipulate metabolites than single peaks (a metabolite which corresponds to an 'object' previously defined consists of different peaks with specific characteristics (location, shape and amplitudes). Since a metabolite already has an analytical and parcimonious description (set of few peaks with estimated specific features (location, shape and amplitudes) expected during detection-alignment step) the metabolite coding step is given and finally the similarity measurement task is simply equivalent to a *metabolite identification* task (*e.g.*, identify the most similar metabolite to a given metabolite). Therefore, we proposed in [3] a new metabolite identification method based on the use of fuzzy set theory which is well appropriate to handle fuzzy situations. Moreover, in order to lead to a more robust, more accurate and efficient similarity measurement method, knowledge on biopsy compositions and peak amplitudes was integrated into the identification scheme. These *a priori* knowledge are available in a metabolites library denoted the "corpus". Indeed, the corpus contains different metabolites of interest within the spectra as well as the characteristics of each metabolites (waited peak locations and peaks amplitudes).

Once these metabolites are identified, we proceed to the metabolite classification for change detection.

C. *Object Classification*

The last step consists in blindly compare/classify the spectrum group query and a given spectrum group defined in the off-line step into two classes for each detected metabolite: class of targets (i.e., no significant changes within biopsies for this metabolite) and the outliers (i.e., biopsies where some metabolite are beyond expected normal variability) using a new version of the SVDD method we proposed in [4] and called Support Vector Data Description with Dependency Handling (SV3DH). Indeed, given two spectrum groups, the SV3DH method is used to detect which metabolites exhibit significant variation. In this case, the feature vector x_{Mi} (which consists of the amplitudes, the locations and the shape of peaks belonging to a given metabolite) is associated to each detected metabolite Mi. The first step consists in selecting a spectrum population and then estimating the hypersphere hyperparameters (ai, Ri), ai is the center and Ri the center of the hypersphere, that model the profile of each metabolite Mi. In order to quantify the metabolite change degree denoted μ_{Mi}, we propose the use of an S-membership function f [4]. The expression of μ_{Mi} is given by:

$$\mu_{Mi} = f(K(xi - ai, xi - ai) / Ri)$$

Where K is the kernel function [4]. Therefore, the more μ_{Mi} is close to 1 the more the metabolite Mi has changed. Thus the metabolites can be ranked according to their change degrees allowing the physicians to select the relevant changed metabolites by a simple threshold and then to control the results. In order to facilitate the threshold setting, we have classified the metabolites according to the change degree into two classes: "weak change" and "high change". To this end, we used a Maximum Likelihood (ML) classifier with respect to the statistical distribution law of metabolite change degrees (cf. Fig.1). Note that μ_{Mi} is assumed to follow a Generalized Gaussian pdf [5].

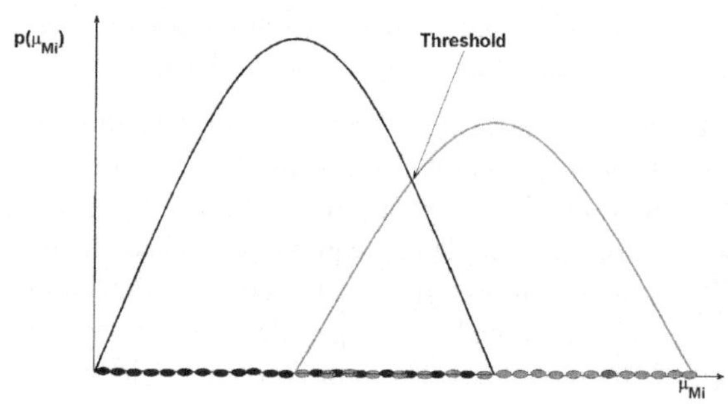

Fig. 1 Threshold setting by Maximum Likelihood classifier.

At the end of this indexing scheme, the changed metabolites between two HSQC spectrum groups (e.g., healthy and pathological group) could be given to the physicians as potential biomarkers.

III. Results

In this section, we present some biomarker identification results obtained on synthetic spectrum database. To model the HRMAS-2D image formation, we consider a 2D spectrum image characterized by its intensity distribution Y, corresponding to the observation of a theoretical 2D-spectrum image X through a nuclear magnetic resonance system. Y is then the convolution product result of X with the Point Spread Function (PSF) of the nuclear magnetic resonance system corrupted by an additive noise. We generate firstly a synthetic 2D spectrum image X_{ref} with fourty metabolites. Then sixty other synthetic theoretical 2D spectrum images are generated from X_{ref} by applying a displacement vector at each peak of X_{ref} . A spatially correlated noise was added to each spectrum to obtain the synthetic spectra used in the simulation. These simulated spectra are then divided into three groups of twenty spectra G_1 , G_2 and G_3 . In each spectrum group, the concentrations of five metabolites belonging to ten spectra were changed. Therefore, each group contains ten spectra modeling the healthy spectra and ten other spectra modeling the pathological ones. Note that the changed metabolites constitute the simulated biomarkers. Table 1, 2 and 3 show the biomarkers identifications. As one can see, all changed metabolites were correctly identified. Note that the estimated threshold by the ML classifier is 0.4

Table 1 Identified biomarker for healthy G_1 vs. pathological G_1 spectra.

Biomarker	μ_{Mi}	Real change/estimated change of concentration
Metabolite 1	0.784	increase/increase
Metabolite 2	0.658	increase/increase
Metabolite 3	0.614	increase/increase
Metabolite 4	0.527	decrease/decrease
Metabolite 5	0.412	decrease/decrease

Table 2 Identified biomarker for healthy G_2 vs. pathological G_2 spectra.

Biomarker	μ_{Mi}	Real change/estimated change of concentration
Metabolite 1	0.980	decrease/decrease
Metabolite 2	0.853	increase/increase
Metabolite 3	0.771	increase/increase
Metabolite 4	0.712	increase/increase
Metabolite 5	0.625	decrease/decrease

Table 3 Identified biomarker for healthy G_3 vs. pathological G_3 spectra.

Biomarker	μ_{Mi}	Real change/estimated change of concentration
Metabolite 1	0.632	increase/increase
Metabolite 2	0.604	increase/increase
Metabolite 3	0.588	decrease/decrease
Metabolite 4	0.491	increase/increase
Metabolite 5	0.412	increase/increase

IV. Conclusions

In this paper, we presented a new content-based object indexing and retrieval scheme for biomarkers detection. The biomarker identification is achieved by comparing 2D NMR MS spectrum groups for metabolite change detection using SV3DH classifier. Results obtained on a large set of raw MS spectrum data base, were presented and we currently work on future development on other biopsy tissues of HSQC spectrum database with physicians from University Hospitals in Strasbourg [6].

Acknowledgment

The authors thank Region Alsace (France) and ARSEP fundation (Fondation pour l'Aide à la Recherche sur la Schlérose en Plaques) for support to this research project. The authors thank Dr Karim Elbayed and Pr Izzie-Jacques Namer (from *Department of Biophysics and Nuclear Medicine, University Hospitals of Strasbourg*) for providing HSQC biopsies spectra and for their validations.

References

1. Smoliar, S.W. and Zhang, H.J, Content based video indexing and retrieval, Multimedia, IEEE Transactions on, vol 1, pages 62-72, 1994.
2. A. Belghith, C. Collet, J.-P. Armspach, A unified framework for peakdetection and alignment: application to HR-MAS 2D NMR spectroscopy , 6th International Conference on Mass Data Analysis of Images and Signals in Medicine, Biotechnology and Chemistry, MDA 2011, pp. 14, New York, USA, septembre 2011.
3. A. Belghith, C. Collet, J.-P. Armspach, A statistical framework for biomarker identification of biopsies using HR-MAS 2D NMR spectroscopy, 8th IEEE International Symposium on Biomedical Imaging, ISBI 2011 pp. 4, Chicago, USA, March 2011.
4. A. Belghith, C. Collet, J.-P. Armspach, Change detection based on Support Vector Data Description handling dependency, 18 th IEEE International Conference on Image Processing, ICIP, pp. 4, Bruxelle, Belgique, septembre 2011
5. Meignen, S. and Meignen, H, On the modeling of small sample distributions with generalized Gaussian density in a maximum likelihood framework, Image Processing, IEEE Transactions on,vol 15, pages 1647-1652, 2006.
6. Akram Belghith, Indexation of HSQC spectra or fMRI images for biomarker identification, PhD Thesis, to appear, http://lsiit-miv.u-strasbg.fr/lsiit/perso/collet, 2012.

Author: Akram BELGHITH
Institute: University of Strasbourg
Street: UDS-ENSPS-LSIIT Boulevard Sébastien Brant - Pôle d'Innovation API, BP 10413
City: Illkirch
Country: France
Email: belghith@unistra.fr

Corresponding author : c.collet@unistra.fr

Theoretical and Experimental Investigation of a Method to Detect an Aortic Aneurysm from Pulse Waves

H. Sato[1], H. Matsuhisa[2], H. Utsuno[3], K. Yamada[2], H. Yoshida[1], and T. Ogura[1]

[1] Omron Healthcare Co., Ltd., Muko, Japan
[2] Department of Mechanical Engineering and Science, Kyoto University, Kyoto, Japan
[3] Faculty of Engineering Science, Kansai University, Suita, Japan

Abstract — **Early identification of aortic aneurysms is important for effective treatment. Although imaging systems have the ability to identify aortic aneurysms, high cost prevents their use for screening. On the other hand, pulse waves can be measured easily, and because pulse wave velocity is affected by the aortic aneurysm, there is a possibility to find an aortic aneurysm by pulse wave measurement. The aim of this study is to propose a method to find an aortic aneurysm from the pulse wave and to develop a measurement device. The aortic aneurysm causes the expansion of artery. The basic theory is based on the wave of a mechanical reactance silencer. The transfer functions of a pulse wave between brachial artery and anterior tibial artery were investigated. The transmission loss is increased by increase of the inner radius of an aortic aneurysm. The interval frequency of local minimum points of the transfer function correlates with the length of an aortic aneurysm. When the stiffness decreases, the characteristic impedance decreases, the pulse wave velocity decreases and the frequencies of the local peaks become low. It is therefore possible to estimate the size of an aortic aneurysm from the transfer function. The method to find an aortic aneurysm from the pulse wave was investigated theoretically and experimentally with silicone tubes. The relation between the transfer function and the pathological arterial change was clarified. This method can be used for diagnosis of aortic aneurysms.**

Keywords — **pulse wave, aortic aneurysm, wave propagation.**

I. INTRODUCTION

The treatment of aortic aneurysms has progressed but early identification is important to save lives. MRI, X-ray CT and ultrasonographic examinations have the ability to find aortic aneurysms but they are mainly used for diagnosis but not for screening because of their high costs. On the other hand, pulse wave can be measured as easily as blood pressure. Since the pulse wave velocity is affected by the aortic aneurysm [1], there is a possibility to find an aortic aneurysm from pulse wave measurement. The aim of this study is to propose a method to find an aortic aneurysm from the pulse wave and to develop a measurement device.

II. THEORETICAL ANALYSIS AND EXPERIMENT

A. *Basic Model and Analysis*

Arterial expansion in an aortic aneurysm causes characteristic impedance to change. Theoretically, this can be modeled as the wave of a mechanical reactance silencer noting that an artery is an elastic tube and a silencer is a rigid tube. In addition to the characteristic impedance, the pulse wave velocity parameter also varies. The transmission loss of the pulse wave in an aortic aneurysm is formulated in consideration with these parameters.

The basic analytical model of a blood vessel with aortic aneurysm is shown in Fig. 1. The expansion area is between points 1 and 2. with inner radius R_A, outer radius R_{Ao}, Young's modulus E_A and length d. The other area is a normal artery of inner radius R_B, outer radius R_{Bo}, Young's modulus E_B and length l_1 and l_3. The boundary condition at point 3 is non-reflection.

When the propagation coefficient between points 0 and 1, γ_B is given by the following function:

$$\gamma_B = j\frac{\omega}{c_B} \tag{1}$$

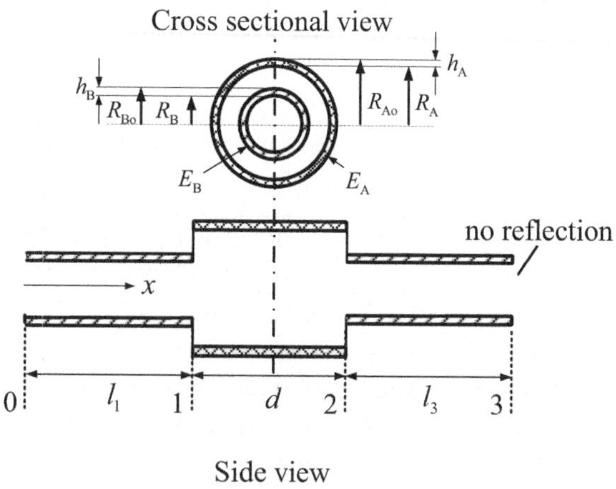

Fig. 1 Theoretical model of aortic aneurysm.

V. Van Toi et al. (Eds.): 4th International Conference on Biomedical Engineering in Vietnam, IFMBE Proceedings 40, pp. 92–96, 2013.
www.springerlink.com

where, ω is the circular frequency and c_B is the propagation velocity.

The pressure $p_{01}(x,t)$ between points 0 and 1, is a function of forward wave p_{f1} and reflection wave p_{r1}.

$$p_{01}(x,t) = p_{f1}e^{-\gamma_B x} + p_{r1}e^{\gamma_B x} \tag{2}$$

In the same area, the volume flow $q_{01}(x,t)$ is given by

$$q_{01}(x,t) = \frac{1}{Z_{0B}}\left(p_{f1}e^{-\gamma_B x} + p_{r1}e^{\gamma_B x}\right) \tag{3}$$

where Z_{0B} is the characteristic impedance. The pressure and the volume flow between points 1 and 2 and points 2 and 3 are expressed by the similar equations.

Since the end at point 3 is non-reflection, the forward wave at point 1 p_{f1} and the forward wave at point 3 p_{f3} have the following relation.

$$\begin{aligned} p_{f1} &= \frac{p_{f3}}{4Z_{0A}Z_{0B}}\left(X^+ + X^-\right) \\ X^+ &= \left(Z_{0A}+Z_{0B}\right)^2 e^{(\gamma_A-\gamma_B)d} \\ X^- &= \left(Z_{0A}-Z_{0B}\right)^2 e^{-(\gamma_A+\gamma_B)d} \end{aligned} \tag{4}$$

Thus, the transmission loss T is given by the following equation.

$$T = 10\log\left|\frac{1}{4Z_{0A}Z_{0B}}\left(X^+ + X^-\right)\right|^2 \tag{5}$$

The characteristic impedance Z_0 is a function of propagation velocity c, blood density ρ, and cross sectional area S.

$$Z_0 = \frac{\rho c}{S} \tag{6}$$

The propagation velocity at the area between 1 and 2, c_A is calculated using the compliance of an elastic thick cylinder.

$$c_A = \sqrt{\frac{E_A}{2(1+\nu)\rho}\cdot\frac{R_{Ao}^2 - R_A^2}{R_A^2(1-2\nu)+R_{Ao}^2}} \tag{7}$$

When the wall thickness is h_A, the equation becomes

$$c_A = \sqrt{\frac{E_A}{2(1+\nu)\rho}\cdot\frac{R_{Ao}^2 - R_A^2}{1+\dfrac{2(1-\nu)}{\dfrac{h_A}{R_A}\left(\dfrac{h_A}{R_A}+2\right)}}} \tag{8}$$

In order to consider the transmission loss, the ratio of wall thickness and inner radius are assumed to be same as

$$\frac{R_{Ao}-R_A}{R_A} = \frac{R_{Bo}-R_B}{R_B} \tag{9}$$

with the ratio of inner radius of the artery and aortic aneurysm as α_R

$$R_A = \alpha_R R_B \tag{10}$$

The maximum transmission loss $T_{\max}$ appears in the case $\gamma_B d = -m\pi$ where m is an integer.

$$T_{\max} = 10\log\left|\frac{\alpha_R^4+1}{2\alpha_R^2}\right|^2 \tag{11}$$

According to these equations, the forward wave from the heart reflects at the end of the aortic aneurysm due to the large radius of aortic aneurysm. The relation between the transmission loss of the pulse wave and size of the aortic aneurysm is as follows:

1) the transmission loss increases when the inner radius of aortic aneurysm increases;
2) the interval frequency decreases when the length of aortic aneurysm increases; and,
3) the transmission loss decreases and the interval frequency of local minimum points decreases when the Young's modulus of aortic aneurysm increases.

B. *Simulations of Whole Body*

The computer simulation model of whole human artery [2] was constructed. The transfer functions of the pulse wave between the brachial artery and the anterior tibial artery were investigated with the inner radius, length and Young's modulus of segment number 34 (thoracic aorta), 50 and 65 (abdominal aorta) changed as shown in Fig. 2.

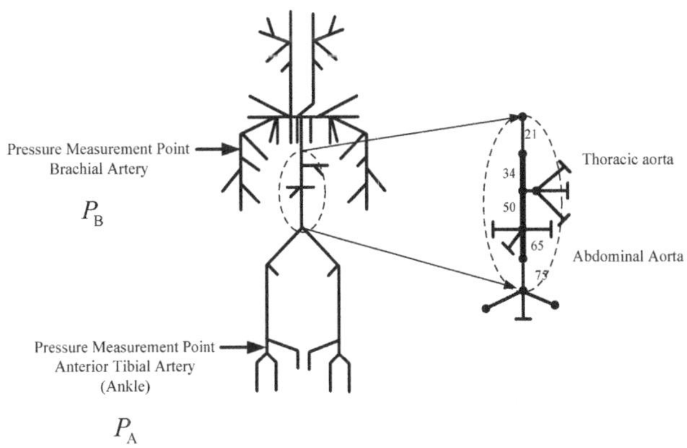

Fig. 2 Schematic diagram of a thoracic aorta and an abdominal aorta and pressure measurement points on the multi-branched model.

C. *Experiment*

The experimental apparatus of a multi-branched aorta was made by silicone tubes as shown in Fig. 3. The dimensions are the same as the computer simulation model. Optical fiber pressure sensors are installed at the ends of tubes to measure pressure P_1 and P_2 at the brachial and tibal arteries to calculate the transfer function by FFT analyzer. The other ends of tubes are closed. A cylinder with an oscillator is connected at the ascending aorta. It is oscillated by sine wave and its frequency is swept from 0.5 to 40 Hz.

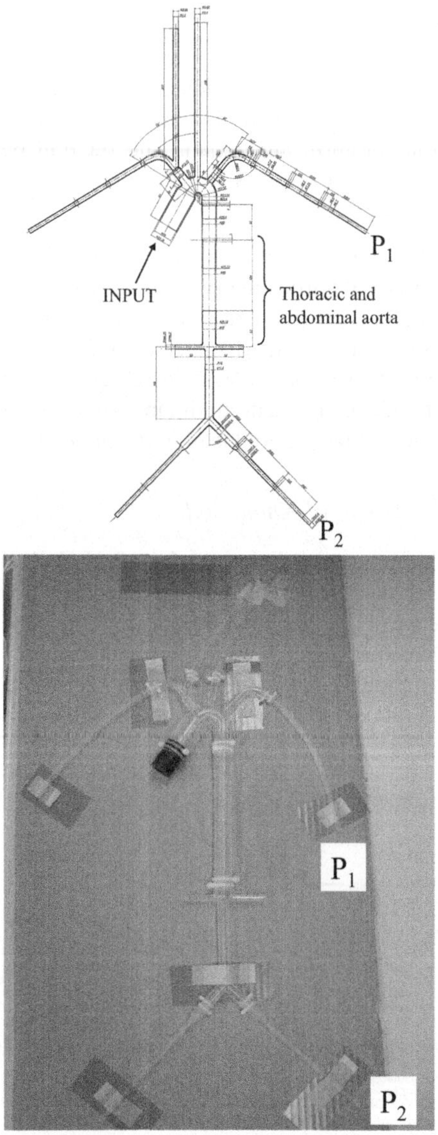

Fig. 3 Schematic diagram and picture of silicon tube model of human artery.

III. Results and discussion

A. *Calculated Result of Basic Theory*

The transmission loss is calculated using equation (5). The transmission loss T is shown in Fig. 4 where the inner radius R_{Ai} is changed from 20 to 40 mm. It shows that the transmission loss increases as inner radius of aortic aneurysm increases but the interval frequency of local minimum points it is not changed. The change of the transmission loss T due to change of the length d from 100 to 300 mm is shown in Fig. 5. When the Young's modulus E_A decreases, the frequencies of the local peaks become lower as shown in Fig. 6. This is caused by the decreases in characteristic impedance and pulse wave velocity. These results show the basic relations between transmission loss and inner radius, length and Young's modulus of an aortic aneurysm.

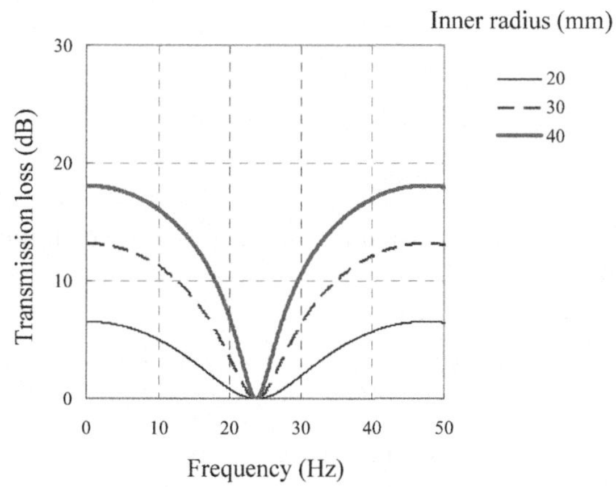

Fig. 4 Transmission loss T of the aortic aneurysm for several inner radii R_A.

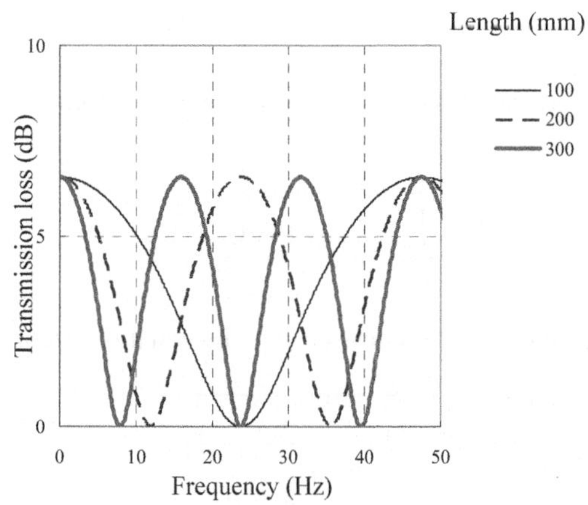

Fig. 5 Transmission loss T of the aortic aneurysm for several lengths d.

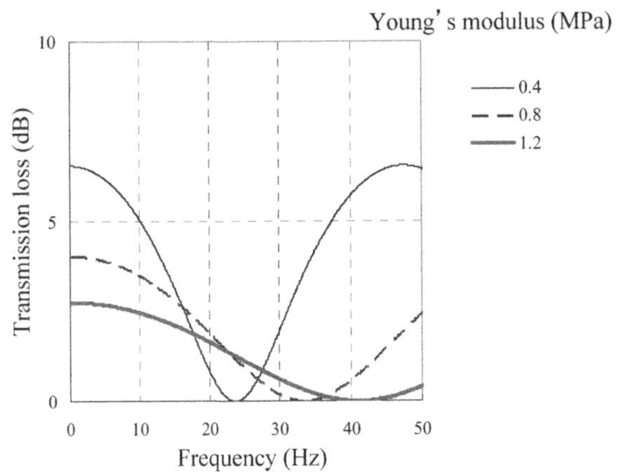

Fig. 6 Transmission loss T of the aortic aneurysm for several Young's moduli E_A.

B. *Calculated Results of Whole Body*

When the radius of aorta increases, the gain of the transfer function decreases as shown in Fig. 7. The interval of peak frequency is not changed much. When the length of expansion artery increases, the interval frequency of the local minimum point decreases as shown in Fig. 8. The gain is not changed in this case. When the Young's modulus of abdominal aortae increase, the interval frequency of local minimum point and gain of transfer function increases.

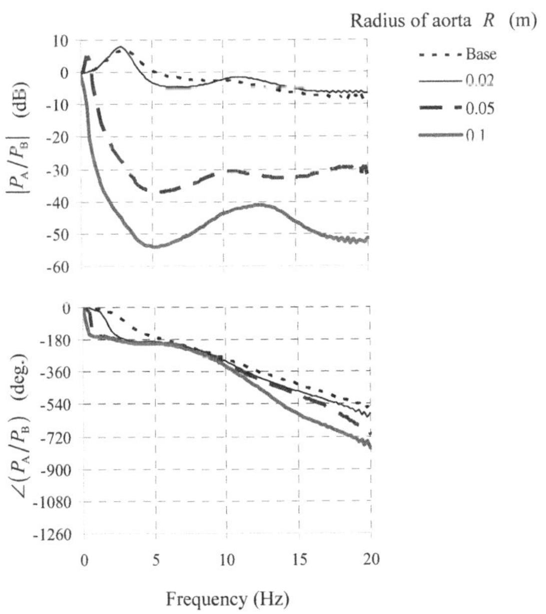

Fig. 7 Simulated transfer functions P_A/P_B between a brachial artery and an anterior tibial artery for several radii R of an abdominal aorta (segment #65).

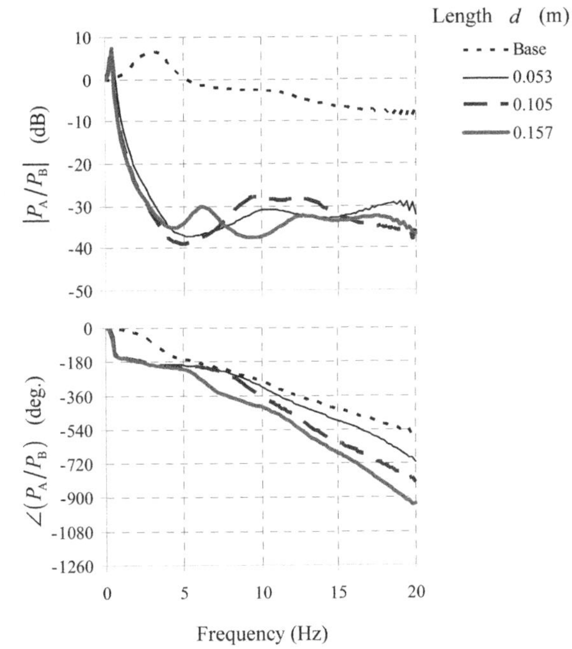

Fig. 8 Simulated transfer functions P_A/P_B between a brachial artery and an anterior tibial artery for several lengths d of aortae (segment #34, 50, 65).

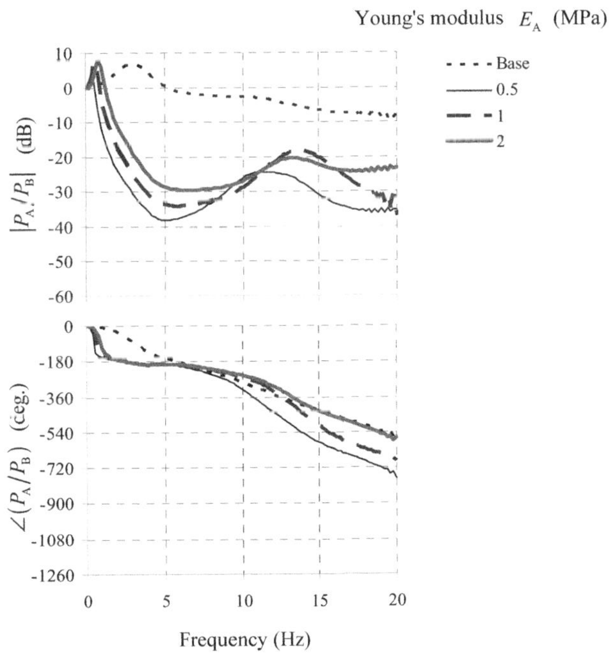

Fig. 9 Simulated transfer functions P_A/P_B between a brachial artery and an anterior tibial artery for several Young's moduli E_A of abdominal aortae (segment #50, 65).

Though those transfer functions are affected from all branches of the artery and reflections, the changes according to the basic theory can be also observed.

C. *Result of Experiment*

The thoracic and abdominal aorta of the experimental model shown in Fig. 3 can be replaced with different aortae. The transfer functions with several inner radii are shown in Fig. 11. The interval frequency of local minimum points are not affected by the inner radius. The gain of transfer functions are not changed because the difference of inner radius is small. The dependence on Young's modulus is shown in Fig. 12. When the Young's modulus of abdominal aorta increases, the interval frequency of local minimum points increases. Though the whole body system include complex factors, these basic characteristics were observed.

The Young's modulus has influences on both the transmission loss and the interval frequency with the latter exhibiting clear changes. The aortic aneurysm should be found in the early stage before changing its Young's modulus at the screening. This study suggests that it is possible to estimate the size of an aortic aneurysm from the transmission loss and the local minimum frequency of pressure transfer function between brachial and anterior tibial arteries. In particular, the length of an aortic aneurysm can be detected easier than the change of the inner radius.

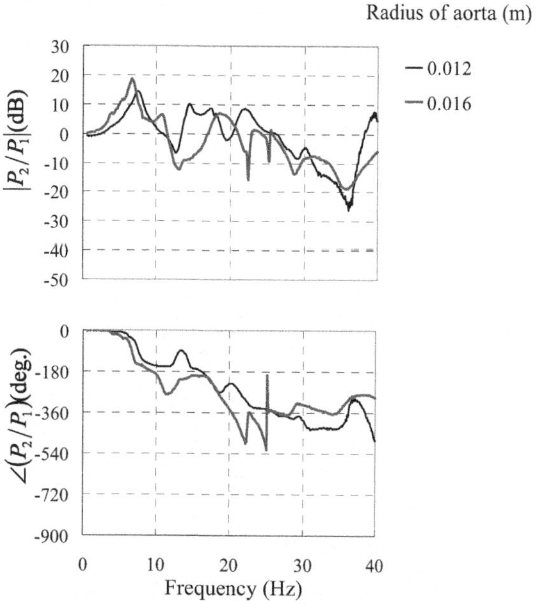

Fig. 10 Experimental results of transfer function P_2/P_1 for several radius R of an abdominal aorta.

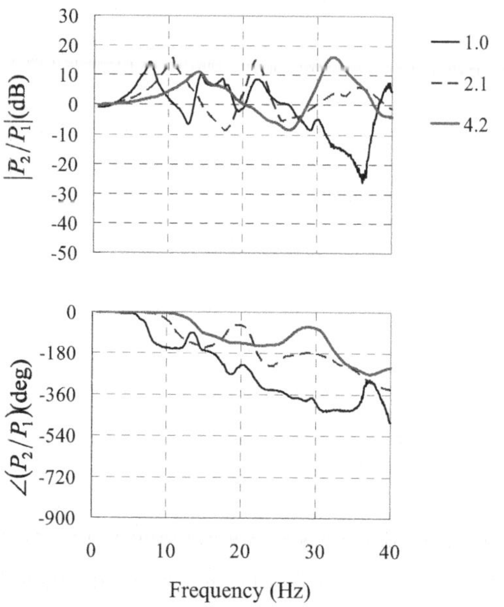

Fig. 11 Experimental results of transfer function P_2/P_1 for several Young's moduli E of an abdominal aorta.

IV. CONCLUSIONS

The method to find the aortic aneurysm from the pulse wave was investigated theoretically and experimentally with the artery system made of silicone tubes. The relation between the transmission function and the pathological change of artery was clarified. This method can be used for diagnosis of aortic aneurysm.

ACKNOWLEDGMENT

This study was supported by Grants for Regional Science and Technology Promotion from the Ministry of Education, Culture, Sports, Science and Technology in Japan.

REFERENCES

1. Lantelme, P., "Effect of abdominal aortic grafts on aortic stiffness and central hemodynamics", Journal of Hypertension, Vol. 27, (2009), pp. 1268-1276.
2. Avolio A. P., "Multi-branched model of human arterial system", Medical and Biological Engineering and Computing, Vol. 18, (1980), pp. 709-718.

Adaptive Medical Image Edge Detection in Contourlet Domain

Nguyen Thanh Binh and Ngoc Minh Le

Faculty of Computer Science and Engineering, Ho Chi Minh City University of Technology, Vietnam

Abstract — **Wavelet transforms and other frequency domain methods are widely used for edge detection, but they suffer from shift and rotation sensitivity. Recently, new x-let multiscale transforms have been developed. This paper describes a method for medical image edge detection in contourlet transform domain that uses cycle spinning. The edge detection is done by performing contourlet transform to the medical image, keeping the coefficients where the signal-to-noise ratio is high, and reducing the coefficients where the signal-to-noise ratio is low. For the contourlet transform to be translation-invariant a 2-D cycle spinning is implemented on subbands Δ_1 and Δ_2. Cycle spinning for edge detection is implemented. The transformed data are shifted, edge detected and unshifted. The method has been compared with earlier edge detection methods that use the discrete wavelet transform and contourlet transform.**

Keywords — **medical image, edge detection, contourlet tranform, cycle spinning.**

I. INTRODUCTION

Medical image edge detection is a fundamental problem in medical diagnosis tasks. In medical imaging diagnosis, medical devices such as ultrasound scan, X-ray... are used to probe and to acquire images inside the patient's body for diagnosis. Therefore, research in diagnostic techniques for exploiting medical images is very important in modern medicine.

For many patients, such as cancer patients, their survival depends on whether the doctor has succeeded or not in diagnosis, e.g., using medical images. These are often blurred, so finding the object boundaries in medical images is a difficult task. For instance, with a blurry image the surgeon may be unable to detect cancer cells. Clearly, medical imaging may help the surgeon to accurately identify the cancer tissue. Several techniques can be used in solving detection problems in medical imaging: hidden Markov models [6], wavelet transform [2],...

Based on the discrete wavelet transform (DWT), tools for an efficient localized multiresolution analysis of images can be provided. However, DWT also has serious drawbacks, namely shift sensitivity [7] and poor directionality [8]. Techniques to reduce these drawbacks have been proposed; for instance, new multiscale transforms have been designed such as the contourlet transform to integrate the concept of directionality in a more useful fashion.

In this paper, we present a method for edge detection in medical images using contourlet transform with cycle spinning. We compare the method with earlier edge detection techniques based on the discrete wavelet transform and on the contourlet transform. The rest of this paper is organized as follows: Section II describes the advantages of contourlet transform for edge detection. Details of the algorithm are given in Section III. In Section IV, the results of the method for edge detection are shown and compared to other methods. We conclude in Section V.

II. THE CONTOURLET TRANSFORM

The contourlet construction presented here is based on the work of Do and Vetterli [1]. Contourlet is easily applied in medical image processing because its representation is a fixed transform. Contourlet allows for a different number of directions at each scale and aspect ratios. This feature allows an efficient contourlet-based approximation of a smooth contour at multiple resolutions. The discrete contourlet transform is a multiscale and directional decomposition using a combination of Laplacian pyramid (LP) and directional filter bank (DFB) [1].

The idea of the contourlet construction [1] is: let $a_0[n]$ be the input image, the output after the LP step is I bandpass images $b_i[n]$, i =1, 2,..., I and a lowpass image $a_I[n]$. Each bandpass image $b_i[n]$ is decomposed by an ℓ_i -level DFB into 2^{ℓ_i} bandpass directional images $c_{i,k}^{(\ell_i)}[n]$, for k = 0, 1,..., $2^{\ell_i} - 1$.

In the discrete contourlet transform, the multiscale and directional decomposition steps are decoupled. So, we have different numbers of directions at different scales.

III. THE METHOD

In this section, we present an approach to edge detection in medical images; the approach is based on the contourlet transform with cycle spinning.

V. Van Toi et al. (Eds.): 4th International Conference on Biomedical Engineering in Vietnam, IFMBE Proceedings 40, pp. 97–100, 2013.
www.springerlink.com

Contourlets constitute a new family of frames that are designed to represent smooth contours in different directions of an image. Contourlet decomposition proceeds through two main steps: first, Laplacian pyramid multiscale decomposition is performed; then directional filter bank decomposition is used to link point discontinuity to linear structures. In more detail, an image is decomposed into a low pass image and bandpass images by the LP decomposition. Each bandpass output is further decomposed by the DFB step. The output of the DFB step consists of smooth contours and directional edges. In this paper, each directional subband at each level consists of 2^n element, where n is a positive integer. For example, if we choose to decompose an image into six levels using $n = 1$ and $n = 2$ then we get 2 and 4 subbands, respectively. Fig 1 shows a contourlet decomposition.

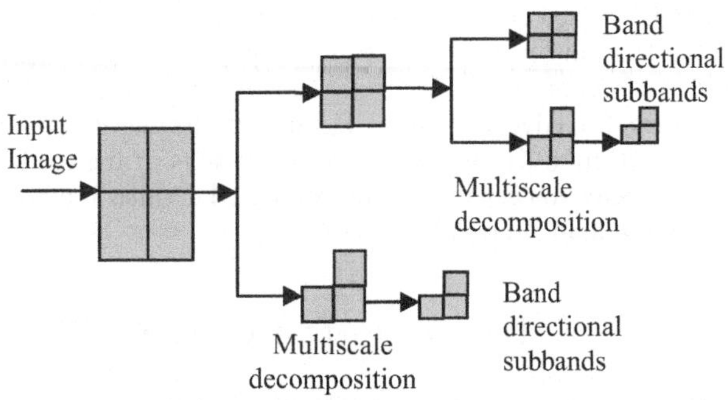

Fig. 1 Contourlet decomposition.

The goal of edge detection is to divide the given medical image into regions that belong to distinct objects in the depicted scene. This process consists of three steps [4], [5], [9]: apply contourlet transform to the image, retain the coefficients where the signal-to-noise ratio is high, and reduce the coefficients where the signal-to-noise ratio is low. A 2-D cycle spinning is implemented on subbands Δ_1 and Δ_2 so that the contourlet transform is translation-invariant. The transformed data are shifted, edge detected, and unshifted. Applying the following procedure,

$$\hat{s} = \frac{1}{K_1 K_2} \sum_{i=1,j=1}^{K_1,K_2} S_{-i,-j}(T^{-1}(\theta[T(S_{i,j}(x))])) \quad (1)$$

where K_1 and K_2 are the maximum number of shifts, we expect an improvement for the estimation $\hat{s}$ compared to the medical image without cycle spinning. For the contourlet transform, if the input image has size $N \times N$, with $N = 2^K$, then after K shifts in each direction, the output becomes repetitive and so the maximum numbers of shifts will be K in each direction. If one decomposes an image of size $N \times N$ using the contourlet transform, then the number of decomposition levels in the Δ_1 stage will be at most K, and therefore, the maximum number of shifts is K in the row and column directions.

The threshold for the contourlet coefficients can be calculated using statistical properties of noise or blur. After thresholding the contourlet coefficients, the image can be reconstructed. Donoho and Johnstone [10] have described a threshold that depends on standard deviation of contourlet coefficients. Here, to compute the threshold value for medical image edge detection, we use a combination of three parameters: the contrast ratio (ratio between standard deviation and mean of contourlet coefficients), the absolute median of contourlet coefficients, and a level dependent parameter.

IV. Experiments and Results

We have applied the procedure described in Section III and observed good performance in our edge detection experiments as briefly demonstrated in this section. Fig. 2 shows the edge detection in medical images with the method presented above.

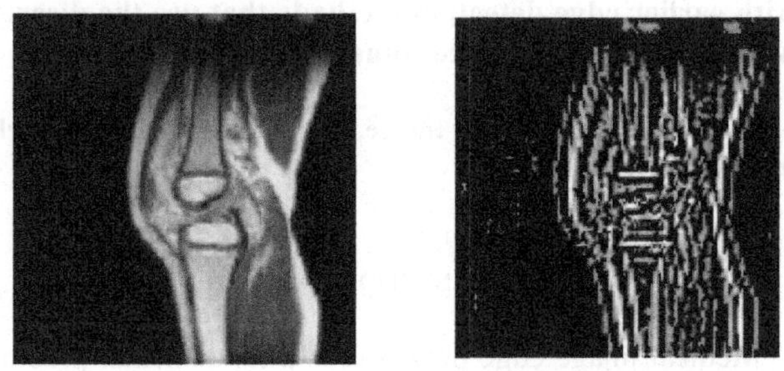

(a) Original image (b) The proposed method

Fig. 2 Edge detection in medical image with our method.

To evaluate the performance of the proposed method using contourlet transform with cycle spinning (CT-CS), we have compared it with earlier edge detection methods using DWT and CT. We use the evaluation function F developed by Liu and Yang [12]:

$$F(I) = \sqrt{N} \times \sum_{i=1}^{N} \frac{e_i^2}{\sqrt{a_i}} \quad (2)$$

where I is the original image, N is the number of regions in the edge-detected image, a_i is the area or the number of pixels of the ith region, and e_i is the sum of the Euclidean distances between the color vectors of the pixels in the regions of the original image and the edge-detected image. Here, F is scaled down by the factor 1/100. If the value of F is smaller, then the edge-detection result is better.

The comparison of results with other methods has been done using our program on the same images and at similar scales. Hard thresholding is applied to the coefficients after decomposition in contourlet domain. In case of contourlet transform with cycle spinning, the input image is of size $N \times N$, with $N = 2^K$. In the case $K = 8$ (i.e., $N = 256$) after 8 shifts in each direction, the transform output repeats and so the maximum number of shifts will be 8 in each direction. We can apply the same approach to the contourlet transform.

To do experiments using the proposed algorithm, we have selected one hundred different images of different sizes: 64×64, 128×128, 256×256, 512×512, and 1024×1024 pixels. Here, we report on some results achieved as shown in Fig. 3, Fig. 4 and Fig. 5.

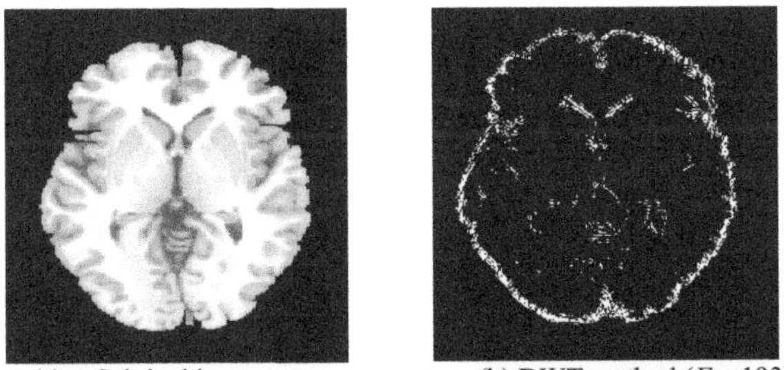

(a) Original image (b) DWT method ($F = 193$)

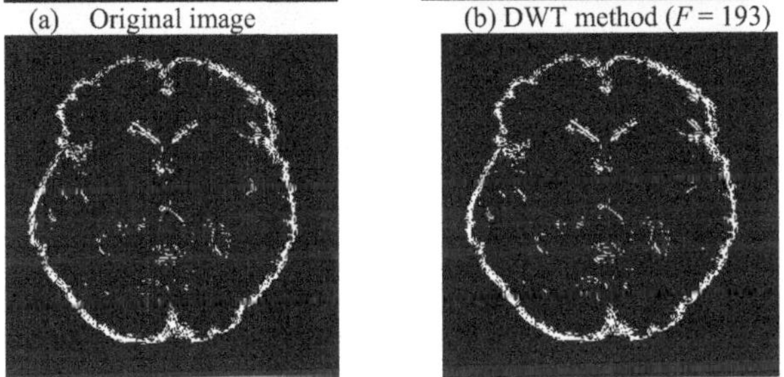

(c) CT method ($F = 124$) (d) The proposed method (CT-CS) ($F = 101$)

Fig. 3 Edge detection results for medical images.

Fig. 3 shows an image corrupted by Gaussian additive noise and its detected edges with two other methods. In this case, the result achieved with CT-CS is better.

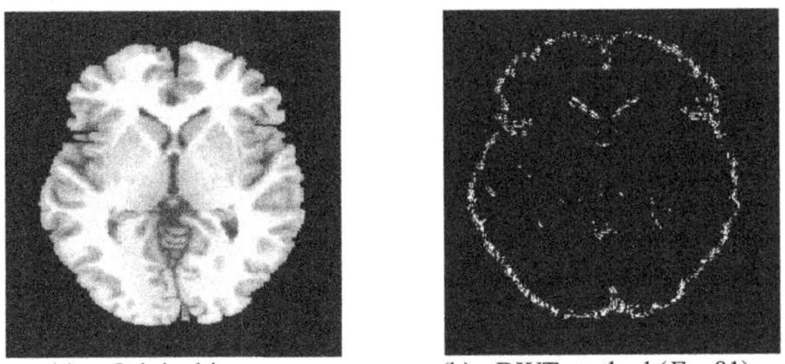

(a) Original image (b) DWT method ($F = 81$)

Fig. 4 Image corrupted with Gaussian additive noise and its edges detected with other methods.

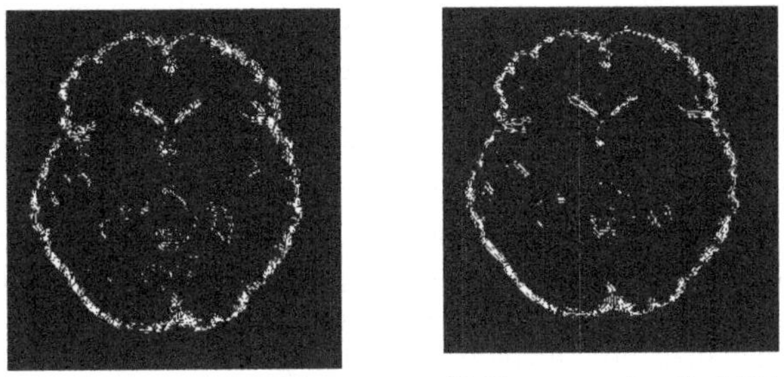

(c) CT method ($F = 65$) (d) The proposed method (CT-CS)($F = 48$)

Fig. 4 *(continued)*

Fig. 5 shows an image corrupted with speckle noise and its detected edges with other methods. In this case, the result using CT-CS is also better.

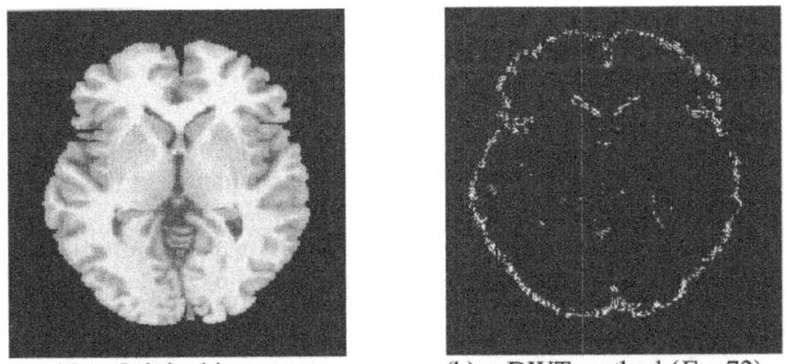

(a) Original image (b) DWT method ($F = 72$)

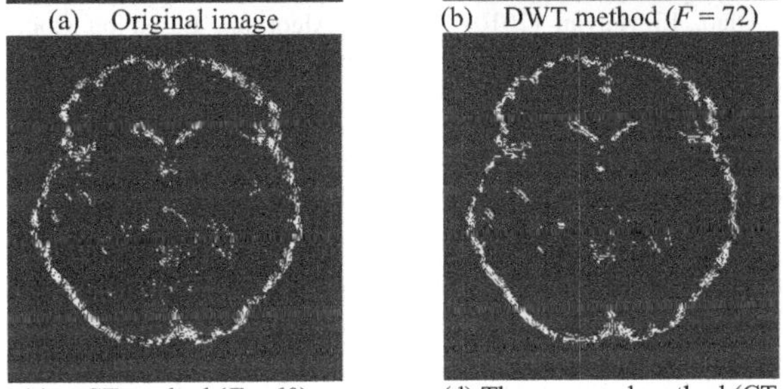

(c) CT method ($F = 60$) (d) The proposed method (CT-CS)($F = 54$)

Fig. 5 Image corrupted with speckle noise and its edges detected with other methods.

In these figures, we observe that the proposed method performs better than the other two methods. Other experiments also show that the proposed method works well and better than the other ones. In addition, clinicians at Children's Hospital, have evaluated these results. In their opinion, the proposed method gets more edges and information from images, and it detects edges efficiently even in the presence of a background, whose intensity is very similar to that of the objects.

V. CONCLUSIONS

Edge detection in medical images is an important step when analyzing medical image data. The main goal of edge detection is to identify an image as a collection of features, each of which has a strong connection with real-world objects.

In this paper we have presented a method for edge detection in medical images using contourlet transform with cycle spinning. The presented method has been compared with earlier methods such as the discrete wavelet transform method and contourlet transform method. The presented edge detection method works very well and better than other methods.

The method is simple and robust. However, in the cases where medical images have complex background and bad quality, they may have effect on the algorithm performance. Most medical images have poor contrasts and complex types of noise. So, we usually denoise and deblur medical images before performing edge detection on them.

References

1. Do M. N., Vetterli M. (2005) The Contourlet Transform: An Efficient Directional Multiresolution Image Representation. IEEE Transactions on Image Processing, vol. 14, pp 2091–2106

Liu H., Chen Z., Chen X., Chen Y. (2005) Multiresolution medical image segmentation based on wavelet transform. Proceedings of the 2005 IEEE Engineering in Medicine and Biology, 27th Annual Conference, Shanghai, China, pp 3418–3421

Li Qiao, Li Haiyun (2010) A Novel Algorithm Based on Contourlet Transform for Medical Image Segmentation. Bioinformatics and Biomedical Engineering (iCBBE), 4th International Conference on, IEEE, pp 1–3

Eslami R., Radha H. (2006) Translation-invariant contourlet transform and its application to image denoising. IEEE Transactions on Image Processing, vol. 15, no. 11, pp 3362–3374

Kun Liu, Lei Guo, Jingsong Chen (2011) Contourlet transform for image fusion using cycle spinning. Journal of Systems Engineering and Electronics, vol. 22, no. 2, pp 353–357

Raghavendra B. S., Subbanna Bhat P. (2005) Contourlet Based Multiresolution Texture Segmentation Using Contextual Hidden Markov Models. Intelligent Information Technology, vol. 3356, pp 336–343

Strang G. (1989) Wavelets and dilation equations: A brief introduction. SIAM Review, 31(4), pp 614–627

Gopinath R. A. (2003) The phaselet transform — An integral redundancy nearly shift-invariant wavelet transform. IEEE Trans. Signal Processing, 51(7), pp 1792–1805

Binh N. T., Khare A. (2009) Medical Image Deblurring in Wavelet Domain. Proceedings of 11th Conference on Science and Technology, University of Technology, HCMC, Vietnam, pp 1027–1032

Donoho D. L., Johnstone I. M. (1994) Ideal spatial adaptation by wavelet shrinkage. Biometrika, vol. 8, pp 425–455

Li Yin-qi, He Ming-yi, Fang Xiao-feng (2007) SAR Image Segmentation Algorithm Using Mean Shift on Contourlet Domain. Computer Engineering, vol. 33, no. 22, pp 48–50

Liu J., Yang Y. H. (1994) Multiresolution color image segmentation. IEEE Trans. Pattern Anal. Mach. Intell. 16(7), pp 689–700

Contourlet Toolbox. Matlab source code available at http://www.ifp.uiuc.edu/~minhdo/software/

Author: Nguyen Thanh Binh
Institute: Ho Chi Minh City University of Technology
Street: 268 Ly Thuong Kiet, Dist 10
City: Ho Chi Minh City
Country: Vietnam
Email: ntbinh@hcmut.edu.vn

Application of Neural Network in Classifying Massive Lesions in Mammogram

Viet Dung Nguyen[1], Duc Thuan Nguyen[1], Tien Dung Nguyen[2], Huu Long Nguyen[1], and Duc Huyen Bui[1]

[1] Department of Biomedical Engineering, School of Electronics and Telecommunications, Hanoi University of Science and Technology, Hanoi, Vietnam

[2] Department of Electronics and Computer Engineering, School of Electronics and Telecommunications, Hanoi University of Science and Technology, Hanoi, Vietnam

Abstract — **Breast cancer is one of the most frequently leading causes of cancer deaths in middle-aged women. Until now, mammography is still the most effective procedure for early diagnosis of the breast cancer. Computer-aided detection (CAD) system can be very helpful for radiologists in identification abnormalities earlier and faster than traditional screening program. Therefore, efforts have been made to develop CAD systems for mammographic interpretation. The identification of massive lesions in mammograms is still challenging because of their irregular shapes and ill-defined margins and faint contrast. Regions-of-Interest (ROIs) that high probably contain massive lesions are first detected. They are then classified as mass region or non-mass region using their extracted features. In this paper we will investigate the application of neural network in classifying massive lesions in digital mammograms.**

Keywords — **mammogram, classification, neural network.**

I. Introduction

Breast cancer is one of the most frequently leading causes of cancer deaths in middle-aged women. The American Cancer Society (ACS) estimates that about 184,450 new breast cancer cases are expected to be diagnosed in 2008 [1]. Early diagnosis and treatment can increase the chances of survival. Screening can be done through mammography, ultrasound, magnetic resonance imaging (MRI) or breast biopsy. Ultrasound produces a good contrast image but it does not contain enough detailed information which can be found in mammography. Due to this reason, ultrasound is not approved by the U.S Food and Drug Administration (FDA) as a screening tool for breast cancer [2]. Although MRI is more sensitive than mammography, its results can also lead to false positive diagnosis which then leads to unnecessary additional tests, biopsies and increased patient anxiety. In addition, the American Cancer Society recommends MRI for women with approximately 20-25% or greater lifetime risk of breast cancer, including women with strong family history of breast or ovarian cancer [3].

Although mammography is proven to be a high effective method, the diagnosis currently is performed based on radiologists' experiences, which becomes a barrier to the diagnosis accuracy especially for less-experienced radiologists. The diagnosis accuracy is also affected by some conditional factors such as stress, distraction, fatigue or low ability of eyes at that time. It is found that during routine screening, a large number of lesions of cancers (about 20%) are missed [4]. The causes of these false negative screening examinations are not clear. Independent double reading by two radiologists has been shown to improve the sensitivity but it may not feasible to routinely perform it due to financial, technical and logistical restrains. Therefore, efforts have been made to develop a computer-aided detection (CAD) system for mammoraphic interpretation [5-6]. CAD can be defined as a diagnosis made to improve radiologists' performance by indicating the sites of potential abnormalities, to reduce the number of missed lesions, and/or by providing quantitative analysis of specific regions in an image to improve diagnosis. CAD systems typically act as "second reader" or "double reading" systems where the final dignosis is made by the radiologist. Typical block diagram of a CAD system is given in fig. 1.

Masses and microcalcification clusters are clear marks of a breast neoplasia. Masses are rather large objects than microcalcification. The identification of massive lesions is still challenging because of their irregular shapes and ill-defined margins and faint contrast.

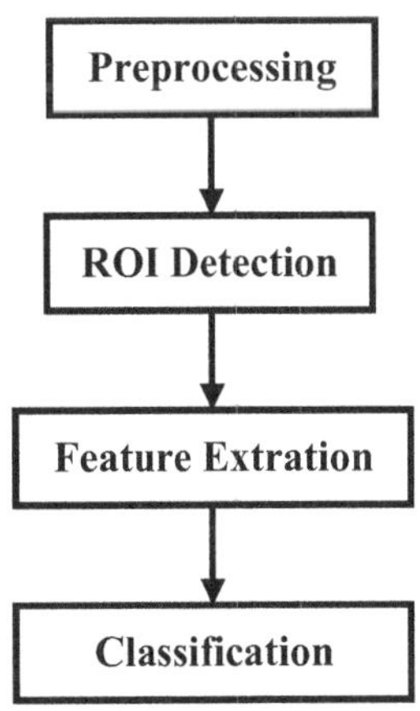

Fig. 1 CAD block diagram.

V. Van Toi et al. (Eds.): 4th International Conference on Biomedical Engineering in Vietnam, IFMBE Proceedings 40, pp. 101–104, 2013.
www.springerlink.com

First of all, mammograms are preprocessed to remove noise and unnecessary parts in order to improve efficiency of ROI (Region Of Interest) detection step. In ROI detection step, regions or areas that highly contain suspicious lesions are detected. Next, each ROI is classified as normal or abnormal one based on extracted features that represent the ROI.

A number of classifiers based on linear discriminant analysis, rule-based methods and artificial neural networks [7-8] have shown effectiveness in detection and diagnostic systems. In this paper, we will investigate the performance of different neural network architectures in classifying massive lesions in mammogram.

II. Methodology

A. Material

In this study, we use mammogram database Mini-MIAS to test the algorithm presented. The database MIAS is created from the Mammographic Image Analysis Society - an organization of United Kingdom research groups. This database includes 322 digital mammograms from 161 patients. Films taken from the United Kingdom National Breast Screening Program have been digitized to 50 micron pixel edge, and presenting each pixel with an 8-bit word. Every image in database always has extra-information or ground truth from the radiologists about characteristic of background tissue, type of abnormality present, severity of abnormality, the coordinates of center and approximate radius (in pixels) of a circle enclosing the abnormality.

Mini-MIAS database is a kind of reduced type of database MIAS, the original MIAS database (digitized at 50 micron pixel edge) has been reduced to 200 micron pixel edge and clipped/padded so that every image has a size of 1024 x 1024 pixels.

B. Method

ROI detection module will extract from background one or more ROI that likely contain a massive lesion. Details of the ROI detection method can be found in our previous work [9].

Features of masses generally present distinct characteristics from those of normal breast tissues. Therefore, before going to the classification step, features of detected ROIs should be calculated or extracted. The extracted features are the inputs to a classifier to determine how pathological the ROI is.

In this study, textural features based on Gray Level Co-occurrence Matrix (GLCM) [10] or Haralick features are calculated. The GLCM $P_{d,\theta}(l_1,l_2)$ represents the probability of occurrence of a pair of gray levels (l_1,l_2) separated by a given distance d at angle θ. GLCMs are constructed by mapping the gray-level co-occurrence counts or probabilities based on the spatial relations of pixels at different angular directions (specified by θ) while scanning the image from left to right and from top to bottom. It may be advantageous to reduce the number of gray levels to a relative small number before computing GLCMs. A reduction in number of gray levels with smoothing can also reduce the effect of noise on the statistics computed from GLCMs. To achieve this, we consider the minimal rectangular fully includes the ROI, specific distance d=1, 2, 3 at quantized angles $\theta=k\pi/4$ with k=0,1,2,3. Smoothing can be done by averaging 4 constructed GLCMs.

There are total of 13 Haralick features. These are listed as follow

- Energy
- Contrast
- Correlation
- Sum of squares
- Inverse difference moment
- Sum average
- Sum variance
- Sum entropy
- Entropy
- Difference variance
- Difference entropy
- Two information-theoretic correlations

Detailed calculation of these features can be found in [11]. Some of above calculated features have values much and much greater than unity, whereas some of the features are far less than unity so the features are normalized, over the dataset, to [0, 1] range.

Artificial neural networks have shown effectiveness in detection and diagnostic systems. Generally, a neural network composes of one input layer, hidden layers and one output layer as shown in fig. 2.

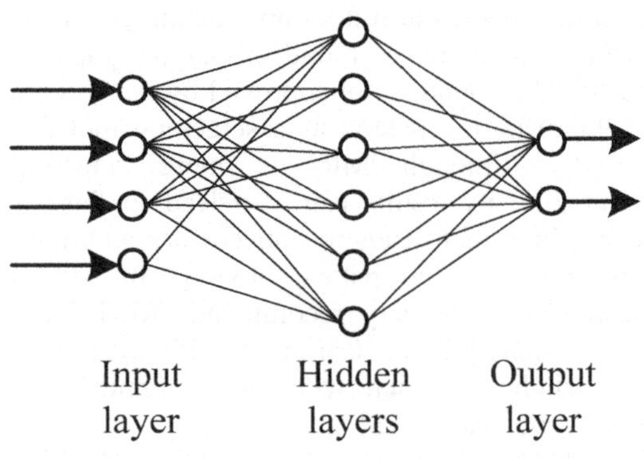

Fig. 2 Typical block diagram of a neural network.

The number of input nodes in input layer equals to number of extracted features. In this study, number of input node is 13. Number of output nodes depends how many categories the ROIs are classified into. In our study, each ROI is classified as mass or non-mass one. Therefore, there is only one output neuron. Values of output will be 0 or 1. Output 1 means the ROI contains massive lesion. Output 0 means the ROI is healthy one. The number of nodes in each hidden layers is changed in order to archive good discrimination results.

In our study, there 2 different neural network architectures are investigated. They are Multilayer Perceptron (MLP) network and Radial Basis Function (RBF) network. We also investigate the combination of MLP and RBF networks so called MLP-RBF with voting scheme. The structure of this network is illustrated in fig. 3.

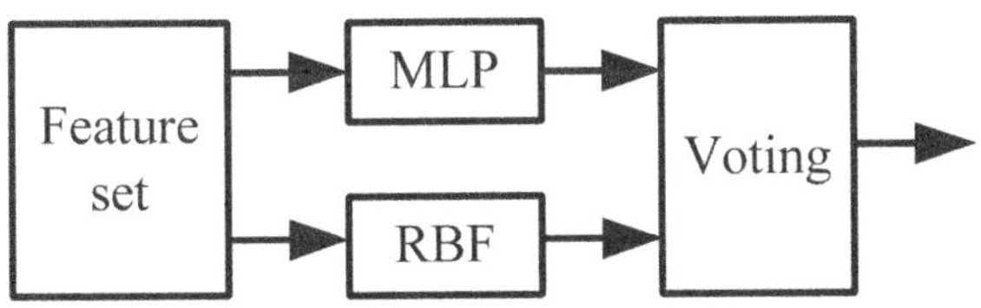

Fig. 3 MLP-RBF with voting network.

In this network, feature set is used for both MLP and RBF. Outputs from MLP and RBF are fed to voting block. Voting scheme is chosen as follow

MLP output=1, RBF output=1 then Output = 1
MLP output=1, RBF output=0 then Output = 1
MLP output=0, RBF output=1 then Output = 1
MLP output=0, RBF output=0 then Output = 0

III. Results

We investigate performance of 3 different neural network architectures mentioned above in classifying massive lesions using 13 GLCM feature set corresponding to d=1,2,3.

To evaluate the performance of a CAD system, a Receiver Operating Characteristics (ROC) is used. A ROC curve is a plotting of sensitivity as a function of specificity at different values of the decision threshold on the neural network output. Higher ROC, approaching the perfection at the upper left hand corner, would indicate greater discrimination capacity.

The ROC curves are provided in fig. 4 to 6. The classification performance is quantified by the area Az under the curve (AUC). Higher A_Z would indicate greater discrimination capacity. If the obtained AUC value lies in the range $0.7 \leq Az \leq 0.9$, the classifier can be considered accurate enough [12].

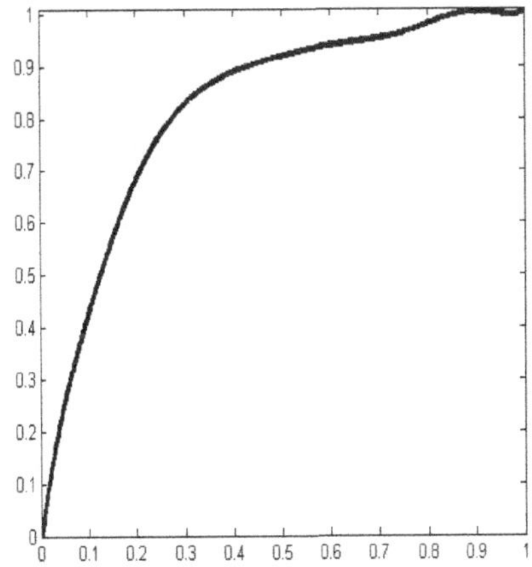

Fig. 4 MLP, d=2.

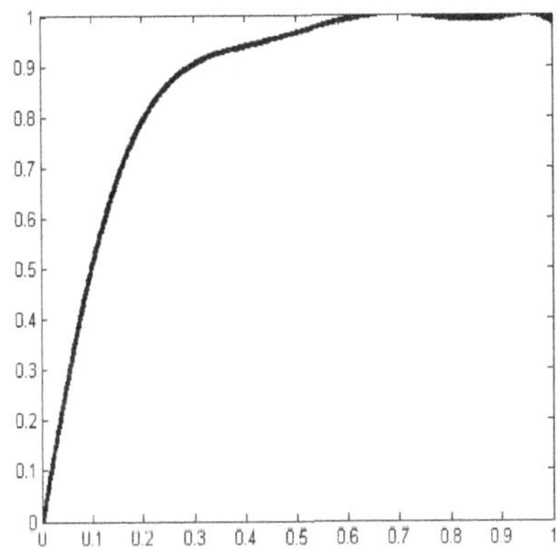

Fig. 5 RBF, d=3.

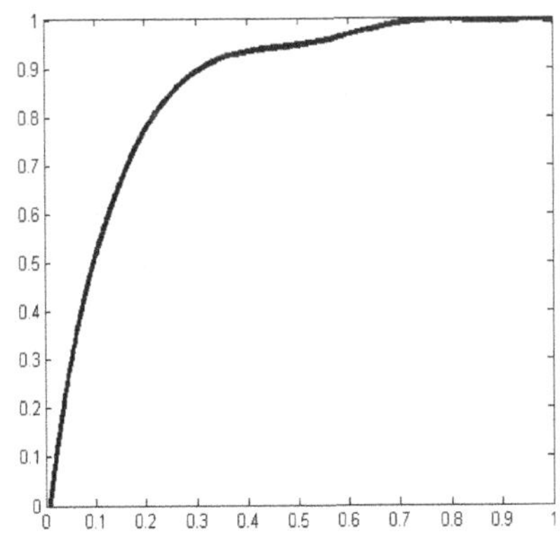

Fig. 6 MLP-RBF, d=2.

Full results are given in table 1.

Table 1 Az with different feature set and network architectures.

	Feature set		
Neural network	d=1	d=2	d=3
MLP	0.838	**0.875**	0.831
RBF	0.804	0.866	**0.867**
MLP-RBF with voting scheme	0.833	**0.871**	0.860

It can be seen that in all cases A_Z lies in range from 0.7 to 0.9 which means classification is considered accurate enough. It is also noted that in most of cases, MLP network gives better result than RBF network. The best A_Z is 0.875 when MLP network and feature set with d=2 are applied. Combination of MLP and RBF with voting scheme doesn't improve classification efficiency.

IV. Conclusions

In this paper, application of neural network to classify massive lesions in digital mammograms is investigated. Satisfactory results indicate that neural network based classification is considered accurate enough. The classification accuracy is not increased when combining different neural networks.

In the future we will investigate other classification methods and feature sets in an attempt to improve the performance.

References

1. American Cancer Society. 2008. Cancer facts and figures 2008. In Statistics for 2008 (available online at http://www.cancer.org/downloads/STT/008-CAFFfinalsecured.pdf)
2. Imaginis, The women's health resource. Ultrasound imaging of the breast (available online at http://www.imaginis.com/breasthealth/ultrasound.asp).
3. Saslow D, Boetes C, Burke W et al (2007) American Cancer Society guidelines for breast screening with MRI as an adjunct to mammography. A Cancer Journal for Clinicians 57 (2007), 79-89.
4. National Institute of Cancer (available at http://www.cancer.gov/cancertopics/factsheet/detection/mammograms).
5. C. Varela, P. G. Tahoces, A. J. M´ endez, M. Souto, J. J. Vidal (2007) Computerized detection of breast masses in digitized mammograms. Comput. Biol. Med. 37, 214–226.
6. R. M. Rangayyan, F. J. Ayres, J. E. Leo Desautels (2007) A review of computer-aided diagnosis of breast cancer: Toward the detection of subtle signs. J. Franklin Inst.344, 312–348.
7. Xu W, Li L, Xu P (2007) A new ANN-based detection algorithm of the masses in digital mammograms. IEEE International Conference on Integration Technology, 26-30
8. Zhang G, Yan P, Zhao H et al (2008) A computer aided diagnosis system in mammography using artificial neural networks. International Conference on BioMedical Engineering and Informatics, 823-826.
9. Nguyen Viet Dung, Pham Van Thanh, Nguyen Tien Dung, Nguyen Duc Thuan (2010) Automatically Contour-based detection of suspicious massive lesions in mammograms. Journal of Science and Technology-Technical Universities No79B, pp. 38-42.
10. R. M. Haralik, K. Shanmugam, I. Dinstein (1973) Textural features for image classification. IEEE Trans. Syst. Man Cyber, 3, 610–621.
11. Rangaraj M. Rangayyan (2005) Biomedical Image Analysis. CRC Press.
12. J. A. Swets (1988) Measuring the accuracy of diagnostic systems. Science, 240, 1285–1293.

Author: Viet Dung Nguyen
Institute: Hanoi University of Science and Technology
Street: 1 Dai Co Viet
City: Hanoi
Country: Vietnam
Email: nvdung-bme@mail.hut.edu.vn

Silicon on Insulator Microring Resonators with Feedback Waveguides for Highly Sensitive Biomedical Sensors

Trung-Thanh Le

Department of Information Technology, Hanoi University of Natural Resources and Environment, Hanoi, Vietnam

Abstract — **A novel sensing structure based on microring resonators with feedback using silicon waveguides is proposed in this paper. The proposed structure can produce a sharp asymmetric Fano resonance line shape, in which the slope between the zero and the unity transmission is greatly enhanced, comparing with that of the conventional microring resonators. As a result, biological sensors using this structure can provide a high sensitivity. In addition, multimode interference (MMI) couplers are employed as coupling elements instead of using directional couplers in the microring resonators. MMI couplers have shown to have relaxed fabrication requirements, simplicity, compactness and are less sensitive to the wavelength. The analytical description of output spectrum response of the proposed sensor is analyzed and derived. The transfer matrix method (TMM) and beam propagation method (BPM) are used to optimally design the sensor structure.**

Keywords — **Biological sensors, chemical sensors, optical microring resonators, multimode interference, transfer matrix method.**

I. INTRODUCTION

Biomedical and chemical detection is an important process in many applications such as biomedical research, healthcare and environmental monitoring [1]. Compared with electrical sensors, optical biosensors have advantages of compact size, multiplexing and immunity to electro-magnetic interference. Typically, detection can be made by the optical absorption of the analytes, optic spectroscopy or the refractive index change [2]. Various configurations of optical biosensors have been reported, including Mach Zehnder interferometers (MZI), surface plasmon sensors and microring resonators (MRR) [3]. Among them, microring resonators are becoming one of the most promising candidates for sensing applications [4]. Microring resonator sensors can provide ultracompact size and easy realization of a sensor array, especially for optical waveguides based on silicon on insulator (SOI) materials due to its high index contrast and compatibility with the existing CMOS fabrication technology.

When detecting target chemicals by using microring resonator sensors, one can use a certain chemical binding on the surface. There are two ways to measure the presence of the target chemicals. One is to measure the shift of the resonant wavelength and the other is to measure the optical intensity with a fixed wavelength.

Recently, it has been suggested that optical Fano resonances have many important applications in resonance line shape sensitive chemical and biological sensing, optical switching, modulating and filtering [2, 5]. However, such structures need a complex fabrication process or complicated configurations. In this study, we propose a novel sensing structure based on microring resonators with feedback using silicon waveguides. The proposed structure can produce a sharp asymmetric Fano resonance line shape, in which the slope between the zero and the unity transmission is greatly enhanced, comparing with that of the conventional microring resonators. As a result, biological sensors using this structure can provide a high sensitivity. Our research is differentiated from prior works in the way that our design is to employ multimode interference (MMI) couplers as coupling elements instead of using directional couplers in microring resonators. MMI couplers have shown to have relaxed fabrication requirements, simplicity, compactness and are less sensitive to the wavelength. The analytical description of output spectrum response of the proposed sensor is analyzed and derived. The transfer matrix method (TMM) and beam propagation method (BPM) are used to optimally design the sensor structure.

II. THEORY

A microring resonator based on multimode interference (MMI) couplers with feedback waveguides is shown in Fig. 1; where 2x2 and 2x1 MMI couplers are used for the coupling elements. In this study, we use three types of MMI couplers based on restricted interference theory, including: 50:50 MMI couplers, 85:15 MMI couplers and 72:28 MMI couplers [6].

The operation of optical MMI coupler is based on the self-imaging principle [7]. Self-imaging is a property of a multimode waveguide by which as an input field is

V. Van Toi et al. (Eds.): 4th International Conference on Biomedical Engineering in Vietnam, IFMBE Proceedings 40, pp. 105–109, 2013.
www.springerlink.com

reproduced in single or multiple images at periodic intervals along the propagation direction of the waveguide. The central structure of the MMI filter is formed by a waveguide designed to support a large number of modes. In the MMI section, the 2-D scalar Helmholtz wave equation is defined as

$$\frac{\partial^2\psi}{\partial x^2}+\frac{\partial^2\psi}{\partial y^2}+\left[\frac{2\pi n(x,y)}{\lambda}\right]^2\psi=\beta^2\psi \tag{1}$$

where $\psi(x,y,z)=\sum_{v=0}^{M-1}c_v\psi_v(x,y)\exp(j(\omega t-\beta_v z))$; x is the lateral dimension; y is the transverse dimension; z is the propagation direction; c_v is the filed excitation coefficient; $\psi_v(x,y)$ is the modal field distribution; n(x,y) is the refractive index profile, $v = 0, 1,\ldots, M-1$ are the mode numbers of the waveguide supporting M modes; λ is the optical wavelength and β is the propagation constant.

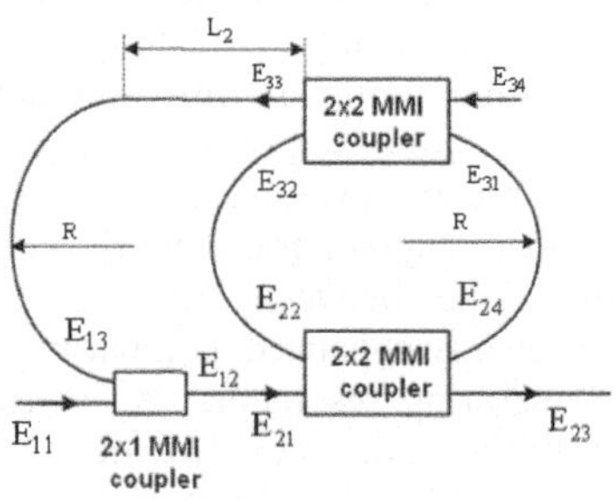

Fig. 1 The sensor structure based on a microring resonator with feedback waveguides.

The MMI coupler can be characterized by a transfer matrix, which describes the relation between the complex amplitudes of the input signals and the complex amplitudes of the output signals. For the first 2x2 MMI coupler, the relations between the input complex amplitudes E_{21}, E_{22} and output complex amplitudes E_{23}, E_{24} of this 2x2 MMI coupler can be expressed as [8]

$$\begin{bmatrix}E_{23}\\E_{24}\end{bmatrix}=\begin{bmatrix}\tau_2 & j\kappa_2\\ j\kappa_2 & \tau_2\end{bmatrix}\begin{bmatrix}E_{21}\\E_{22}\end{bmatrix} \tag{2}$$

where κ_2 and τ_2 are the cross coupling coefficient and transmission coupling coefficient of the coupler. The relations between the input complex amplitudes and output complex amplitudes of the second 2x2 MMI coupler can be written by

$$\begin{bmatrix}E_{33}\\E_{32}\end{bmatrix}=\begin{bmatrix}\tau_3 & j\kappa_3\\ j\kappa_3 & \tau_3\end{bmatrix}\begin{bmatrix}E_{34}\\E_{31}\end{bmatrix} \tag{3}$$

The complex amplitude fields E_{31}, E_{24} , E_{13}, E_{33} , E_{12}, E_{21} and E_{32}, E_{22} are related by

$$E_{21}=\alpha_2\exp(j\phi_2)E_{12}, \tag{4}$$

$$E_{31}=\alpha_1\exp(j\phi_1)E_{24}, \tag{5}$$

$$E_{23}=\alpha_1\exp(j\phi_1)E_{32} \tag{6}$$

$$E_{13}=\alpha_1\alpha_2\exp(j(\phi_1+\phi_2))E_{33} \tag{7}$$

where α_1, α_2 are the attenuation coefficients of the ring waveguide and straight waveguide (with a length of L_2); R is the radius of the ring waveguide and $L_1=\pi R$; ϕ_1 and ϕ_2 are phase factors, $\phi_1=\frac{2\pi n_{eff}}{\lambda}L_1$, $\phi_2=\frac{2\pi n_{eff}}{\lambda}L_2$; here n_{eff} is the effective refractive index and λ is the wavelength.

From the above equations, the output power intensity of the device can be obtained by

$$I=\left|\frac{E_{22}}{E_{11}}\right|^2=\left|\frac{\tau_1(\tau_2-\tau_3\alpha_1e^{j2\phi_1})e^{j\phi_2}}{1-\tau_2\tau_3\alpha_1e^{j\phi_1}-j\tau_1\tau_3\kappa_2\alpha_1\alpha_2e^{j2(\phi_1+\phi_2)}}\right|^2 \tag{8}$$

Here τ_1 is the transmission coefficient of the 2x1 MMI coupler of Fig. 1. The optical biosensor based on microring resonators can be optimized from two aspects: sensitivity and detection limit optimization. For detection limit optimization, the accuracy of measurement for a given response should be optimized, which can be obtained by optimizing design process. We will show later that our proposed sensor structure can greatly improve the sensitivity compared with sensors based on conventional microring resonators. The sensitivity of the sensor is [3]: $S=\partial I/\partial n_a=(\partial I/\partial n_{eff})(\partial n_{eff}/\partial n_a)$; where I is the normalized output intensity, n_a is the refractive index of the analytes and n_{eff} is the effective refractive index. The intensity change ∂I originates in the shift of shape of the resonance due to the change in effective index ∂n_{eff} at a fixed wavelength. The term $\partial n_{eff}/\partial n_a$ depends on the waveguide design and is a constant for a given waveguide structure. Thus the sensitivity can be enhanced by increasing the term $\partial I/\partial n_{eff}$; which can be obtained by using the proposed structure having a steeper slope in the transmission characteristic.

IV. Simulation Results and Disscusion

It is shown that the proposed sensor can be realized using silicon nanowire waveguides. The cross-section of the waveguide is shown in Fig. 2 [9]. The core thickness of the silicon waveguide is h_{co}=220nm. The access waveguide is tapered to a width of 0.8 μm to improve device performance. Because the transverse magnetic (TM) polarization is more sensitive to the change in refractive index of the analyte, we use the TM polarization in the following simulations. It is assumed that the central optical wavelength is $\lambda = 1550\text{nm}$.

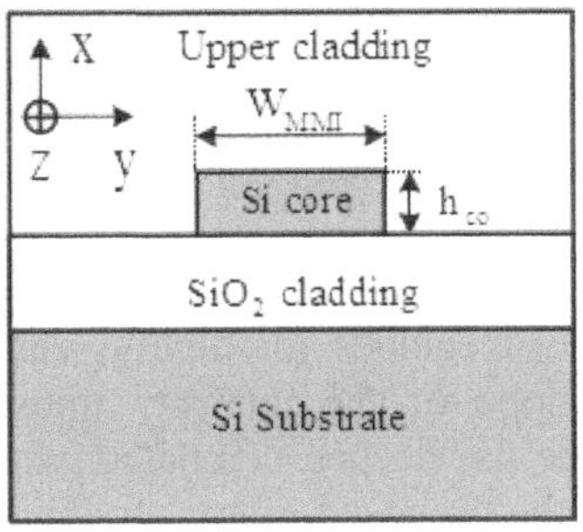

Fig. 2 Cross-section of the waveguide used in the designs of the proposed sensor.

For 2x2 MMI couplers having rectangular shape, the 2x2 coupler are theoretically capable of providing cross-coupling ratios $|\kappa|^2$ (defined as the cross output power divided by the sum of bar and cross output powers) of 1, 0.85, 0.72, 0.5, 0.27, 0.15, and 0. In this study, three types of 2x2 MMI couplers having coupling splitting ratios of $|\kappa|^2 = 0.5$, 0.85 and 0.72 are used in the sensor structure. Therefore, designs of the three types of these MMI couplers in order to achieve the desired power splitting ratios are considered first.

Design of 50:50 MMI couplers: In order to realize a 50:50 2x2 MMI coupler, the MMI width is chosen to be $W_{MMI} = 3(s + W_a)$; where W_a is the width of the input and output waveguides and s is the separation between two parallel input waveguides. Using the mode propagation analysis method (MPA) and beam propagation method (BPM) [9], it is shown that the width of the MMI coupler is $W_{MMI} = 4\mu m$ and the optimal length of the MMI is $L_{MMI} = 20\mu m$. In addition, the positions of the input and output waveguides must be carefully chosen in order to achieve the desired splitting ratio. One can show that the location of the input and output waveguides are at $W_{MMI}/6$ and $-W_{MMI}/6$. Fig. 3 shows the BPM simulation of the field propagation through the 50:50 MMI coupler having an optimal width of $W_{MMI} = 4\mu m$ and an optimal length of $L_{MMI} = 20\mu m$.

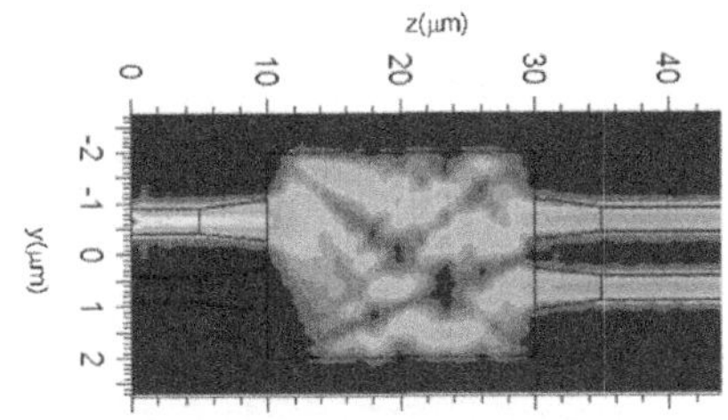

Fig. 3 Field propagation through a 50:50 MMI coupler.

Design of 85:15 MMI couplers: In order to realize a 85:15 2x2 MMI coupler, the MMI width is chosen to be $W_{MMI} = 4\mu m$. Using the mode propagation method (MPA), the length of the MMI coupler should be $L_{MMI} = 3L_\pi / 4$, where L_π is the beat length of the MMI coupler. Similarly, the positions of the input and output waveguides must be carefully chosen in order to achieve the desired splitting ratio. One can show that the location of the input and output waveguides are at $W_{MMI}/4$. Using BPM simulation, the optimal length of the 85:15 MMI coupler is found to be $L_{MMI} = 33\mu m$. Fig. 4 shows the BPM simulation of the field propagation through the 85:15 MMI coupler having an optimal width of $W_{MMI} = 4\mu m$ and an optimal length of $L_{MMI} = 33\mu m$.

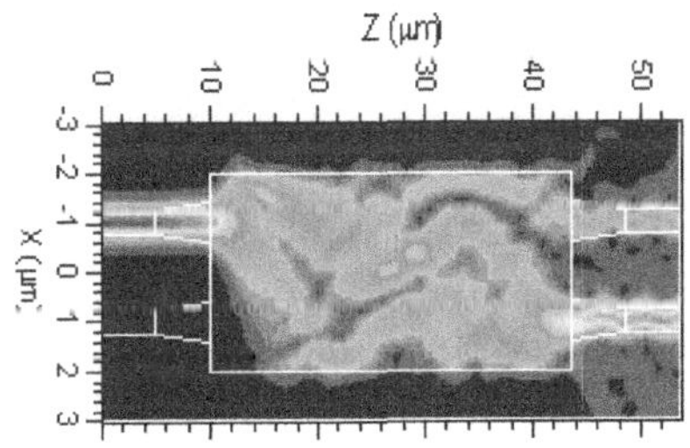

Fig. 4 Field propagation through an 85:15 MMI coupler

Design of 72:28 MMI couplers: In order to realize a 72:28 2x2 MMI coupler, the MMI width is still chosen to be $W_{MMI} = 4\mu m$, but the length of the MMI coupler should be $L_{MMI} = 3L_\pi / 5$. In this case, the positions of the input and output waveguides must be carefully chosen in order to achieve the desired splitting ratio (73:28). One can show that the locations of the input waveguides are at $W_{MMI}/5$ and $2W_{MMI}/5$. Using BPM simulation, the optimal length of the 72:28 MMI coupler is found to be $L_{MMI} = 26.2\mu m$. Fig. 5 shows the BPM simulation of the field propagation through the 72:28 MMI coupler having an optimal width of $W_{MMI} = 4\mu m$ and an optimal length of $L_{MMI} = 26.2\mu m$.

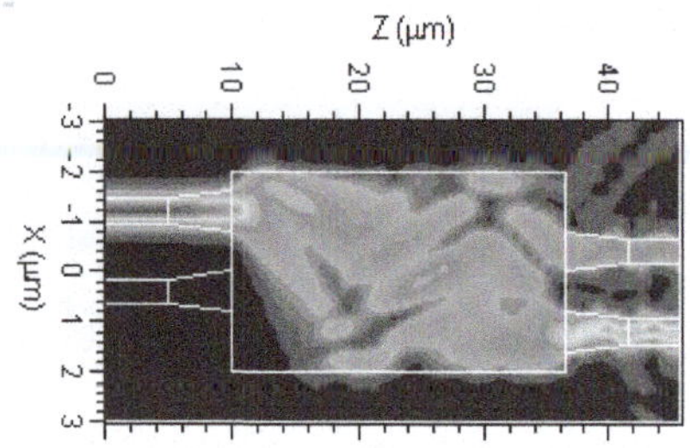

Fig. 5 Field propagation through an optimal 72:28 MMI coupler.

Finally, the 2x1 MMI coupler needs to be designed. Using the MPA method, the length of the 2x1 MMI coupler is to be $L_{MMI} = 3L_\pi / 8$. Using the BPM simulation, the optimal length of the 2x1 MMI coupler is found to be $L_{MMI} = 26\mu m$. Fig. 6 shows the BPM simulation of the field propagation through the 2x1 MMI coupler having a width of $W_{MMI} = 4\mu m$ and an optimal length of $L_{MMI} = 26\mu m$.

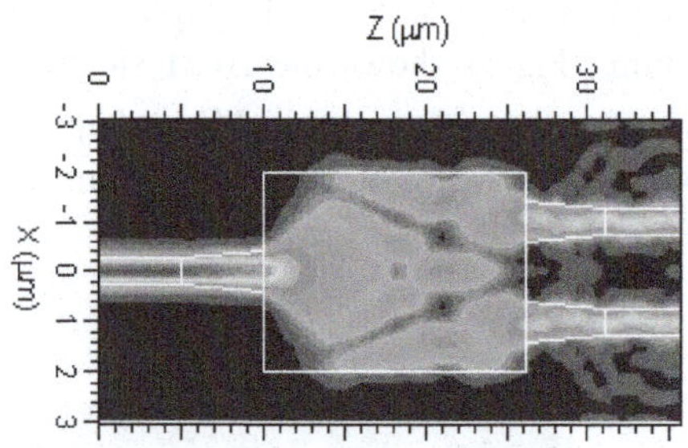

Fig. 6 Field propagation through an optimal 2x1 MMI coupler.

Now the transmission of the proposed sensor using three types of the MMI couplers are investigated. Fig. 7 shows

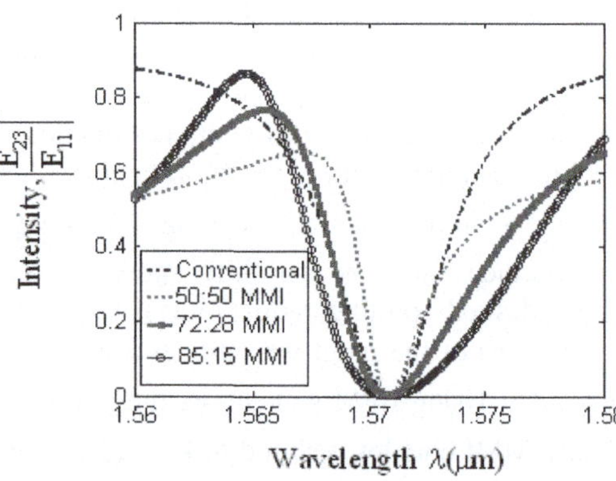

Fig. 7 The output intensity transmission for conventional sensors using conventional microring resonator, 50:50 MMI couplers, 85:15 MMI couplers and 72:28 MMI couplers.

the normalized intensity transmission spectra $I = |E_{23} / E_{11}|^2$ of the proposed sensor structure for four cases: sensors based on conventional microring resonators [10], sensors based on 50:50 MMI couplers, 85:15 MMI couplers and 72:28 MMI couplers. The simulation results show that the steeper slope of the proposed structure can be obtained. As a result, such a sensor structure can provide a higher sensitivity compared with the sensors based on conventional microring resonators.

V. Conclusions

We have proposed and analyzed a microring resonator with feedback waveguides for highly sensitive biological sensing. The slope of the intensity transmission of the proposed device is steeper compared with the reported structure; thus the proposed sensor structure can provide a higher sensitivity. The design of the proposed sensor using silicon waveguide has been investigated. The transfer matrix method and beam propagation method have been used to design the proposed sensor structure.

References

1. Vo DT (2003) Biomedical Photonics Handbook: CRC Press
2. Yi H, Citrin DS, and Zhou Z (2010), Highly sensitive silicon microring sensor with sharp asymmet-rical resonance, Optics Express, vol. 18, pp. 2967-2972.
3. Passaro V, Dell'Olio F, Casamassima B. and Leonardis F. D. (2007), Guided-Wave Optical Biosensors, Sensors, vol. 7, pp. 508-536.
4. Chao CY, Fung W and Guo L (2006), Polymer Microring Resonators for Biochemical. Sensing Applications, IEEE Journal of Selected Topics in Quantum Electronics, vol. 12, pp. 134 - 142.
5. Hon KY and Poon A, "Silica polygonal micropillar resonators: Fano line shapes tuning by using a Mach -Zehnder interferometer," presented at Proceedings of SPIE Vol. 6101, Photonics West 2006, Laser Resonators and Beam Control IX, San Jose, California, USA, 25-26 January, 2006.
6. Le TT (2010), Multimode Interference Structures for Photonic Signal Processing: Modeling and De-sign: Lambert Academic Publishing, Germany.
7. Bachmann M, Besse PA, and Melchior H (1994), General self-imaging properties in N x N multi-mode interference couplers including phase relations, Applied Optics, vol. 33, pp. 3905-.
8. Yariv A, Universal relations for coupling of optical power between microresonators and dielec-tric waveguides (2000), Electronics Letters, vol. 36, pp. 321–322.
9. Le TT and Cahill LW (2009), The Design of 4×4 Multimode Interference Coupler Based Microring Resonators on an SOI Platform, Journal of Telecommunications and Information Technology, Poland, pp. 98-102.

10. Shi L, Xu Y, Tan W, and Chen X (2007), Simulation of Optical Microfiber Loop Resonators for Ambient Refractive Index Sensing, Sensors, vol. 7, pp. 689-696.

The address of the corresponding author:

Author: Dr Trung-Thanh Le
Institute: Hanoi University of Natural Resources and Environment
Street: 41A, K1, Phu Dien road, Tu Liem
City: Hanoi
Country: Vietnam
Email: thanh.le@hunre.edu.vn

Measurement of Humidity Measurement in Adverse Conditions

Zuoqing Guo[1] and Yaodong Gu[2]

[1] The International College, Xiamen University, Xiamen, China
[2] Faculty of Sport Science, Ningbo University, Ningbo, China

Abstract — **Humidity is defined as the amount of moisture containing in the air. The most common way used to describe it is relative humidity. The function of some of breathing devices is used to humidify air for medical purpose, and the method of humidification is involved with water droplets being produced. The small water droplets sometimes attach to the sensor and affect the readings from the humidity measuring tools. Therefore this project was aiming to measure humidity accurately and quickly in a presence of water droplets.**

Keywords — **Humidity, eliminator, sensor, droplets, transducer.**

I. Introduction

Humidity is defined as the amount of moisture containing in the air. Relative humidity, specific humidity and absolutely humidity are three ways for qualifying humidity [1, 2]. Among these, Relative humidity is the most common way to use. It is the ratio of actual amount of water vapour presenting in the air comparing to the maximum amount of water vapour that would be presented in the air at saturation.

Humidity is an important factor to human. If humidity is very high, there is a lot of water vapour in the air, and people are more difficult to sweat [3]. Hence people feel hot. Oppositely, humidity in the air is very low, there is little water vapour containing in the air, people would feel cold, that is due to lots of moisture move out from skin much quicker [4].

The small water droplets sometimes attach to the sensor and affect the reading of the sensor. The primary aim of this project is to measure humidity accurately and quickly in a presence of water droplets. The suitable experimental work is needed to analysis the accuracy and response time for measuring humidity in a presence of droplets.

II. Methods

Even though from theory calculation, it is known that the mesh eliminator is not suitable to be used for extremely low velocity. However it was necessary to check from experiments and compared the results with theory prediction. Materials was shown in Fig.1.

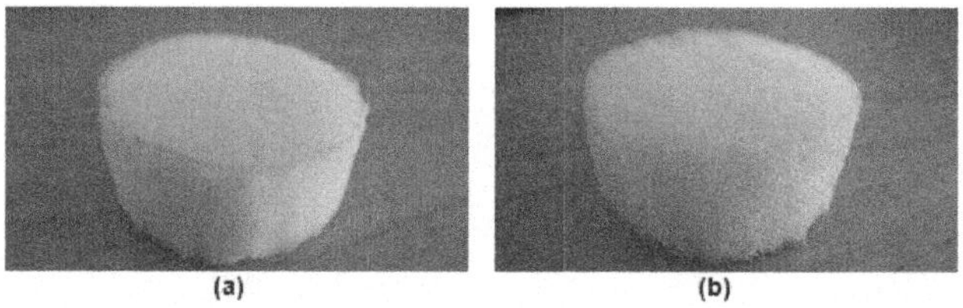

Fig. 1 (a) More dense mesh eliminator and (b) Less dense mesh eliminator.

Set up the droplets generator, regulating the frequency to 1.46MHz and amplitude to 2VPP by a signal generator. Increase the gain to the point where was at one sixth of one turn on the amplifier device. Secondly, Connecting the compressed air pipe. Then, lace the eliminator in between the sensor and the droplets generator, switch on amplifier and increase the air flow to 0.019m/s, and finally, Start measuring humidity by the sensor (Fig. 2).

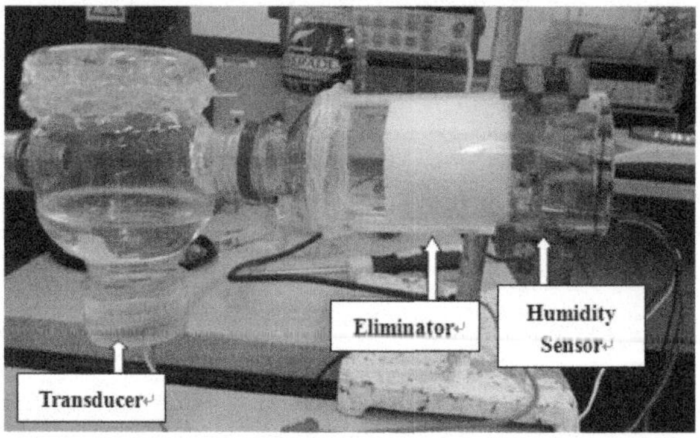

Fig. 2 Experiments setup with eliminator.

III. Results

The picture in Fig. 3 showed that the droplets continued to flow through the eliminator and out of the tube at that low air velocity. The readings in Fig. 4 and Fig.5 showed almost same curve compared to the one without presence of eliminator. From theory calculation, the droplets were predicted to still flow through the eliminator, and the

V. Van Toi et al. (Eds.): 4th International Conference on Biomedical Engineering in Vietnam, IFMBE Proceedings 40, pp. 110–112, 2013.
www.springerlink.com

experiment result just proved it. However, the droplets distribution in Fig.6 and Fig.7 showed that the eliminator did catch some relative big droplets at this low velocity. The sufficient momentum of those relatively big droplets kept the droplets to flow in a straight path. Therefore, they were caught by the eliminator.

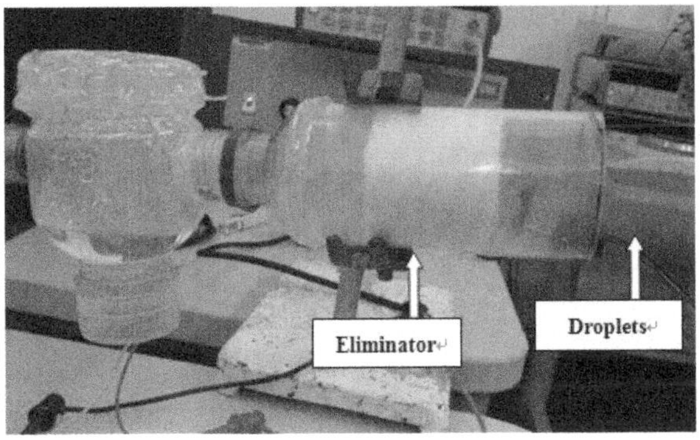

Fig. 3 Relatively small droplets flow easily through the eliminator.

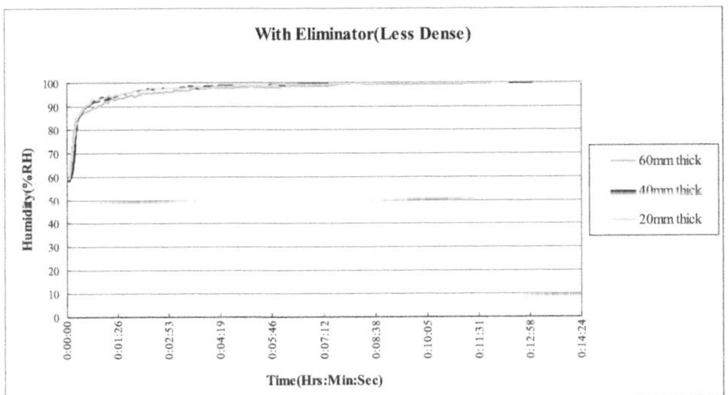

Fig. 4 Reading from SHT15 in the presence of less dense eliminator.

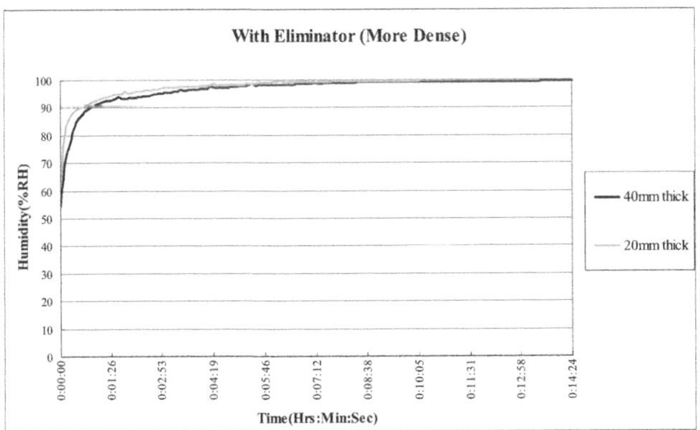

Fig. 5 Reading from SHT15 in the presence of more dense eliminator.

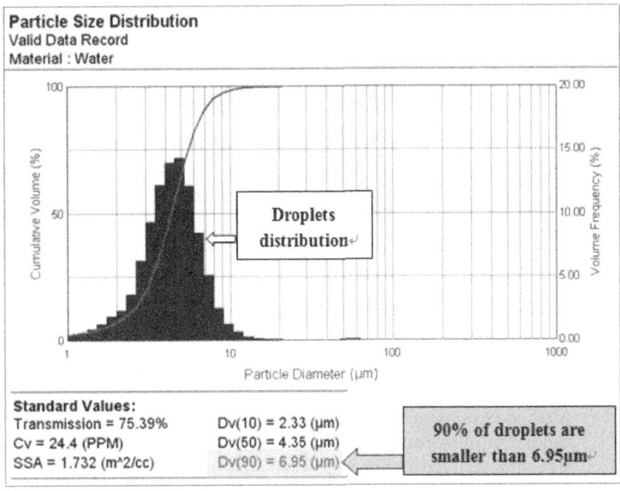

Fig. 6 Droplets distribution without eliminator.

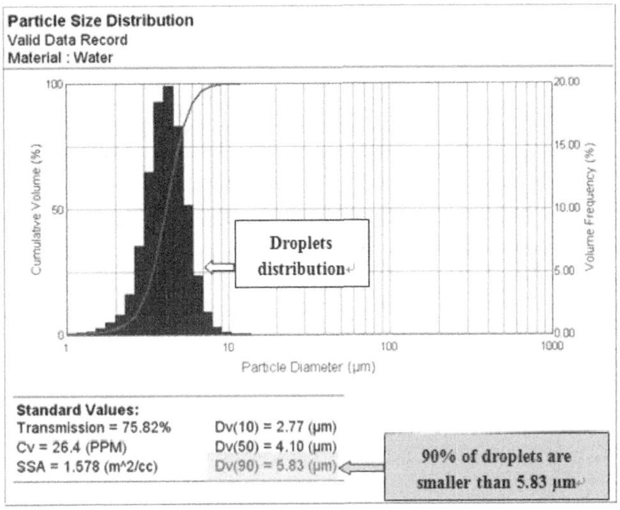

Fig. 7 Droplets distribution with eliminator.

The eliminator was also good to prevent the sensor from corrosion. Even though those relatively small droplets attached to the sensor, but they were too small and they evaporate very fast. Therefore, there would be no droplets appeared on the sensor as it was taken out from the tube.

IV. Discussion

The sensor can be used in many different areas and purpose [5]. In some cases with presence of droplets, the sensor is strongly suggested to be faced to the air flow direction if the response time is the most important factor to users. However this may cause decreasing in sensor's life time. If users want to measure humidity accurately without decreasing the sensor's lifetime, it is recommended to put the sensor facing parallel to the air flow direction.

According to the function of eliminator, it is good to be placed in front of the sensor resulting in protection of sensor. As the deposition of droplets on the eliminator, this may increase the environment humidity and the reading would be relatively higher. So this phenomenon should be noticed.

REFERENCES

1. Rouadi P, Baroody FM, Abbott D et al. (1999) A technique to measure the ability of the human nose to warm and humidify air. J. Applied Physiology 87: 400-406.
2. James PW, Azzopardi BJ, Wang Y et al. (2005) A model for liquid film flow and separation in a wave-plate mist elimniator. Chemical Engineering Research and Design 83: 469-477.
3. Galletti C, Brunazzi E, Tognotti L (2008) A numerical model for gas flow and droplet motion in wave-plate mist eliminators with drainage channels. Chemical Engineering Science 63: 5639-5652.
4. Setekleiv AE, Helsor T, Svendsen H (2010). Liquid holdup in wire-mesh pads. Chemical Engineering Research and Design 88: 1523-1531.
5. Fernicola V, Banfo M, Rosso L et al. (2008). Investigation of capacitive-based relative humidity sensors and their stability at high temperature. International Journal of Thermophysics 29: 1668-1677.

Author: Y.D. Gu
Institute: Ningbo University
Street: No.818 Fenghua Road
City: Ningbo
Country: China
Email: guyaodong@nbu.edu.cn

Cell Classification Using Shape Analysis

Thao X. Hoang, Hoan Nguyen, Son Tran, and Nguyen Dinh Thuc

Dept. of Knowledge Engineering, Faculty of Information Technology,
University of Science, Ho Chi Minh City, Vietnam

Abstract — **In this paper, we present a new method of cell classification using geometrical features of shape analysis. Firstly, the contour of cell is detected and approximated by one polygon. Secondly, by using the bounded polygon, the shape of cell is normalized. Then, the cell is classified into three kinds of cells: bull sperm cell (B-SC), human fibroblast cell (H-FC), and human mesenchymal stem cell (H-MSC) by using the structure of bounded polygon. Our approach overcomes the difficulty caused by the similarity between three kinds of cells when their shapes are unpredictably deformed in microscope image. We have applied our proposed method into data collected by one microscope with 40x eyepiece lens.**

Keywords — **Binary image, Otsu method, contour, Douglas Peuker algorithm.**

I. INTRODUCTION

Cell image analysis is one of the main functions in the lab for examining cell by microscopy. Through achievements in computer and semiconductor, a technician can take hundreds of high-resolution images of cells by using one microscope. In practice, there are two approaches of cell analysis. One is based on physical characteristics such as method of flow cytometry for cell counting. The other is based on image analysis. The number of cells existing in each image changes from a dozen to a few thousand of items. Therefore, it is not easy to count and classify cells by eye. Much as it is a boring and time-consuming process, it serves as an indispensable part of biology research projects. Until now, although many companies such as Nikon Instrument Inc., BMG Labtech, GE Healthcare, Nova Biomedical, Meta System have provided both software and hardware solutions to cell analysis, the price of one solution, ranging from tens to hundreds of thousands U.S dollar, the educational budget could hardly bear the whole cost of it. Especially, in these solutions, software takes a more important role than hardware - microscope, mechanical parts, and video card.

In practice, analyzing microscope images is a two-phase process. The first phase is the segmentation process to extract one microscope image including multiple cell images to single box containing one cell image. The second one is the classification process to label each cell in a single box. Most published researches are about segmentation or classification only. In this paper, we deal with the classification process problem. Much work has explored cell classification by using supervised learning [1][2][3] or shape matching [4][5].

In this paper, we present a new method of cell classification using geometrical features of shape analysis. In the manner of bio-image processing, this approach is robust and has low computational cost. In the proposed method, our assumption is that the region of interest is manually selected by the end-user so that there is only one cell image in the input picture.

This paper is organized as follows. Section II describes the workflow and structure of our method, presents the application of Douglas Peuker algorithm and our development in detecting the bounded polygon. Section III shows data and results of our proposed method. Finally, some conclusions are drawn in Section IV.

II. PROPOSED METHOD

The following workflow in Figure 1 outlines our proposed method. An image has to go through 5 steps including Pre-Processing, Approximating the bounded polygon, Approximating the shape, Extracting features and Classifying in order to decide which type of cell it is.

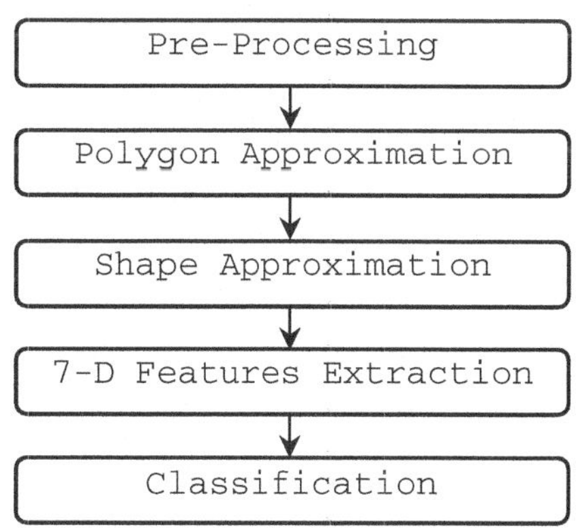

Fig. 1 Method Workflow.

The following parts of Section II will describe the contents of this workflow in detail.

V. Van Toi et al. (Eds.): 4th International Conference on Biomedical Engineering in Vietnam, IFMBE Proceedings 40, pp. 113–117, 2013.
www.springerlink.com

A. Preprocessing Based on Contour

In order to capture, enhance and remove noises from the cell image, we propose a multi-step preprocessing method. First, the edge image is received by applying the Sobel Operator [6]. Second, a simple threshold background removing method is used, which calculates the most frequent color in the image. Third, Otsu threshold selection method [7] is used to convert the image to binary. Next, we refine the image by using Median Filter to remove noises and Morphological Transforms (dilate, erode, closing and opening) to enhance the details of the cell. Finally, the contour of the cell is captured using the algorithm of Suzuki [8]. In this stage, small contours known as noises are removed and the biggest extreme outer contour (the cell) is remained.

B. Bounded Polygon and Geometrical Feature

Approximating the Bounded Polygon

The Douglas-Peucker Algorithm [9] is applied to approximate the polygon bounding the cell contour. Inputting a piecewise linear curve, a set of points between the first and the last point V_i and V_j, the algorithm makes it smooth by eliminating all the points in the set if the furthest point V_f from the line segment with V_i and V_j (the two end points) is not far enough. Otherwise, the algorithm will be recurred twice, using V_i and V_f; V_f and V_j as new first and last points. V_i, V_f and V_f is marked to be kept as part of the approximate polygon. In order to speed up the recursive algorithm, we use the Speed-up technique [10] that integrates with the Convex Hull Algorithm of Melkman [11] to find the furthest point mentioned above.

The simple pseudo code of Douglas-Peuker Algorithm is displayed below [10]

```
Given an array of vertices V, the call
DPbasic(V, i, j) simplifies the sub-
chain from Vi to Vj.
  Procedure DBbasic(V, i, j)
 1. Find the vertex Vf farthest from the
    line ViVj.
    a. Let dis be its distance
 2. If dis > ε then
    a. DBbasic(V, i, f)  /* Split at Vf
       and               approximate */
    b. DBbasic(V, f, j)  /*
       Recursively */
 3. Else
    a. Output(ViVj)         /*
       Use ViVj in the approximation */
```

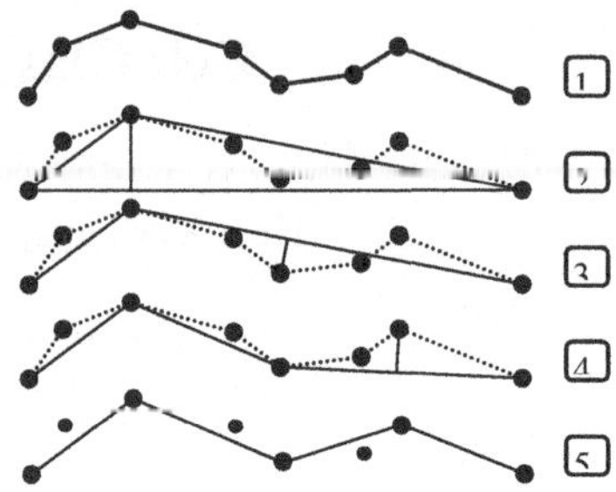

Fig. 2 Example: Douglas-Peucker Algorithm.

Figure 2 is an example of the Douglas-Peucker algorithm. In this example, the original polyline having 8 points becomes a 5-points one after the process. The advantages of this method are that it not only approximates the bounding polygon but also reduces the number of points of the contour. This gives great effort to the afterward steps of the procedure in both manner of speed and complexity.

C. Approximating the Shape

After applying the Douglas-Peucker algorithm to get the approximate polygon, there are excess parts that are not the main features of the cell. We propose a method to get rid of these unwanted curves.

An odd curve is a curve that the start vertex is near enough to the end vertex in both case of Euclidean Distance and Path Distance:

$$IsOddCurve(Curve[i,j]) \leftrightarrow \begin{cases} EDis[i,j] < \varepsilon 1 \\ PDis[i,j] < \varepsilon 2 \\ j - i > 1 \end{cases}$$

$EDis[i,j]$ stands for the Euclidean Distance between vertex i and vertex j. The $PDis[i,j]$ stands for the Path Distance between the two vertexes mentioned which is defined as

$$PDis[i,j] = EDis[i,i+1] + \ldots + EDis[j-1,j].$$

To speed up the procedure, we build a Path Distance Matrix $PDisM[n \times n]$, which contains all the Path Distance of the polygon.

$$PDisM[i,j] = \begin{cases} 0, i = j \\ PDisM[i,j-1] + EDis[j-1,j], i > j \\ PDisM[j,i], i > j \end{cases}$$

The mission is to detect all the odd curves in the polygon and find an appropriate solution to remove them. We build an Odd Curve Matrix $OddCurveM[n \times n]$ as follows:

$$OddCurveM[i,j] = \begin{cases} false, (i = j) \\ IsOddCurve(Curve[i,j]), (i < j) \\ false, (i < j \\ \wedge \exists\, OddCurveM[i,j'] = false \\ \wedge\ PDisM[i,j'] > \varepsilon 1) \\ false, (i > j) \end{cases}$$

Starting from vertex V_i, if the curve $C_{i,j}$ is not an odd curve because of too long path distance, all the curve after will be marked as normal curve. The purpose of building the Odd Curve Matrix is to help processing the odd curves more easily when it represents an adjacency matrix of a directed graph that has n nodes. There is an edge from node i to node j if $Curve[i,j]$ is an odd curve.

Each vertex has a set of odd curves starting from it and the Main Odd Curve of the set is the longest curve:

$$MainOddCurve[i,j] \leftrightarrow \forall k, i < k < n,$$
$$OddCurve[i,k].PDis < MainOddCurve[i,j].PDis$$

Starting from the first vertex, its Main Odd Curve is evaluated as isolated or intersecting as follows:

$$IsoMainOddCurve[i,j] \leftrightarrow \nexists\, MainOddCurve[k,l],$$
$$i < k < j < l \text{ or } k < i < j < l$$

So as to delete an Isolated Main Odd Curve, we remove all vertexes between the start and the end vertex. However, in case of two $InterMainOddCurve[i,j]$ and $InterMainOddCurve[k,l]$, we find a Partition Set of odd curves by looking for a route from node i to node j in the graph.

$$PartSetOC[m] of\ [i,l] \leftrightarrow \forall i < m,$$
$$\begin{cases} PartSetOC[i].end = PartSetOC[i+1].start \\ PartSetOC[0].start = i \\ PartSetOC[m-1].end = l \end{cases}$$

The most efficient route will be chosen as the Partition Set of overlapped odd curves. If there is no route from node i to node j, a partition set of odd curves between them is non-existent. In this case, a greedy approach is used to remove as many odd curve lengths as possible: the first odd curve is removed, then the available odd curve that is right after the first one is chosen to be evaluated.

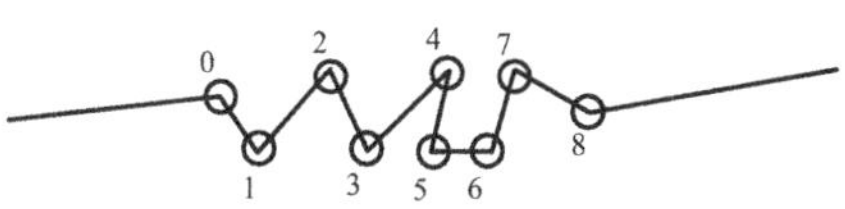

Fig. 3 Odd Curves.

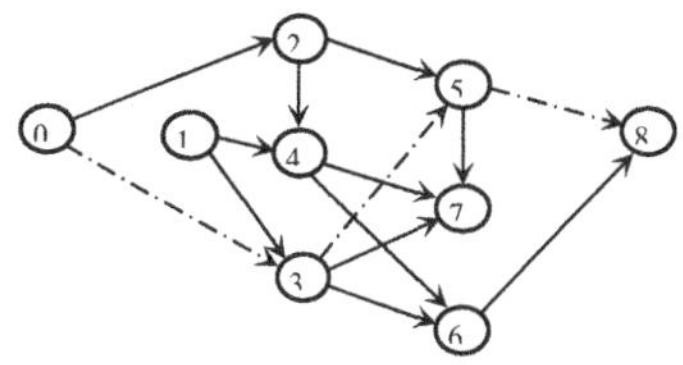

Fig. 4 Directed Graph of Odd Curves.

Fig. 5 Smooth Curve.

Figure 4 is a directed graph built from overlapped odd curves in Figure 3. A route from V_0 to V_8 through V_3 and V_5 is available, so the Partition Set is $\{V_0V_3|V_3V_5|V_5V_8\}$. And Figure 5 is the result of the removing odd curve process. Figure 6 is a completed example of this Shape Approximation process. In this example, there are 3 IsoMainOddCurves and 5 InterMainOddCurves.

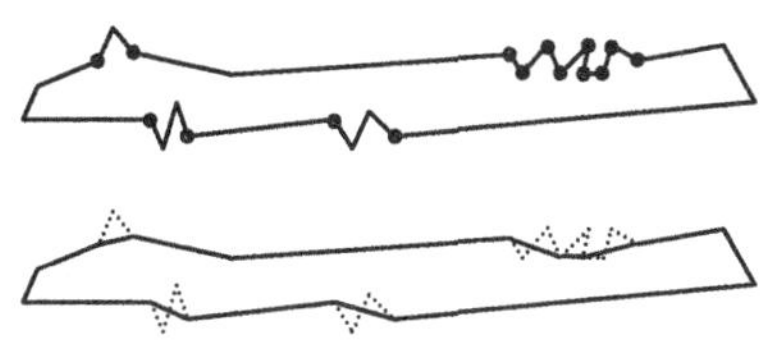

Fig. 6 Shape Approximation.

D. Extracting Geometrical Feature

To get features of the cell, we rotate the approximate polygon by making the line composed of two furthest poles parallel to the vertical axis. The polygon is then bounded by a rectangle. All the features are extracted from this rectangle containing the rotated approximate polygon. Since the resolutions of different types of cell are the same, these features below reflect the exact information about the cells. There are 7 features extracted, which are Height, Width, MeanWidth, Perimeter, Number of Approximate Points (NAP), Scale between Height & MeanWidth (Scale HMW), and Area.

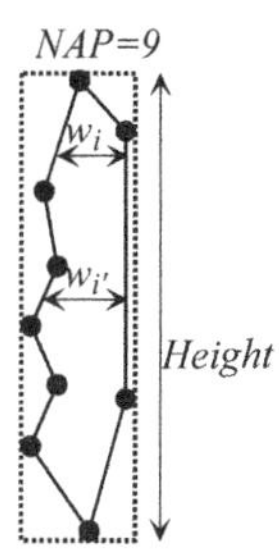

Fig. 7 Height, W_i and NAP.

Figure 7 describes the Height, width (w_i) and NAP of an approximate polygon. Width is the furthest distance between the left and the right point of the approximate polygon that have the same vertical coordinate and MeanWidth is the mean of all width. Here are details of extracted features:

$$Width = \max_{i:\,0\rightarrow Height-1}(w_i)$$

$$MeanWidth = {}^{1}/_{Height} \times \sum_{i=0}^{Height-1} w_i$$

$$Perimeter = \sum_{i=0}^{NAP-1} Distance(p_i, p_{(i+1)\%NAP})$$

$$Area = \sum_{i=0}^{Height-1} w_i$$

$$Scale\ HMW = {}^{Height}/_{MeanWidth}$$

E. Classification

Each feature extracted is a weak filter of these three cell types that distinguishes one type from the others. Through the experiment of a set of 50 images of each type, weak filters classifying H-FC from the others are Height, Perimeter, NAP and Scale HMW. And those that distinguish B-SC and H-MSC are Width, MeanWidth and Area. Thresholds of these filters are also achieved through the experiment. Each set of weak filters above generates a strong filter used in the final classification. The H-FC Strong Filter is passed when 3/4 first weak filters are satisfied. Otherwise, three other filters are used to determine whether the cell is a BSC or a H-MSC. If 2/3 last weak filters are satisfied, the BSC strong filter is passed.

III. Experiments

A. Data

The microscope used to take images has its total magnification ranging from 10X to 40X. Under the microscope, the shape of cell and noise are both magnified. Sometimes the shape of noise is as large as that of cell, which causes mistake in results of cell detection and classification. Figure 8 presents the structure of a counting cell chamber. Cells are usually counted in one of the large chambers {#1, #2, #4, #5, or #3}. The order of scanning and counting cell on the large chambers is complicated and easy to cause mistakes. To overcome it, we take advantages of image processing and pattern recognition techniques to define the boundary of chamber and detect cells in the image.

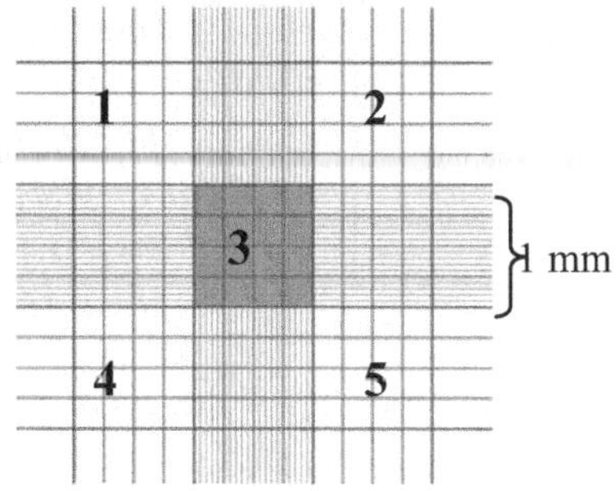

Fig. 8 The structure of one chamber.

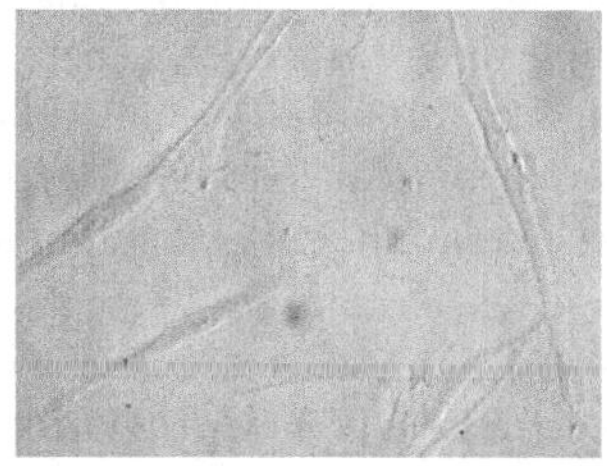

Fig. 9 H-FC Sample.

Figure 9 is a sample of H-FC cell that has a size of 1380x1034. The H-MSC samples have the size as H-FC. However, the B-SC samples' size is 4140x3096, which is different from the other two. Hence, we scale all the samples to the same size before applying the procedure on them.

B. Result

The experiment was tested on a computer having Windows 7 x64 OS, Core i5 2.3Ghz processor, 4GB DDR3 RAM. The testing program was built on .Net Framework 4.0 Environment using C# programming language.

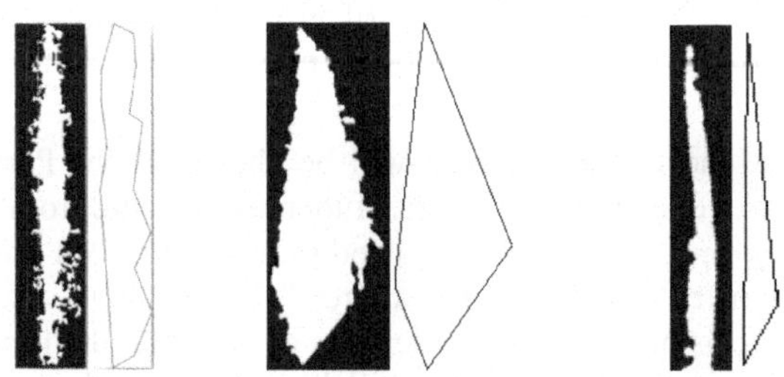

Fig. 10 Samples of our methods.

We have applied our algorithm to 30 testing images for each kind of cell. All images are collected by one microscope with 40x eyepiece lens. Table 1 illustrates the classification results. Figure 10 presents samples of our method. In each part of Figure 10, the left image is the binary image of cell and the right image is the shape approximation.

Table 1 Classification Result

	Successful Rate (%)	Average Time (ms)
H-FC	93.33	84.9
H-MSC	90	30.6
B-SC	93.33	21.6

IV. Conclusions

We have proposed a new method of cell classification by utilizing the geometrical feature in image analysis. The proposed method can get 90 percent of average accuracy for three kinds of cells. The calculating time is less than 100 ms for each sample. Moreover, the proposed method can be easy to implement in many other kinds of cells.

Acknowledgment

The authors are greatly in debt to the Laboratory of Stem Cell, University of Science for providing excellent data that makes this work possible.

References

1 A. Chebira, et.al., "Multiresolution Techniques for the Classification of Bioimage and Biometric Datasets," in Proceedings of the SPIE, Volume 6701, pp. 67010G (2007).
2 K. Huang and R. Murphy, "Boosting accuracy of automated classifi-cation of fluorescence microscope images for location proteomics," BMC Bioinformatics 5(78), 2004.
3 L.P. Coelho, et.al., "Principles of Bioimage Informatics: Focus on Machine Learning of Cell Patterns," Springer LNCS 2010, Vol. 6004/2010, 8-18.
4 R.F. Murphy, M. Velliste and G. Porreca (2002) Robust Classification of Subcellular Location Patterns in Fluorescence Microscope Images. Proc 2002 IEEE Intl Workshop Neural Networks Signal Processing (NNSP 12), pp. 67-76.
5 Z. Pincus Z, JA. Theriot, "Comparison of quantitative methods for cell-shape analysis," J. Microsc. 2007; 227 (Pt 2): 140-56
6 Rafael C. Gonzalez and Richard E. Woods, "Digital Image Processing 2nd Edition," pp. 572-585, 2002
7 N. Otsu, "A Threshold Selection Method from Gray-Level Histograms," in IEEE Transactions on Systems, Man and Cybernetics, vol. 9, pp. 62-66, January 1979.
8 Satoshi Suzuki and Keiichi Abe, "Topological Structural Analysis of Digitized Binary Images by Border Following," in Computer Vision, Graphics, And Image Processing, vol. 30, pp. 32-46, 1985.
9 D.H.Douglas and T.K Peucker, "Algorithms for the Reduction of The Number of Points Required to Represent a Digitized Line or Its Caricature," in The Canadian Cartographer, vol. 10, pp. 112-122, December 1973.
10 John Hershberger and Jack Snoeyink, "Speeding Up the Douglas-Peucker Line-Simplification Algorithm," in International Symposium on Spatial Data Handling, vol. 5, pp. 134-143, 1992.
11 Avraham A. Melkman, "On-line construction of the convex hull of a simple polyline," in Information Processing Letters, vol. 25, pp. 11-12, 1987.

Foot Loading Character under Inner Heel Height Increasing Way

Qi Hao, W.W. Shen, J.B. Ma, Jianshe Li, and Yaodong Gu

Faculty of Sport Science, Ningbo University, Zhejiang 315211, China

Abstract — **This study aims to observe the silica gel insoles for adult females foot plantar pressure changes thus to find any other factors that cause female foot pain further. Six female subjects were participated in this study, the foot pressure parameters was recorded by Novel emed pressure measurement system. Each person had walked three times under bare foot state, and with 2 layers, 4 layers, 6 layers of silica gel insoles under the heel separately. Multiple comparison results showed that different layers of silica gel pad used in experiment was sensitive to plantar pressure change. With simulated inner heel height increased, the peak pressure reduced significantly in the heel. The experimental silica gel insoles, as its special material different from common insoles, all subjects feel soft and comfortable in the experimental process, which may indicates the insoles have a cushion to foot pressure load increase.**

Keywords — **plantar pressure, heel height, biomechanics.**

I. Introduction

As people's living standards improve, their health conscious is more and more intense. High-heeled shoes is essential to women groups in the daily work and life, as we all know, a large number of studies have demonstrated that long-term wearing the high-heeled shoes of a certain height have adverse impact on foot. Previous studies have demonstrated that walking in high-heeled shoes alters lower-extremity joint function [1], put too much load on the medial plantar metatarsal area, causing metatarsal injuries and pain, and even lead to foot deformation [2]. Walking in high heels change the gait kinematic and kinetic parameters of human for that heel elevation making the plantar pressure redistribution and force situation of lower limb even the whole body skeletal systems is different with wearing flat shoes. Appropriate height can make the body weight well-distributed on the whole foot, while too high and thin heel and a narrow front, changing the foot status that bearing weight uniform. Walking with medium-hard-soled shoes and hard-soled shoes, the foot pressure center path length are both longer than bare feet, while track length of soft-soled shoes become shorten. Walking with soft-soled shoes and medium-hard-soled shoes, the total pressure are both reduce compared with bare feet , and their walking phase pressure are both reduced. Walking with hard-soled shoes, the total pressure has increased significantly compared with bare feet walking, also increased each walking phase pressure. Metatarsal pad has been found to reduce forefoot pressure and transfer weight bearing to the longitudinal and metatarsal arch [3].

This paper aims at that only the heel height increase can influence the plantar pressure parameters or not. The experiment used the common silica gel insoles in market, with different insole layers simulating different heel height. The study aims to observe the silica gel insoles for adult females foot plantar pressure changes, and compare with the results of the study at ordinary high-heeled shoes by the corresponding biomechanical parameters, thus to find any other factors that cause female foot pain further. Then we can optimize the design of the high-heeled shoes, so that people can have a comfortable and beautiful condition, with a pair of healthy feet.

II. Methods

Six female subjects joined in the experiment. Participants are well-balanced, healthy body, without legs, ankle injury. Written consent was obtained from each subject before commencement of the experiment.

The silica gel insoles used in experimental study was organic silicon material, it is non-toxic, tasteless, physical inertia. By cross-linking system choice, the filler choice and used with the base polymer, it can get dimensionally stable, high precision and smooth surface characteristics of organic silicon material. The experiment used a total of 6 silicone pads, the thickness of each block are the same, it is 0.8cm. The experiment is divided into different experimental conditions by different silicone heel pad layers simulating different heel height. Specific for bare foot (without heel pad), 2-layer silica gel pad, 4-layer silica gel pad and 6-layer silica gel pad. Subjects started experiment in the four states respectively, and then collect experimental data to compare and analyze.

The experimental apparatus used in this study is the pressure distribution measurement plate. It is a Novel emed measurement system. The first step is to have environmental adaptation training on subjects who are required to wear uniform socks for walking exercise. They need to

V. Van Toi et al. (Eds.): 4th International Conference on Biomedical Engineering in Vietnam, IFMBE Proceedings 40, pp. 118–120, 2013.
www.springerlink.com

guarantee that their right foot contact force measurement flat when past it in the natural walking. The walking direction each time is the same. Then we began formal experiment after all subjects have been adapted. Clicking the computer to start test and then issue the order to subjects. They walk through the force measurement flat in natural speed, which calls that only their right foot contact force measurement flat each time trial, and walking direction is one-way consistent (Fig. 1). Everyone walked three times (as bare foot state). Then the subjects put silicone insoles in the right foot of the sock, fix them on the foot heel position, and through force measurement version by the same line. Interval of each experiment is half a minute to let subjects have a fully rest. According to the different layers of silica gel insoles, the whole experiment is divided into four parts: (1) bare feet (without silicone insoles); (2) 2-layers of silica gel insole (heel height of 1.6cm);(3) 4-layers of silica gel insole(heel height of 3.2cm); (4) 6-layers of silica gel insole (heel height of 4.8cm). Each part of the experiment is tested three times at least. Collect and record data of dynamic foot pressure distribution with instruments for each participant in four different state.

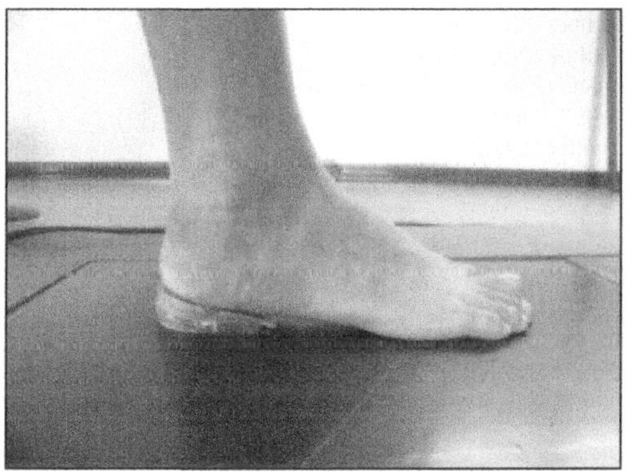

Fig. 1 Inner heel height increasing method.

III. Results

Use system software for data analysis after experiment testing. Firstly, the foot plantar is divided into five regions through the computer manually partition: big toe (BT), medial metatarsal (MM), lateral metatarsal (LM), arch (A) and heel (H). Export the experimental data used in this experiment such as peak pressure, mean pressure and pressure time integral values. Calculate the average value of each participant's three times test in each experiment state. Then summarize the experimental data of six subjects in bare feet, 2 layers, 4 layers and 6 layers of silica gel insoles state. Finally, compare the different plantar pressure distribution under four states and the difference of biomechanical parameters.

According to the measured dynamic plantar pressure data of each subject, we can obtain corresponding value changes of plantar regions under different test conditions. Among them, select peak pressure (PP), mean pressure (MP) and pressure time impulse (PTI) for comparison and analysis. Peak pressure is the maximum pressure value in five plantar areas and maximum ground reaction force suffered per unit in specific area. Roy has pointed out that the largest plantar pressure often has a high correlation with forefoot lesions; Mean pressure is the average value of pressure in each region. Comparison of mean pressure in different plantar regions may reflect the distribution of plantar pressure; Pressure time impulse is the function degree of pressure in unit time which reflects the interaction intensity between foot and ground.

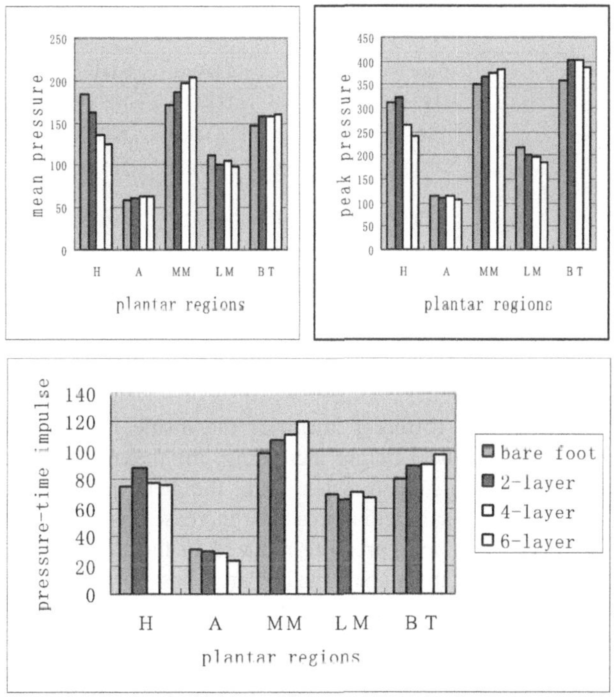

Fig. 2 MP, PP and PTI comparison in four states.

Fig.2 illustrates the comparisons of three plantar biomechanical parameters in different foot regions and Table1 lists the comparisons of mean pressures for each test condition. ANOVA results indicated a significant heel height effect on peak pressure of the heel regions, and effect on mean pressure of the heel, the metatarsal bone inside area.

Multiple comparison results showed that different layers of silica gel pad used in experiment is meaningful for foot pressure change (Table 1). In the heel area, peak pressure in naked feet and 6 layers of insoles exist significant statistically difference ($p<0.05$), 2 layers and 6 layers of silica gel insoles also exist significant statistically difference ($p <0.01$). The mean pressure value in heel area there are much more differences. Results showed that in the heel area, the bare feet and 4 layers ($p <0.01$), bare feet and 6 layers ($p <0.01$), 2 layers and 4 layers ($p <0.05$), 2 layers and 6 layers ($p<0.01$) of silica gel insoles all exist significant statistically differences; mean pressure in metatarsal medial area also exists significant differences. It shows that in the metatarsal medial area, MP value in bare feet and 4 layers ($p<0.01$), MP value in bare feet and 6 layers ($p<0.01$) of silicone pads exist statistically differences; the remaining three regions (arch, metatarsal lateral and thumb area) does not exist significant difference by the statistical analysis software ($p> 0.05$).

Table 1 Comparisons of mean pressures for each test condition (KPa).

	Bare foot	2 layers	4 layers	6 layers
Heel	183±19	162±8	135±16	125±15
Arch	59±11	60±12	63±8	64±14
Medial Metatarsal	171±10	186±12	196±9	203±14
Lateral Metatarsal	111±21	101±24	106±21	98±22
Big Toe	146±35	157±67	157±65	159±58

IV. Conclusions

After placed different layers of the insole in the heel, with simulated heel height increasing, the maximum value of average pressure transfer from the heel part to the forefoot metatarsal medial area. The increase has significant difference. The peak pressure reduced in the heel, but the increase in the metatarsal medial area has not significant statistically difference. The pressure time impulse in the medial metatarsal and the thumb area both has an increasing trend, but they don't exist statistically difference. The experimental silica gel insoles, as its special material different from common insoles, all subjects feel soft and comfortable in the experimental process, which may indicates the insoles have a cushion to foot pressure load increase. Thus as the pressure in the heel decreased, it only add small load relatively to the metatarsal bone inside area. At the same time, it will also reduce injury risk which the high-heeled add to the forefoot metatarsal area. Previous research about women foot injuries caused by high heels generally believes that it is due to excessive height of shoes. The study found that not only the high heel, but the structure of shoes such as the width of shoes, narrow shoes head, the hard bottom shoes and so on also are important factors that cause female foot pain even the toe deformation. Given the article is a preliminary study for heel height simulation made by silica gel insoles, it need to be further in-depth understanding of the precise mechanism that high-heeled shoes act on women. One day people can have a comfortable and beautiful condition, with a pair of healthy feet.

Acknowledgment

This study is supported by Ningbo University's talent project and Anta Sports Products Limited (HS11085).

References

1. Wu J., Li J.S. (2003) Biomechanics research on flat walking gait of young women with high heels. Sports Research. 24(3): 9-11

Li J.S., Gu Y.D. Ren X.J. et al (2010) Biomechanical effects of foam inserts on forefoot load during the high-heeled gait: a pilot study. Journal of mechanics in medicine and biology 10: 667-674.

Esenyel M., Walsh K., Walden J.G. et al. (2003) Kinetics of high-heeled gait, J Am Podiatr Med Assoc 93: 27-32

Author: Yaodong Gu
Institute: Ningbo University
Street: No.818, Fenghua Road
City: Ningbo
Country: China
Email: guyaodong@nbu.edu.cn

Orthopedic-Based Biomechanics for Undergraduate Vietnamese Students

B.S. Kelley[1], R.M. Unruh[1], B.R. Rigby[1], H.D. Vu[2], and Tới V. Võ[3]

[1] School of Engineering and Computer Science, Baylor University, Waco, Texas, USA
[2] Department of Biomedical Engineering, Hanoi University of Science and Technology, Hanoi, Vietnam
[3]Department of Biomedical Engineering, VNU-HCMC-International University, Ho Chi Minh City, Vietnam

Abstract — **Biomedical Engineering is among the emerging technological disciplines in Vietnam. This is evidenced by the identification of undergraduate BME as one of the *Advanced* programs supported by the Vietnamese Ministry of Education and Training (MOET) and hosted by the new Biomedical Engineering Department within the School of Electronics and Telecommunications at the Hanoi University of Science and Technology (HUST). Furthermore, a second BME department has emerged at the Vietnam National University- Hoc Chi Minh City- International University (IU). The predominant focus of both of these BME departments and curriculums is in biomedical instrumentation, which historically has been the most frequent field of employment for biomedical engineers in Vietnam. Because the intent of both of these BME programs is to educate their students to be competitive internationally for graduate programs and employment, there is an emphasis to expand their curricula to include broader BME topics such as biomechanics and biomaterials. With the award of a grant from the Vietnam Education Foundation (VEF) as a VEF U.S. Faculty Scholar, the lead author taught an engineering biomechanics course for Vietnamese undergraduate students for three consecutive years. Two of these offerings were at HUST and one was at IU. The pedagogical format of the course changed each year and ranged from a mixture of onsite and internet distance learning, to onsite and video conferencing, to fully onsite. Each year the course was centered around the design of an orthopedic fixation appliance...a bone plate. Course materials include PowerPoint slides, supplemental readings, and assignment descriptions, and were customized specifically for this course. Each offering presented unique challenges relating to instructional technology, course pace, and cultural and language differences. Notwithstanding, each of the course offerings was considered a success based on student learning, mastery and performance, and student course evaluations.**

Keywords — **biomechanics, engineering education, bone plate, Vietnam.**

I. Introduction

The genesis of this effort derives from the Vietnam Education Foundation's (VEF) U.S. Faculty Scholar Program. Established by the U.S. Congress in 2000, the purpose of VEF is to establish educational exchange activities to build the bilateral relationship relating to science and technology between the United States and Vietnam. Since 2003 nearly 400 Vietnamese have pursued graduate degrees in the U.S. as VEF Fellows. Since 2007 over 30 Vietnamese with doctorates have further developed their professional capabilities through additional study and research in the U.S.

In VEF's newest program, American faculty members are selected through a competitive grant process to teach at a Vietnamese university, either fully onsite or by also using video conferencing from the United States. During the past three years eleven VEF U.S. Faculty Scholars have taught at ten different Vietnamese universities. It was this program that sponsored the majority of first two years of the efforts described in this paper.

The main goals of the VEF U.S. Faculty Scholar Program are:

1. To help build capacity in Vietnam in science, mathematics, medicine, engineering and technology through teaching and related research activities.
2. To help Vietnamese institutions address specific educational needs including curriculum development, integration of research opportunities, identification of student learning outcomes and evaluation.
3. To build positive relationships between U.S. and Vietnamese higher education institutions and faculty members leading to sustainable partnerships. [1]

II. Background

There are several prominent reports written on comparative studies of engineering and science education between Vietnam and the United States. [2, 3, 4, 5] Companion efforts have also contributed to raising the quality and profile of biomedical engineering research within Vietnam, [6, 7, 8, 9] as well as on aspects of teaching biomedical engineering in Vietnam. [10, 11, 12, 13]

To better understand the contemporary practices of engineering education in Vietnam, VEF, MOET, and the U.S. National Academies partnered to conduct a multiple casestudy quantitative research project. These efforts centered on four Vietnamese universities and were conducted by American authorities in assessment and instructional design in the selected fields of computer science, electrical engineering, and physics. Follow-on studies also concentrated on undergraduate agricultural education. [3, 4] The results of the site visits identified five critical areas of Vietnam higher education in need of change. [2] They are:

1. *Undergraduate teaching and learning*: ineffective teaching methods and inadequate facilities and resources.
2. *Undergraduate curriculum and courses*: too many courses and requirements; out-of-date content; an imbalance between theoretical concepts and applied experiences; lack of professional skills; and absence of expected student learning outcomes.
3. *Instructors*: lack of qualified teachers, low level of academic preparation; lack of modern teaching practices; overworked and underpaid faculty.
4. *Graduate education and research*: little opportunity to pursue research and apply modern teaching methods; academic inbreeding; separation of teaching departments from research activities.
5. *Assessment of student learning and institutional effectiveness*: lack of articulated outcomes; institutional effectiveness and course quality not evaluated in terms of student learning; lack of institutional research infrastructure.

It was these types of issues, and specifically the classroom experience, that VEF asked the inaugural group of U.S. Faculty Scholars to seek to affect.

Initial efforts associated with this project were to determine the best candidate university and the most appropriate course to offer. In 2008 when preparations were being made, the most active undergraduate BME program in Vietnam was at the Hanoi University of (Science and) Technology (HUST). As was traditional in Vietnam, many of the biomedical engineering graduates were employed by hospitals to forecast and purchase and maintain the facilities electronic equipment and instrumentation. Because of this HUST's (and IU's) biomedical engineering curriculum has a strong biomedical electronics component. During this same period the Vietnam Ministry of Education and Training (MOET) approved and funded HUST's proposal to raise the quality and stature of BME to the level of an *Advanced* program. With that came also the need for interactions with more U.S. universities and to extend the curriculum to include a broader spectrum of BME topics.

The lead author has with some success taught an undergraduate engineering biomechanics course on numerous occasions [14, 15]. The content and scope of that course was not fully transferable to the Vietnamese context, nor did it have the specific intent to address more broadly the concerns indentified in the VEF/MOET/National Academics study [2]. Therefore, adjustments were made to the course design, and new course materials that were easily transferrable to the Vietnamese students were prepared. These included PowerPoint slides, homework assignments, and supplemental readings that were all combined into a single PDF document. This document was later modified with the intent to enact improvements between the subsequent offerings of the course.

The engineering biomechanics course has four learning objectives. By the end of the course the students should…

1. Have a basic understanding of the function of the musculoskeletal system
2. Be able to apply principles of materials mechanics to evaluate stresses in bones
3. Have an awareness of the materials and design of orthopedic implants
4. Be able to apply the engineering design process for an orthopedic device

An assumption that permeated the decision-making process was that it is not realistic to expect American/ foreign professors to regularly be able to spend an entire semester teaching in Vietnam, and therefore a portion of the course must be offered using a distance learning technology from the United States.

III. Engineering biomechanics course

A. Course Design

The central focus of the biomechanics course is the design of a bone plate to stabilize and reduce long-bone fractures. A design project was chosen as the center point so that students could directly apply the knowledge they learned in the class in a practical application, and so that they could also gain experience in technical presentations and writing that is common and valuable in American engineering courses. The semester topics of the course included review principles in mechanics with biomedical examples that also served to confirm a common nomenclature and vernacular. The specific bone plate problem was selected because though the engineering analysis leading to the design can be complex, the visualization of a metal plate attached with screws to a bone is relatively straight forward. The course's specific engineering topics supported and culminated in the bone plate design. These included bending and moment diagrams, neutral axis identification, parallel axis theorem, and material properties of engineering materials and hard and soft biological tissues. Physiological systems and anatomical considerations were necessarily woven throughout the course content. Other course topics address technical presentations and writing, group dynamics and responsibilities, the engineering design process, and medical device regulations.

B. *Teaching Environment*

The engineering biomechanics course was taught for three consecutive years from 2009 thru 2011. However, from an experimental design perspective, few if any of the variable parameters remained constant during that period. The course materials were improved each year. The location and delivery method changed each year. The background of the students changed each year. For all three environments the students were in their third year and receptive and open to being taught in English by a foreign professor. For all scenarios the rapport between the instructor and the students was strong and productive. The student course evaluation process used by Baylor University was completed by the students at the end of each course. Course enrollment ranged from 14 to 22 students.

Year 1- 2009- Hanoi University of Technology: The first course offering was at HUT, which is an established institution located near the center of the city. The course duration was fifteen weeks. The initial and final two weeks were taught onsite and the intervening eleven weeks were delivered using Elluminate Live® via the internet. The course materials were printed in the U.S. and given in book format to the students. The onsite teaching components were delivered at twice the normal pace, i.e., two lectures per day, three days each week. The students were pre-screened by the department based on their English communication skills. There were 14 students enrolled in the course.

The onsite pace of the course provided for conducive learning. There were shortcomings in the course relating to the portion delivered over the internet. The particular software platform, Elluminate Live®, worked well and reliably. The PowerPoint slides displayed well, audio quality was acceptable, and the students could view a web cam image of the instructor. The shortcomings related to the technology available to the students in Vietnam, and the pace of the course. Bandwidth limitations prevented the students from using the HUT computer classroom. Then up to six computers in a laboratory were used with up to three students sharing a monitor. Later during the semester some of the students chose to log in from a different location, and likewise some decided not to log in at all. The single 1.5-2 hour session per week was not frequent enough to keep the students engaged and active. The time difference between Texas and Vietnam necessitated the course being delivered at 7:00 a.m. in Hanoi, which also contributed to student apathy.

Year 2- 2010- VNU-HCMC-International University: IU is a relatively young institution, and the BME department is even younger. The entire institution is currently housed in a single modern seven-story building located in a developing region on the outskirts of the city. All of the third- and fourth-year IU BME students enrolled in the biomechanics course, in addition to several lecturers and graduate students (not for credit) from the HCMC University of Technology. Because of the young age of the department, none of IU BME students entered IU as BME students, but rather transferred into the BME program from other engineering departments.

The overall format of the course was similar to that of the previous year, with several notable exceptions. While the initial and final two weeks of the course were conducted onsite, the intervening weeks were shortened from eleven to five, with a corresponding increase in weekly sessions. The delivery vehicle was video conferencing instead of the internet. The class materials were emailed to IU in advance of the course. The in-country printing was in black-and-white only and the quality poorer than anticipated and was sometimes difficult to read. Circumstances with student schedules also lead to an irregular class schedule and meeting location, and a week-long break between the initial onsite component and beginning the video conferencing portion. Thirteen IU undergraduate BME students, plus four HCMCUT instructors and graduate students enrolled in the course. The addition of the HCMCUT representatives added positively to the environment of the class.

The video conferencing systems offered several advantages over the previous year's internet protocol. The video feed was richer, including a much larger field of view, and the audio quality was often superb. However, although several advance trial sessions were conducted, there were continuing obstacles in getting the video conference systems from Baylor and IU to consistently communicate with each other. And because of the time difference between Texas and Vietnam (7:00 p.m.–7:00 a.m.), it was not usual for there to be class-time technician help readily available from either end. Instead of near-high-definition video quality from both ends, the viewability was relatively poor and the students could view the instructor but not separately or simultaneously the PowerPoint slides.

Year 3- 2011- Hanoi University of Science and Technology: The third course offering moved back to the site of the first, and though the classroom was in the same building, it was not identically the same room. Twenty-two *Advanced* third year BME students were enrolled in the course. This meant that they had a full year of intensive English instruction during their first year at HUST. Like the IU students, instruction in all of their courses was in English. The course materials were again emailed in advance, and the PDF file was distributed to the students. A good portion of the students elected to have print shops prepare nicely bound copies, while others worked solely with the digital version.

The greatest difference between this version of the biomechanics course offering and the other two was that it was offered completely onsite. This occurred over a compressed three-week schedule during the summer. While the onsite schedule eliminated most of the distance-wise technology difficulties (notwithstanding projector challenges), the accelerated schedule introduced a new and different set of obstacles. The class met in double morning and double or single afternoon lecture sessions. The exception was Friday afternoons which were not scheduled until the final exam. Completing and grading homework assignments was a continuous activity. Assigning student design teams occurred at the beginning of the second week of the course, and the draft of the introduction to their design report occurred prior to several critical lectures involving the design's engineering analysis. An advantage of the fully onsite approach was that the students remained more engaged and developed a deeper rapport with the instructor.

IV. Course Evaluation

To better understand the perspective of the student participating in the biomechanics learning environment, a student class evaluation was conducted near the end of each course offering. The Baylor course-evaluation system was used, and consists of student responses to each of fifteen statements. These statements are divided into seven groups and are as listed below.

Communication

1. The instructor explained material clearly.
2. The instructor had an effective style presentation.

Attitude Toward Students

3. The instructor treated students with respect.
4. The instructor was concerned that students learn the material of the course.

Attitude Toward Subject

5. The instructor appeared interested in the subject material.
6. The instructor stimulated my interest in the subject.
7. The instructor stimulated my thinking.

Preparation and Organization

8. The instructor made effective use of class time.
9. The instructor was well prepared for each class.
10. The course was well organized.

Policy and Evaluation

11. The requirements of the class were clearly explained.
12. The exams were a good measure of my knowledge.

Text and Instructional Material

13. The assignments contributed to my understanding of course content.

Overall

14. I learned a great deal from this course.
15. The instructor used methods conducive to learning.

Figure 1 shows the percentage of student who agree or strongly agree with each of the fifteen statements. One can examine the specific course format to help explain some of the variation in ratings. For example, responses to statement №2, dealing with effective presentation style, were lowest when there were challenges with the video conferencing compatibility. Student responses to statements six and seven relating to stimulating interest and thinking were higher when the course was taught only onsite. Students were also encouraged to provide open-ended comments, and included among these were the desire for laboratory exercises and an attendance/participation grade.

Overall these student course evaluations are higher than the average for comparable courses at Baylor, suggesting that the Vietnamese students enjoyed and believe they benefited from the course. An evaluation of student success on achieving the course learning objectives, and overall student grades, supports that students enrolled in the class largely accomplished each of the four learning objectives.

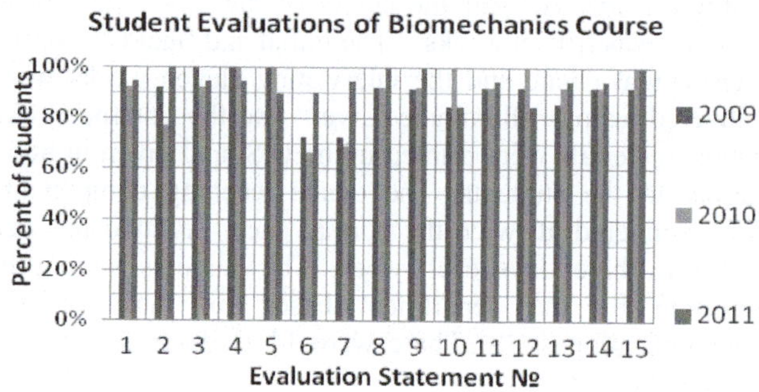

Fig. 1 Student course evaluation summary.

V. Conclusions

During the past three years Vietnamese students from two different Vietnamese universities have demonstrated the capacity to study, learn, and master biomechanics topics when taught in English by a foreign professor. This occurred regardless of the delivery modality, ranging from distance learning to accelerated onsite instruction. Each of these formats possesses limiting challenges.

Acknowledgment

This work was supported under a U.S. Faculty Scholar grant from the Vietnam Education Foundation, Baylor

University, the Hanoi University of Science and Technology, and VNU-HCMC-International University.

References

1. McNamara, L, et. al. (2010) VEF Achievements and Activities: Strengthening the Relationship between the United States and Vietnam through Educational Exchanges… - 2003-1010.
2. Director, S, Doughty, P Gray, P Hopcroft, et al. (2006) Observations on undergraduate education in computer science, electrical engineering, and physics at select universities in Vietnam, Washington DC.
3. Alfen, N Angle, J Gamble, H, Hashimoto, A Wang, J McNamara, L Nguyen, P (2007) Observations on the current status of education in the agricultural sciences in Vietnam, Vietnam Education Foundation and the National Academies of the United States, Washington DC
4. Petrochenkov, M (2007) Opportunities for enhancing STEM education in Vietnam, Vietnam Education Foundation and the National Academies of the United states, August, Hanoi, Vietnam
5. Michalak, M., Luan, P. (2009) U.S.-Vietnam Education Task Force-Final Report, U.S. Dept of State and the Vietnam MOET
6. Vo, T (2004) "Needs Assessment for Biomedical Engineering in Vietnam," A report submitted to the NSF, 38 pps
7. Webster, J, "Biomedical Engineering in Vietnam Today," IEEE Engr in Medicine and Biology Magazine, pp 7-8, May/June 2005
8. Vo, T (2005) The First International Conf on the Development of Biomedical Engineering in Vietnam, NSF (INT-0243803).
9. Hoang, C, Phan, D, Vo, T., Phan, R, (2007) BME 2007, Proceedings of the Second International Conf on the Development of Biomedical Engineering in Viet Nam, Bach Khoa Publishing House, 359 pps
10. Kelley, B, Rigby, B Vu, H (2009) Teaching engineering biomechanics in Vietnam ASEE GSW Proc. Waco, Texas, 2003 TD1-3
11. Kelley, B, (2009) Teaching Engineering Biomechanics at the Hanoi University of Technology, VEF Final Report, 31 pps
12. Kelley, B, Rigby, B, Vu, H (2010) VEF-Sponsored HUT Biomechanics Course, BME 2010 in Vietnam, in: New Frontiers in Biomedical Engineering, IFMBE Proc Vol 7403, pp 246-249.
13. Kelley, B (2010) Teaching Engineering Biomechanics at the VNU-HCMC-International University, VEF Final Report, 26 pps.
14. Werk, S (2009) Young Alumni Making a Difference: Rajesh Pandey '93, Mercerian, Fall 2009, pp 16-17.
15. Kelley, B, Dittenbar, D, Ayers, S, Gonzalez, R, Goldberg, J., Malkin, R. (2008) Innovative Biomedical Design Competitions and Scenarios Exploring Underserved Populations. Proc. 2008 National Collegiate Inventors & Innovators Alliance Conf, Dallas, Tex, pp 125-131

Author: Benjamin S. Kelley
Institute: Baylor University
Street: One Bear Place #97356
City: Waco, Texas 76798
Country: United States
Email: Ben_Kelley@Baylor.edu

Comparative Study of the Different Outsole Effect on Badminton Movement

S.D. Li[1], Y.D. Gu[1], Y.C. Lu[2], and J.S. Li[1]

[1] Faculty of Sports Science, Ningbo University, Ningbo, China
[2] Institute of Physical Education, Zhejiang Wanli University, Ningbo, China

Abstract — **This study focuses on an typical of the most commonly used footwork of pushing to catch a ball across the front, and analyze impact, friction force during touchdown. Meanwhile, different sport shoes' biomechanical features were compared, and the effect of badminton's biomechanical characteristics of the lower limbs on the badminton shoes and its possible mechanism was further explored. The results found that comparing with ordinary shoes, professional badminton shoes have better ground grip ability, but the cushioning property are in the same level. Although this is a pilot study, professional badminton shoes still reflecting certain advantages in improving performance.**

Keywords — **Outsole, badminton, biomechanics.**

I. Introduction

Badminton is very popular in all age groups as a physical exercise and has an increasing trend years by years [1]. To further improve the badminton's athletic ability and training level and prevent injury of badminton to ensure its sustained development, we must do biomechanical research of badminton and understand body characters helping guidance of daily training. The characters of high intensity and long motion bumping of badminton lead to that it is easy injured in the lower limbs of badminton players. Yu et al. [2] analyzed the injured occasions of 56 players of 3 badminton's teams in Guangzhou area, finding that foot and ankle injury ranked second. If merging the thigh, shank, knee, foot and ankle's injury, the lower limbs' injury is as high as 47.16% [2]. Although this investigation has certain regional, there is still certain representativeness. Besides, there are lots reports about the effect of sport shoes on motion [3-5]. Some studies detailed analyzed the evolution and development of sport shoes' study, and pointed out that sport shoes can really change the motion's technological features. The study about sport shoes also showed that sport shoes have enormous influence on the lower extremity's load and kinematics [6].

Because foot is always being up-stair load condition in a badminton match and it has high risk of being injured, it asks players to buy and use special shoes which can increase athletic level and prevent the own injury and professional disease. Simth and Bunch [7] consider that special shoe must possess two important properties of preventing injury and improving performance. To sum up, sport shoes' property is placed major emphasis on two aspects. Firstly, suffice the functional requirement of the motion. Secondly, player's physical consumption should be decreased while the prevention to foot's skeleton and tendon.

Based on the present situation and developing trend, this study focuses on an example of the most commonly used footwork of pushing to catch a ball across the front, and applies modern biomechanics to analyze the biomechanical properties of foot reaction at touchdown. Meanwhile, different sport shoes' biomechanical features were compared, and the effect of badminton's biomechanical characteristics of the lower limbs on the badminton shoes and its possible mechanism were further explored. And we hope to provide theoretical basis for research of badminton special shoes.

II. Methods

The six subjects participated in this study are badminton players from sports department of Ningbo University. Testing shoe A is professional badminton shoe. Testing shoe B is common sports shoe (Fig.1). This study focuses on a testing motion of the most commonly used footwork of pushing to catch a ball across the front (Fig.2). Experimental instrument uses Kistler force platform (Kistler, Germany). The subjects' front foot was asked to be located in the center of force platform in the process of data collection. When everything is prepared, the subject gives commands and instructions, subjects try their best to step on the force platform actively with the leg behind stepping to right front to pick up ball in front of the net. The testing order of two kinds of sport shoes is random. The study mainly examine the landing foot after taking a step initiatively at full stretch and explore the plantar vertical force and crosswise friction of the initiative front foot in the movement. Data derived from Kistler dynamometric system is analyzed by SPSS.

V. Van Toi et al. (Eds.): 4th International Conference on Biomedical Engineering in Vietnam, IFMBE Proceedings 40, pp. 126–128, 2013.
www.springerlink.com

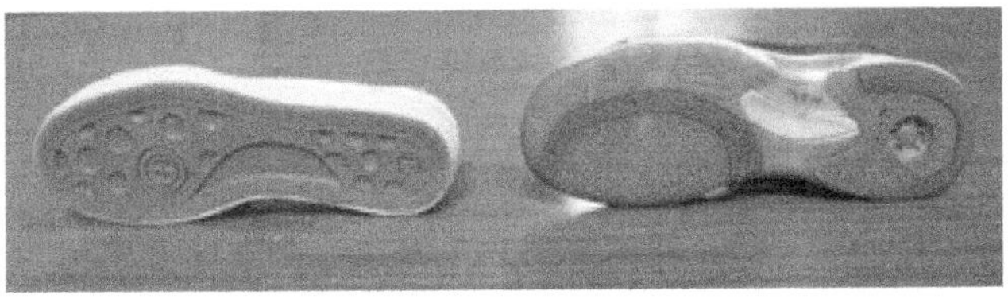

Fig. 1 The outsole patterns of two different shoes (left: ordinary; right: professional badminton shoes)

Fig. 2 Illustration of push forward step in badminton. (major sub-steps are: initial contact, heel contact, and whole foot contact).

III. Results

Fig.3 showed that testing shoe A's vertical reaction peak force turned up later than testing shoe B, among the tested data, while foot vertical peak force is smaller than testing shoe B, but not exist significant difference. The contact time of testing shoe A's movement completion is 0.782s, and testing shoe B is 0.651s, also the testing shoe A is significantly greater than testing shoe B. It is concluded that testing shoe A spends more time in undertaking relatively smaller force while testing shoe B is absolutely opposite, but can't demonstrates testing shoe A's cushion in vertical direction is better than testing shoe B's.

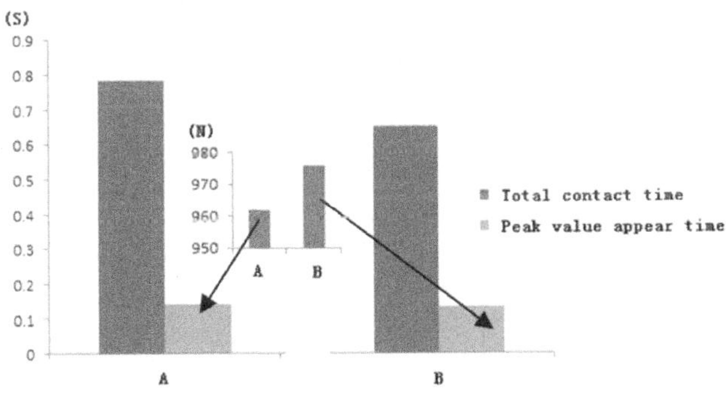

Fig. 3 The peak vertical reaction force and appearing time during the whole movement completion.

Crosswise friction is the force generated in horizontal direction at contact when doing the footwork of pushing to catch a ball cross the front. Fig.4 showed that testing shoe A's crosswise friction is significantly larger than testing shoe B's, which means shoe A is able to provide more ground grip ability.

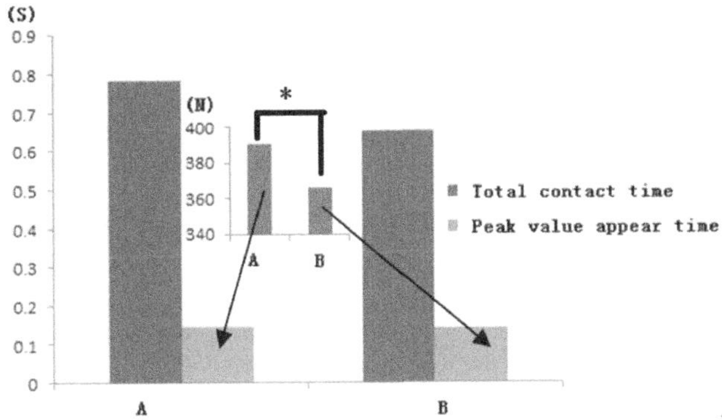

Fig. 4 The peak horizontal friction force and appearing time during the whole movement completion (*p<0.01).

IV. Conclusions

With the gradual maturity of badminton's technology, the special shoe's effect on match's performance is becoming more and more important. Although there is no direct physical contact, the badminton's complexity, emergentness and persistence are all very strong. Because the player should do sundry complex technical motion constantly on the field, such as a sudden stop, starting, reversing, taking off, quick running forward, left, right and back and torso extension, lateral bending, turning around, antexion, winding up to hit a ball and so on, the player should possess the technology of flexible moving footwork, good sudden stops and taking off. However, this technological motion's truth is based on the interaction between sport shoe and ground.

Badminton's cushion is an important index. Shiang [3] thinks that the most important factor of influence on comfort is the shock absorber performance and foot cushion. However, from the experimental results finding in this piloting study, the professional shoes' cushioning property not distinctively better than ordinary training shoes. Special shoe can complete various kinds of technological motion better in a drastic match, and make the touchdown's time of ground reaction force longer which contributes to relatively mitigate the impact force to foot to prevent the lower limbs to be injured.

The results suggest that professional badminton's shoe has certain advantages in preventing injury and improving performance through the good ground grip ability. The cause is likely that material and stripe pattern in the professional badminton's shoe sole can better increase friction between shoe sole and ground.

Acknowledgment

This study is supported by Ningbo University's SRIP project.

References

1. Wei Y. (2009) Shoes on metatarsophalangeal joint and rearfoot motion in typical badminton footwork. China sport science 29: 16-18.
2. Yu X. (2005) The study of athletic injuries for badminton athlete in Guangzhou. Journal of PLA institute of physical education 24:100-102.
3. Shiang T.Y., (1997) The nonlinear finite element analysis and plantar pressure measurement for various shoe soles in heel region. Proc Nat Sci Counc Repub China B 21: 168-174.
4. Robbins S.E., Gouw G.J. (1990) Athletic footwear and chronic overloading, A brief review. Sports Med 9:76-85
5. Li J.S., Gu Y.D. Ren X.J. et al (2010) Biomechanical effects of foam inserts on forefoot load during the high-heeled gait: a pilot study. Journal of mechanics in medicine and biology 10: 667 671.
6. Esenyel M., Walsh K., Walden J.G. et al. (2003) Kinetics of high-heeled gait, J Am Podiatr Med Assoc 93: 27-32
7. Smith L.S., Bunch R. (1986), Athletic footwear. Clin Podiatr Med Surg. 3:637-647.

Author: Jianshe Li
Institute: Ningbo University
Street: No.818, Fenghua Road
City: Ningbo
Country: China
Email: lijianshenbu@yahoo.com.cn

Development and Validation of a Gas Exclusion Technique to Allow Sterile Procedures in Non-sterile Environment

M.S. Salehi Dashtbayaz, B. Pingguan-Murphy, A. Ahmady, and E.K. Moo

Department of Biomedical Engineering, Faculty of Engineering, University of Malaya, Kuala Lumpur, Malaysia

Abstract — **Tissue Engineering is one the most encouraging form of treatment for patients with advanced forms of articular cartilage degeneration today, and remains under active development. The extraction of required samples for effective TE research is complicated by the absolute need for sterility during the biopsy procedure. However, there are situations in which a standard bio-safety hood is not able to provide sterility for a required operation, particularly in the case of large subject matter which will not fit within the hood without disturbing laminar hood flow or in the case where large pieces of equipment such as powered tools are required. This gives rise to the necessity of an alternative method of sterile biopsy. The main aim of this study is the development and evaluation of a technique based on using positive pressure nitrogen gas in order to provide a locally and temporary sterile environment, which can be suitable in a general laboratory or clinical setting and also may be applicable in emergency applications, outdoor surgery and so on.**

Keywords — **Tissue Engineering, Articular cartilage, Nitrogen, Sterility, Bio-safety hood.**

I. Introduction

A Synovial joint is a complex unit which provides painless stable movement for living things, with a low level of friction, and is usually able to function for a lifetime. The extraordinary functional capacities in this unit particularly derive from articular cartilage, the tissue that forms the bearing surface of every synovial joint . Unlike other kinds of tissue, cartilage tissue has a low level of metabolic activity; therefore, it has very little response to any change caused during loading or injury. However, when articular cartilage is damaged, not only does it not heal, but it also often degenerates further; a degeneration that will lead to pain and loss of function of the joint. Despite many advances in treatment for patients with advanced forms of articular cartilage degeneration, the reconstruction of a state equivalent to that of healthy undamaged cartilage remains to be achieved. The most promising form of treatment under active development today is the Tissue Engineering solution, essentially comprising the biopsy of cartilage, the isolation and expansion of the cells therein, the placing of those cells within a scaffold material, and the re-implantation of the scaffold to affect a fully biological repair.

Tissue engineering has been employed as an alternative method for the repair of defective articular cartilage . Several surgical procedures for cartilage treatment, some of which involve tissue transplantation, have been performed by clinics in recent years. One of these procedures is termed autologous chondrocyte implantation (ACI), a process made possible by the isolation of autologus articular chondrocytes by biopsy from excised non loading-bearing tissue and subsequent in vitro expansion in an appropriate culture medium.

There are three main constituents which generally form the basis of tissue engineering for cartilage: A cell source; A signaling strategy (both biochemical and/or mechanical) and appropriate scaffold material.

Unfortunately, the extraction of the required samples for effective research is complicated by the absolute need for sterility during the biopsy procedure. Explants should be dissected using an appropriate protocol to develop and avoid contamination and retain cell survival. The level of sterility needed is somewhat in excess of that required for routine surgical operations, because in this case biopsy must survive incubation in a high growth medium for some time, in the absence of the immune system. Even a trace of bacterial, viral or fungal pathogen is likely to grow unabated until the biopsy must be disposed of. For this reason it is usual that, in the research context, the biopsy is undertaken within a sterile laboratory environment, usually by means of a class II biosafety hood equipped with microbiological filtering and laminar flow system.

Regrettably there are situations in which such a biosafety hood is not able to accommodate the required operation, particularly in the case of large subject matter which will not fit within the hood without disturbing laminar hood flow, but also in the case where large pieces of equipment such as powered tools are required for specific forms of biopsy procedure.

V. Van Toi et al. (Eds.): 4th International Conference on Biomedical Engineering in Vietnam, IFMBE Proceedings 40, pp. 129–131, 2013.
www.springerlink.com

Accordingly, the main aim of this study is the development and evaluation of a technique based on using positive pressure nitrogen gas in order to provide a localised and temporary sterile environment, which can be employed in a general laboratory or clinical setting, and also may find future application applicable in emergency situations, outdoor surgery, agricultural veterinary care and so on.

Material and Methods: The basic concept to be developed is to use a pre-existing source of sterile nitrogen gas to produce an enclosed zone of positive pressure at the region of dissection, in a way that any pathogens from the region are carried away from the incision and any pathogens in the surrounding area are prevented from entering via diffusion by means of a dominant counter-active convection. Nitrogen gas was selected for this study because it is an a diatomic molecular gas (N2) which is not very reactive.

This method presents the series of procedures used for the preparation of articular cartilage explants. Fresh articular cartilage were be used for this study in order to obtain more viable cells. Therefore, bovine feet were obtained on day of slaughter from a local abattoir. Feet were washed and then were put into alcohol (Ethanol C_2H_5OH) 70 % (v/v), for at least 15 minutes. Each foot was put in a stainless steel tray and sprayed with alcohol, then transferred to a sterile laminar hood (class II) in order to removing skin of the foot by using scalpel and expose the articular surfaces .

The exposed metatarsus was wrapped with sterilized aluminum foil. The joint was placed in an autoclaved plastic bag. Before tying the plastic bag, one end of a sterile plastic tube was placed inside the plastic bag while another end was left free (Figure 1). The open end of the pipe will be used to allow the transfer of nitrogen gas to the joint. The assembly was then transferred from the clean laminar hood to the open air.

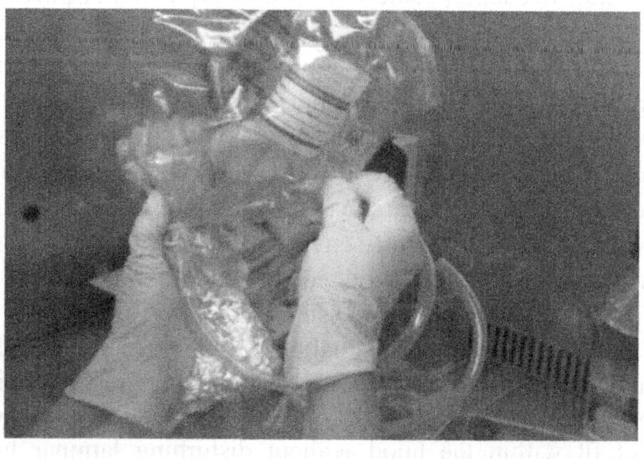

Fig. 1 Metacarpal-phalangeal joint, sterile universal container and sterile plastic pipe was placed inside the autoclaveble plastic. A cable tie was used to tighten the bag. One end of the pipe was left free.

A custom-built glass bubble column including two coaxial glass columns was used as a gas delivering system in this research. The rehydration of the nitrogen gas was the main reason why the nitrogen gas delivering system was built in this form.

The metacarpal-phalangeal joint, which was wrapped in an autoclaved plastic sheet (Figure 2), was placed in a clamp (size 6) and tightly secured for more fixation. The loose end of the pipe from the bag was connected to the gas delivering system. The purified Nitrogen gas was allowed to flow through the nitrogen gas delivery system and autoclaved plastic was inflated with humidified nitrogen gas which caused the expansion of the bag. A small hole was created in the plastic bag and by using a punch and a hammer the expants were derived from the the metacarpal-phalangel joint surface.

Fig. 2 The hammer was placed on top the punch in order to obtain an articular cartilage explant.

All necessary equipment, such as a hammer, punch, scissors and forceps were autoclaved prior to use. The articular cartilage explant was retrieved by using sterile forceps which were already in the bag, and the explant was put into the sterile container (also already within the bag). About 4 explants were extracted from each joint. At the end of the study, the container cover was tightened and transferred to the laminar hood. This was followed by a series of sterility tests in order to confirm that the proposed nitrogen gas delivery system was bacteria free.

In order to allow the microorganisms to grow, the growth media or agar was inoculated with a sample and then put in an incubator. Four different kinds of agar were used as culture media in this study. After inoculation, these were incubated at 37° C with enhanced Co_2 for at least 48 hours in order to allow the microorganisms to growth. The

Chocolate Agar + horse Blood, MacConkey, Sabouraud Dextrose, and Columbia Blood Agar with 5% Horse Blood agar plates were removed from the incubator after 48 hours to check for colony formation.

Results and Discussion: The plates were checked visually for colony formation. These experiments were repeated two times per week during a period of three weeks. In total 48 samples which were extracted from 12 synosival bovine joints (each joint yielding four samples). The results of the sterility test show no colony formation on any of the 48 agar plates, which means there is no observed infection in this technique at all.

As a control group, a series of articular cartilage explants were extracted in the open air without the use of nitrogen exclusion. Explants were placed in a sterile container and washed in PBS. After inoculation, the 4 types of agar were left inside the incubator as before, in order to allow the microorganisms to grow. 3 plates out of at total of 4 inoculated showed colony formation, revealing infection acquired during extraction of the explant in the open air.

Conclusion: Because of the importance of explants for tissue engineering research and articular cartilage treatment a simple and easily accessible extraction method is of research value. In this study of the effectiveness of a gas exclusion technique, results indicate a the use of Nitrogen gas to provide a convenient means to ensure sterility within a working environment outside of the controlled environment of a biosafety hood.

References

1. Buckwalter, J., et al., Articular cartilage: composition and structure, in Injury and Repair of the Musculoskeletal Soft Tissues, S. Woo and J. Buckwalter, Editors. 1988, IL: American Academy of Orthopaedic, Surgeons: Park Ridge. p. 405-425.
2. Buckwalter, J., L. Rosenberg, and E. Hunziker, Articular cartilage: composition, structure, response to injury, and methods of facilitation repair, in Articular Cartilage and Knee Joint Function: Basic Science and Arthroscopy, E. JW, Editor. 1990, Raven Press: New York. p. 19-56.
3. Buckwalter, J. and H. Mankin, Tissue design and chondrocyte- matrix interactions. Bone Joint Surg, 1997. 79A(4): p. 600-611.
4. Gevork, L.E., The tissue-engineering of functional osteochondral constructs for cartilage repair. 2008, Columbia University. p. 296.
5. Lohmander, S., Tissue Engineering of Crtilage and Bone. Tissue Engineering of Cartilage : do we need it, can we do it , is it good and can we prove it? 2003, West Sussex: John Wiley & Sons Ltd.
6. Brittberg, M., Autologous chondrocyte implantation-technique and long term follow-up. Injury International Journal of the Care of the Injured, 2008. 39(Supp!): p. S40-S49.
7. Dario, O. and M. Fauza, Tissue Engineering:Current State of Clinical Apllication. Current Opinion in Paediatrics, 2003. 15(2): p. 67-71.

Author: Massoumeh-Sepideh Salehi Dashtbayaz
Institute: Department of Biomedical Engineering, Faculty of Engineering, University of Malaya
City: Kuala Lumpur
Country: Malaysia
Email: sbiomecial.um@gmail.com

Development of a Computer-Controlled Tactile Stimulator with Simultaneous Visual and Auditory Stimulation

Hyung-Sik Kim, Mi-Hyun Choi, Ji-Hye Kim, Hong-Won Yeon, Jeong-Han Yi, and Soon-Cheol Chung*

Department of Biomedical Engineering, College of Biomedical and Health Science, Research Institute of Biomedical Engineering, Konkuk University, Chungju-si, South Korea

Abstract — **In this study, a computer-controlled tactile stimulator was developed, which can stimulate visual and auditory senses simultaneously by using the E-Prime software. The newly developed system consists of three units: a control unit, a drive unit and a vibration unit. Since the developed system is a small, lightweight with low electrical consumption, a maximum of 35 stimulation channels and various visual and auditory stimulation combinations. The system was designed to stimulate the fingers. The stimulator is expected to be highly practical due to a diverse combination of stimuli, such as tactile–visual, tactile–auditory, visual–auditory and tactile–visual–auditory stimulation with computer-control.**

Keywords — **Computer-control, Tactile stimulator, Simultaneous stimulation, Tactile-Visual-Auditory.**

I. INTRODUCTION

To study the human five senses such as sight, hearing, touch, smell and taste, numerous stimulators were developed last years. Due to these developments, the auditory and visual stimulation fields achieved good results, but it has been difficult for the tactile stimulation field to achieve effective and quantitative stimulation control. For tactile stimulation, mechanical, electrical, and thermal methods commonly are used [1]. The mechanical method uses vibrations from moving-coil actuators [2-4], vibration motors [5-6], RC servomotors [7] and piezoelectric element [8]. The electrical method uses electrostatic and electrical stimulators [9]. The thermal stimulation system uses the peltier elements [10] and Shape memory alloy [11]. A previously developed tactile stimulator had differences and several disadvantages: 1) Problems related to system size, weight, complexity of structure, availability, electricity consumption, number of stimulation channels, and cost and problems related to simultaneous stimulation with visual and auditory stimuli; 2) problems related to stimulation-control, such as the intensity, duration, and frequency of stimulation; and 3) several additional problems, such as pain, discomfort, and stability to the human body.

* Corresponding author.

Therefore, this study strove to develop a tactile stimulator that would correct these systematic, stimulation control, and additional problems. It was also developed to control and generate simultaneous visual, auditory, and tactile sensations using E-Prime software.

II. MATERIALS AND METHODS

Figure 1 shows the proposed computer-controlled tactile stimulation system. The system is operated using E-Prime software. Therefore it can provide various combinations of visual and auditory stimulation with simultaneous tactile stimulation.

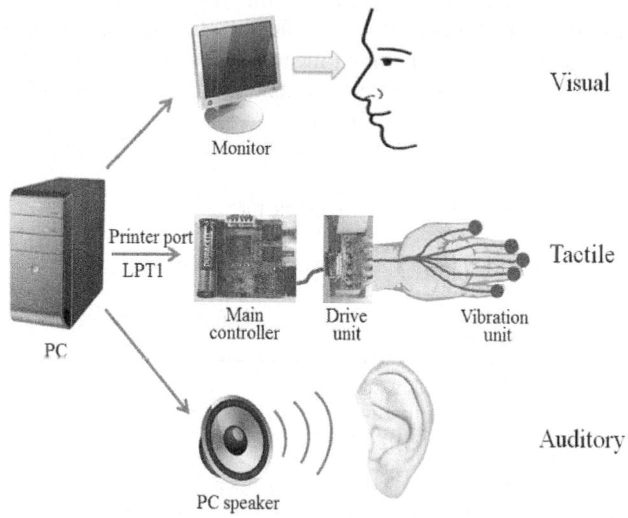

Fig. 1 Computer-controlled tactile stimulation system.

This system consusts of three parts: the main controller, drive unit, and vibration unit.

The main controller uses an AVR-type ATMEGA128 processor (Atmel, USA), which is a general-purpose 8-bit microprocessor. This processor works at a 4.5–5.5V operating voltage, and it has several capabilities, such as an 8-bit and 16-bit timer counters.

V. Van Toi et al. (Eds.): 4th International Conference on Biomedical Engineering in Vietnam, IFMBE Proceedings 40, pp. 132–134, 2013.
www.springerlink.com

The main controller and PC are connected to the printer port (LPT1), and through this channel the stimulation command is received from the E-Prime software. The frequency of the stimulation signal can be changed from 100 to 400 Hz and the level of stimulation was controlled by manipulating the duty rate of pulse width modulation (PWM), which was generated by microprocessor, using the software. The operator can handle the duration of vibration and stimulation time using digital switch. The duration of vibration can be changed from 0.5 to 5 s with 0.5 s intervals. The strength of stimulation can be controlled up to 10 levels (Figure 2).

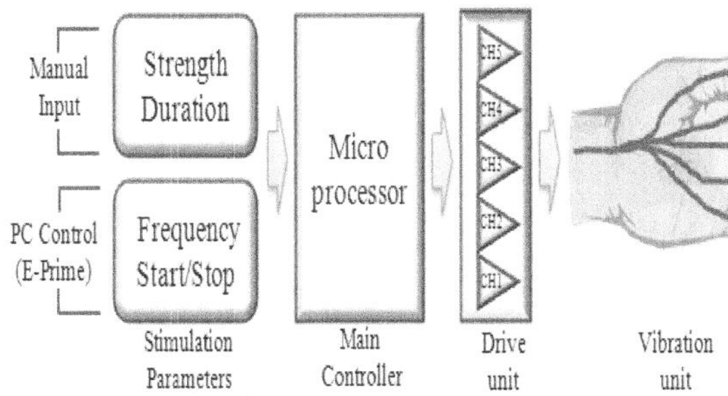

Fig. 2 Block diagram of a tactile stimulator.

The drive unit operates the vibration motor in response to commands from the main controller. For stimulate five fingers, it was designed to have a five stimulation channels and it is possible to extend the unit up to 35 channels.

The vibration unit uses a small, coin-type vibration motor (LG Innotek Co., Ltd., Korea; 10.5mm in diameter). Each vibration unit is attached to the tip of the finger using Velcro.

III. Results

The entire system operates with a single AAA (1.5 V) battery and the main circuit operates with a +5V source voltage using a step-up circuit. No discomfort during the experiment was reported because of compact size of the main controller.

Simultaneous tactile stimulation with visual and auditory stimulation was verified experimentally. Tactile and auditory stimulation was measured from the starting time of its response, and visual stimulation was measured using an oscilloscope from the transition point from low signal to high signal using a cadmium sulfide (CDS) cell, optic sensor and transistor switch. Figure 3 showed that it was possible to apply simultaneous (Figure 3(a)) or non-simultaneous (Figure 3(b)) tactile, visual and auditory stimulations.

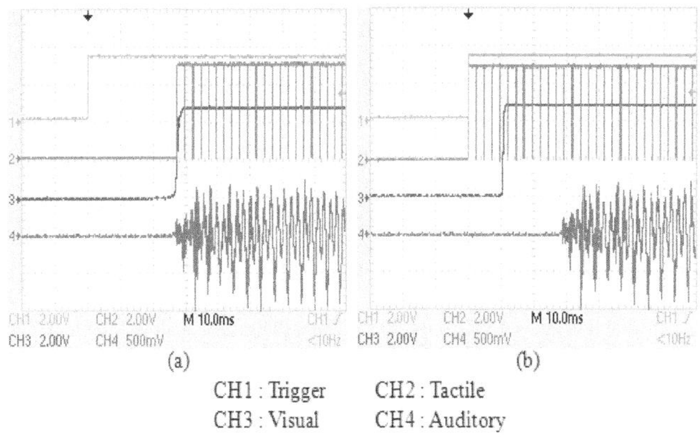

Fig. 3 Simultaneous and non-simultaneous tactile, visual and auditory stimulation.

IV. Discussion and conclusions

In this study, a tactile stimulator was developed that can simultaneously stimulate visual and auditory senses using the E-Prime software. This novel tactile stimulator has several advantages related to systematic and stimulation control: the compact and lightweight device can be carried easily and the simple structure attaches to the subject easily. With widely used, low-cost elements, an optimal design and algorithm make it possible to operate with low power consumption. Additionally, the number of stimulation channels can be extended up to 35. A diverse combination of tactile–visual, tactile–auditory and tactile–visual–auditory stimulation is possible. The strength and duration of stimulation can easily be controllable up to 10 levels using a digital switch and the range of operating frequency (100–400 Hz) increases its applicability.

Since the developed tactile stimulator was based on E-Prime, which is the most widely used software for visual and auditory stimulation, highly practical applicability and accessibility is expected due to the diverse combination of tactile–visual, tactile–auditory and tactile–visual–auditory stimulation. Tactile stimulation at the toe was possible by adjusting the length of the Velcro, and by using adhesive tape, stimulation of any part of the body was possible.

Thus, it can be widely used for various tactile and sensory studies. For further study, a much simpler system can be made with a wireless connection between a PC and a tactile stimulator.

Acknowledgment

This research was supported by the Pioneer Research Center Program through the National Research Foundation of Korea funded by the Ministry of Education, Science and Technology (2011-0027920).

References

1. Kyung KU, Son SW, Kwon DS (2004) Design of an integrated tactile display system, Proc IEEE Int Conf Robot Autom. 776-781
2. Pawluk DTV, Buskirk CP, Killebrew JH, Hsiao SS, Johnson KO (1998) Control and pattern specification for high density tactile display, Proc ASME Dyn Syst Control Division ASME Int Mech Eng Congr Expo. 64:97-102.
3. Graham SJ, Staines WR, Nelson A, Plewes DB, Mcilroy WE (2001) New devices to deliver somatosensory stimuli during functional MRI, Magn Reson Med. 46:436-442
4. Khoudja MB, Hafez M, Alexandre JM, Kheddar A, Moreau V (2004) VITAL: A VIbroTActiLe interface with thermal feedback, Proc IEEE Int Conf Robot Autom
5. Wagner CR, Lederman SJ, Howe RD (2002) A tactile shape display using RC servomotors, 10th Symp Haptic Interface Virtual Env Teleoperator Syst. 354-356
6. Drewing K, Fritschi M, Zopf R, Ernst MO, Buss M (2005) First evaluation of a novel tactile display exerting shear force via lateral displacement, ACM Trans Appl Percept. 2:118-131
7. Stetten G, Klatzky R, Nichol B, Galeotti J, Rockot K, Zawrotny K, Weiser D, Sendgikoski N, Horvath S (2007) Fingersight: Fingertip visual haptic sensing and control. IEEE Int Workshop on Haptic Audio Vis Env Appl. 80-83
8. Han JM, Adriaens TA, de Koning WL, Banning R (2000) Modeling piezoelectric actuators, IEEE/ASME Trans Mechatron. 5:331-341
9. Jungmann M, Schlaak HF (2002) Miniaturized electrostatic tactile display with high structural compliance, Proc Eurohaptics Int Conf. 12-17
10. Ino S, Shimizu S, Odagawa T, Sato M, Takahashi M, Izumi T, Ifukube T (1993) A tactile display for presenting quality of materials by changing the temperature of skin surface, Proc IEEE Int Workshop Robot Human Commun. 220-224
11. Velazquez R, Pissaloux EE, Wiertlewski M (2006) A compact tactile display for the blind with shape memory alloys, Proc 2006 IEEE Int Conf Robot Autom. 3905-3910

Author: Soon-Cheol Chung
Institute: Research Institute of Biomedical Engineering, Konkuk University
City: Chungju-si
Country: South Korea
Email: scchung@kku.ac.kr

Associations between Cultural Characteristics and Antibacterial Activities of Streptomyces Isolates: A Preliminary Analysis

Nor Asmara Tasrip[1], Mohd Nasir Mohd Desa[1,3], Cheah Yoke Kqueen[1], Chong Pei Pei[1], and Mariana Nor Shamsudin[2,4]

[1] Dep. of Biomedical Sciences and [2] Dep. of Medical Microbiology & Parasitology,
Faculty of Medicine & Health Sciences, Universiti Putra Malaysia, Selangor, Malaysia
[3] Halal Products Research Institute, Universiti Putra Malaysia, Selangor, Malaysia
[4] Institute of Bioscience, Universiti Putra Malaysia, Selangor, Malaysia

Abstract — **40 *Streptomyces* isolates were recovered from soils taken from various sites around Universiti Putra Malaysia (UPM) and nearby peripheral areas, including 4 from a far-off location. Identification was done by conventional and molecular approaches (*16S rRNA* and *rpoB* genes sequence homology). All isolates were subjected to antibacterial assay against 2 pathogenic bacteria of gram positive and gram negative (*Escherichia coli* ATCC 25922 & *Staphylococcus aureus* ATCC 25923), by perpendicular streak method using MHA and TSA as the test media. With MHA as the test media in the antibacterial assay, > 50 % of the isolates showed antibacterial activity to at least one of the test bacteria as compared by using TSA as the test media. Color series of aerial mycelium was described based on growth on ISP-2 medium; there was a significant association between grey and non-grey color series of the aerial mycelium in association with the antibacterial activity of the isolates on MHA (89% vs. 42% respectively, P = 0.013) and TSA (78% vs. 35% respectively, P = 0.025). As a conclusion, different chemical composition of the test media could have an effect on the outcome of the antibacterial assay (MHA vs. TSA). The observed relationship between aerial mass-color (grey vs. non-grey) and antibacterial activity warrant further investigation in a larger sample of isolates.**

Keywords — ***Streptomyces*, antibacterial activity, aerial mycelium, color.**

I. Introduction

Streptomycetes are filamentous gram-positive bacteria which are responsible for the musty odor of soil. They are well known as the producer of various bioactive metabolites of commercial value particularly the antibiotics. Given the diversity of *Streptomyces* spp., identification of phenotypic markers associated with antibacterial activities may help screen for useful isolates with potential novel antibacterial properties.

This study was conducted to examine the antibacterial activities of *Streptomyces* isolates from local soil and potential linkage with their cultural characteristics.

II. Materials and Methods

40 *Streptomyces* isolates were recovered from soils by standard serial dilution method using ISP-2 medium. Soils were taken from various sites around Universiti Putra Malaysia (UPM) and nearby peripheral areas, including 4 from a far-off location. Identification was done by conventional **[1,2]** and molecular approaches (*16S rRNA* and *rpoB* genes sequence homology) **[2, 3]**.

All isolates were subjected to antibacterial assay against 2 pathogenic bacteria of gram positive and gram negative (*Escherichia coli* ATCC 25922 & *Staphylococcus aureus* ATCC 25923), by perpendicular streak method using Mueller Hinton Agar (MHA) (Merck, Darmstadt, Germany) and Tripticase Soy Agar (TSA) (Merck, Darmstadt, Germany) as the test media.

Antibacterial activities of each isolate on MHA and TSA were analyzed accordingly and compared. Chi-square (χ^2) analysis in a 2 x 2 contingency table was then performed to compare isolates with and without antibacterial activity on the respective test media in association with their aerial mycelium color series, substrate mycelium, and spore production ($P < 0.05$ for significant level), using the online tool at *http://www.graphpad.com/quickcalcs/contingency1.cfm.*

III. Results and Discussion

With MHA as the test media in the antibacterial assay, > 50 % of the isolates showed antibacterial activity to at least one of the test bacteria. However, that on TSA was < 50%. Based on growth on ISP-2 medium, the series of aerial mass-color were yellow (n = 15), grey (n = 9), brown (n = 7), white (n = 6), black (n = 1), orange (n = 1) and peach

V. Van Toi et al. (Eds.): 4th International Conference on Biomedical Engineering in Vietnam, IFMBE Proceedings 40, pp. 135–136, 2013.
www.springerlink.com

(n = 1). There was a significant association between grey and non-grey series of the aerial mycelium in association with the antibacterial activity of the isolates on MHA (89% vs. 42% respectively, P = 0.013) and TSA (78% vs. 35% respectively, P = 0.025), while those of the other categorical variables were negligible. This study was in line with a finding that showed *Streptomyces* isolates with grey-color series exhibiting more antibacterial activities [4].

Table 1 Antibacterial activity of *Streptomyces* isolates on MHA.

Colour P = 0.013	Both[1]/either one[2] (N = 21)	None[3] (N = 19)
Grey (N = 9)	8 (89%)	1 (11%)
Others (N = 31)	13 (42%)	18 (58%)

Table 2 Antibacterial activity of *Streptomyces* isolates on TSA.

Colour P = 0.025	Both[1]/either one[2] (N = 18)	None[3] (N = 22)
Grey (N = 9)	7 (78%)	2 (22%)
Others (N = 31)	11 (35%)	20 (65%)

[1] Both = inhibition of both test bacteria.
[2] Either one = inhibition of either one of the two test bacteria.
[3] None = no inhibition.

IV. Conclusions

Different chemical composition of the test media could have an effect on the outcome of the antibacterial assay (MHA vs. TSA). The observed relationship between aerial mass-color (grey vs. non-grey) and antibacterial activity warrant further investigation in a larger sample of isolates.

References

1. Shirling EB, Gottlieb D (1966) Int J Syst Bacteriol 16: 313-340.
2. Duraipandiyan V, Sasi AH, Islam VIH et al. (2010) J Med Mycol 20: 15-20.
3. Kim BJ, Kim CJ, Chun J et al. (2004) Int J Syst Evol Microbiol 54: 593-598.
4. Malibari AA (1991) J King Abdul Aziz Univ: Sci 3: 31-42.

Corresponding author:

Mohd Nasir Mohd Desa, PhD
Department of Biomedical Sciences
Faculty of Medicine & Health Sciences
Universiti Putra Malaysia
43400 UPM Serdang, Selangor
Malaysia
Email: nasirdes@medic.upm.edu.my

Effect of Defatted Dabai (Canarium Odontophyllum Miq.) Pulp Ingestion on Lipid Peroxidation and Antioxidant Status of Hypercholesterolemic-Induced Rabbits

Azrina Azlan[1,2], Amin Ismail[1,2], Zulkhairi Amom[3], and Faridah Hanim Shakirin[1]

[1] Department of Nutrition and Dietetics, Faculty of Medicine and Health Sciences, 43400 UPM Serdang, Selangor, Malaysia
[2] Laboratory of Halal Science Research, Halal Products Research Institute, Universiti Putra Malaysia, 43400 UPM Serdang, Selangor, Malaysia
[3] Department of Human Anatomy, Faculty of Medicine and Health Sciences, 43400 UPM Serdang, Selangor, Malaysia

Abstract — **Antioxidant activity of Dabai (*Canarium odontophyllum* Miq.) pulp polyphenols was observed in animal model to investigate its potential in reducing *in vivo* lipid peroxidation. Dried fullfat and defatted Dabai pulp were prepared and administered daily to hypercholesterolemic-induced rabbits (0.5% dietary cholesterol) at 5% of purified diet, besides a group receiving statin (10mg/kg/day) for 56 days. All groups were induced with hypercholesterolemia except in negative control. At the end of experimental period, the *in vivo* antioxidative capacities of Dabai pulp were evaluated using assays; TBARS, SOD, GPx and TAS in whole blood and plasma samples of animals. Results indicated that defatted- and fullfat-Dabai pulp significantly reduced 36% and 15% plasma MDA (relative to without treatment [PC]), respectively while only 4% insignificant reduction was due to statin. On the contrary, plasma TAS levels were 76% and 66.2%, significantly higher (than without treatment) attributable to defatted and fullfat Dabai parts supplementation, respectively as compared to 9% insignificant reduction in statin group. Both pulps also caused marked increments in activity of enzymes especially SOD thus suggesting the strong *in vivo* antioxidant capacities of Dabai pulp (defatted, especially) as future active pharmaceutical ingredient.**

Keywords — **Dabai, hypercholesterolemia, rabbit, artherosclerosis, polyphenols.**

I. Introduction

Canarium odontophyllum Miq. (CO) or Dabai is an indigenous fruit of Sarawak, Malaysia. The fruit is consumed by communities in Sarawak and categorized as underutilized. The fruit is highly nutritious and rich in vitamins and minerals [1]. Flesh of CO fruit is also high in phenolics with substantial antioxidant activities as shown by Shakirin et al. and Chew et al. using various assays; 1,1-diphenyl-2-picrylhydrazyl radical (DPPH), β-carotene bleaching, ferric reducing ability (FRAP) and trolox equivalent antioxidant capacity (TEAC) [2,3].

Previous studies also indicated that the high antioxidant activities were highly correlated with total phenolic and flavonoid contents of the fruit pulp. Chew et al. also showed CO pulp consist of appreciable levels of phenolic compounds such as catechin, epigalocatechin gallate and epicathecin [3]. Therefore, this paper aims to explore the *in vivo* antioxidative capacities of dabai polyphenols against CHD in hypercholesterolemic-induced rabbits. The data is useful in revealing the potential of dabai polyphenols and protecting human against negative effect of free radicals in the body.

II. Materials and Methods

A. Materials

Fresh fruits of CO were obtained from Agriculture Research Centre, Semongok, Sarawak, Malaysia. Chloroform, methanol were purchased from Merck (Germany), Cholesterol was from Sigma Chemical Co (USA), Simvastatin from Pharmaniaga Logistic (Malaysia), total antioxidant status (TAS), superoxide dismutase (SOD), glutathione peroxidase (GPx) kits were from Randox Laboratory Ltd (United Kingdom).

B. Preparation of Fruit

The fruit pulp was prepared by separating it from pit manually. Then, the pulp was freeze-dried, ground and sieved to obtain homogenized particles. Later, sample was separated for fullfat pulp and defatted pulp portion. The defatted pulp was prepared by soaking powdered pulp in chloroform-methanol (2:1 v/v) for overnight.

C. Study Design and Biochemical Analyses

Thirty five male New Zealand white rabbits weighing 1.5-1.7 kg of 8-10 weeks old were purchased from East

V. Van Toi et al. (Eds.): 4th International Conference on Biomedical Engineering in Vietnam, IFMBE Proceedings 40, pp. 137–140, 2013.
www.springerlink.com

Asia Company, Malaysia. The animals were placed in individual cages at 25ºC under 12h light/dark cycle and water was provided *ad libitum*. Food intake was measured daily. Body weight was measured at baseline and week 8 of the study. After 2 weeks of acclimatization, the rabbits were divided into 5 groups of seven rabbits each (n=7). Rabbits in group 1 was assigned as negative control (NC); consumed normal diet; group 2 was positive control; consumed normal diet containing 0.5% cholesterol (PC); group 3 (HS), 4 (HF) and 5 (HD) were the supplemented groups which consumed cholesterol-enriched diet containing simvastatin (10 mg/kg/day), fullfat pulp powder (50g kg^{-1}) and defatted pulp powder (50g $kg^{-1}$of CO fruit), respectively. All groups received experimental diet prepared within three days before feeding. Blood was drawn from marginal ear of rabbits at week 0 and 8 of the study. Samples for plasma MDA was collected in EDTA-containing tubes while samples for superoxide dismutase (SOD), glutathione peroxidase (GPx) and total antioxidant status (TAS) were collected in lithium heparin-containing tubes. All experimental protocols were approved by the Animal Care and Use Committee (ACUC) of Faculty of Medicine and Health Sciences, Universiti Putra Malaysia (UPM) Serdang, Selangor. Plasma was obtained by centrifugation at 3000 rpm for 10 minutes at 25 ºC. Plasma MDA level was measured by using thiobarbituric acid (TBA) assay as described by Beuge and Aust [4] while activity of enzymes (SOD, GPx) and TAS were determined on Clinical Chemistry Analyzer (Vitalab Selectra E, Germany).

D. Statistical Analysis

The data are presented as group means ± SEM. Analysis of Variance (ANOVA) using Statistical Package Social Sciences (SPSS) version 16.0 was used for data analysis. Statistical differences between treatment and control group were determined using one way ANOVA. Turkey pos hoc tests were used for multiple group comparison. The significance value was set at $p < 0.05$.

III. RESULTS

The effects of dabai pulp supplementation on plasma MDA and TAS are summarized in Table 1. At the beginning of the study, plasma MDA and TAS were not significantly different between groups. Feeding high cholesterol diet (PC group) resulted to steady increased of plasma MDA levels and simultaneous reduction in plasma TAS levels. Among the supplemented groups, defatted dabai-containing diet resulted to reduction in plasma MDA and increment in plasma TAS, both significantly ($p<0.05$) at 8 weeks of treatment. Fullfat dabai-containing diet however showed similar trend as group receiving statin with marked increment of MDA level at 8 weeks of treatment. The plasma TAS of animals supplemented with statin was reduced with increased supplementation period and was not significantly different with PC and NC groups at week 8 of the study. On contrary, the supplementation of fullfat dabai to animals favoured the increment in plasma TAS of animals similar to effect of defatted dabai pulp. The effects of dabai pulp supplementations on antioxidant enzymes and status were also depicts in Table 2. Generally from week 0 to 8, erythrocyte SOD activities were increased significantly ($p<0.05$) in both dabai supplemented groups (HF and HD) but decreased significantly with statin treatment ($p<0.05$). On the other hand, erythrocyte GPx at week 8 were reduced significantly ($p<0.05$) in all groups from baseline with most reduction observed in the statin supplemented animals.

Table 1 Changes in plasma MDA and TAS, erythrocyte SOD and GPx in experimental rabbits

Group		Week 0	Week 8
NC	MDA(mmol/L)	$2.60 \pm 0.04^{a,1}$	$3.74 \pm 0.15^{b,2}$
	TAS (mmol/L)	$0.814 \pm 0.01^{b,1}$	$0.660 \pm 0.01^{a,2}$
	SOD (U/ml)	$4.84 \pm 0.03^{a,1}$	$2.74 \pm 0.04^{a,2}$
	GPx (U/L)	$756 \pm 4.77^{bc,1}$	$729 \pm 20.16^{b,1}$
PC	MDA(mmol/L)	$2.92 \pm 0.32^{a,1}$	$4.37 \pm 0.18^{c,2}$
	TAS (mmol/L)	$0.814 \pm 0.02^{ab,1}$	$0.634 \pm 0.03^{a,2}$
	SOD (U/ml)	$4.73 \pm 0.02^{a,1}$	$5.33 \pm 0.17^{b,2}$
	GPx (U/L)	$827 \pm 12.5^{c,1}$	$565 \pm 14.8^{a,2}$

Table 1 (*continued*)

HS	MDA(mmol/L)	2.80 ± 0.10[a,1]	4.17 ± 0.17[bc,2]
(Simvastatin)	TAS (mmol/L)	0.712 ±0.07[ab,1]	0.692 ±0.00[a,1]
	SOD (U/ml)	4.79 ±0.09[a,1]	2.93 ±0.04[a,2]
	GPx (U/L)	801 ±13.6[c,1]	682 ±53.43[ab,2]
HF	MDA(mmol/L)	3.06 ± 0.12[a,1]	3.70 ± 0.09[b,2]
(Fullfatdabai)	TAS (mmol/L)	0.826 ±0.01[b,1]	1.054 ±0.05[b,2]
	SOD (U/ml)	4.80 ±0.05[a,1]	6.45 ±0.05[d,2]
	GPx (U/L)	685 ±016.89[a,1]	631 ±12.69[ab,1]
HD	MDA(mmol/L)	2.92 ± 0.18[a,1]	2.81 ± 0.04[a,1]
(Defatted	TAS (mmol/L)	0.780 ±0.05[ab,1]	1.114 ±0.03[b,2]
dabai)	SOD (U/ml)	4.91 ±0.08[a,1]	6.53 ±0.06[d,2]
	GPx (U/L)	696 ±36.25[b,1]	587 ±20.12[a,2]

NC: normal diet group, PC: positive control, HS, HF, and HD: hypercholesterolemic group treated with 10 mg/kg/day simvastatin, 5% full-fat pulp, 5% defatted pulp, respectively.
Values within week followed by same superscript letter are not significantly different at $p<0.05$.
Values within group followed by same superscript number are not significantly different at $p<0.05$.

IV. Discussion

Findings of this study has shown that feeding high cholesterol diet to rabbits result in production of high free radical condition in the *in vivo* system [5] thus enhanced the lipid peroxidation process. In this study, supplementation of dabai parts, the defatted part especially has stimulated the best effect better than expected from statin. Statin is a well accepted cholesterol-lowering drug with potent hypolipidemic effect known to inhibit HMG-CoA reductase in cholesterol biosynthesis [6]. The reason for this favourable effect of defatted dabai could be due to antioxidative properties of the fruit pulp that is rich in water soluble antioxidants especially phenolics.

Previously, Azrina et al., have revealed the potency of antioxidants available in the fruit pulp especially skin. This observation also is in agreement with report by Hock Eng et al., who identified catechin, epigallocatechin, elargic acid as well as anthocyanidin as the main polyphenolics in the defatted dabai [1, 7]. By calculation, daily diet of rabbits containing defatted dabai supplementation contained about 30.8mg polyphenols that favourably improved both antioxidant status and reduced *in vivo* lipid peroxidation of the animals. Results of this study also have shown that activities of antioxidant enzymes were stimulated by the defatted dabai supplementation. Generally, administration of high cholesterol diet results to increased activities of SOD in animals. The increased of SOD is related to excessive oxidative stress induced by the hypercholesterolemic condition (0.5% dietary cholesterol) as a mean to neutralize toxic effect of free radicals. Glutathione peroxidise (GPx) and superoxide dismutase (SOD) are important antioxidant enzymes that protect organisms from oxidative damage. SOD is very efficient in catalyzing dismutation of superoxide anion radical (O2•) that cause oxidative damage in the *in vivo* system and produces hydrogen peroxide. The hydrogen peroxide later will be transformed to water by GPx thus preventing formation of other free radicals such as hydroxyl (•OH) and hydrogen (•H) [8]. Increment in both antioxidant enzymes constitutes a protective effect against radicals in the *in vivo* system. In this study, significant increased in the enzyme activities of SOD due to fullfat- (Δ=1.65 U/ml) and defatted-dabai (Δ=1.62 U/ml) supplementations were observed while the enzyme activities of GPx were reduced (in all groups) indicating the possibility of insufficient SOD activities to convert superoxide radicals to hydrogen peroxide to be broken down by the GPx. Both dabai supplemented diets (HD group, especially) also caused significant increment in TAS value suggesting the enhanced capacities of both enzymatic and non-enzymatic antioxidant defences that exist within the *in vivo* system of the model used [9,10] following intake of dabai supplementation.

V. Conclusion

This study has demonstrated that fullfat- and defatted-dabai pulps possess strong antioxidant properties related to the presence of polyphenolic compounds in the fruit pulp. Defatted dabai has more promising effect better than commercial drug, statin. Further studies are under way to separate and purify the polyphenols from the fruit fractions as well as both *in vitro* and *in vivo* analyses of the polyphenol rich fraction.

Acknowledgement

The financial support of science fund from Ministry of Agricluture (MOA) , vote number 5450555 (Project number 05-01-04-SF1137) is gratefully acknowledged. The authors also acknowledged the assistance of laboratory staffs from the Department of Nutrition and Dietetics, UPM throughout the research project. They also extend their thanks to Universiti Putra Malaysia for the use of laboratory facilities.

References

1. Azrina A, Nurul N, Amin I. (2010). Antioxidant properties of methanolic extract of Canarium odontophyllum fruit. International Food Research of Journal 2010;17: 319-326
2. Shakirin FH, Prasad KN, Ismail A, Yuon LC, Azlan A. (2010) Antioxidant capacity of underutilized Malaysia Canarium odontophyllum (dabai) Miq. fruit. Journal of Food Composition and Analysis 23:777-781.
3. Chew L, Khoo H, Amin I, Azrina A, Lau C. (2011). Analysis of Phenolic Compounds of Dabai (*Canarium odontophyllum* Miq.) Fruits by High-Performance Liquid Chromatography. Food Analytical Methods 1-12. DOI: 10.1007/s12161-011-9217-1 (In press)
4. Beuge, J.A. and Aust, S.D. (1978). Microsomal lipid peroxidation. Methods Enzymology 52: 302-10.
5. Prasad, K. (2007). Regression of hypercholesterolemic atherosclerosis in rabbits by secoisolariciresinol diglucoside isolated from flaxseed. Atherosclerosis. 197: 34-42.
6. Shin, M., Chung, N., Lee, J.H., Jang, Y., Park, E., Jeon, K., Chung, J.H., Seo, B. (2007). Effects of simvastatin on plasma antioxidant status and vitamins in hyper cholesterolemic patients. International Journal of Cardiology 118: 173-177.
7. Khoo Hock Eng, Azrina Azlan, Amin Ismail, Faridah Abas. (2011). Influence of different extraction media on phenolic contents and antioxidant capacity of defatted Dabai (*Canarium odontophyllum)* fruits. Food Analytical Methods. DOI 10.1007/s12161-011 9250-0 (In press)
8. Switala, J. and Loewen, P. (2002). Diversity of properties among catalases. *Arch. Biochem. Biophys*. 401: 145-154
9. Yu, L., Scott, H., Jonathan, P., Mary, H., John, W. and Ming, Q. (2002). Free radical scavenging properties of wheat extracts. Journal of Agricultural and Food Chemistry 50 (6): 1619-1624.
10. Anagnostopoulou, M. A., Kefalas, P., Papageorgiou, V. P., Assimepoulou, A. N. and Boskou, D. (2006). Radical scavenging activity of various extracts and fractions of sweet orange peel (*Citrus sinensis*). Food Chemistry 94: 19-25.

Sibu Olive Inhibits Artherosclerosis by Cholesterol Lowering Effect in Cholesterol Fed-Rabbit

M.H. Nurulhuda[1], Azrina Azlan[1,2], Amin Ismail[1,2], Zulkhairi Amom[3], and Faridah Hanim Shakirin[1]

[1] Department of Nutrition and Dietetics, Faculty of Medicine and Health Sciences, 43400 UPM Serdang, Selangor, Malaysia
[2] Laboratory of Halal Science Research, Halal Products Research Institute, Universiti Putra Malaysia, 43400 UPM Serdang, Selangor, Malaysia
[3] Department of Human Anatomy, Faculty of Medicine and Health Sciences, 43400 UPM Serdang, Selangor, Malaysia

Abstract — **'Sibu olive' is known botanically as "*Canarium Odontophyllum* Miq." The present study was designed to assess the effect of Sibu olive (SO) consumption on cholesterol level and the development of atherosclerosis of mild hypercholesterolemia rabbits. A total of thirty five rabbits were randomly divided into 5 groups (n=7); NC (normal diet), PC (normal diet + 0.5% cholesterol), HS (hypercholestrolemic diet + 10 mg/kg/day simvastatin), HF (hypercholsterolemic diet + 5% fullfat SO) and HD group (hypercholesterolemic diet + 5% defatted SO). Main outcome measures were body weight changes, changes in total cholesterol, low-density lipoprotein cholesterol (LDL-C), high-density lipoprotein cholesterol (HDL-C), and atherosclerotic plaque formation. After 8 weeks, there was increment in body weight changes (1.75–2.24 kg) of animals. The HD group showed significant reduction in total cholesterol (96.3%), LDL-C (26.5%) and HDL-C increment (46.8%) relative to PC. Futhermore, the atherosclerotic plaque formation in HD group diminished by 79.4% compared to PC. The polyphenols like phenolic acids, flavonoids, and anthocyanidins) that promote the lipid lowering potential of the SO fruit. Hence, in the future, SO fruit may serve as potent alternative treatment in the prevention of atherosclerosis.**

Keywords — **Sibu olive, hypercholesterolemia, rabbit, artherosclerosis, polyphenols**

I. INTRODUCTION

'Sibu olive' (SO) is known botanically as *Canarium odontophyllum* Miq which is categorized in the genus of canarium, subfamily of *Burseraceae* [1]. The morphology of the fruit is similar to olive fruit as well as some physicochemical properties such as in terms of total fat and total phenolic contents, but differs in fatty acid content as reported by Azrina et.al [2]. Previous studies claimed that the fruit is nutritious due to its richness essential minerals such as potassium (352 mg/100 g), phosphorus (65 mg/100 g), calcium (200 mg/100 g) and magnesium (106 mg/100 g [1]. Moreover, the potential health benefits of the fruit have also been reported due to the strong antioxidant capacity of the different SO parts [4]. Therefore, the purpose of the present study was to investigate the cholesterol lowering as well as inhibition of atherosclerosis formation potentials of the SO parts in hypercholesterolemic-induced rabbits.

II. MATERIAL AND METHODS

A. Preparation of Fruit Extract

Fresh SO fruits were obtained from the Agriculture Research Centre, Semongok, Sarawak, Malaysia. The pulps were separated and freeze-dried and ground using a dry grinder. The powder is known as full-fat pulp. The defatted pulps were prepared by extracting the oil from the dried powder sample using chloroform: methanol (2:1 v/v) at a ratio of 1:5 (w/v). The residue after oil extraction known as defatted pulp was dried in oven at 40 °C.

B. Experimental Design

Thirty five male New Zealand white rabbits were purchased from East Asia Company, Malaysia. After acclimatization for about two weeks, the animals were randomly distributed into five groups (n=7) of normal and hypercholesterolemic groups. Normal group was named NC and received normal diet while the hypercholesterolemic groups (PC, HS, HF and HD) received hypercholesterolemic diet (0.5% of cholesterol). The hypercholesterolemic groups: PC received only the hypercholesterolemic diet; HS received hypercholesterolemic diet + 10mg/kg/day simvastatin; HF received hypercholesterolemic diet + full-fat pulp of SO and HD received hypercholesterolemic diet + defatted pulp of SO. All experimental protocols involving animals were approved by the Animal Care and Use Committee of the Faculty of Medicine and Health Sciences, Universiti Putra Malaysia (UPM), Serdang, Selangor. (Approval no.: UPM/PADS/BR-UUH/00238).

V. Van Toi et al. (Eds.): 4th International Conference on Biomedical Engineering in Vietnam, IFMBE Proceedings 40, pp. 141–144, 2013.
www.springerlink.com

C. Preparation of Animal Diets

The experimental diets were prepared based on the NC diet, which is a normal basal diet. The normal basal diet contained soybean (15%), corn (30%), starch (10%), molasses (2%), corn oil (2%), vitamin mixture (0.3%), mineral mixture (3.5%), DL-methionine (0.2%), calcium carbonate (0.5%) and calcium hydrogen phosphate (0.5 %). Meanwhile for the HF and HD diets, 5% of SO full-fat pulp and SO defatted pulp was added, respectively.

D. Preparation of 0.5% Cholesterol Diet

The cholesterol diets were prepared by dissolving the, 0.6 g cholesterol in 2 ml chloroform then was sprayed on the 20 g of normal basal diet for each animal's diet. The pellets were dried in an oven at 40°C overnight to allow the evaporation of chloroform [5].

E. Lipid Profile Analysis

Plasma total cholesterol, LDL-c, HDL-c were determined automatically on a chemistry analyzer (Hitachi 902, Japan).

F. Atheroma Plaques Scoring

At the end of week 8, the rabbits were sacrificed. The aorta preparation and Sudan IV staining was adapted from Prasad, K. [6]. The macroscopy study was conducted to quantify the atherosclerotic lesion area adapted from Bocanet. et al. [7].

G. Statistical analysis

Data were analyzed using SPSS 16.0 (SPSS Corporation, Chicago, IL) and presented as group means ± SEM. Turkey pos hoc tests were used for multiple group comparison. The significance value was set at $p < 0.05$.

Table 1 Changes in body weights and lipid profile (Total cholesterol, LDL-c, HDL-c) in experimental rabbits.

Group		Week 0	Week 8
NC	Body Weight (kg)	$1.87 \pm 0.03^{a,1}$	2.09 ± 0.06ab,2
	Total Cholesterol(mmol/L)	$1.79 \pm 0.06^{d,1}$	$1.94 \pm 0.02^{a,1}$
	LDL-c (mmol/L)	$0.66 \pm 0.03^{b,1}$	$0.66 \pm 0.05^{a,1}$
	HDL-c (mmol/L)	$0.59 \pm 0.04^{c,1}$	$0.57 \pm 0.04^{a,1}$
PC	Body Weight(kg)	$1.82 \pm 0.09^{a,}$	$2.34 \pm 0.04^{b,2}$
	Total Cholesterol(mmol/L)	$1.54 \pm 0.06^{c,1}$	$34.42 \pm 0.84^{c,3}$
	LDL-c (mmol/L)	$0.62 \pm 0.03^{b,1}$	$14.84 \pm 1.26^{c,2}$
	HDL-c (mmol/L)	$0.36 \pm 0.02^{ab,1}$	$3.59 \pm 0.19^{d,3}$
HS	Body Weight(kg)	$1.86 \pm 0.02^{a,1}$	$2.22 \pm 0.09^{ab,2}$
(Simvastatin)	Total Cholesterol(mmol/L)	$1.58 \pm 0.09^{cd,1}$	$3.74 \pm 0.11^{a,3}$
	LDL-c (mmol/L)	$0.65 \pm 0.02^{b,1}$	$3.99 \pm 0.31^{a,3}$
	HDL-c (mmol/L)	$0.49 \pm 0.03^{bc,1}$	$1.29 \pm 0.13^{a,3}$
HF	Body Weight(kg)	$1.84 \pm 0.08^{a,11}$	$2.02 \pm 0.01^{a,1}$
(Fullfat SO)	Total Cholesterol(mmol/L)	$1.23 \pm 0.01^{b,1}$	$33.61 \pm 0.99^{c,3}$
	LDL-c (mmol/L)	$0.58 \pm 0.04^{b,1}$	$17.97 \pm 1.18^{c,2}$
	HDL-c (mmol/L)	$0.55 \pm 0.04^{c,1}$	$4.12 \pm 0.24^{e,3}$
HD	Body Weight(kg)	$1.75 \pm 0.00^{a,1}$	$2.24 \pm 0.01^{ab.3}$
(Defatted SO)	Total Cholesterol(mmol/L)	$0.89 \pm 0.02^{a,1}$	$1.29 \pm 0.02^{a,1}$
	LDL-c (mmol/L)	$0.29 \pm 0.02^{a,1}$	$10.90 \pm 0.59^{b,3}$
	HDL-c (mmol/L)	$0.30 \pm 0.03^{a,1}$	$1.91 \pm 0.11^{b,2}$

NC: normal diet group, PC: positive control, HS, HF, and HD: hypercholesterolemic group treated with 10 mg/kg/day simvastatin, 5% full-fat pulp, 5% defatted pulp respectively.

Values are expressed as mean ± SEM.

Values within week followed by the same superscript letter are not significantly different at $p<0.05$.

Values within group followed by the same superscript number are not significantly different at $p<0.05$.

III. Results

In this study, the incorporation of 0.5% cholesterol (PC) resulted in the highest body weight changes compared to other groups. Moreover, the PC group exhibited 94.3% greater in total cholesterol level in plasma compared to normal diet. The HD group showed significant reduction in total cholesterol by 96.3% compared to PC. Generally, remarkable changes in lipid profiles were observed after 8 weeks of the study (Table 1). The treatment with defatted pulp in HD group exhibited significant reduction in LDL-c (26.5%) concentration in plasma compared to control group.

In the groups with high cholesterol diet (HS, HF and HD), the general trend is an increase in HDL-c levels from 0 to 8 weeks with the highest increment observed in the HF group (73.3%). The least increment in HDL-c level at 8 week was observed in the HS group (26%). Despite of that, there was a significant different in HDL-c increment (46.8%) in the HD group relative to PC. The atherosclerotic plaques were more severe in animals fed with high cholesterol diet (PC) (22.08%) than in animals supplemented with full-fat pulp (HF) and defatted pulp (HD). The plaque areas of the aortas in HF and HD were significantly lower ($p<0.05$) by 43.39% and 79.4%, respectively compared to PC. Thus, among the treated groups, HD exhibited the greatest reduction in atherosclerotic plaque formation.

IV. Discussion

The aim of this study was to explore the cholesterol-lowering potential of SO in an animal model of hypercholesterolemia. Firstly, the changes of body weight among the groups showed that hypercholesterolemic diet exhibited the highest body weight gain. The remarkable physical changes indicate the hypercholesterolemic model have successfully established. In addition, the increment of body weight of animals receiving a high cholesterol diet (PC) might be due to lipid depositions [8]. Based on the lipid parameters, animals fed on high cholesterol diet (PC) showed increased TC, LDL-c and HDL-c as compared to normal animals. These differences were statistically significant ($p<0.05$).

The present study showed that supplementation of defatted pulp of SO exhibited a positive hypocholesterol emic effect which was manisfated by remarkable reduction in TC and LDL-c levels, ($p<0.05$) (Table 1). When feeding hypercholesterolemic diet, the total cholesterol of PC group showed an increment of 96% compared to normal diet. These changes are associated with a phenomenon that excessive load of cholesterol to the liver above the acceptable level of its normal process causes the system to be unable in metabolising the lipids resulting in high cholesterol return in the circulating blood [9]. Treatment of high cholesterol fed animals concomitantly with SO defatted pulp however showed a remarkable reduction of lipid profiles towards the level of the control group. The reduction of plasma cholesterol levels could be due to the action of polyphenol compounds in the defatted pulp, especially in the skin part. The polyphenol content was found at 68±1.2 mg GAE/g in skin extracts, which was found the highest compared to other parts [10]. Previous study has revealed the existence of polyphenols in SO such as phenolic acids (ellagic and vanillic acids), flavonoids (catechin, epicatechin,epicatechin gallate, epigallocatechin gallate, apigenin), anthocyanidins (cyanidin, pelargonidin and delphinidin) and anthocyanins (malvidin-3,5-di-O-glucoside, cyanidin-3-O-glucoside, cyanidin-3-O-rutinoside and peonidin-3-O-glucoside) [11]. Moreover, polyphenol compounds are believed to play a significant role in lowering plasma cholesterol by binding to bile acids which in turn increased faecal loss [12].

Apart from that, cholesterol-lowering properties of defatted SO may also attributed to the presence of dietary fiber. The SO defatted pulp consisted of more than 50% total dietary fiber (TDF), characterized by 73% and 27% of insoluble dietary fiber (IDF) and soluble dietary fiber (SDF), respectively [13]. The capacity of dietary fiber to lower serum cholesterol by hindering the digestion and absorption of dietary fat, modifying bile acid absorption and metabolism, forming a short chain fatty acid which can inhibit cholesterol and fatty acid synthesis in the liver, and altering the concentration of insulin and hormones [14]. Apart from that, we also found that supplementation of SO defatted pulp in high cholesterol fed animals could inhibit the progression of atherosclerosis. The formation of atheromatous plaque in the HD group was found to be significantly lower compared to that of the non-treated group (PC), might be attributable partly to its hypo-cholesterolemic property, since decrease of TC level is capable to improve endothelial function [15].

V. Conclusion

Among the fruit parts, the defatted pulp of SO exhibited the remarkable reduction in lipid profile and greatest reduction of atherosclerotic plaque formation. Therefore, defatted pulp may be beneficial in preventing hyper-cholesterolemic, atherosclerosis and reduce the risk factors for coronary artery disease.

Acknowledgement

The financial support of science fund from Ministry of Agricluture (MOA), Vote no 5450555 (Project no. 05-01-04-SF1137) is gratefully acknowledged. The authors also acknowledged the assistance of laboratory staffs from the Department of Nutritionand Dietetics, UPM throughout the research project. They alsoextend their thanks to Universiti Putra Malaysia for the use of laboratory facilities.

REFERENCES

1. L.Y. Chew, K. Nagendra Prasad, I. Amin, A. Azrina, C.Y. Lau, (2011), Nutritional composition and antioxidant properties of Canarium odontophyllum Miq. (dabai) fruits, Journal of Food Composition and Analysis, 24: 4-5, 0889-1575: 670-67710.1016/j.jfca.2011.01.006. (In press)
2. A. Azlan, K. N. Prasad, H. E. Khoo, N. Abdul-Aziz, A. Mohamad, A. Ismail, Z. Amom (2010). Comparison of fatty acids, vitamin E and physicochemical properties of Canarium odontophyllum Miq. (dabai), olive and palm oils, Journal of Food Composition and Analysis, 23: 8, 0889-1575,772-776.
3. Faridah Hanim Shakirin, K. Nagendra Prasad, Amin Ismail, Lau Cheng Yuon, Azrina Azlan,(2010), Antioxidant capacity of underutilized Malaysian Canarium odontophyllum (dabai) Miq. fruit, Journal of Food Composition and Analysis,. 23:8, 777-781.
4. Azrina A., Nurul Nadiah M.N., Amin I. (2010).Antioxidant properties of methanolic extract of Canarium Odontophyllum fruit. International Food Research Journal 17: 319-326.
5. Shimizu, T., Nakai, K., Morimoto, Y., Ishihara, M., Oishi, H., Kikuchi, M and Arai, H..(2009). Simple rabbit model of vulnerable atherosclerotic plaque. Neurologica Medico- Chirurgica 49: 327-332.
6. Prasad, K., 2007. Regression of hypercholesterolemic atherosclerosis in rabbits by secoisolariciresinol diglucoside isolated from flaxseed. Atherosclerosis 197: 34-42.
7. Bocan, T.M.A., Mueller, S.B., Mazur, M.J., Uhlendorf, P.D., Brown, E.Q. and Kieft, K.A.(1993). The relationship between the degree of dietary-induced hypercholesterolemia in the rabbit and athersclerotic lesion formation. Atherosclerosis 102: 9-22.
8. Rauch, U., Osende, J.J., Chasebro, J.H., Fuster, V., Vorchheimer, D.A., Harris, K., Harris, P., Sandler, D.A., Fallon, J.T., Jayaraman, S. and Badimon, J.J.(2002).. Statin and cardiovascular diseases: the multiple effects of lipid-lowering therapy by statin. Atherosclerosis 153: 181-189.
9. Ramos, S., Moulay, L., Granada-Serrano, A.B., Vilanova, O., Muguerza, B. and Bravo, L (2008). Hypolipidemic effect in cholesterol-fed rats of a soluble fiber-rich product obtained from Cocoa husks. Journal of Agriculture of Food Chemistry 56: 6985-6993.
10. K. Nagendra Prasad, L.Y. Chew, K. W. Kong,A. Azrina, I. Amin (2010). Antioxidant Capacities of Peel, Pulp and Seed Fractions of Canarium odontophyllum Miq. Journal of Biomedicine and Biotechnology. Volume 2010: 871379. DOI: 10.1155/2010/871379
11. L.Y. Chew, H. E. Khoo, I. Amin, A. Azrina, C.Y. Lau (2011). Analysis of Phenolic Compounds of Dabai (Canarium odontophyllum) Fruits by High-Performance Liquid Chromatography. Food Analytical Method DOI: 10.1007/s12161-011-9217-1 (In press)
12. Zunft H. J. F., Luder, W., Harde, W., Haber, B., Graubaum, H.J., Koebnick, C. and Grunwald, J. (2003). Carob pulp preparation rich in insoluble fibre lowers total and LDL cholesterol in hypercholesterolemic patients. European Journal of Nutrition 42: 235–242.
13. Goldstein, J.L. and Brown, M.S. (2009). The LDL receptor. Arteriosclerosis Thrombosis Vascular Biology 29; 431-438.
14. Kang, M., Kawai, Y., Naito, M., Osawa, T. (1999). Dietary defatted sesame flour decreases susceptibility to oxidative stress in hypercholesterolemic rabbits. Journal of Nutrition 129: 1885-1890.
15. Valgimigli, M., Merli, E., Malagutti, P. (2003). Endothelial dysfunction in acuteand chronic coronary syndromes: evidence for a pathogenetic role of oxidativestress. Archives of Biochemistry and Biophysics 420, 255–258

Analysis of Liquid Meniscus Shape in Various Microchannels for Biomedical Diagnostic Device

Truong Dinh Sy, In Kyum Park, Kwang Won Hong, Hyun Chul Bang, Young Ho Seo*, and Byeong Hee Kim

Medical & Bio-Materials Research Center and Dept. of Mechanical and Mechatronics Engineering
Kangwon National University, Chuncheon 200-701, South Korea

Abstract — **Biomedical diagnostic devices always have crucial aspects on development of biomedical technology. Blood sampling process through microchannels is required in biomedical diagnostic devices. In this paper, we have investigated advancing meniscus of biomedical sample in various side-wall shaped microchannels. Advancing meniscus of liquid mainly depends on side-wall shape of microchannel, even though dynamic contact angle is sustained as same value regardless of side-wall shape.**

Keywords — **Meniscus, Microchannel, PDMS chip, Capillary force, Biomedical diagnostic device.**

I. INTRODUCTION

For biomedical applications of microfluidic devices, filling process of biological liquid through microchannels is important issue for accurate sensing of target protein. Because air trapping during filling process of biological liquid causes sensing error, shape of advancing meniscus of biological liquid should be controlled to avoid air trapping. In additions, advancing meniscus shape is also important, because antibody-antigen reaction rate depends on resident time of bio-fluids. That is advancing meniscus control becomes critical in almost of microfluidic equipment to reach stable antigen-antibody reaction. Figure 1 shows the general advancing meniscus shape of bio-fluid in capillary microchannel, provided a general concave meniscus formation. Concave advancing meniscus causes the difference resident time of flow between center and edge of the microchannel. Therefore flat shape of advancing meniscus is more suitable in biomedical microfluidic devices as shown in Fig.2. This paper presents how to control the meniscus in microscale.

Recent researches have demonstrated of capillary and the aspects have effected on flow rate such as surface tension, contact angle, viscosity force. The theoretical fluidics had been presented [1]. Meniscus such as 'W' formation in microchannels from thin polymer films investigated and Laplace pressure is considered to explain the capillary rise of the liquid [2]. Viscosity of liquid as well as channels' surface characteristic directly influences on capillary. Interplay between surface forces and inter-molecular demonstrated as complicated meniscus. [3] Furthermore, micro-channels filling process was caused by surface tension influences [4] as well as simulation of filling process were calculated in recent research [5].

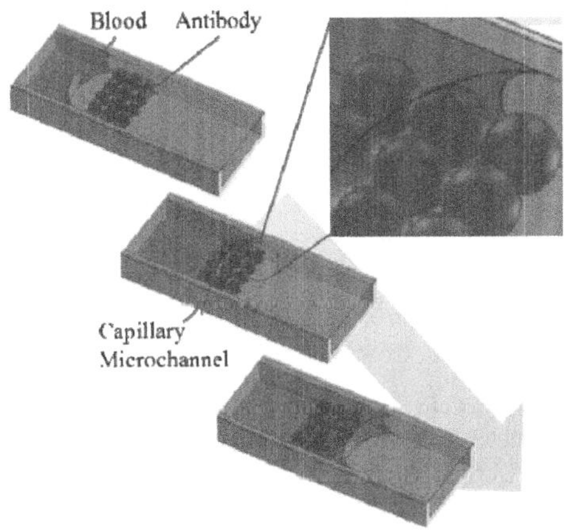

Fig. 1 Advancing meniscus of general capillary microchannel.

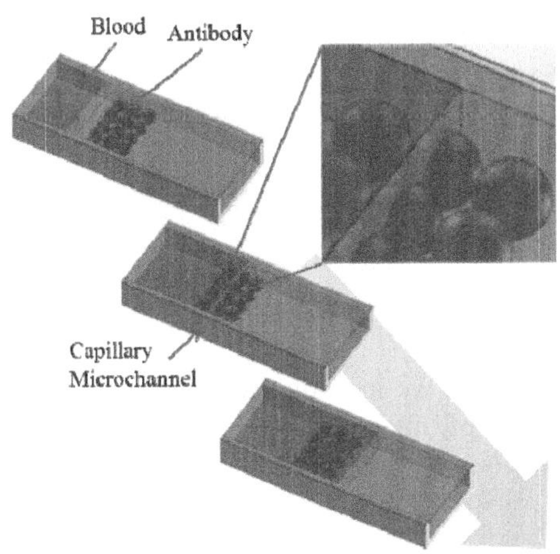

Fig. 2 Requirement shape of meniscus in biomedical diagnostic devices.

* Corresponding author.

V. Van Toi et al. (Eds.): 4th International Conference on Biomedical Engineering in Vietnam, IFMBE Proceedings 40, pp. 145–148, 2013.
www.springerlink.com

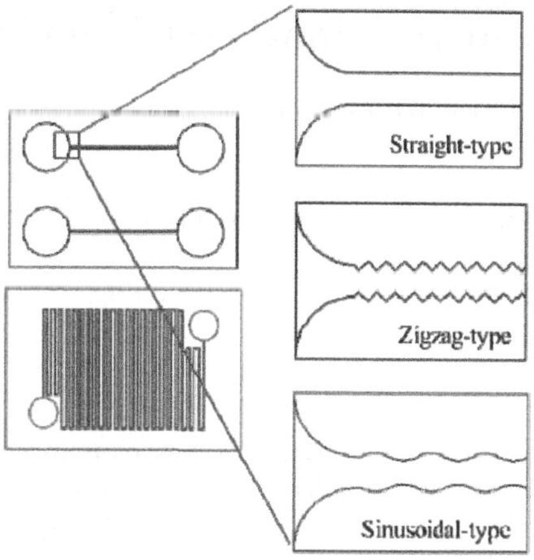

Fig. 3 Design of side-wall shape of microchannel.

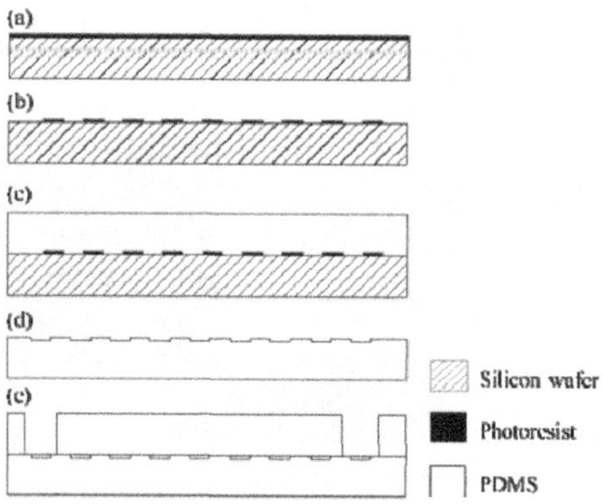

Fig. 4. Design of side-wall shape of microchannel. Fig.4 PDMS chip fabrication process: (a) SU-8 photoresist (PR) spin coating; (b) Preparing PR mold by lithography; (c) PDMS casting on PR mold; (d) PDMS peeling; (e) Bonding with upper PDMS plate.

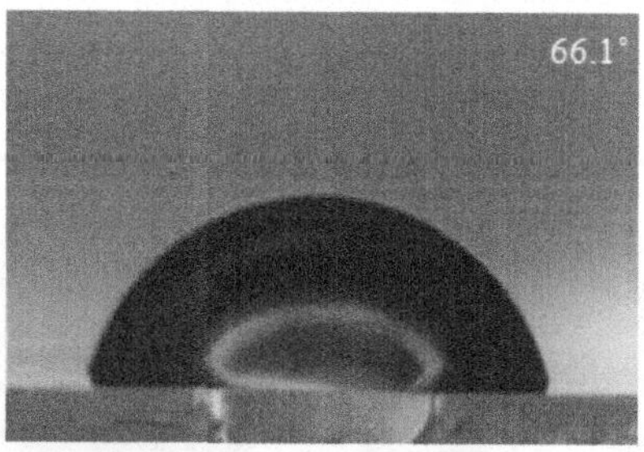

Fig. 5 Static contact angle of hydrophilic PDMS.

In this paper, we investigated the relation between side-wall shapes of microchannels and advancing meniscus. Side-wall shape of microchannel was designed as three different shapes including straight, triangular waveform (zigzag) and sinusoidal waveform. Figure 3 presents the side-wall shape of each microchannel. Width and depth of microchannel were designed as 200μm and 60μm, respectively. The paper compared shape of advancing meniscus with different side-wall shaped microchannels.

II. Materials and Methods

Microchannels were fabricated by polydimethylsiloxane (PDMS, Dow Corning) replication process. There were a number of stages to fabricate the microchannels. The fabrication procedure was presented in Fig.4. Firstly, SU-8 photoresist (Microchem Co.) was coated on silicon wafer to reach a target thickness of 60μm before using the available mask in the contact photolithography stage. After UV exposure step, unexposed area of SU-8 was developed using SU-8 developer, and then rinsed in IPA (Isopropyl Alcohol).The wafer was put on hot plate for 150-200^{0}C for improvement of adhesion between silicon wafer and SU-8 polymer. SU-8 pattern mold was treated with silane (Trichlorosilane, Sigma-Aldrich Inc.) in order to reduce the surface energy of the mold. Silane was deposited by vacuum coating method in desiccator for 1hr. Because PDMS was hydrophobic material and static contact angle was higher than 100°, there was no capillary action in PDMS microchannel. In order to make capillary action in PDMS microchannel, surfactant of 1wt% silwet L-77 was added into PDMS prepolymer. Finally, upper and lower hydrophilic PDMS replicas were bonded together at 200°C for 2 hours. Static contact angle of hydrophilic PDMS was measured as 66.1±1.6° as shown in Fig.5.

After hydrophilic PDMS microchannels were completely fabricated, deionized water with red ink whose viscosity was similar with blood plasma was poured into the microchannels. Advancing meniscus of deionized water was observed using inverted optical microscope and CCD. Due to rapid speed of flow, it might be hard to take clear capture images in experiments. Thus, additional tube was connected at the outlet port in order to increase hydraulic resistance that reduces flow speed.

III. Results and Dicussion

Shape of advancing meniscus of deionized water with red ink through straight-wall shaped microchannels was almost flat as shown in Fig.6. Interfacial angle between tangential direction of side-wall and front of advancing meniscus was approximately 90°, even though static contact angle was 66.1±1.6°. For straight-wall shaped microchannels, four different widths of microchannels including 50μm, 90μm, 150μm and 200μm were used. The meniscus formations were almost perpendicular to the side-wall direction for all these channels as shown in Fig.5. Dynamic contact angle which means contact angle of moving fluid.

For triangular (zigzag) and sinusoidal waveform shaped side-walls, microchannel had repeating structures of diverging and converging regions.

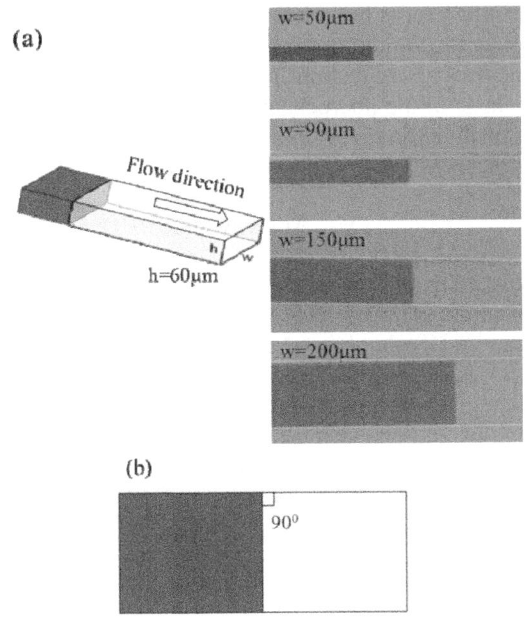

Fig. 6 Flat side-wall microchannels and meniscus shape.

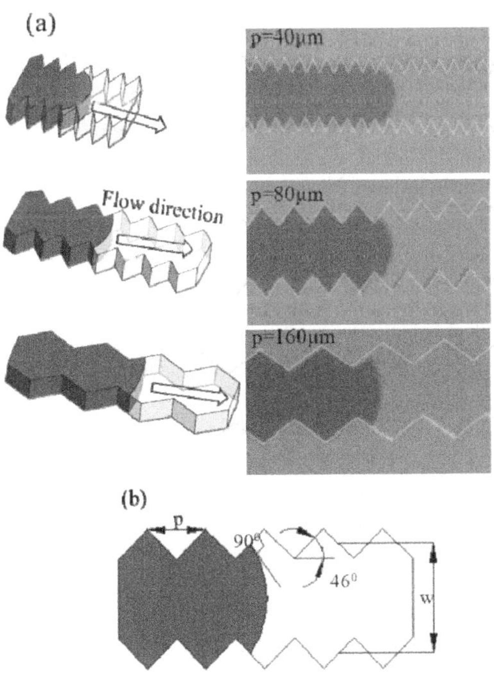

Fig. 7 Triangular side-wall microchannels and meniscus shape.

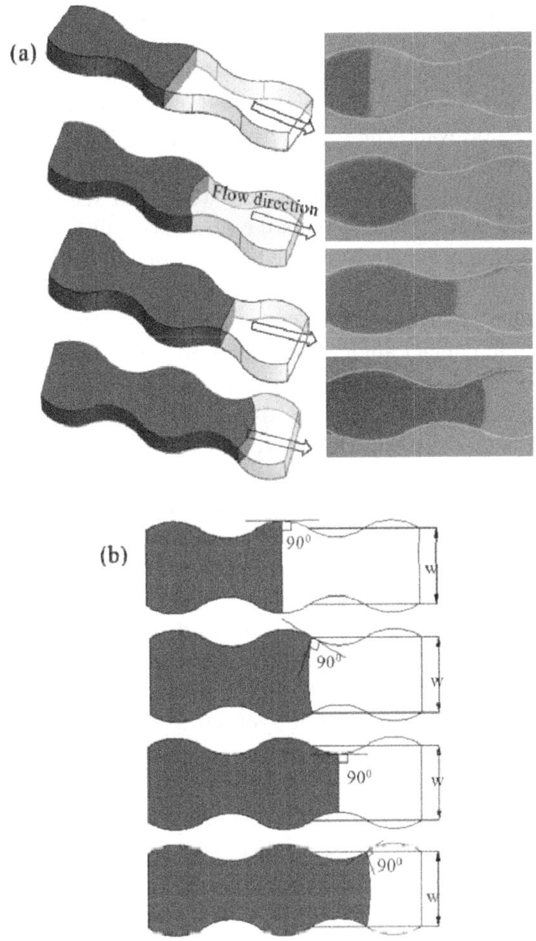

Fig. 8 Sinusoidal waveform side-wall microchannels and meniscus shape.

In a diverging region, shape of advancing meniscus was convex shape, while concave shaped advancing meniscus was observed in a converging region. At any moment, interfacial angle between tangential direction of side-wall and front of advancing meniscus was maintained as 90° as shown in Fig.7 and 8.

When diverging and converging angles were than 45°, shape of advancing meniscus was maintained as convex shape. On the other hand, if that angle was lower than 45°, shape of advancing meniscus alternated convex and concave shapes as shown in fig. 7. Apparent shape of advancing meniscus was dependent on side-wall shape of micro-channel, but interfacial angle between tangential direction of side-wall and advancing meniscus was similar with 90°.

In particular, the dynamic shape meniscus in sinusoidal wall shaped microchannel was more complicated than triangular side-wall microchannel. Figur 8 illustrated that there were four meniscus formations of liquid flow in sinusoidal waveform microchannels. Shape of menisucs was concave in converging region, convex in diverging region, and flat in maximum and minum width.

The meniscus distinction caused geometry characteristic of the side-wall. The results have presented that it is posible to control meniscus shape of liquid flow in microchannels for improving sensitivity. Hence, we obtain stable antigen-antibody reaction as well as the flow rate will be indentical at the center and the edge of microchannels. The experimental results illustrated a significant difference between dynamic meniscus and static meniscus. The dynamic contact angle was around 90°, while static contact angle was slightly higher than 66°. In order to be understanding about this difference, there were three dimensionless parameters including dimensionless pressure, capillary number and Weber number that investigated in [4]. In this paper, the other aspect influenced on dynamic contact angle is the velocity of liquid flow, that to be caused by extremely narrow microchannels.

In experimental study, filling speed was also measured. Figure 8 compares the speed of fluid flow in the straight-wall microchannels. Three different microchannels was used. Width, depth and length of three different microchannels were 50μm × 60μm × 1088mm, 100μm × 60μm × 900mm, and 200μm × 60μm × 665mm, respectively. From Fig.9, it can be seen that there were a gradually decrease of velocity during the filling process. Because hydraulic resistance depends on microchannel size, as the width of microchannel is wider, filling speed is faster.

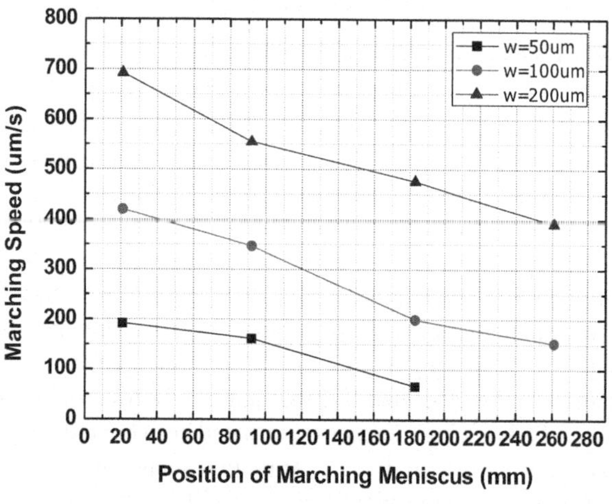

Fig. 9 Velocity of the flow versus position.

IV. Conclusion

Shape of advancing meniscus in the microchannel depends on surface energy as well as inner wall shape of patterns. This paper investigated the concave or convex meniscus correspondingly wall shape of microchannels. Interesting results provided design guideline of microfluidic channels for the biomedical diagnostic devices. The results of this paper have demonstrated that we could modify side-wall shape of the microchannel in order to control flow rate and sensitivity in microfluidic diagnostic devices.

V. Acknowledgment

This study was supported by the Regional Core Research Program funded by the Korea Ministry of Education, Science and Technology (Medical & Bio-material Research Center). Also, this research was financially supported by the Ministry of Education, Science Technology (MEST) and Korea Institute for Advancement of Technology (KIAT) through the Human Resource Training Project for Regional Innovation.

References

1. Henrik B Theoretical Microfluidics, MIC- Department of Micro and Nanotechnology Technical University of Denmark
2. Kahp Y S, Seungwoo C, Hong H L (2004) W-shaped meniscus from thin polymer films in microchannels. Journal of Microchannels and Microengineering, 2004, pp 1185-1189
3. Pilnam K, Ho-Young K, Jae Kwan K et al (2009) Multi-curvature liquid meniscus in a nanochannel: Evidence of interplay between intermolecular and surface forces. Lab chip, 2009, pp 3255-3260
4. Dong Sung K, Kwang-Cheol L, Tai Hun K et al (2002) Micro-channel filling flow considering surface tension effect. Journal of Microchannels and Microengineering, 2002, pp 236-246
5. Donggang Y, Byung K (2002) Simulation of the filling process in microchannels for polymeric materials. Journal of Microchannels and Microengineering, 2002, pp 604-610
6. Auro A S, Sushanta K M (2009) Effect of dynamic contact angle in a volume of fluid (VOF) model a mircofluidic capillary flow. Journal of Colloid and Interface Science, 2009, pp 461-480

Author: Young Ho Seo
Institute: Kangwon National University ENG.BLDG 6-308
Street: 1 Kangwondeahakgil
City: Chuncheon
Country: South of Korea
Email: mems@kangwon.ac.kr

Microfluidic Blood Cuvette for Hemoglobin Measurement

J.K. Son[1], H.J. Kim[1], W.K. Jang[1], Y.H. Seo[1], B.H. Kim[1,*], W.H. Lee[2], K.T. Park[2], K.B. Nahm[2], and E.Y. Choi[2]

[1] Medical & Bio-Materials Research Center and Dept. of Mechanical and Mechatronics Engineering
Kangwon National University, Chuncheon 200-701, South Korea
[2] Boditech Med Inc., G-tech Village, Chuncheon, South Korea

Abstract — **This paper presents hemoglobin measurement based on microfluidic blood cuvettes. Blood sample filled MBC by capillary action, and hemoglobin in the blood sample was measured by absorbance of 530nm-wavelenth light. Microfluidic blood cuvette was fabricated by poly-methylmethacrylate (PMMA) substrate and dry film photoresist (DFR). Two different MBC including dual-side wall and sing-side wall microchannels were designed and fabricated. During blood filling process, air was trapped in microfluidic blood cuvette of dual-side wall microchannel, while there was no air trapping in microfluidic blood cuvettes of single-side wall microchannel. Sensitivity of hemoglobin measurement was more linear in 105μm-depth microchannel rather than 35μm-depth microchannel.**

Keywords — **microfluidic blood cuvette, hemoglobin, blood-flow, capillary force, dry film photoresist (DFR).**

I. INTRODUCTION

Blood is supplying oxygen and nutrients at the cells of the body. Also blood is transporting carbon dioxide and the waste caused by the metabolism. The blood is composed plasma, red blood cells, white blood cells, platelets and various cells. By examining the various components of blood can diagnose various diseases. As an example, such as acquired immune deficiency syndrome (AIDS), malaria, anemia, and cancers, leukemia can be diagnosed. In particular, hemoglobin occupied approximately 33% of red blood cells volume, hemoglobin concentration analysis is necessary anemia or diagnosis of postoperative bleeding or blood dialysis of renal disease patients.

Iron deficiency anemia is an important public health problem resulting in considerable morbidity and mortality. It is a problem of serious public health significance, given its impact on psychological and physical development, behavior and work performance [1]. Anemia is primarily caused by impaired hematopoiesis with aging, and secondarily caused by various conditions associated with chronic diseases [2,3]. The World Health Organization(WHO) has recommended cutoff points of 13.0g/dl for males and 12.0g/dl for females. Hemoglobin measurement is the primary method for anemia screening and to study the prevalence in populations [4,5]. Measuring hemoglobin concentration is relatively easy and inexpensive, and this measurement is frequently used as a proxy indicator of iron deficiency. However, anemia can be caused by factors other than iron deficiency. In addition, in populations where the prevalence of inherited haemo-globinopathies is high, the mean level of Hemoglobin concentration may be lowered. This underlines that the etiology of anemia should be interpreted with caution if the only indicator used is Hemoglobin concentration. The main objective for assessing anemia is decision-makers on the type of measures to be taken to prevent and control anemia [6]. Estimate of the prevalence of anemia in turn depends upon the methods used for assessing hemoglobin concentration and the cut-off point applied [7,8]. There are various methods which have been recommended for estimation of hemoglobin for assessment of anemia. Each method has its advantages and limitations. The most widely used quantitative methods are Sahli's method, Cyan-methemoglobin method [9], and HemoCue methods [10].

Because sampling of accurate amount of blood and uniform distribution of blood cells are important in hemoglobin measurement, microfluidic chip is used [10,11]. Air tapping and blood cell aggregation in microchannel must be avoided during blood sampling into microchannels. We suggested microfluidic blood cuvette based on single-side wall microchannel to prevent air trapping, and controlled depth of microchannel to prevent blood cell aggregation. We also measured hemoglobin concentration based on Cyanmethemoglobin method with semiconductor laser of 530nm wavelength.

II. METERIALS AND METGIDS

A. *Hemoglobin Measurement Method*

In order to measure hemoglobin concentration based on cyanmethemoglobin method, blood cuvette, laser diode and photodiode are required. Cyanmethemoglobin has a maximum absorbance at 540 nm [12]. Color intensity is proportional to total hemoglobin concentration. Generally, Cyanmethemoglobin method uses a 520~550nm wavelength

V. Van Toi et al. (Eds.): 4th International Conference on Biomedical Engineering in Vietnam, IFMBE Proceedings 40, pp. 149–152, 2013.
www.springerlink.com

range. In this study, semiconductor lasers of 530nm wavelength were used. Figure 1 shows the hemoglobin measurement principle. Laser was exposed onto the blood contain chip. Cyanmethemoglobin absorbs 530nm-laser and then remaining laser was transmitted to photodiode. The higher hemoglobin concentration, the lower laser is transmitted.

B. Contact Angle Measurements of the Material

Polymer material of the PMMA substrate was used to fabricate microchannel and the wall of the microchannel was fabricated using the DFR film [13,14]. Blood sample was filled with microfluidic blood cuvette by capillary action, thus contact angle of the materials is important. Figure 2 shows the contact angle of DRF film and PMMA substrate. Contact angles of DRF and PMMA were measured 41° and 73°. These materials could generate capillary force.

C. Micro-channel Design and Simulation

Major requirement of microfluidic blood cuvette is prevention of air-trapping during blooding filling process.

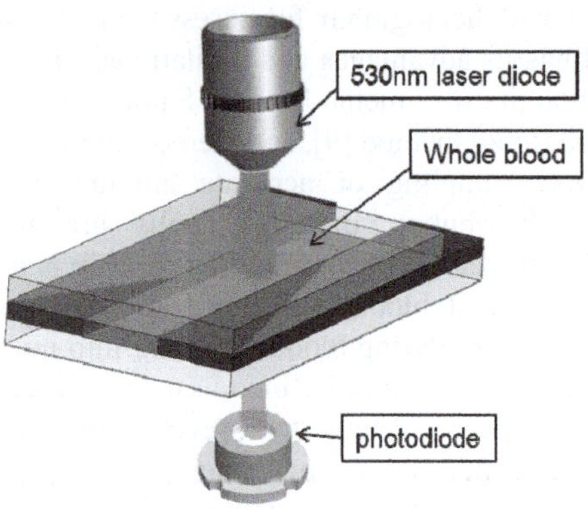

Fig. 1 Hemoglobin measurement principles: Cyanmethemoglobin absorbs 530nm-laser.

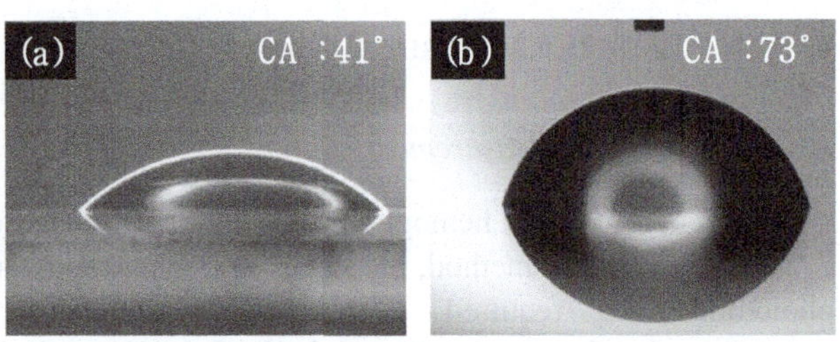

Fig. 2 Contact angel of DFR film and PMMA substrate; (a) Contact angle of DFR film, (b) Contact angle of PMMA substrate.

In this paper, two different microfluidic blood cuvettes were designed including dual-side wall and single-side wall cuvettes. Figure 3(a) and 3(b) show the dual-side wall and single-side wall microfluidic blood cuvette, respectively. Each type of microfluidic blood cuvette has three-different depth of microchannel. Totally six-different microfluidic blood cuvettes were fabricated and tested. Length and width of microchannels were 25mm and 5mm, respectively. Depth of microchannels was designed as 35, 70 and 105μm.

Firstly we estimated the fluid flow in microfluidic blood cuvette using computational fluid dynamics tool of CFD-ACE+. Figure 4(a) and 4(b) show 2-dimensional simulation results of advanced meniscus in dual-side wall and single-side wall microfluidic blood cuvettes, respectively. From the simulation results, dual-side wall microfluidic blood cuvette had possibility of air-trapping at the center line of microchannel.

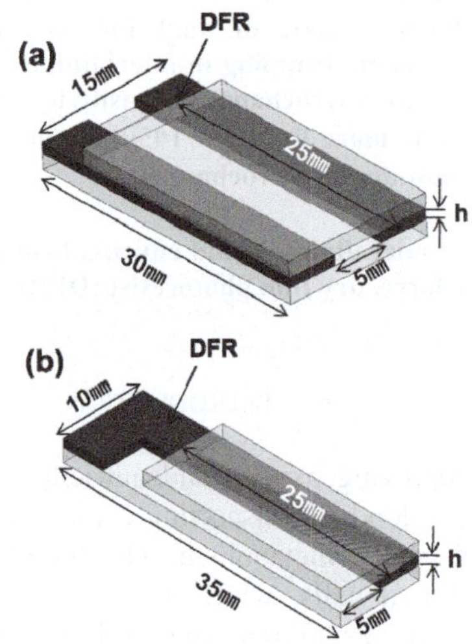

Fig. 3 Design of PMMA chip using DFR; (a) Microfluidic blood cuvette based on dual-side wall microchannel, (b) Microfluidic blood cuvette based on single-side wall microchannel.

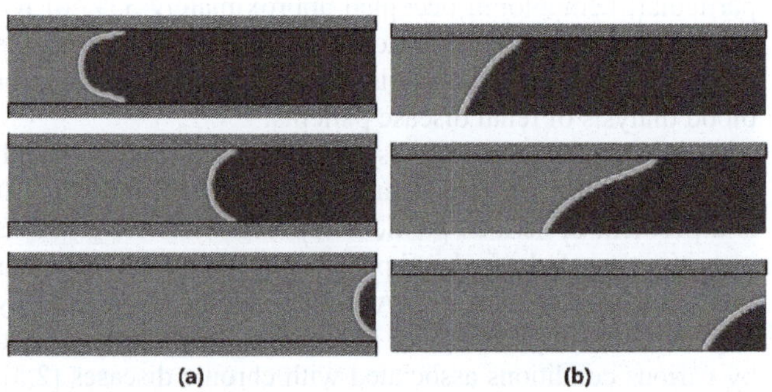

Fig. 4 2D simulation results of the blood flow characteristics; (a) MBC based on dual-side wall microchannel, (b) MBC based on single-side wall microchannel.

D. Fabrication of Microfluidic Blood Cuvette

Microfluidic blood cuvettes were fabricated by lamination process. Dry film resist (DFR) was laminated on PMMA substrate. Since thickness of DFR was approximately 35μm, 105μm-depth microchannel could be fabricated by three-times of lamination of DFR. Figure 5 shows the fabrication process steps. Firstly, DFR was bonded with PMMA substrate using a laminator at 90°C. After that, DFR was patterned by typical photolithography process. UV exposed DRF was developed by well-known developer of 1wt% of Na_2Co_3 solution during 10min. Figure 6 and 7 shows fabricated microfluidic blood cuvettes based on dual-side wall and single-side wall microchannels, respectively.

III. RESULTS AND DISSCUSSION

A. Blood Filling through Microfluidic Blood Cuvette

Using fabricated cuvettes, blood filling characteristics was observed. Blood of 20 μl was dropped at inlet channels and then blood moved into microchannel due to capillary action. Figure 7 shows blood filling characteristics of dual-side wall microfluidic blood cuvette. Figure 6(a), 6(b), and 7(c) blood filling process into 35μm-depth, 70μm-depth and 105μm-depth microchannels, respectively. In microfluidic blood cuvette based on dual-side wall microchannel, blood quickly moved at side wall region rather than center region of microchannel, because contact angle of DFR (41°) lower than contact angle of PMMA substrate (73°). Filling process was completed just in 10 sec. As shown in Figure 6(a), 6(b), and 6(c), air-trapping took place at the center of microchannels. Blood filling speed in 35μm-depth microchannel was approximately 1.7 mm/s, and that of in 105μm-depth microchannel was about 2.7 mm/s. Because hydraulic resistance was inversely proportional to microchannel depth, blood filling speed in 105μm-depth microchannel was faster than that in 35μm-depth microchannel. Moreover, blood flow mark which might be caused by blood aggregation was severely occurred in 35μm-depth microchannel.

Secondly, blood filling characteristics in single-side wall microfluidic blood cuvette was measured. As the first experiment, blood of 20μl was used. Dimension of microchannels in single-side wall type was same as that of dual-side wall type. Figure 7 shows the blooding filling characteristics through single-side wall microchannel. There was no air-trapping. These results were very well agreement with simulation results in Fig. 4(b).

Blood filling speed was faster as increasing depth of the microchannel. Blood filling speed through 35μm-depth microchannel was 1.0 mm/s, and that in 105μm-depth microchannel was 2.3 mm/s. Blood filling speed in single-side wall microchannels was slightly slower than that in dual-side wall microchannels. In single-side wall micro-channel, severe flow mark of blood also observed in 35μm-depth microchannel.

B. Hemoglobin Measurement Method

From the blooding filling experiments, air-trapping center of microchannels was generated in dual-side wall microchannels, and blood flow mark due to blood aggregation was observed in 35μm-depth microchannel. Therefore, measurement test of hemoglobin concentration, we used single-side wall microfluidic blood cuvette based on 105μm-depth microchannel, and compared with 35μm-depth microchannel.

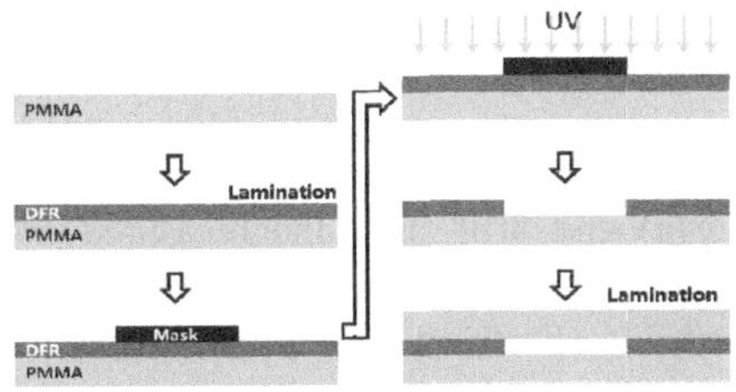

Fig. 5 Schematics of the fabrication process of chip.

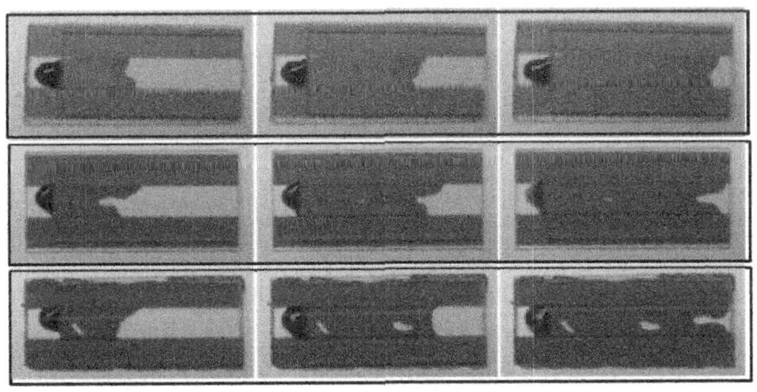

(a)
(b)
(c)

Fig. 6 Blood filling through different depth of MBC based on dual-side wall microchannel; (a) 35 μm, (b) 70μm, (c) 105 μm.

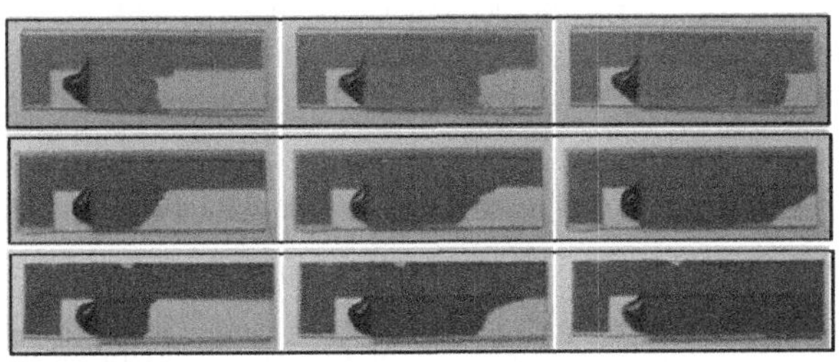

(a)
(b)
(c)

Fig. 7 Blood filling through different depth of MBC based on single-side wall microchannel; (a) 35 μm, (b) 70 μm, (c) 105 μm.

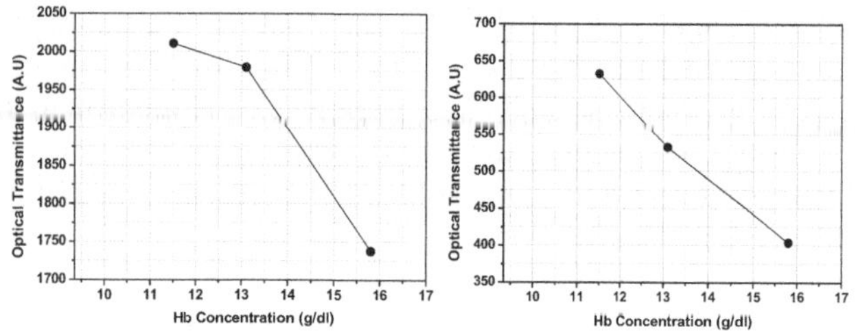

(a) (b)

Fig. 8 (a) Transmittance measurement of 35μm-depth microchannel (b) Transmittance measurement of 105μm-depth microchannel.

Because normal level of hemoglobin concentration for women was about 12 ~ 18 g/dl, we used three different hemoglobin concentration of 11.5, 13.1, 15.8 g/dl. Also we exposed 530nm-laser diode onto blood filled microfluidic cuvette and then measured transmitted intensity of 520nm-laser. Before experiment, hemoglobin was converted to cyanmethemoglobin(HiCN) by addition of the reagents potassium cyanide and potassium ferricyanide, and light transmittance at 530nm was measured.

Figure 8(a) and 8(b) show the measured optical transmittance using 35μm-depth and 105μm-depth microfluidic blood cuvettes, respectively. As depth of microchannel was increased, containing blood volume was increased, thus light transmittance was decreased. Conversely, light absorbance was increased. With increasing hemoglobin concentrations, light transmittance was decreased. In 105μm-depth microchannel, light transmittance (absorbance) was more linearly changed.

IV. Conclusions

Microfluidic blood cuvette for hemoglobin concentration was designed and fabricated. Using microfluidic blood cuvette based on single-side wall microchannel, we successfully avoid air-trapping and light transmittance was more linearly changed against hemoglobin concentration in 105μm-depth microfluidic blood cuvette. Consequently, single-side wall microchannel was most suitable to prevent air-trapping in blood cuvette, and microchannel depth must be larger than 100 μm to eliminate blood flow mark due to blood aggregation.

Acknowledgment

This study was supported by the Regional Core Research Program funded by the Korea Ministry of Education, Science and Technology (Medical & Bio-material Research Center). Also, this research was financially supported by the Ministry of Education, Science Technology (MEST) and National Research Foundation of Korea (NRF) through the Human Resource Training Project for Regional Innovation.

References

1. De Maeyer EM, Adiels-Tegman M. (1985) The prevalence of anaemia in the world. World Health Stat 1985, pp. 302-316.
2. Lipschitz DA, Udupa KB, Milton KY et al. (1984) Effect of age on hematopoiesis in man, Blood, 1984, pp. 502–509.
3. Baldwin JG, (1989) True anemia: Incidence and significance in the elderly, Geriatri 1989, pp. 33–36.
4. WHO Technical Report Series. Nutrition(1971): Eighth report of the joint FAO/WHO Expert Committee 1971 No. 477
5. DeMaeyer EM, Dallman P, Gurney JM, Hallberg L, Sood Sk, Srikantia SG, (1989), Preventing and controlling iron deficiency anemia through primary health care. A guide for health administrators and programme managers. Geneva, World Health Organization, 1989
6. Bruno de Benoist, Erin McLean, Ines Egli, Mary Cogswell, (2008), Worlfwide prevalence of anemia 1993-2005, 2008,
7. Stoltzfus RJ, (1997), Rethinking anemia surveillance, Lancet 1997, pp. 1764-1766.
8. Yip R, (1994), Iron Deficiency: contemporary scientific issues and international programmatic issues, J Nutr 1994, pp. 1479-1490.
9. Program for Appropriate Technology in Health (Path), (1996), Anaemia Detection in Health Services Guideline for Program Managers, U.S. Agency for International Development 1996, pp 37-38.
10. Von Schenk, H Falkensson M, Lundberg B, (1986), Evaluation of "HemoCue", a new device for determining hemoglobin. Clin Chem 1986, pp. 526-529.
11. Chakraborty, S. (2005) Dynamics of capillary flow of blood into a microfluidic channel, Vol. 5, Lab on a chip, 2005, pp. 421~430.
12. Tietz, N. W., Fundamentals of Clinical Chemistry, 2nd ed., W.B. Saunders Co., Philadelphia, p.411, (1976).
13. Vulto, P. Glade, N. Altomare, L. Bablet, J. Medoro, G. Leonardi, A. Romani, A. Chartier, I. Manaresi, N. Tartagni, M. Guerrieri, R. (2004) "Dry film resist for fast fluidic prototyping", 8th International Conference on Miniaturized Systems in Chemistry and Life Sciences - Micro Total Analysis System. 2004, pp. 43~45
14. Yuan-Chien Tsai, Hsiu-Ping Jen, Kuan-Wen Lin, You-Zung Hsieh, (2006). Fabrication of microfluidic devices using dry film photoresist for microchip capillary electrophoresis, Vol. 1111, Journal of chromatography A. 2006, pp. 267~271

Use macro [author address] to enter the address of the corresponding author:

Author: Byeong Hee Kim
Institute: Kangwon National University ENG.BLDG.6
Street: 1 Kangwondaehakgil
City: Chuncheon-si
Country: South of korea
Email: kbh@kangwon.ac.kr

Detection of Platelet Activation in Blood Tubing Line by Observing Ultra-Weak Luminescence

Ryoma Kakegawa and Toshia Fujisato

Major in Biomedical Engineering, Graduate School of Engineering, Osaka Institute of Technology, Osaka, Japan

Abstract — **We have applied the luciferin-luciferase reaction and are developing a technique to detect the released ATP from platelets in their aggregation. A photomultiplier tube (PMT) was used as a photodetector and a serial counting device was connected to count the output signal from the PMT. Using this apparatus, the luminous reaction of rat heparinized blood was observed in two types of blood circuits; one was totally composed of poly vinyl chloride (PVC) tube and the other was inserted a glass tube section into the PVC circuit. As a result, after the removal of residual ATPs in plasma by luciferin, emission intensity progressively increased in the both circuits. In the circuit without a glass section, the emission intensity was remarkably lower.**

Keywords — **platelet, thrombus formation, luciferin, in vitro.**

I. INTRODUCTION

The thrombus formation in blood pump used in extracorporeal circulation and artificial heart is still a major problem. The thrombus formation occurs in association with contact with the pump material. It is also affected by the flow-field in blood circulation such as high-shear or stagnation [1][2][3][4][5]. It is necessary for the development of blood pump to investigate the association between flow-field and thrombus formation. Usually the thrombus formation in blood pump is evaluated by both in vivo (long-term animal experiment, e.g.) and in vitro (Lee-White test, e.g.) experiments [6][7][8]. However, it is difficult to observe its process in real time by these methods. There are ethical and economic problems, too.

Initial phase of thrombus formation in platelet aggregation, adenosine tri-phosphates (ATPs) are released from activated platelets. These ATPs can be detected safe by luminescence measurement using the luciferin-luciferase reaction (LL reaction, Formula 1) [9].

In this study, an apparatus to detect platelet aggregation at a high sensitivity by using LL reaction was reported.

$$\mathbf{ATP + Luciferin}$$
$$\downarrow \mathbf{Luciferase, O_2, Mg^{2+}}$$
$$\mathbf{Oxyluciferin + CO_2 + Pyrophosphate + Light} \quad (1)$$

II. MATERIALS AND METHODS

A. Detection Apparatus

The LL reaction radiates very weak luminescence in this system since the amount of ATPs in platelet is very small. A photomultiplier tube (PMT; H7360-01, Hamamatsu Photonics) was used because it has very high sensitivity for detecting a single photon. A serial counting device was connected to the PMT to detect the output signal (Fig.1).

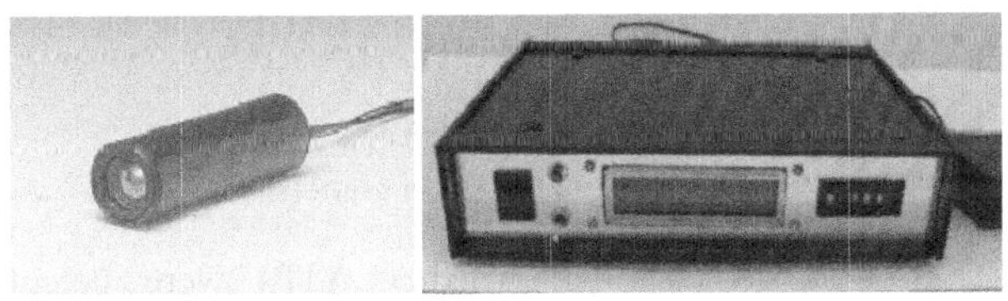

Fig. 1 PMT (left) and counting device (right).

B. Experimental System

The LL reaction was observed in a dark room. A glass test tube was set with holding fixture on a side of the apparatus box. The PMT was put on the bottom of the box and it was 10cm away from the test tube. One ml syringe filled with reagent solution was located on the top of the test tube and the reagent in syringe was injected into the test tube by push bench. The output signals from PMT were counted by the serial counting device connected. The data was corrected to PC (Fig.2).

V. Van Toi et al. (Eds.): 4th International Conference on Biomedical Engineering in Vietnam, IFMBE Proceedings 40, pp. 153–156, 2013.
www.springerlink.com

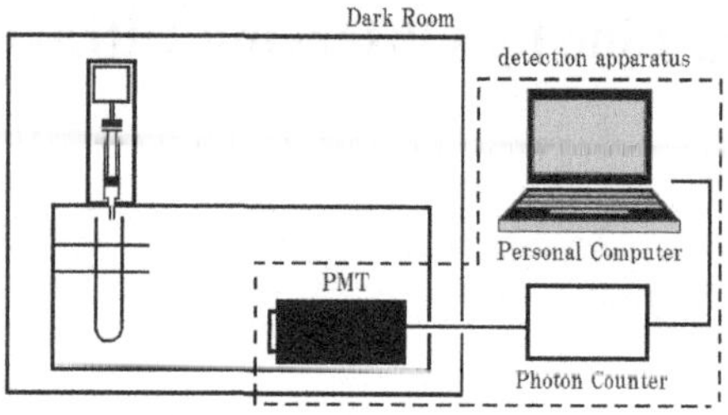

Fig. 2 Experimental System.

C. *Blood Coagulation Test*

The coagulation time of blood was measured by Lee-White method [8]. One ml of rat venous blood was injected into two glass test tube having diameter of 11mm and length of 75mm. After 3 minutes from blood collection, the first test tube was mixed every 30 seconds. If the blood was coagulated in the first test tube, the second test tube was mixed every 30 seconds. The blood coagulation time was measured from blood collection in the second test tube (Fig.3). All animal experiments were performed with the approval of the Animal Experiment Committee of the institution.

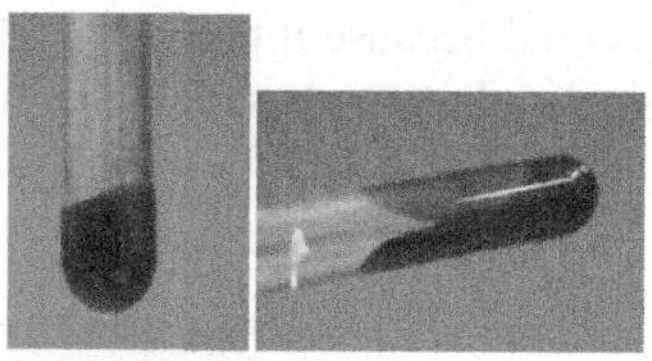

Fig. 3 Lee-White test.

D. *Static Emission*

At first, to establish whether ATPs were detected quantitatively by luminous reaction using the apparatus, LL reaction was observed between ATP and luciferin reagents. After glass test tube was filled with 500μL of ATP reagent, emission intensity was observed by 100μL luciferin reagent injection (Fig.4).

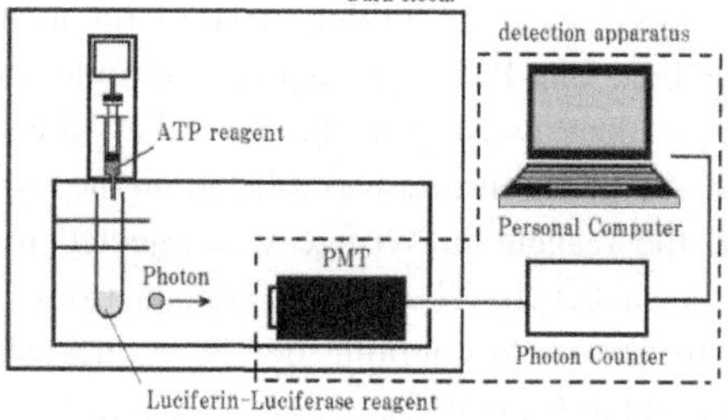

Fig. 4 LL reaction using luciferin and ATP reagents in test tube.

Second, to establish whether this method was applied to thrombus formation, LL reaction was observed between heparinized blood and luciferin reagents in glass test tube. After glass test tube was filled with 500μL of rat heparinized blood, 100μL of luciferin and protamine sulfate reagent was injected into the tube. Then the rat heparinized blood was reacted with protamine sulfate and blood coagulation was started by the foreign-body reaction at glass surface. The amount of light emission was detected by the detection apparatus (Fig.5).

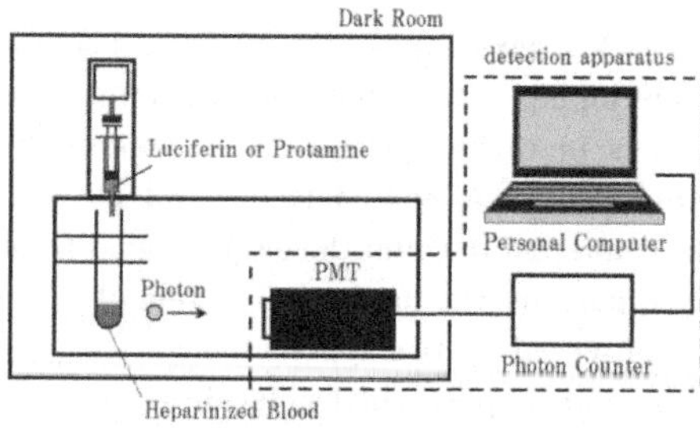

Fig. 5 LL reaction using blood in test tube.

E. *Dynamic Emission*

The thrombus formation in blood circuit was observed using this detection method. The amount of thrombus formation was compared with two different circuit materials; one was totally composed of poly vinyl chloride (PVC) tube (type A circuit) and the other was inserted a glass tube (length of 6cm) section into the PVC circuit (type B circuit). Blood circulation was driven by a roller pump. After the each blood circuit was filled with 4mL of rat heparinized blood, luciferin and protamine sulfate reagent was injected into the blood circuit by 1ml syringe at the same time. The amount of light emission was detected by the detection apparatus (Fig.6).

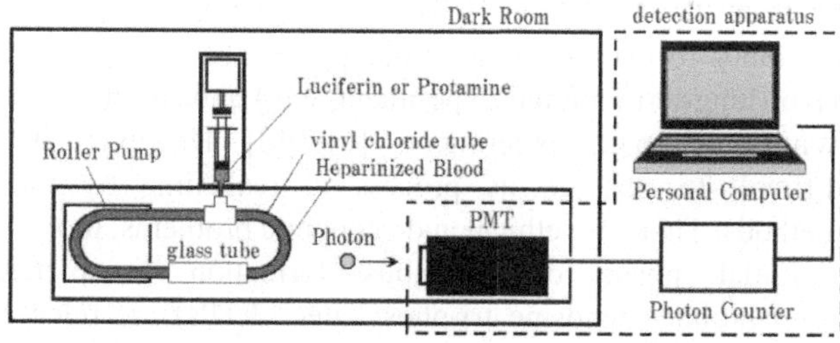

Fig. 6 LL reaction using blood in blood circuit.

III. RESULTS

Fig.7 shows the intensity change of the light emission after luciferin reagent was injected into the glass test tube filled with ATP reagent. This result shows a linear relationship between ATP reagent concentration and emission intensity. The coagulation time using Lee-White method was about 8 minutes (Fig.8).

Fig.9 shows the intensity change of the light emission after ATP and luciferin reagent were injected into the glass test tube filled with rat heparinized blood. Immediately after the luciferin injection, luciferin reacted with residual ATPs in plasma. After the amount of light emission was lower and protamine sulfate was injected, the emission intensity increased according to the reaction with ATPs released from the activated platelets on the glass surface. The peak emission intensity was observed about 8 minutes after protamine sulfate injection.

Fig.10 shows the intensity change of the light emission after reagent was injected into the each blood circuits. Dynamic emission experiment showed similar findings to the static emission experiment. After protamine sulfate injection, the emission intensity increased. The peak emission intensity in both circuits was observed about 7 minutes after protamine sulfate injection. After dynamic emission measurement, thrombus formation on the tube surface was higher in type B circuit than type A circuit (Figs.11 and 12).

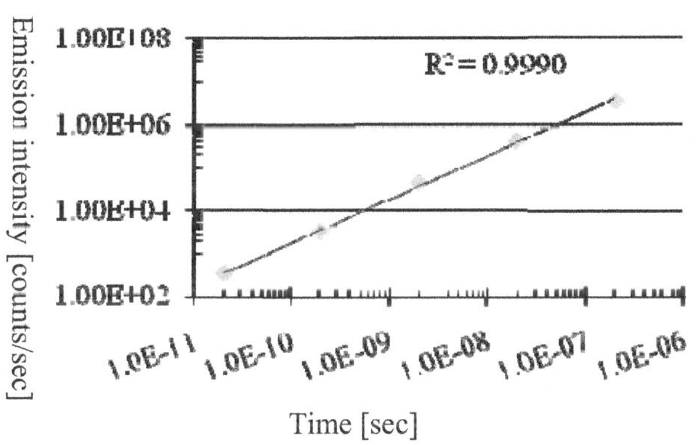

Fig. 7 Chenge of emission intensity by LL reaction.

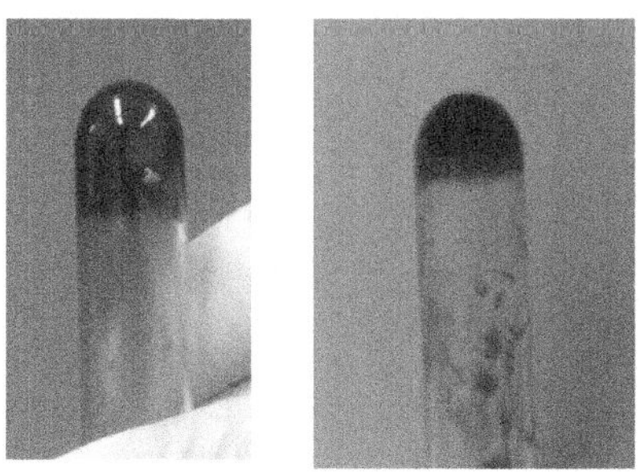

Fig. 8 The blood coagulation test using Lee-White method.

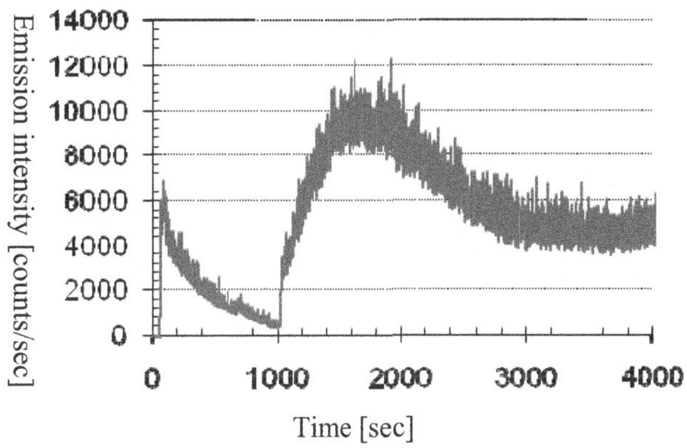

Fig. 9 Chenge of emission intensity by LL reaction in test tube.

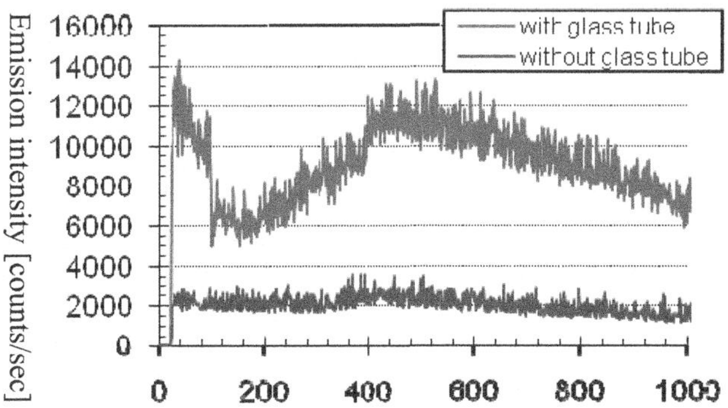

Fig. 10 Chenge of emission intensity by LL reaction in blood circuit.

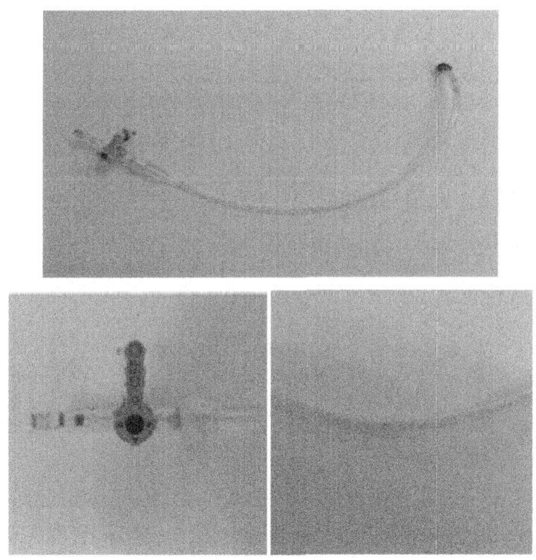

Fig. 11 Thrombus formation in type A circuit; whole circuit is top, reagent inlet is left down, and blood path is right down.

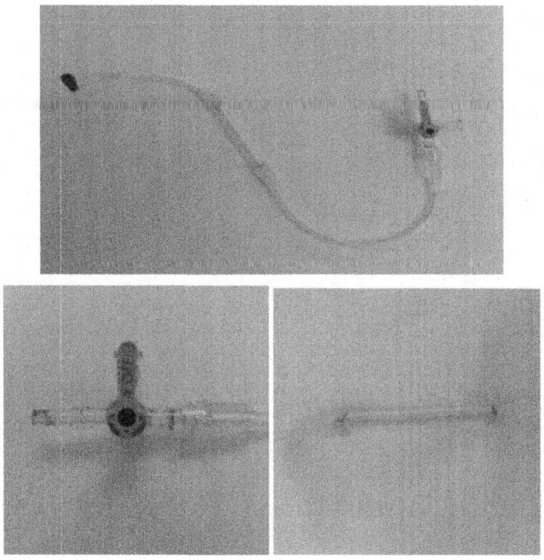

Fig. 12 Thrombus formation in type B circuit; whole circuit is top, reagent inlet is left down, and blood path is right down.

IV. Discussion

After luciferin injection, a sudden increase of emission intensity was observed by the reaction with residual ATPs in plasma. Then after protamine sulfate injection, emission intensity increased according to the reaction with ATPs released from platelets activated on the glass surface.

We think it is still necessary to optimize experimental condition. Also, since this apparatus could not get locational information of blood coagulation, additional CCD cameras may be introduced in this system.

V. Conclusions

The thrombus formation in blood circuit may be evaluated in real time by detecting luminescence from LL reaction.

References

1. Denton A. Cooley (2003) The total artificial heart. Nature Medicine 9(1): 108-111

2. Kozaburo Hayashi. (1977) Artificial Heart and its Materials. TheJapan Society of Materials Science 26 (286): 617-629.

3. Masahiro Toyoda et al. (2001) Geometric Optimization for Nonthrombogenicity of a Centrifugal Blood Pump through Flow Visualization. The Japan Society of Mechanical Engineers. 67 (664): 3127-3133.

4. Yosuke Tomari et al. (2006) Simple in vitro anti-thrombogenic testing for a centrifugal blood pump. The Japan Society of Mechanical Engineers. 5 (66): 353-354.

5. Shahrokh Falati et al. (2002) Real-time in vivo imaging of platelets, tissue factor and fibrin during arterial thrombus formation in the mouse. Nature Medicine. 8(10): 1175-1182.

6. Tetsumei Urano. (2008) Intra vita real time imaging analysis of thrombus formation. The Japan Scociety on Thrombosis and Hemostasis. 19 (6): 808-813.

7. Katsuya Fukushima et al. (2007) Visualization of Thrombus Formation Process and Flow Analysis on Shear Blood Flow. Japanese Society for Medical and Biological Engineering. 45(1): 99-105.

8. Roger I. Lee, Paul D. White. (1913) A clinical study of the coagulation time of blood. American J. Med. Sciences 145 (4): 495-503

9. Kenichiro Nakashima. (2000) Analytical methods utilizing a chemiluminescence measurement. The Japan Society for Analytical Chemistry 49 (3): 135-159.

Surface Nano-texturing of Stainless Steel for Biocompatibility Enhancement of Implantable Medical Devices

M. Kang, Y.M. Park, H.C. Yang, Y.H. Seo, and B.H. Kim*

Medical & Bio-Materials Research Center and Dept. of Mechanical and Mechatronics Engineering, Kangwon National University, Chuncheon 200-701, South Korea

Abstract — **Stainless steel (SUS) has been used in many industrial fields such as bio and medical parts. Especially, implanted medical device (IMD) has been widely researched for enhancing a human life from biological treatment using those devices. These medical devices require some essential requirements, such as biocompatible, non-toxic and human safety. For these reason, many researchers have been suggest a method of surface treatment to decrease a biological damages due to implanted in vivo. However, these proposed methods were had some disadvantages such as semi-permanent and require additional processes. To overcome these disadvantages, our research proposes a surface texturing method that is anodic oxidation based on electro-chemical reaction. Firstly, 1μm aluminum thin layer was deposited on the 4inch SUS-316L wafer. At this point, for the comparison of SUS-316L wafer with different surface roughness conditions, we have prepared polished and unpolished surface wafer. Using suggested method, porous aluminum oxide layer was formed and then we have removed oxide layer using alumina chemical etchant. Finally, we have fabricated various shapes of irregular nano/micro dimple structures with 100nm to 50μm according to surface roughness conditions.***

Keywords — **Stainless steel-316L, Surface texturing, Micro/nano dimple structure, Biocompatibility, aluminum anodic oxidation, Dielectric breakdown.**

I. Introduction

Recently, as quality of life is improving and aging population increasing, medical system has increased the need for qualitative improvement. In particular, implanted medical devices (IMD) such as stent, total hip, spinal rod and spine are one of the most profitable parts of the modern healthcare industry[1-3]. These IMD systems have been researched for replacement of body part that has lost function and for assist in healing. Furthermore, Bio-materials have been treated as an important part of IMD because of there should be non-toxic and biocompatibility to human. Actually, the human body has been formed harsh environment conditions, adverse reactions and immune responses were caused by implantable device [4]. Also, when in direct contact with tissues or cells in the affected area, that materials mechanical properties and functions were changed by reaction on the surface. As an example the case of stent, surface treatment for reducing the blood vessel inflammations reduced the restenosis. [5, 6] In additions, the case of medical needle, the research has been reported about minimize the pain and damage to normal tissues from the surface treatment. [7] Therefore, IMD should be improving the biocompatibility through the chemical and mechanical surface treatment. Conventional method for improve the biocompatibility such as choose the bio-materials and chemical surface coating or mechanical surface texturing have been researched, though [8, 9] these methods have a disadvantages such as required for high cost, long process time, limits of material and additional processes. For these reason, our research propose the surface texturing method using electro-chemical reaction based anodic oxidation process on the 4inch SUS-316L wafer that have a biocompatibility. Using that method we have performed surface texturing on polished and unpolished surface for the comparison with different surface roughness conditions and then we have fabricated dimple structures. Finally, various shapes of micro/nano dimple structures with 100nm to 50μm were fabricated on all surfaces and we have verified a structure formation mechanisms and differences according to surface roughness.

II. Fabrication of micro/nano dimple structures

A. *Aluminum Anodic Oxidation Process*

Anodic oxidation is the process of adding a layer of metal oxide with porous structures to specific metal surface and aluminum is the most commonly anodized metal because it is cheap and oxidized very easily. That process is reacted based on electro-chemical reaction on the boundary surface of aluminum and electrolyte. The hydrogen liberated from the electrolysis forms in the electrolyte at the cathode and the oxygen forms on the surface of the aluminum part at the anode. Oxygen readily combines with

* Corresponding author.

V. Van Toi et al. (Eds.): 4th International Conference on Biomedical Engineering in Vietnam, IFMBE Proceedings 40, pp. 157–160, 2013.
www.springerlink.com

the aluminum to form a layer of aluminum oxide. Since the verified about the formation of porous nanostructures by anodic oxidation process, various application researches have been reported. [10] in this paper, we have also carried out aluminum oxidation process using homemade process equipment. Figure 1 shows the schematics of experimental. As shown in Fig.1, temperature of target material was directly controlled during the process. Because temperature directly influences on chemical reaction rate, temperature of electrolyte is most significant process parameter. Anodic oxidation equipment is consisted of DC power supply, counter electrode of titanium, electrolyte bath, and cooling plate. Using that equipment, we have fabricated various shapes of dimple structures on that surface.

Typically, pure aluminum plate or thin film aluminum deposited by vacuum system were used in anodic oxidation process. In order to fabricate nano-porous alumina, thin film aluminum was generally deposited on silicon and glass substrates which had high electrical resistance or insulated material. When the underlying layer of oxidized aluminum was aluminum, scallop-shape bottom-barrier oxide formed. [11] On the other hand, when aluminum layer was deposited on noble metals or high electric resistive materials, inverted scallop-shape barrier oxide formed at the interface between aluminum and underlying material.[11]

However, when underlying layer of aluminum is high electric conductive material, formed alumina is non-porous layer not porous layer. In this case, non-porous alumina acts as dielectric layer. Due to irregular surface morphology, electric field is concentrated on a specific point, and then stainless steel which is subjacent materials of non-porous alumina is locally damaged by electro-chemical reaction. At the same time, dielectric breakdown is occurred on non-porous alumina.

B. Surface Nano- texturing

In order to fabricate a porous dimple structures on the SUS-316L wafer, anodic oxidation process with aluminum was carried out. In generally, anodic oxidation process has been performed in the acidic electrolyte such as sulfuric acid, oxalic acid, and phosphoric acid. Size of nano dimple structures depends on electrolyte and voltage. Oxalic acid and DC 10V were used in this study. Figure 2 show the schematic diagram of process steps of anodic oxidation. In the experimental study, we prepared two types of 4inch SUS-316L wafer including polished and unpolished wafers for investigation of surface roughness effect on electro-chemical reactions (Fig.2(a)). 1μm-thick aluminum layer was sputtered on the wafer (Fig.2(b)), and anodic oxidation was carried out in 0.3M oxalic acid solution at 2°C, 10V for 2hr. (Fig.2(c)). Because stainless steel which is underlying layer of aluminum is electric conductive material, non-porous alumina was formed as shown in Fig.2(d). Due to irregular surface morphology of alumina, electric field was locally concentrated, and then stainless steel was damaged. Damaged region was eventually etched away. (Fig.2.(e)) Finally surface textured stainless steel with micro- and nano-dimple structures was fabricated after removing alumina layer. (Fig.2(f)) As the process time of anodic oxidation increased, the number and size of formed micro- and nano-dimple structures increased.

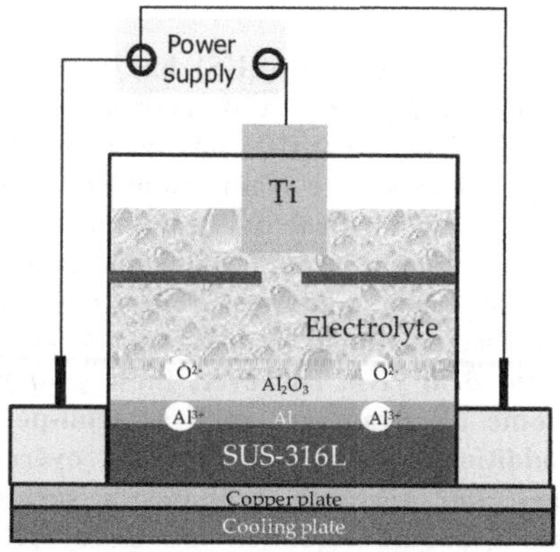

Fig. 2 Schematic diagram of anodizing equipment and that including principle of anodizing phenomenon; Process equipment configuration: counter electrode, bath, copper plate, cooling plate, DC power supply and electrolyte

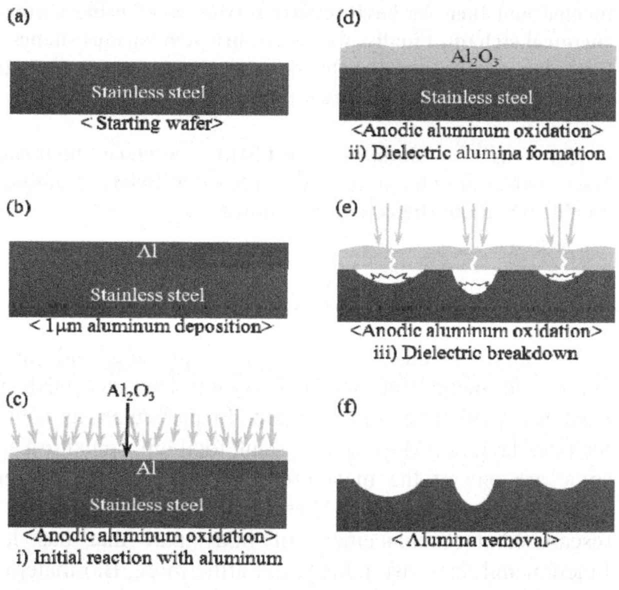

Fig. 2 Schematic diagrams of fabrication procedure with micro- and nano dimple structures on the 4inch SUS-316L wafer using anodizing process: (a) Preparation of polished SUS-316L wafer; (b) 1μm aluminum deposition on the wafer for anodizing process; (c) Initiation of aluminum anodic oxidation (0.3M oxalic acid, 2°C, 10V, 30min); (d) Completion of dielectric alumina formation; (e) Stainless steel was damaged by locally concentrated electric field and dielectric breakdown of alumina; (f) Alumina removal.

Non-porous alumina layer was removed using alumina etch solution of 1.8 wt.% chromic acid and 6 wt.% phosphoric acid at 65°C for 2h. Dimple structures were fabricated on the entire surface of SUS-316L wafer.

C. *Experiment Results*

Micro- and nano-dimple structures were fabricated on the SUS-316L surface using suggested anodizing process. For investigation of surface roughness effect, polished and unpolished wafers were prepared. Figure 3(a) and 3(b) show the SEM images of unpolished and polished SUS-316L surface, respectively. Surface roughness was improved by polishing process and measured surface roughness of unpolished and polished wafers were 20.59nm and 0.97nm, respectively. Anodic oxidation process was carried out on those two different wafers.

Figure 4(a) and 4(b) show anodized aluminum surface on unpolished and polished wafer, respectively. Dielectric alumina was still remained on the stainless steel. Size of damaged region due to electric field concentration on unpolished SUS-316L wafer was smaller than that on polished SUS-316L wafer. On unpolished wafer, size of damaged region was approximately 100 ~ 200nm, while that on polished wafer was about 10 ~ 50 μm. The more surface was rough, the smaller region was damaged. Alumina layer which underwent dielectric breakdown was observed in Fig.4(b)

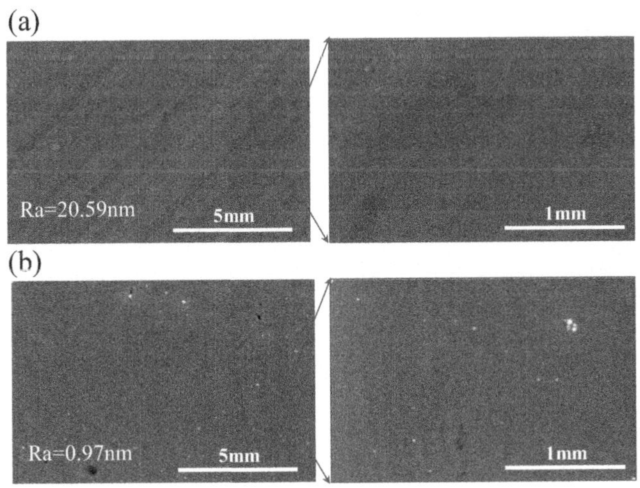

Fig. 3 SEM images of SUS-316L wafer surface with different surface roughness conditions; (a) Unpolished (Ra=20.59nm), (b) Polished (Ra= 0.97nm).

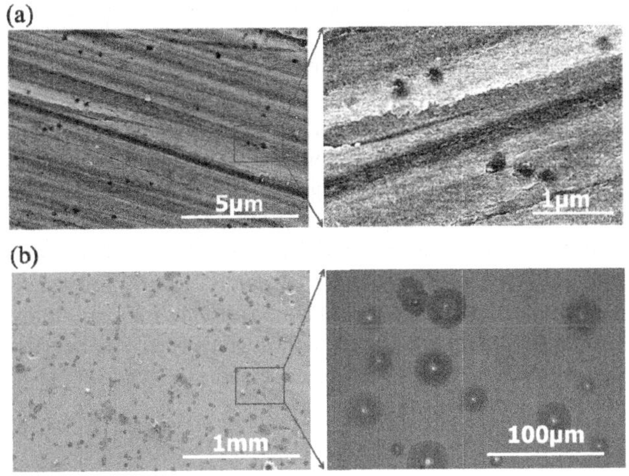

Fig. 4 Fabrication results of anodizing process using 1μm aluminum deposited SUS-316L wafer. All surface covered with irregularly formed concave micro/nano structures; (a) base material: unpolished SUS-316L wafer, (b) base material: polished SUS-316L wafer

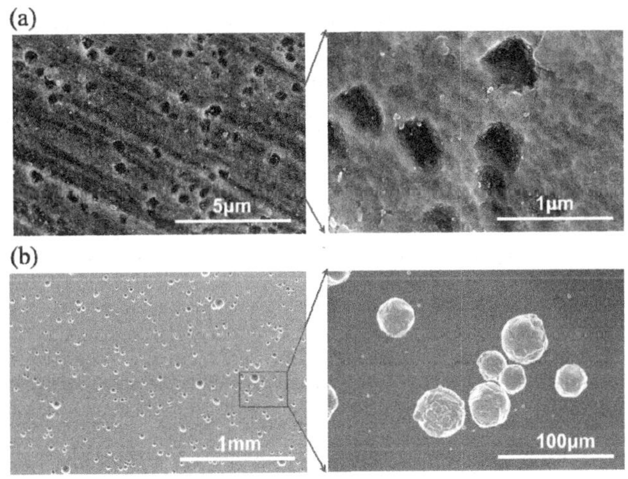

Fig. 5 SEM images of micro/nano concave structured SUS-316L surface after selectively removing the aluminum oxide layer by chemical etchant; (a) using a unpolished surface, (b) using a polished surface.

Figure 5(a) and 5(b) show micro- and nano-dimples structures on the SUS-316L surface after selectively removing the dielectric alumina layer by chemical etchant. Unpolished SUS-316L surface was covered with irregularly formed nano-dimple structures of 100 ~ 200nm, and polished wafer was covered by several tens micrometers size dimples. These dimple structures were formed by local concentration of electric field and their location was just underneath of the dielectrically breakdowned alumina.

III. Conclusions

We have suggested surface texturing method of stainless steel which based on electro-chemical reaction. In order to forming micro- and nano-dimple structures on the surface of stainless steel, modified aluminum anodic oxidation process was used. Irregular nano-dimple structures of 100 ~ 200nm were formed on unpolished SUS-316L surface, while polished wafer was covered by several tens micrometers size dimples. These dimple structures were formed by localized electro-chemical damage effect. Size of dimple structure formed on unpolished surface was smaller than that on polished surface. The more surface was rough, the smaller dimple was formed.

Acknowledgment

This research was supported by a grant (06K1401-00511) from the Center for Nanoscale Mechatronics & Manufacturing, one of the 21st Century Frontier Research Programs, which are supported by Ministry of Education, Science and Technology, Korea. Also, This study (work) was supported by the Regional Core Research Program funded by the Korea Ministry of Education, Science and Technology (Medical & Bio-material Research Center).

References

1. Leiming Gao, PeiranYang, IanDymond, JohnFisher, ZhongminJin, (2010) Effect ofsurfacetexturingontheelastohydrodynamiclubricationanalysis of metal-on-metal hip implants,vol. 43, TribologyInternational, 20010, pp 1851-1860
2. L. Le Gu´ehennec, A. Soueidan, P. Layrolle , Y. Amouriq, (2007) Surface treatments of titanium dental implants for rapid osseointegration, vol. 23, dental materials, 2007, pp 844-854
3. Paul K. Chu, (2007) Plasma surface treatment of artificial orthopedic and cardiovascular biomaterials, vol. 201, Surface & Coatings Technology, 2007, pp 5601-5606
4. Ho-Yeon Song, Young-Hee Kim and Byong-Taek Lee, (2009) P Biocompatibility of Biomedical Materials, vol. 22, Trends in metals & materials engineering, 2009, pp 21-28
5. Joong Gwun Park, Tae Won Kang, Kee Sung Lee, and Tae-Woo Kim, (2007) Material Design and Analysis of Coronary Artery Stents,vol. 7, Journal of the Korean Ceramic Society, 2007, pp 362-367
6. E. Schampaert, E. a Cohen, M. Schlüter, F. Reeves, M. Traboulsi, L.M. Title, R.E. Kuntz, and J.J. Popma, (2004) The Canadian study of the sirolimus-eluting stent in the treatment of patients with long de novo lesions in small native coronary arteries (C-SIRIUS),vol. 43, Journal of the American College of Cardiology, 2004, pp 1110-1115
7. S.M. Cho, H.J. Chung, M.T. Kim, and K.Y. Rhee, (2009) A Study on the Ceramic Coating of Biopsy Needle,vol.26, Precision Engineering, 2009, pp 121-126
8. I.-bu Sohn, M.-jin Ko, Y.S. Kim, and Y.-chul Noh, (2009) Femtosecond Laser Lithography for Mas kless PR Patterning,vol. 26, Precision Engineering, 2009, pp 36-41
9. Jeongwon Han, Jun-Hyuk Choi, Yeong-Eun Yoo, Byeong Hee Kim, Jaesook Lee and Shinill Kang, (2009) Nano Replication Technology of Nano Patterns and Application Fields, vol. 26, Precision Engineering, 2009, pp 30-35
10. Z. Su and W. Zhou, (2008) Porous Anodic Metal Oxides, vol.16, Science Foundation in China, 2008, pp 36-53
11. Jihun Oh and Carl V. Thompson, (2008) Selective Barrier Perforation in Porous Alumina Anodized on Substrates, vol.20, Adv. Mater, 2008, pp 1368-1372

Use macro [author address] to enter the address of the corresponding author:

Author: Byeong Hee Kim
Institute: Kangwon National Univesity
Street: 1 Knagwondaehakgil Eng. Bldg.6-111
City: Chuncheon
Country: South Korea
Email: kbh@kangwon.ac.kr

Blood Separation Device Based on Microstructures in Microchannels

Kang Il Byun, Hyun Chul Bang, Hyung Jin Kim, Eui Don Han, Young Ho Seo*, and Byeong Hee Kim

Medical & Bio-Materials Research Center and Dept. of Mechanical and Mechatronics Engineering
Kangwon National University, Chuncheon 200-701, South Korea

Abstract — **In this paper, plasma extraction device based on microstructure in microchannels was presented. Also, we analyzed the blood flow according to blood transportation methods for application of blood separation. The micro-channels for blood separation were designed blood cell channel, plasma channel and separation channel. The separation channel to separate the plasma was performed the role of micro filter based. The overall shape of microchannels was designed to spiral shape. Spiral microchannel was designed for prevention of blood cell cross-over from blood cell channel to plasma channel and clogging. Microchannels were made of polymethylsiloxane which is optically transparent. Blood sample was diluted with phosphate buffered saline at the ratio of 1:3. Characteristics of blood cell flow through microchannel, also, were observed with respect to two different blood transportation methods including pushing by syringe pump and pulling by vacuum pump. As results, blood flow driven by vacuum pump (pulling) was slower than that driven by syringe pump (pushing), but almost no blood cell was crossover from blood cell channel to plasma channel in pulling transportation method.**

Keywords — **Spiral microchannel, Microstructure, Blood separation device, Plasma extraction.**

I. Introduction

Point-of-care testing devices can enable medical system to monitor individual patient continuously, inexpensively and easily. So patient can be self-diagnosed in the home. Also, biologists have developed many effective antibodies for heart disease, diabetes, prostate cancer and so on. However, automation and miniaturization of sample preparation techniques are difficult. A blood based medical diagnosis generally uses a specific protein in blood, thus blood cell was separated and removed by centrifugal separation system. A diagnostic chip based on whole blood hardly provides the quantitative information of pathogens, because of the blood cell. Therefore, plasma extraction from whole blood is first step in a diagnostic chip.

There were so many methods of plasma extraction from whole blood [1-6]. The first is method using the inertia of the fluid particle. When the fluid is flow in the microchannels, fluid occur lift and drag because fluid speed is different in the microchannel wall and center. Then fluid particle is stop in equilibrium position of forces. As such, blood is can be separated using the induced dean vortices [1, 2]. The second is method using the fluid dynamics. Generally fluid is laminar flow in the microchannel. In the laminar flow, if width of microchannel is suddenly wider, particle is flow along flow streamline. There is a zone where particles (blood cells) are not flowing in the microchannels. When you create microchannel in this zone is capable of separating plasma. The particles can be separated using this zone [3, 4]. The third method uses the microstructure as microfilters. Xing Chen was introduced the separation method using microstructure according to particle size of blood such as red blood cell, white blood cell, plasma and so on [5, 6].

Microfiltration based on microstructures such as micropillar arrays, microweir, and membrane with holes could be classified into dead-end filtration and crossflow filtration. But most of the current separation microchips are based on dead-end filtration which might lead to clogging or jamming. For dead-end filtration, the fluidic flow is perpendicular to microfilters so that smaller particles pass through the microfilters with the fluid, while larger particles are stopped and clogging or jamming take place [7, 8].

To overcome clogging of dead-end filtration, crossflow filtration, which allows the bigger particles to stay in a suspended state instead of being deposited after being stopped, can be used for separation [9, 10]. For crossflow filtration, the fluidic flow is parallel with to microfilters. In plasma extraction from whole blood, blood cells are clogging into microfilters or crossover the microfilter due to deformability of the blood cells.

In this paper, we suggested microfilters in spiral micro-channels in order to prevent blood cell crossover through microfilters. Also, we compared blood transportation methods including pulling and pushing methods.

II. Blood separation device design

Plasma extraction device consists of two spiral micro-channels and microfilters which are connected two spiral channels. One of two spiral channels is blood cell channel and the other is plasma channel where extracted plasma flows. When fluid flows through spiral microchannel,

* Corresponding author.

secondary flow of dean vortex is generated across the microchannel [11]. Rotational flow across the microchannel due to dean vortex in the spiral micro-channels might help to remove the clogged blood cell in the microfilters.

Figure 1 shows the schematic view of the plasma extraction devices and roles of each microchannel. When blood sample is dropped in the blood inlet chamber, blood is moved into blood cell channel, and then only plasma is extracted into plasma channel. Size of the microfilters is smaller than red blood cell.

Dimensions of microchannels are presented in Fig.2. For the spiral microchannels, the starting radius was 3mm, and increment of radius was 140μm every turn. Cross-sectional dimension of blood cell channel and plasma channel were 50 × 20μm and 50 × 1.5μm, respectively. Array of micro-dimple of 2μm-diameter were located in plasma channel to prevent adhesion of upper and lower polymethylsiloxane (PDMS) plates, because width was large in comparison with height. Length of blood cell and plasma channels was approximately 108mm. Cross-sectional dimension of microfilter was 2.0 × 1.5μm, and each length was 20μm as shown in Fig.2.

III. FABRICATED AND EXPERIMENT

A. Fabrication of Plasma Extraction Device

Most of laboratory-scale experiments of blood separation used PDMS microchannel, because microchannels were easily fabricated by PDMS casting process, PDMS was biocompatible material and PDMS was optically transparent. We also fabricated microchannels using PDMS casting process. Silicon mold fabricated by typical MEMS process was used as mold for PDMS casting process. Figure 3(a) and 3(b) show fabrication process of silicon mold and PDMS casting process, respectively.

In order to fabricate different height of channels, thermally growth SiO_2 and silicon deep etching process were used. Firstly, 1.5μm-thick SiO_2 was thermally on the 4inch silicon wafer for the plasma channel and microfilter. Secondly, the plasma channel and microfilter were patterned by dry etching process of SiO_2, and then, the blood cell channel was fabricated by silicon deep etching process. Etched depth was 20 μm. The fabricated silicon pattern was used as the mold of PDMS casting process.

PDMS casting process was performed two times to fabricate the same pattern of silicon mold. In order to prevent adhere between mold and replica, Teflon AF (Teflon AF-77, 3M Inc.) was used [12]. The Teflon was coasted on mold at 4000rpm for 60 second, and then mold was treated at 175℃ for 15 min. Generally PDMS was hydrophobic material. Surfactant of Silwet-L77 was added in prepolymer of PDMS to convert hydrophilic PDMS. Finally hydrophilic PDMS microchannel was bonded with flat hydrophilic PDMS substrate. Figure 4(a) and 4(b) show photograph and SEM image of the fabricated plasma extraction device, respectively.

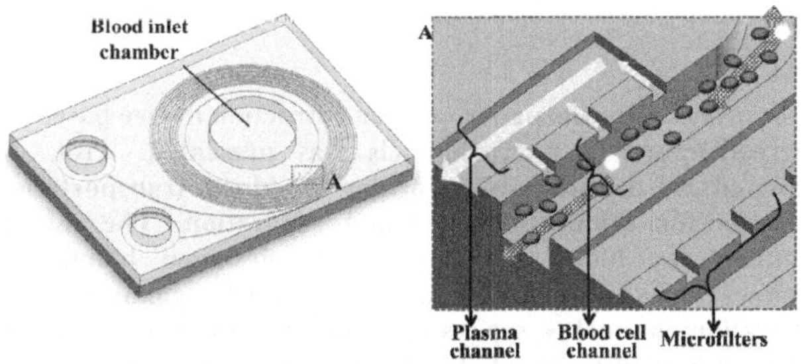

Fig. 1 Schematic view and working principle of the plasma extraction based on microfilters and spiral microchannel.

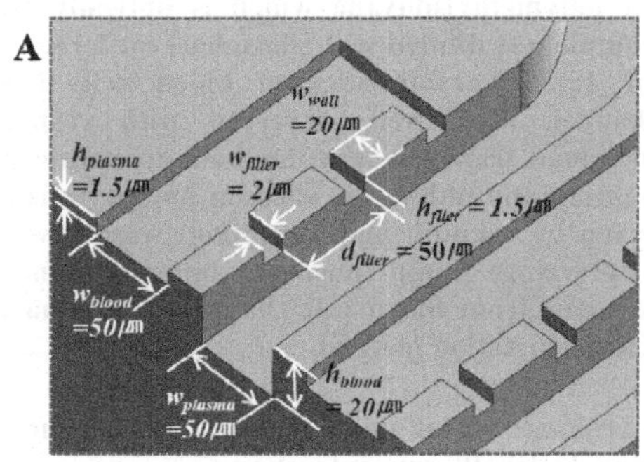

Fig. 2 Dimension of designed blood separation device

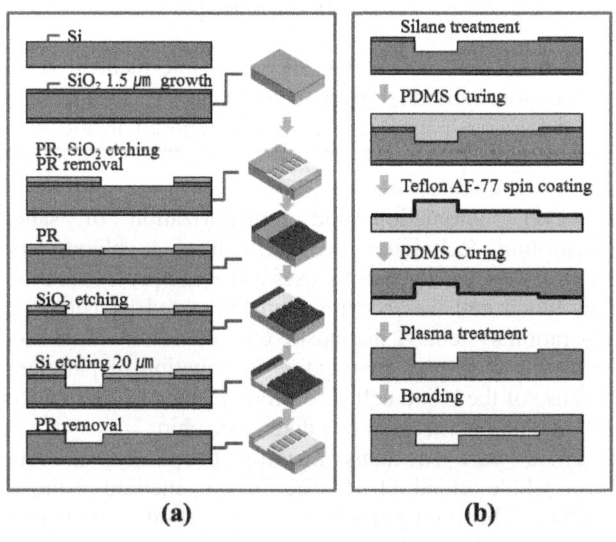

Fig. 3 Fabrication process; (a) fabrication process of silicon mold of plasma extraction device based on microfilter, (b) PDMS casting process.

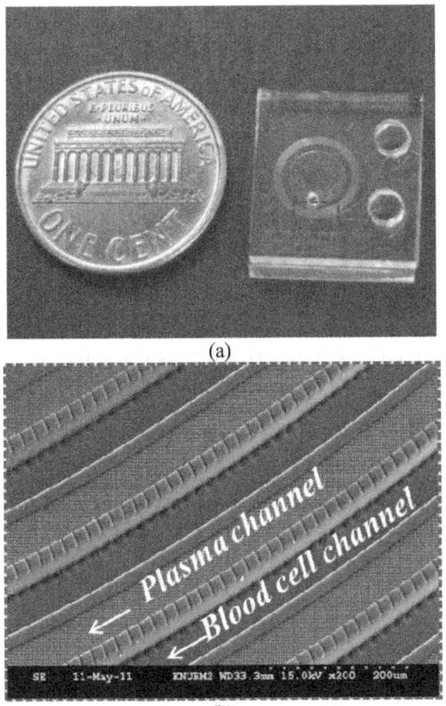

(a)

(b)

Fig. 4 Fabricated PDMS plasma extraction device; (a) photograph of fabricated PDMS device, (b) SEM image of PDMS device.

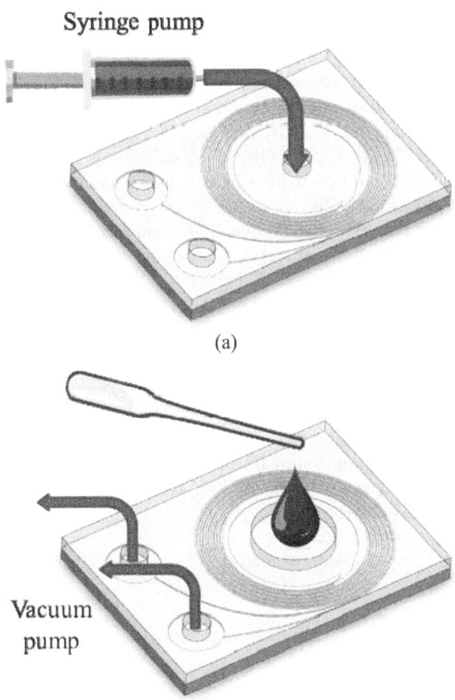

(a)

(b)

Fig. 5 Blood transportation methods; (a) Pushing transportation by syringe pump, (b) Pulling transportation by vacuum pump.

B. *Experiment of Plasma Extraction*

In medical diagnosis based on blood, generally blood was diluted with phosphate buffered saline (PBS). Dilution ratio of blood and PBS was from 1:1 to 1:5. The used blood in the experiment was diluted with PBS (10×concentrate, Sigma Inc.). Dilution ratio was 1:3.

In plasma extraction experiment, capillary force was not enough to transport diluted blood. So blood was transported by two different ways of pulling and pushing methods. In pushing transportation, blood was pushed by syringe pump at constant flow rate of 5ul/min from inlet port. On the other hand, blood was pulled by vacuum pump in pulling method at constant suction pressure of 45kPa. Diluted blood was sucked from outlet port. Figure 5 shows these two methods. The all of experiments were captured by the invert microscope (TS-100, Nikon).

Figure 6 shows experimental result of plasma extraction using syringe pump, i.e., pushing transportation. In pushing experiment, average flow rate of 1:3 diluted blood in spiral microchannel was 0.05ul/min. In the first turn of spiral microchannel, blood cells only were located in blood cell channel. However, after the first turn, a lot of blood cells were crossover from blood cell channel to plasma channel due to high back pressure and high deformability of blood cells.

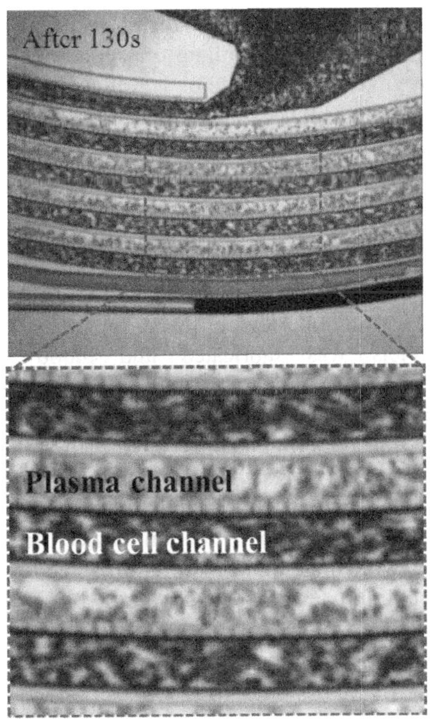

Fig. 6 Experiment result of plasma extraction device syringe pump at constant flow rate of 5ul/min.

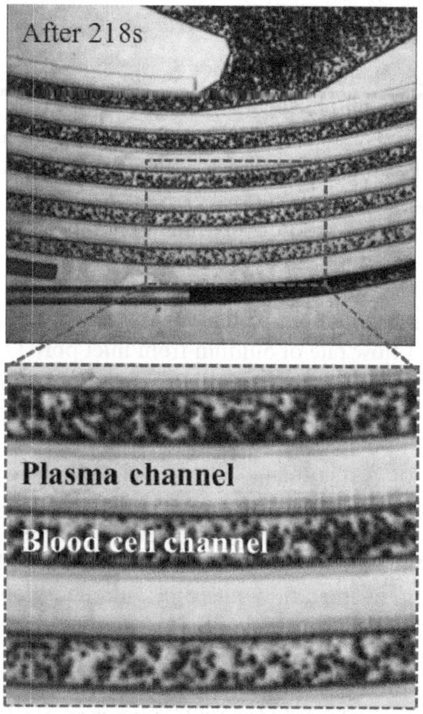

Fig. 7 Experiment result of plasma extraction device vacuum pump at constant suction pressure 45kPa.

Figure 7 shows experimental result of plasma extraction using vacuum pump, i.e., pulling transportation. In pushing experiment, average flow rate of 1:3 diluted blood in spiral micro channel was 0.03ul/min which was 60% of flow rate in pushing transportation. However, there was no blood cell crossover into plasma channel as shown in Fig.7.

IV. Conclusions

Plasma extraction device based microfilters in spiral microchannels was fabricated and blood flow was characterized. The more blood cells were crossover from blood cell channel into plasma channel through microfilter in pushing transportation method rather than pulling method due to high back pressure generation. Therefore, it is necessary to use vacuum suction as the way of blood transportation in plasma extraction device. In the future work, we will integrate microfluidic vacuum pump in plasma extraction device to transport blood.

Acknowledgment

This study was supported by the Regional Core Research Program funded by the Korea Ministry of Education, Science and Technology (Medical & Bio-material Research Center). Also, this research was financially supported by the Ministry of Education, Science Technology (MEST) and National Research Foundation of Korea (NRF) through the Human Resource Training Project for Regional Innovation

References

1. Jens D, Stefan H, Thilo B, Thomas G, Roland Z, (2006) Patterning of flow and mixing in rotating radial microchannels,vol. 2, Microfluid Nanofluid, 2006, pp 97-105
2. Di C. D, Edd J. F, Irimia D, et al. (2008) Equilibrium separation and filtration of particles using differential inertial focusing. Vol. 80, Analytical chemistry, 2008, pp 2204-2211
3. Sollier E, Rostaing H, Pouteau P, Fouillet Y, Achard J. (2009) Passive microfluidic devices for plasma extraction from whole human blood, vol. 141, Sensors and Actuators, 2009, pp 617-624
4. Elodie S, Myriam C, et al. (2010) Fast and continuous plasma extraction from whole human blood based on expanding cell0free layer devices, vol. 12, Biomed Microdevices, 2010, pp 485-497
5. Xing C, Da F C, Chang C L, Hui L, (2008) Microfluidic device for blood cell separation and collection based on crossflow filtration, vol. B130, Sensors and Actuators, 2008, pp 216–221
6. Hong Miao J, Victor S, Yu C, Chew Kiat H, Tit Meng L, Levent Y, (2008) Silicon-based microfilters for whole blood cell separation, vol. 10, Biomed Microdevices, 2008, pp 251–257
7. P. Wilding, L.J. Kricka, J. Cheng, G. Hvichia, M.A. Shoffner, P. Fortina, (1998) Integrated cell isolation and polymerase chain reaction analysis using silicon microfilter chambers, vol. 257, Anal. Biochem. 1998, pp 95–100
8. P. Bacchin, M. Meireles, P. Aimar, (2002) Modelling of filtration: from the polarised layer to deposit formation and compaction, vol. 145, Desalination, 2002, pp 139–146
9. B. Keskinler, E. Yildiz, E. Erhan, M. Dogru, Y.K. Bayhan, G. Akay, (2004) Crossflow microfiltration of low concentration-nonliving yeast suspensions, vol. 233, J. Membr. Sci. 2004, pp 59–69
10. Y. Lee, M.M. Clark, (1998) Modeling of flux decline during crossflow ultrafiltration of colloidal suspensions, vol. 149, J. Membr. Sci. 1998, pp 181–202
11. Ali Asgar S. B, Sathyakumar S. K. and Ian P, (2008) Continuous particle separation in spiral microchannels using dean flows and differential migration, vol. 8, Lab chip, 2008, pp 1906-1914
12. Pollack M. G., Chenderov A. D. and Fair R. B., (2002) Electrowetting-based actuation of droplets for integrated microfluidics, vol. 2, Lab chip, 2002, pp 96-101

Author: Young Ho Seo
Institute: Kangwon National University
Street: 1 Knagwondaehakgil Eng. Bldg.6-111
City: Chuncheon
Country: South Korea
Email: mems@kangwon.ac.kr

Molecular Cloning of Gene Encoding for Lipoprotein in *Streptomyces Lividans*

Nguyen H.K. Tu

School of Biotechnology, Hochiminh City International University, Hochiminh City, Vietnam

Abstract — **A lipoprotein contains both proteins and lipids water-bound to the proteins including enzymes, transporters, structural proteins, antigens, adhesins and toxins. In the present study, a LpnS gene encoding lipoprotein from Streptomyces lividans 66 was cloned by using polymerase chain reaction, southern blot and hybridization. After sequencing, the LpnS gene had 1235 bps coding 415 amino acids that had 100% homology to lipoprotein of Streptomyces lividans TK 24 and 99 % of Streptomyces coelicolor A3(2) and 71 % Streptomyces flavogriseus ATCC 3331 and 71% of ABC transporter of Streptomyces griseoflavus. This gave more data about lipoprotein in Streptomyces lividans 66.**

Keywords — **lipoprotein, molecular cloning, *Streptomyces lividans* 66.**

I. Introduction

A lipoprotein contains both proteins and lipids water-bound to the proteins including enzymes, transporters, structural proteins, antigens, adhesins and toxins. There are four major classes of lipoproteins. Low-density lipoproteins are also called "bad cholesterol," form 60 to 70 percent of a person's total cholesterol while high-density lipoproteins also called "good cholesterol," removes excess cholesterol from tissues in the body [2]. Examples include the high density (HDL) and low density (LDL) lipoproteins which enable fats to be carried in the blood stream as the transmembrane proteins of the mitochondrion and the chloroplast, and bacterial lipoproteins. All cells use and rely on fats and cholesterol as building blocks to create the multiple membranes which cells use to both control internal water content, internal water soluble elements and to organize their internal structure and protein enzymatic systems. However, lipoprotein in *Streptomyces lividans* 66 may be play role a as an antibiotics or in a distinct class of bacterial membrane proteins that are translocated across the cytoplasmic membrane primarily by the Sec general secretory pathway. *Streptomycetes* supplied many products for human [5]. However, the function of the other components was still unknown. To clarify more function of *Streptomycetes*, the study exploit to obtain more lipoprotein from *Streptomyces lividans* 66 by applying the molecular cloning techniques that will contribute to the lipoprotein source for life sciences.

II. Materials and Methods

A. Bacterial Strains, Plasmids, and Growth Conditions

E. coli JM 109 and a plasmid pUC19 were used for the cloning and sequencing experiments. *E. coli* was grown in LB medium [4] at 37°C. If necessary, ampicillin (100 μg/ml) was added to the media. *Streptomyces lividans* 66 and *Streptomyces lividans* TK24 were grown in YEME medium [1].

B. DNA Manipulations

DNA manipulation were applied for *E. coli* following the protocols described by Hanahan [3]. The plasmid DNA in *E. coli* was isolated using the Wizard Plus Minipreps DNA Purification System (Promega).

C. Southern Hybridization

Southern hybridization was performed by using a Hybond-N+ (Amersham Biosciences) membrane. Probe labeling, hybridization and detection were performed with AlkPhos Direct Labeling and Detection System (Amersham Biosciences) according to the protocol supplied by the manufacturer.

D. Polymerase Chain Reaction

PCR amplification was performed by using the S. lividans TK24 genomic DNA as a template in association with a sense primer (5'-GGCGACCGACCCGGAGGGCC GGACGACGCT-3') and an anti-sense primer (5'-GCTG CCGGCCCGGTGGCGAAAGGCCGCGGC-3').

PCR was done under the following conditions: an initial 5 min at 96°C; then, 29 cycles of 1 min at 96°C followed by 1 min at 65°C; and, finally, an extension period of 1 min at 72°C.

V. Van Toi et al. (Eds.): 4th International Conference on Biomedical Engineering in Vietnam, IFMBE Proceedings 40, pp. 165–167, 2013.
www.springerlink.com

E. DNA Sequencing

DNA sequencing was performed with the ABI PRIZM 310 genetic analyzer (Applied Biosystems) by using the protocol of the BigDye terminator cycle sequencing ready reaction kit ver.1.1.

The open reading frames (*orfs*) were predicted by using the FramePlot ver.2.3.2 analysis program (Wright and Bibb, 1992). A homology search was conducted by using the FASTA and BLAST programs at the web site.

Comparison of the LpnS protein from *S. lividans* 66 with the lipoprotein-like proteins was performed by using the CLUSTAL W program at the web site. Hydropathy analysis was performed as described previously.

III. RESULTS AND DISCUSSION

A. Cloning of the DNA fragment containing lpnS in S. lividans 66

The chromosomal DNA of *S. lividans* 66 which was digested by KpnI/PvuII, was fractionated on 0.8% agarose-gel electrophoresis and transferred to a Hybond-N+ membrane by using the standard protocol [4]. Southern hybridization analysis was performed by using a putative gene (400-bp) of *S. lividans* TK24 as a probe.

To obtain the probe DNA,

The probe DNA was hybridized to the KpnI/PvuII-digested *S. lividans* 66 genomic DNA. Southern hybridization were performed at 65°C, for 16 h. After detection of the hybridized complex, the 2.5–3.5 kb genomic DNAs of *S. lividans* 66 digested with KpnI/PvuII were extracted from the agarose gel. The DNA fragments were purified, ligated to KpnI/HincII-digested pUC19, and then introduced into *E. coli* JM109. The resulting genomic libraries were screened by using the colony hybridization technique [4]. The candidates carrying the putative *L* gene were obtained among 2000 clones by using a putative *lipoprotein* gene (400-bp) of *S. lividans* TK24 as a probe. The clone gene was confirmed by the determination of DNA sequence.

B. Identification of lpnS Gene

Using a putative *lipoprotein* gene (400-bp) of *S. lividans* TK24 as a probe to detect a positive clone. The positive clone had a 2.9-kb DNA fragment from *S. lividans 66*. In the DNA fragment, one *orf was* complete (Fig. 1). The FASTA and BLAST search suggest that *orf* consists of 1245-bp (designated *lpnS*) and the gene product (designated LpnS). A protein designated LpnS consists of 415 aa that shows a 100% identity to lipoprotein of *S. lividans TK24*, 99 % of *Streptomyces coelicolor* A3(2) and 71 % *Streptomyces flavogriseus* ATCC 3331 and 71% of ABC transporter of *Streptomyces griseoflavus*..

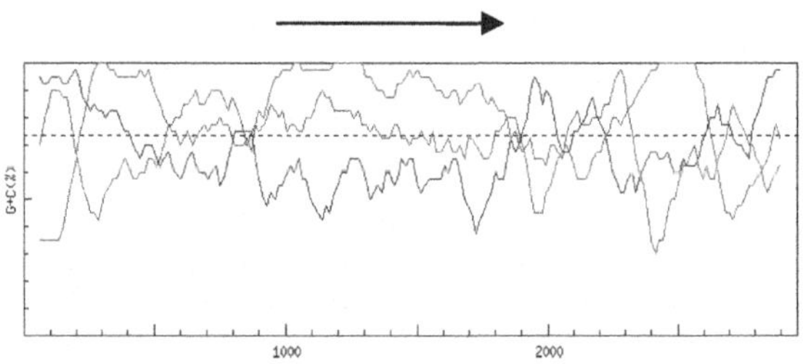

Fig. 1 Frame analysis of DNA sequence of *lpnS* from *S. lividans 66.*

Comparison of the LpnS protein from *S. lividans* 66 with the lipoprotein-like proteins was performed by using the CLUSTAL W program at the web site (Fig. 2).

```
Lipo_STK24
MEPTVRTHPTRRLPTALVLTAALLATGCSEQSDDGREPAATESSGTPHGYVEGAREAA
Lipo_Sco
MEPTVRTHPTRRLPTALVLTAALLATGCSEQSDDGREPAATESSGTPHGYVEGAREAA
Lipo_S66
MEPTVRTHPTRRLPTALVLTAALLATGCSEQSDDGREPAATESSGTPHGYVEGAREAA
Lipo_Sfla
MAIAPTVRRNVTPWASAALALTALLTTSCAGKHSDAPAPQKNSSATTPHGYVEGAKEAA
ABC_
MSLFVRRNVTAGAPAALLTLALLTAGCTGEGGNPSTPKAGSATPTAPHGHVEGAREAA
          :    ** : *    .:**    * * :    .    ..     .:.:***:****:***

Lipo_STK24
EEQSRLILGDPGTGASRVLDLVTGKVHESAPVTGADGLSTDGRFGFFHTDRGTHVVDGGA
Lipo_Sco
EEQSRLILGDPGTGASRVLDLVTGKVHESEPVTGADGLSTDGRFGFFHTDRGTHVVDGGA
Lipo_S66
EEQSRLILGDPGTGASRVLDLVTGKVHESAPVTGADGLSTDGRFGFFHTDRGTHVVDGGA
Lipo_Sfla
EQQSRLLLNDPGTGDSRVLDLITGKVRSTAQVAGAEQLETDGRFGFFHTARGTHVLDSGA
ABC_
EEQSRLLLNDPTTGDTRVLDLTTGKAHETSPRAGAVELTTDGRFGYFRTADGIRVLDGGT
        *:****:*.** ** :***** ***.:.:    :**   * ******:*:*   * :*:*.*:

Lipo_STK24
WTVDHGDHVHYYRAAIRHVGDLPGGRGAEIRSDAAVTAATDPEGRTTLYDRTALEKGEG
Lipo_Sco
WTVDHGDHVHYYRAAIRHVGDLPGGRGAEIRSDAAVTAATDPEGRTTLYDRTALEKGEG
Lipo_S66
WTVDHGDHVHYYRAAIRHVGDLPGGRGAEIRSDAAVTAATDPEGRTTLYDRTALEKGEG
Lipo_Sfla
WTVDHGDHVHYYRAAIRDIGVLPAGADAQIRSDAAVTAATGKDGRTGLYDRTELEKGSIC
ABC
WTVDHGDHVHYYRAGIRDVGRLPADKGAHVRGDTAVTAVTGEDGRAVTYRRTAFEKHG
**************.**.:* **..*.:*.*:****.*..:**:   * ** :**   :.

Lipo_STK24
PARTLDGVHAGAVVPYEEHLVALAGDDESAEVAVYDRSGRRVASPDATCERPAGDVR
Lipo_Sco
 PARTLDGVHAGAVVPYEEHLVALAGDDESAEVAVYDRSGRRVASPDATCERPAGDAVTR
Lipo_S66
 PARTLDGVHAGAVVPYEEHLVALAGDDESAEVAVYDRSGRRVASPDATCERPAGDVR
Lipo_Sfla
SIRTLPGPHTGAVIPYAEHLLALVERDGTAKVSAYDRAGRPVASPDAECDEPRGDAVR

ABC
```

Fig. 2 Alignment of the amino acid sequence of LpnS from *S. lividans* 66 with those from various bacteria.

```
PARPLPGKHAGAVVPYAEHLLTLTDDGTGTPEVTVLARDGARAASTDATCEDPKGDAVTR
 .*.*.* *.***.** ***.:.*.       ..:*:.   * *   .**.** *: * ******

   Lipo_STK24
RGVVLGCADGALLVSAEDGTFTAERIPYGQDVPEKERAAAFRHRAGSDTLTAPAGERAVW
   Lipo_Sco
RGVVLGCADGALLVSAEDGTFTAERIPYGQDVPEKERATAFRHRAGSDTLTAPAGERAVW
   Lipo_S66
RGVVLGCADGALLVSAEDGTFTAERIPYGQDVPEKERAAAFRHRAGSDTLTAPAGERAVW
   Lipo_Sfla
RGVVLGCADGALVVGEKDGVLTADRIRYGRDVPAAERATAFRHRPGSDTLTAPAGDRAW
   ABC_
RGVVLGCADGALLISEEDGHFTAERIPYAGDVPADERAVSFRHRAGSDTLTAHAGDRAVW
****************:*.:* *. ***   ***.:****.******* **:****

   Lipo_STK24
VLDVTDRDWTRVRTTGPVLAANTAGEGSPLLVLESDGALHGYDIATGRHTARTEPLLTRT
   Lipo_Sco
VLDVTDRDWTRVRTTGPVLAANTAGEGSPLLVLESDGALHGYDIATGRHTARTEPLLTRT
   Lipo_S66
VLDVTDRDWTRVRTTGPVLAANTAGEGSPLLVLESDGALHGYDIATGRHTARTEPLLTRT
   Lipo_Sfla
VLDVTDRAWKRVSTGPVVATGTAGEGSALVLETDGALHGYDIDTGEEVARTEPLLS
   ABC
VLDVTERTWTRVETGPVVAANTAGEGSPLLVLETDGALHGYDVATGEETARTERLLTGE
*****:* *.** *   **:*:.******.*****:********: **...**** ***

   Lipo_STK24
AKTPAGAAAPVVEVDRSRAYVNDRTGKRVYEIDYNDALRVARSFDLDIAPGLMAETGR
   Lipo_Sco
AKTPAGAAAPVVEVDRSRAYVNDRTGKRVYEIDYNDALRVARSFDLDIAPGLMAETGR
   Lipo_S66
AKTPAGAAAPVVEVDRSRAYVNDRTGKRVYEIDYNDALRVARSFDLDIAPGLMAETGR
   Lipo_Sfla
QRTGTSTSGSTIEVDRSRAYVNDPKGKRVYEIDYNDDLRVARTFDVDIEPHLMTETGR
   ABC
HRAGAGGGGPVVENDRSRAYVNAPKDRRVYEIDYNDGLRVARTFDLDMEPRLMVETGR
          :: :. ....:* ********   ...:********* *****:**:*: * **.****
```

Fig. 2 (*continued*)

Alignment shows the comparison of deduced amino acid sequences of *lpnS* and *lpnS*-like genes from *S. lividans TK24* (Stk24), *Streptomyces coelicolor* A3(2) (Scoe), *Streptomyces flavogriseus* ATCC 3331 (Sfla) and ABC transporter of *Streptomyces griseoflavus* (ABC). The asterisks and dot showed the identities of those amino acid sequences. LpnS was a putative protein that function was not studied yet. However, after alignment and homology search, LpnS showed high identity to ABC transporter sequence of *Streptomyces griseoflavus*. ABC transporter is an ATP binding cassette protein that has many function in antibiotic resistance and transporter in human. This might be a base to direct the functional study of LpnS to improve the way of treatment for human.

In order to understand the structure of LpnS, we used the Kyte-Doolittle hydropathy with window 9 to analyze the hydrophobicity with a minium score of -3.0 and highest score of 3.0. The protein existed some regions at over 1.6 that was supposed as membrane related protein.

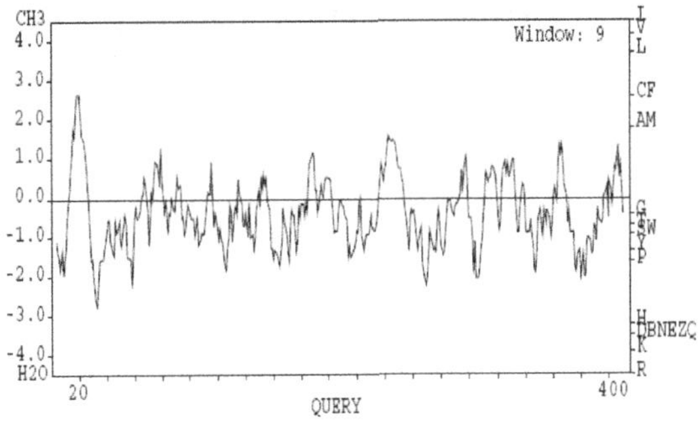

Fig. 3 Hydopathy data of *lpnS* in *Streptomyces lividans* 66.

IV. Conclusions

The study supplied the data about *lpnS* encoding LpnS protein in *Streptomyces lividans* 66. By analysis the gene and protein structures with different programs LpnS was related to membrane of the cells. The function of *lpnS* does not known. This work demonstrates a gene coding lipopeptide that contained regions of hydrophobic relating to membrane. Based on the advantages of lipoprotein [2] [5], *lpnS* gene will be studied so far.

References

1. Bibb, M. (1996) The regulation of antibiotic production in *Streptomyces coelicolor* A3(2) *Microbiology* 142: 1335-1344.
2. Brown M.S., Goldstein J.L. (1986) A receptor –mediated pathway for cholesterol homeostasis *Science* 232: 34-47
3. Hanahan, D. (1983) Studies on transformation of Escherichia coli with plasmids. J Mol Biol 166: 557-580
4. Sambrook, J., Fritsch, E.F., Maniatis, T. (1989) Molecular Cloning: a Laboratory Manual. Cold Spring Harbor, NY: Cold Spring Harbor Laboratory
5. Vivian M., Marie-Francois C., Kien Nguyen, Paul B., Julia P.enn,2 A. Whiting, Jonathan S. et al. (2006) Genetic Engineering in Streptomyces roseosporus *Chemistry & Biology* 13: 269–276

Author: Nguyen Hoang Khuc Tu

Institute: Hochiminh city International University
Street: Quarter 6, Linh Trung ward, Thu Duc Dist.
City: Ho Chi Minh
Country: Vietnam
Email: nhktu@hcmiu.edu.vn

Effects of Rice-Washing Water on the Hyaluronic Acid Production of *Streptococcus Thermophilus*

Nguyen H.K. Tu and Phi T.T. Trang

School of Biotechnology, Hochiminh City International University, Hochiminh City, Vietnam

Abstract — **Hyaluronic acid (HA), one of the most potential biological materials, is increasingly widely used in pharmaceutical and cosmetics nowadays. Hyaluronic acid was produced commercially from rooster combs and the fermentation of Streptococci groups A and C. In order to contribute to the safety and cheap hyaluronic acid for human, we used rice-washing water as medium to culture Streptococcus thermophilus MR10 which was detected as hyaluronic acid production source isolated in fermented foods in Vietnam. This is the first report pointing the the effects of cheap rice-washing water on the HA production of Streptococcus thermophilus. In the medium containing rice-washing water, Streptococcus thermophilus produced HA better than the usual MRS medium for Streptococcus thermophilus growth.**

Keywords — **hyaluronic acid, rice-washing water, *Streptococcus thermophilus.***

I. Introduction

Hyaluronic acid (HA) is a polymer of disaccharides, composing of D-glucuronic acid and N-acetyl-D-glucosamine. It plays a key role in body structures due to it acts to hold water in connective tissue, help to lubricate the cartilage of joints, be a primary component of synovial fluid and healthy skin in which about 50% of the structure. When human is going to be old, production and content of HA in human body slow down, resulting in a loss of viscoelasticity in the tissue that manifests visibly in wrinkles and physically in joint pain and also other connective tissue complications. Reflecting its variety of natural functions, HA has found a number of applications in pharmaceutical, cosmetics and special foods. HA production can be extracted from animal sources such as rooster combs and bovine's eyes, and microbial production through the use of large scale fermenters in certain bacterial species. Traditionally only rooster combs and bovine eyes have been used as sources for commercially available HA. However the collection of rooster combs and the extraction and purification procedures of HA from these tissues are time consuming and labor intensive, making hyaluronan production very costly [4]. In fact, in animal tissues hyaluronan is complex with proteoglycans and often contaminated with HA degrading Hence, since 1980s microbial production is gradually replacing extraction from animal tissues [3]. Consequently, microbial producing HA has been a preferred source with hope to obtain HA having advantages of being derivable from a microbiological source at relatively low costs, in high yields, and with low immunity and being substantially free protein and nucleic acid impurities. Bacteria known to be capable of the synthesis of HA are *Streptococci* of groups A and C. HA is a major component of the capsule of some pathogenic and non-pathogenic. HA from group A hemolytic *Streptococci*, for instance strains of the *Streptococcus pyogenes* a human pathogen was first isolated by Kedall, Heidelberger and Dawson by using acetate and ethanol. Their HA is chemically identical to that found in the connective tissue and biological fluids of mammals. However, the safety problems of using pathogenic bacteria for the production of HA remain controversial problems due to toxicity to human and exotoxin contamination. There for more approaches on non-pathogenic bacteria have been studied. It has been reported that streptococcus thermophilus which is a generally recognized safe bacterium produces useful exopolysaccharides (EPSs) including HA from traditional dairy food products [2].

There were *S. thermophilus* used as generalized as safe (GRAS) for HA production. In Viet Nam, there was no studies on HA production while *S. thermophilus* was used widely for industrial production of fermented dairy products such as milk, cheese and yogurt. The purpose of this study is to use *S. thermophilus*, and cheap nutrient sources or suitable temperature changes for HA production in Vietnam conditions. The nutrient source used was the waste obtained from rice washing before cooking, but that contains carbohydrates and vitamins sources.

II. Materials and Methods

A. *Materials*

Tested hyaluronan were collected from *Streptococcus thermophilus* MR10 while cetylpyridium chloride (CPC), carbazole, sulfuric acid, sodium borate, n-butanol, formic acid, hydrochloric acid, absolute ethanol (Merck), MRS broth were analyzed chemicals.

V. Van Toi et al. (Eds.): 4th International Conference on Biomedical Engineering in Vietnam, IFMBE Proceedings 40, pp. 168–170, 2013.
www.springerlink.com

B. *Extraction of Hyaluronic Acid from Streptococcus*

After *Streptococcus* was cultured, the cells were removed. The protein in supernatant was removed by chloroform. The extraction of exopolysaccharide was treated with 1% cetylpyridinium chloride (CPC) and later centrifuged. The precipitate obtained was re-suspended by adding 0.9 M calcium chloride solution to dissociate the HA precipitation complex. The suspension was treated with 25% v/v aqueous ethanol and the nucleic acids were precipitated and separated by centrifugation. The resulting suspension was deproteinized and defatted four times by chloroform solvent extraction (10% of the total volume). HA was precipitated by first adding 75% v/v aqueous ethanol and then acetone.

Residue protein was eliminated by shaking it with chloroform four times and then with a chloroform-amyl alcohol mixture several times until a gel no longer formed. The final solution was dialyzed. Following acidification to pH 4.0, the solution was precipitated with ethyl alcohol and the precipitate was desiccated.

C. *Determination of Uronic Acid*

Determine the glucuronic acid content by reaction with carbazole based on the British Pharmacopeia, 2007, monograph 1472. Reaction changing from no color into purple red indicated the positive result of uronic acid when mixing 200µl samples with 1ml of sulfuric acid containing borate that was heat for 10minutes at 100 °C. The reaction was let on ice to cool the reaction. Then, the mixture was added with 40µl carbazole and heat for 12minutes at 100 °C.

D. *Optimization*

The optimization was performed in MRS broth at 34, 37, 40 °C. The rice washing water was prepared by washing 200g rice for three times with 100ml of distilled water. The collection of 80ml of supernatant rice water was steriled at 121 °C for 15minutes.

E. *High Pressure Liquid Chromatography (HPLC)*

All samples were crystallized and re-suspended in the solution of 1M NaCl, centrifuge and pass through 0.22um filter membrane. The after-treatment solutions were analyzed by the HPLC machine Agilent, with C18 column, refractive index detector along with the commercial hyaluronan as standard. The mobile phase was solution of KH2PO4 0,5M pH 2.5 with current speed of 1ml/minute. Column temperature was maintained at 30°C.

III. Results and discussion

A. *High Pressure Liquid Chromatography (HPLC)*

Via result from HPLC analysis, obviously, retention time of exopolysaccaride standard and HA extraction from *S. thermophilus* is similar. The result was showed as in Fig.1.

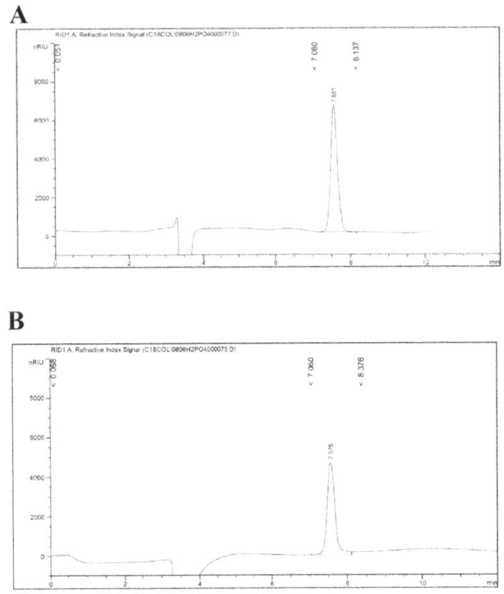

Fig. 1 High pressure liquid chromatography of exopolysaccharide in *Streptococcus* (A) and standard hyaluronic acid (B).

Using Refractive Index Detector, which is specific for polysacharide analysis also bring more convenience than using UV detector. As we can see in the chromatogram, only one peak appear, mean that extracted sample was not along with other glycosaminoglycan. In the Fig.1B, the retention time of the standard hyaluronan was 7.575. The retention time of extraction from *Streptococcus* was 7.561 (Fig.1A). Therefore, the extraction from *Streptococcus* contained hyaluronic acid and did not contain other polysaccharide from rice-washing water that might destroy when treated at 121°C for 15minutes. Actually, polysaccharides are destroyed at high temperature [3].

B. *Determination of Uronic Acid*

After reacting with carbazole, the color changes into purple, the uronic acid was determined in the extraction. The results were showed in Fig.2. The hyaluronic acid was produced and gave the purple solution while the medium did not.

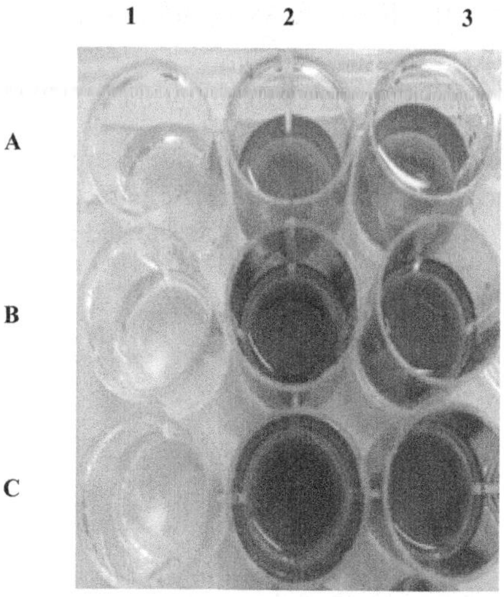

Fig. 2 The reactions of extracted samples from *Streptococcus* after incubated at 34, 37, 40 °C. From left to right: 1-media containing rice-washing water, 2-extracted hyaluronic acid in media containing rice-washing water, 3-2-extracted hyaluronic acid in MRS broth. From up to down: A- extracted hyaluronic acid at 34 °C, B- extracted hyaluronic acid at 37 °C, C- extracted hyaluronic acid at 40 °C.

C. *Optimization of Hyaluronic Acid Production in Streptococcus*

Based on the colorimetric mesurement of the reaction between hyaluronic acid and carbazole. Also, the hyaluronic acid produced in rice-washing water at 37 °C was the highest. The hyaluronic acid produced in rice-washing water at 34 °C was higher than at 40 °C.

The results of the production ability of Streptococcus.

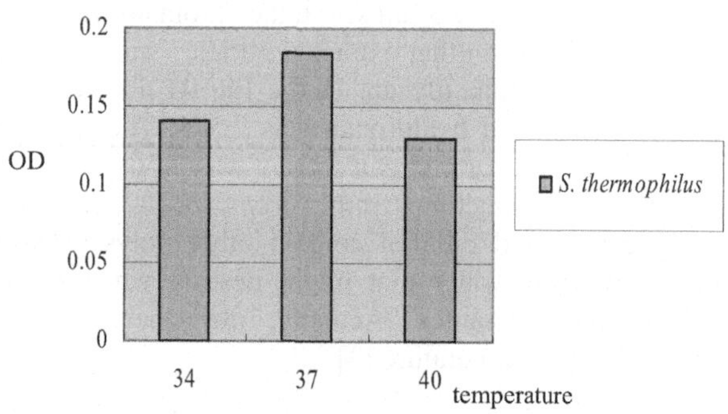

Fig. 3 Effects of temperature on hyaluronic acid production in *Streptococcus.*

After cultured in MRS broth and rice-washing water, the hyaluronic acid production was obtained basing on the optical density as shown in Fig.4. The hyaluronic acid yielded in rice-washing water was higher than in MRS broth. The procedure was performed and measured for 3 days. The third day gave the highest yield. The study did not let the cultures more long time because the microorganism did not growth well after this time. In Fig.4, the hyaluronic acid in third day of culture seemed to increase twice.

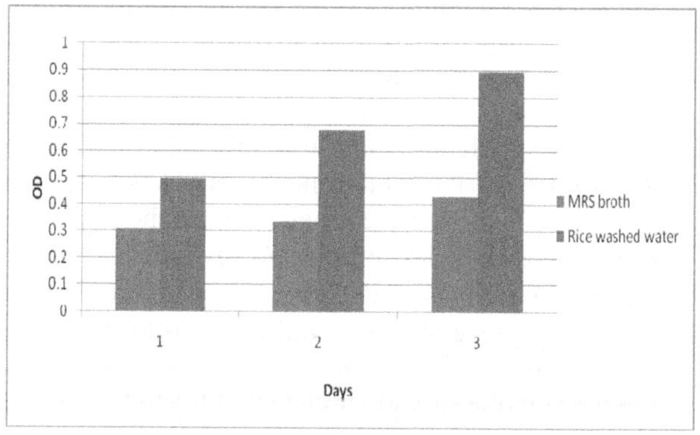

Fig. 4 The comparision of hyaluronic acid in Streptococcus in MRS broth and rice-washing water based on optical density.

IV. Conclusions

Rice-washing water used to tbe thrown away. However, this residue gave the high yield in hyaluronic acid. With HPLC, the hyaluronic acid production was affected by components of rice-washing water after autoclaved that was not any polysaccharide or starch from rice. That was will be studied in future to find out the component which gave the high yield in hyaluronic acid production. Also, the temperature of 37 °C was optimal condition for HA production n *Streptococcus thermophilus.*

References

1. Eriksen, PB., Knudsen PJ., Fenger M., Florentz K. (1980) High-Performance liquid chromatography of hyaluronic acid and oligosaccharides produced by bovine testes hyaluronidase. J Chromatogr A 187: 373–379.
2. Izawa N., Hanamizu T., Iizuka R., Sone T., Mizukoshi H., Kimura K., Chiba K (2009) Streptococcus thermophilus produces exopolysaccharides including hyaluronic acid. J Biosci Bioeng 107(2) :119-123
3. Lago G., Oruna L., Cremata JA., Pérez C., Coto G., Lauzan E., Kennedy JF. (2005) Carbohydr Polym 62(4): 321–326
4. Meyer K., Palmer J. (1934) J Biol Chem 107: 629–634

Comparision of Antimicrobial Activities and Antibiotic Resistance of Lactobacillus Rhamnosus and Lactobacillus Acidophilus before and after Co-cultured

Nguyen H.K. Tu and Nguyen T.N. An

School of Biotechnology, Hochiminh City International University, Hochiminh City, Vietnam

Abstract — **Lactic acid bacteria (LAB) have the important role in general health, providing a beneficial micro flora in the intestinal tract. The tested Lactobacillus acidophilus** VTCC-B-702 **and Lactobacillus rhamnosus** ATCC 11443 **were grown individually and co-cultured together in the same environment. Agar-well diffusion test was performed to determine the antibiotic susceptibility and antimicrobial activity. Two tested LAB exhibited the sensitivity to ampicilin and tetracycline at 5x10-3 to 1x10-2 (mg/ml), but resistant at 5x10-4 (mg/ml). L. acidophilus was resistant to erythromycin at all concentration, but L. rhamnosus was sensitivity to erythromycin at 5x10-3 to 1x10-2 (mg/ml), but resistant at 5x10-4(mg/ml). After co-cultured, both of these microrganisms were resistant to erythromycin. This action showed there were erythromycin resistant determinant from L. acidophilus to L. rhamnosus. The antimicrobial activities were also tested on Escherichia coli, Staphylococcus aureus, Salmonella typhi, Pseudomonas aeruginosa. The effects showed the synergy except on Escherichia coli.**

Keywords — **antibiotic susceptibility, antimicrobial activity, co-culture, *Lactobacillus rhamnosus, Lactobacillus acidophilus.***

I. INTRODUCTION

Lactic acid bacteria (LAB) refer to a large group of beneficial bacteria that take part in natural ecosystems such as vagina or the gastro-intestinal tract of humans and other animals [1]. In these environments, LAB have the ability in the inhibition of pathogenic microorganisms. For example, *Lactobacilli* and *bifidobacteria* (*Lb. casei, Lb. rhamonsus, Lb. acidophilus, Bf. Bifidum)* used for treatment of diarrhea, inflammatory disease of digestive tract and *L. acidophilus* can inhibit the growth of *Candida albicans*, the fungus caused the diseases in mouth, esophagus or vagina [2]. LAB also have the important role in industrial fermentation such as fermented dairy product, pickling of vegetables, baking, winemaking, curing fish, meats and sausages [4]. Based on these reasons, this research was focused on the interactions between *L. acidophilus and L. rhamnosus*. Each species of LAB was grown individual and co-cultured with another species in order to test the effects of associating these LAB species on the antibiotic resistance, antimicrobial activity with different kinds of pathogens.

II. MATERIALS AND METHODS

A. Materials

Ampicilin and tetracycline at 5x10-3 to 1x10-2 (mg/ml), but resistant at 5x10-4 (mg/ml). *Lactobacillus rhamnosus* ATCC 11443, *Lactobacillus acidophilus* VTCC-B-702 were obtained from VTCC. *Escherichia coli, Staphylococcus aureus, Salmonella typhi, Pseudomonas aeruginosa were supplied by IDC.* Man-Rogosa-Sharpe (MRS) and Luria-Bertani (LB) media were in order used to cultivate LAB and pathogens. Gram stain kit was performed to test the LAB morphology under microscope.

B. Antibiotic Susceptibility

The antibiotic solutions were prepared to reach the concentration at 500 mg/ml, by using erythromycin, ampicillin, and tetracycline (500mg) which were popularly sold in drugstores. LAB were grown overnight in MRS broth. The inoculums density of liquid cultures was adjusted to an absorbance reading at 600nm of 0.1±0.02 (approximately $1x10^8$ cfu ml^{-1}) [3]. The aliquots of 50μl of LAB were introduced onto the surface of sterile MRS agar plate, and distributed the inoculums by using a sterile spreader. The plates were left for about thirty minutes to allow the test organisms fully embedded before wells (d=5mm) of equal distant were dug. Different concentrations of antibiotics: $5x10^{-4}$mg/ml, $5x10^{-3}$mg/ml, $1x10^{-2}$mg/ml were transferred into the wells by using sterile pipette. The plates were incubated for 24 hours, and the diameters of zone of inhibition were recorded.

C. Antimicrobial Activity

One week LAB broths were centrifuged at 10000 rpm in 15 minutes at 4^0C. The bacteria indicators which had OD=0.05, were spread on LB plates by using cotton swap. The plates were left for about 15 minutes, and the wells were dug. The wells were filled with the LAB supernatants.

V. Van Toi et al. (Eds.): 4th International Conference on Biomedical Engineering in Vietnam, IFMBE Proceedings 40, pp. 171–173, 2013.
www.springerlink.com

The plates were incubated at room temperature for 12-18 hours to record the diameter zones of inhibition.

D. Co-culturing the Microorganism

L. rhamnosus and *L. acidophillus* were cultured separately for 1 day at 30°C. Then, two cultures were mixed together and continously incubated for 2 days at 30°C. The mixed culture was spreaded on MRS agar and incubated. The suspected colonies were chosen and tested for morphology and biochemical tests.

III. RESULTS AND DISCUSSION

A. Antibiotic Susceptivity

Before co-cultured, *L. rhamnosus* were susceptible to ampicilin, tetracyline, and erythromycin (22mm-31mm) at 5×10^{-3} mg/ml (Table 1), while *L. acidophilus* resisted to erythromycin at this concentration (Table 1). At 5×10^{-4} mg/ml of tested antibiotics, these strains were resistant (Table 1). Fig. 1 and Fig.2 showed the well signed with the numbers contained different concentration of ampicillin and tetracycline. The well 1 pointed the resistance at 5×10^{-4} mg/ml of tested antibiotics. The results of erythromycin on these strains were measured by the changing of the zone size (Table 1) and the resistant effect of erythromycin was similar and not shown.

However, after co-cultured, these LAB had some changes in resistance to antibiotic when they lived in the same environment based on the decrease in the diameter of inhibition zones with the diameter of zone inhibition ranged from 13.7mm-27.8mm (P>0.05). At all the concentration of erythromycin, the co-cultured strains were resistant. In case of ampicillin and tetracycline, the resistance became higher (Tabe 2, Fig.1, Fig. 2). The possible reason for these phenomena can be the horizontal gene transfer. The positive effect when co-cultured different strains of LAB can provide the useful way to protect the probiotics when the patients use ampicillin, tetracycline, erythromycin.

Table 1 Diameter zones of inhibition when LAB were grown individually.

	Concentration (mg/ml)	Ampicillin	Tetracycline	Erythromycin
S1	5×10^{-4}	-	-	-
	5×10^{-3}	22.2±1.1	26.4±1.9	20.2±1.6
	1×10^{-2}	30±1.1	30.3±3.2	23.7±2.3
S2	5×10^{-4}	-		-
	5×10^{-3}	25.7±3.4	31.7±1.3	-
	1×10^{-2}	30.6±1.9	36.2±5.4	-

Mean±SD, (-): resistance, parameter: mm, *L. rhamnosus (S1), L. acidophillus (S2).*

Table 2 Diameter zones of inhibition when co-cultured.

	Concentration (mg/ml)	Ampicillin	Tetracycline	Erythromycin
S1+S2	5×10^{-4}	-	-	-
	5×10^{-3}	21.2±0.7	23±1.7	-
	1×10^{-2}	27.8±1.4	27.7±2.5	-

Mean±SD, (-): resistance, parameter: mm, *L. helvetica (S1), L. acidophillus (S2), S. thermophillus (A1), L. lactis (A2)*

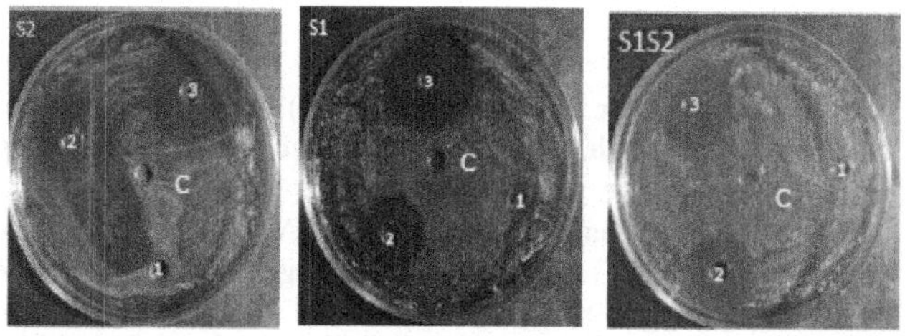

Fig. 1 The resistance to tetracycline of *L. rhamnosus* and *L. acidophilus* before and after co-cultured.

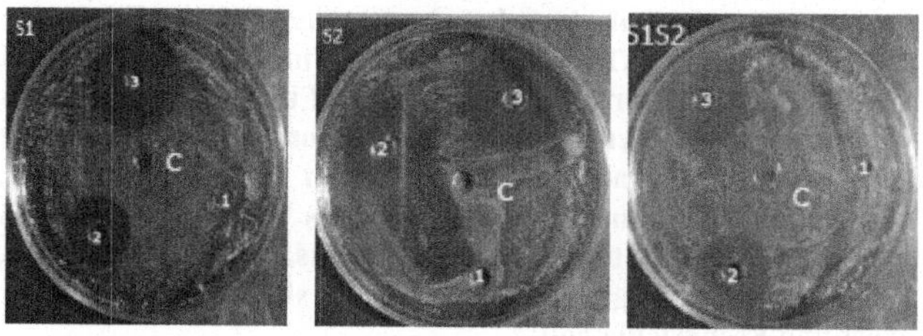

Fig. 2 The resistance to ampicillin of *L. rhamnosus* and *L. acidophilus* before and after co-cultured.

B. Antimicrobial Activities

The agar-well diffusion assay revealed that all strains of LAB before co-cultured were inhibitory to the indicator bacteria which cause many popular human diseases. The pathogens were *E. coli, S. typhi, P. aeruginosa, S. aureus* which were inhibited by *L. rhamnosus* and *L. acidophilus* before co-cultured (Table 3). After co-cultured, the inhibition of *L. rhamnosus* and *L. acidophilus* showed the changes in diameter of zones. These co-cultured LAB were not only showed the antimicrobial activity but also the symbiotic relationships based on the increases of diameter zones of inhibition to *S. aureus, S. typhi, P. aeruginosa.* The antimicrobial activity of LAB after co-culture increased. The result showed that the positive effects were found when two different LAB were grown in the same media. The increasing in antimicrobial activities based on the combinations of different strains of LAB will open the new ways to treat the popular diseases that are caused by many harmful bacteria which become resistant to many antibiotic drugs. In addition, these combinations are safety methods to protect the patients who are infected with the above pathogens and are sensitive to antibiotic drugs as ampicillin, tetracycline and erythromycin. On the other hand, the effect in *E. coli* of the co-cultured *L. rhamnosus* and *L. acidophilus* had nothing (Table 4). The mechanism of this effect on *E. coli* will be studied so far.

Table 3 Antimicrobial activity of LAB when were grown individually.

Samples	S1	S2
S. aureus	10.5±0.6	11.5±0.6
E. coli	15.5±0.8	14.5±0.7
S. typhi	12±0.7	11
P. aeuginosa	14.5±0.5	15.5±0.5

Mean±SD, Parameter: mm, *L. rhamnosus (S1), L. acidophilus (S2).*

Table 4 Antimicrobial activity of LAB after co-cultured.

Samples	S1	S2
S. aureus	14.3±0.5	14.3±0.5
E. coli	13.3±0.5	12.7±0.5
S. typhi	11.6±0.5	11.4±0.5
P. aeruginosa	16±0.7	16±0.7

Mean±SD, Parameter: mm, *L. rhamnosus (S1), L. acidophilus (S2).*

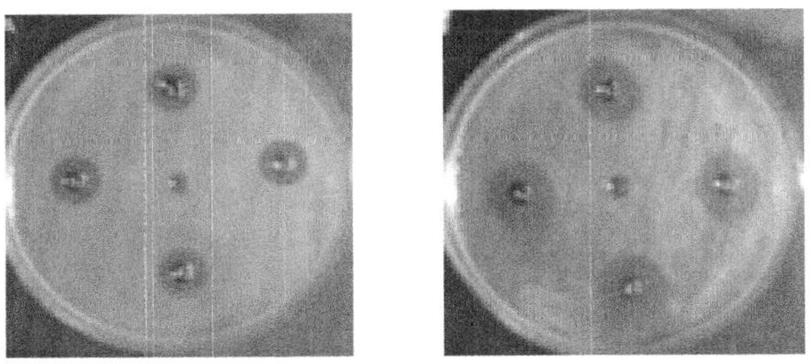

Fig.3 The antimicrobial activities on *S. aureus* before and after co-culturing *L. rhamnosus* and *L. acidophilus*. The left showed the results before co-cultured. The right showed the results after co-cultured.

The LAB with inhibitory affinity over the bacteria indicators exhibited various degree of inhibitory zones. The mechanism of these changes can be explained by the changes in cell wall after co-culture to provide the condition for conjugation. The changes in the pores that increase the secretion bacteriocin or hydrogen peroxide to the environment. Finally, the antimicrobial activity of LAB goes up. However, the mechanism of these results was not well-known. The answer for these interactions needs more chemical and biological tests.

IV. CONCLUSIONS

In conclusion, the present study showed the positive effects between *L. acidophilus* to *L. rhamnosus* when they were co-cultured in the same environment. The increasing in antibiotic resistance for ampicillin, tetracycline and erythromycin. The antimicrobial activities proved the good impacts of interaction between lactic acid bacteria on *Escherichia coli*, *Staphylococcus aureus*, *Salmonella typhi*, *Pseudomonas aeruginosa*. This study pointed out the way to formulate the probiotics as well as the best use in the human health or food industry.

REFERENCES

1. Bernt-commard M. F., Leivin V., Brassant D., Neeser J.R., Sewin A.L., Hudnait S. (1997) The human *Lactobacillus acidophilus* strain LAI secretes and nonbacteriocin antibacterial substance activity in-vitro and in-viro. Applied Environment and Microbiology 63(27): -2753
2. De Roos N. M., Katan M.B., (2000) Effects of probiotic bateria on diarrhea, lipid metabolism, and carcinogenesis: a review of papers published between 1988 and 1998. The American journal of clinical nutrition 71: 405 411
3. Huys G., D'Haene K., Swings S. (2002) Influence of the culture medium on antibiotic susceptibility testing of food-associated lactic acid bacteria with the agar overlay disc diffusion method. Appl Microbiol 34: 402-406
4. Patel A. K., Michaud M., Singhania R.R., Soccol C.R., Pandey A. (2010) Polysaccharides form probiotics: New developments as food additives. Food technol biotechnol 48(4): 451-463

Author: Nguyen Hoang Khue Tu

Institute: Hochiminh city International University
Street: Quarter 6, Linh Trung ward, Thu Duc Dist.
City: Ho Chi Minh
Country: Vietnam
Email: nhktu@hcmiu.edu.vn

Detection of N-Acetyl-D-Glucosamine in Hyaluronan by Thin Layer Chromatography

Nguyen H.K. Tu and Pham V.M. Thien

School of Biotechnology, Hochiminh City International University, Hochiminh City, Vietnam

Abstract — **Various reactions that produce color are known to be selective for certain types of carbohydrates. It was difficult to identify N-acetyl-glucosamine in hyaluronan because the hyaluronan was produced very low in microorgansims. Besides, in the traditional methods of Morgan-Elson, using alkali borate to lysis hyaluronan to give N-acety-D-glucosamine caused the unspecific products that could not react to di-methyl-benzaldehyde (DMAB). The study developed the method to detect N-acetyl-D-glucosamine by thin layer chromatography (TLC) that showed the stable purple traces on TLC than in liquid. The study was confirmed by applying high pressure liquid chromatography (HPLC).**

Keywords — **hyaluronan, N-acetyl-d-glucosamine, thin layer chromatography.**

I. Introduction

Highly reactive intermediates including 0_2- , H_20_2, -OC1, chloramines and HO ° are thought to be released from activated leukocytes both in vitro and during acute and chronic inflammatory conditions in vivo [4]. A number of previous studies have examined the reaction of HO ° with hyaluronan (a free, linear, A-BA-B copolymer of/3-D-glucuronic acid and N-acetyl-Dglucosamine with the two monomers linked 1 -3 and the disaccharide units linked 1-4 (Fig. 1) and related materials (such as heparin sulfate) in order to determine the site(s) of attack and the mechanism(s) of strand cleavage. Such studies have either studied, indirectly, the intermediate radicals (e.g. by pulse radiolysis 19 with optical or conductivity detection), or the products of the reaction (by, e.g., viscosity, end group analysis, oxygen consumption, peroxide formation, CO_2 release, molecular weight measurement)[3]. The study are however limited in the sense that neither can give direct information as to either the identity of the radical species formed as a result of HO" attack on the sugar rings, or the subsequent reactions of these substrate-derived radicals (and hence the mechanism of strand cleavage and polymer alteration).

In many projects involving a transformation or isolation of carbohydrates, it is important to quantify the amount of product obtained. It is sometimes difficult to measure the content of a specific component in a mixture. Knowledge of the properties of the desired product can suggest aspecific method of analysis, several of which involve colorimetric analysis. With colorimetric analyses, a selective chemical transformation will produce color. The absorbance or transmittance of the colored sample is directly correlated with the presence of acertain functional group or product in the mixture. Various reactions that produce color are known to be selective for certain types of carbohydrates containing N-acetyl-glucosamine. In the traditional methods of Morgan-Elson [1], using alkali borate to lysis hyaluronan to give N-acety-D-glucosamine caused the unspecific products that could not react to di-methyl-benzaldehyde (DMAB). The reaction was performed in solution. The color was also unstable. In this study, we tried to modify the detection of N-acety-D-glucosamine by using hyaluronidase and thin layer chromatography [2] to find out the way for determining hyaluronan in commercial products fastly.

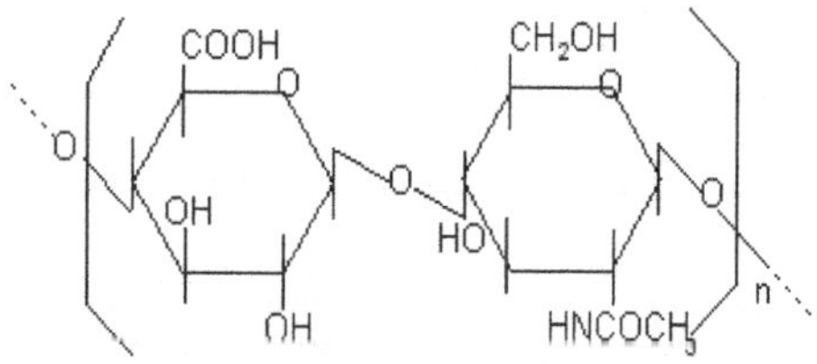

Fig. 1 Chemical structure of hyaluronan

II. Materials and Methods

A. Materials

Tested hyaluronan were collected from umbilical cord, rooster comb and *Streptococcus thermophilus* MR10 while sulfuric acid, sodium borate, n-butanol, formic acid, hydrochloric acid, absolute ethanol, p-dimethylamin-obenzaldehyde (DMAB) (Merck), hyaluronidase (Sigma) were analyzed chemicals.

V. Van Toi et al. (Eds.): 4th International Conference on Biomedical Engineering in Vietnam, IFMBE Proceedings 40, pp. 174–177, 2013.
www.springerlink.com

B. *Extraction of Exopolysaccharide from Streptococcus, Human Umbilical Cord, Rooster Combs*

After *Streptococcus* was cultured, the cells were removed. The protein in supernatant was removed by chloroform. The extraction of exopolysaccharide was followed as in umbilical cord and rooster comb sections.

Human umbilical cord residuals were treated with 4 l of 0.2% sodium chloride solution and then filtered. The supernatant was treated with 300–600 ml of 1% cetylpyridinium chloride (CPC) and later centrifuged. The precipitate obtained was re-suspended by adding 0.9 M calcium chloride solution to dissociate the HA precipitation complex. The suspension was treated with 25% v/v aqueous ethanol and the nucleic acids were precipitated and separated by centrifugation. The resulting suspension was deproteinized and defatted four times by chloroform solvent extraction (10% of the total volume). HA was precipitated by first adding 75% v/v aqueous ethanol and then acetone.

Adult cock's combs were defatted by cold acetone, then, the defatted material were treated with 5% sodium acetate solution as solvent for HA. Absolute ethanol was added to precipitate HA. Precipitation was collected and dissolved in 5% sodium acetate.

Residue protein was eliminated by shaking it with chloroform four times and then with a chloroform-amyl alcohol mixture several times until a gel no longer formed. The final solution was dialyzed. Following acidification to pH 4.0, the solution was precipitated with ethyl alcohol and the precipitate was desiccated.

C. *Determination of Uronic Acid*

Determine the glucuronic acid content by reaction with carbazole based on the British Pharmacopeia, 2007, monograph 1472. Reaction changing from no color into purple red indicated the positive result of uronic acid.

D. *Determination of N-Acetyl-Glucosamine*

Detection of N-acetylhexosamine was based on Morgan-Elson method [3] [4]. The changing of color from yellow into pink pointed the positive reaction. Extraction from Umbilical cord, rooster comb, bacteria and commercial hyaluronan were treated with bovine testicular hyaluronidase in 37°C for 24h. The reactions were stopped and undergo alkaline treatment by 0.8M potassium borate, pH 9.1 at 100°C in 15 minutes. All the samples were then developed on the thin-layer-chromatography with the solvent system comprised by n-butanol / formic acid / water (4:8:1, v/v/v). Thin layer chromatography after developed adequately will be dry in ambient temperature. Reaction traces were detected by aerosol DMAM solution. DMAB solution was prepared as described. Dissolved 5 gram of crystallization DMAB in 5 ml of concentrated HCl and 50 ml of glacial acetic acid. This solution was kept chilled and only be diluted ten times with glacial acetic acid, prior to use.

E. *High Pressure Liquid Chromatography (HPLC)*

All samples were crystallized and re-suspended in the solution of 1M NaCl, centrifuge and pass through 0.22um filter membrane. The after-treatment solutions were analyzed by the HPLC machine Agilent, with C18 column, refractive index detector along with the commercial hyaluronan as standard. The mobile phase was solution of KH2PO4 0,5M pH 2.5 with current speed of 1ml/minute. Column temperature was maintained at 30°C.

III. Results and discussion

A. *High Pressure Liquid Chromatography (HPLC)*

To have samples containing N-acetyl-glucosamine, the study isolated exopolysaccharide from the different sources containing hyaluronan studied previously. Via result from HPLC analysis, obviously, retention time of exopolysaccaride standard and HA extraction form other sources is similar. Minor different in retention time can be explained by different degree of polymerization.

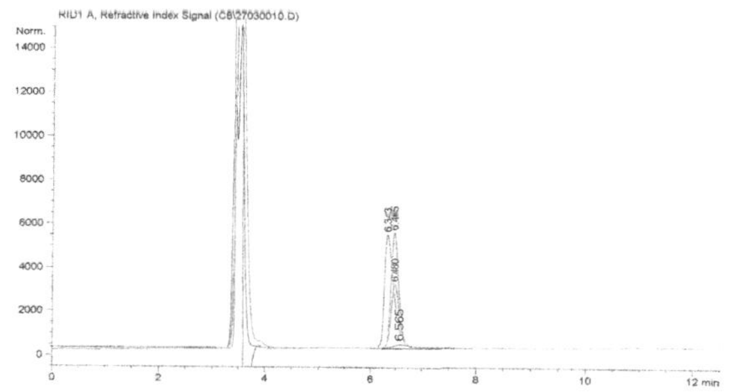

Fig. 2 High pressure liquid chromatography of umbilical cord, rooster comb and Streptococcus (retention time of umbilical cord, rooster comb and Streptococcus were 6.455, 6.430, 6.565, respectively).

Using Refractive Index Detector, which is specific for polysacharide analysis also bring more convenience than using UV detector. As we can see in the chromatogram, only one peak appear, mean that extracted is not go along with other glycosaminoglycan. In the Fig.2, the retention time of the standard hyaluronan was 6.343 while the retention time of umbilical cord, rooster comb and *Streptococcus* were 6.455, 6.430, 6.565, respectively.

A. Determination of Uronic Acid

After reacting with carbazole, the color changes into purple, the uronic acid was determined in the extraction. After measure the absorbance, based on the equation of Y= 0.0065x-0.0068 (R2 = 0.9994). The concentration of umbilical cord, rooster comb and Streptococcus was 500, 600, 180 mg/ml. With the data, the study used to determine the other monomer in extracted exopolysaccharide. That was N-acetyl-glucosamine.

B. Determination of N-Acetyl-D-Glucosamine

Based on Morgan Elson, the N acetyl glucosamine was detected by reacting with DMAB (Fig 3). However, the hyaluronan was lysed with bovine testicular hyaluronidase, the reaction was detected on thin layer chromatography. The results were showed in fig 3. Remarkably, the umbilical cord, rooster comb were the hyaluronan sources. With the detection, the N-acetyl-glucosamine in these sources reacted with DMAB and gave the stable traces for two weeks (Fig 4).

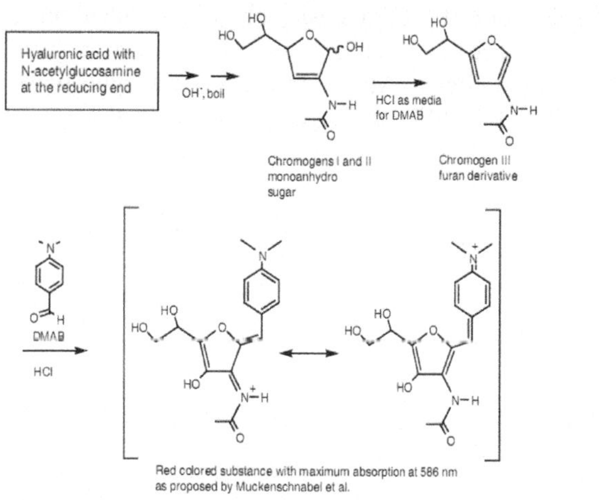

Fig. 3 Colorimetric reaction of N-acetyl-glucosamine and DMAB.

By using bovine testicular hyaluronidase, this enzyme digested hyaluronan at the specific sites as in Fig. 5 and formed non-reducing N-acetyl-glucosamine that helped the reaction performance when leach hyaluronidase formed non-reducing glucuronic acid [2]. Therefore, the study applied the bovine testicular hyaluronidase to lyse hyaluronan.

The study could detect HA in both products and the detectable level was 10 mg.

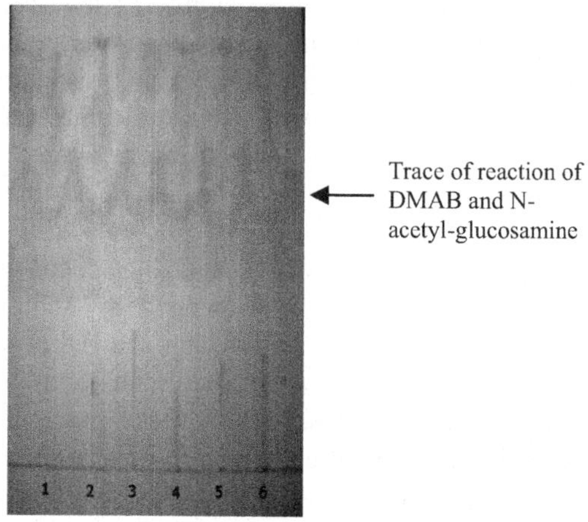

Fig. 4 Thin-layer Chromatography of N-Acetyl-Glucosamine (1-4-6. Standard HA, 2. Umbilical cord HA, 3. Rooster Comb HA, 5. non-pathogenic *Streptococci* HA).

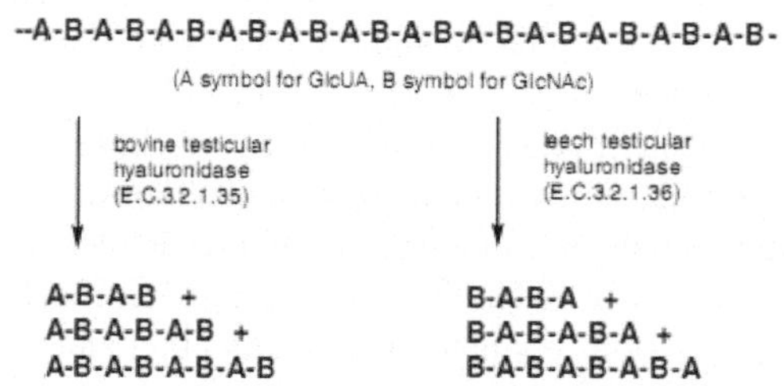

Fig. 5 Products after digested with bovine testicular hyaluronidase and leach testicular hyaluronidase.

IV. Conclusions

The study supplied the better and reliable test for N-acetyl-glucosamine monomer in hyaluronan because of the specificity of bovine testicular hyaluronidase on hyaluronan. This test will help us detect hyaluronan in medicine or foods, cosmetics rapidly and confidently.

References

1. Elson LA., Morgan WT. (1933) Biochem J 27:1824-1828
2. Eriksen, PB., Knudsen PJ., Fenger M., Florentz K. (1980) High-Performance liquid chromatography of hyaluronic acid and oligosaccharides produced by bovine testes hyaluronidase. J Chromatogr A 187: 373–379.
3. Lago G., Oruna L., Cremata JA., Pérez C., Coto G., Lauzan E., Kennedy JF. (2005) Carbohydr Polym 62(4): 321–326
4. Meyer K., Palmer J. (1934) J Biol Chem 107: 629–634

Author: Nguyen Hoang Khue Tu

Institute: Hochiminh city, International University
Street: Quarter 6, Linh Trung ward, Thu Duc dist.
City: Ho Chi Minh
Country: Vietnam
Email: nhktu@hcmiu.edu.vn

A DNA Extraction Method Applied for Living Seahorses

Nguyen Thi Hue and Nguyen Thi Thanh Tran

School of Biotechnology, International University, VNU, HCM City

Abstract — **The DNA extraction is the first step provided DNA samples for further genetic studies. As a small animal the DNA extraction from seahorse means killing the living specimens. This study is investigated to estimate the tissue sample size which can be used for extraction of enough DNA for further studies without killing living specimens and find out the suitable method for extraction of DNA from seahorse. Tissue samples were obtained by taking small tail rings at the end of the tails or the fins of living seahorses. Two methods were applied for extraction of DNA: Phenol method and salt method. The extracted DNA was concentrated by Nanodrop system. The quality of DNA was measured by agarose gel and PCR assay. While DNA from fin tissue gave no result in PCR, all DNA from tail tissue with about 20ng/ul gave the good PCR result even extracted by phenol or salt method. With the tail tissue, high concentration of DNA was achieved in both methods. 41% samples and 57.4% samples gave DNA concentration higher than 20ng/ul by phenol and salt method, respectively. The purity of DNA yield extracted by phenol is lower than by salt method (18% in phenol method compared to 44.5% in salt method). Thus, salt method gave higher efficiency than phenol method. The salt method gave higher percentage of sample with high concentration and high purity of DNA than phenol method. The salt method is considered as the suitable method. With the weight of tail tissue around 20mg total amount of DNA at least is 3μg which adequate for further genetic analysis. Beside the over success of salt method compare to phenol method, the salt method is lower cost and less hazardous chemicals as in salt method phenol, an expensive chemical and toxic, is not required.**

Keywords — **Seahorse, DNA extraction, Phenol method, salt method.**

I. INTRODUCTION

DNA extraction is a routine essential step in any biological studies about DNA. This step is widely utilized in a variety of genomic and molecular studies as well as other fields of medical examinations, clinical diagnostics, and forensic investigation. Therefore, many methods have been established to extract DNA molecules from biological materials [1].

Various effects can be obtained on DNA extraction by using different methods [2, 3]. An ideal isolation technique should optimize DNA yield, minimize DNA degradation, and be efficient in terms of cost, time, labor, and supplies. It must also be suitable for extracting multiple samples and generate minimal hazardous waste [3].

The Phenol and Salt methods are commonly used for DNA extraction from diverse organisms. In principal DNA is first released by lysing the cells with an anionic detergent in the presence of a DNA stabilizer. Proteins and other contaminants are removed by salt precipitation. Finally, DNA precipitated with absolute ethanol or isopropanol from its aqueous solution. There are tremendous variations in volume of ethanol or isopropanol (1- 2.5x volume of supernatant with DNA), incubation temperature (-80°C-25°C) and time (1- 24 hours) used for DNA precipitation. DNA molecules then stored in sterilized water or TE solution [2, 3].

Seahorses (genus *Hippocampus*) are in the family Syngnathidae, which also includes pipefish, pipehorses, and seadragons. Seahorses have a worldwide distribution inhabiting both temperate and tropical seas, with 33 species identified to date [4, 5]. Some species are reported in Vietnam: the spotted seahorse (*H. kuda*), three-spotted seahorse (*H. trimaculatus*), Japanese seahorse (*H. mohnikei*), and tiger tail seahorse (*H. comes*) [5]. Seahorse populations are in decline due to habitat destruction and overexploitation [6]. Monitoring the status of seahorses is possible via an analysis of the mitochondrial (mt)DNA control region. The mtDNA control region is sufficiently variable to use at the population level [7, 8] [9-11]. The results from an mtDNA analysis can subsequently be applied to the development of seahorse conservation and management strategies, such as increasing natural seahorse abundances while avoiding impacts on genetic variations [12].

Hippocampus kuda, also known as spotted seahorse. Spotted seahorse is the main objective in the study due to their status as vulnerable species needs to be protected. Genetic studies in these seahorses are necessary to understand more about the genetic diversity which may support the breeding selection for the new generation. The DNA extraction is the first step provided DNA samples for further genetic studies. With small animal like seahorse the DNA extraction from tissue means killing the living specimens. For the spotted seahorse, an animal need to be protected, genetic studies need to be performed but killing

V. Van Toi et al. (Eds.): 4th International Conference on Biomedical Engineering in Vietnam, IFMBE Proceedings 40, pp. 178–183, 2013.
www.springerlink.com

the seahorses must be avoided [13]. The same situation, with other small animals, extracting DNA from small tissues and do not scarify the living specimens is necessary.

This study is investigated in comparison the efficiency of two methods phenol and salt with the limitation of weight of tissue and kind of tissue based on the DNA concentration, purity and cost also.

II. MATERIALS AND METHODS

A. Materials

The spotted seahorse *Hippocampus kuda* was used as the model sample. Seahorses were collected from Khanh Hoa province, Vietnam in September 2010. Tissue samples were obtained nondestructively by taking small tail rings from the end of the tails or the fins of living seahorses. They were simply taken by a decisive cut down at the tail to get small parts of the tail or fin and did not harm the seahorses. Tissue samples collected from living seahorse were then stored in -20°C freezer for later use.

B. Methods

DNA extraction. Two methods were applied for extraction of DNA: Phenol method and salt method. Phenol method: Sample tissue is incubated overnight with proteinase K in the buffer TNES pH 8.0, the combination of 1M Tris, 0.5M Na_2EDTA, SDS 20% and 2M NaCl. The mixture of phenol: chloroform: isoamylalcohol (25:24:1) is then added to the incubated samples, and then 3M Sodium Acetate is added. The mixture is then centrifuged at 4°C, 14000 rpm for 10 mins. The supernatant is then collected for the next precipitation step. Cold absolute ethanol is then added to the collected supernatant for DNA precipitation. The centrifugation at 4°C, 14000rpm, for 5 mins is applied after DNA precipitation step to collect the DNA pellet. DNA pellet is then washed two time by ethanol absolute and 70%. The final DNA pellet is then dried at room temperature for at least 1h. Dried DNA pellet is subsequently suspended in 150µl of sterilized distilled water. The DNA solution is then ready for further analysis [14].

Salt method: Sample tissue is treated with proteinase K overnight as the same way with phenol method. The process DNA extraction is performed the same with phenol method except the mixture phenol: chloroform: isoamylalcohol is replaced by chloroform and 6M sodium chloride [14].

DNA quantification. DNA concentration was measured by Nanodrop Spectrophotometer ND- 1000. DNA concentration peak was observed by 260nm light.

DNA qualitative. The quality of DNA was measure by observing the band of DNA on agarose gel and the successful in PCR amplification using a set of primers (forward primer: 5'-aacggttcttgatggtcagg- 3', reverse primer: 5'- ccgtgtgcactctgaaatgt- 3') to amplify a fragment of 247 nucleotides of mtDNA control region.

The PCR assay was performed with total volume of 25µl contained 40ng of genomic DNA, 12.5µl of Toptaq DNA polymerase, 25mM of primers. The PCR profile consisted of an initial denaturation step (3 min at 94°C), followed by 30 cycles of denaturation at 94°C for 30s), then annealing at 61°C for 30s, and elongation at 72°C for 1min, and a final extension step at 72°C for 5min). PCR products were then analyzed by agarose gel 1.5% and visualized by UV and taken photograph by GelDoc- It System from UVP Company. To control the size of PCR product the first well in each gel was loaded by the 100kb marker.

III. RESULTS

76 samples with the range of weight from 9.2mg to 68.7mg were colleted from tail rings of seahorse. Table 1 showed the summary of collected samples and the distribution of samples in two DNA extraction methods: phenol and salt. 22 ring samples were extracted by phenol and 54 samples were extracted by salt (table 1). 9 other samples were collected from fin of seahorse. The size of fin sample is range from 0.5mm^2 to 2mm^2. 4 fin sample were extracted by phenol and 5 samples were extracted by salt method (table 2).

Table 1 The collected ring tissue samples with the range of weight and distribution of samples into two DNA extraction methods.

Method	No samples	Ring tisue sample weight (mg)		
	n=76	<15	15 - 40	>40
Phenol	4	8	7	7
Salt	5	11	26	17

The extracted DNA was checked on agarose gel. Figure 1 showed the presence of DNA in the extracted solution. Most of DNA extracted by salt method showed the clear and bright band on agarose gel while DNA extracted by Phenol method showed no band, smear or not clear band (figure 1). The DNA extracted from fin showed very light smear on agarose gel (figure 1: lane 7, 8).

Table 2 The collected fin tissue samples with the range of size and distribution of samples into two DNA extraction methods.

Method	No samples	Fin tisue sample size (mm^2)		
	n=9	0.5	1	2
Phenol	3	0	2	1
Salt	6	1	3	2

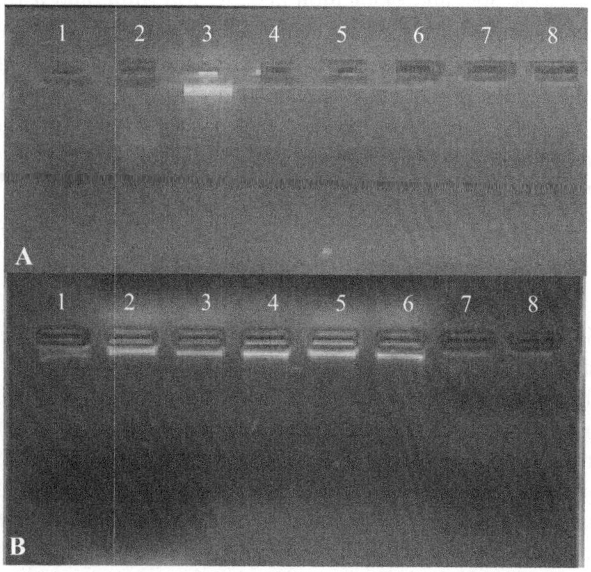

Fig. 1 Extracted DNA on 1% agarose gel. A: DNA extracted by Phenol method. B: DNA extracted by salt method. Lane 1, 2, 3, 4, 5, 6: ring tissue samples. Lane 7, 8: fin tissue samples.

Comparing the successful rate of DNA extraction on tail tissue between two methods by checking the DNA band on agarose gel, the result showed that 27.3% samples extracted by phenol method showed the DNA band, while 46.3% samples extracted by salt method showed the DNA band (table 3). With fin tissue it is not clear in comparison through agarose gel.

Table 3 Success rates of phenol and salt method based on DNA band on agarose gel.

DNA extraction method	No samples	Sample showed DNA band	% success
Phenol	22	6	27.3
Salt	54	25	46.3

DNA quantitation was compared between DNA extracted from tail tissue by phenol and salt methods, the result showed that DNA extracted by salt showed the higher purity (44.5%) than by phenol method (18%). Percentage of sample with DNA concentration higher than 20ng/µl in salt method is 57.4% compared to those in phenol method is 41% (table 4). Within the fin tissue samples, 3 samples extracted by phenol give concentration of DNA lower than 10ng/µl with low purity. 2/6 samples extracted by salt method give concentration of DNA at 15 and 16ng/µl with low purity and 4 other samples give low concentration of DNA at lower than 10ng/µl with low purity also.

Table 4 Quality and quantity of DNA through UV light measurement.

Method	No Samples	DNA Concentration (150µl)		Purity
		C> 20ng/µl	C< 20ng/µl	(1.7<A260/280<1.9)
Phenol	22	9 (41%)	13 (59%)	4 (18%)
Salt	54	31 (57.4%)	23 (42.6%)	24 (44.5%)

DNA extracted samples were testing for quality by PCR assay. The PCR assay was carried out using 40ng DNA for each assay. The DNA samples extracted from tail tissue showed the amplified PCR products in both phenol and salt method while DNA extracted from fin samples gave no PCR result.

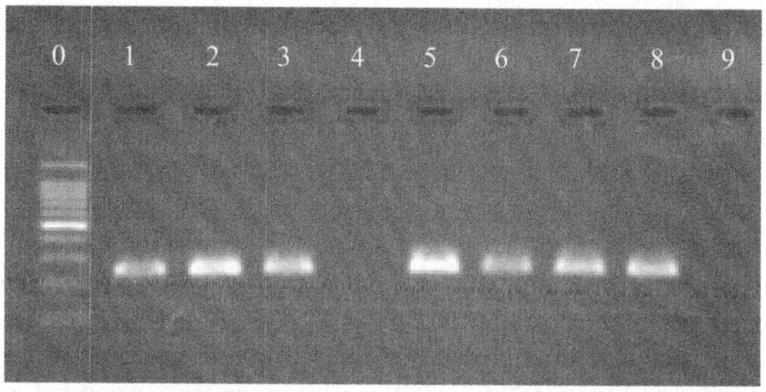

Fig. 2 Results of testing DNA quality by PCR assay. PCR product is showed by the bright band on agarose gel at 247bp. Lane 0: 100bp DNA molecular weight. Lane 4, 9: DNA isolated from fin tissue. Lane 1, 2, 3, 5, 6, 7, 8: DNA extracted from tail tissue. Lane 1, 2, 3, 4: DNA samples extracted by phenol method. Lane 5, 6, 7, 8, 9: DNA samples extracted by salt method.

Low weight or small size of tissue sample is result in low concentration of DNA. In both method of DNA extraction, the weight of tail tissue lower than 15mg gave too low DNA concentration, lower than 10ng/µl, while weight of tail sample higher than 15mg gave concentration of DNA higher than 19ng/µl (table 5). It results in total DNA amount in average higher than 2850ng. This amount of DNA can be used to perform 71 PCR assays with 40ngDNA/PCR assay. With the tissue weight higher than 40mg the high amount of DNA is easy to get. The average amount of DNA for 40mg tissue sample is 69ng/µl x 150µl = 10350ng (table 5). It is plenty DNA provide for further genetic analysis.

Table 5 The weight of tail tissue sample affected to DNA yield concentration.

Method	Phenol		Salt		
Weight (mg)	No samples	DNA conc	No samples	DNA conc	P value
<15	8	5.05	11	7.4	0.932
15<w<40	7	19.3	26	19.05	0.011
>40	7	96.7	17	69.6	0.008
	22	38.7	54	32.6	

The small size of fin tissue, smaller than $1mm^2$ gave inadequate amount of DNA (0.4ng/µl) (table 6). However, with the size of over $1mm^2$ a sufficient amount of DNA can be achieved. The minimum average amount of DNA got from $1mm^2$ fin tissue is 5.14ng/µl x 150µl = 771ng (table 6). It can be used for 19 PCR assays with 40ng/PCR assay.

Table 6 The size of fin tissue sample affected to DNA yield concentration.

Method	Phenol		Salt		
size (mm^2)	No samples	DNA conc	No samples	DNA conc	P value
0.5	0	0	1	0.41	NA
1	2	6.8	3	5.14	0.524
2	1	9.2	2	15.6	0.898
	3	7.6	6	7.8	

The efficiency between phenol and salt method was compared based on concentration of DNA in each group of sample weight and size (table 5, table 6). There is not significant different efficiency between two methods in fin samples (table 6). The significant different in efficiency of two method was observed in tail samples with high weight of tissue (table 5).

IV. Discussion

This study is conducted to evaluate the efficiency of two DNA extraction methods which may help to select a suitable method for extraction of DNA from seahorses without scarifying them and provide an adequate amount of DNA for further genetic studies that need a quite large amount of DNA for many PCR assays. The finding from this study can be developed more to produce a KIT for molecular genetic studies in seahorses.

Phenol method and Salt method were carried out in parallel space and time in the study. DNA concentration and it purity are the most important things that must be concerned. As the results of previous studies, DNA concentration is influenced by sevearl factors: tissue material, the weight or size of tissue materials, precipitation time and the volume of absolute ethanol or isopropanol [15].

Tissue material in the report is mainly tissue which cut off from various parts of living seahorses without scarifying them. The tissue can originate from the tail or fins. For a seahorse's safety, tissue should be taken out from the last tail rings by sterile equipments. Many studies about genetics of seahorses suggested for DNA extraction of seahorses, the genetic source should be fins (dorsal fins or pectoral fins) in small sizes (1-2mm^2) [4, 7, 12], and using the phenol method for DNA extraction. However, the amount of DNA is not adequate for deeply study in genetic as many PCR assays need to be performed. In this study, 9 samples were being worked on fins with the similar sizes by both of Phenol and Salt methods; however none of the samples had given the result of high concentration of DNA. With the low concentration the volume of DNA put into PCR assay need to be increased. This could also bring more contaminant or inhibitor into to the PCR assay. So that it could be the reason of no PCR products are collected from these samples even an equally amount of DNA was used for each PCR assay. The DNA can be extracted from fin tissue, however in both of the methods cannot achieve with high concentration or large amount of DNA when the material is a very small. The fin tissue cannot be collected with a bigger size without killing the seahorse. Thus, fin tissue is not the sufficient target for DNA extraction support for genetic studies even use phenol or salt method. There is no significant different in efficiency of two these methods applied in fin tissue (table 6).

Notably, while DNA sample extracted from fin tissue gave no result in PCR, all sample extracted from tail tissue with higher than 20ng/ul concentration gave the good PCR result even extracted by phenol or salt method (figure 2). Thus, extraction of DNA from tail tissue may give higher efficiency than fin tissue. This can be explained as the fin tissue is too small but cannot increase the size of fin tissue without affecting to living seahorses.

With the tail tissue, high concentration of DNA was achieved in both methods. 41% samples and 57.4% samples gave DNA concentration higher than 20ng/ul by phenol and salt method, respectively. The purity of DNA yield extracted by phenol is lower than by salt method (18% in phenol method compared to 44.5% in salt method). Thus, salt method gave higher efficiency than phenol method. The salt method gave higher percentage of sample with high concentration and high purity of DNA than phenol method.

Another factor has been mentioned is the weight of tissue, the more weight of tissue or the higher quantity of DNA can be yielded. With the weight of tail tissue around 50mg, DNA yield was in the range of 50- 120ng/µl (data not shown), while the amounts of were rapidly decreased with the weight of tissue of 30mg, 20mg, 10mg and 5mg. However, with the weight of tissue around 20-30 mg the concentration of DNA is around 20ng/ul (data not shown). This will be provided enough for an experiment and do not kill living specimens. With lower weight of tissue (lower than 15mg) the concentration of DNA extracted by both two method is not significant different. With the higher weight of tissue (higher than 20mg) the concentration of DNA extracted by phenol is higher than salt method (table 5). However, the purity of DNA was compared the result showed that the percentage of high purity sample in salt method is higher than in phenol method (table 4). This is an important factor which may affecting directly to the successful of PCR assays. In addition, with the weight of tissue around 20 - 40mg the difference in DNA concentration between two methods is not very significant (P = 0.011). Thus, the salt method is considered as an advantage method compare to phenol method when using the pieces of tissue which not affecting much to the living specimens.

In addition, compared with Phenol method, Salt method using the same chemicals, but without Phenol solution which is usually hazardous to technicians and toxic for PCR assay also, therefore salt method is less of toxicity. In term of cost, the price of each DNA extraction assay by salt method is significant less than a commercial DNA extraction KIT (Qiagen KIT). The salt method requires cheap chemical compared to phenol method in term of using phenol, a toxic and expensive chemical. Only a simple chemical preparation is required in salt method which can be made easily in any laboratory.

In conclusion, the salt method is the suitable method for DNA extraction applied for tail tissue of seahorses to produce an adequate amount of DNA support to genetic studies. This method can also be applied for other small animals which some modify in tissue used.

A. *Salt Method DNA Extraction Protocol for Tail Tissue of Seahorse*

1. 20 mg of tail tissue (first ring of the tail) is collected and transferred into a 1.5µl microcentifuge tube. Pipet 20µl Protease (or 10µl Proteinase K). Add 300µl of TNES buffer. Continue with step 3
2. Mix thoroughly by vortexting and incubate the mixture in the tube at 56° for 5- 24 hours (longer incubation higher DNA yield will be have at the end)
3. After incubation time, invert the tube slowly 3-5 times to recognize the fragmentation of tissue done completely.
4. Add 300µl of Chloroform and 80µl of NaCl 6M, invert the tube for 1-3 time, centrifuge maximum speed 12000- 14000rpm for 10 mins.
5. Transfer the supernatant to a fresh tube. Adding an equally cold absolute ethanol in. Invert the tube gently 3 times.
6. Optional step - Place the tube into a -20°C freezer in 24 hours.
7. Centrifuge the tube for 15 minutes, at 4°C, and at maximum speed 12000- 14000rpm.
8. Discard the supernatant; do not disturb the pellet.
9. Wash the pellet with cold ethanol by adding 500ul of 70% ethanol into the pellet and centrifuge for 5 minutes, at 4°C, at maximum speed 12000- 14000 rpm. Remove the ethanol, keep the pellets inside. Repeat this step for one more time.
10. Dry the DNA pellet at room temperature for at least 30 mins. Make sure no ethanol remain in the pellet.
11. 50µl distilled water is added and suspended the DNA pellet. Keep DNA solution in 4°C or -20°C freezer for later use.

ACKNOWLEDGMENT

Dr. Hoang Tung (Biotechnology School of International University- VNU, Vietnam) for providing living seahorses. This research was supported by The School of Biotechnology - International University- VNU HCMC, Vietnam.

REFERENCES

1. Milligan BG, Total DNA isolation. 2nd Edition. Oxford, New York, Tokyo: Oxford University Press, 1998. pp 29–64.
2. Waldschmidt AM, Extraction of genomic DNA from Melipona quadrifasciata (Hymenoptera: Apidae, Meliponinae). Braz. J. Genet., 1997. (20) 421– 423.
3. Chen M, Z.Y., Tao J, Luo Y., Methodological comparison of DNA extraction from Holcocerrus hippophaecolus (Lepidoptera: Cossidae) for AFLP analysis. Forestry Studies in China, 2008. (10): 189–192.
4. Lourie SA, Fin-clipping procedure for seahorses. Fisheries Centre, Univ. of British Columbia: Project Seahorse Technical Bulletin 3., 2003. (vers. 1.1), pp 1-4.

5. Lourie SA, A.V., HJ Hall, Seahorse: an identification guide to the world's species and their their conservation. London: Project Seahorse, 1999.
6. IUCN, IUCN red list of threatened species. 2004.
7. P. R. TESKE, M.I.C.a.C.A.M., Population genetics of the endangered Knysna seahorse, Hippocampus capensis. Molecular Ecology, 2003. 12: 1703- 1715.
8. Zhou JF, Q.W., YZ Ye, JG Tong, Genetic divergence between Cyprinus carpio carpio and Cyprinus carpio haematopterus as assessed by mitochondrial DNA analysis, with emphasis on origin of European domestic carp. Genetica, 2003. 119: 93-97.
9. Faria PJ, N.G., C Yamashita, P Martuscelli, CY Miyaki, Genetic variation and population structure of the endangered Hyacinth Macaw (Anodorhynchus hyacinthinus): implications for conservation. Biodivers. Conserv, 2008. 17: 765-779.
10. Guillen AKA, G.B., O Takenaka, Genetic diversity among African great apes based on mitochondrial DNA sequences. Biodivers. Conserv, 2005. 14: 2221-2233.
11. Liao TY, T.W., HD Lin, SC Shen, CS Tzeng, Phylogeography of the endangeredspecies, Sinogastromyzon puliensis (Cypriniformes: Balitoridae), in southwestern Taiwan based on mtDNA. Zool. Stud, 2008. 47: 383-392.
12. Thadsin Panithanarak, R.K., Uthairat Na-Nakorn, and Thuy Thi Thu Nguyen., Population Genetics of the Spotted Seahorse (Hippocampus kuda) in Thai Waters: Implications for Conservation. Zoological Studies, 2010. 49. 4: 564- 576.
13. Shan-Rong Shi, R.J.C., Lin Wu, Cheng Liu, Ram Datar, Yan Shi, Dongxin Liu, Hyoeun Lim, and Clive R. Taylor., DNA Extraction from Archival Formalin-fixed, Paraffin-embedded Tissue Sections Based on the Antigen Retrieval Principle: Heating Under the Influence of pH. The Journal of Histochemistry & Cytochemistry, 2002. 50(8): 1005- 1011.
14. Russell, S.a., Molecular Cloning: A Laboratory Manual. Cold Spring Harbor Laboratory Press., 1989. 11 (4): 133- 145.
15. Zeugin JA, H.J., *Ethanol Precipitation of DNA*. Focus, 1985. **7: 1–2**.

Detection of Toxic Aspergillus Species in Food by a Multiplex PCR

Nguyen Thi Hue[1], Nguyen Huu Nhien[1], Pham Minh Thong[1], Chu Nguyen Thanh[2], and Pham Van An[2]

[1] School of Biotechnology, International University – VNU
[2] Microbiology – GMO Testing Laboratory, Quality Assurance and Testing Centre 3

Abstract — ***Aspergillus parasiticus*** **and *flavus* are known as the main producers carcinogen aflatoxins. The presence of these fungus and aflatoxins in food is a serious risk for human and animal health. The identification of these fungi is not straightforward due to its similarities with closely related species. In this study, a multiplex PCR assay have been developed which overcame the disadvantages in detection of *Aspergillus parasiticus* and *flavus* by conventional morphological method. Twelve isolates of filamentous fungi, representing six *Aspergillus* and *Aspergillus*-related species were used. DNA was extracted from mycelium following SDS method modified from Plaza's method. The extracted DNA was used for the PCR method to identify the presence of selected fungi using specific primer sets. The result of identification was analysed by agarose gel electrophoresis. The sensitivity and specificity of the detection were checked by result of a gradient PCR assays. Two specific set of primers were successful designed based on Aflatoxin biosynthesis gene cluster to detect A. *parasiticus* and A. *flavus* species, with the limitation of detection (LOD) for each set of primer is 0.005ng and 0.008ng. A multiplex PCR assay set up with two these sets of primers was also successful in detection of both targets of *A. parasiticus* and *A. flavus* at the annealing temperature of 65°C with high sensitive and specificity. The limitation of detection (LOD) is 0.5ng. A highly specific multiplex PCR assay have been developed to detect the presence of *Aspergillus parasiticus* and *flavus* in food. This will permit prediction of the presence of aflatoxins type G and aflatoxins type B, the most potent natural carcinogens.**

Keywords — ***Aspergillus flavus, Aspergillus parasiticus,*** **aflatoxin, aflatoxin biosynthesis gene clusters, *norB-cypA* genes.**

I. Introduction

Aspergillus species are among the most ubiquitously found mold fungi throughout the world of high importance in medicine, agriculture and biotechnology. Besides causing direct pathogenesis, they also produce various types of toxic secondary metabolites, mycotoxins and cause non-contagious mycotoxicoses [1]. Out of all different types of mycotoxins, aflatoxins are the most potent natural carcinogens known, have hepatoxic and immunosuppressive properties which cause acute liver damage, liver cirrhosis, tumor induction and teratogenesis [2]. They recognized as a possible human carcinogen by International Agency of Research on Cancer [3]. These mycotoxins are produced primarily by *Aspergillus parasiticus* and *A. flavus* which may invade agricultural products during plant growth, during harvest and afterwards in storage, results in significant economic losses [4].

More than 300 mycotoxins have been discovered up to date, but the most toxic group is aflatoxin. Aflatoxins are both acutely and chronically toxic to animals, including human and are categorized into 4 major kinds, aflatoxin B1 (AFB1), aflatoxin B2 (AFB2), aflatoxin G1 (AFG1), and aflatoxin G2 (AFG2). It has been demonstrated aflatoxin B1 (AFB1), in a variety of animal species, was the most potent liver carcinogen known. Aflatoxin B1 has been linked to human hepatocellular carcinoma in several regions of Africa and Southeast Asia. M_1 and M_2, compounds of lower toxicity, are found to be excreted in milk, when AFB_1 and AFB_2 are ingested by lactating cows. Children that consume large quantities of milk are susceptible to the long-term exposure to aflatoxins. In developing countries, where detection and decontamination policies are impractical, aflatoxin contamination is predominately a food safety issue. In nature, aflatoxins are produced by *Aspergillus flavus, A. parasiticus* and other *Aspergillus* species.

Aspergillus flavus belongs to the Genus *Aspergillus*, and is able to grow on various nutrient sources in nature. Despite playing an important factor in nature, the quest for knowledge about potential aflatoxin problems made *A. flavus* became the most widely reported foodborne fungus [5]; [6]. In many reports, this mold is determined as a plant pathogen; particular in developing seeds: corn, peanuts, cottonseed, and tree nuts [7]; [8]; [5]. To animal including human, aspergillosis is a disease caused by *Aspergillus* species, especially *A. flavus* and *A. parasiticus* [9]. *A. parasiticus*, not same as *A. flavus* only produces aflatoxin B, may produce both aflatoxins B and G. Discrimination of these two species from each other and from closely related species is difficult when using conventional methods, based mainly on morphological or immunological features including culturing the fungus in suitable inducing media, extracting aflatoxins with other solvents, and monitoring their presence by chromatographic and ELISA techniques [10];[11]. The limits of these methods are time-consuming and require considerable expertise. Furthermore,

V. Van Toi et al. (Eds.): 4th International Conference on Biomedical Engineering in Vietnam, IFMBE Proceedings 40, pp. 184–189, 2013.
www.springerlink.com

the complexity in the presence of aflatoxinogenic and non-aflatoxinogenic *Aspergillus* species which can cause false negative results in detection [12]; [13].

The application of DNA-based techniques, particularly polymerase chain reaction (PCR), permits rapid, sensitive and specific detection, that overcome the disadvantages of conventional method, necessary to devise strategies to control and reduce fungal mass and toxin production at early and critical stages of the food chain. Up to date, there have been few studies for detection of *A. parasiticus* by using molecular methods. RT-PCR with target genes aflD, aflO, aflP [14], PCR-RFLP with target gene aflR , multiplex-PCR with target genes nor-1, ver-1, omt-1 and apa-2 (Chen *et al.*, 2002), real-time PCR with primers and probe designed on multicopy ITS2 rDNA target sequence were used [15], however these methods were established to detect both *A. parasiticus* and *A. flavus* for aim of monitoring contamination of food and feedstuffs by aflatoxin-producing *Aspergillus*. A recent study for detection of unique *A. parasiticus* using real-time PCR assay with primers designed on the multicopy internal transcribed region of the rDNA unit (ITS1-5.8S-ITS2 rDNA) was done by Sardinas N. in 2010 [15]. This study obtained successfully in accurate detection and quantification of *A. parasiticus* at spore concentrations equal or higher than 10^6 spores/g. Nevertheless, this method requires an expensive equipped-laboratories and consumable which may increase the cost for the detection of. In this study, a cheap, accurate, sensitive and specific assay is developed to detect A. *flavus* and *A. parasiticus*.

II. Materials and Methods

A. *Materials*

12 isolates of filamentous fungi, representing six *Aspergillus* and *Aspergillus*-related species were used in this study (table 1). All isolates of *A. flavus* (VTCC-F-160, VTCC-F-898), *A. oryzae* (VTC-F-910, VTCC-F-912), and *A. parasiticus* (VTCC-F-1130, VTCC-F-1132, ATCC-16404) were obtained from Institute of Microbiology and Biotechnology (IMBT), Vietnam National University, Hanoi; *A. flavus* (AF.IV26.1) and *A. niger* (ATCC-16404) were obtained from Pasteur Institute, HCM city, and American type culture collection, respectively; *A. flavus* (TN1), *A. candidus*, and *P. aethiopicum* (TNTT) were obtained from Quatest3. Fungal strains were maintained on potato dextrose agar (Downes *et al.*, 2001; Wehr *et al.*, 2004; FDA, 2005).

Fungal strains were incubated in Blackeslee's malt extract broth (2% malt extract, 2% dextrose, 0.1% peptone) at 25°C, 120rpm for 3 days before DNA were extracted.

Table 1 List of selected Aspergillus species and Aspergillus-related species.

Species	Code	Provider
A. parasiticus	VTCC-F-1130 VTCC-F-1132 VTCC-F-1159	Vietnam Type Culture Collection
A. flavus	VTCC-F-160	Vietnam Type Culture Collection
	VTCC-F-898	Vietnam Type Culture Collection
	AF.IV26.1	Pasteur Institute HCM city
	TN1	Isolated from maize – Quatest3
A. oryzae	VTCC-F-910 VTCC-F-912	Vietnam Type Culture Collection
A. niger	ATCC-16404	American Type Culture Collection
A. candidus		Isolated from food – Quatest3
P. aethiopicum	TNTT	Quatest3

B. *Methods*

Primer design. Aflatoxin biosynthesis gene cluster of A. flavus (Isolate AF13, AF36, AF70); A. oryzae (Isolate RIB40); A. parasiticus; A. niger were obtained from Genebank (National Center for Biotechnology Information, NCBI, http,//www.ncbi.nlm.nih.gov/nuccore/) and aligned with Bioedit software (Version 7.0.5.3). Primers were designed based on nucleotides that are unique to A. flavus and A. parasiticus. Primers to be designed were compared against all other oligonucleotides available online with the basic local alignment search tool algorithm (BLAST, NCBI) [16].

DNA extraction. The protocol was adopted with some modifications from Plaza's method [17]. Mycelium of *Aspergillus* species from broth culture were collected and washed with 1.5ml NaCl 0.85% before being air-dried at 65°C. Mycelium were kept at -70°C in 10 minutes and ground well in mortar before process the DNA extraction.

PCR set up. PCR assays were performed in a 10μl reaction mixture containing Buffer 1X, 2.5mM Mg^{2+}, 0.8mM dNTP mix, 7.2mM primer mix, 0.5 unit Taq polymerase, 1μl of DNA template. The thermal cycle including 5mins initializtion at 95°C, 35 cycles of 30 seconds at 95°C for denaturation, 30 seconds at optimal annealing temperature for primer annealing, and 30 seconds at 72°C for extension was used in the Applied Biosystems® Veriti® 96-Well Fast Thermal Cycler. PCR products were analyzed by electrophoresis (Midi-Ultra - Nyx Technik, Inc.) containing 2% agarose gel in TBE buffer 0.5X (Promega) stained with bromide.

Singplex PCR. The annealing temperature for gradient PCR amplification ranged from 54°C to 69°C for the set of primer used in detection of A.*flavus*. The range of temperature from 58 - 66°C was set for detection of A. *parasiticus*. The best annealing temperature for primer set will be selected based on the presence of PCR product on agarose gel. The specificity of detection was checked by testing the ability of amplification target within 12 different species of *Aspergillus*. The specificity assays were performed with 70ng DNA from individual strains or DNA mixture of all 12 strains. To assess the detection limit of detection of the assay, the dilution series of factor 10 was prepared for individual DNA concentration; and for mixture of DNA from 12 *Aspergillus*, and PCR assays were performed in each dilution.

Multiplex PCR. Two set of primers which successful for detection of A.*flavus* and A.*parasiticus* in singplex PCR were used in one PCR assay, and tested again the best annealing temperature for both. The specificity and LOD were also tested in the same method as singplex PCR with the target is the mixture of DNA from 12 species.

III. Results

A. Singplex PCR for Detection of A. flavus

A. *flavus* primers. BLAST result for forward primer (AFLA-F-5' – GGTGGTGAAGAAGTCTATCTAAGG – 3') showed that all 24 nucleotides of AFLA-F matches oligonucleotide of *A. flavus* (Identities = 24/24) and no gap was found (Gaps = 0/24). No other 100% matches and no gap was found on the other species. On the contrary, BLAST result for reverse primer (AFLA-R-5' – AAGGCATAAAGGGTGTGGAG – 3') showed that all 20 nucleotides of AFLA-R matches oligonucleotide of both *A. flavus* and *A. oryzae* (Identities = 20/20) with no gap was found (Gaps = 0/20). This indicates the designed forward primer is specific for *A. flavus* only. The PCR product is a 413bp sequence.

Figure 1 shows the result of checking PCR products from the gradient PCR assay on agarose gel, the presence of the band at 413bp indicated the expected PCR product. The annealing temperature of 54 °C, 57 °C, 60 °C and 63°C revealed sharp amplified products; especially a greater intensity appears at 60°C. When the temperature increases the PCR product become faint and then no PCR product was observed at 69°C (figure 1). Between 54°C to 66°C, any point of temperature can be used for further study, but 65°C was selected as it is high enough to avoid the appearing of extra PCR product due to low temperature in some cases.

Specificity of A. *flavus* primers. The result for testing the specificity of the designed primer is shown in figure 2, this indicated that only DNA from *A. flavus* could be amplified (lane 2 to 5), no matter what the presence of other species (lane 1). No other fungal species (non-flavus *Aspergillus* and *Penicillium*) PCR product was obtained (lane 6 to 13). Only *A. flavus* species is detected by PCR using the designed primer set. Thus the primer set is specific for detection of *A. flavus*.

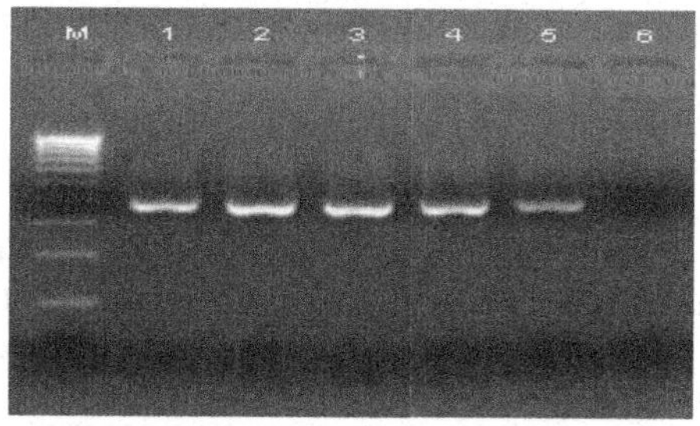

Fig. 1 Checking annealing temperature of primers. Lane 1 to 6: 54°C, 57°C, 60°C, 63°C, 66°C, 69°C.

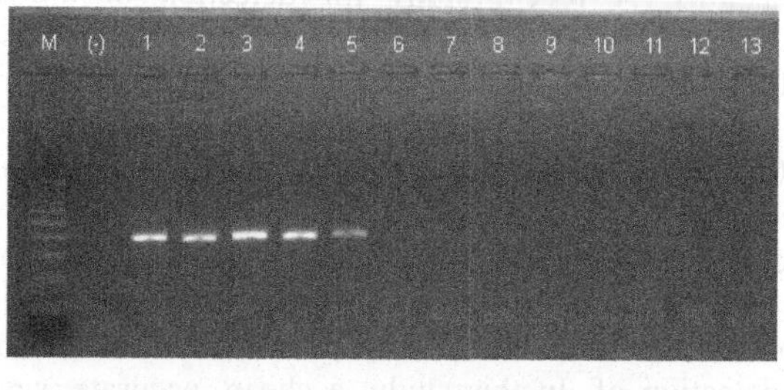

Fig. 2 The specificity of *A.flavus* primers. M: 100-bp DNA molecular weight ladder; (-): negative control; Lane 1: DNA mixture; Lane 2: *A. flavus* VTCC-F-160; Lane 3: *A. flavus* VTCC-F-898; Lane 4: *A. flavus* AF.IV26.1; Lane 5: *A. flavus* TN1; Lane 6: *A. oryzae* VTCC-F-910; Lane 7: *A. oryzae* VTCC-F-912; Lane 8: *A. parasiticus* VTCC-F-1130; Lane 9: *A. parasiticus* VTCC-F-1132; Lane 10: *A. parasiticus* VTCC-F-1159; Lane 11: *A. niger* ATCC-16404; Lane 12: *A. candidus*; Lane 13: *P. aethiopicum* TNTT.

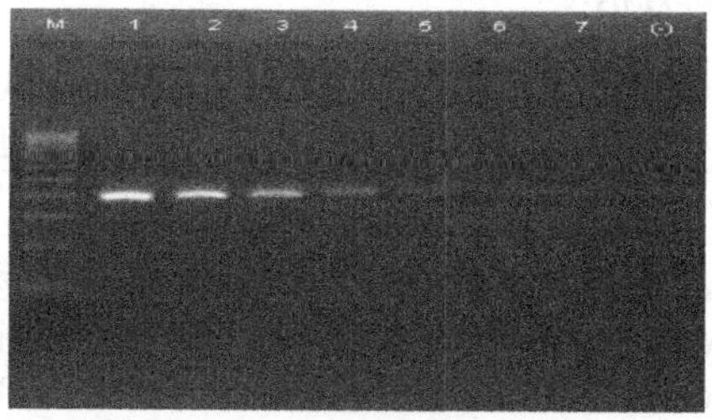

Fig. 3 Sensitivity of A. *flavus* primers for A. flavus DNA only. M: 100-bp DNA molecular weight ladder; (-): negative control; Lane 1 to 7: 5ng, 1ng, 0.2ng, 0.04ng, 0.008ng, 0.0016ng, 0.00032ng.

Sensitivity of A. *flavus* primers. DNA solution of *A. flavus* VTCC-F-898 was diluted to have 7 dilutions; value of LOD is the dilution at which the PCR product gives the band on agarose gel. For individual dilution of DNA from A. flavus only, figure 3 showed that when the DNA amount is reduced from 5ng to 0.00032ng the appearance of PCR product on agarose gel is also reduced and gave no product since 0.0016ng. Thus, the limit of detection was established at 0.008ng (LOD=0.008ng) (figure 3).

With the mixture DNA from 12 species, the LOD is found at 0.016ng (figure 4).

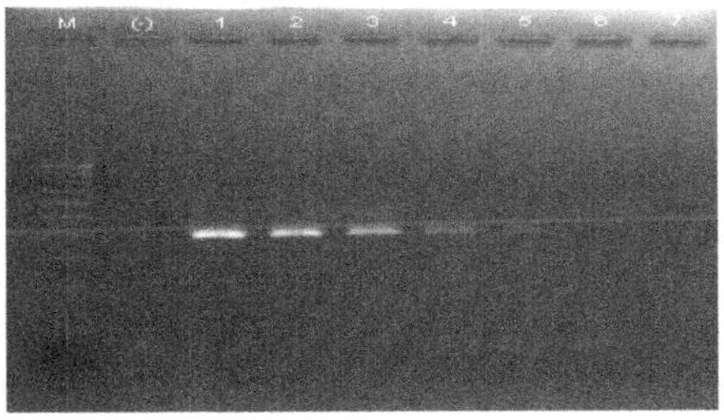

Fig. 4 Sensitivity of A. *flavus* primers for DNA mixture. M: 100-bp DNA molecular weight ladder; (-); negative control; 1, Lane 1 to 7 in order: 10, 2, 0.4, 0.08, 0.016, 0.0032, 0.00064ng.

B. Singplex PCR for Detection of A.parasiticus

A. *parasiticus* primer. Primers were designed on the basis of sequence alignments of the *norB-cypA* genes, the farthest upstream portion of the aflatoxin biosynthesis gene cluster, which were evaluated by Ehrlich [18]. The primers APA1 (5' GGATTCGTGAGTGTCTTTAGGG 3') and APA2 (5' GGTAAATGCTCCGCACAGTC 3') were picked up as shown the highly matching and identity as A. *flavus* primer above.

The gradient temperature PCR was performed to evaluate the annealing temperature of *A. parasiticus* primers on the A. *parasiticus* target only. It result in the best and be selected temperature for further analysis for A. parasiticus primers is 62°C (data not shown).

Specificity of A. parasiticus primers. The test on DNA of 12 aspergillus species showed that only A. parasiticus is be detected among several species under A. parasiticus primers (figure 5). The specificity of the test also demonstrated when A. parasiticus was detected in the mixture of the DNA solution of all tested species (lane 1, figure 5). This indicated that the A. parasiticus primers is specific for detection of A. parasiticus only whenever the presence of other species.

Sensitivity of A. *parasiticus* primers. The test was performed on DNA of A. *parasiticus* (VTCC-F-1132) only (figure 6) and DNA mixture of all selected species (figure 7). Figure 6 showed that the LOD of the test on A. *parasiticus* only is 0.005ng (lane 10, figure 6). The figure 7 showed that LOD of detection A. parasiticus in the DNA mixture is 0.005 ng also (lane 10, figure 7).

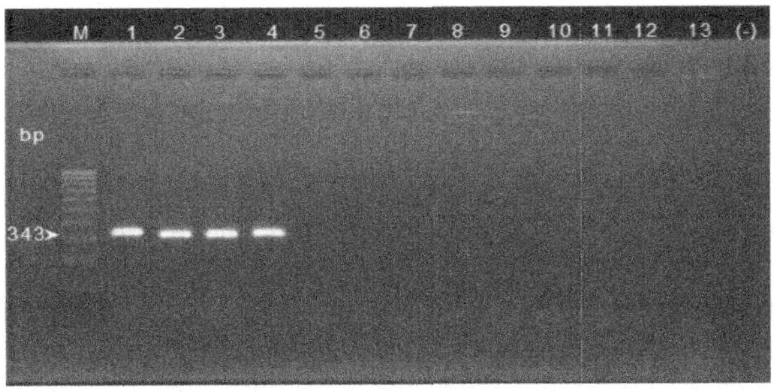

Fig. 5 Specificity testing for A. *parasiticus* primers. M: 100-bp DNA molecular weight ladder, (-): negative control. Lane 1: DNA mixture of selected species; Lane 2: *A. parasiticus* VTCC-F-1130; Lane 3: *A. parasiticus* VTCC-F-1132; Lane 4: *A. parasiticus* VTCC-F-1159; Lane 5: *A. flavus* VTCC-F-160; Lane 6: *A. flavus* VTCC-F-898; Lane 7: *A. flavus* AF.IV26.1; Lane 8: *A. flavus* TN1; Lane 9: *A. oryzae* VTCC-F-910; Lane 10: *A. oryzae* VTCC-F-912; Lane 11: *A. niger* ATCC-16404; Lane 12: *A. candidus*; Lane 13: *P. aethiopicum*.

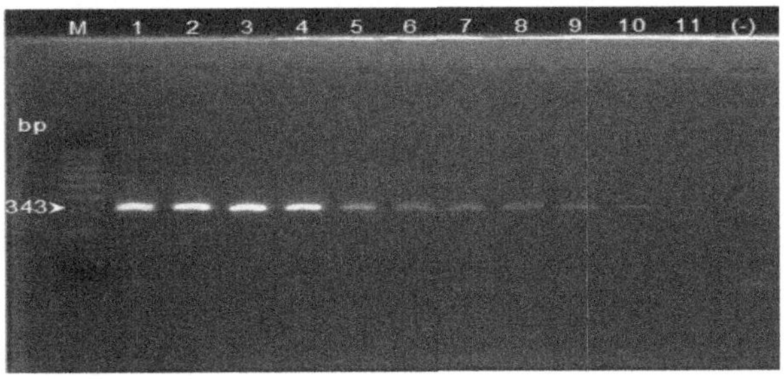

Fig. 6 Sensitivity of A. *parasiticus* primers. M: 100-bp DNA molecular weight ladder, (-): negative control. Lane 1 to 11 on order: 100, 50, 10, 5, 1, 0.5, 0.1, 0.05, 0.01, 0.005, 0.001ng.

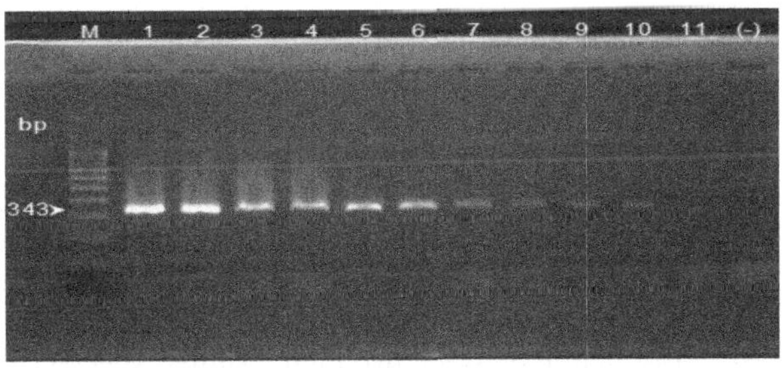

Fig. 7 Gel electrophoresis of sensitivity testing on mixture of all selected species. M: 100-bp.DNA molecular weight ladder, (-): negative control. Lane 1 to 11 in order: 100, 50, 10, 5, 1, 0.5, 0.1, 0.05, 0.01, 0.005, 0.001ng.

C. Multiplex PCR for Detection of A.flavus and A.parasiticus

From the singplex PCR result the combination of two set of primers for detection of A.*flavus* and A.*parasiticus* was used to test the ability of detection two these kinds of fungi in the same PCR assay. Put on together, two couples of primers were performed in a single PCR assay, multiplex PCR, for a simple detection of *Aspergillus* species.

DNA of *A. parasiticus* and *A. flavus* were tested both in individual samples and in mixtures of many different DNA from other *Aspergillus* species (*A. flavus, A. parasiticus, A. oryzae, A. niger, A. candidus, and P. aethiopicum*).

Figure 8 indicates the specificity using two pairs of primer in multiplex PCR. The presence of *A. flavus* DNA was identified by one band appearing at 413bp; the band appears 343bp represented the appearance of *A. parasiticus* DNA. Only one band can be observed in individual samples (lane 1 and 2, figure 8). Two separated bands are observed in the presence of both *A. parasiticus* and *A. flavus* in a mixture. The LOD of the multiplex PCR assay in detection of two these fungi is approximate 0.5ng of mixture DNA (figure 8).

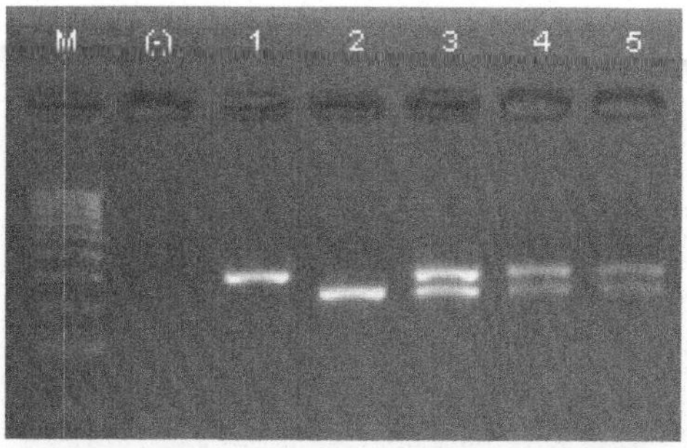

Fig. 8 Specificity of multiplex PCR in detection both A. *flavus* and A. *parasiticus*. M: 100-bp.DNA molecular weight ladder, (-): negative control; Lane 1: A. *flavus* only - 70ng; Lane 2: A. *parasiticus* only – 70ng; Lane 3-5: mixtures of *Aspergillus* DNA in dilution series of 50ng, 5ng, and 0.5ng.

IV. Discussion

This study focuses on the search of new primers for identification of *A. flavus* and *A. parasiticus*, two most popular causes of aflatoxin contamination in food, in only one PCR reaction. In nature, both species often share substrates such as corns, nuts and cereals. This enhances the quest of distinction between two highly similar species. The presence of both aflatoxins type B (B1 and B2) and type G (G1 and G2) can be predicted using this assay.

Applied conventional PCR assays, the specificity and sensitivity of two couples of primer were tested individually. The specificity was confirmed in a wide selection of strains, even in the mixture of many strains. The sensitivity was tested using a gradient concentration of target DNA and a gradient concentration of DNAs mixtures. Due to some correlations of both PCR assays, a multiplex PCR can be performed simultaneously reducing time of analysis.

With the successful of singplex PCR for each A. *flavus* and A. *parasiticus* in a very high LOD (0.005ng and 0.008ng) the multiplex could also very high sensitivity. The specific of detection using the designed primers have been demonstrated by the non-appearance of PCR in other species targets while it is appeared very clearly in the A. *flavus* with the A. *flavus* primers and the A. *parasiticus* with A. *parasiticus* primers. This indicated the successful in designing primer. Moreover, in the multiplex PCR two set of primer didn't show any conflict or dimer or hair pin or cross binding between these primers. This again demonstrated the ability of using the designed primers in the specific detection of two these target species among several other species.

The multiplex PCR set up here could be used to confirm and identify the presence A. *flavus* and A. *parasiticus* in food while the morphology method is not easy to identify or take long time to do. However, in comparison with previous quantitative PCR assays [15], the success of this work is the beginning of a new method, which means further studies are required before being applied in real. The sensitivity of the multiplex PCR must be further tested. In addition, to detect false negative results, internal amplification controls are necessary. The false negative results are usually caused by PCR reagents defaults, thermal cycling machines or the inhibitors from cultures or substrates. The later studies should concern the research on food samples for the applications in real life.

Acknowledgment

Authors would like to thank to Microbiology – GMO Testing Laboratory, Quality Assurance and Testing Centre, where the study is carried out.

References

1. Beck-Sague, C., Jarvis, W. R., Secular Trends in The Epidemiology of Nosocomial Fungal Infections in The United States, 1980-1990. National Nosocomial Infections Serveillance System. The Journal of Infectious Diseases, 1993. 167(1274-1251).
2. JECFA, Evaluation of Certain Food Addictives and Contaminants. Forty-sixth Report of the Joint FAO/WHO Expert Committee on Food Addictives 1996, in WHO Technical Report Series, WHO, Editor. 1997: Geneva.
3. IARC, International Agency for Reaearch on Cancer: Monographs on the Evaluation of Carcinogenic Risks to Huaman, IARC, Editor. 1993: Lyon, France. p. 257-263.
4. Bennett, J.W., Klich, M., Mycotoxins. Pathogenesis. 2009: Elsevier Inc. 559-565.
5. Pitt, J.I., Hocking, A. D., Fungi and Food Spoilage. 3 ed. 2009: Springer Science, Business Media. 275-337.
6. Kwon, C.K.J., Bennett, J. E., Medical mycology. 2 ed. 1992: Lea & Febiger.

7. Samson, R.A., Hoekstra, E. S., Frisvad J. C., Filtenborg, O., Introduction to food and airborne fungi. Centraalbureau voor Schimmelcultures: Utrecht, the Netherlands., 2000.
8. Goldman, G.H., Osmani, S. A., Mycology. The Aspergilli genomics, medical aspects, biotechnology, and research methods, ed. J.W. Bennett. Vol. 26. 2008, London: CRC Press.
9. Stevens, D.A., Kan, V. L., Judson, M. A., Morrison, V. A., Dummer, S., Denning, D. W., Bennett, J. E., Walsh, T. J., Patterson, T. F., Pankey, G. A., Practice guidelines for diseases caused by Aspergillus. Infectious Diseases Society of America. Infectious Diseases Society of America, 2000. 30(4): p. 696-709.
10. Lin, Y.C., J.C. Ayres, and P.E. Koehler, Influence of temperature cycling on the production of aflatoxins B1 and G1 by Aspergillus parasiticus. Appl Environ Microbiol, 1980. 40(2): p. 333-6.
11. Yan, Z.Y., et al., Detection of Aflatoxin Producing Moulds in Korean Fermented Foods and Grains by Multiplex PCR. J. Food Protect., 2004. 67(11): p. 2622-2626.
12. Hamed, K.A., Sheir, W. T., Horn, B. W., Weaver, M. A., Cultural Methods for Aflatoxin Detection. Journal of Toxicology. Toxin Reviews, 2004. 23(2-3): p. 295-315.
13. Rashid, M., Khalil, S., Ayub, N., Ahmed, W., Khan, A. G., Categorization of Aspergillus flavus and Aspergillus parasiticus Isolates of Stored Wheat Grains in to Aflatoxinogenics and Non-Aflatoxinogenics. Pak. J. Bot., 2008. 40(5): p. 2177-2192.
14. Scherm, B., et al., Detection of transcripts of the aflatoxin genes aflD, aflO, and aflP by reverse transcription-polymerase chain reaction allows differentiation of aflatoxin-producing and non-producing isolates of Aspergillus flavus and Aspergillus parasiticus. Int J Food Microbiol, 2005. 98(2): p. 201-10.
15. Sardinas, N., et al., Specific detection and quantification of Aspergillus flavus and Aspergillus parasiticus in wheat flour by SYBR(R) Green quantitative PCR. Int J Food Microbiol, 2010.
16. Altschul, S.F., Basic local alignment search tool. J Mol Biol, 1990. 215(3): p. 403-410.
17. Plaza, G.A., Upchurch, R., Brigmom, R. L., Whitman, W. B., Rapid DNA extraction for screening soil filamentous fungi using PCR amplification. Polish journal of Environmental Studies, 2003. 13(3): p. 315-318.
18. Ehrlich, K.C. and P.J. Cotty, An isolate of Aspergillus flavus used to reduce aflatoxin contamination in cottonseed has a defective polyketide synthase gene. Appl Microbiol Biotechnol, 2004. 65(4): p. 473-8.

Evaluation of the Human Genomic DNA Extraction from Hair Root Method

Nguyen Thi Hue and Nguyen Thi Thao Linh

School of Biotechnology, International University, VNU, HCM City

Abstract — **Extraction of DNA from specimens is a required step which provides DNA for further studies in molecular genetics. In human genetic study, blood is a prefer specimen as they provide adequate DNA. However, blood specimen not always easy to collect while hair root is easier. This research is carried out to select an efficient DNA extraction method using hair root specimens. Totally 200 hair roots were collected from 20 volunteers and divided into 2 groups different in period time: fresh or 1 week, for evaluation of 2 extraction methods: phenol and salting out. The comparison will performed to evaluate the extraction method for each group based on the quantity and quality of DNA extracted. Quantity of DNA was measured by Nanodrop system. Quality of DNA was measured by the DNA on agarose gel and the product of a PCR assay. The phenol method is considered as a better method for extraction of DNA from hair root which give a higher quantity and quality of DNA. With 5 fresh hair roots, extraction by phenol method result in >1000ng of DNA and the average purity of DNA at 1.7, while extraction by salting out method result in about 800ng of DNA and the average purity at 1.0. Storing hair root leading to reducing the quantity of DNA extracted. PCR assay testing on DNA extracted showed the result that either salting out method or phenol method with the same amount of DNA in PCR assay they showed the successful PCR products on agarose gel. The phenol method is an efficient method for extraction of DNA from hair roots. 5 hair roots can be achieved in around 1μg of DNA which is enough for at least 10 PCR assays in human molecular genetic studies. Fresh hair root is recommended for extraction of DNA.**

Keywords — **Hair roots, DNA, extraction, efficient, PCR assay.**

I. Introduction

Efficient extraction of nucleic acids from clinical samples is highly determinant for the reliability and performance of any molecular diagnostic assay. The quality and quantity of template DNA influence the success of subsequent experiments such as polymerase chain reaction (PCR), enzyme digestion and recombinant selection [1]. Because the efficacy of PCR amplification may be affected by the presence of a number of inhibitors of Taq polymerase, such as phenol [2], polyamines [2], polysaccharides [3], DNA extraction is required before amplification [4], etc. The optimal choice of samples for commercial tests however depends on the ease of collecting, costs of transporting, costs of processing and reliability. Blood is the most commonly used sample because it provides enough and high quality DNA for further genetic studies. However, sampling blood sometimes is not easy. Compared with sampling blood, sampling hair root is relatively easier. Hair root is the preferred sample for genetic testing because it is also the most cost efficient sample-type to collect and process. However, hair roots contain extremely small amount of DNA [5] [6] and it is difficult to extract DNA of sufficient quality and quantity which is normally required in human genetic studies. An efficient method of DNA extraction that yields pure and high-quality DNA is important for conducting PCR and sequencing reactions [7]. The phenol method and salting-out method have a potential of being developed into suitable methods apply for hair roots which may result in high quality and quantity of DNA [8]. In addition, the phenol method and salting-out have lower cost than other methods, therefore, they can be applied in various laboratories [8]. It is quite clear that the extraction methods have to be adapted in such a way that they can efficiently purify DNA from fresh samples or has been stored samples . In this study, the phenol and salting-out methods will be performed and compared for extraction of human genomic DNA in different groups of hair storage time to identify the most suitable storage time and the best method which may give high quality and quantity of DNA.

II. Materials and Methods

A. Materials

200 hair roots have been collected from volunteers and divided into two groups for two experiments. Group 1 contained 100 fresh hair roots (10 hair roots per person), group 2 contained 100 hair roots which is stored in one week (10 hair roots per person). Each experiment is performed to compare the efficiency of two extraction methods: phenolic and salting-out.

B. Methods

With phenolic extraction method, the hair roots (5 hair roots/assay) were treated with lysis buffer, proteinase K, and then by phenol: chloroform: Isoamylalcohol (25:24:1).

V. Van Toi et al. (Eds.): 4th International Conference on Biomedical Engineering in Vietnam, IFMBE Proceedings 40, pp. 190–194, 2013.
www.springerlink.com

Phenol solution was prepared by dissolving crystal phenol in 10X TE and left saturating for 1 week before use. Lysis buffer I (Tris HCl 10mM, $MgCl_2$ 5mM, Triton X100 (1%v/v)) was put in the hair root samples, and mixed well in 30s and incubated at 85°C for 10 minutes. Let the tube cool down at room temperature then add protease, vortex 30s and incubate at 56°C for 1 hour. After that, 100 μl lysis buffer II (TrisHCl 30mM, EDTA 20mM, SDS 3%, pH 8.0) was used for cell lysis. The buffer phenol: chloroform: izoamylalcohol (25:24:1) was added with equal volume, then inverted continuously and gently to avoid DNA damages. Centrifugation was applied at 10,000 rpm for 4 minutes at room temperature. The upper phase was collected into a new eppendorf. The DNA precipitation was performed by adding Sodium Acetate (3 M) pH 5.2 and an equal volume of Iso-Propanol. Centrifugation at 10,000 rpm for 4 minutes at room temperature was then applied after few times of inverting. The removal of supernatant was performed to keep the pellet, which was originally washed with absolute ethanol then Ethanol 70%, followed by centrifugation at 12,000 rpm for two minutes at room temperature. Let the pellet dry at room temperature until the ethanol completely evaporated; DNA pellet then was resuspended by 20μl dH_2O.

The extraction of genomic DNA by salting – out was performed to compare the efficiency with the phenolic extraction method. The two protocols of the two investigated methods are almost identical except for in salting – out method, phenol: chloroform: Isoamylalcohol (25:24:1) step was not prepared.

Concentration and purity of DNA was measured by NanoDrop ND 1000 spectrophotometer. Quality of DNA extracted was checked by gel electrophoresis and PCR assay using a set of two human growth hormone primers hGH1 5'- GCCTTCCCAACCATTCCCTTA-3' and hGH2 5'- CAAGGATTTCTGTTGTGTTTC-3' which were designed to amplify a 400bp DNA fragment.

The PCR assay was performed using the QIAGEN TopTaq Master Mix Kit. 1x concentration of master mix, 0.5μM of each primer and 50ng of DNA was used in the total volume of 25μl. The PCR was carried out using Eppendorf® PCR system with the thermal cycle of 40 cycles of 94°C for 30secs, 63°C for 30secs and 72°C for 1min, and the final extention is 72°C for 5 mins. The PCR products were then analysed by electrophoresis with 1.5% agarose gel, 100 voltage, in 30 mins.

III. Results

The overall average concentration and purity of DNA extracted from phenolic and salting-out methods are shown in table 1 and 2. The highest DNA concentration is achieved in phenolic extraction protocol. Moreover, the purity of extracted DNA of phenolic method is better than the salting-out extraction method in both fresh hair samples and stored one-week-old samples. Therefore, the phenol method is an efficient method for extraction of DNA from hair roots.

Table 1 Comparison of DNA concentration extracted from fresh hair roots by different methods.

	Extraction Method	
	Phenol	Salting - out
[Conc.] (ng/μl)	87	42
[Purity]	1.7	1

Table 2 Comparison of DNA concentration extracted from stored one-week-old hair roots by different methods.

	Extraction Method	
	Phenol	Salting - out
[Conc.] (ng/μl)	17	8
[Purity]	1.7	1

Analyse data for each assay in the phenolic protocol (table 3); the results showed that there were 60% of samples of which concentrations of extracted DNA reached over 50ng/μl and the purity of DNA is 1.7, in the range of highly pure DNA (1.7 -1.9). This is a considerably good result. The phenolic method was applied in stored one-week-old hair roots which gave the results of 50% assay with DNA concentration less than 20 ng/μl and the purity of DNA is

Table 3 Concentrations of extracted DNA by phenolic method.

Conc. (ng/μl)	Assay (%)	
	Fresh hair roots	One-week hair roots
>50	60	0
20-50	10	50
<20	30	50
Purity	1.7	1.7

1.7 as well. With salting-out method (table 4), the DNA extraction from fresh hair roots also took the concentration of DNA with 30% samples which give the concentrations of extracted DNA over 50ng/μl but the purity is lower than 1.7. However, when applying salting-out method to stored one-week-old hair roots, the results showed that 80% extracted DNA have concentrations that are lower than 20ng/μl and the purity is not good either (less than 1.7).

Table 4 Concentrations of extracted DNA by salting - out method.

Conc. (ng/μl)	Assay (%)	
	Fresh hair roots	One-week hair roots
>50	30	0
20-50	30	20
<20	40	80
Purity	1	1

Comparing results of the genomic DNA extracted from both stored one-week-old hair roots, and fresh hair roots, phenolic extraction protocol showed higher level in both purity and extracted DNA concentration (ng/μl). In salting-out method, the results were the same: DNA extracted from fresh hair had higher concentration than DNA extracted from hair roots which had been stored for one week before DNA extraction. It implied that in order to increase the effeciency of the extraction process, fresh samples were highly recommended. However, hair roots stored in 1 week are still considered sufficient enough for extraction of DNA by using phenolic extraction method.

Gel electrophoresis was applied to check the quality of DNA products. The results showed that with the same amount of DNA (10μl/well), all DNA products from fresh hair roots extracted by using phenolic method and salting-out method could be observed whereas, the extracted DNA from stored one-week-old hair roots could be seen clearly when phenolic method was applied.

PCR assays were performed to check the ability to be amplified for further genetic studies. All fresh hairs samples of two methods gave a good result in PCR assay as the presence of PCR product at 469bp on the agarose gel (figure 3). With stored one-week-old hair roots (figure 4), some of the samples could not presence in the gel because of the PCR inhibitors.

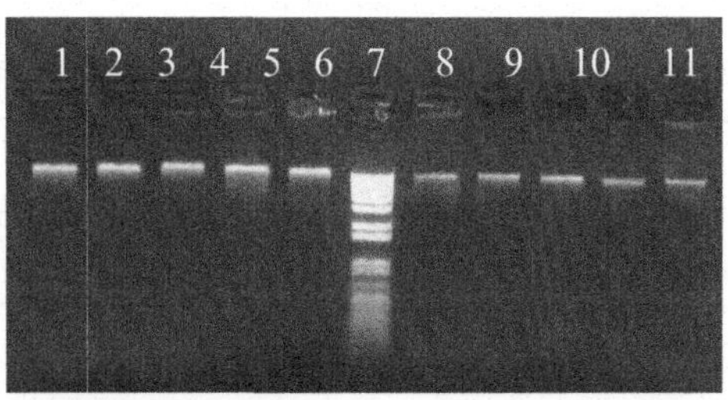

Fig. 1 Quality of DNA from fresh hair roots checked by agarose gel electrophoresis. Lane 1,2,3,4,5: DNA is extracted by phenolic method. Lane 7,8,9,10,11: DNA is extracted by salting-out method. Lane 6: ladder 1kb.

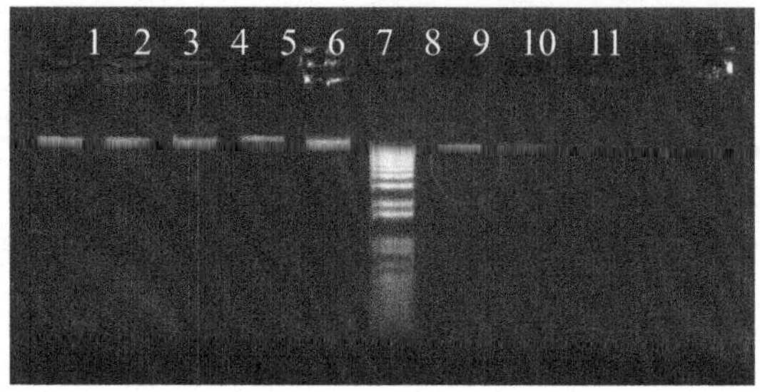

Fig. 2 Quality of DNA from stored one-week-old hair roots by gel electrophoresis. Lane 1,2,3,4,5: DNA is extracted by phenolic method. Lane 7,8,9,10,11: DNA is extracted by salting-out method. Lane 6: ladder 1kb.

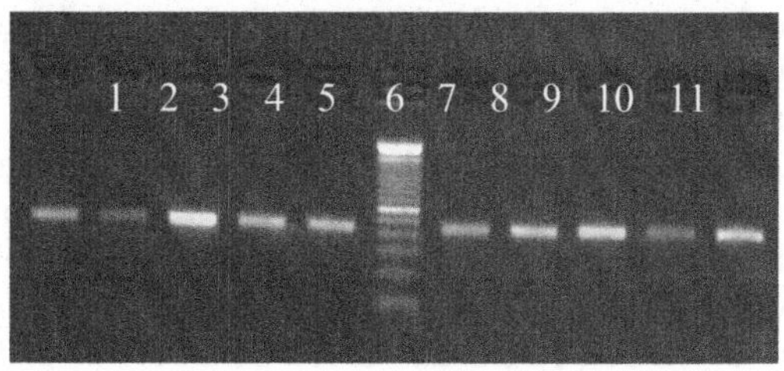

Fig. 3 Quality of PCR products from fresh hair by gel electrophoresis. Lane 1,2,3,4,5: PCR products by phenolic method. Lane 7,8,9,10,11: PCR products by salting – out method. Lane 6: ladder 100 bp.

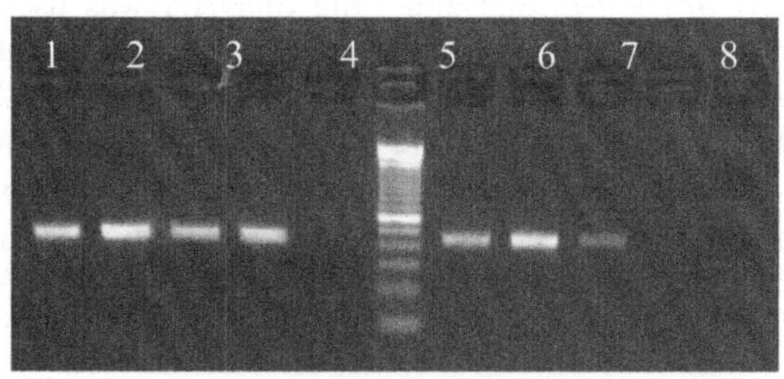

Fig. 4 Quality of PCR products from stored one-week-old hair by gel electrophoresis. Lane 1,2,3,4,5: PCR products by phenolic method. Lane 7,8,9,10,11: PCR products by salting – out method. Lane 6: ladder 100 bp.

IV. Discussion

Salting-out is a method of separating proteins based on the principle that proteins are less soluble at high salt concentrations. When trying to apply an efficient method for extracting and purifying genomic DNA from hair roots, salting – out method was used and gave mediocre results. The DNA concentration at the end is too low and the purity is not high enough for using in PCR assay. The contamination could have occurred during the process of extracting DNA from hair roots because of the salts or the inactivation of protease. Moreover, the PCR assay needed the most purified extracted DNA to get good results. Another problem is that with low concentration, a large volume of DNA solution is needed; this also increases the chance of contamination, which will in turn affect PCR assay results. Thus, unsuccessful PCR assays on DNA samples extracted by salting-out method can be explained as either there was too much contamination in the samples or a large volume of DNA solution was used for low concentration of DNA in each PCR assay.

Notably, the concentration of DNA solution extracted by phenolic method is significantly greater than that of salting - out method. This could be explained that the phenolic method contained phenol: chloroform: isoamylalcohol step which helps remove unwanted protein and precipitate DNA from the cell lysate. The phenol solution could be considered as the key factor of DNA extraction. Still, contamination cannot be avoided completely in phenolic method; however, using a small volume of DNA solution with high concentration of DNA for each PCR assay may overcome this obstacle. With lower concentration and lower purity, the DNA samples from the salting-out method give no result in PCR assay compared to DNA samples from the phenolic method. This also indicates that the quality of DNA extracted by phenolic method is higher than in salting-out method. The purity and concentration of DNA is believed to directly affect the PCR results.

In addition, the extracted DNA from two extraction methods can be observed by using gel electrophoresis. However, some of extracted DNA was smeared. This indicated that the extracted DNA was broken down during the extraction process. In phenolic method, oxidized phenol can result in DNA damages and cannot be used; thus, to decrease the damage of the extracted DNA, redistilled phenol is recommended for purification procedures. The purity and pH of phenol solutions used in biochemical work is very important. The phenol solution needs to be stored in good conditions to avoid the oxidation of phenol. Moreover, gentle manipulations may not always be possible during the isolation process. Grinding, shaking, stirring, and other disruptive procedures may cause cleavage (shearing or scission) of the DNA chains. This usually does not cause damage to the secondary structure of the DNA, but it does reduce the length of the molecules.

The most significant problem presented by the hair samples is the low amount of DNA available. The success of extraction protocols were depended on the strategy of collecting and the natural characteristic of the hair roots. Hair has a natural color-producing factory, simply by varying the proportion of only two melanin pigments; it can produce a surprising diversity of shades. Only two melanin pigments give hair its color: Eumelanin and Pheomelanin. In all, there isn't more than 1% of melanin in each hair, even in the darkest hair. While we know that melanin is what gives hair its color, the range of natural variation remains astounding. The wide range of natural hair colors is determined by the total amount of melanin (either Eumelanin or Pheomelanin, or a mixture of the two) present in the cortex of the hair fiber. However, melanin was considered as PCR inhibitions, especially in phenol and salting-out method, by which DNA was separated from other components based upon the difference in their solubility.

In economic aspect, the price for one DNA extraction assay by phenolic method is twice cheaper than using QIAamp® DNA Mini Kit (2.74$ USD). Comparing to Chelex method which also applied for genomic DNA extraction from hair root [9], the phenolic method also take lower costs. With simple protocol, low cost, high concentration and exceptional purity, the phenolic method is a promising DNA extraction method can be used in any molecular genetic laboratories.

Acknowledgment

Authors would like to thank to all volunteers who provided hair root for this study. Thank to Pediatric Hospital – Quang Nam Province where we collect samples for this study. This study is funded by NAFOSTED and VNU-HCMC

References

1. Wang, T.Y., et al., A simplified universal genomic DNA extraction protocol suitable for PCR. Genet Mol Res, 2011. 10(1): p. 519-25.
2. Katcher, H.L. and I. Schwartz, A distinctive property of Tth DNA polymerase: enzymatic amplification in the presence of phenol. Biotechniques, 1994. 16(1): p. 84-92.
3. Shioda, M. and K. Murakami-Murofushi, Selective inhibition of DNA polymerase alpha by a polysaccharide purified from slime of Physarum polycephalum. Biochem Biophys Res Commun, 1987. 146(1): p. 61-6.

4. Khaja, M.N., C.D. Poduri, and C.M. Habibullah, Modified protocol for hepatitis B virus DNA isolation and detection. Anal Biochem, 2003. 314(1): p. 142-3.
5. Higuchi, R., et al., DNA typing from single hairs. Nature, 1988. 332(6164): p. 543-6.
6. Forslind, B. and G. Swanbeck, Keratin formation in the hair follicle. I. An ultranstructural investigation. Exp Cell Res, 1966. 43(1): p. 191-209.
7. McOrist, A.L., M. Jackson, and A.R. Bird, A comparison of five methods for extraction of bacterial DNA from human faecal samples. J Microbiol Methods, 2002. 50(2): p. 131-9.
8. Miller, S.A., D.D. Dykes, and H.F. Polesky, A simple salting out procedure for extracting DNA from human nucleated cells. Nucleic Acids Res, 1988. 16(3): p. 1215.
9. Vigilant, L., An evaluation of techniques for the extraction and amplification of DNA from naturally shed hairs. Biol Chem, 1999. 380(11): p. 1329-31.

Study of Lactobacillus Acidophilus by Restriction Fragment Length Polymorphism (RFLP) Analysis

H.K. Tu Nguyen[1], D. Ly Ha[2], T.T. Vinh Doan[1], T. Tu Quach[1], and H. Kim Nguyen[3]

[1] School of Biotechnology, International University, Hochiminh City National University
[2] Department of Reference Substances, Institute for Drug Quality Control, Hochiminh City
[3] The Retina Centers of Washington, VA

Abstract — ***Lactobacillus acidophilus*** **strain is one of the most commonly used probiotics. *L. acidophilus* is linked to decreased instances of vaginal yeast infection, gastrointestinal dysfunction and even boosting immune function. Common food sources of *L. acidophilus* include yogurt and fermented soy products, such as miso and tempeh. There are many species of *Lactobacillus acidophilus* bacteria that are found in a variety of environments, from dairy products to the human gastrointestinal tract (GI). Identification of *L. acidophilus* species from different sources is important that will supply their different products to contibute in medicinal and life sciences. Therefore, this study was to find the DNA profiles in *Lactobacillus acidophilus* ATCC 4356 and *Lactobacillus acidophilus* VTCC-B-871 by restriction fragment length polymorphism (RFLP) analysis. Besides, using a probe having 400 base pairs (bp) of *minD* gene from *Lactobacillus rhamnosus* JCM 1553, Southern blotting and hybridization were performed. There were a difference between two tested *Lactobacillus acidophilus*.**

Keywords — ***Lactobacillus acidophilus*** **ATCC 4356, *Lactobacillus acidophilus* VTCC-B-871, restriction fragment length polymorphism (RFLP).**

I. INTRODUCTION

Chemotherapeutic drug therapy in cancer is seriously hampered by severe toxicity primarily due to indiscriminate drug distribution and consequent collateral damage to normal cells. Molecularly targeted drugs such as cell cycle inhibitors are being developed to achieve a higher degree of tumor cell specificity and reduce toxic side effects. Unfortunately, relative to the cytotoxics, many of the molecularly targeted drugs are less potent and the target protein is expressed only at certain stages of the cell cycle thus necessitating regimens like continuous infusion therapy to arrest a significant number of tumor cells in a heterogeneous tumor mass [2] [3]. Therefore, trend to produce minicells from non-pathogen bacteria is essential. *Lactobacillus acidophilus* is used as the common probiotic. The minicell production is more benefit than the other bacterial source. The minicells were purified from the parents [2] [3]. Therefore, the purity of minicell will not have high quality as minicell production by gene expression.

Beisdes, the research on the cell division of *Lactobacillus acidophilus* will give the approach in the survival of *Lactobacillus acidophilus* in medicinal products. Lactobacillus acidophilus probiotics were controlled but the survival of *Lactobacillus acidophilus* was usually lower than the amounts noted in probiotic lablel [4]. The improvement of Lactobacillus acidophilus with the suitable properties in Vietnam condition is necessary. To get this improvement, the research and function analysis of *minD* the genes coding for cell division [1] will be done to understand. *Lactobacillus acidophilus* in pharmaceutical product clearly.

For the above reasons, the proposal will focus the *minD* gene from *Lactobacillus acidophilus* to detect the diversity in *Lactobacillus acidophilus* that will supply more information for study in *Lactobacillus acidophilus.*

II. MATERIALS AND METHODS

A. Bacterial strains

The *Lactobacillus rhamnosus*, *Lactobacillus acidophilus* ATCC 4356 and *Lactobacillus acidophilus* VTCC-B-871 supplied by Vietnam type culture collection were used in the study.

B. Culture for Lactobacillus and Genomic DNA Isolation

The samples were incubated in MRS broth at 30°C in 48-72 hours. The white colonies with the milky smell were picked up and transferred onto MRS agar. The incubation condition was at 37°C in 24-48 hours aerobically. The genomic DNA was isolated according Sambrook protocol [5].

C. Design of Synthetic Oligonucleotide Probe

According to the information from *minD* of *Lactobacillus rhamnosus*, DDBJ/GeneBank/EMBL data base, and the

V. Van Toi et al. (Eds.): 4th International Conference on Biomedical Engineering in Vietnam, IFMBE Proceedings 40, pp. 195–197, 2013.
www.springerlink.com

supplier of Invitrogene company, using PCR technique, DNA sequencing, an oligonucleotide probe was synthesized to clone the desired gene from *Lactobacillus rhamnosus*.

D. Polymerase Chain Reaction

PCR amplification was performed by using the *Lactobacillus rhamnosus* genomic DNA as a template in association with a sense primer OMR1(5'-GAATGCGACC GGGGCGGCTGACGGTGCGA-3') and an anti-sense primer OMR2 (5'-TCAACGGCACGCTATCACCTAGTA ACCGGC-3').

PCR was done under the following conditions: an initial 4 min at 95°C; then, 29 cycles of 1 min at 95°C followed by 30 secs at 65°C; and 30 secs at 72°C, finally, an extension period of 30secs at 72°C.

E. Restriction Digestion

The genomic DNA from Lactobacillus acidophilus strains were digested with restriction enzymes supplied by Takara (Japan). The digestion was followed as instructions of the company.

F. Southern Hybridization

Southern hybridization was performed by using a Hybond-N+ (Amersham Biosciences) membrane. Probe labeling, hybridization and detection were performed with AlkPhos Direct Labeling and Detection System (Amersham Biosciences) according to the protocol supplied by the manufacturer.

G. Agarose Gel Electrophoresis

Agarose (Promega) was dissolved in TAE buffer by heating in a microwave oven. When the agarose had cooled to 60°C, ethidium bromide was added to a final concentration of 0.5 μg/ml and the molten gel poured into the gel maker and the come implaced. For electrophoresis, the comb was removed and the gel placed in the apparatus. Samples were applied by micropipette after mixing with 1/6 volumes of loading buffer and ran at 100 V. Gels were visualized with a 360 nm UV transilluminator.

III. Results and discussion

A. Oligonucleotide Probe Preparation

According to the information of *Lactobacillus rhamnosus*, DDBJ/GeneBank/EMBL data base, a DNA fragment for *minD* structural gene which may control differentiation of *Lactobacillus rhamnosus* was determined. From this sequence, primers were indicated as OMR1 and OMR2. Then, a 400bp-DNA fragment from this *minD* structural gene was prepared by the PCR technique and visualized on 0.8 % agarose gel electrophoresis (Fig. 1)

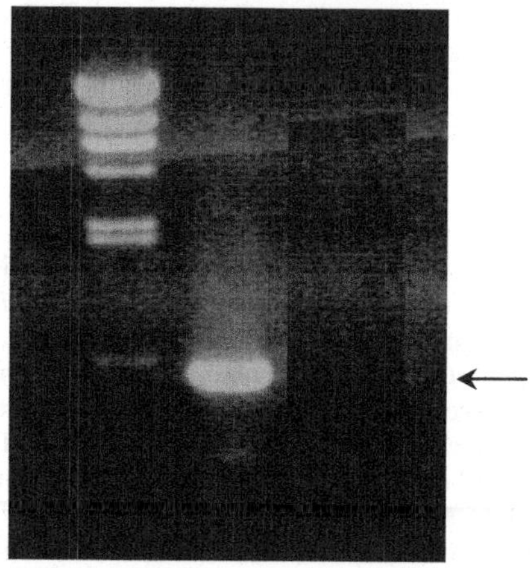

Fig. 1 PCR product on agarose electrophoresis
Lanes 1, λ/HindIII marker; 2, PCR product. An arrow shows a 400 bp size.

B. Isolation of Genomic DNA of Lactobacillus Acidophilus

After isolation genomic DNA from *Lactobacillus acidophilus* and agarose electrophoresis, the size of total chromosomal DNA is about 30 kb (Fig. 2).

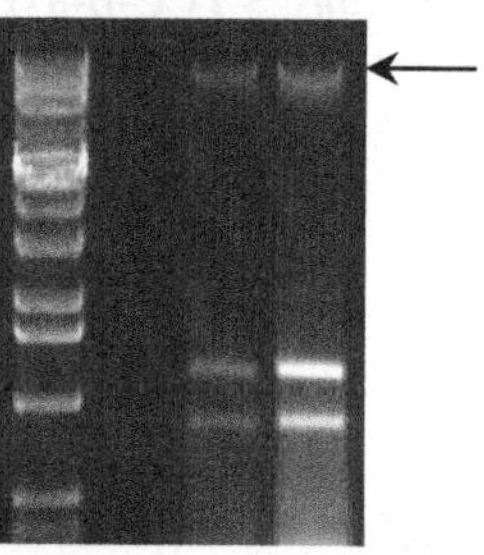

Fig. 2 Genomic DNA of *Lactobacillus acidophilus* on agarose electrophoresis Lanes 1: extension 1-kb marker; 3, 4: genomic DNA. An arrow shows 30 kb size.

After isolation of genomic DNA, there were also two bands in lower pattern that were suspected plasmids conferring for the antibiotic resistance in *Lactobacillus acidophilus*. At this step, it seemed that the DNA systems in *Lactobacillus acidophilus* ATCC 4356 and *Lactobacillus acidophilus* VTCC-B-871 were the same (Fig. 2). The genomic DNA of these strains were digested by restriction enzymes.

C. *Analysis of Genomic DNA of Lactobacillus Acidophilus Strains*

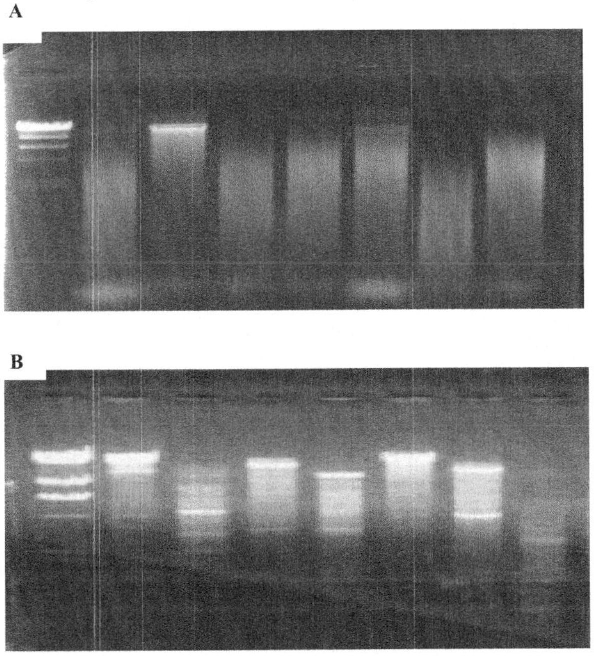

Fig. 3 Restriction fragment length pattern analysis of genomic PCR product on agarose electrophoresis
Lanes 1, λ/HindIII marker; 2: KpnI, 3: EcoRI, 4: HincII, 5: EcoRV, 6: SacI, 7: HindIII, 8: BglII.

(A) Lactobacillus acidophilus ATCC 4356; (B) Lactobacillus acidophilus VTCC-B-871

After digested with KpnI, EcoRI, HincII, EcoRV, SacI, HindIII, BglII, the DNA reactions were checked on electrophoresis, the patterns in two strains were different (Fig. 3). In order to make confirmation, southern blot and hybridization were performed.

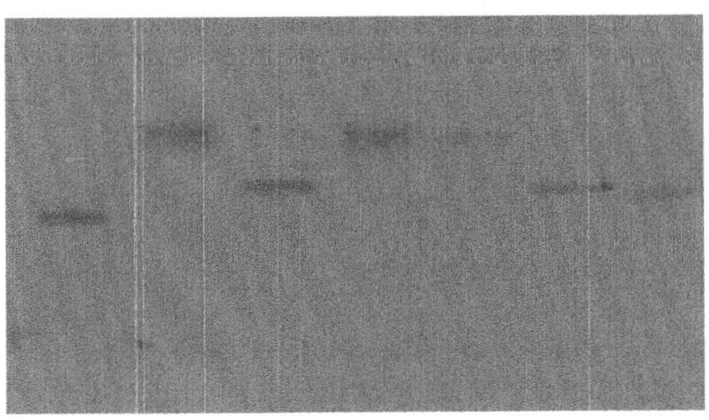

Fig. 4 Southern blot analysis for genomic DNA of *Lactobacillus acidophilus* ATCC 4356
Lanes 1, λ/HindIII marker; 2: KpnI, 3: EcoRI, 4: HincII, 5: EcoRV, 6: SacI, 7: HindIII, 8: BglII.

Conditions for hybridization were determined in consideration of the buffer concentration and Tm value between the probe and its complementary DNA sequence. After exposure, the results showed as in Fig. 4. The genomic DNA of *Lactobacillus acidophilus* VTCC-B-871 was not actually digested with EcoRI, EcoRV, SacI but genomic DNA of *Lactobacillus acidophilus* ATCC 4356 was digested with EcoRI, EcoRV, SacI. Those results were explained by the higher signals of the non-digested patterns (Fig. 4). During the genomic DNA isolation, there were the existence of plasmids in extraction. Therefore, the southern blot and hybridization contributed to the difference in genomic DNA digestion, not only because of plasmids. In Fig. 3, the patterns of digestion in both *Lactobacillus acidophilus* strains were different.

IV. Conclusions

By using RFLP analysis and Southern blotting along with hybridization, genomic DNA profiles of *Lactobacillus acidophilus* ATCC 4356 and *Lactobacillus acidophilus* VTCC-B-871 were different.

Using *minD* gene encoding a cell division protein in the study showed there might have a different division in these strains that will be study so far. Even the same strain of *Lactobacillus acidophilus*, the living sources are different that will affect on the genetic diversity in microorganisms.

Acknowledgements

Thanks to National Foundation for Science and Technology development (NAFOSTED) of Vietnam for supporting us to fulfill the study.

References

1. De Boer P. and Crossley R. (1989) A division inhibitor and a topological specificity factor coded for by the minicell locus determine proper placement of the division septum in Escherichia coli. Cell 56: 641-649.
2. Jennifer, A., Jocelyn C., Nancy B. et al. (2007) Bacterially- derived nanocells for tumor targeted delivery of chemotherapeutics and cell cycle inhibitors.
3. Jennifer, A., Nancy B., Jocelyn C. et al. (2007) Bacterially derived 400nm particle for encapsulation and cancer cell targeting of chemotherapeutics. Cancer cell. 11: 431-445.
4. Kaur, I.P., Chopra, K. and Saini, A. (2001). Probiotics: potential pharmaceutical applications. Eur J Pharm Sci 15: 1-9.
5. Sambrook, J., Fritsch, E.F., Maniatis, T. (1989) Molecular Cloning: a Laboratory Manual. Cold Spring Harbor, NY: Cold Spring Harbor Laboratory

Author: Nguyen Hoang Khue Tu
Institute: Hochiminh city International University
Street: Quarter 6, Linh Trung ward, Thu Duc Dist.
City: Ho Chi Minh
Country: Vietnam
Email: nhktu@hcmiu.edu.vn

Study of Tetrodotoxin Detection in Cultured Medium

Quach T. Tu, Doan N. Hieu, and Nguyen H.K. Tu

School of Biotechnology, Hochiminh City International University, Hochiminh City, Vietnam

Abstract — **Detection of TTX produced in cultured medium containing peptone, yeast extract that was made from animal or other biological sources caused the false positive results when using thin layer chromatgraphy method. With such results, it was a lot of mistakes when detect tetrodotoxin in food as a toxin or in medicine as a drug. Therefore, it was needed to look up a modification in TTX detection in cultured medium. In the study, the tested samples were designed and the mobile phases for thin layer chromatgraphy were optimized. The results showed the suitable system for TTX detection was pyridine: ethylacetate: acetic acid: chloroform with the ratio as 15:5:5:2 or butanol: acetic acid: distilled water (15:3:6). The tetrodotoxin level was detectable at 10 ng.**

Keywords — **detection, media, thin layer chromatography, tetrodotoxin.**

I. INTRODUCTION

Tetrodotoxin (TTX) is the main causative agent of thousands of intoxication cases around the world due to consuming the buffer fish. In the other hands, TTX could be applied in pharmaceutical industry to create new drug for reducing the malignant cancer pain without addiction whereas current drug such as paracetamol that reduces pain slightly and morphine which cause the drug dependence. Besides, this compound was also proven for heroin withdrawal capacity. TTX is a potent toxin which commonly found in many marine species such as puffer fish [1], blue-ring octopus crab [2] and newt - Triturus spp [3]. TTX had been classified as a natural product which is a non-proteinous compound, its molecular weight was about 319 U and has a very low LD50 of 10 µg kg-1 in mice. TTX acts by binding to the voltage-gate sodium channels, preventing Na+ flow through the channels and consequently blocking the action potential generation. Symptoms of intoxication appear within minutes to hours and may progress from mouth shortly after ingestion to respiratory failure, hypotension and even death [4]. In Vietnam, there were many studies about extraction of TTX in tetryodonniae [5] and the serological effects of TTX on mice [6]. To date, detection of TTX produced in cultured medium containing peptone, yeast extract that was made from animal or other biological sources caused the false positive results when using thin layer chromatgraphy method. With such results, it will be a lot of mistakes when detect tetrodotoxin in food as a toxin or in medicine as a drug. Therefore, it was needed to look up a modification in TTX detection in cultured medium. In the study, the tested samples were designed and the mobile phases for thin layer chromatgraphy were optimised.

II. MATERIALS AND METHODS

A. Materials

Tetrodotoxin was perchased from Tocris company while pyridine, chloroform, ethyl acetate, yeast extract, chloride sodium, acetic acid, peptone, sodium sulfite, sodium citrate, iron citrate, butanol were analyzed chemicals. Puffer fish genus of *Lagocephalus* was collected from Nha Trang Thin layer plates were from Merck. The UV light with wavelength of 254 nm.

B. Liver Extraction

Pufferfish Lagocephalus (weight 400g) was dissected to isolate the organs such as liver, intestine, ovary and skin. These organs were extracted using 0.9% NaCl, 1.5% K2HPO4, and distilled water in 24 hours.

C. Medium Preparation

In order to detect tetrodotoxin in media, the study focus on the effects of yeast, peptone, sodium chloride, sodium sulfite, sodium citrate. Therefore, the Luria broth (LB) and TCBS were prepared according to the instructions of the manufacturing companies.

D. Thin Layer Chromatography(TLC)

The mobile phase was prepared by mixing chloroform, ethyl acetate, acetic acid, distilled water or by mixing butanol, acid acetic, distilled water with different ratios. The samples applied on thin layer chromatography were prepared by mixing standard or liver extraction with different medium. The detection was visible under UV light at 265nm.

V .Van Toi et al. (Eds.): 4th International Conference on Biomedical Engineering in Vietnam, IFMBE Proceedings 40, pp. 198–200, 2013.
www.springerlink.com

III. Results and discussion

A. Detection of Tetrodotoxin by Using the Mobile Phase Containing Pyridine, Chloroform, Acetic Acid and Ethyl Acetate

The samples prepared by mixing the standard tetrodotoxin with LB medium or peptone, yeast, extract, sodium chloride, TCBS medium. After applied on TLC plates, the mobile phase conditions were optimized. The mobile phase contained methanol, chloroform, acetic acid and distilled water but the ratios between them were changed. By screening the mobile phase, the obtained ratio for pyridine, ethyl acetate, acetic acid, chloroform was 10:5:5:2. Data was shown in Fig.1, Fig.2.

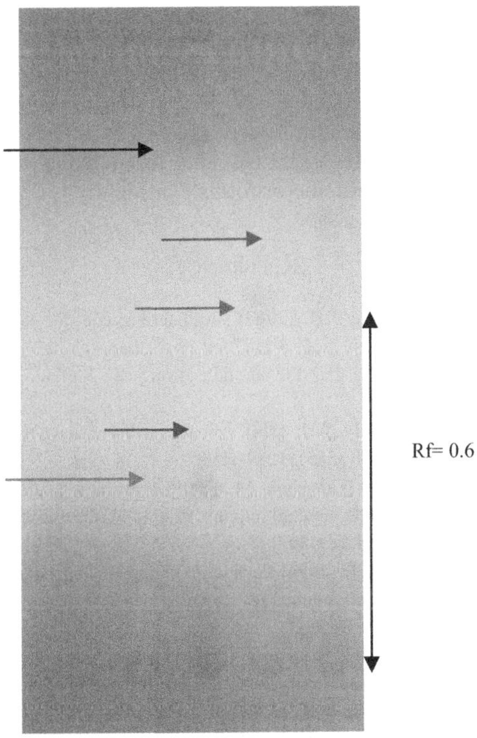

Fig. 1 Thin layer chromatography with mobile phase is pyridine: ethyl acetate:acetic acid: chloroform (10:5:5:2). From left to right: Peptone; Yeast extract; sodium chloride; liver extract; standard. The arrow showed the position of tested component.

In Fig. 1, with this system, the different components gave the Rf differently. There could be the useful way to identify tetrodotoxin in medium. To confirm this system, the study was done in the other thin layer chromatography steps with different samples. The results were showed in Fig 2.

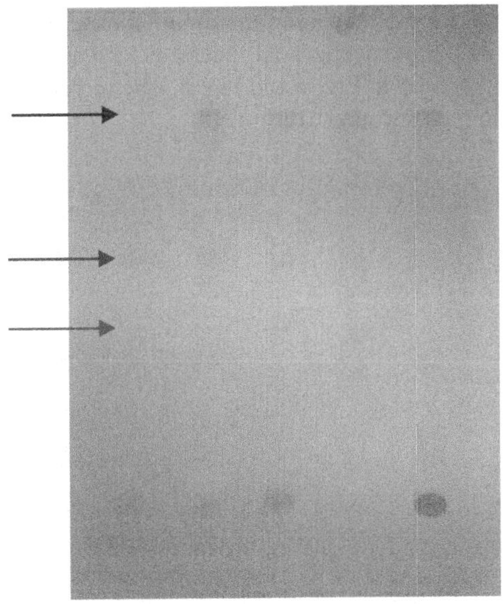

Fig. 2 Thin layer chromatography with mobile phase is pyridine: ethyl acetate:acetic acid: chloroform (10:5:5:2). From left to right: Standard, liver sample, TCBS with TTX1, TCBS with TTX2, TCBS. The red arrow showed the position of TTX.

In both running, the standard tetrodotoxin gave the Rf = 0.6. This Rf value was determined similarly in Fig.3.

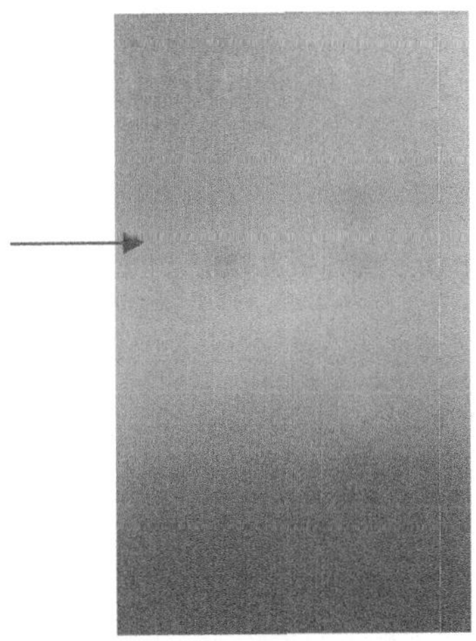

Fig. 3 Thin layer chromatography with mobile phase is pyridine: ethyl acetate:acetic acid: chloroform (10:5:5:2). From left to right: Standard, tested sample.

B. Detection of Tetrodotoxin by Using the Mobile Phase Containing Butanol, Acetic Acid and Distilled Water

After applied on TLC plates, the mobile phase conditions were optimized. The mobile phase contained buthanol, acetic acid and distilled water but the ratios between them were changed. By screening the mobile phase, the obtained ratio for butanol, acetic acid, acetic acid was 15:3:6. The data was showed in Fig. 4 and Fig. 5. The red arrow pointed the position of tetrodotoxin.

Fig. 3 Thin layer chromatography with mobile phase is butanol: acetic acid: distilled water (15:3:6). From left to right: liver extraction, LB, LB containing sample, LB containing TTX, TTX. The red arrow showed TTX.

Fig. 4 Thin layer chromatography with mobile phase is butanol: acetic acid: distilled water (15:3:6). From left to right: liver extraction, TTX. The red arrow showed TTX.

Comparision of two mobile phases, the components run with different Rfs clearly. The other conditions were also experimented. However, there was a conflict result when running the same solvent system for the negative sample. There were some reasons stated in this situation. Firstly, sample had not been purified yet leading to the compounds in the sample could not separate from each other. Secondly, low quality of chemical was reasonable concerned to the wrong positive result. As a result, modification of the solvent system was required. The detectable TTX applied was 10 ng.

IV. Conclusions

The study supplied the better and reliable test for tetrodotoxin. By doing the experiment for optimization of the solvent system for TLC, better results were achieved. Solvent were changed gradually to achieve the best. The most remarkable outcome for TTX separation from the medium was conducted via the system pyridine: ethyl acetate: acetic acid: chloroform (10:5:5:2) or butanol: acetic acid: distilled water (15:3:6). This test will help us detect tetrodotoxin in medicine or foods confidently.

References

1. Đoàn S.N., Hoàng T.H. (2009) Ảnh hưởng của độc tố cá nóc (S. spadiceus rich) lên một số chỉ tiêu huyết học của chuột nhắt trắng Swiss và khả năng thải độc của nấm hoàng chi. Journal of science 5: 33-40.
2. Huey R.B., Moody W.J. (2002) Snake sodium channel resist TTX arrest. Science 297(5585): 1289-1290.
3. Jang J. and Yotsu-Yamashita M. (2006) Distribution of tetrodotoxin, saxitoxin, and their analogs among tissues of the puffer fish Fugu pardilis. Toxicon 48(8): 80-987.
4. Lê Q.H. và Lê X.T. (1994) Tách chiết và tinh chế tetrodotoxin từ một số loài cá nóc (Tetryodontidae) tại vùng biển miền Trung Việt Nam. Journal of biology 16(3): 38-41.
5. Tsai Y.H., Ho P.H., Hwang C.C.,Hwang P.A.,Cheng C.A., Hwang D.F(2006) Tetrodotoxin in several species of xanthid crabs in southern Taiwan. Food Chem 95(2): 205-212.
6. Yotsu-Yamashita M., Merbs D., Kwet A., Schneider M. (2007) Tetrodotoxin and its analogue 6-epitetrodotoxin in newts (Triturus spp; Urodela, Salamandridae) from southern Germany. Toxicon 50(2): 306-309.

Author: Nguyen Hoang Khue Tu

Institute: Hochiminh city International University
Street: Quarter 6, Linh Trung ward, Thu Duc Dist.
City: Ho Chi Minh
Country: Vietnam
Email: nhktu@hcmiu.edu.vn

Investigation of the Effects of the Histone Deacetylase Inhibitor Saha on the Medulloblastoma Cell Line Daoy Using Gel-Based Proteomics

Thu Thuy Pham[1,2], Christian Scharf[2], Elke Hammer[2], M. Gesell[2], J. Sonnemann[3], J. Beck[3], and Uwe Völker[2]

[1] Institute of Biotechnology and Environment, Nha Trang University, Nha Trang, Vietnam
[2] Interfaculty Institute of Genetics and Functional Genomics, Ernst-Moritz-Arndt University, Greifswald, Germany
[3] Pediatric Hospital, University Medicine, Greifswald, Germany

Abstract — **Suberoylanilide hydroxamic acid (SAHA) belongs to the class of histone deacetylase inhibitors emerging as potent antitumor agents. In this study, effects of SAHA on DAOY cells as model cell line for medulloblastoma, one of the most common malignant pediatric brain tumors, were analysed at the protein level by using gel-based proteomics. A 2D proteome map of DAOY cells in the pH range of 4-7 was created using 2D electrophoresis combined with MALDI-TOF/ TOF-MS. The 2D reference map of control and SAHA treated DAOY cells comprises 1196 identified protein spots of 778 distinct proteins. This is the first report of a 2D proteome map of SAHA treated DAOY cells and the largest catalogue of proteins identified in DAOY cells so far. Quantitative analysis of SAHA induced effects on the cells was performed using 2D-DIGE technique. After treatment with 10 µM SAHA, alterations were observed for 220 protein spots. Especially, SAHA induced fragmentation of many known caspase substrates. Higher levels of mitochondrial proteins indicated apoptosis via the mitochondrial pathway. Altogether, these findings point to induction of various antitumor activities by SAHA.**

Keywords — **DAOY, gel-based proteomics, medulloblastoma, SAHA.**

I. Introduction

Medulloblastoma is one of the most common malignant childhood brain tumors with a very low survival rate. Therefore, there is an urgent need to develop novel treatment approaches that can increase survival and reduce long term side effects of patients. Histone deacetylase inhibitors (HDIs) have emerged as a promising new class of antineoplastic agents in cancer therapy. Among them, suberoylanilide hydroxamic acid (SAHA, vorinostat, Zolinza®) has been approved for treatment of cutaneous T-cell lymphoma and is currently used for treatment of various tumor cell types both *in vivo* and *in vitro*. In a previous study, it was reported that SAHA induced apoptosis and cell cycle arrest of DAOY cells, the most widely used model cell line for investigation of medulloblastoma [1]. However, molecular mechanisms of this antitumor activity are still not clear.

Proteomics can provide a genome-wide perspective of the mode of action of anticancer drugs and display changes in gene expression. Current proteomic approaches employing cancer cells focus on systematic identification of expressed proteins and quantification of changes in protein levels during drug treatment [2]. This provides insights into tumor genesis, mode of action, toxicity, side effects of drugs and allows identification of biomarkers for early diagnosis [3]. Gel-based proteomic approaches rely on separation of protein mixtures by 2D gel electrophoresis (2-DE) followed by mass spectrometry for protein identification [4]. Excellent quantitative information can be obtained using 2D-differential in gel electrophoresis (2D-DIGE), where an internal standard is added to each gel. Also, by separation of different species of a single protein on the gel, knowledge on protein degradation and post-translational modifications (PTMs) can be gained, which is especially important when an agent inducing PTM like SAHA is under investigation.

II. Materials and Methods

A. Cell Line

DAOY cells (ATCC HTB-186, LGC Promochem, Wesel, Germany) were grown in DMEM medium (Biochrom, Berlin, Germany) and incubated with SAHA as described previously [1]. Cell death was determined by cytofluorometric analysis of propidium iodide uptake.

B. Protein Preparation and Quantification

Total protein extracts were prepared by freezing and thawing for 6 cycles using lysis buffer containing 8M urea, 2M thiourea and 2% CHAPS. Cell debris was discarded after a centrifugation at 21,000 g for 1 hour at 4°C. Protein concentration was determined by Bradford method.

C. Gel-Based Proteomic Approach

2DE and 2D-DIGE techniques were performed as described previously [5, 6]. For the construction of 2D

V. Van Toi et al. (Eds.): 4th International Conference on Biomedical Engineering in Vietnam, IFMBE Proceedings 40, pp. 201–204, 2013.
www.springerlink.com

proteome map, total protein extracts of 600 µg were used. Gels were then stained with CBB-R250 for protein spot visualization. Image analysis was performed using Delta2D® 3.6 (Decodon, Greifswald, Germany) and GeneSpring® 7.2 (Agilent Technologies, Waldbronn, Germany). Protein spots were further excised manually for protein identification. After tryptic digestion, peptide mixtures were subjected to MALDI-TOF/TOF-MS on a 4800 MALDI TOF/TOF™ Analyzer (Applied Biosystems, Darmstadt, Germany). Protein identification was performed using MASCOT search engine 2.0 (Matrix Science, London, UK) embedded into the GPS Explorer 3.6 software (Applied Biosystems, Foster, CA) with a fully tryptic human database (IPI human 3.23).

D. Protein Functional Classification

Identified proteins were translated into corresponding gene names of the NCBI database using Ingenuity Pathway Analysis (IPA, Ingenuity Systems, Inc.). Functional assessment was performed by using the PANTHER classification system (*http://www.pantherdb.org*).

III. Results

A. SAHA Induces Cell Death of DAOY Cells

After 24h incubation with 10µM SAHA, around 20% to 35% of DAOY cells were non-viable while no remarkable changes in cell viability were detected after treatment with a lower dose of 2µM SAHA. A dose-dependent increase in dead cells was observed in both bioreplicates tested (Fig. 1).

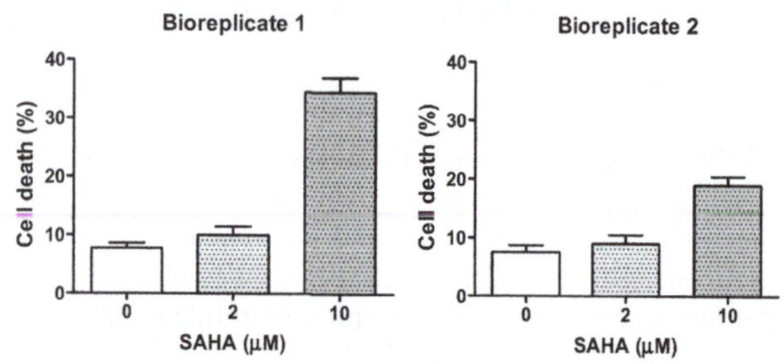

Fig. 1 SAHA induces cell death of DAOY cells after 24h incubation.

B. 2D-Proteome Map of DAOY Cells

Initially, several pH-ranges were comparatively investigated for their coverage of the proteome of DAOY cells. This test established that majority of proteins were visualized in the pI range 4-7 and that only of 13% of the detected spots were allocated outside this range. Thus, all subsequent experiments were conducted with IPG-strips covering the pI range 4-7. A 2D reference map of non-treated control and 10 µM SAHA treated DAOY cells has been created, which covers 1700 spots. For 1196 spots, the corresponding proteins could be identified, resulting in 778 distinct proteins (or protein sequences, TrEMBLs). This is the first report of a 2D proteome map of SAHA treated DAOY cells. A majority of protein spots (89%) were identified as unique proteins indicating a good protein separation while the residual (11%) contained two or more proteins due to the co-migration of different proteins, a drawback of 2-DE.

Identified proteins were assigned to the following categories: protein metabolism and modification (19%); nucleotide and nucleic acid metabolism (12%), signal transduction, cell structure and motility, and intracellular protein traffic (6.0-6.5% each); cell cycle, immunity and defense, carbohydrate metabolism (4.5-5% each); development processes, transport (3% each); apoptosis and oncogenesis, cell proliferation and differentiation (2% each). Proteins belonging to other processes accounted for 15.6 % while about 10% of identified proteins are still poorly characterized.

C. Quantitative Proteomic Analysis by 2D-DIGE

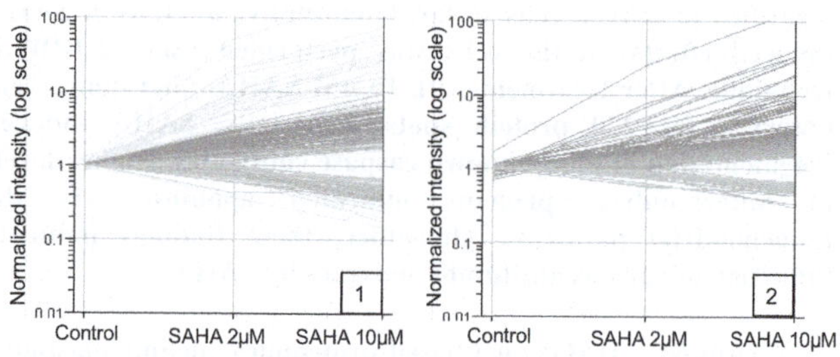

Fig. 2 Intensity profiles of protein species with altered levels in DAOY cells after SAHA treatment from two bioreplicates 1 and 2 (up – orange, down – green).

Table 1 List of proteins displayed aberrant migration on the 2D gel

Protein	Protein modification and degradation	
	Posttranslational modification	Enzyme responsible for degradation
MARCS	Myristoylation	
TOR3A	Glycosylation	
DREB, NASP	Unknown	
ACTB, ACTG, DUT, SPTA2, SSRP1, TBB2C		Caspase-3
LMNA		Caspase-6
VIME		Caspase-3,6
MYH9		Caspase-3,7
HS90B		Granzyme B
CLH1, DNAJC14, EF1D, EF2, ESYT1, FAS, FLIP1, FLNA, GIPC1, HEXB, TRAD1		Unknown

Quantitative data were obtained from two independent bioreplicates of three conditions (Fig. 2) analysed in four technical replicates each. Changes were considered as significant when fold change of spot intensities exceeded 1.5 and a p-value ≤ 0.05 (ANOVA, Welch t-test) in both bioreplicates. Among 1700 protein spots detected on the DIGE gels, 220 spots were recognized as changed spots after treatment with 10 μM SAHA, of which 158 and 62 spots displayed higher or lower intensities, respectively (Fig. 2). Also, trends of changes were in the same direction in both bioreplicates, which indicate a dose dependent effect.

Of 220 altered spots, 162 spots were identified whereas 58 spots still escaped identification due to low protein levels or different modifications. The 2D gel image of those 162 regulated protein spots is given in Fig. 3.

Among the up- and down-regulated proteins were structural proteins (e.g. calreticulin, calumenin, lamin B1); heat shock proteins (e.g. HSP71, CH60, GRP78); transcriptional regulators (e.g. EF2, CSTF1); brain signal transductors (e.g. 1433T, 1433E, CRK, MARCS); mitochondrial protein VDAC1; proteins involved in microtubule stabilization (FLIP1), redox system (e.g. BIEA, TXND4, TXND5), nucleosome remodeling (e.g. RUVB1, RUVB2) and mRNA splicing (e.g. HNRPK, HNRPC). A series of proteins were found to be subjected to protein degradation after SAHA treatment, including mainly cytoskeleton proteins. Many of those proteins are known as substrates for caspases (Table 1). Besides, some protein spots displayed a slower migration than expected and so could be targets for various post-translational modifications.

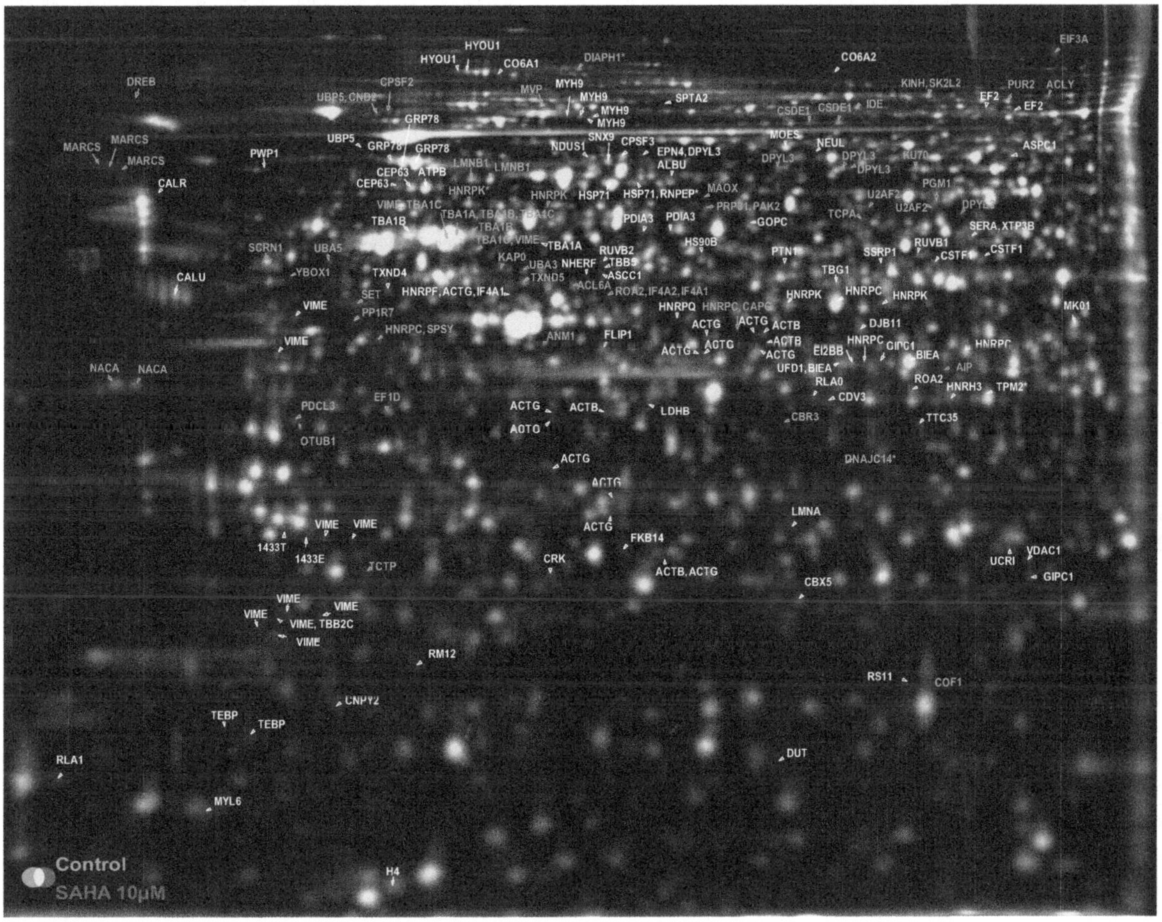

Fig. 3 Monitoring of changes in protein levels of DAOY cells after treatment with 10 μM SAHA. Of 220 altered spots, proteins from 162 spots were identified by mass spectrometry. Identified spots were labelled by corresponding abbreviated Swiss Prot entry names or gene names (*). Spot colour indicates the effect of SAHA on protein abundance: red = increased; green = decreased; yellow = no change. Increased and decreased spot intensities can be also distinguished by white and green colored, respectively.

IV. Discussion

Proteomic technologies provide an excellent tool to identify novel targets, validate efficacy and toxicity, and define the mode of action of drugs. In this study, a gel-based proteomic approach was applied to study the response of DAOY cells to SAHA. Using 2-DE combined with MALDI-TOF/TOF-MS analysis, the first 2D proteome reference map of SAHA treated DAOY cells with 1196 identified protein spots was created. This result contributes to basic and clinical research involving DAOY cells. Compared to the previous 2D proteome map of DAOY cells with only 229 proteins [7], this 2D proteome map covers a much higher number of identified proteins. For the unidentified spots, further high resolution LC-MS should be employed.

Histone deacetylase inhibitors represent a promising new class of therapeutic drugs. Among them, SAHA, a member of second generation of synthetic HDIs, is one of the most potential candidates. In a previous study, SAHA was found to remarkably inhibit cell proliferation, induce apoptosis, and harm mitochondria of DAOY cells [1].

In this study, a series of proteins were found to be subjected to protein degradation, reported as apoptotic events in brain cells and other tumor cells such as lamin A/C, α-spectrin, myosin-9 and SSRP1 [8].

Our proteomic analysis revealed a total of 220 differentially expressed protein spots. Many alterations in protein expression are compatible with known antitumor activities of SAHA reported for other cancer cells such as the up-regulation of heat shock proteins, prostaglandin synthase 3, ubiquinol cytochrome c reductase or the down-regulation of MARCS protein [9, 10, 11]. Lastly, the mitochondrial protein VDAC1 was upregulated by SAHA, which indicate mitochondrial perturbation.

SAHA can cause transcription mediated effects and exerts transcription independent effects on tumor cells [12]. Consistent with that, in our study treatment of DAOY cells with SAHA induced hyperacetylation of histones as the primary target (our unpublished data) and changes in protein level of some transcriptional factors and cofactors as shown above, which indicate transcription mediated effects. Besides, treatment with SAHA has been shown to induce cell cycle arrest of DAOY cells at G1/S and G2/M transitions as a transcription-independent mechanism [1].

V. Conclusions

This study presents a proteomic analysis of the response of DAOY cells to SAHA using gel-based approach. A 2D proteome reference map of DAOY cells in the pH range of 4-7 with around 1200 protein spots was established. This is the first proteome map of SAHA treated DAOY cells. Proteomic analysis revealed extensive protein degradation which is indicative of typical early apoptotic events after SAHA treatment. Many protein alterations found in this study are in agreement with known antitumor activities of SAHA established in previous studies. The mitochondrial protein VDAC1 which plays an important role in cell death of tumor cells can be a potential target for further investigation.

References

1. Sonnemann J, Kumar K S, Heesch S et al. (2006) Histone deacetylase inhibitors induce cell death and enhance the susceptibility to ionizing radiation, etoposide, and TRAIL in medulloblastoma cells. Int J Oncol 28: 755-766
2. Chen E I, Yates J R (2007) Cancer Proteomics by Quantitative Shotgun Proteomics. Mol Oncol 1: 144-159
3. Chuthapisith S, Layfield R, Kerr I D et al. (2007) Principles of proteomics and its applications in cancer. Surgeon 5: 14-22
4. Wolff S, Antelmann H, Albrecht D et al. (2007) Towards the entire proteome of the model bacterium Bacillus subtilis by gel-based and gel-free approaches. J Chromatogr B Analyt Technol Biomed Life Sci 849: 129-140
5. Hammer E, Phong T Q, Steil L et al. (2010) Viral myocarditis induced by Coxsackievirus B3 in A.BY/SnJ mice: Analysis of changes in the myocardial proteome. Proteomics 10: 1802–1818
6. Thiele T, Steil L, Gebhard S et al. (2007) Profiling of alterations in platelet proteins during storage of platelet concentrates. Transfusion 47:1221-1233
7. Spiller S E, Ravanpay A C, Hahn A W et al. (2006) Suberoylanilide hydroxamic acid is effective in preclinical studies of medulloblastoma. J Neurooncol 79(3): 259-270
8. Luthi A U, Martin S J (2007) The CASBAH: a searchable database of caspase substrates. Cell Death Differ 14: 641-650
9. Tong A, Zhang H, Li Z et al. (2008) Proteomic analysis of liver cancer cells treated with suberonylanilide hydroxamic acid. Cancer Chemother Pharmacol 61: 791-802
10. Furchert S E, Lanvers-Kaminsky C, Jürgens H et al. (2007) Inhibitors of histone deacetylases as potential therapeutic tools for high-risk embryonal tumors of the nervous system of childhood. Int J Cancer 120(8): 1787-1794
11. He J, Huang C, Tong A et al (2008) Proteomic analysis of cervical cancer cells treated with suberonylanilide hydroxamic acid. J Biosci 33: 715-721
12. Mehnert J M, Kelly W K (2007) Histone deacetylase inhibitors: biology and mechanism of action. Cancer J 13(1): 23-29

Author: Thu Thuy Pham
Institute: Institute of Biotechnology and Environment, Nha Trang University
Street: 02 Nguyen Dinh Chieu
City: Nha Trang, Vietnam
Country: Vietnam
Email: thuypt@ntu.edu.vn

Detection of Cholesterol Esterase in Leuconostoc Dextranicum Isolated in Foods

Nguyen Hoang Khue Tu and Tran Chu Minh Hoang

School of Biotechnology, Hochiminh City International University, Hochiminh City, Vietnam

Abstract — **Streptococcus thermophilus was isolated from ferment foods in Vietnam. This strain was determined by API 50 CHL (BioMerieux). Interestingly, we have detected and characterized the enzyme called cholesterol esterase that will be used in cholesterol diagnostic and treating the cholesterol ester storage disease in women. After culturing, the free cell supernatant was precipitated with 60% ammonium sulfate and resuspended in distilled water. After dialysis, the supernatant had been titrated with 1M sodium hydroxide and detected by thin layer chromatography (TLC). The detection was based on the determination the free fatty acid that was the product of cholesterol ester hydrolyzation by cholesterol esterase. The color changed in titration and the spot appearance compared with the standard in TLC showed the presence of cholesterol esterase. The cholesterol esterase production was also optimized in different conditions containing vitamin D, and palmitic acid as the substrates of cholesterol. The study showed that Streptococcus thermophilus should be supplied in ferment food as yogurt in order to improve the health for human.**

Keywords — **cholesterol esterase, *Streptococcus thermophilus*, thin layer chromatography, titration.**

I. Introduction

In the body, cholesterol works with high-density lipoproteins (HDL), often referred to as "good cholesterol". High-density lipoproteins carry cholesterol from the body's tissues to the liver for removal. When the cholesterol is attached to a fatty acid, it is a cholesterol ester. Normally, the cholesterol esters are broken down by lysosomal acid lipase called cholesterol esterase into cholesterol and a fatty acid and then excreted or used by the body as nutrients. Cholesterol esterase is acid lipase; an enzyme that catalyzes the hydrolytic cleavage of cholesterol and other sterol esters and triglycerides. This enzyme is found in the lysosome (compartments that digest and recycle materials in the cell) [4] or some bacteria as *Streptomyces lavendulae* [2], [3]. In enzymology, cholesterol esterase is known as a sterol esterase is an enzyme that catalyzes the chemical reaction:

$$\text{Cholesterol ester} + H2O \xrightarrow{\text{cholesterol esterase}} \text{cholesterol} + \text{fatty acid}$$

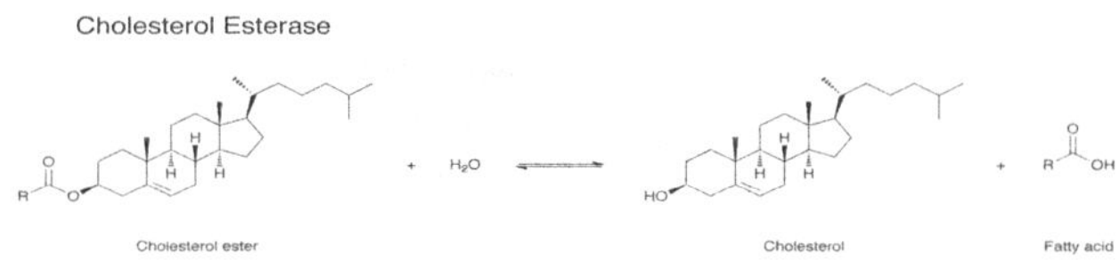

Fig. 1 Reaction of cholesterol ester hydrolyzation of cholesterol esterase.

In this study, we research and detect cholesterol esterase in *Streptococcus thermophilus* isolated from food. This strain is commonly used in the pharmaceutical and food products.

For healthcare, cholesterol causes the big problem in human circulation system. cholesterol esterase is the main key to solve it. The medicine used in cholesterol treating is bile acid resins, ezetimibe (zetia), fibric acid, niacin, commonly. Up to now, there was no information about study of cholesterol esterase in *Streptococcus thermophilus* while these strains used as biological products for human. This project is the first step to detect cholesterol esterase from the *Streptococcus thermophilus*. From the result of this project, we will investigate cholesterol esterase the from *Streptococcus thermophilus* that will be useful for health care so far.

II. Materials and Methods

A. Materials

Foods as yourgurts suspected to contain *Streptococcus thermophilus* were selected for study. *Streptococus thermophilus* isolated from yourgurts and identified was used as a cholesterol esterase-producing microorganism. MRS agar and MRS broth were from Merck. Kit API 50CHL and software were supplied by BioMérieux. Palmitic acid, oleic acid, vitaminD, ethanol, cholesteryl oleate, triton X-10, ammonium sulfate were in purified form.

B. Streptococcus Isolation

The isolation on yourgurts and fermented products was performed. The samples were incubated in MRS broth at 45°C in 24-48 hours anaerobically, then spreaded onto MRS

V. Van Toi et al. (Eds.): 4th International Conference on Biomedical Engineering in Vietnam, IFMBE Proceedings 40, pp. 205–207, 2013.
www.springerlink.com

agar. The white colonies with the milky smell were picked up and transferred onto MRS agar. The incubation condition was at 37 °C in 24-48 hours aerobically. The desired colonies were stained and observed under microscope. The suspected colony was identified by the API 50CHL kit (BioMérieux).

C. Culture for Cholessterol Esterase Production

Streptococcus thermophilus was grown in MRS Agar at 37 °C for 2 days in the aerobic condition. Then, they was transferred into MRS broth and incubated at 37 °C on the aerobic condition. The palmitic acid and vitamin D were added in the MRS broth. The palmitic acid added in the MRS broth to get the final concentration was 0.012% while the vitamin D amount was added in medium to get the final concentration of 7.5 % for optimizing the cholesterol esterase. All the samples were incubated 3 days at 37 °C aerobically.

D. Partially Purification of Cholesterol Esterase

The culture broths were centrifuged at 12000 rpm for 15 min and were carried out at 4°C with 60% ammonium sulfate. The precipitate was collected, dialized and dissolved in 100 ml distilled water by prepared for experiment.

E. Detection of Free Fatty Acid by Titration

Preparation sample for experiment: 1ml mixture of precipitate dissolved in distilled water was added with 0.5ul cholesteryl oleate and 0.5 % Triton X-100. Then, this mixture was incubated at 37°C for 20min. The detection of free fatty acid by titration was performed according to Arti nigam. 1 to 2 drops of phenolphtalein solution were added to the sample and then titrated against 1M NaOH till a persistent faint pink color develops. The amount of 1M NaOH was recorded and calculated.

F. Thin Layer Chromatography

Thin layer chromatography was performed according to Arti Nigam [1]. The mobile phase was prepared by mixing chloroform: methanol: acetic acid: distilled water with the ratio 5:15:4:2. The standard oleic acid and cholesterol were prepared with 1% and 2%, respectively. For the spraying solution, the 20% percholric acid in aqueous base was used. 5 ul of each sample was applied. After running, the TLC plate was taken out, air dried, and detected with the help of percholric spray. The spots of the resolved lipid were located and compared with the spots of standard.

III. Results and Discussion

A. Microorganism Identification

3 isolated microorganisms were cocci, positive gram (Fig.2). After identification of them by using the API 50 CHL kit (BioMérieux) and analyzing with the API software, there were only one microorganism identified as Streptococcus thermophilus with Id: 100% and T: 0,5 (Fig. 3).

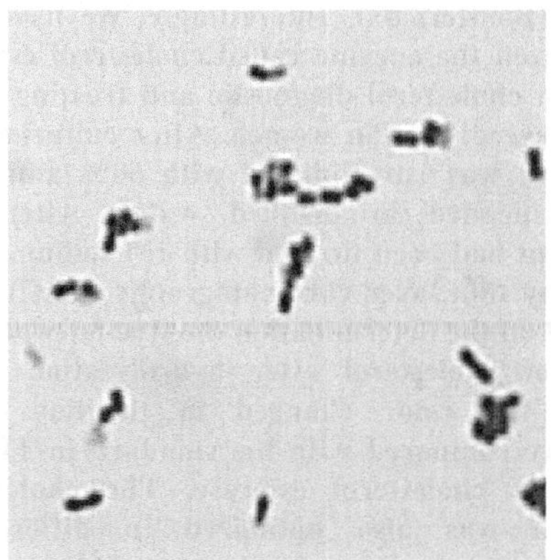

Fig. 2 Morphology of *Streptococcus thermophilus* under microscope after Gram stain.

B. Titration to Detect of Fatty Acid

After precipitation with 60% ammonium sulfate, the obtained precipitant was dissolved in distilled water and used to react with cholesterol oleate for 20 minutes at 37° C. The solutions were titrated with 1M NaOH and by using phenolphtalein as an indicator till the permanent red color appeared (Fig.3).

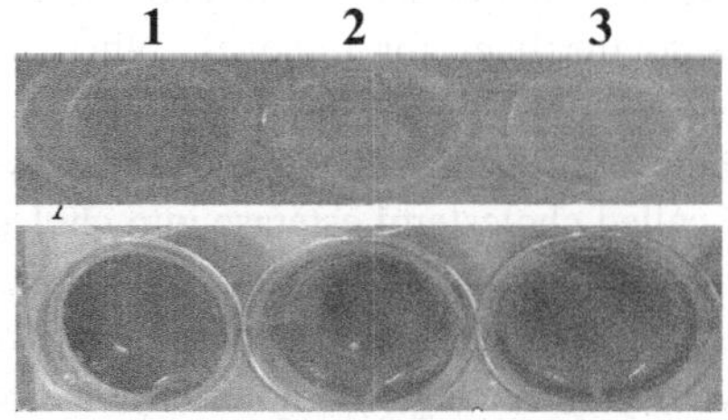

Fig. 3 The solution before (A) and after titration (B) (From left to right of A and B: cultured MRS broth (1), added palmitic acid (2), added vitamin D (3)).

The volume of 1 M NaOH was recorded as in table 1. With the same amount of 1M NaOH, the color of sample under vitamin D condition was weaker than in MRS broth or under palmitic acid. With the same amount of biomass produced in MRS broth or under palmitic acid, the color was red-orange

(Fig.4B-3). Therefore, vitamin D may interfere the cholesterol esterase production. The mechanism for cholesteryl esterase production should be studied so far. Remarkably, with this procedure of precipitation and dialysis, the palmitic acid and vitamin D were removed out of the samples.The obtained samples might contain any acidic products due to there were the change color (Fig. 5B). After adding phenolphthalein as indicator and titration with 1M NaOH, all solutions had changed color. And the volume of NaOH made color change is less than 100 ul. It was demonstrated having the reaction between NaOH and acid. To make sure the products formed after hydrolyzation, thin layer chromatography was performed to detect which is fatty acid.

Table 1 The volume of 1M NaOH (ul)

Strain	Cultured MRS	Palmitic acid	Vitamin D
Streptococcus thermophilus	100	100	100
	Dark red	Dark red	Red-Orange

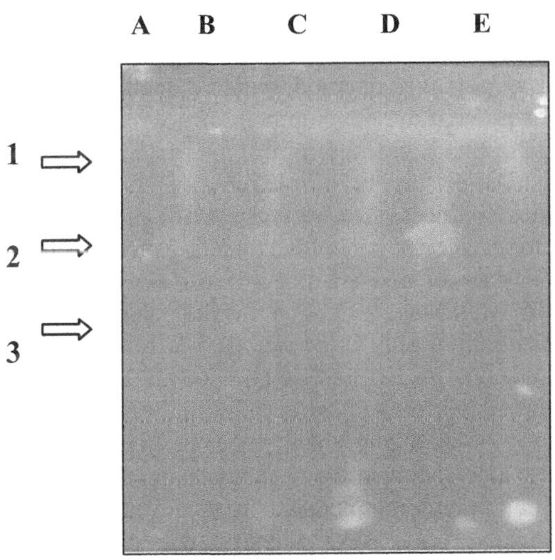

Fig. 6 Thin layer chromatography for cholesterol detection in *Streptococcus thermophilus*. From left to right: (A) sample from MRS,(B) sample from MRS with palmitic acid, (C)sample from MRS with vitamin D, (D) oleic acid, (E) cholesterol. From up to down: (1) cholesterol pattern, (2) oleic acid pattern, (3) others in tested samples.

C. *Thin Layer Chromatography*

The tested samples were reacted with cholesteryl oleate and then were run on TLC with solvent system. Then the spots were detected by spraying with 20% perchloric acid and then dried. Finally, the test was observed under UV (Fig.5). The image showed the spots had equal Rfs with the oleic acid and cholesterol. So, it make be confirmed that have ester hydrolyzation was acted. From which, in the precipitate, they have cholesteryl esterase. However, there were also the existence of spots in the lowest pattern (Fig.5-3 A, B, C) showing that more products were formed after hydrolyzation. With the such detection, the other fatty acids will be formed. These spots will be analyzed so far.

IV. Conclusions

The precipitation of cholesterol esterase with 60% ammonium sulfate was used to detect the activity. Thin layer chromatography was used to determine the free fatty acid that is the product of cholesterol ester hydroxylation. After culture the microorganism, the media were used to detect the cholesteryl esterase. The positive results showed that the cholesteryl esterase was produced extracellularly. The location of this enzyme in the *Streptococcus thermophilus* will be done in next step. In this research, *Streptococcus themophilus* grew up in MRS without or with palmitic acid, vitamin D having the cholesterol esterase. This is the basic step for next research about the cholesterase esterase in lactic acid bacteria. This is the primary detection of cholesterol esterase in lactic acid bacteria in foods. In order to understand and collect the purified enzyme for treatment and diagnosis, more purified steps and chemical reaction as well as structure analysis should be studied. With the purified cholesterol esterase, the genes encoding for this enzymes will be identified so far to understand more the mechanisms of cholesterol reduction of lactic acid bacteria when using in human health care.

References

1. Nigam, A. (2001) Lab manual in Biochemistry,Immunology and Biotechnology, McGraw-Hill Book Co.,New Dehil 110 008: 73-107
2. Nishimura, M. and Sugiyama, M. (1994) Appl MicrobiolBiotechnol 41: 419-424
3. Nishimura, M., Inouye S. (2000) J Biosci Bioeng 90(5): 564-566
4. Stokke, K. T. (1972) Biochim Biophys Acta 270 : 156-199

Author: Nguyen Hoang Khuc Tu

Institute: Hochiminh city International University
Street: Quarter 6, Linh Trung ward, Thu Duc Dist.
City: Ho Chi Minh
Country: Vietnam
Email: nhktu@hcmiu.edu.vn

Comparing the Effect of Ficus Benjamina Extract and Pereskia Saecnarosa Extract on the Level of Micro and Macro Minerals in Normal and Induced Liver Cancer Rats

Asmah Rahmat, Fahmida Parveen Saib, and Nurul Amira Buslima

Department of Nutrition and Dietetics, Faculty of Medicine and Health Sciences, Universiti Putra Malaysia, 43400, Serdang, Selangor, Malaysia
asmah@medic.upm.edu.my

Abstract — **Hepatocellular carcinoma (HCC) is the fifth most common malignancy in the world. It is much prevalent in many of the developing countries than in the industrialized world. The effect of *Ficus Benjamina* extract and *Pereskia Saecnarosa* extract was studied on the level of micro and macro minerals in normal and induced liver cancer rats. The extracts of *Ficus Benjamina* and *Pereskia Saecnarosa* were prepared and were given to the rats through *ad libitum*. Forty-four male rats from the species *Sprague Dawley* were used as a model. They were acclimatized for duration of one week and further divided into 6 groups at random. The 6 groups were: control, cancer; normal rats supplemented with *Ficus Benjamina,* cancer rats supplemented with *Ficus Benjamina*, normal rats supplemented with *Pereskia saecnarosa* and cancer rats supplemented with *Pereskia saecnarosa*. Twelve weeks later, the blood of each rat was taken for the analysis of calcium, iron and zinc and body weight was taken before and after the treatment. The very obvious result was the highest weight gained by the normal rats supplemented with *Ficus benjamina* extract. However, the weight gain was not significant to the control group but it was significant to the cancerous group. In all the 6 groups irrespective of their treatment, the level of minerals was almost the same. From the statistical analysis, it was found that there was a significant relationship in the level of calcium between the cancerous rats and the cancerous rats given *Pereskia saecnarosa* ($p<0.05$). Moreover, a significant relationship in the level of iron was found between the cancerous rats and the cancerous rats supplemented with *Pereskia saecnarosa* ($p<0.05$). However the cancerous rats supplemented with *Pereskia saecnarosa* showed a lower level of ferrum in their blood serum. Furthermore a significant relationship was found in the level of zinc between the control rats and normal rats supplemented with *Pereskia saecnarosa* ($p<0.05$). In conclusion, there is no significant relationship between the effect of *Ficus Benjamina* and *Pereskia saecnarosa* extract and the level of calcium, iron and zinc.**

Keywords — ***Ficus Benjamina*, *Pereskia Saecnarosa*, hepatocarcinogenesis, micro and macro minerals.**

I. INTRODUCTION

Since ancient time, plants have been used an exemplary sources of medicine. Ayurveda and other Indian literature have mentioned the use of plants in the treatment of various ailments [1]. Indeed, most of the new applications of plant secondary metabolites and their derivatives over the half century have been applied towards combating cancer [2]. Since ancient times, more than 3000 plant species were reported to be used in the treatment of cancer [3]. Since the year 1980's, more than 60% of cancer approved drugs developed are of natural products origin [4] and 25% are directly derived from plants [5]. Plants products are regarded as potential chemopreventive agents act to block reverse or prevent the development of invasive cancers [6]. Recently, tropical and subtropical plants are perceived to produce a large variety of phytochemicals or secondary metabolites and possess a wide range of cancer preventive. Lots of importance is given to cancer as cancer is one of the leading diseases of new era and it is also one of the greatest killers. In Malaysia, it is one of the leading causes of morbidity and mortality [7].

Antioxidants are defined as compounds that inhibit the oxidation of other molecules by inhibiting the initiation the propagation of oxidizing chain reactions [8]. are two basic of antioxidants; synthetic and natural. Synthethis antioxidants are compounds with phenolic structures of various degrees of alkyl substitution, whereas natural antioxidants can be phenolic compounds (tocopherols, flavonoids and phenolic acid). Nitrogen compounds alkaloids, chlorophyll derivatives, amino acids and amines) or carotenoids as well as ascorbic acid [9].

Many of natural antioxidants, especially flavonoids, exhibit a wide range of biological effects, including antibacterial, antiviral, anti-inflammatory,a ntiallergic, anti-thrombotic and vasodilatory actions. Antioxidants activity is the fundamental property important for life. Many of the biological functions, such as antimutagenicity, anti-carcinogenocity and antiaging, among others, originate from

V. Van Toi et al. (Eds.): 4th International Conference on Biomedical Engineering in Vietnam, IFMBE Proceedings 40, pp. 208–212, 2013.
www.springerlink.com

this property [10]. The antioxidantactivity of several plant materials has cerently been reported.

Two plant extracts were used in this research, namely *Ficus benjamina* extract and *Pereskia saecnarosa* extract. According to Wagner, [11], *Ficus benjamina* is a plant from the family Moracacea (Mullberry family) that are native to large area including India, southern China, Southeast Asia, Malaysia, Filippines, northern Australia, and the island of South Pacific [12]. Whereas, *Preskia saecnarosa* is categorized as a primitive cactus because of its leaves and one of an edible herb in Malaysia, though it is not very common. It is known as "jarum tujuh bilah" among Malaysian and usually they are made into juice after boiled and drank it every morning for the purpose of health practice and body refreshing.

In the Ficus species, there have been some studies carried out on *Ficus elastica* as in possessing inflammatory properties and on *Ficus citrifolia* as a therapeutic agent against cancer therapy [13]. However, so far, there has been no study carried out of *Ficus benjamina* as in possessing any anti-cancerous properties. As for *Pereskia saecnarosa* is concerned, there has been few researchers carried out to see its potential in cancer therephy. The leaves of *Pereskia saecnarosa* is believed to possess a dietary component in reducing the severe level in the incidence of hepatocarcinogenesis.

II. Materials and Methods

2-acetylaminofluorene, diethylnitrosamine, acetone, rat basal diet, ether, centrifuge apparatus, spectrAA-40 atomic absorption spectrometer, calcium hollow cathode lamp, iron hollow cathode lamp, zinc hollow cathode lamp.

III. Methodology

A. *Sample Preparation and Extraction*

The leaves of *Ficus benjamina* and *Pereskia saecnarosa* were randomly plucked from Universiti Agricultural Park, Universiti Putra Malaysia. The leaves were cleaned, weighed and further grounded into powder form. A volume of 2000ml of distilled water was added to the ground leaves to make a concentration of 5.0% (w/v). The solution were then filtered and given to the group of rats which was concerned. The same procedures were done to prepare *Pereskia saecnarosa* extract.

B. *Cancer Inducer Preparation*

Diethylnitrosamine (DEN) inducer is prepared by dissolving 1.0 ml DEN in 3.33 ml corn oil, which is equivalent to 200 mg DEN / kg body weight of the rats. About 0.1 ml of the solution was injected to each rat. Besides that, 1.0 g of 2-acetylaminofluorene (AAF) is dissolved into 50.0 ml acetone. 1.5 ml of this solution is then dropped to 150 g rat chow to obtain the final concentration at 0.02 % (w/w) AAF in the diet. The acetone is then dried in vacuum at 15 mmHg for an hour.

C. *Induction of Hepatocarcinogenesis*

The cancerous groups were induced with hepatocarcinogenesis using the method of [14]. which can induce liver cancer is injected into the rats intraperitoneally with necrogenic doses of 200 mg/kg body weight to initiate hepatocarcinogenesis. Following a two weeks period, as described by method of Cameroon, [15], the rats were then fed with 0.02 % (w/w) AAF treated rat chow for another 2 weeks without partial hepatectomy to promote hepatocarcinogenesis.

Group	Treatment		
Normal control (N)	Basal diet + water		
Cancer induced (C)	Basal diet + water	AAF + water	Basal diet + water
Normal + FB (NF)	Basal diet + water	Basal diet + FB	
Cancer + FB (CF)	Basal diet + water	AAF + FB	Basal diet + FB
Normal + PS (NP)	Basal diet + water	Basal diet + PS	
Cancer + PS (CP)	Basal diet + water	AAF + PS	Basal diet + PS
Week	0	2	4 12

DEN ↑ (week 0)

Fig. 1 Protocol to study the effect of *Ficus benjamina* and *Pereskia saecnarosa* extract during rat hepatocarconogenesis.

D. *Animals Management*

In this research, a total of 44 male rats from the species of *Sprague dawley* were used as a model. They were housed in a separate cage in the animal house at the Faculty of Medicine and Health Sciences of Universiti Putra Malaysia with woodchip bedding in a well ventilated room with equal periods of daylight and darkness with temperature 32 ± 2°C. Hygienic conditions are maintained by twice-weekly changes of the woodchip beds and daily changes of water and treatment. Each rats initially weighing between 150 to 200 g (7-8 weeks old) were acclimatized for at least 1 week before use, on basal diet and water *ad libitum*. All rats are randomly distributed into 6 groups comprising groups of normal control (N), cancer-induced control (C), normal with 0.1ml of *Ficus benjamina* (NF), cancer with 0.1ml of *Ficus benjamina* (CF), normal with 0.1ml of *Pereskia saecnarosa* (NP) and cancer with 0.1ml of *Pereskia saecnarosa* (CP.) A total of 5-8 rats are allocated for each group. The rats in normal control group had only a basal diet and water throughout the experiment. For the groups of treatment and cancer-induced + treatment, about 0.1ml samples will be given everyday using force feeding. For two weeks period, groups of cancer-induced are feed with a treated AAF chow pellet in their diet as substitute to basal diet in the 3rd week of experiment following two weeks of DEN injection to promote hepatocarcinogenesis. On the 5th week, feeding of treated AAF chow pellet is terminated and the rats were then given back basal diet until the end of the experiment.

E. Experiment Termination

All rats are killed after 12 weeks periods of experimental protocol. The rats were starved for 24 hours before being sacrificed. At the termination of the experiment, the animals are weighed and immediately before death, blood is collected by venipuncture and placed into EDTA tubes. The samples were centrifuged at 3,000 rpm with bench centrifuge for 10 minutes and the plasma was collected to perform analysis of macro minerals.

F. Macro Minerals Analysis

The atomic absorption spectroscopy (AAS) was used to determine the following mineral level; calcium, iron and zinc in the blood plasma of the 44 male rats after their respective treatment. A SpectrAA-40 atomic absorption spectrometer, PSC-56 programmable sample changer, Epson LX-80 printer and calcium, ferum and zinc hollow cathode lamps from Varian were used in the procedure. The main instrument parameters for the analysis were as Table 1.

Table 1 Parameters for the selected macro minerals analysis.

Parameters	Zinc	Ferum	Calcium
Wavelength (nm)	213.9	248.3	422.7
Slit width (nm)	1.0	0.2	0.5
Lamp current (mA)	4	8	3
Burner height (nm)	10	10	4
Acetylene flow (L/min)	2.00	2.50	2.50
Air flow (L/min)	13.5	13.5	13.5
Standard concentration(µg/ml)	0.1-0.5	0.1-0.5	4.0-20.0

As for reagents concerned, the water used were doubly distilled and deionized. All the reagents used for the preparation of standard solutions and anti-interference reagents were ultra pure grade. The standard solutions were prepared in glass bottles, previously washed with diluted nitric acid and rinsed with deionized water. One milliliter of serum was placed in a test tube diluted to 10 milliliter with deionized water. The test tube was placed in the auto sampler carousel after mild shaking. The measurement was automatically carried out and the results were the average of two replicated according to the standard addition calibration method. The final result was presented as unit of mg/100g sample.

G. Data Analysis

All data were presented as mean ± standard deviation of duplicates and statistically analyzed using Statistical Package for Social Sciences (SPSS) version 19.0. One-way analysis of variance (ANOVA) was used to compare means of different groups. A significant difference was considered at the level of $p<0.05$.

IV. Results and Discussion

A. Comparison of Body Weight

The average final weights gained of the rats were shown in Table 2. The table clearly shown that the group of normal rats supplemented with *Ficus benjamina* extract was the highest weight gained among all groups. Actually the group of normal rats supplemented with *Ficus benjamina* extract has gained weight significantly ($p<0.05$) compared with the DEN/AAF treated group, but the weight gain is not significant to the other groups. The small sample size may be an explanation of this effect. The presence of some bioactive compounds in *Ficus benjamina* might have lead to the weight gained, where it might have increased the appetite of the rats.

Table 2 Effect of *Ficus benjamina* extract, *Pereskia saecnarosa* extract and DEN/AAF on the final body weight gained.

Group	Body weight gained (g)
Normal control (N)	178.94 ± 8.40
Cancer induced (C)	149.26 ± 4.56[a]
Normal + FB (NF)	213.28 ± 19.12[a]
Cancer + FB (CF)	176.74 ± 4.10
Normal + PS (NP)	180.81 ± 8.46
Cancer + PS (CP)	150.28 ± 14.55[b]

Values are the mean ± standard deviation.
Letter a: significantly different from the control group ($p<0.05$).
Letter b: significantly different from the control cancer ($p<0.05$).

There was also a significant reduction on weight in the DEN/AAF treated group to the control group. The cancerous group had lost a significant amount of weight. The significant reduction in body weight by the cancerous rats might be because of low food consumption and hence loss of weight. The reduction in body weight by the cancerous rats showed that the effect of DEN/AAF has been met. A research carried out by Hosaka, [16] also showed the same pattern in weight gained.

Furthermore, the body weight gained of the DEN/AAF group also differs significantly ($p<0.05$) with the DEN/AAF group supplemented with the *Pereskia saecnarosa* extract group. Cancerous rats supplemented with *Pereskia saecnarosa* extract have higher body weight gained. Cancerous rats do not have appetite but *Pereskia saecnarosa* extract might have increased the appetite of the rats, resulting in a weight gained.

B. *Plasma Mineral Analysis Using Atomic Absorption Spectroscopy (AAS)*

From the result shown in Table 3, all the plant extracts did not influence the level of minerals in the rat's plasma. In all the 6 groups irrespective of their treatment, the levels of minerals were almost the same. This may be because the rats have been consuming the plant extract for a too short period of time. For a herb to be effective, it must be consumed over a long period of time.

Table 3 Selected mineral level in mg/100g of sample in each group.

Group	Calcium	Iron	Zinc
Normal control (N)	1.66 ± 0.475	0.58 ± 0.292	0.010 ± 0.387
Cancer induced (C)	1.89 ± 0.203	0.71 ± 0.029	0.63 ± 0.335
Normal + FB (NF)	1.44 ± 0.575	0.07 ± 0.026	0.64 ± 0.391
Cancer + FB (CF)	1.88 + 0.142	0.06 + 0.028	0.79 + 0.441
Normal + PS (NP)	1.43 ± 0.648	0.08 ± 0.028	0.83 ± 0.416[b]
Cancer + PS (CP)	1.48 ± 0.503[a]	0.07 ± 0.017[a]	0.10 ± 0.068

Values are the mean ± standard deviation.
Letter a: significantly different from the control group ($p<0.05$).
Letter b: significantly different from the control cancer ($p<0.05$).

From the statistical analysis, there was a significant relationship in the level of calcium between the cancerous rats and the cancerous rats given *Pereskia saecnarosa*. It is know that calcium can prevent precancerous cells from becoming cancerous [17]. But vitamin D is essential for the absorption and metabolism of calcium. If calcium is taken but the level of vitamin D is too low, the calcium will not be absorbed and will result in high level of calcium in the blood.

Moreover, a significant relationship in the level of iron was found between the cancerous rats and the cancerous rats supplemented with *Pereskia saecnarosa*. However, the cancerous rats supplemented with *Pereskia saecnarosa* showed a low level of iron in their blood plasma. There might be substances in *Pereskia saecnarosa* which inhibit the absorption of iron. Iron plays an important role in the prevention of cancer but should not be taken in excess due to the initiation of production of radicals by iron complexes. According to Charles, [18], excess stores of iron can increase the risk of cancer in man and probably women as well. Furthermore, a significant relationship was found in the level of zinc between the control rats and normal rats supplemented with *Pereskia saecnarosa*.

In a research carried out by undergraduate student from the Faculty of Medicine and Health Sciences, Universiti Putra Malaysia, the level of 4 minerals were analyzed from the leaves of the plant *Ficus benjamina*. The minerals of interest were selenium, calcium, copper and iron. The result showed that calcium (9.12mg/100g) was of the highest amount, followed by iron (0.64mg/100g), selenium (0.46mg/100g) and copper (0.06mg/100g). The nutrients that were present in *Ficus benjamina* leaves have a great potential to be used as a source of anticancer medicine.

V. Conclusions

Results of the body weight gain by the normal group of rats supplemented with *Ficus benjamina* and by the normal group of rats supplemented with *Pereskia saecnarosa* showed that both plants extract have the potential to increase appetite. Moreover, the low body weight of cancerous rats indicates that the effect of Diethylnitrosamine and 2-acetyl-aminofluorene have been successfull. As for the levels of minerals were concerned, tehre were almsot no variation among the groups. This may due to the short period of taking the plant's extract.

Acknowledgment

A warm appreciation dedicated to the Department of Nutrition and Dietetics, Universiti Putra Malaysia for the support, help and contribution in this research project.

References

1. Grover J.K., Yadav S. & Vats S. (2002). Medicinal plants of India with anti-diabetic potential. Journal of Ethnopharmacology. 81:1-100.
2. Newman D.J. and Cragg G.M. (2007). Natural Products as Sources of New Drugs over the Last 25 Years. J. Nat. Prod. (70) pg 461-477.
3. Cragg G.M., Newman D.J. (2001). Medicinals for the millennia: The historical record. Ann NY Acad Sc. (953): pg 3-25.
4. Cragg G.M., Newman D.J., Snader K.M. (1997). Natural products in drug discovery and development. J Nat Prod. (60) pg 52-60.
5. Amin A., Gali-Muhtasib H., Ocker M. and Schneider-Stock R. (2009). Overview of Major Classes of Plant-Derived Anticancer Drugs. International Journal of Biomedical Science. (5):1 pg 1-11.

6. Reddy L., Odhav B. and Bhoola K.D. (2003). Natural products for cancer prevention: a global perspective. Pharmacology and Therapeutics. 99;(1):1-13.
7. Malaysia Department of Statistics. (2010). Vital Statistics Peninsular Malaysia 2010. National Printing Department, Kuala Lumpur.
8. Velioglu Y.S., Mazza G., Gao L. and Oomah B.D. (1998). Antioxidant activity and total phenolic of selected fruits, vegetables and grains products. Journal of Agricultural and Food Chemistry 46: 4113-4117.
9. Larson R.A. (1998). The antioxidant of higher plants. Phytochemistry. 27:969-978.
10. Huang M.T., Ho C.T. & Lee C.Y. (1992). Phenolic compounds in food and their effects on health. Antioxidants and cancer prevention: ACS Symposium Series 507. American Chemical Society: Washington DC.
11. Wagner W.L., Herbts D.R. & Sohmer S.H. (1999). Manual of the flowering plants of Hawaii. Bishop Museum Special Publication 83, University of Hawaii and Bishop Museum Press, Honolulu, HI.
12. Riffle R.L. (1998). The tropical look. Timber Press Inc., Portland, Oregon.
14. Solt and Farber (1976). Solt, D. B. and Farber, E. (1976). A new principle of the analysis of chemical carcinogenesis. Nature 263: 701-703
15. Cameron, R., Kellen, J., Kolin, A., Malkin, A. and Farber, E. (1978). Gamma glutamyltransferase in putative premalignant Liver Cell Populations during Hepatocarcinogenesis. Cancer Research 38: 823-829.
16. Hosaka S., Tawa S., Aoki Y., Tanaka E., Yoshizawa K., Karasawa Y. & Kiyosawa K. (2001). Hepatocarcinogenesis inhibition by caffeine in ACI rats treated with 2-acetylaminofluorene.
17. Balch J.F. & Balch P.A. (1992). Prescription for Nutrition Healing, Second Edition, pp121.

Fuzzy Determination of the Human's Level of Psycho-Emotional

Nikolay Korenevskiy[1], Riad Taha Al-Kasasbeh[2], Florin Ionescouc[3], Mohdi Alshamasin[2], E. Alkasasbeh[4], and Andrew P. Smith[5]

[1] Faculty Biomedical Engineering, South West State University, Russia
[2] Faculty of Engineering Technology, Al-balqa Applied University, Amman, Jordan
[3] Department of Mechatronics, University of Applied Sciences (HTWG), Konstanz, Germany
[4] Faculty of Karak, Al-balqa Applied University, Karak, Jordan
[5] Faculty of Computer Science, University of Applied Sciences, Konstanz, Germany

Abstract — **Determination of psycho-emotional tension on humans is performed by means of psychological tests, the electrical characteristics of biologically active points connected with the psycho-emotional sphere, and on indicators characterizing the state of a subject's visual attention. The level of such tension is determined as an aggregate of these three components with the rules of fuzzy logic. It is shown that the resulting method can be used as an informative feature for prediction and medical diagnosis of conditions such as stomach and heart disease. The use of dissimilar parameters characterizing different areas of psychology takes into account the systemic nature of mental phenomena such as psycho-emotional tension. Fuzzy logic decision-making can correctly account for uncertainties in a parameter under consideration, combined with incomplete understanding of its properties in practice. Additionally, such an assessment of psycho-emotional tension allows one to integrate this indicator into fuzzy output systems for a broad range of tasks in the fields of medical and psychological diagnosis, including vocational suitability.**

Keywords — **fuzzy logics, psychoemotional tension, certainty factors.**

I. Introduction

There are a significant number of works that convincingly show the role of psycho-emotional tension (PET) in the occurrence and development of a variety of psychosomatic diseases.

There is a large arsenal of methods and facilities for the identification of PET according to various systems of psychological, psychophysiological and physiological factors. However the problem of estimation of the PET level remains far from solved. It is possible to raise the reliability of PET diagnostics by using a combination of the indicators characterizing various occurrences of apparent activity, whose quantity and structure is optimized by means of modern mathematical methods and information technologies.

In this paper, we propose to make an assessment of PET using the set of diverse, heterogeneous indicators. The first group of indicators relies on traditional tests such as Spielberger-Hanin, where there is a scale reflecting the degree of emotional stress. The second group of indicators evaluates PET according to the largest deviation in the electrical resistance of biologically active points (acupuncture points) from their nominal values, for points which are "connected" with the psycho-emotional sphere (R8, VB20, P9, F3, VG24). The third group of indicators characterize properties such as switchable attention (the ability to switch attention) (SWA), concentration (ability to concentrate) (CNA), stability of visual attention (STA) and parameters determining the state of memory [2,3,4,10].

II. Methods

The experience in solving various problems of medical and psychological diagnosis, including the task of assessing the psycho-emotional, sphere shows that the measured parameters (factors) are often incomplete and unclear. The structure of the classes for which the decision is made (especially for prediction, early diagnosis, classification of functional states, determining the levels of psycho-emotional tension, etc.) exhibits fuzzy bounds with crossing zones passing from class to class. Under these conditions, according to recommendations [6,7,8,9] for the synthesis of congruent solving rules, it is reasonable to use the theory of fuzzy logic for decision-making. From this theory we chose two variants. The first variant is based on the membership functions for the investigated disease class ω_ℓ and was described in L. Zadeh's works [6]. The second variant is based on certainty factors in the hypothesis (prognosis, diagnosis, etc.) ω_ℓ and was described in E. Shortliffe's works [7].

Taking into account the fuzzy nature of the conception and classification "level of PET" the interval [0,…,1] is defined as the scale for PET measuring. The concept "very

V. Van Toi et al. (Eds.): 4th International Conference on Biomedical Engineering in Vietnam, IFMBE Proceedings 40, pp. 213–216, 2013.
www.springerlink.com

high level of PET" corresponds to the upper limit (1). This value experts can define as the level of psycho-emotional tension which leads to impairment of the body's normal functioning, to the occurrence and development of psychosomatic diseases, failure of the person to perform properly in the work environment, etc.

The parameters that are used in this work in any case reflect the current level of the psycho-emotional tension. The different values of these parameters correspond to different level of the psycho-emotional tension. If the selected groups of factors are used as the basic membership functions to the conception "high level of PET" then calculated for the basic variable membership function will define the current level of the psycho-emotional tension. The experts believe that the conception "high level of PET" can't be defined by one factor. So the limiting value of the membership functions of one factor is always less that 1.

On the other hand during the researches it was established that each of used factors makes a "contribution" in conception "level of PET" of all organism. It allows one to estimate the level of PET by the expression that is similar to the iterative rule of the fuzzy output by E. Shortliffe [8].

$$LT(j+1)=LT(j)+\mu_p(S^*_{j+1})[1-LT(j)] \qquad (1)$$

where $\mu_p(S^*_{j+1})$ - numerical value of the membership functions to conception "very high level of PET" calculated in concrete point S^*_{j+1} (S – basic variable with number j+1); $LT(1)=\mu_p(S^*_1)$

Due to the fact that the basic variables S_i for different groups of factors are different, rule (1) should be sorted as heterogeneous rule for fuzzy estimation of the level of PET. Formula (1) defines the level of PET in the interval [0,1]. Value 1 means very high level of PET.

III. Results

To determine the first component of the general formula for definition of the PET level the scales of situational anxiety and personal anxiety from the Spilberger-Hanin test can be used [3]. The scale of situational anxiety (S_1) allows one to reliably measure various degrees of PET corresponding to psychological tension. On this scale the following membership function corresponds to the concept "very high level of PET":

$$\mu_p(S_1)=\begin{cases}0, if\ S_1<35;\\0,013\cdot S_1-0,58, if\ S_1\geq 35.\end{cases}$$

The scale of personal anxiety (S_2) is directed to the diagnosis of anxiety-related occurrences in behavior in relation to various situations. The diagnostics of the level of human's anxiety is very important because it influences on the work activity. High level of the personal anxiety can be result of the accumulation of effects of the overstrain and stress. On the other hand, high level of PET can show that the human "complicate" problems.

The level of PET on the scale S_2 is defined by expression:

$$\mu_p(S_2)=\begin{cases}0,\ if\ S_2<40;\\0,012\cdot S_2-0,5,\ if\ S_2\geq 40,\end{cases}$$

Taking into account (1) the level of PET is defined by the Spilberger-Hanin test.

$$LT^T=\mu_p(S_1)+\mu_p(S_2)[1-\mu_p(S_1)] \qquad (2)$$

The upper index "T" means that the level of PET is defined by the group of factors that were derived during the subjective testing.

Experience of using of the energy performance of acupuncture points (AP) for a variety of practical problems has shown that the quality of diagnosis increases considerably if for the choice of informative points and the synthesis of the corresponding decision rules we take into account the features of "output" of information from the body's internal structures to these points.

In work [4] the main feature of the information output is shown - the majority of AP do not have completely coincident lists of diagnoses, syndromes and symptoms. These lists we call situations.

This research has shown that if the situations do not coincide, it is possible to choose AP combinations, analysis of which allows one to confirm an investigated situation (diagnosis) and refute the situations which are indicated by AP output but absent at the patient. These combinations are called diagnostically important points (DIP) [4].

In work [4] the algorithm of the DIP search is shown. Also when the electrical resistance is used, it's reasonable to use the following structure of the fuzzy decision rule as the informative factor.

IF [for all AP from the list DIP for class ω_ℓ the resistance deviation R_j is larger than the selected threshold $\delta R_{j,treshold}$] then [calculate the measuring parameters according to (1)] else [AP is mot used in solving of the problem]. (3)

The level of PET is defied by this expression
IF [(δR$_{VB20}$>10%) and (δR$_{P9}$>10%) or (δR$_{R8}$>10%)]
THEN $\{LT^A(j+1) = LT^A(j) + \mu_p(\delta R_{j+1})[1 - LT^A(j)]\}$
ELSE $LT^A = 0$ (4)

where $\mu_p(\delta R_6)$ - membership function to "very high level of PET" with basic variable δR_6 , S=1,2,3; $LT^A(1) = \mu_p(\delta R_{VB20})$;

$\delta R_1 = \delta R_{R8}; \delta R_2 = \delta R_{F3}; \delta R_3 = \delta R_{P9}$;

During the specific research the experts using the method Delphi built the corresponding graphics of the membership functions.

$$\mu_{PT}(\delta R_{R8}) = \begin{cases} 0, & if \quad \delta R_{R8} < 10\%; \\ 0{,}004\,\delta R_{R8} - 0{,}04, & if \quad 10\% \le \delta R_{R8} < 60\%; \\ 0{,}65, & if \quad \delta R_{R8} \ge 60\%, \end{cases}$$

$$\mu_{PT}(\delta R_{VB20}) = \begin{cases} 0, & \text{if} \quad \delta R_{VB20} < 10\%; \\ 0{,}0025\,\delta R_{VB20} - 0{,}025, & \text{if} \quad 10\% \le \delta R_{VB20} < 70\%; \\ 0{,}15, & \text{if} \quad \delta R_{VB20} \ge 70\%, \end{cases}$$

$$\mu_{PH}(\delta R_{P9}) = \begin{cases} 0, & \text{if} \quad \delta R_{P9} < 10\%; \\ 0{,}004\,\delta R_{P9} - 0{,}04, & \text{if} \quad 10\% \le \delta R_{P9} < 60\%; \\ 0{,}2, & \text{if} \quad \delta R_{P9} \ge 60\%, \end{cases}$$

To estimate the level of PET the results from [3,10] are used. According to this the following expressions are obtained:

IF $[(Y1 > 10) and (Y2 > 0)]$
THEN $[LT^A = F(Y1, Y2)]$
ELSE $(LT^A = 0)$ (5)

where $Y1 = SWA - SWA_0 + CNA - CNA_0$; $Y2 = STA - STA_0$; SWA and SWA_0 are the switchable attentions (current and during the rest); CNA and CNA_0 are he corresponding concentrations of attention; STA and STA_0 - corresponding stability of visual attention.

For calculation LT^A the membership functions are obtained:

$$\mu_{PT}(Y1) = \begin{cases} 0, & \text{if} \quad Y1 < 10; \\ 0{,}0086 \cdot Y1, & \text{if} \quad 10 \le Y1 < 80; \\ 0{,}6, & \text{if} \quad Y1 \ge 80, \end{cases}$$

$$\mu_{PT}(Y2) = \begin{cases} 0, & \text{if} \quad Y2 < 10; \\ 0{,}0086 \cdot Y2, & \text{if} \quad 10 \le Y2 < 80; \\ 0{,}6, & \text{if} \quad Y2 \ge 80, \end{cases}$$

Taking into account the membership functions for (5) $F(,Y1,Y2) = \max\{\mu_{PT}(Y1), \mu_{PT}(Y2)\}$.

Taking into account (1) the general formula for calculation of the level of PET based on the three groups of factors is:

$$\begin{aligned} LT = LT^T + LT^A + LT^{AM} - LT^T LT^A - LT^T LT^{AM} - \\ -LT^A LT^{AM} + LT^T LT^A LT^{AM} \end{aligned} \qquad (6)$$

IV. WRKING MODEL

In the course of the research we studied the informative value of indicators characterizing the level of psycho-emotional tension in solving problems of early diagnosis for diseases such as diseases of the nervous system (class ω_N), digestive system (class ω_D), skin diseases (class ω_S), etc.

Graphs of the functions belonging to the class of early stages of these diseases are shown in Fig. 1.

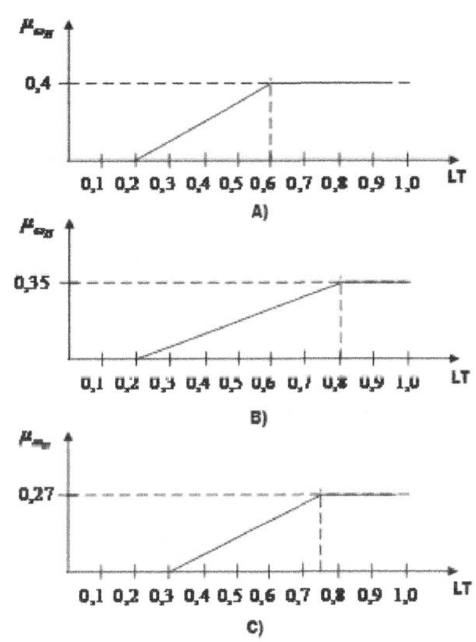

Fig. 2 Functions belonging to the class of early stage of diseases of :a) the nervous system ($\mu_{\omega N}$) ; b) digestive digestive system ($\mu_{\omega D}$); c) skin ($\mu_{\omega S}$) with base variable LT.

Analysis of the graphs shows that it is impossible to judge with certainty based only on the level of PET the existence of early stages of the investigated classes of diseases. However, as shown by the results of the studies of other informative features, confidence in the decisions made can be given at the level of 0.85 or higher, which is quite acceptable for use in medical practice.

The description of the combined decision rules for different diseases that use the level of PET as the informative factor s given in [4].

V. Conclusions

The work of many scientists has clearly shown that for a number of diseases, one of the significant risk factors is long-term psycho-emotional tension (PET). The same factor has a significant impact on the quality of the performance of a human operator of complex machinery.

There is a considerable amount of work to determine the level of PET using psychological tests and to construct the devices for monitoring physiological and psycho-physiological parameters, but credible and generally accepted quantitative estimates of the psychological properties are still pending. There is no conformity in the precise definition of this concept by various specialists dealing with the phenomenon of mental and emotional stresses from various viewpoints (physiology, clinicians, psychologists, etc.).

The use of dissimilar parameters characterizing different areas of psychology can take into account the systemic nature of mental properties such as psycho-emotional tension. The use of fuzzy logic decision-making can correctly take into account uncertainties in the parameter under consideration, combined with incomplete under-standing of its properties in practice. Additionally, fuzzy assessment of psycho-emotional tension allows one to integrate this indicator into fuzzy output systems, to address a broad range of tasks in the fields of medical and psycho-logical diagnosis, choice of occupation and vocational guidance, etc.

Acknowledgment

We gratefully acknowledge funding from the DFG (German Research Council) and additional support from Al-Balqa Applied University in Jordan, South West State University in Russia, HTWG (University of Applied Sciences) Konstanz in Germany and the Higher Council for Science and Technology of Jordan.

References

1. Plotnikov V , Korenevskii N , Zabrodin U(1989) Automation of methods of psychological investigations. Orel: Publishing house of Psychology Institute, Russia
2. Sergeenko S., Bodrov V, Pisarenko U(2003) Training in engineering psychology and ergonomics. Publishing house "Academy", Moscow
3. Korenevsky N, Skopin D, Al Kasasbeh R, Kuz'min A(2010) System for Studying Specific Features of Attention and Memory. Biomedical Engineering Journal 44(1): 32-35
4. Korenevsky N, Krupchatnikov R, Seregin P (2010)Theoretical fundamentals biophysics acupuncture with applications in biology and medicine and ecology using fuzzy logic. Monograph, Kursk State Technical University press, Kursk, ISBN 978-5-7277-0530-8.
5. Zhao J, Zheng L, Zhuang T et al (2003) Represention of acupoints based on the the visible Human. Engineering in Medicine and Biology Society, vol.18, Proceedings of the 25th Annual International Conference of the IEEE, Cancun, 2003 , pp. 525 – 528
6. Klir GJ, Yuan B (1996) Fuzzy sets, fuzzy logic, and fuzzy systems. selected papers by L.A. Zadeh. Singapore: World Scientific. 826 p. Singapore.linkinghub.elsevier.com/retrieve/pii/ S0165011402003524.
7. Buchanan B, Shortliffe E (1984) Rule-Based Expert Systems: The MYCIN Experiments of the Stanford Heuristic Programming Project. Addison-Wesley Publishing Company. Reading, Massachusetts,p. 742, ISBN 0-201-10172-6.
8. Al-Kasasbeh R, Korenevskiiy N, Ionescu F, Kuzmin A (2009) Synthesis of the Combined Fuzzy Rules for Medical Applications with Using Tools of Exploration Analysis. Proc. 4th IAFA Intern, Conference Interdisciplinary Approaches in Fractal Analysis, Bucharest, Romania, May 26-29, 2009 ISSN 2066-4451, pp. 71-78.
9. Al-Kasasbeh R, Ionescu F, Korenevskiy N, Alshamasin M (2011) Prediction and Prenosological Diagnostics of Gastrointestinal Tract Diseases Based on Energy Characteristic of Acupuncture Points and Fuzzy Logic. ICBBT 2011, Proc. 3rd International Conference on Bioinformatics and Biomedical Technology, Sanya, China, March 25-27, 2011, pp.307-3011
10. Korenevskiy N, Filatova O, Lukashov M, Krupchatnikov R(2009) Complex Estimation of Level of the Psychoemotional Pressure. Biomedical electronics vol. 1(5):4-9

The Relationship between Jaw Imbalance and Arm Strength Loss through Fractal Analysis of Surface EMG Signals

Le Minh Hoa, Truong Quang Dang Khoa, Nguyen Van Hoa, Dao Tien Tuan*, and Vo Van Toi

Biomedical Engineering Department, International University of Vietnam National University, Ho Chi Minh City, Viet Nam
* Biomechanics and Biomedical Engineering Department, University of Technology of Compiègne, France

Abstract — **It has been thought that imbalance in the jaws can cause loss in arm strength. In this study, EMG will be used to determine the EMG-Force relationship of the jaw, neck and arms muscles. This will give us useful insights on the arms strength loss due to the biomechanical effects of the imbalance in the jaw mechanism. Three male subjects (average age: 20.67±2.89 years old) participated in this investigation. The subject were asked to stand with their arms and legs extended in frontal plan in such a way that the ratio subject intra-feet distance/ subject height equal 0.25. The experimental protocol consisted of two paradigms that (1) From the initial position (0 mm, 5 kg), external loading weight was added incrementally (6kg, 7kg) until the subject can hold the weight in 10 seconds. Check EMG signal quality and data storage; (2) Perform variation in spacer thickness from 2mm to 3mm and execute step 1 for each variation. Myoelectric signals of the masseter, trapezoid, deltoid and brachiodiolis muscles were measured. Higuchi fractal dimension was used to analyze the complexity of surface EMG signal of these four muscles. The results showed that (1) The activities of arm muscles varied according to the applied pull -down force; (2) The masseter, trapezoid and deltoid muscles displayed the greatest amount of activity in this study. There was also a significant difference in the fractal dimension values of the surface EMG signal before and after the loss of arm strength.**

Keywords — **jaw imbalance, arm strength, higughi fractal dimension.**

I. INTRODUCTION

It has been reported that there is a strong relationship between the oral maxillofacial muscle, temporomandibular joint, and neck and arm muscles. Linderholm et al. [1] who investigated the relationship between bite force magnitude and forces produced by several skeletal muscle groups in children suggested that there was a relationship between bite forces and other forces, and the maximal bite force correlated with elbow flexion force and hand-grip force. Raadsheer et al. [2] found that the size of the jaw muscles was significantly related to the size of the limb muscles; however, maximal voluntary bite force moments were not significantly related to the moments of the arm flexion and leg extension forces.

In a study of finding a relationship between the height of bite plates and the strength of deltoid muscles, Chakfa et al. [3] performed experiments with 20 female subjects using the bite plates of adjusted heights of 2, 4, 6 and 12 mm. Each subject was seated on a dental chair with the arms extended to the side at the shoulder level and parallel to the floor. With or without introducing the bite plate in the subject's mouth, the examiner applied a downward force on the wrist of the extended arm of the subject with a strain gauge while applying a stabilizing force on the contra lateral shoulder of the subject until the subject's arm could no longer resist to the downward force. These researchers found that while increasing the height of the plate, the arm strength increased (from the strength without the plate of 6.9 kg on the right arm and 6.4 kg on the left arm) to a maximum (of 8.6 kg on either side of the arms) then decreased to 6.5kg to the right and 6.3 kg to the left arm. Although these investigators did not indicate the height of the plate when the maximum strength occurred and simply reported that it depended on the subject, we deemed that they were either 4 or 8mm.

In a study to determine the effects of the imbalance in the jaw to the strength of the arms, we [4] conducted experiments with a pool of 34 (age: 19.8±0.9 years old) and healthy subjects of both genders. The subjects stood straight with both arms extended and feet distanced apart. They resisted to a pull-down force applied on one hand at a time while firmly holding in one side of the jaws a firm spacer using essentially their molar and pre-molar teeth until they could no longer resist to the force. We found that when the subjects held the spacer on one side of the jaw the contra lateral arm lost its strength. This loss varied linearly with respect to the thickness of the spacer and it stopped when the thickness of the spacer reached about 2mm.

V. Van Toi et al. (Eds.): 4th International Conference on Biomedical Engineering in Vietnam, IFMBE Proceedings 40, pp. 217–221, 2013.
www.springerlink.com

In literature, the quantification of the jaw healthy and pain muscle activities during different jaw movements (opening, chewing, closing, etc.) were performed [5-8]. These studies showed a strong influence of the pain jaw muscles on the jaw movement patterns. Moreover, it has been showed that SEMG technique is a good solution to measure and analyze, non-invasively, the jaw muscle activities.

In experiment, Ohtsuki [9] showed a decreasing voluntary isometric strength of the human arm under the condition of simultaneous bilateral exertion of ontralateral corresponding muscles as compared to the unilateral condition. The decrease ratio of strength was 24.6 and 18.8% for the right and left for extension and 6.3 and 7.6% for the right and left for flexion. Integrated EMG showed the tendency comparable to the strength and high correlation was found between strength and integrated EMG. Recently, SEMG was used to predict the applied forces on human hands [10]. An artificial neural network was developed using frequency content of the SEMG signals. The investigators found a significant performance of the proposed model for predicting the loading forces based on the spectral features of the EMG signals.

According to the best of our knowledge, there are no studies related directly to the quantification of muscle activities in relation to the bite and the anisometric arms strength by using EMG technique. In this study, EMG was used to determine the EMG-Force relationship of the jaw, neck and arms muscles. This gave us useful insights on the arms strength loss due to the biomechanical effects of the imbalance in the jaw mechanism.

II. Materials and Methods

Three male subjects (average age: 20.67±2.89 years old) participated in this investigation. All participants were healthy and showed no musculoskeletal or neurological restrictions or diseases. They had a complete dentition, i.e. no premolar or molar was missing in a quadrant. There were no signs of severe malocclusions or facial malformations. All subjects were not on any medication and had no complain of any kinds of muscle pains throughout the experiments.

Before the experiments, each subject filled out a questionnaire, which was kept confidential and included patient's identification, age and gender. The tenets of the Declaration of Helsinki were followed; local Institutional Review Board approved the study and informed consent was obtained for all subjects.

The goal of the experiments was to determine the relationship between the bite and arm strength based on EMG signal. The subject were asked to stand with their arms and legs extended in frontal plan in such a way that the ratio subject intra-feet distance/ subject height equal 0.25. This ratio was chosen based on the fact that many subjects felt comfortable and stable. The experimental protocol started with asking the subject hold the handle of a load which was at rest. Next at the warning signal the load was released at once; this generated a pull-down force on the subject's arm. The subject must hold the load for a maximum of ten seconds without moving his body. The pull-down force consisted of a load in kilogram that can be adjusted in the increment of 100g and then converted into Newton. In these paradigms, the subject was asked to bite on a spacer using his premolar and molar teeth. The spacers had different heights of 2mm and 3mm and were applied in random order. The spacers were made out of a firm material to assume a good occlusal support. While the subject was biting on a spacer pull -down force was applied on the subject's arm as aforementioned.

Myoelectric signals of jaw (masseter), neck (trapezoid) and arms (deltoid, brachiodiolis,) muscles were measured. The masseter was selected due to its main active role during biting at the temporomandibular (TM) joint. The concentric and eccentric contractions of the masseter create the biting motions as well as control the bite's force. The trapezoid was selected due to its essential loading contribution in supporting the arm during persistent extension position. Arm extensors such as deltoid, brachiodiolis or extensor digitorum muscles were selected according to their role of support for maintaining the arm in extension under external loading conditions.

Paradigm 1: From the initial position (0 mm, 5 kg), external loading weight was added incrementally (6kg, 7kg) until the subject can hold the weight in 10 seconds. Check EMG signal quality and data storage

Paradigm 2: Perform variation in spacer thickness from 2mm to 3mm and execute step 1 for each variation.

Higuchi is an efficient algorithm for measuring the fractal dimension of discrete time sequences [11]. Higuchi's algorithm calculates the fractal dimension from time series. Higuchi fractal dimension has already been used to analyze the complexity of surface EMG signal of the biceps [12]

Given a one dimensional time series $X = x[1], x[2], \ldots, x[N]$, the algorithm to compute the HFD can be described as follows [11]:

Form k new time series X_k^m is defined

$$X_k^m = \{x[m], x[m+k], x[m+2k], \ldots, x\left[m + \text{int}\left(\frac{N-m}{k}\right) \times k\right]\}$$

by where k and m are integers, and int(•) is the integer part of •. k indicates the discrete time interval between points, whereas $m = 1, 2, \ldots, k$ represents the initial time value.

The length of each new time series can be defined as follows:

$$\frac{\left(\sum_{i=1}^{\text{int}((N-m)/k)} \left| x[m+ik] - x[m+(i-1)\times k] \right| \right) [(N-1)/(\text{int}((N-m)/k)]}{k}$$

where N is length of the original time series X and $(N-1)/\{\text{int}\ [(N-m)/k] \times k\}$ is a normalization factor.

Then, the length of the curve for the time interval k is defined as the average of the k values $L(m,k)$, for m= 1, 2, …, k: $L(k) = \frac{1}{k} \times \sum_{m=1}^{k} L(m,k)$

Finally, when $L(k)$ is plotted against $1/k$ on a double logarithmic scale, with k = 1, 2, …, k_{max}, the data is expected to fall on a straight line, with a slope equals to the fractal of X. Thus, Higuchi fractal dimension is defined as the slope of the line that fits the pairs $\{\ln[L(k)], \ln(1/k)\}$ in a least-squares sense. A value of k_{max} = 10 was chosen for our study.

III. Results

Surface EMG was successfully measured from four kinds of muscles during different tasks in ten seconds .There was an increase in the activity of the trapezoid and deltoid muscles according to the time when the pull down force applied on the subject was 50 N (Figure 1). The fractal dimension calculated for surface EMG of deltoid and trapezoid muscles were highest at ninth second. Moreover, with higher pull - down force (63N) applied on the subject's arm, there was a rise activity of all four muscles (Figure 2). When the subjects clenched their teeth to carry the heavy weight, it led to the movement in the masseter muscle. And the branchiodiolis muscle also took place in the activity, therefore the fractal dimension of the EMG signal from this branchiodiolis muscle also increased. The deltoid and trapezoid muscles had a significant amount of activity in the arm.

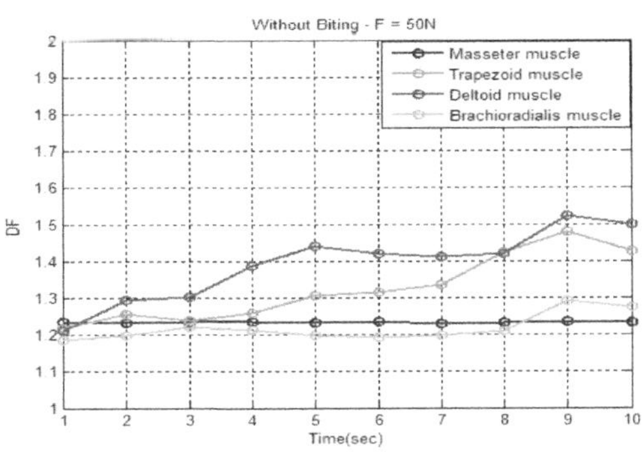

Fig. 1 The activity of four muscles when 50N pull- down force was applied on the subject's arm.

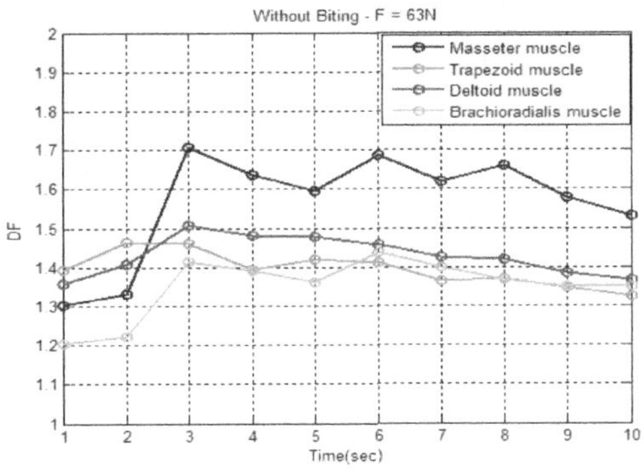

Fig. 2 The activity of four muscles when 63N pull- down force was applied on the subject's arm

The activity of the trapezoid and deltoid muscles varied according to the applied force (Figure 3 & 4). The fractal dimension of these muscles increased slightly during ten second activity until it reached the saturation point with the small applied force. With a bigger applied force, the muscles strongly contracted in respect to the fractal dimension; it reached the saturation point in the very beginning of the ten seconds. For 50N, fractal dimension reached the saturation point at ninth second, for 60N it was fifth second and for 63N it second.

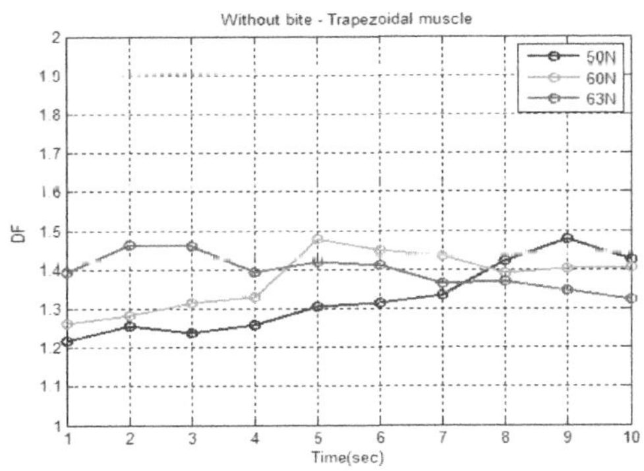

Fig. 3 The activity of the trapezoid muscle in paradigm 1.

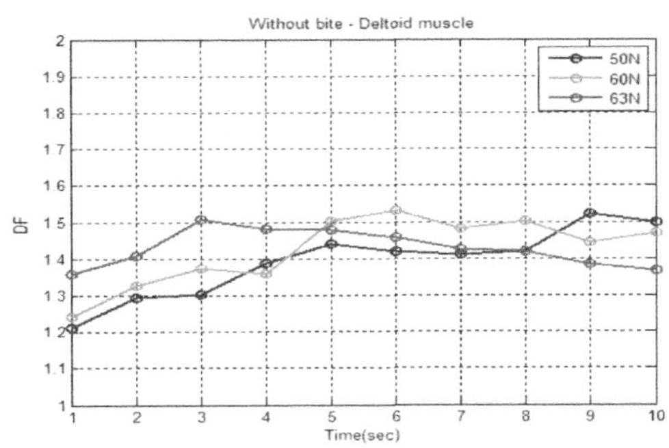

Fig. 4 The activity of the deltoid muscle in paradigm 1.

In the second paradigm, when the subject bit on the variation of thickness spacer, there was a change in the activity of these four muscles. In particular, the masseter muscle contracted in different way according to the thickness variation of the spacer (Figure 5). With the increasing thickness, the masseter muscle contracted strongly at the first three seconds until it reached the saturation point. The subject bit on the spacer at the constant force during the period of time so the fractal dimension was almost stable.

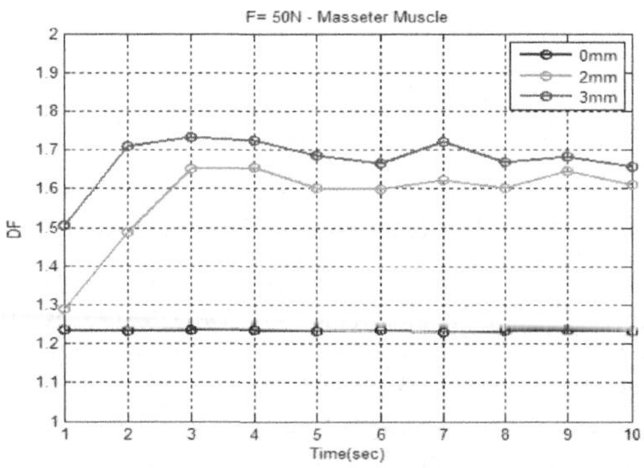

Fig. 5 The activity of the masster muscle during biting on the variation of spacer thickness.

There was a difference in the activity of the trapezoid muscle before and after the jaw imbalance (Figure 6).As our knowledge, imbalance in the jaw can cause loss in arm strength. In this situation, the fractal dimension of surface EMG of the trapezoid muscle has demonstrated the following: before biting, the fractal dimension of EMG signal of the trapezoid muscle increased slightly from first to ninth second where it reached the saturation point. But after biting, once the imbalance of the jaw occurred, the fractal dimension started at a higher level and quickly reached the saturation point even though the same pull - down force was applied on his arm. The fractal dimension of the trapezoid muscle slightly went down as a result of muscle fatigue.

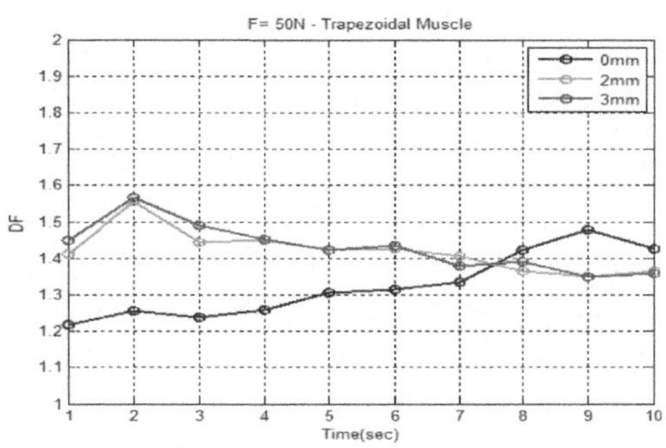

Fig. 6 The activity of the trapezoid muscle during biting on the variation of spacer thickness.

The deltoid muscle also behaved similarity to the trapezoid muscle (Figure 7). The fractal dimension was successfully demonstrated in the loss of arm strength due to the jaw imbalance.

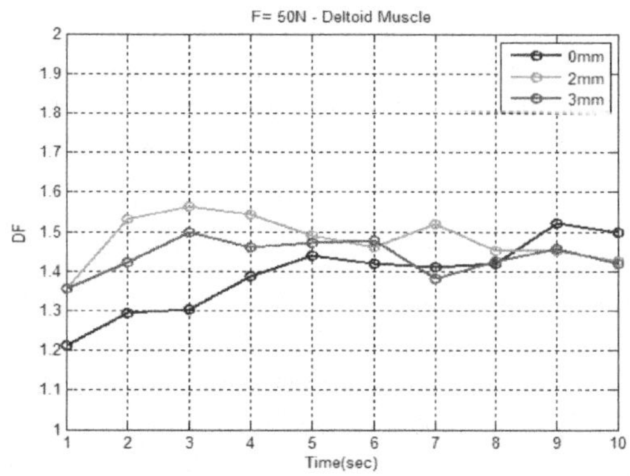

Fig. 7 The activity of the deltoid muscle during biting on the variation of spacer thickness.

The activity of the branchiodiolis muscle was arised respectively to the increasing of spacer thickness (Figure 8). It displayed similar activity as the masseter muscle, increasing at the very beginning and quickly reaching the saturation point.

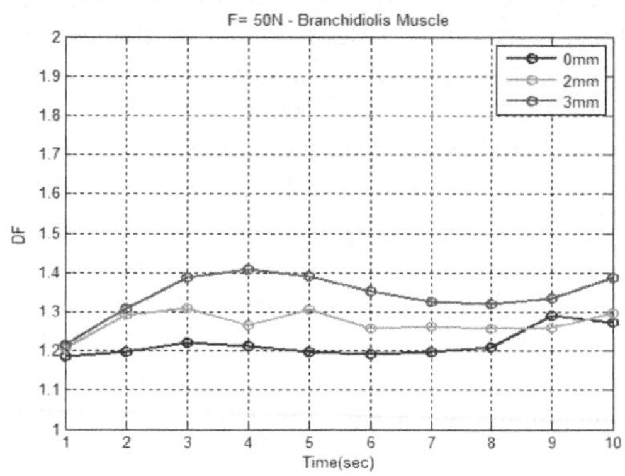

Fig. 8 The activity of branchiodiolis muscle during biting on the variation of spacer thickness.

IV. CONCLUSIONS

The fractal dimension of EMG signal of the masseter, trapezoid, deltoid and branchiodiolis muscles has successfully demonstrated the activity of jaw and arm muscles during the loss of arm strength due to the jaw imbalance. The activities of arm muscles varied according to the applied pull -down force. Muscles contracted with increasing applied force. In particular, the masseter, trapezoid and deltoid muscles displayed the greatest amount of activity in this study. There was also a significant difference in the fractal dimension values of the surface EMG signal before and after the loss of arm strength.

Acknowledgment

We would like to thank Vietnam National Foundation for Science and Technology Development-NAFOSTED for supporting research grant No. 106.99-2010.11. Furthermore research was partly supported by a research fund from the Vietnam National University in Ho Chi Minh City. We also would like to thank Associate Professor, Doctor Nguyen Thi Doan Huong, Ho Chi Minh City Medicine and Pharmacy University for her valuable information about human anatomy and physiology. Last but not least, we are deeply grateful for the support we have been receiving from our volunteers, families and friends.

References

1. Linderholm H, Lindqvist B, Ringqvist M, Wennström A. (1971) Isometric bite force and its relation to body build and general muscle force. Acta Odont Scand; 29: 563–568.
2. Raadsheer M.C., van Eijden T.M.G.J., van Ginkel F.C., Prahl-Andersen B. (2004) Human jaw muscle strength and size in relation to limb muscle strength and size. Eur J Oral Sci; 112: 398–405.
3. Chakfa. A.M., Mehta. N.R., Forgione A.G., et al (2002). The effect of step-wise increases in vertical dimension of occlusion on isometricstrength of cervical flexors and deltoid muscles in nonsymptomatic females. Cranio; 20(4):264-273
4. Hoa LM, Huan DN, Thao NH, Hoan NT, Khoa TQ, Tam NH and Vo Van Toi (2010): Relationship between Dental Occlusion and Arm Strength, International Federation of Medical and Biological Engineering (IFMBE) Proceedings Series, Springer, Vol. 27, p. 265-268
5. Lee. D.S., Wanigaratne. K., Whittle. T., Peck. C.C and Murray. G.M (2006). A method for studying jaw muscle activity during standardized jaw movements under experimental jaw muscle pain. Journal of Neuroscience Methods; 157: 285–293.
6. Wang. K., Lobbezoo. F., Svensson. P and Nielsen. L. A (2007). Influence of jaw gape on EMG of jaw muscles and jaw-stretch reflexes. Archives of Oral Biology; 52: 562-570.
7. Uchida. S., Iwasaki. L. R., Marx. D.B.,Yotsui. Y., Inoue. H and Nickel J.C (2008). Variations in activities of human jaw muscles depend on tooth-tipping moments. Archives of Oral Biology; 53:199–205.
8. Hansen. L.B., Hara. S., Marumo. Y., Miles T and Svensson. P (2009). Effect of experimental pain on EMG-activity in human jaw-closing muscles in different jaw positions. Archives of Oral Biology; 54: 32–39.
9. Ohtsuki.T(1983). Decrease in human voluntary isometric arm strength induced by simultaneous bilateral exertion. Behavioural Brain Research; 7(2): 165-178.
10. Arslan. Y.Z., Adli. M.A., Akan.A and Baslo. M.B (2010). Prediction of externally applied forces to human hands using frequency content of surface EMG signals. Computer Methods and Programs in Biomedicine; 98: 36–44.
11. Higuchi T (1988). Approach to an irregular time series on the basis of the fractal theory. Physica D; 31 :277-83.
12. Vineet .G., Srikanth .S ., Narender.P.R. (1997). Fractal analysis of surface EMG signals from the biceps. International Journal of Medical Informatics ; 45 : 185- 92.

Author: Truong Quang Dang Khoa
Institute: Biomedical Engineering Department – Ho Chi Minh City International University
Street: Quarter 6, Linh Trung, Thu Duc Dist.
City: Ho Chi Minh City
Country: Viet Nam
Email: tqdkhoa@hcmiu.edu.vn

A Portable Respiratory Monitor Using Respiratory Inductive Plethysmography

H.T. Ngo[1,2], C.V. Nguyen[1], T.M.H. Nguyen[1], and Toi Van Vo[1]

[1] Biomedical Engineering Department, International University, Quarter 6, Linh Trung, Thu Duc Dist., Ho Chi Minh City, Vietnam
[2] Biomedical Engineering Department, Pratt School of Engineering, Duke University, Durham, North Carolina, USA

Abstract — **Monitoring respiratory activity provides important information about patient's well-being. There are many respiratory activity monitoring techniques introduced so far. This paper presents a portable respiratory monitor using Respiratory Inductive Plethysmography (RIP) technology. RIP sensor's characteristics are also studied. Experimental results have shown that RIP sensor's sensitivity and linearity can be improved by designing the RIP sensor properly. In addition, the monitor's accuracy in term of respiratory rate measurement is verified by experiment on 10 subjects. Experimental results have shown that our monitor can measure respiratory rate with error ±1 bpm. Furthermore, it can transfer data to PC (via Bluetooth) where chest and abdomen movement efforts are plotted. These plots are useful for diagnosing and monitoring several respiratory disorders.**

Keywords — **Respiratory Monitor, Respiratory Inductive Plethysmography.**

I. Introduction

From a clinical point of view, 'respiratory activity' includes such descriptors as respiratory rate and depth, as well as quantifiable information about the degree of gas exchange actually taking place. Continuous monitoring of respiratory activity is important for identifying or predicting high-risk situations, making appropriate monitoring techniques potentially life-saving [1]. Many respiratory monitoring devices and methods have been introduced so far. They are classified into three categories which are listed in Table 1.

Despite the large number of reported methods and devices for respiratory monitoring, convincing evidence of their clinical usefulness is still lacking. Inherent limitations regarding the detection principles, non-specific response, poor adaptation to clinical conditions and difficulties in clinical evaluation are identified as the primary causes [1]. Therefore, developing a new respiratory monitor is needed.

In this work, Respiratory Inductance Plethysmography (RIP) technology is used. There are hundreds of publications on RIP. Investigators suggest that RIP may be less prone to some of the artifacts caused by cardiac activity, motion of the subject or sensor and changes in external variables such as temperature and sound [2]. This technology is also recommended by American Academy of Sleep Medicine for measuring respiratory effort [3]. Nowadays, RIP is widely used at sleep centers across the world [4].

The principle of RIP technology, its qualities and applications have been introduced in [3], but simply put, the RIP sensor is an inductive coil made of insulated wires woven or sewn in a sine wave or zig-zag pattern on an expandable belt. One belt is placed around the rib cage and the other around the abdomen. During respiration, the length of the belts change and, therefore, self-inductance of the coils change proportionally. By measuring variation in the coils' self-inductance, we can detect variation of rib cage's and abdomen's circumference or cross-sectional area. An important quality of RIP is linearity. In addition, because there is no electrical current passing through the patient and only a weak magnetic field is created, this technology is safe. Some RIP applications include distinguishing central apnea from obstructive apnea during sleep studies, detecting paradoxical breathing or slight phase shifts, and measuring the actual volume of airflow to create a "flow-volume loop" [3].

However, we found no detailed technical description of RIP sensor. The question is that: how pattern of wire woven or sewn along the belt and its parameters affect RIP sensor characteristics? In this work, we investigate effect of one parameter, step size, of sine wave pattern to sensitivity and linearity of RIP sensor. Sine wave pattern is chosen due to its proven effectiveness and ease of construction. In addition, we present a complete design of a respiratory monitor that is accurate, portable, inexpensive, and simple to use. Because RIP sensor is integrated into belt, the term sensor and belt are used interchangeably in this paper.

V. Van Toi et al. (Eds.): 4th International Conference on Biomedical Engineering in Vietnam, IFMBE Proceedings 40, pp. 222–225, 2013.
www.springerlink.com

Table 1 Categories of sensing principles for respiratory monitoring devices and methods (table from [1]).

Category	Typical measured quantity	Typical sensor position
Movement, volume and tissue composition detection	Electromyography, abdomen and thoracic circumference, impedance or blood volume	Abdomen and chest wall
Airflow sensing	Respiratory gas flow	Nasal/oral area
Blood gas measurement	Arterial gas concentration	Peripheral organ or Nasal/oral area

II. EFFECTS OF SINE WAVE'S STEP SIZE TO RIP SENSOR CHARACTERISTICS

To investigate effect of step size of sine wave pattern to characteristics of RIP sensor, we construct three RIP sensors with the same pattern of wire (sine wave), height (2.2 cm), and length (30 cm) but different step sizes as illustrated in Fig. 1 and Fig. 2.

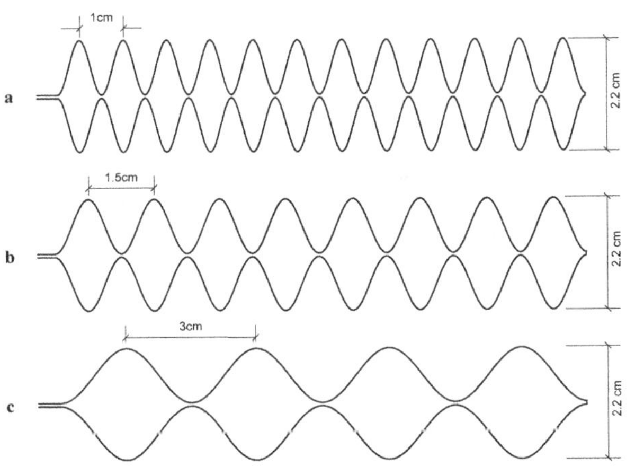

Fig. 1 Three RIP sensors with the same sine wave pattern of wire, height, and length but different step sizes: 1 cm (a), 1.5 cm (b), 3 cm (c).

Fig. 2 Three RIP sensors depicted in Fig. 1 are constructed.

We then measure inductance of each sensor versus stretched length of the sensor. The Agilent 4285A Precision LCR Meter is used. The measured data are plot in Fig. 3, Fig. 4, and Fig. 5. In these figures, in addition to plotting measured data scatterly, trend-line based on linear regression is also plotted. Slope of the trend-line gives a measure of sensitivity of sensor. R-squared value gives a measure of linearity of sensor (values close to 1 indicate excellent linearity). Sensitivity and linearity of sensors with different step sizes are depicted in Fig. 6. The figure clearly shows that the smaller step size, the higher sensitivity. In term of linearity, there is slightly difference between 1 cm and 1.5 cm step size. However, for 3 cm step size, linearity decreases remarkably. Therefore, step size should be kept small to achieve good linearity.

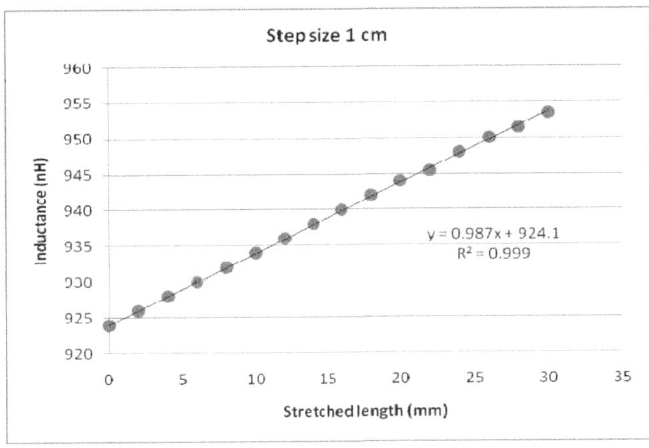

Fig. 3 Inductance vs. stretched length of RIP sensor with step size 1 cm.

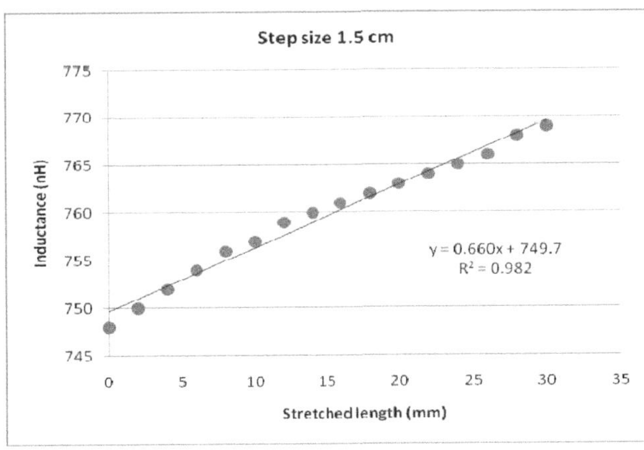

Fig. 4 Inductance vs. stretched length of RIP sensor with step size 1.5 cm.

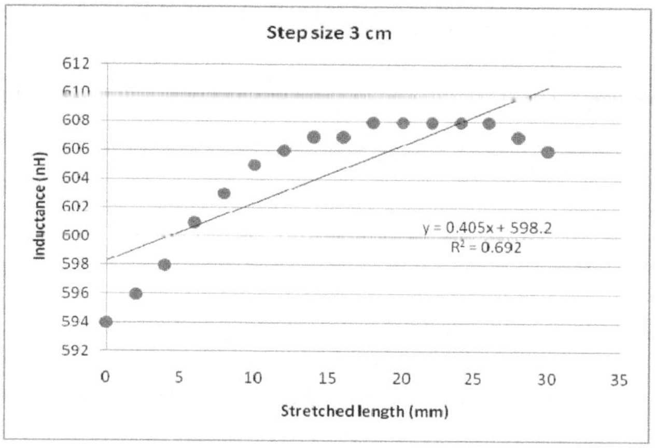

Fig. 5 Inductance vs. stretched length of RIP sensor with step size 3 cm.

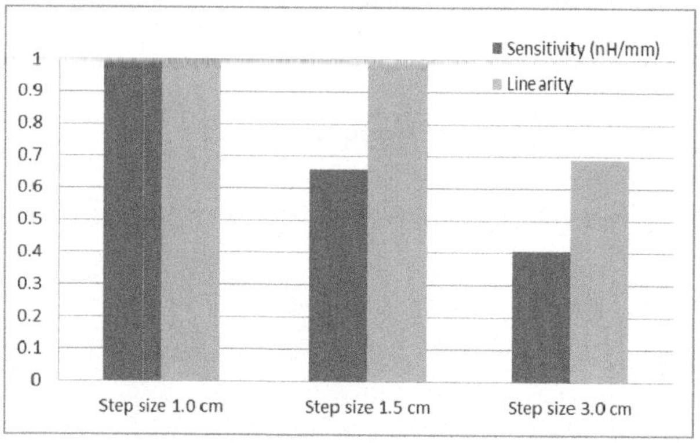

Fig. 6 Sensitivity and linearity of RIP sensors with different step sizes.

III. DESIGN OF A PORTABLE RESPIRATORY MONITOR

A portable respiratory monitor has been built. Two RIP belts are worn around chest and abdomen of subject. The belts are connected to Colpitts oscillators (Fig. 7). Variation in chest's and abdomen's cross-sectional area results in variation of inductance of the belts and therefore of frequency of the oscillators' sinusoidal output. An FM demodulation circuit using Phase-locked loop is used to convert the variation in frequency into voltage. The voltage signal is then filtered and amplified before being fed to a microcontroller with built in 10-bit ADC. Here, the signal is digitized and is transmitted to computer wirelessly via Bluetooth module. In addition, an algorithm for the microcontroller has been developed to calculate breathing rate and display on a 16 characters x 2 lines LCD. Whenever breathing rate drops lower than a predefined threshold, an alarm buzzer is turned on. On the computer side, we have developed a LabVIEW program to display breathing rate, plot chest's and abdomen's movement, and Konno-Mead loop (Lissajous). The monitor is powered by a lithium-ion cell phone battery which can be recharged. The monitor is hence portable. We decided to use this type of battery due to its commercial availability, rechargeable ability, lightweight, and high capacity. Total cost of the monitor (Fig. 8) is less than one-hundred USD.

To test the monitor's accuracy in term of respiratory rate measurement, we do experiment with 10 subjects (7 male, 3 female, age from 18 to 21), each subject do experiment two times (Fig. 9). We measure their respiratory rate by two approaches simultaneously; one using our monitor and one asking subjects to count their own respiratory rate. Each experiment lasts one minute. During the experiments subjects seat comfortably without changing body's posture. They were instructed to focus on counting their respiratory rate and not allowed to talk. They were also not allowed to look at the monitor's LCD display and computer's screen. Student T-test shows that there is no statistically significant difference between the two approaches with level of significance 5%. In addition, maximum error between two approaches is ±1 respiratory beat per minute.

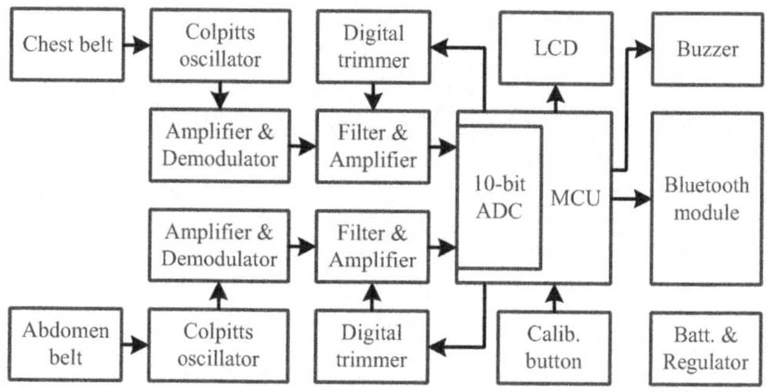

Fig. 7 Portable respiratory monitor's block diagram.

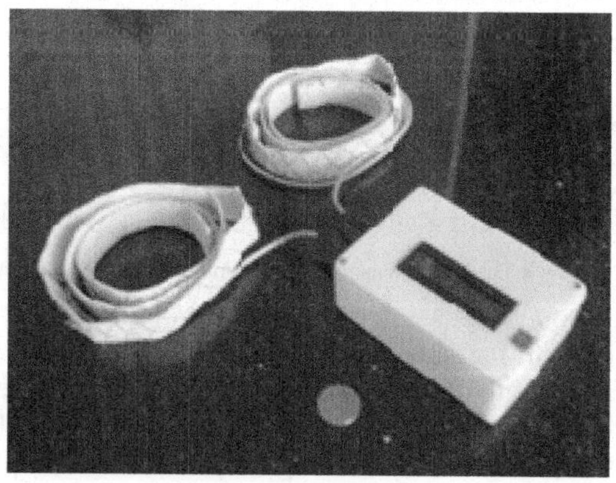

Fig. 8 Portable respiratory monitor with two RIP belts.

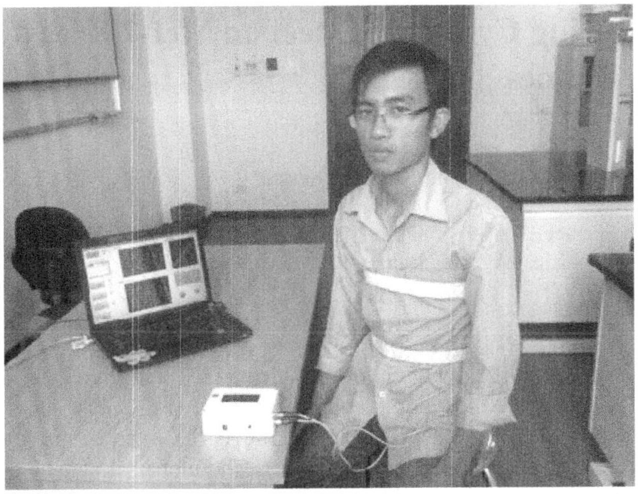

Fig. 9 Our monitor can transmit data to laptop wirelessly via Bluetooth. On laptop side, chest & abdomen movement efforts and Konno-Mead loop are plotted. Breathing rate is displayed on both monitor and laptop.

IV. Conclusions

Two parts of our work are presented. In the first part, we investigate the effect of step size of sine wave pattern to RIP sensor characteristic. Three RIP sensors with different step sizes are built and their inductance are measured. Experimental results have shown that the smaller step size, the better sensitivity and linearity. In the second part, we present a complete design of a portable respiratory monitor. We then verify the monitor's performance in term of respiratory rate measurement by doing experiment on 10 subjects. Experimental results have shown that the presented monitor can measure respiratory rate with high accuracy. Furthermore, the monitor can transfer data to PC (via Bluetooth) where chest & abdomen movement efforts and Konno-Mead loop are plotted. These plots are useful for diagnosing and monitoring several respiratory disorders.

Acknowledgment

The authors are grateful to Associate Professor M.D./Ph.D. Le Thi Tuyet Lan, the University of Medicine Hospital. This study was funded by the International University, VNU-HCM.

References

1. M. Folke, L. Cernerud, M. Ekstrom, B. Hok, Critical review of non-invasive respiratory monitoring in medical care, Medical & Biological Engineering & Computing, Vol. 41, 377-383, 2003.
2. Kevin P Cohen, Dorin Panescu, John H Booske, John G Webster and Willis J Tompkins, Design of an inductive plethysmograph for ventilation measurement, Physiol. Meas. 15, 217-229, 1994.
3. J. Scott Cardozo, New AASM Recommendations for Sensors: A Simple Guide for the Sleep Technologist, Sleep Diagnosis and Therapy, Vol 3 No 5, September-October, 2008.
4. Jahan Naghshin, Patrick J. Strollo, Respiratory Monitoring Equipment and Detection of Respiratory Events, Sleep Med Clin 4, 353–360, 2009.

Author: H. T. Ngo
Institute: Duke University
Street:
City: Durham, NC 27708
Country: USA
Email: hoan.ngo@duke.edu

Nanoparticle-Encapsulated Tamoxifen Inducing Cytotoxic Effect on Mcf-7 Breast Cancer Cell Lines

A.M. Akim[1], E.E. Tung[1], P.P. Chong[1], M.Y. Hamzah[2], and K.Z.M. Dahlan[2]

[1] Department of Biomedical Science, Faculty of Medicine and Health Sciences, University Putra Malaysia, 43400, Selangor, Malaysia
[2] General Research and Technology Development, Malaysian Nuclear Agency, Bangi 43000 Selangor, Malaysia

Abstract — **Nanoparticle in drug delivery offers major improvements to those limitations by providing longer circulation times, greater specificity, and enhances anticancer effects. Thus, the objective of this study is to synthesize a polymeric nanoparticle encapsulating TMX, and investigate the *in vitro* characteristics on breast cancer cell line, MCF-7.** Materials and Methods. **The nanoparticles were synthesized through gamma irradiation polymerization between N-isopropylacrylamide (NIPAAM), N-vinyl-2-pyrrolidone (VP), poly(ethyleneglycol) diacrylate (PEG-DA). TMX was directly loaded into the nanoparticles by random physical entrapment. Photon Cross Correlation Spectroscopy Nanophox (PCCS-Nanophox) and Transmission Electron Microscopcy (TEM) were used to observe size and morphology of the nanoparticles. To study the *in vitro* effects, MCF-7 cells were treated with TMX-nanoparticles and its cytotoxicity effects were investigated through MTT assay.** Results and Discussion. **From the results, it was confirmed that nanoparticles at 0.055M subjected to 8kGy of gamma irradiation posses a consistent size distribution of 49.89 nm SD ±1.82. This was further confirmed through pictures taken with TEM in which the nanoparticles had a clearly distinguishable core-shell structure, with a spherical or elliptical shape with smooth surface. Through the cytotoxicity assay, it showed that TMX-nanoparticles had a greater inhibitory effect compared to TMX alone at 48 hrs and 72 hrs. There was also no evidence of toxicity by void polymeric nanoparticles.**

Keywords — **Polymeric nanoparticle, gamma irradiation, Tamoxifen, TMX-nanoparticles, MCF-7, breast cancer.**

I. INTRODUCTION

Many new materials and devices were created in a wide range of applications. One of the emerging strategies that have been receiving a lot of attention is the use of nanoparticle systems in the treatment of cancer. Through prior studies, it was suggested that these nanoparticles in drug delivery offers major improvements to those limitations by providing longer circulation times, greater specificity, and enhances anticancer effects[1][2]. Hence, nanoparticles conjugated anticancer drugs can be a promising and potential strategy in the treatment of breast cancer. Tamoxifen (TMX) is one of the most common anti-cancer drug in breast cancer treatment. On breast cancer tissue, it acts as an anti-estrogen. However, upon other tissue such as the uterus, it acts as an estrogen [3]. Other side effects include liver cancer, blood clotting, retinopathy and corneal opacities which is dependent to be dose dependent. This suggests the use of a lower dose of TMX and a better delivery system to enhance its efficiency in breast cancer treatment.

II. MATERIALS AND METHODS

A. Reagent

NIPAAM, VP, PEG-DA, 3-[4,5-dimethylthiazol-2-yl]-2,5-diphenul tetrazolium bromide] (MTT) and tamoxifen were purchased from Sigma, USA.

B. Cell Line

MCF-7 breast cancer cell line was purchased from American Type Culture Collection (ATCC), Rockville, Maryland, USA. The cell lines were maintained in DMEM 4.5g/l Glucose culture medium supplemented with 10% FBS and 1% of antibiotics (100 IU/ml penicillin and 100 µg/ml streptomycin). The cells were grown in tissue culture flasks in the incubator at humidified atmosphere containing 5% CO_2 and 95% air, at 37°C.

C. Synthesis of Polymeric Nanoparticles

Polymeric nanoparticles were prepared from NIPAAM, VP and PEG-DA at 90:5:5 molar ratios ranging from 0.01M-0.08M. The samples were subjected to gamma irradiation at doses ranging from 2kGy to 50kGy. The size and stability of the nanoparticles were detected through Photon Cross Correlation Spectroscopy Nanophox (PCCS-Nanophox).

V. Van Toi et al. (Eds.): 4th International Conference on Biomedical Engineering in Vietnam, IFMBE Proceedings 40, pp. 226–228, 2013.
www.springerlink.com

D. Transmission Eletron Microscopy (TEM)

TEM pictures of the nanoparticles sample were taken under JEM 2100 TEM instrument operating at magnification of 100kV. A drop of aqueous solution of the sample was placed on a membrane grid surface. Excess fluid was removed using a filter paper. The grid surface was air dried at room temperature before loaded into the microscope.

E. Loading of Drug into Nanoparticles

TMX were loaded in the polymeric nanoparticles by using a post-polymerization method where TMX is directly loaded into the nanoparticles by random physical entrapment through constant vortexing or mild sonication.

F. Cell Viability Assays

Growth inhibition of cells was measured through MTT assay, which relies on the conversion of MTT, 3-[4,5-dimethylthiazol-2-yl]-2,5-diphenul tetrazolium bromide], to a purple coloured formazan product by the activity of living cells. 1 x10^5 cells/ml of MCF-7 cells were plated in 96 well plates. The cells were treated with void nanoparticles, TMX and TMX-nanoparticles at concentrations ranging from 0 to 60 µM for 24, 48 and 72 hours. MTT assay was performed by adding 50ul of MTT (2mg/ml) into each well. After 4 hours incubation, the treatment were removed and 100ul of dimentyl sulphoxide (DMSO) were added to solubilize the purple formazen formed. The readings were measured by microplate reader. From the results, the relative growth inhibition was determined.

III. RESULTS AND DISCUSSIONS

The particle size and stability of nanoparticles measured with PCCS-Nanophox as shown in Fig. 1. At a concentration of 0.055M of NIPAAM-VP-PEGDA at 8kGy of gamma irradiation, the nanoparticles that were formed have a very consistent size with average of 49.89nm SD± 1.82. This is further confirmed through TEM of the nanoparticles is illustrated in Fig. 2. From the picture, it was demonstrated that the nanoparticles synthesized have a spherical morphology, with clear distinguishable core-shell structure. It has a smooth surface with an approximate size of 47nm core and 86nm outer shell.

It is important that a drug delivery platform must be biodegradable, biocompatible and non toxic towards cells. Therefore, the toxicity of void nanoparticles was also studied. From the results showed in Fig. 3, 4 and 5 there was no evidence of toxicity associated with the void nanoparticles across the dose range. On the other hand, both TMX and TMX-nanoparticles effectively inhibited the growth of MCF-7 cells when compared to untreated cells and cells exposed to void nanoparticles. At 48 hours and 72 hours, the effect of TMX-nanoparticles was more pronounced than TMX alone, giving a lower IC50

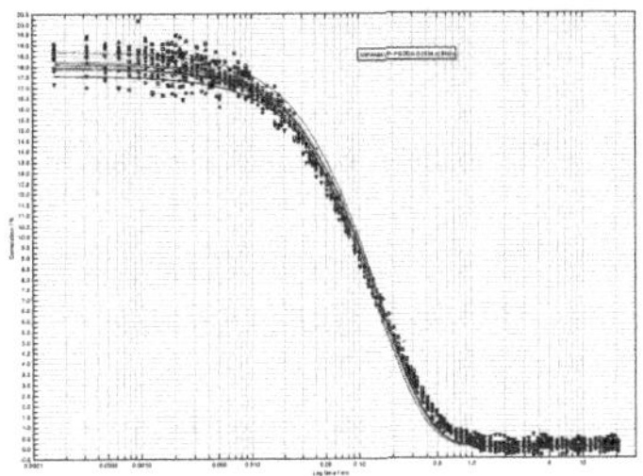

Fig. 1 Size particles and stability graph of void nanoparticles from Photon Cross Correlation Spectroscopy-Nanophox.

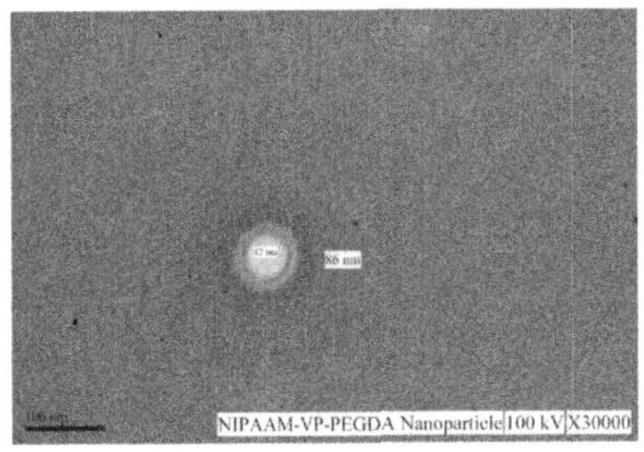

Fig. 2 Picture of void nanoparticle taken from Tranmission Electron Microscope.

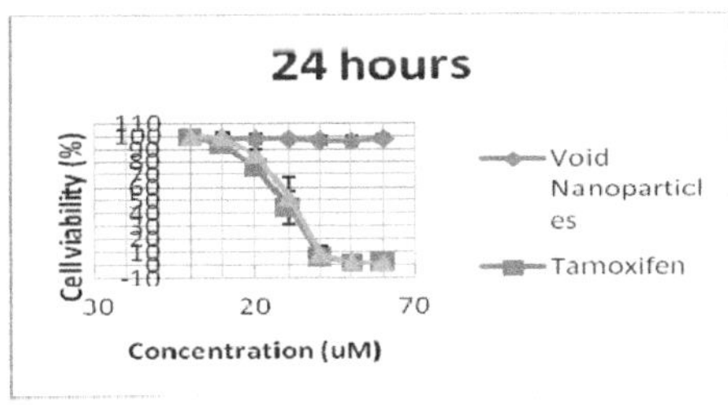

Fig. 3 MTT result of nanoparticle after 24 hours.

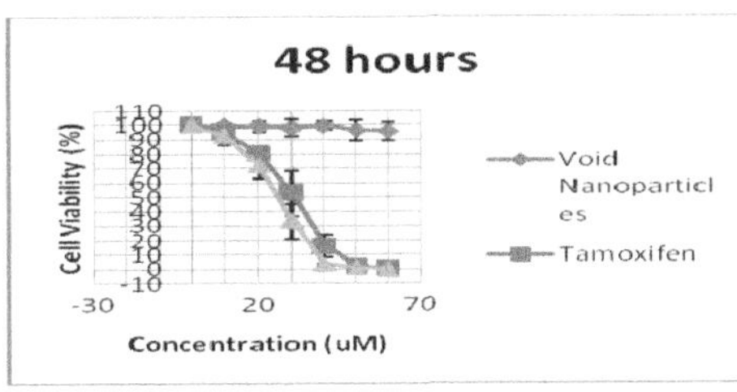

Fig. 4 MTT result of nanoparticle after 48 hours.

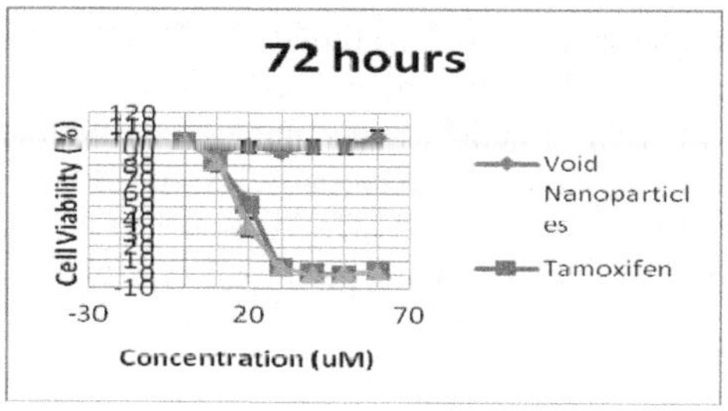

Fig. 5 MTT result of nanoparticle after 72 hours.

IV. Conclusions

The nanoparticles are non-toxic ideal as a drug delivery platform and provide an opportunity for other diseases and drugs.

Acknowledgment

The author would like to acknowledge Universiti Putra Malaysia and Agensi Nuklear Malaysia for the facilities that they provide for this research.

References

1. Byrne, J. D., Betancourt, T., and Peppas, L. B. (2008). Active targeting schemes for nanoparticle systems in cancer therapeutics. Advanced Drug Delivery Review 60, 1615-1626.
2. Wang, A. and Li, S. (2008). Hydroxycamptothecin-loaded nanoparticles enhance target drug delivery and anticancer effect. BMC Biotechnology 8 (46).
3. Cohen, I. 2004. Endometrial pathologies associated with postmenopausal tamoxifen treatment, Gynecologic Oncology 94,256–266.

Author: Abdah Md Akim
Institute: Universiti Putra Malaysia
Street: Faculty of Medicine and Health Sciences
City: Serdang
Country: Malaysia
Email: abdah@medic.upm.edu.my

Proposed Implementable Model of Computer Interfaced Electrocardiogram (ECG) Module

Ijlal Shahrukh Ateeq[1], KamranHameed Muzammil Khan[1], Junaid Khalid[1], and Muhammad Ali[2]

[1] Sir Syed University of Engineering and Technology, Department of Biomedical Engineering, Karachi, Pakistan
[2] Karachi Institute of Radiotherapy and Nuclear Medicine, Department of Biomedical Engineering, Karachi, Pakistan

Abstract — **Electrocardiogram measurement plays a fundamental role for the finding and monitoring of a variety of cardiac problems. In this investigational study we considered three lead computer interface ECG module with PIC Controller, based on the Einthoven's triangle technique. Our developed ECG module operates on bipolar limb lead arrangement and a particular gold plated switch is used for leads selection with the intention to reduce contact noises. LAB WINDOWS software is used to display a real time ECG signal that supports us to evaluate heart rate and PQRST amplitudes. Our investigational outcome illustrates 80% – 90 % precision as compared to conventional ECG machine, therefore it allowing an efficient ECG measurement regardless of artifacts.**

Keywords — **Electrocardiogram, Einthoven Triangle method, bipolar limb lead, Heart rate, PQRST amplitudes.**

I. INTRODUCTION

The heart is composed of cardiac muscle, an involuntary muscle. The signal that formulates the heart's muscle fibers derives from the sinoatrial node, which is the natural pacemaker of the heart [1]. The recording of changes in electric currents formed by the contracting heart muscles is termed as electrocardiogram (ECG) [2], its records are usually shown on a piece graph of paper or a display device and is an essential tool in assisting many problems such as heart's rhythm, diagnosing poor blood flow to the heart muscle (ischemia), diagnosing a heart attack and evaluate certain abnormalities [3]. Hence determination and analysis of true ECG signal is important for a variety of procedures particularly in biological system in which a minor difference can prove to be fatal.

The ECG signal is dependent upon the mechanical movement of the cardiac muscles of the patient whose ECG is to be determined; the ECG wave is consists of five essential points P, Q, R, S and T. These points are grouped together to represent important electrical events, and are represented by 3 discrete waves as shown in "Fig. 1".

The P wave represents atrial depolarization it is typically smooth and positive followed by atrial contraction The QRS complex represents the time it takes for depolarization of the ventricles, represent the period of ventricular muscle contraction before repolarization followed by ventricular expulsion The T wave due to ventricular repolarization The T wave is normally rounded and positive [4].

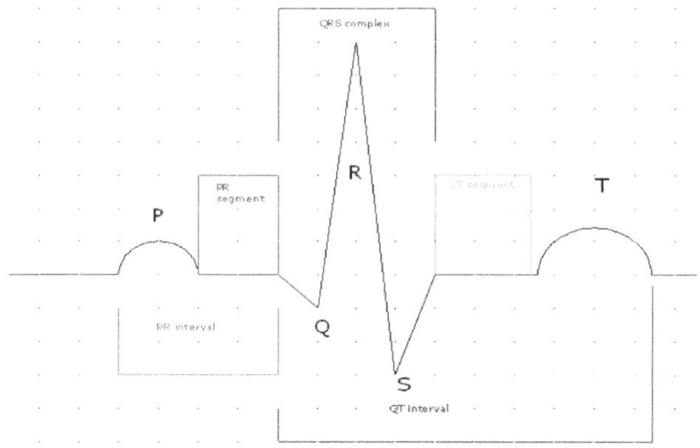

Fig. 1 Standard ECG Wavelet Representation.

II. METHODOLOGY

An ECG signal formed by one cardiac cycle, is consists of 1mv. To the clinician the shape and duration of each feature of the ECG are significant. The waveform, however, depends greatly upon the lead configuration used; here the method of three lead techniques is used. This method uses the three limbs leads I, II, III based on Einthoven triangle which is shows in "Fig. 2" and it determine the alliance of the QRS [5]. In general the cardiologist looks critically at the various time intervals, polarities, and amplitudes to arrive at his diagnosis [6].

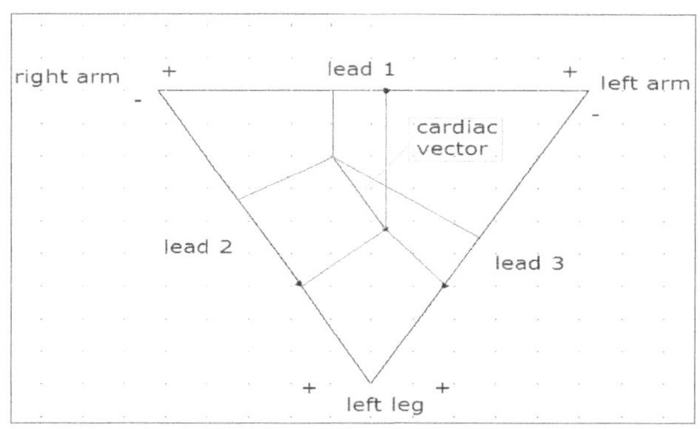

Fig. 2 Representation of Einthoven triangle.

V. Van Toi et al. (Eds.): 4th International Conference on Biomedical Engineering in Vietnam, IFMBE Proceedings 40, pp. 229–232, 2013.
www.springerlink.com

The method proposed in this research to compensate for noise variation includes an instrumentation amplifier, filters and a microcontroller with a software algorithm to analyze, calculate and display an ECG wavelet.

A. Hardware

The proposed module of ECG device is shown in Fig. 3. The hardware includes an ECG cable consists of three leads made up of stainless steel. The three leads of ECG are connected to the lead selector switch, signal from the ECG lead is pre-amplified through INA114, it is an instrumentation amplifier having a high common mode rejection ratio and also provide a flexibility to set the gain [7], the preamplifier is followed by the isolation amplifier PS2506, it is used to isolate the patient body from the reverse current [8], the isolation amplifier is followed by the band pass filter constructed using F353 IC, the IC is used to obtain the specific frequency for diagnostic ECG from 0.38 to 150 Hz [9], the band pass filter is followed by the notch filter LMF 90, it is used to eliminate the line frequency of 50 Hz [10], the notch filter is followed by the post amplifier and summing circuit using AD822 IC, AD822 is a dual instrumentation amplifier and its first stage is worn for the post amplifier for impedance matching[11].

B. Software

The ECG wave is display in the computer screen by the determination of various enable points and these enable points are also depend upon the machine cycle of the microcontroller. This approach is useful in designing and implementation of ECG software algorithm. The designs metrics of the software is shown in "Fig. 4" coded in Turbo C language and then programmed in the PIC16F877A a 10 bit ADC and then it is interface with the LAB WINDOWS /CVI 8.0 from National Instruments for the final display result [12].The heart rate and PQRST amplitudes of the patient can be calculated from the following relation shown in figure "(1)" and "(2)"

H= [60/(X2-X1)*10ms] sec **(1)**

Where,

H is the heart rate of the patient, the X2 is distance of 2nd wave, X1 is the distance of 1st wave of ECG, 60 is the time period for heart rate/minute, 10ms is the one machine cycle of Microcontroller.

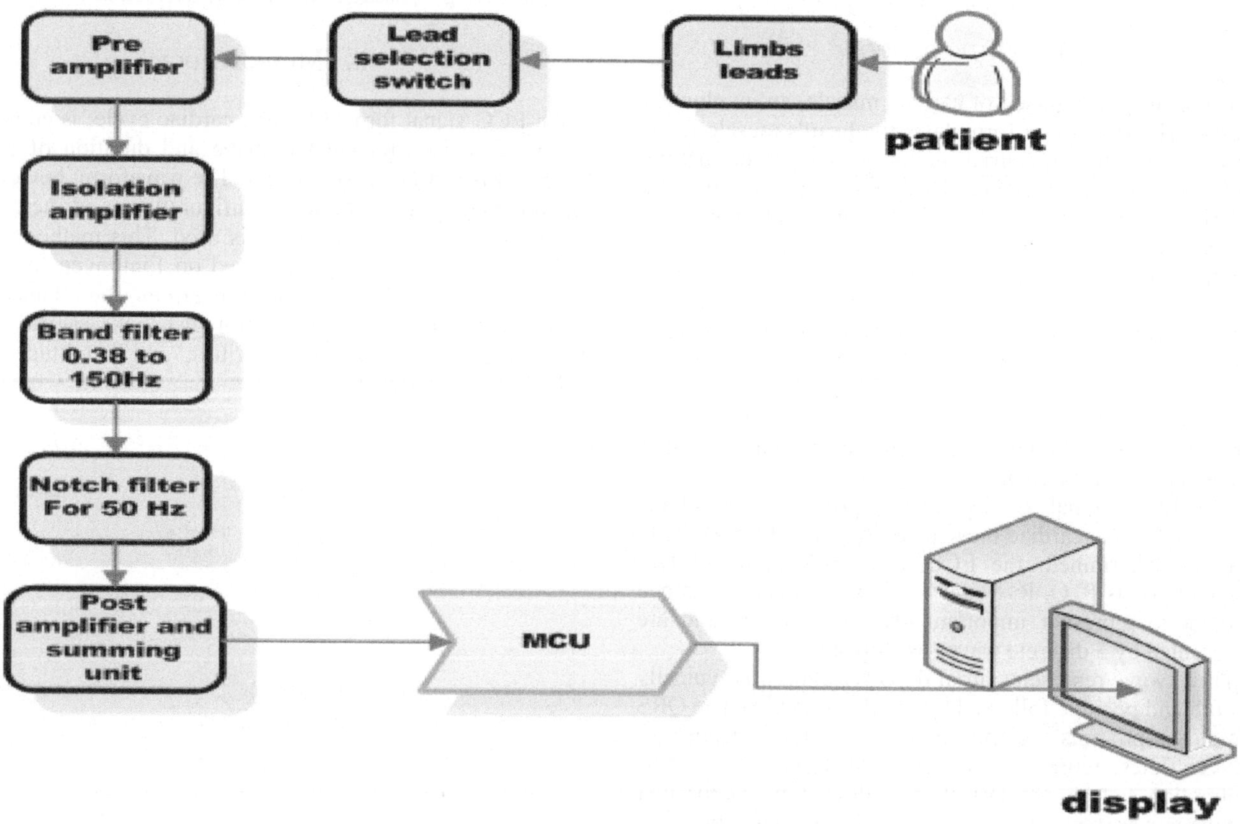

Fig. 3 Block diagram of the hardware.

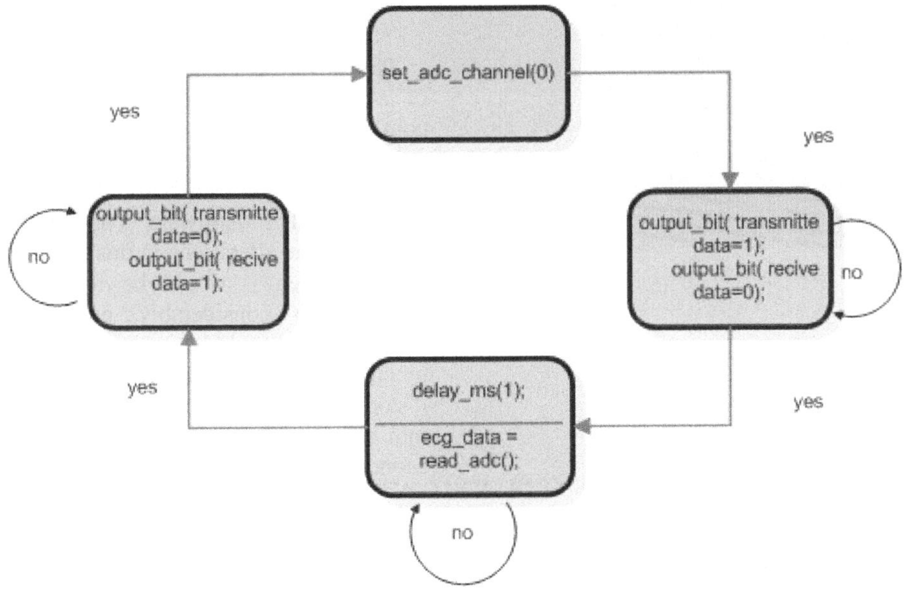

Fig. 4 Proposed metrics of the software.

Y=a*(5/1024) volts **(2)**

Where,

Y is the formula for calculating amplitudes of ECG 1024 is the maximum sampling rate of the ADC, 5 volts is the maximum peak amplitude of the ECG is the peak point of any amplitude of the ECG:

C. Discussion and Results

"Fig. 5," gives a good indication of the effectiveness of the designed circuit. While "Fig. 6," demonstrate the clear assessment of the final signals with the other conventional ECG traces which shows that all traces are acceptable & enhanced. A software algorithm designed on LAB WINDOWS/CVI 8.0 is used with microcontroller PIC16F877A, the microcontroller with the built in ADC eliminates the need of synchronizing micro controller with an external ADC if used separately, as a result increased efficiency, reduced hardware circuitry and cost effectiveness is achieved. In hardware section one percent tolerance resistance is used which provide the 99% efficient and effective response as a result the roll of rate of the filter increases and the gain of preamplifier is stabilized [13], the ground of the circuit is connected with the shielding wire of ECG probe, the connection compensates for motion artifacts [14]. This is because motion artifacts having frequency band from 0.01Hz to around 30Hz will distort the ECG signals. On comparison with standard waveforms it is obtained that the results are 70%-80% accurate. The major flexibility of this work is that it is universal serial bus (USB) compatible with the computer.

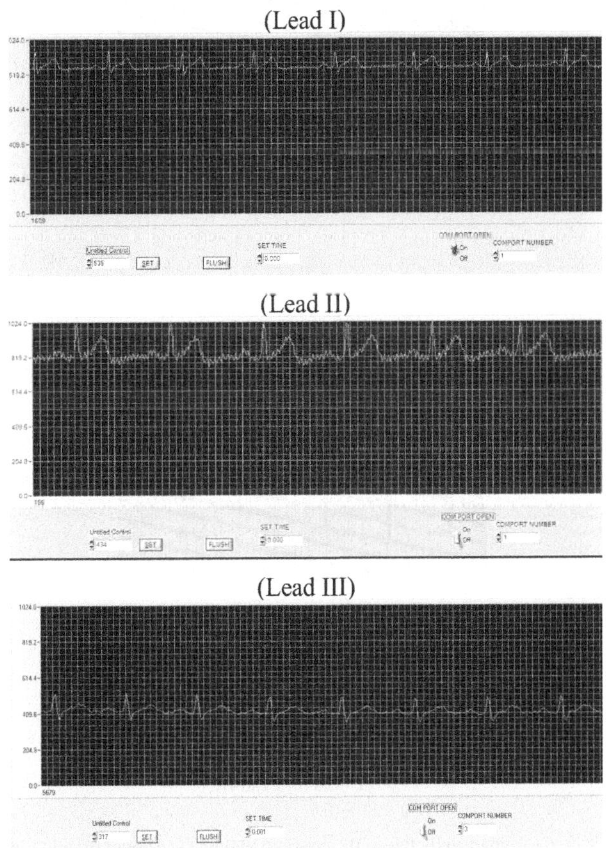

Fig. 5 Interfaced ECG signal of lead I, II & III of running software in LAB WINDOWS/CVI 8.0.

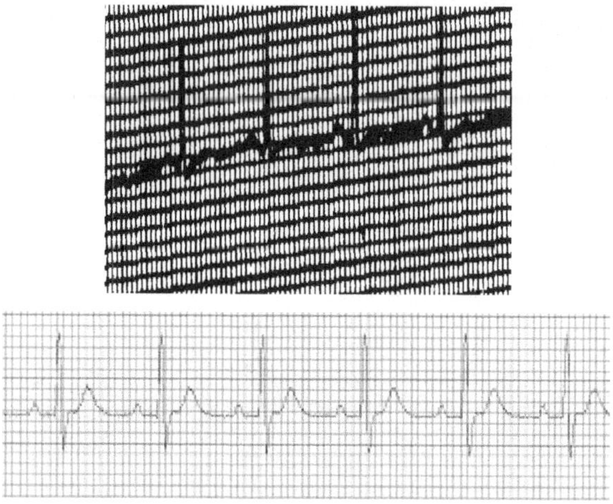

Fig. 6 (a) Lead I.

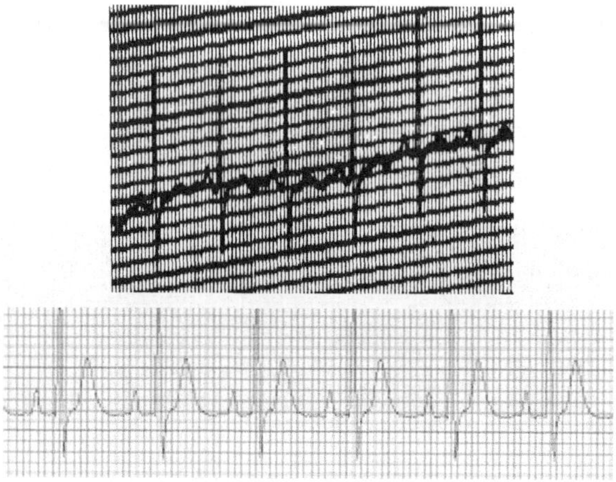

Fig. 6 (b) Lead II.

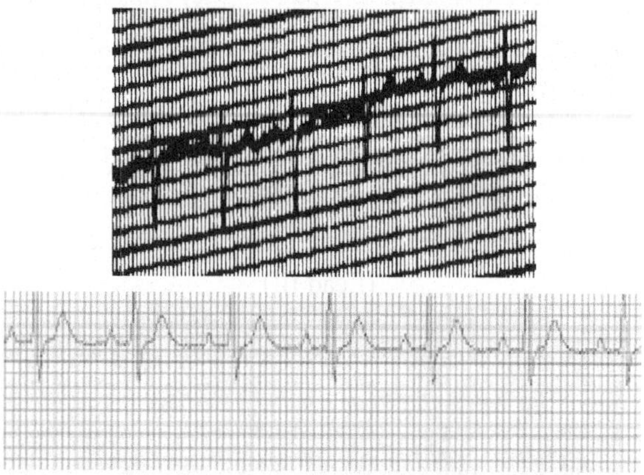

Fig. 6 (c) Lead III.

Fig. 6 Comparison between the conventional ECG (Above) [6] and our designed project (Below) (6a) shows the output of lead I (6b) shows the output of lead II (6c) shows the output of lead III.

III. CONCLUSION

The waveforms discussed above, demonstrate the results of a subject connected in Einthoven Lead I, II & III configurations. As can be viewed good results are attained regardless of the simplicity of the hardware. The research reveals that significant improvements can be achieved with unsophisticated hardware combined with attention to software. The alterations in this case are by no means at the optimal level; it would be feasible for an enthusiastic designer to considerably get better results. Further enhancements could be interfacing of data in a hospital or a much broader global domain through telemedicine domain, if code with other particular characteristics were to be put into practice. An eventual outcome could be a compact economical ECG for a potentially huge volume market.

ACKNOWLEDGMENT

We sincerely acknowledge Engr. Ashfaque Ahmed (Manager Electronics Dept, SUPARCO) for his efforts, time and support on this project.

REFERENCES

1. Van De Graaff, Kent M. Human Anatomy. McGraw Hill Publishing, Burr Ridge, IL.2002, pp.148-151
2. Memmler, R.; Cohen, B. & Wood, D., The Human Body in Health & Disease, Lippincott Williams & Wilkins, 1996
3. Dr Carl J Brandt, GP, Article reviewed at netdoctor website.
4. Francis Musyoki, McGill University, Class 2003, Molson Medical Informatics Students Project, "Electrocardiology and Cardiac Arrythmias".
5. Harvey Connor., Basic of 12 Lead ECG, Paramedic, I/C. Paramedic, Oklahoma City Community College, website, File "Vector_Basics_of_12_Lead_ECG_s.ppt"
6. eslie Cromwell, Fred J. Weibell & Erich A. Pfeiffer, Biomedical Instrumentation and Measurements, 2nd ed, Prentice-Hall, Inc ,1980, pp. 105-117
7. Texas instrument website, INA114, "ina114.pdf" datasheet.
8. NEC website, PS2506, " PS2506L-2-E4.pdf" datsheet.
9. Fairchild semiconductor website , "LF353.pdf" datasheet.
10. National Semiconductor website, "LMF 90.pdf" datasheet.
11. Analog Modules website,"AD822.pdf "datasheet.
12. Microchip technology website, "PIC16F877A.pdf " datasheet.
13. National Semiconductor website, LAB WINDOWS /CVI 8.0, File "00lwlist.pdf"
14. Rolf schunman & Mac E.van valkenburg, Design of analog filters, Oxford, pp. 26.

Automatic Spine Extraction in Posteroanterior Radiographs Based on Spine's Features

Phan Minh Khanh, Tran Giang Nam, and Pham The Bao

Faculty of Mathematics and Computer Science, University of Science, Ho Chi Minh City, Vietnam

Abstract — **Nowadays, radiographs are used more widely in clinical diagnostics. Therefore, automatic detection and extraction of spine in radiographic image are problems which are paid more attention to quantify the severity of spinal diseases. This will help reduce the number of clinical evaluation by a doctor and will allow studies based on large database. The purpose of this paper is to present a new approach to automatic detection of the spine in posteroanterio radiographic image. First, we define a region of interest (ROI) based on a set of radiographs. This region contains sufficient information and data of the vertebra. Filters with threshold based on the brightness and the features of images and spine, are used to remove redundant information. Then a mask is brought in to save the position of the spine. Based on the determined position of the spine, a method is applied to smooth the edge and to add parts of the spinal disability. In addition, the mask processing will detect the direction of curvature of the spine to facilitate other processes. The result which is obtained from the steps above for good quality X-ray image is almost entirely derived spine and less noise, although the poor quality images still have high noise surrounding the bone.**

Keywords — **Spine detection, X-ray, radiograph, spine extraction.**

I. INTRODUCTION

A spine is a column composed of many vertebrae, is responsible for supporting the body, moving and protecting spinal marrow. Many diseases relatıng to the spine are found accidentally when taking a body X-ray photo for diagnosing other diseases. The estimation of the severity of the disease relating to the spine is usually performed by using 2D spine X-ray image from front to back. In the past and even in present, spinal deformation is almost depended entirely on observing and diagnosing of orthopedic surgeons. However, a doctor can not undertake all tasks such as reviewing, measuring, analysing and classifying, especially with a large number of X-ray images. The automatic system detects and extracts spine from X-ray image is the first step to help doctors diagnose more accurately. This system is the foundation of other applications to detect diseases relating to the spine. Spinal fractures, osteoporosis, gout in the spine, scoliosis, spondylolisthesis... are examples. In addition, the results of this system can also be used to model backbone from 2D into 3D.

Many applications and solutions have been proposed to automatically detect one or more parts, organs (such as lung, kidney, bone ...) from the radiograph. To identify lung from X-ray images, the aim of two methods after is to identify the common features and thus does not pay much attention to the impact of noise on the image. Specifically, McNitt-Gray [1] proposed classification approach using a linear discriminant to determine the ROI of X-ray images. Luo [2] classified automatically the direction of the X-ray images based on anatomical features to ensure accurate labeling, and therefore he stores position points of radiograph exactly. In addition, active models (Active Appearance Model - AAM [3] and Active Shape Model - ASM [4]) and their corresponding statistics [5] have been launched with the same basic idea, both of them try to search on the surrounding of points of new pattern that fit to the model of the sample (obtained by training data). Smyth [6, 7] used the quantitative morphometry (QM) to build ASM of the lumbar vertebrae in the Dual-energy X-ray absorptiometry (DEXA) image to meansure the shape. In this method, ASM is initialised by placing three anchor points, ie, the ASM fixed at three locations and the rest of the points in the model are free to move in the restricted area of model. Robert [8] used AAM which is characterized by a parameter model of the object. ASM and AMM are the effective techniques in high quality image, but may be not applied correctly for the spine X-ray image; because the spine radiograph has many overlapping structures and high noise level which increase the technical complexity for this automatic detection. Furthermore, the lumbar and cervical spine has different contrast and noise than the thoracic spine. These studies are not successful in determining the spinal continuity and spinal deformity of X-ray image, while these results will be useful for clinical diagnosis of diseases relate to the spine, especially the curves of scoliosis. Luc Duong [9] gave a technique based on anisotropic filtering is introduced to enhance the noisy spine radiographs, and then the filtering process is followed by a detection step based on pattern classification devised to recognize spinal features to detect the spinal curve, this method bring to us a good result.

V. Van Toi et al. (Eds.): 4th International Conference on Biomedical Engineering in Vietnam, IFMBE Proceedings 40, pp. 233–237, 2013.
www.springerlink.com

In this paper, we propose a new approach based on the characteristics of X-ray image and the spine. This method is presented in three steps. First, identify ROI in the X-ray images, including the scoliosis [10], and then creating the dynamic threshold based on the characteristics of those areas for the spine region is shown more clearly. Then we use a mask to scan and record the position of the spine in image. Finally, based on the position of the spine was obtained; we proceed to adjust the spine correctly.

II. METHOLOGY

A. General System

We perform extraction spine of posteroanterio radiograph of spine, person who is shot, must be in a normal upright position, figure 1.

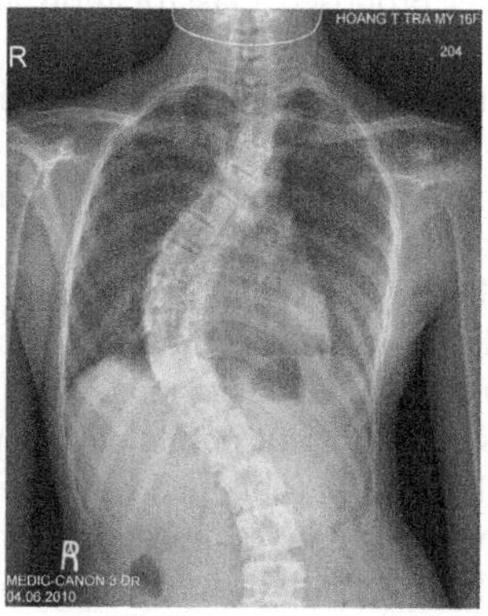

Fig. 1 Input image.

To proceed to extract spine from spinal X-ray image, we perform the following steps. First, we classify threshold input image to highlight the spine. Then we segment image based on classify - threshold - image.

B. Preprocessing

Due the fact that X-ray image changes only in brightness between black and white, we change input image to gray image for subsequent processing steps easier.

A normal posteroanterio radiograph of the spine has two different parts to another. In the upper half of image, the bright scale of background and spine is lower than the lower haft of image. So, as a convenience for image segmentation, we split the image into two parts vertically and conduct classification threshold base on the threshold of each parts.

Selecting threshold for parts bases on comments on upper half of image, the highlights in this region are ribs and backbone. The ribs are well defined and have big distances, so they should not interfere much in the classification threshold to highlight the backbone. About the lower half of image, this region has so many bones in abdomen closed to the spine, so it is easy to causes interference. Besides, the spine is always the most prominent part of the whole image.

Through out empiric statistics for feedback characteristic X-ray of the spine in multiple images, we set each threshold based on the the brightness behaviour of each image. Then we calculate the average brightness of the pixels having nonzero value (black color) and also find the highest value of pixel of an image. Although each part of each image have different thresholds, formulas 1, *ave* is average brightness of the pixels having nonzero value of image, *max* is the highest value of pixel of an image, α is coefficient of variation, and $0 < \alpha < 1$.

$$threshold = ave + (\max - ave) \times \alpha \quad (1)$$

Thus, when we consider an image, the threshold formulas differ only on the coefficient α. Basing on comments on input image, coefficient of the upper half of image will certainly be lass than the lower half of image, with α=1/3 for the upper half of image, we will obtain the best output for the next steps. For convenience, from here we sign $\alpha 1$ is the coefficient for the upper half of the image and $\alpha 2$ is the coefficient for the lower half of the image. We have the result shown in figure 2.

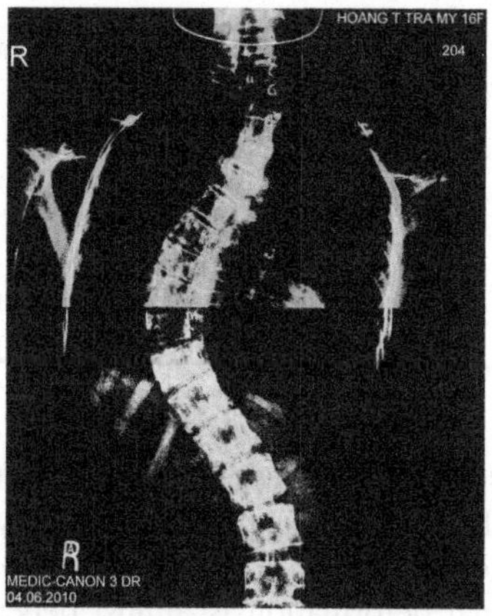

Fig. 2 Result after threholding.

C. Segmentation

After the image is classified threshold, the spine will be shown clearly. However, there are still many areas of interference.

From this step, we change gray image to binary image for convenience of handling.

In the human life, the middle part of the spine is the most pressure, so the bones in the middle spine are very close together and and apare gradually to two sides. Therefore, first, we will detect the bones in the middle spine.

On the binary image after segmenting, figure 2, we draw a line in middle images. We consider the longest line segments with the whitest pixels passing through areas are the backbone. However, the problem which is in the image where the white pixels are not continuous and fragment into sections that affects finding the longest line segment. To handle this situation, we consider the pixels in an error span, the pixels in the error span is considered to be continuous with each other. Since bone in middle spine is quite prominent and has high occlusion when comparing with the surrounding area, selecting the longest line segment is easy and this line is also part of the spine. We build the algorithm to solve it, algorithm 1.

Algorithm 1
Input: An image (binary classified-threshold-image).
Output: left vertex coordinates and the length of the line segment.

I. Step 1: Select a row in middle image, this row calls mid.
II. Step 2: Set j=1.
III. Step 3: iteration.
   o If (j < (horizontal image)), go to step 4.
   o Else, go to step 10.
IV. Step 4:
   o If (A(mid,j)=0), go to step 5.
   o Else, go to step 6.
V. Step 5: Assign j=j+1. Turn back to step 3.
VI. Step 6: Assign lj=j. Go to step 7.
VII. Step 7:
   o If (A(mid,j+t)=1, with (1 < t < error_span)), go to step 8.
   o Else, go to step 9.
VIII. Step 8: Assign j=j+t, turn back to step 7.
IX. Step 9: Save result which has lagest value of (j-li), turn back to step 3.
X. Step 10: End algorithm.

Based on the above results we obtain the line segment in the spine. We will move this line segment to upper row to continue determine the spine. The determination process is executed as follows algorithm 2:

Algorithm 2
Input: line segment L is found above.
Output: spinal region.

XI. Step 1: Put (x, y1) and (x, y2) which are two vertexcoordinates of line segment.
XII. Step 2: Assign x=x+1.
XIII. Step 3: Increasing or decreasing y1 and y2 to contract and stretch line segment to take full upper line segment.

Similarly, we will move to detect the spine in the lower half of image. However, another problem occurs, that is the mark is lost when line segment moves down, figure 2 is an example. To solve this problem, we use algorithm 3.

Algorithm 3

XIV. Step 1: Find the row in the input image corresponding to the lost-mark-row on the classified-threshold-image.
XV. Step 2: Classify threshold again and overlap this row into classified - threshold - image.
XVI. Step 3: Move line segment down by algorthim above.
XVII. Step 4:
   o If the next row have no mark, go to step 1.
   o Else, go to step 3.

Cofficient α is used in step 2 determined by using α equal to $\alpha 1$ or α less than $\alpha 2$. In this case, we use $\alpha 1$ to classify threhold. Figure 3 illustrates that process.

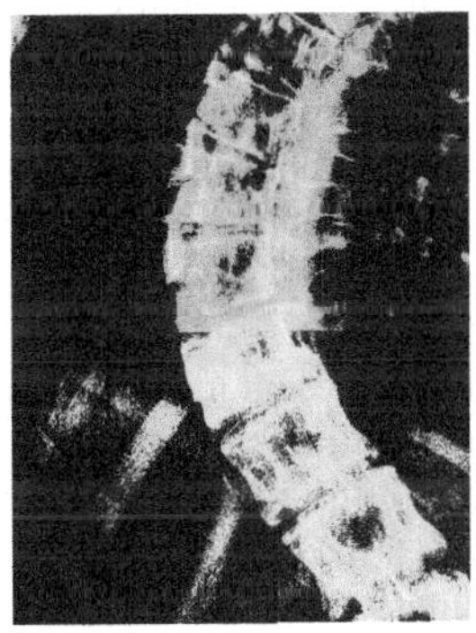

Fig. 3 Region of middle spine after thresholding again

In some cases, when the input image's quatily is not good, a problem is occurs in the lower half of image, that is middle part of image is bright but bottom part is dark. Therefore, we used the third threshold $\alpha 3$ to decrease noise. In other words, we decompose the bottom half of image in two parts and do similarly as above.

In addition, coefficients $\alpha 1$ and $\alpha 2$ can be changed to keep track of the spine and discard abundant details. When the width of the spine increases abnormally, coefficient will increase. On the contrary, when width of the spine decrease suddenly or loss, cofficient will decrease.

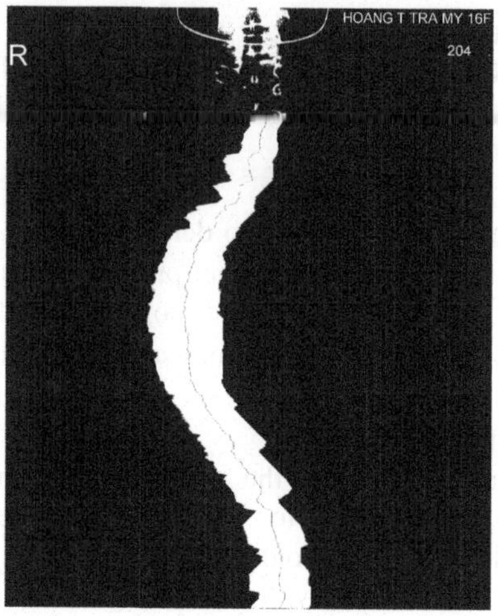

Fig. 4 Result after segmentation.

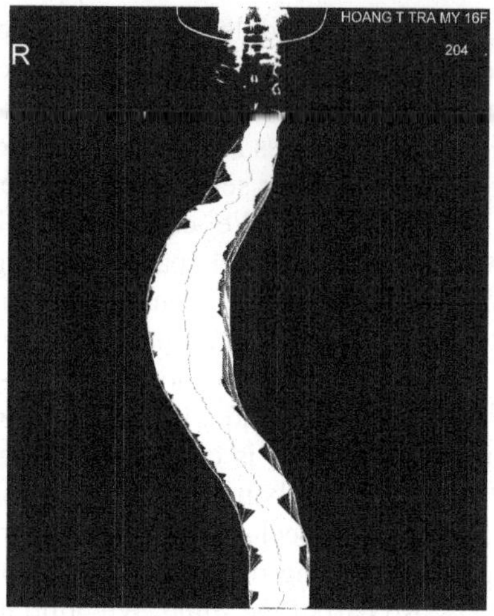

Fig. 5 New spine boundary.

During implementation, the midpoint of the line segment will also be saved to serve to determinating the direction of curvature of the spine.

D. Post Processing

After segmenting, an image will be highlighted the parts that the lines segment run up and down draw. These highlighted parts are called spinal case. To process easier in next steps, we remove all the pixels outside the spinal case.

Although the obtained spinal case contains most of spine, the border is folded much. This problem leads to deduction of some bones. In order to this problem, we conducte a smooth edge.

We smooth border line with the idea of filling up bottom land. In this part, we present method to fill up the left border line; the right border line is similarly. At the any width of a spine, we determine the pixel on the left boder which has minimum abscissa (maximum abscissa for right border). This pixel combines with another pixel on left border create a new border line and all pixel in bottom land will be filled up (change value to 1). If the distance between those two pixels is so long, the spine is rough after processing. On the contrary, if that distance is so short, effect is not good. In our opinion, that distance is approximately equal to 1/3 the width of the spine, that will give a good result. The smotth method is described as follows:

From a fixed point, line segment will scan an angle from left to right (right to left for right border) until line segment contact a point on border line of the spine. If the line segment from fixed point to contact point is shorter than defaul length of line segment, we will cut abundant part. Do similarly for all points on boundary.

Now, boundary of the spine is relatively smooth. Then we will extract the spine from the input image corresponding to the points which are on and in the boundary which is detected newly from the image above.

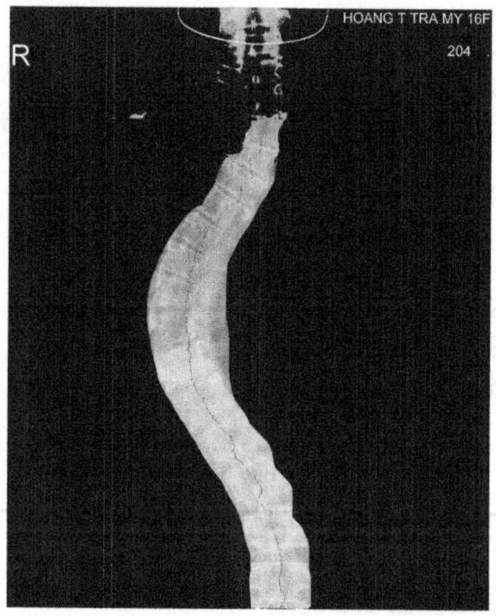

Fig. 6 Spine extraction.

III. RESULT

We implemented our research on Laptop: chip Intel Core 2 Duo, CPU 2x2.0GHz, 3GB RAM, windows 7 Professional by Matlab.

Standard data used to illustrate the article, are spine X-ray images and is posrteroanterior radiographs of Hoang Thi Tra My, female, 16 years old, scoliosis patient.

The rest data are from internet, so they have large differences between the size, and many images are not good quality.

Table 1 Result of nine test samples.

Item	Size (pixel)	Recall	Time	Ratio of spine detection	Noise magnitude
1	2208x2688	10	40 mins	100%	very little
2	599x1741	10	20 mins	65%	big
3	1231x2420	10	30 mins	80%	medium
4	768x1024	10	5 mins	60%	big
5	1600x1600	10	10 mins	75%	very little
6	630x630	10	10 s	90%	little
7	250x683	10	7 s	100%	no
8	335x582	10	7 s	70%	medium
9	238x583	10	7 s	90%	medium

Most images are extracted at least 60% of spine (excluding cervical vertebrae), but some results are still have a lot of noise in the lumbar vertebrae. The average accuracy is high. In terms of program execution time, the small image will process fastly; but the bigger the image is, the more running time is increased.

IV. Conclusions

The method that we proposed not only has good results, but also it's simple and effective. In addition, by saving the midpoint of the line, we can determine the direction of curvature of the bone easily.

A few areas do not handled well. Therefore the program also needs to be studied more to improve the accuracy of the images. Smoothing frame algorithm is not good, we are testing a new algorithm to hope for better results and lower the processing time.

Due to the fact that the length of the spine is usually around 1/5 to the horizontal image, so it can improve the method that finds first line of bone by selecting the limitative length of the segment.

Acknowledgment

The authors wish to thank Dr. Truong Quang Dang Khoa and acknowledge Biomedical Engineering Department of International University, Vietnam National University – Ho Chi Minh City for valueble input images on this research.

References

1. M. F. McNitt-Gray, H. K. Huang, and J. W. Sayre (1995) Feature selection in the pattern classification problem of digital chest radiograph seg mentation. IEEE Transactions on Medical Imaging, vol. 14, no. 3, pp. 537–547.
2. H. Luo, W. Hao, D. H. Foos, and C. W. Cornelius (2006) Automatic image hanging protocol for chest radiographs in PACS. IEEE Trans. Inf. Technol. Biomed., vol. 10, no. 2, pp. 302–311.
3. Ginneken B. (2002) Active Shape Model Segmentation with Optimal Features. IEEE Transactions on Medical Imaging, vol. 21, pp. 924-933.
4. T.F.Cootes, G.J. Edwards and C.J.Taylor (2001) Active Appearance Models. IEEE Transactions on Pattern Analysis and Machine Intelligence, Vol.23, No.6, pp.681-685.
5. T. F. Cootes and C. J. Taylor (2004) Anatomical statistical models and their role in feature extraction. Br. J. Radiol., vol. 77, Spec no 2, pp. S133–S139.
6. P. P. Smyth, C. J. Taylor, J. E. Adams (1997) "Automatic Measurement of Vertebral Shape using Active Shape Models," Image Vision Comput. 15, pp. 575-581.
7. P. P. Smyth, C. J. Taylor, J. E. Adams (1999) Vertebral Shape: Automatic Measurement with Active Shape Models. Radiology 211, pp. 571-578.
8. M. Roberts, T. Cootes, and J. Adams (2003) Linking sequences of active appearance submodels via constraints: an application in automated vertebral morphometry. in Proc. 14th Br. Mach . Vis. Conf., p. 10.
9. Luc Duong, F. Cheriet and H.Labelle (2010) Automatic Detection of Scoliotic Curves in Posteroanterior Radiographs. IEEE Transactions on Biomedical Engineering, vol. 57, no. 5, pp. 1143-1150.
10. L. R. Long and G. R . Thoma (2001) "Landmarking and feature localization in spine x -rays," J. Electron. Imaging, vol. 10, pp. 939–956.

Use macro [author address] to enter the address of the corresponding author.

Author: Pham The Bao
Institute: University of Science
Street: 227 Nguyen Van Cu
City: Ho Chi Minh
Country: Vietnam
Email: ptbao@hcmus.edu.vn

Betulinic Acid Inhibits Lps-Induced Mmp-9 Expression by Suppressing Nf-Kb Activation in Bv2 Microglial Cells

Sung-Soo Kim

Department of Pharmacology, School of Medicine, Kangwon National University, Chuncheon, Korea

Abstract — **The objective of the present study was to evaluate anti-inflammatory effects of betulinic acid on lipopolysaccharide (LPS)-stimulated BV2 microglial cells. Pretreatment of betulinic acid significantly attenuated LPS-induced NO production and protein expression of iNOS. Betulinic acid also significantly suppressed LPS-induced release and expression of cytokines such as TNF-α and IL-1β. Furthermore, betulinic acid significantly suppressed LPS-induced MMP-9 expression, which has been suggested to play an important role in the migration of activated microglia. In order to understand the possible mechanism by which betulinic acid suppresses LPS-induced cytokine production and migration of microglia, the role of NF-kB, a major pro-inflammatory transcription factor, was examined. Betulinic acid significantly suppressed LPS-induced degradation of IkB, which retains NF-kB in the cytoplasm. Therefore, nuclear translocation of NF-kB upon LPS stimulation was significantly suppressed bybetulinic acid. Taken together, the present study for the first time demonstrates thatbetulinic acid possesses anti-inflammatory activity and inhibition of microglial migration through the suppression of nuclear translocation of NF-kB in BV2 microglial cells.**

Keywords — **Betulinic acid, MMP-9, Lipopolysaccharide, Cytokines, iNOS.**

I. INTRODUCTION

Microglia, resident macrophages and immune surveillance cells in the central nervous system, have been reported to play a critical role in host defense and tissue repair in brain [1]. It has been demonstrated that microglia present in remote areas migrate into the injured sites and clear the cell debris via the process called phagocytosis [2]. However, excessive activation of microglia has also been proposed to play a pathogenic role in neuro-inflammatory conditions such as ischemia [3], multiple sclerosis [4], Parkinson's disease [5], Alzheimer's disease [6]and HIV-associated dementia [7]. Under these immunologically mediated CNS diseases, activated microglia have been suggested to contribute to neurodegenerative process by producing excessive amounts of biologically active molecules such as nitric oxide (NO) and several cytokines [8,9,10,11].

Matrix metalloproteinases (MMPs) are zinc-dependent proteolytic enzymes that are involved in the remodeling of the extracellular matrix (ECM) in numerous physiological and pathological processes [12,13]. MMPs have been reported to play a key role in the microglial migration through the regulation of cellular adhesion, movement, and chemotaxis[14]. Furthermore, it has been reported and that MMPs could regulate chemotactic properties of chemokines by cleaving or inactivating chemokines[15]and facilitate the release of chemokines from proteins or other molecules [16]. Especially, increased activity of MMP-9 has been reported to significantly contribute to the microglial migration[17].

Betulinic acid is naturally occurringpentacyclictriterpene that is widely distributed in the plant kingdom and it has been reported that considerable amount of betulinic acid is available in the outer bark of many tree species such as white-barked birch trees [18]. Although betulinic acid has been reported to possess a variety of biological activities including anti-malarial, anti-inflammatory, anthelminthic, anti-tumor, and anti-retrovial properties [18,19], the exact mechanism for its biological activities has not been clearly examined.

In the present study, in order to bring forth a novel pharmacological agent that could suppress abnormally activated microglia for the treatment of inflammation-related CNS diseases, the anti-inflammatory effects of betulinic acid and its underlying mechanism were examined in LPS-stimulated BV2 microglial cells.

II. MATERIALS AND METHODS

A. *Reagents and Cell Culture*

Bacterial lipopolysaccharide (LPS, E. coli serotype 055:B5) was purchased from Sigma-Aldrich (St. Louis, MO, USA). Betulinic acid (Fig. 1) was isolated and identified from *Forsythia koreana*. Briefly, the dried *Forsythia koreana* was extracted with methanol and then partitioned with n-hexane, dichlormethane, ethyl acetate, and butanol, successively. Betulinic acid was isolated from an-hexane fraction using silica gel column chromatography. The chemical structure of the compound was determined by

V. Van Toi et al. (Eds.): 4th International Conference on Biomedical Engineering in Vietnam, IFMBE Proceedings 40, pp. 238–243, 2013.
www.springerlink.com

means of ^{1}H-NMR (600 MHz) and EI-MS and identified to betulinc acid by comparison with spectral data from the literature [20,21]. The compound was dissolved in ethanol and added to the cell culture at the desired concentrations. The BV2 microglia cell line was maintained in Dulbecco's modified Eagle's medium (DMEM; Gibco BRL, Grand Island, NY, USA) supplemented with 5% heat-inactivated fetal bovine serum (FBS; Gibco BRL) and 50 μg/ml gentamicin (Sigma, St. Louis, MO, USA). Cells were maintained in a humidified incubator at 37℃ with 5% CO_2. In all experiments, cells were incubated in the presence of the indicated concentration of betulinic acid before the addition of LPS (200 ng/ml).

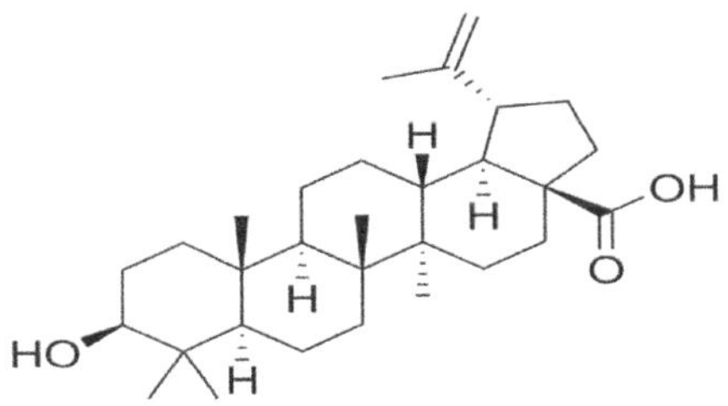

Fig. 1. Chemical structure ofbetulinic acid.

B. *Cell Viability*

Cell viability was determined by 3-(4,5-dimethylthiazol-2-yl)-2,5-diphenyltetrazolium bromide (MTT) assay. The BV2 microglial cells were plated into 96-well plates at a density of 5×10^4 cells per well for 24 hr. BV2 microglial cells incubated with various concentrations of BA for 24 hr. MTT (0.5 mg/ml in PBS) was added to each well, and the cells were incubated for 3 hr at 37 ℃. The resulting formazan crystals were dissolved in dimethyl sulfoxied (DMSO). The optical density was measured at 540 nm. The results were expressed as a percentage of surviving cells over control cells

C. *Nitrite Quantification Assay*

The production of NO was estimated by measuring the amount of nitrite, a major stable metabolite of NO, using the Griess reagent (0.1% naphthylethylenediaminedi-hydrochloride and 1% sulfanilamide in 5% phosphoric acid). BA-pretreated BV2 microglial cells were stimulated with LPS in 12-well plates for 24 hr, and then 100 μL of each culture medium was mixed with an equal volume of Griess reagent. The absorbance at 540 nm was measured on a microplatereader. The results were expressed as percentages of released NO from LPS-stimulated BV2 cells. To prepare a standard curve, sodium nitrite was diluted in culture medium to concentrations.

D. *TNF-α and IL-1β ELISA*

The BV2 microglia cells were treated with BA in the absence or presence of LPS. After 24 hr incubation, TNF-α and IL-1β levels in culture media were quantified using monoclonal anti-IL-1β or anti-TNF-α antibody according to the manufacturer's instruction (R &D Systems)

E. *Western Blot Analysis*

The BV2 microglial cells were washed with PBS two times and lysed with lysis buffer [50 mMTris-HCl, pH 8.0, 150 mMNaCl, 0.02% sodium azide, 0.1% sodium dodecyl sulfate (SDS), 1% NP-40, 0.5% sodium deoxycholate, 1 mM phenyl methylsulfonyl fluoride]. Equal amounts of protein were separated on 10% SDS-polyacrylamide gel. Proteins were transferred to Hypond PVDF membrane (Amersham Biosciences, Piscataway, NJ, USA). The membranes were blocked in 5% skim milk in TBST for 1 hr at room temperature and sequentially incubated with an appropriate antibody; iNOS (BD Bioscience, Franklin Lakes, NJ), IkB-α(Santa Cruz Biotechnology Inc., Eugene, OR, USA),MMP-9 (Affinity BioReagents, Inc., Golden, CO, USA) or β-actin(Sigma, St. Louis, MO, USA). After thoroughly washing with TBST, the membranes were then washed three times with TBST and incubated with HRP-conjugated goat anti-rabbit IgG (Jackson ImmunoResearch) or HRP-conjugated goat anti-mouse IgG (Jackson ImmunoResearch) for 2 hr at room temperature. The blots were developed by the enhanced chemiluminescence detection kit (ECL Amersham Biosciences, Piscataway, NJ, USA).

F. *Immunocytochemistry*

The effect of betulinic acid on the nuclear translocation of p65 subunit of NF-kB was examined by immunofluorescence assay using confocal microscopy. BV2 microglial cells were cultured in sterile coverslips and pretreated with betulinic acid for 1 hr and stimulated with LPS. At 30 min after the LPS treatment, the cells were fixed in 4% paraformaldehyde for 20 min at room temperature. The fixed cells were then permeabilized with 0.1% Triton X-100 in PBS and blocked with 3% BSA. Afterwards, the cells were sequentially incubated with rabbit p65 antibody (Santa Cruz Biotechnology Inc., Eugene, OR, USA) at room temperature and FITC-conjugated donkey anti-rabbit IgG (Jackson ImmunoResearch Laboratories, Inc., West Grove, PA, USA) at room temperature for 1 hr. After washing with PBS, the sample were mounted and observed by means of confocal microscopy.

G. RNA Isolation and RT-PCR

Total RNA was isolated using the Total RNA Extraction Kit (iNtRON Biotechnology, Inc,USA) according to the manufacturer's instruction. Total RNA (2 μg) was obtained from cells was reverse-transcribed using oligo-(dT)15 primers (Promega, Madison, WI, USA). The following PCR conditions were applied: β-actin, 24 cycles of denaturation at 94°C for 30 sec, annealing at 55°C for 30 sec and extended at 72°C for 30 sec; MMP-9, 25 cycles of denaturation at 94°C for 30 sec, annealing at 61°C for 30 sec and extended at 72°C for 30 sec. The β-actin was used as an internal control to evaluate relative expression of MMP-9. PCR primers were as follow: MMP-9, 5'-GAG CTG TGC GTC TTC CCC TTC -3' (forward) and 5'-GGA ATG ATC TAA GCC CAG TGC -3' (reverse) and β-actin, 5'- ATC CTG AAA GAC CTC TAT GC-3' (forward) and 5'- AAC GCA GCT CAG TAA CAG TC-3' (reverse). After amplification, PCR reactions were electrophoresed in an agarose gel.

H. Statistical Analysis

All values shown in the figures are expressed as the mean±SD obtained from at least three independent experiments. Statistical analysis was carried out by one-way analysis of variance (ANOVA) with Tukey's post-hoc test using SPSS software 12K (SPSS, Chicago, IL, USA). A value of P<0.05 was considered as statistically significant.

III. RESULTS

A. Betulinic Acid Inhibits NO Production and iNOS Expression in LPS-Stimulated BV2 Cells

The effects of betulinic acid on NO production and iNOS expression were examined in LPS-stimulated BV2 microglial cells. Although higher concentrations of betulinic acid over 25 μM appeared to be cytotoxic, no significant cell death was observed in the concentration range of betulinic acid used in the present study (Fig. 2A). Treatment of LPS resulted in the excessive production of NO and up-regulation of iNOS in protein levels (Fig. 2B & C). Pretreatment of betulinic acid significantly suppressed the LPS-induced NO production in a concentration-dependent manner (Fig. 2B). In accordance with the suppression of NO production, betulinic acid also significantly attenuated LPS-induced up-regulation of iNOS expression in protein levels (Fig. 2C). These results demonstrate thatbetulinic acid exhibits inhibitory effects on LPS-induced iNOS expression and activity in BV2 microglial cells.

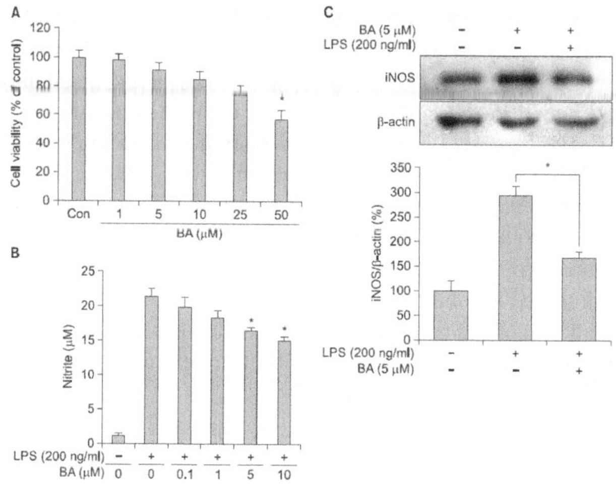

Fig. 2 Effects of betulinic acid on NO production and iNOSexpression in LPS-stimulated BV2 microglial cells.(A) Effect of betulinicacid(BA) on the viability of BV2 microglial cells. No significant cell death was observed up to 10 μM. (B) Concentration-dependent suppression of LPS-induced NO production by betulinic acid. (C) Suppression of LPS-induced iNOS protein expression by betulinic acid: top, representative immunoblot of iNOS. β-Actin was used as an internal control; bottom, quantitative analysis of immunoblots; Quantitative data represent three independent experiments and are expressed as mean±SD. *p<0.05 indicate statistically significant difference with LPS alone. BA stands for betulinic acid.

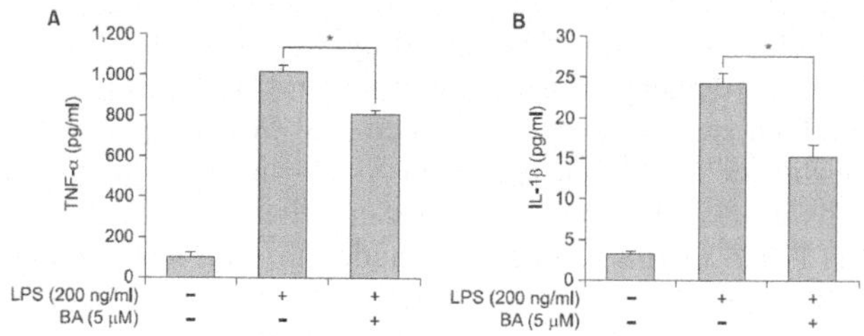

Fig. 3. Inhibitory effects of betulinic acid on the LPS-induced release of TNF-α (A) and IL-1β(B)in BV2 microglial cells. BV2 microglia cells were incubated with 200 ng/ml of LPS in the presence or absence of indicated concentrations of betulinicacid(BA) for 24 hr. Cell culture media were collected and subjected to TNF-α and IL-1β sandwich ELISA. Data represent three independent experiments in triplicate and are expressed as mean±SD. *p<0.05 and **p<0.01 indicate statistically significant difference with LPS alone. BA stands for betulinic acid.

B. Betulinic Acid Attenuates the Expression of Pro-inflammatory Cytokines in LPS-Stimulated BV2 Cells

To determine the effect of betulinic acid on the expression of pro-inflammatory cytokines such as TNF-α and IL-1β, extracellular release of these cytokines was examined using ELISA assay. Betulinic acid significantly suppressed LPS-induced extracellular release of TNF-α and IL-1β in BV2 microglial cells (Fig. 3). In accordance with the attenuation of extracellular cytokine release, betulinic acid significantly suppressed LPS-induced up-regulation of these cytokine genes (data not shown).

C. Betulinic Acid Suppresses LPS-Induced MMP-9 Gene Expression

Based on the previous report that MMP-9 might play a significant role in the migration of activated microglia to remote places [17], the expression of MMP-9 gene and its extracellular cellular release were examined in the present study. The mRNA expression of MMP-9 was significantly increased with LPS stimulation (Fig. 4A). However, pretreatment of betulinic acid significantly attenuated LPS-stimulated gene expression of MMP-9 in BV2 microglial cells (Fig. 4A). Furthermore, betulinic acid significantly attenuated LPS-stimulated gene expression of MMP-9 as shown in a representative western blotting (Fig. 4B). These results strongly suggest that the primary effect of betulinic acid on MMP-9 might be through the suppression of gene expression rather than the inhibition of extracellular release of MMP-9

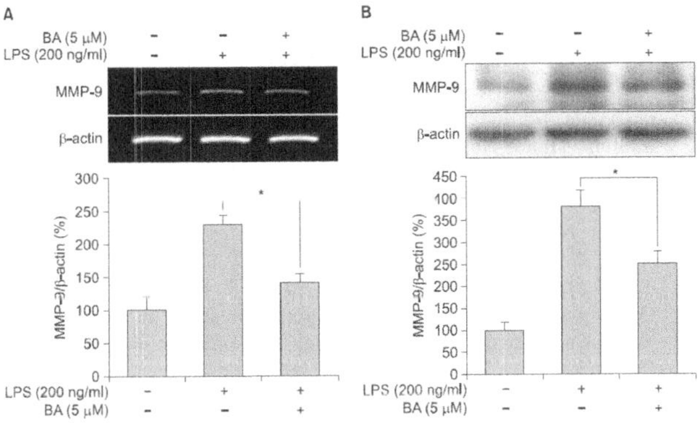

Fig. 4 Effect of betulinic acid on MMP-9 expression in LPS-stimulated BV2 microglia cells. (A) BV2 cells were treated 200 ng/ml LPS in the absence or presence of betulinic acid(BA), and total RNA was isolated at 6 h after LPS treatment. The mRNA expression of MMP-9 was measured by RT-PCR. (B) Western blots was performed using the cell lysates of BV2 cells with LPS in the presence of betulinic acid for 24 h. Quantitative data represent three independent experiments and are expressed as mean±SD. *p<0.05 and **p<0.01 indicate statistically significant difference with LPS alone. BA stands for betulinic acid.

D. Betulinic Acid Suppresses LPS-Induced Degradation of IkB and Nuclear Translocation of NF-kB

The transcription factor NF-kB has been extensively reported to play a predominant role in the expression of numerous inflammatory cytokine genes and IkB has been demonstrated to inhibit the nuclear translocation of NF-kB by retaining it in the cytoplasm [22,23]. Therefore, in the present study, effects of betulinic acid on LPS-induced nuclear

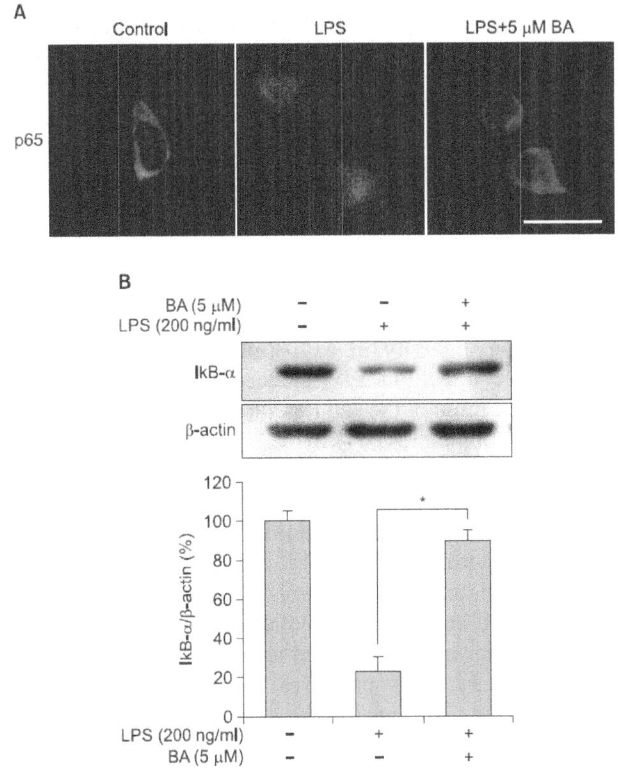

Fig. 5 Blockade of nuclear translocation of p65 subunit of NF-kB in LPS-stimulated BV2 microglial cells by betulinic acid(BA). (A) Localization of NF-kB p65 subunit was determined using a p65 antibody and an FITC-conjugated donkey anti-rabbit IgG antibody. In basal condition, immunostaining of p65 subunit was diffuse throughout the cytoplasm. However, LPS stimulation resulted in the apparent translocation of p65 subunit into the nucleus. Pretreatment of betulinic acid appeared to attenuate LPS-induced nuclear translocation of p65 subunit. Scale bar, 20 μm. (B) Total cell lysates obtained 15 min after the LPS stimulation were subjected to Western blotting to assess the levels of IKB-α proteins (top). Quantification of IKB-α degradation was performed by densitometric analysis (bottom). β-Actin was used as an internal control. Quantitative data represent three independent experiments and are expressed as mean±SD. *p<0.05 indicate statistically significant difference with LPS alone. BA stands for betulinic acid.

translocation of NF-kB and degradation of cytosolic IkB were examined. To determine the nuclear translocation of NF-kB, the location of nuclear NF-kB was examined using confocal microscopy. Immunostaining of RelA (p65) subunit of NF-kB was present predominantly in the cytoplasm in basal condition (Fig. 5A). LPS stimulation resulted in apparent nuclear translocation of p65 subunit of NF-kB in BV2 microglial cells. However, pretreatment of betulinic acid appeared to attenuate the LPS-induced nuclear translocation of p65 subunit of NF-kB (Fig. 5A). To examine the effect of betulinic acid on IkB, intracellular level of IkBwas examined upon LPS stimulation using immunobloting assay. LPS challenge resulted in a significantly decreased intracellular level of IkB in BV2 microglial cells (Fig. 5B). However, pretreatment of betulinic acid showed a significant attenuation of LPS-induced degradation of IKB (Fig. 5B).

IV. Discussion

The present study demonstrated thatbetulinic acid, isolated from *Forsythia koreana*, possesses an anti-inflammatory activity in LPS-stimulated microglia and a suppressive effect on LPS-induced microglial migration, presumably through the suppression of nuclear translocation of transcription factor NF-kB.

Betulinic acid, a pentacyclicterterpene, and its derivatives have been reported to possess a variety of biological properties [18] including inhibition of human immuno-deficiency virus (HIV) [24], anti-bacterial, anti-cancer [19,25], and anti-inflammatory activities [26]. Although anti-HIV and anti-cancer activities of betulinic acid and its derivatives have been reported to be through the inhibition of HIV entry, HIV-proteases or reverse transcriptase [18,27], and through increased degradation of the transcription factors specificity protein 1 (Sp1), Sp3, and Sp4[25], respectively, the underlying mechanism of anti-inflammatory property of betulinic acid has not been clearly demonstrated[26]. The present results evidently demonstrate thatbetulinic acid significantly suppresses the activity of transcription factor NF-kB, which has been reported to play a key role in the inflammatory responses, through the inhibition of LPS-induced ikB degradation in the cytoplasm [28].

Aberrant activation of microglia has been reported to be involved in several pathologic conditions by releasing pro-inflammatory cytokines and oxidants such as TNF-α, interleukin (IL)-6, IL-lβ, IL-6, IL-10, IL-12, interferon-γ, and NO [9,29]. Accordingly, suppression of microglial activation and subsequent release of cytokines have been suggested to protect neuronal damage in several inflammation-related CNS diseases [3,30,31,32]. In the present study, betulinic acid exerts anti-inflammatory property through the significant suppression on LPS-induced expression and activity of iNOS, and the release of TNF-α and IL-1β.

It has been previously reported that MMP-9 plays an important role in the regulation of chemotaxic migration of leukocyte[33] as well as monocyte[34]. Recently, it has been reported that increased expression and activity of MMP-9 are involved in microglial activation and that the inhibition of MMP-9 suppresses LPS-induced expression of pro-inflammatory cytokines in microglia[35]. In the present study, betulinic acid significantly attenuated the LPS-induced MMP-9 expression.

Aberrant activation of a pro-inflammatory transcription factor NF-kB has been associated with various inflammatory conditions such as autoimmune diseases [36,37]. NF-kB-dependent microglial activation has been reported to be a crucial contributor to ischemia [38]. In the present study, betulinic acid significantly attenuated LPS-induced nuclear translocation of NF-kB and subsequent extracellular releases of pro-inflammatory cytokines and MMP-9, through the inhibition of LPS-induced IkB degradation.

In conclusion, the present results strongly demonstrate that betulinic acid exertsthe anti-inflammatory activity and the inhibition of microglial migration by suppressing the production of pro-inflammatory mediators and MMP-9 through the inhibition of the nuclear translocation of NF-kBinLPS-stimulated BV2 microglial cells. The present study suggests thatbetulinic acid might be a valuable therapeutic agent for the treatment of inflammatory diseases in CNS caused by activated microglia.

References

1. Perry VH, Gordon S (1988) Macrophages and microglia in the nervous system. Trends Neurosci 11, 273-277.
2. Kato H, Walz W (2000) The initiation of the microglial response. Brain Pathol 10, 137-143.
3. Hailer NP (2008) Immunosuppression after traumatic or ischemic CNS damage: it is neuroprotective and illuminates the role of microglial cells. ProgNeurobiol 84, 211-233.
4. Matsumoto H (1992) [Some markers reflecting the pathology and disease activity of multiple sclerosis]. No ToShinkei 44, 95-102.
5. McGeer PL, McGeer EG (1998) Glial cell reactions in neurodegenerative diseases: pathophysiology and therapeutic interventions. Alzheimer Dis AssocDisord 12 Suppl 2, S1-6.
6. Itagaki S, McGeer PL, Akiyama H, Zhu S, Selkoe D (1989) Relationship of microglia and astrocytes to amyloid deposits of Alzheimer disease. J Neuroimmunol 24, 173-182.
7. Merrill JE, Chen IS (1991) HIV-1, macrophages, glial cells, and cytokines in AIDS nervous system disease.FASEB J 5, 2391-2397.
8. Chao CC (1992) Cytokine release from microglia: differential inhibition by pentoxifylline and dexamethasone. J Infect Dis 166, 847-853.
9. Graber MB, Streit WJ (2010) Microglia: biology and pathology. ActaNeuropathol 119, 89-105.
10. Merrill JE, Benveniste EN (1996) Cytokines in inflammatory brain lesions: helpful and harmful. Trends Neurosci 19, 331-338.
11. Tuttolomondo A, Di Raimondo D, di Sciacca R, Pinto A, Licata G (2008) Inflammatory cytokines in acute ischemic stroke. Curr Pharm Des 14, 3574-3589.
12. Yong VW, Krekoski CA, Forsyth PA, Bell R, Edwards DR (1998) Matrix metalloproteinases and diseases of the CNS. Trends Neurosci 21, 75-80.
13. Cuzner ML, Opdenakker G (1999) Plasminogen activators and matrix metalloproteases, mediators of extracellular proteolysis in inflammatory demyelination of the central nervous system. J Neuroimmunol 94, 1-14.

14. Zuo J, Ferguson TA, Hernandez YJ, Stetler-Stevenson WG, Muir D (1998) Neuronal matrix metalloproteinase-2 degrades and inactivates a neurite-inhibiting chondroitin sulfate proteoglycan. J Neurosci 18, 5203-5211.
15. McQuibban GA, Gong JH, Tam EM, McCulloch CA, Clark-Lewis I, et al. (2000) Inflammation dampened by gelatinase A cleavage of monocyte chemoattractant protein-3. Science 289, 1202-1206.
16. Li Q, Park PW, Wilson CL, Parks WC (2002) Matrilysin shedding of syndecan-1 regulates chemokine mobilization and transepithelial efflux of neutrophils in acute lung injury. Cell 111, 635-646.
17. Choi MS, Cho KS, Shin SM, Ko HM, Kwon KJ, et al. ATP induced microglial cell migration through non-transcriptional activation of matrix metalloproteinase-9. Arch Pharm Res 33, 257-265.
18. Yogeeswari P, Sriram D (2005) Betulinic acid and its derivatives: a review on their biological properties. Curr Med Chem 12, 657-666.
19. Fulda S (2008) Betulinic Acid for cancer treatment and prevention. Int J MolSci 9, 1096-1107.
20. Do JC, Chai JY, Son KH (1991) Studies on the components of Lycopuslucidus (I). Kor J Pharacogn 22, 162-165.
21. Patra A, Chaudhuri KS, Panda KS (1988) Betulun-3-cafferate from Quercussuber 13C-NMNR spectra of some lupense. J Nat Prod 51, 217-220.
22. Ghosh S, May MJ, Kopp EB (1998) NF-kappa B and Rel proteins: evolutionarily conserved mediators of immune responses. Annu Rev Immunol 16, 225-260.
23. Bonizzi G, Karin M (2004) The two NF-kappaB activation pathways and their role in innate and adaptive immunity. Trends Immunol 25, 280-288.
24. Soler F, Poujade C, Evers M, Carry JC, Henin Y, et al. (1996) Betulinic acid derivatives: a new class of specific inhibitors of human immunodeficiency virus type 1 entry. J Med Chem 39, 1069-1083.
25. Chintharlapalli S, Papineni S, Ramaiah SK, Safe S (2007) Betulinic acid inhibits prostate cancer growth through inhibition of specificity protein transcription factors. Cancer Res 67, 2816-2823.
26. Recio MC, Giner RM, Manez S, Gueho J, Julien HR, et al. (1995) Investigations on the steroidal anti-inflammatory activity of triterpenoids from Diospyrosleucomelas. Planta Med 61, 9-12.
27. Fujioka T, Kashiwada Y, Kilkuskie RE, Cosentino LM, Ballas LM, et al. (1994) Anti-AIDS agents, 11. Betulinic acid and platanic acid as anti-HIV principles from Syzigiumclaviflorum, and the anti-HIV activity of structurally related triterpenoids. J Nat Prod 57, 243-247.
28. Lee JW, Cheong IY, Kim HS, Lee JJ, Lee YS, et al. Anti-inflammatory Activity of 1-docosanoyl Cafferate Isolated from Rhusverniciflua in LPS-stimulated BV2 Microglial Cells. Korean J PhysiolPharmacol 15, 9-15.
29. Kreutzberg GW (1996) Microglia: a sensor for pathological events in the CNS. Trends Neurosci 19, 312-318.
30. Aquilano K, Baldelli S, Rotilio G, Ciriolo MR (2008) Role of nitric oxide synthases in Parkinson's disease: a review on the antioxidant and anti-inflammatory activity of polyphenols. Neurochem Res 33, 2416-2426.
31. Hashioka S, McGeer PL, Monji A, Kanba S (2009) Anti-inflammatory effects of antidepressants: possibilities for preventives against Alzheimer's disease. Cent NervSyst Agents Med Chem 9, 12-19.
32. Ray B, Lahiri DK (2009) Neuroinflammation in Alzheimer's disease: different molecular targets and potential therapeutic agents including curcumin. CurrOpinPharmacol 9, 434-444.
33. Sellebjerg F, Sorensen TL (2003) Chemokines and matrix metalloproteinase-9 in leukocyte recruitment to the central nervous system. Brain Res Bull 61, 347-355.
34. Kappert K, Meyborg H, Clemenz M, Graf K, Fleck E, et al. (2008) Insulin facilitates monocyte migration: a possible link to tissue inflammation in insulin-resistance. BiochemBiophys Res Commun 365, 503-508.
35. Woo MS, Park JS, Choi IY, Kim WK, Kim HS (2008) Inhibition of MMP-3 or -9 suppresses lipopolysaccharide-induced expression of proinflammatory cytokines and iNOS in microglia. J Neurochem 106, 770-780.
36. Li Q, Verma IM (2002) NF-kappaB regulation in the immune system. Nat Rev Immunol 2, 725-734.
37. Karin M, Takahashi T, Kapahi P, Delhase M, Chen Y, et al. (2001) Oxidative stress and gene expression: theAP-1 and NF-kappaB connections. Biofactors 15, 87-89.
38. Cho IH, Hong J, Suh EC, Kim JH, Lee H, et al. (2008) Role of microglial IKKbeta in kainic acid-induced hippocampal neuronal cell death. Brain 131, 3019-3033

Author: Sung-Soo Kim, M.D., Ph.D.
Institute: Department of Pharmacology, School of Medicine, Kangwon National University
Street: 1 Kangwondaehak-gil
City: Chuncheon
Country: South Korea
Email: ksslsy@kangwon.ac.kr

The Effects of Thrombin on Migration and Metalloproteinase Expression in C6 Glioma Cell Line, and Their Related Intracellular Signaling

Sang-Hoon Cha

Department of Systems Immunology, College of Biomedical Science, Kangwon National University, Kangwon 200-701, South Korea

Abstract — **Glioblastoma multiforme is one of the most common and aggressive tumors in central nervous system. It often possesses characteristic necrotic lesions with hemorrhages, which increase the chances of exposure to thrombin. Thrombin has been known as a regulator of MMP-9 expression and cancer cell migration. However, the effects of thrombin on glioma cells have not been clearly understood. In the present study, influences of thrombin on glioma cell migration were examined using Boyden chamber migration assay and thrombin-induced changes in MMP-9 expression were measured using zymography, semi-quantitative RT-PCR, and Western blotting. Furthermore, underlying signaling pathways by which thrombin induces MMP-9 expression were examined. Thrombin-induced migration and MMP-9 expression were significantly potentiated in the presence of wortmannin, a PI3K inhibitor, whereas MAPK inhibitors suppressed thrombin-induced migration and MMP-9 expression in C6 glioma cells. The present data strongly demonstrate that MAPK and PI3K pathways evidently regulate thrombin-induced migration and MMP-9 expression of C6 glioma cells. Therefore, the control of these pathways might be a beneficial therapeutic strategy for treatment of invasive glioblastoma multiforme.**

Keywords — **Thrombin, Migration, MMP, Glioma.**

I. INTRODUCTION

Gliomas constitute nearly 60% of primary brain tumors and usually infiltrate normal brain parenchyma. Gliomas are categorized as their histologic features, and the major type of gliomas are astrocytomas [1]. Glioblastoma multiforme (GBM) is the most common type of astrocytomas arising in adults and clinically very aggressive [2]. These tumors typically show necrosis and hemorrhage. Patients diagnosed as GBM survive less than 1 year due to diffusely infiltrating nature of GBM. Glioma cells migrate away from the primary tumor site so that surgical resection of a tumor mass is not effective for remission [3].

Matrix metalloproteinases (MMPs) are critically involved in many physiological and pathological processes such as tissue remodeling and cancer progression [4]. Especially, MMP-2 and MMP-9 (known as gelatinases) have been reported to play crucial roles in tumor angiogenesis and metastasis. It has been also reported that enhanced gelatinase expression in various types of cancers corresponds to increased invasion and metastasis [5].

Thrombin, a trypsin-like serine protease, is the most copious enzyme participating in the coagulation cascade. In addition to its hemostatic roles, thrombin modulates a diversity of pathophysiological functions such as inflammation, cell migration and proliferation, apoptosis and so on. There are many evidences that thrombin regulates MMP-9 expression [6-8] and cancer cell migration [9-11]. Nevertheless, the effect of thrombin on glioma cell migration is not fully understood even though thrombin released by hemorrhagic site of GBM may have a important role in invasion.

In the present study, we focused on demonstrating the effects of thrombin on the migration of C6 rat glioma cells, which are morphologically similar to GBM [12], and possible underlying signaling pathways that thrombin induces MMP-9-mediated migration.

II. MATERIALS AND METHODS

A. Reagents

Thrombin and wortmannin were obtained from Sigma-Aldrich (St. Louis, MO, USA) and Calbiochem (Darmstadt, Germany), respectively. SB203580, SP600125, U0126 and MMP-9 inhibitor, batimastat, were purchased from Tocris Bioscience (Ellisville, MO, USA).

B. Cell Culture and Thrombin Treatment

The C6 rat glioma cell line was cultured in Dulbecco's modified Eagle's medium (DMEM; GIBCO BRL, Grand Island, NY, USA) in the presence of 10% fetal bovine serum (FBS; Hyclone Laboratories, Inc.), 100 U/ml of penicillin, 100 mg/ml of streptomycin and 2 mM of glutamine (GIBCO BRL, Grand Island, NY, USA). Cells were maintained at 37℃ in a humidified atmosphere with 5% CO_2. All inhibitors were pretreated in serum-deprived culture medium for 1 hr followed by thrombin treatment as indicated.

V. Van Toi et al. (Eds.): 4th International Conference on Biomedical Engineering in Vietnam, IFMBE Proceedings 40, pp. 244–249, 2013.
www.springerlink.com

C. *Cell Migration Assay*

Cell migration assay was performed by BD FalconTM Cell Culture Inserts (BD Bioscience, Bedford, MA; pore size, 8-μm) in 24-well culture plates. Cell culture inserts were coated with fibronectin (Sigma, St. Louis, MO, USA; 3 μg/ml) both upper and lower surfaces. Approximately 0.7 $\times 10^5$ cells in 400 μl of serum-free medium were seeded in the upper chamber and 400 μl of the same medium was placed in the lower chamber. When seeded, cells were respectively pretreated with appropriate concentration of inhibitors; SB203580, SP600125, U0126, and wortmannin. After 1 hr of pretreatment, thrombin was added to the lower chamber and the plates were incubated for 6 hr at 37℃ in 5% CO_2 incubator. Subsequently, cells were fixed in 3.7% formaldehyde solution for 10 min and permeabilized by 0.2% Triton X-100 for 10 min. Then, cells on the lower surface were stained with 0.2% crystal violet and rinsed with PBS several times. Non-migrating cells in the upper chamber were attentively removed by cotton swabs. The migrating cells were counted by light microscopy.

D. *Western Blot Analysis*

After removed the culture media, the C6 rat glioma cells were harvested and lysed with PRO-PREPTM Protein Extraction Solution (iNtRON Biotechnology, Inc.). The obtained protein samples were quantified by the BCA protein assay and the same amounts of total cellular proteins were resolved on 10% SDS-polyacrylamide gels. After SDSPAGE, proteins were transferred to polyvinylidene difluoride (PVDF) membranes (Amersham Biosciences, Piscataway, NJ, USA). The membranes were incubated in blocking solution (5% skim milk in TBST) for 1 hr at room temperature and then probed overnight with suitable antibodies at 4oC; anti-MMP-9 (Affinity BioReagents, Golden, CO, USA; 1 : 1,000), p-Akt, Akt (Cell Signaling Technology, Inc.; 1 : 1,000), or β-actin (Sigma-Aldrich, St. Louis, MO, USA; 1 : 5,000). After three washes with TBST, the membranes were incubated with either horseradish peroxidase (HRP)-conjugated goat anti-rabbit or donkey anti-mouse secondary antibody (Jackson ImmunoResearch) for 2 hr at room temperature. The blots were detected by enhanced chemiluminescence using Kodak® BioMaxTM XAR Film.

E. *RNA Isolation and Semi-quantitative RT-PCR Analysis*

Isolation of total cellular RNA from C6 glioma cells was performed using the Total RNA Extraction Kit (iNtRON Biotechnology, Inc.) according to the manufacturer's instructions. Total RNA was quantified by NanoDrop spectrophotometer (Thermo Fischer Scientific Inc., Wilmington, DE, USA) and 2 μg of total RNA was immediately reverse-transcribed by Omniscript® RT Kit (QIAGEN GmbH, Hilden, Germany) with oligo-$(dT)_{15}$ primers (Promega, Madison, WI, USA). The cDNA was amplified by polymerase chain reaction (PCR) with appropriate primers (MMP-9 forward, TCCAGTAGAC-AATCCTTGCA and reverse, CTCCGTGATTCGAG-AACTTC; GAPDH forward, GACAACTTTGGCAT-CGTGGA and reverse, ATGCAGGGATGATGTTCTGG). 30 cycles of PCR was performed by n*Taq* polymerase under the following conditions: denaturing at 94℃ for 30 sec, annealing at 62℃ for 45 sec, and elongation at 72℃for 30 sec. The PCR products (MMP-9, 110 bp; GAPDH, 133 bp) were electrophoresed on a 2% agarose gel containing ethidium bromide and visualized by UV transillumination.

F. *Gelatin Zymography*

The gelatin zymography was carried out in 8% polyacrylamide gels containing 1 mg/ml of gelatin. C6 glioma cells were plated in 60 mm culture dishes and incubated overnight in serum- free medium. Cell culture supernatants were collected and concentrated by Amicon® Ultra Centrifugal Filter (Millipore, Bedford, MA) subsequent to 24 hr-thrombin treatment. These samples were mixed with non-reducing 5× loading buffer without heating and electrophoresed on the gel as mentioned above. After PAGE, gels were thoroughly washed with enzyme renaturing buffer (2.5% Triton X-100 solution) at room temperature for 4 times every 15 min. The gels were then transferred into incubation buffer (50 mM Tris-HCl, 0.15 M NaCl, 10 mM CaCl2, and 0.02% NaN_3) and placed in a 37℃ shaking incubator for 24 hr followed by Coomassie brilliant blue R-250 staining and destaining procedures.

G. *Statistical Analysis*

All data are expressed as the mean ± standard deviation (SD). Statistical significance was analyzed by two-tailed Student's t-test. Data with values of $p < 0.05$ were considered as statistically significant. Single (* and #) and double (** and ##) marks represent statistical significance in $p < 0.05$ and $p < 0.01$, respectively.

III. RESULTS

A. *Thrombin Induced Migration of C6 Glioma Cells*

In the previous studies, it has been demonstrated that thrombin induces cell motility in many cell lines [9,13,14]. To investigate the effect of thrombin on the migration of C6 glioma cells, we used the Boyden chamber migration assay.

Approximately 0.7 × 10^5 cells were plated in the upper chamber of cell culture inserts and incubated with or without 25 U/ml of thrombin. After 6 hr of incubation, the cells migrating through the membrane toward thrombin containing serum-free media were stained with crystal violet and the number of migrated cells was counted (Fig. 1). The number of migrated cells was significantly increased in the presence of thrombin compared to control, suggesting that thrombin has a chemotactic effect on C6 glioma cells.

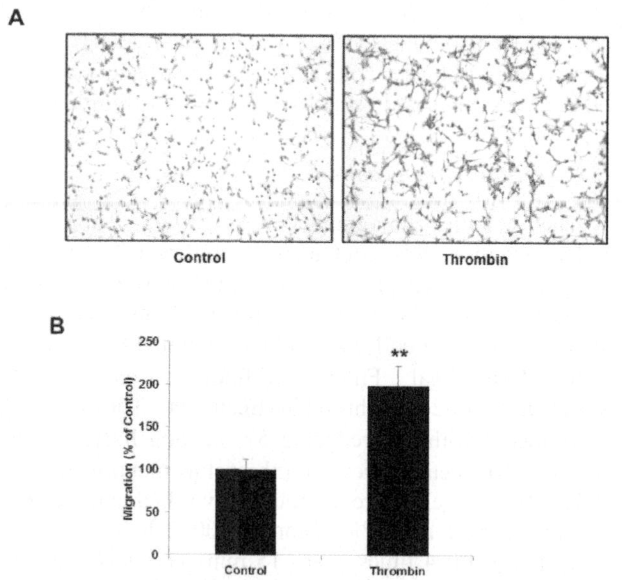

Fig. 1 Thrombin stimulates C6 glioma cell migration. (A) Representative images of C6 cell migration assay. Thrombin (25 U/ml) treatment for 6 hr exhibited approximately 2-fold increase in number of migrating cells. Migrating cells and unmigrating cells were separated by polyethylene terephthalate (PET) membranes. Microscopy images were detected the migrating cells on the lower surface of the membrane. (B) Quantitative analysis of migration assay. The cell migration was quantified by counting the cells that migrating through the membranes. Cell counting data were expressed as a percentage in comparison with control group. The experiments were conducted in triplicate and data are shown as the means ± SD. **$p < 0.01$ indicates statistically, significant difference with control.

B. MAPK Inhibitors Had an Inhibitory Effect on Thrombin-Induced Migration of C6 Glioma Cells

Involvement of MAPK pathway in the regulation of cancer cell migration has been investigated by many studies [15-17]. In the present study, to determine the underlying mechanism for chemotactic action of thrombin, C6 cells were treated with some MAPK pathway inhibitors: SB203580 (20 μM), a specific inhibitor of p38-MAPK; SP600125 (20 μM), a selective inhibitor of c-Jun N-terminal kinase (JNK); U0126 (10 μM), a selective inhibitor of MEK/ERK. All inhibitors were pretreated for 1 hr in serum free media with indicated concentrations and thrombin was subsequently treated for 6 hr. As shown in Fig. 2, thrombin-induced migrating cells were reduced with pretreatment of the MAPK inhibitors. Inhibition of p38 with SB203580 showed significant inhibition of C6 migration (Fig. 2). Especially, inhibition of JNK and ERK with SP600125 and U0126, respectively, exerted such a strong inhibition that the number of migrated cells was reduced almost to control level (Fig. 2). These results suggest that MAPK pathway may regulate thrombin-induced migration of C6 glioma cells.

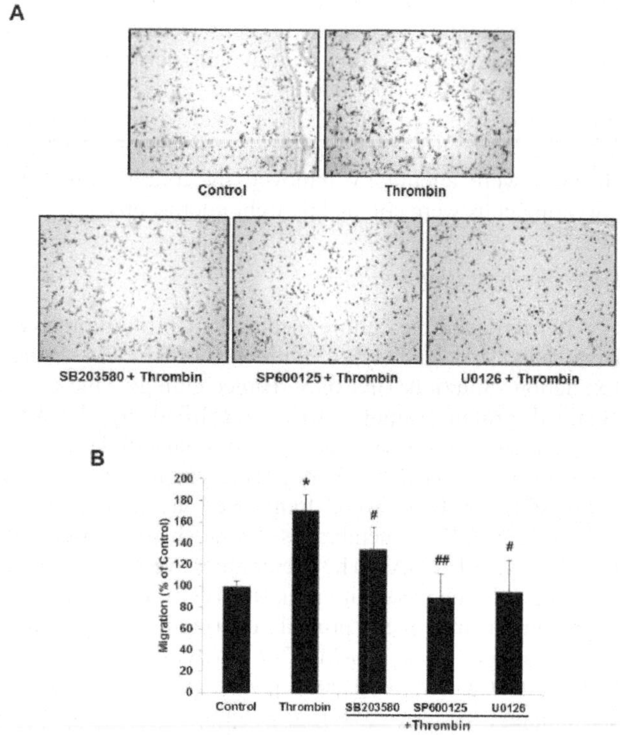

Fig. 2 MAPK inhibitors (MAPKIs) inhibit thrombin-induced C6 glioma cell migration. (A) Representative images of the MAPKI effects on thrombin-induced migration. SB203580, SP600125, and U0126 were pretreated for 1 hr at the concentrations of 20, 20, and 10 μM, respectively, followed by thrombin treatment for 6 hr. (B) Quantitative analysis of migration assay. The cell migration was quantified by counting the cells that migrating through the membranes. All data were obtained from three independent experiments and presented as the means ± SD.

C. Thrombin-Induced C6 Cell Migration Was Significantly Increased with Wortmannin, a PI3K Inhibitor

The previous studies have been shown that PI3K pathway is also related to the cancer cell motility [18,19]. On this account, we investigated the effect of the PI3K pathway on thrombin-induced C6 cell migration with a PI3K inhibitor wortmannin, an antifungal antibiotic. 100 nM of wortmannin was used for migration assay and other experimental procedures were the same as mentioned above.

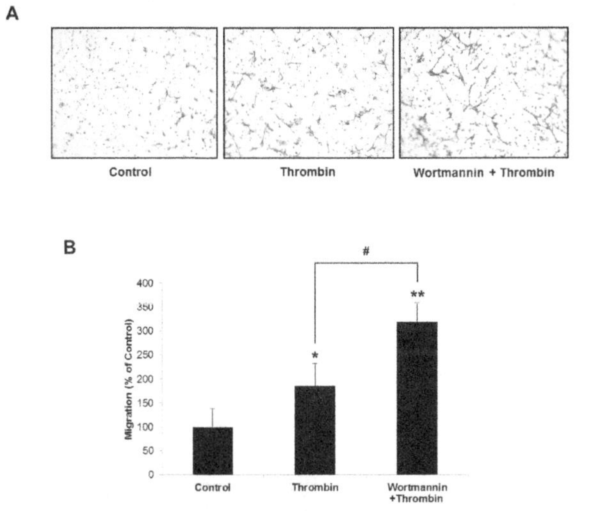

Fig. 3 Thrombin-induced chemotactic migration was markedly increased by PI3K inhibitor, wortmannin. (A) Potentiating effects were observed in the group that treated thrombin for 6 hr subsequent to preincubated with wortmannin for 1 hr. (B) Quantitative analysis of migration assay. The cell migration was quantified by counting the cells that migrating through the membranes. All data were obtained from three independent experiments and presented as the means ± SD. *$p < 0.05$ and **$p < 0.01$ indicate statistically significant difference with control. #$p < 0.05$ indicates statistically significant difference between thrombin and wortmannin thrombin.

Wortmannin significantly potentiated the migratory activity induced by thrombin (Fig. 3). This result may suggest that the PI3K pathway negatively regulates thrombin-induced C6 cell migration.

D. MMP 9 Expression Was Correlated with C6 Cell Migration Propensity

To elucidate whether the thrombin-induced migration and MMP-9 expression are directly related, we used an MMP-9 inhibitor (MMPI), batimastat, for migration assay. The suppressive effect of MMPI is represented in Fig. 6. It clearly showed that MMP-9 is directly involved in cell migration stimulated by thrombin. To determine whether MMP-9 is regulated by MAPK and PI3K pathways in the thrombin-induced migration of C6 glioma cells, we examined the changes of MMP-9 expression in multiple stages such as mRNA, protein, and extracellular levels. C6 cells were pre-incubated with the MAPK or PI3K inhibitors for 1 hr and thrombin was subsequently treated for 24 hr in serum-free media. After incubation, the conditioned media were collected and concentrated for zymography and the cells were harvested and lysed for western blot analysis. Gelatinolytic activities of MMP-9 were visualized by zymography. As shown in Fig. 4A, thrombin induces MMP-9 secretion and it is suppressed by MAPK inhibitors whereas thrombin-induced MMP-9 secretion was potently increased by wortmannin (Fig. 4A). These results closely correspond to migration assay data. Western blot analysis also presented the same tendency to zymographic data. To further confirm, we measured MMP-9 expression in mRNA level using semi-quantitative RT-PCR assay. After pretreated the inhibitors for 1 hr, C6 cells were incubated with thrombin for 4 hr. MMP-9 mRNA was expressed in the same manner as MMP-9 protein expression (Fig. 5). These data strongly support that thrombin-mediated MMP-9 expression is regulated by MAPK and PI3K pathways.

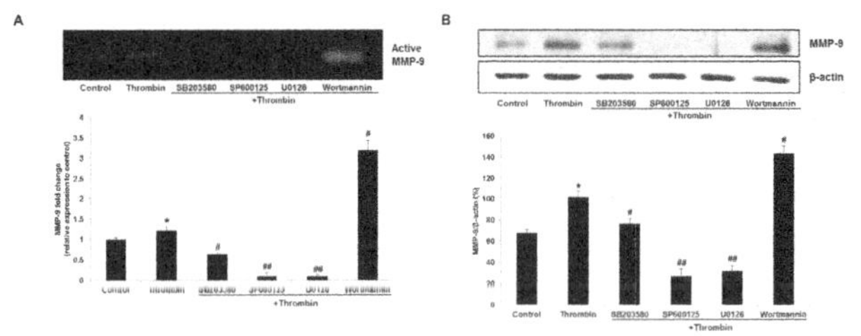

Fig. 4 Effects of thrombin and MAPK or PI3K inhibitors treatment on MMP-9 expression in protein level. All inhibitors were pretreated for 1 hr at the indicated concentrations followed by 24 hr of thrombin (5 U/ml) treatment: SB203580 and SP600125, 20 μM; U0126 and wortmannin, 10 μM. C6 cells were incubated with inhibitors in the serum free media. (A) MMP-9 activities were analyzed by zymography. Thrombin-induced MMP-9 expression was blocked by MAPK inhibitors and increased by PI3K inhibitor. (B) The cellular level of MMP-9 was examined by western blot analysis. β-actin was used as an internal control. *Indicates statistically significant difference with control. #$p < 0.05$ and ##$p < 0.01$ indicates statistically significant difference with thrombin.

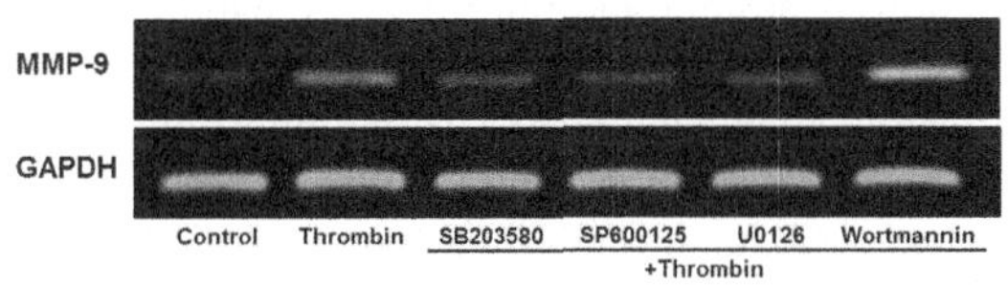

Fig. 5 Effects of thrombin and MAPK or PI3K inhibitors treatment on MMP-9 expression in mRNA level. C6 cells were pretreated with inhibitors for 1 hr, followed by thrombin (5 U/ml) treatment for 4 hr. Subsequently, RNA samples were isolated and subjected to semi-quantitative RT-PCR for MMP-9 with a housekeeping gene, GAPDH, as an internal control. MMP-9 mRNA expression patterns were similar to western blot and zymographic data.

IV. Discussion

Glioblastoma multiforme (GBM) is one of the most aggressive CNS tumors with very poor prognosis because of its highly-infiltrative nature [20]. GBM has a histologically characteristic feature which forms a serpentine pattern of necrosis and vascular or endothelial cell proliferation. These necrotic lesions often contain hemorrhagic components [2]. These conditions lead glioma cells to have more chances of exposure to thrombin. Previous studies have been revealed that thrombin is involved in cancer cell migration and invasion [9,10,21]. In the present study, we found that thrombin enhances migration of C6 rat glioma cells by Boyden chamber migration assay. To determine the mechanism of the thrombin effect, we considered MAPK and PI3K pathways as pivotal factors based on the former studies [22,23]. Thrombin-induced migratory ability was significantly blocked by MAPK inhibitors. Wortmannin, a specific inhibitor of the PI3K pathway, on the other hand, surprisingly reinforced the thrombin-mediated cell migration. We used 100 nM and 1 μM of wortmannin for migration assay, but 1 μM was less potent than 100 nM of wortmannin in inducing migratory ability (data not shown). These results might be caused by excessive PI3K pathway inhibition in 1 μM of wortmannin treatment. Because PI3K pathway is critical in cell survival, its inordinate suppression subjects to cellular damage. The C6 cells treated with 1 μM of wortmannin presented slight morphological changes and with 10 μM showed more severe changes in migration assay. As MMPs are key components of cell migration and invasion [5,24], we investigated the relationship between C6 cell migration and MMP-9. Thrombin-induced migration was effectively suppressed by an MMP-9 inhibitor, batimastat. It supports that MMP-9 is directly involved in cell migration stimulated by thrombin. However, batimastat alone could not totally block the thrombin-induced migration. This implies that there are more factors responsible for thrombin-induced cell migration other than MMP-9. Thrombin induced MMP-9 expression in protein and mRNA level, parallel to increase in C6 cell migration. Induced MMP-9 expression was significantly attenuated by MAPK inhibitors and augmented by the PI3K inhibitor. Our results indicate that these pathways regulate thrombin-mediated MMP-9 expression and it affects migratory activity of C6 cells. That is, the activator of PI3K pathway, such as the platelet-derived growth factor (PDGF), or MAPK inhibitors might inhibit the thrombin-induced MMP-9 expression and migration. A previous study actually demonstrated that PDGF inhibits MMP-9 activity in C6 cells [25]. Wortmannin, for itself, did not increase both MMP-9 expression and cell migration (Data not shown). In conclusion, the present study suggests that thrombin mediates C6 glioma cell migration by increasing expression of MMP-9. In addition, MAPK and PI3K pathways might be involved in thrombin-induced cell migration and MMP-9 expression. The migratory ability of cells is an important factor for cancer metastasis. Together with the previous studies, this study suggests that inhibition of MAPK pathway or potentiation of PI3K pathway could be a valuable therapeutic target for the treatment of metastatic glioblastoma.

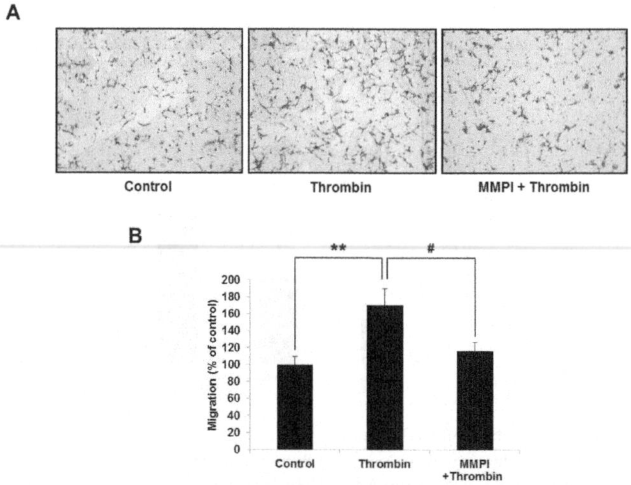

Fig. 6 Thrombin-induced migration was directly blocked by batimastat, an MMP-9 inhibitor (MMPI). (A) Representative images of the effect of MMPI on thrombin-induced C6 cell migration. 0.5μM of batimastat was pretreated for 1 hr and thrombin was subsequently treated for 6 hr. (B) Quantitative analysis of migration assay. The cell migration was quantified by counting the cells that migrating through the membranes. All data were obtained from three independent experiments and presented as the means ± SD. $^{**}p<0.01$ and $\#p<0.05$ indicate statistically significant difference between control and thrombin, and thrombin and MMPI thrombin respectively.

References

1. Louis DN, Ohgaki H, Wiestler OD, Cavenee WK, Burger PC, Jouvet A, Scheithauer BW, Kleihues P. The 2007 WHO classification of tumours of the central nervous system. Acta Neuropathol. 2007;114:97-109.
2. Lefranc F, Brotchi J, Kiss R. Possible future issues in the treatment of glioblastomas: special emphasis on cell migration and the resistance of migrating glioblastoma cells to apoptosis. J Clin Oncol. 2005;23:2411-2422.
3. Holland EC. Glioblastoma multiforme: the terminator. Proc Natl Acad Sci USA. 2000;97:6242-6244. 4. Kessenbrock K, Plaks V, Werb Z. Matrix metalloproteinases: regulators of the tumor microenvironment. Cell. 2010;141: 52-67.
4. Page-McCaw A, Edward AJ, Werb Z. Matrix metalloproteinases and the regulation of tissue remodeling. Nat Rev Mol Cell Biol. 2007;8:221-233.
5. Björklund M, Koivunen E. Gelatinase-mediated migration and invasion of cancer cells. Biochim Biophys Acta. 2005;1755: 37-69.

6. Kawakita K, Kawai N, Kuroda Y, Yasashita S, Nagao S. Expression of matrix metalloproteinase-9 in thrombin-induced brain edema formation in rats. J Stroke Cerebrovasc Dis. 2006;15:88-95.
7. Choi MS, Kim YE, Lee WJ, Choi JW, Park GH, Kim SD, Jeon SJ, Go HS, Shin SM, Kim WK, Shin CY, Ko KH. Activation of protease-activated receptor1 mediates induction of matrix metalloproteinase-9 by thrombin in rat primary astrocytes. Brain Res Bull. 2008;76:368-375.
8. Radjabi AR, Sawada K, Jagadeeswaran S, Eichbichler A, Kenny HA, Montag A, Bruno K, Lengyel E. Thrombin induces tumor invasion through the induction and association of matrix metalloproteinase-9 and beta1-integrin on the cell surface. J Biol Chem. 2008;283:2822-2834.
9. Chiang HS, Yang RS, Huang TF. Thrombin enhances the adhesion and migration of human colon adenocarcinoma cells via increased beta 3-integrin expression on the tumour cell surface and their inhibition by the snake venom peptide, rhodostomin. Br J Cancer. 1996;73:902-908.
10. Kaufmann R, Junker U, Junker K, Nuske K, Ranke C, Zieger M, Scheele J. The serine proteinase thrombin promotes migration of human renal carcinoma cells by a PKA-dependent mechanism. Cancer Lett. 2002;180:183-190.
11. Heider I, Schulze B, Oswald E, Henklein P, Scheele J, Kaufmann R. PAR1-type thrombin receptor stimulates migration and matrix adhesion of human colon carcinoma cells by a PKCepsilon-dependent mechanism. Oncol Res. 2004;14: 475-482.
12. Grobben B, De Deyn PP, Slegers H. Rat C6 glioma as experimental model system for the study of glioblastoma growth and invasion. Cell Tissue Res. 2002;310:257-270.
13. Bizios R, Lai L, Fenton JW 2nd, Malik AB. Thrombin-induced chemotaxis and aggregation of neutrophils. J Cell Physiol. 1986;128:485-490.
14. Pankonin G, Teuscher E. Stimulation of endothelial cell migration by thrombin. Biomed Biochim Acta. 1991;50:1073-1078.
15. Reddy KB, Nabha SM, Atanaskova N. Role of MAP kinase in tumor progression and invasion. Cancer Metastasis Rev. 2003;22:395-403.Jadeski LC, Chakraborty C, Lala PK. Nitric oxide-mediated promotion of mammary tumour cell migration requires sequential activation of nitric oxide synthase, guanylate cyclase and mitogen-activated protein kinase. Int J Cancer. 2003;106: 496-504.
16. Jadeski LC, Chakraborty C, Lala PK. Nitric oxide-mediated promotion of mammary tumour cell migration requires sequential activation of nitric oxide synthase, guanylate cyclase and mitogen-activated protein kinase. Int J Cancer. 2003;106: 496-504.
17. Frankenberry KA, Somasundar P, McFadden DW, Vona-Davis LC. Leptin induces cell migration and the expression of growth factors in human prostate cancer cells. Am J Surg. 2004; 188:560-5.
18. Barber MA, Welch HC. PI3K and RAC signalling in leukocyte and cancer cell migration. Bull Cancer. 2006;93:E44-52.
19. Vasko VV, Saji M. Molecular mechanisms involved in differentiated thyroid cancer invasion and metastasis. Curr Opin Oncol. 2007;19:11-7.
20. Kanu OO, Hughes B, Di C, Lin N, Fu J, Bigner DD, Yan H, Adamson C. Glioblastoma Multiforme Oncogenomics and Signaling Pathways. Clin Med Oncol. 2009;3:39-52.
21. Chen HT, Tsou HK, Tsai CH, Kuo CC, Chiang YK, Chang CH, Fong YC, Tang CH. Thrombin enhanced migration and MMPs expression of human chondrosarcoma cells involves PAR receptor signaling pathway. J Cell Physiol. 2010;223:737-745.
22. Wang Z, Castresana MR, Newman WH. Reactive oxygen species-sensitive p38 MAPK controls thrombin-induced migration of vascular smooth muscle cells. J Mol Cell Cardiol. 2004; 36:49-56.
23. Wang L, Luo J, He S. Induction of MMP-9 release from human dermal fibroblasts by thrombin: involvement of JAK/STAT3 signaling pathway in MMP-9 release. BMC Cell Biol. 2007;8:14.
24. Watanabe H. Extracellular matrix--regulation of cancer invasion and metastasis. Gan To Kagaku Ryoho. 2010;37:2058-2061.
25. Estève PO, Robledo O, Potworowski EF, St-Pierre Y. Induced expression of MMP-9 in C6 glioma cells is inhibited by PDGF via a PI 3-kinase-dependent pathway. Biochem Biophys Res Commun. 2002;296:864-869.

Author: Sanghoon Cha
Institute: Department of Systems Immunology, College of Biomedical Science, Kangwon National University
Street: 192-1, Hyoja 2-dong
City: Chuncheon
Country: South Korea
Email: chash@kangwon.ac.kr

Diffuse Optical Tomography and Biophysical Modeling of the Aging Brain

F. Lesage[1,2], C. Bonnery[1], and P. Pouliot[1,2]

[1] Institute of Biomedical Engineering, École Polytechnique de Montréal, Montreal, Canada
[2] Montreal Heart Institute, Montreal, Canada

Abstract — **Diffuse optical imaging (DOI) of the brain has emerged as a new neuroimaging technique that is seeing growing applications in brain research. The technique's simplicity, lower cost and applicability to children have contributed to its popularity. However, poor quantification of local hemodynamic changes based on optical signals remains a major hurdle and significant efforts are still being pursued to better understand how to recover accurate images of functional activation, generate statistical maps and recognize the limitations of DOI. In the context of aging, further confounds arise due to anatomical and physiological changes that must be accounted for in the signal modeling. In this work, DOI signals are revisited in this context. In particular we investigate the diffusion path length factor and its modifications with age.**

Keywords — **Diffuse Optical Imaging, Aging.**

I. Introduction

Cognitive decline of the elderly severely affects quality of life for millions of people, highlighting the need for a better understanding of the mechanisms of aging in the brain. Aging is normally associated with various systemic and specific vascular changes. In large arteries, possible changes in diameter, smooth muscle vascularisation, elastin and collagen content, wall stiffness and stress, lead to increased flow pulsatility that could threaten micro-circulation particularly in organs which are highly perfused, such as the brain.

Since many components affecting the hemodynamic response seem to be modified with age, the interpretation of imaging studies of age is difficult. To untangle the contributions of the different components from measured data, biophysical models describing their interaction might provide a useful tool. In particular, biomechanical models of cerebral vasculature offer an opportunity to explore this aspect of aging in the absence of alternative direct methods for probing micro-vascular mechanics in vivo.

Diffuse optical imaging (DOI) of the brain is a relatively new functional imaging modality using near-infrared (NIR) light propagation (650-900 nm) in biological tissues [1]. Since hemoglobin is the main absorber in this range, DOI has enabled researchers [2–5] to image hemodynamics *in vivo* through centimeters of biological tissues. By acquiring data at two or more wavelengths, near-infrared spectroscopy (NIRS) can distinguish and record simultaneous changes in deoxyhemoglobin (HbR) and oxyhemoglobin (HbO_2) [6]. NIRS has seen growing interest in neuroscience due to the recent widespread availability of commercial instruments. A non-exhaustive list of other applications includes bedside monitoring [7], imaging newborns [8], and epilepsy [9]. In this work, we summarize our investigation of the capabilities of optical imaging and NIRS to characterize changes in vascular health with age.

II. Background

A. Principles of Optical Imaging

In the wavelength range 650–900 nm light travels in tissues with reduced attenuation. At these wavelengths, oxy (HbO_2) and deoxyhemoglobin (HbR) dominate light absorption while other chromophores such as water, cytochrome oxydase and lipids have a marginal contribution to absorption. The amplitude of DOI signals are modulated by many factors, all of which need to be accounted for quantification. Amongst static factors, the optical properties of tissues, absorption and scattering, the anatomy and the so-called partial volume effect all modulate the signal. A second set of challenges in quantifying the brain response can be attributed to dynamical factors: distinguishing spatially and functionally the origin of hemodynamic changes is required to differentiate (surface) physiology from neuronal induced signals.

In a typical functional imaging experiment, the signature of interest is the response evoked by a controlled stimulus. In this pre-processing step, significant differences emerge with functional magnetic resonance imaging (fMRI). The acquisitions are done with higher temporal sampling (10-100 Hz) and both oxy and deoxyhemoglobin can be measured. As a result, physiology appears very prominently rather than being aliased, as the acquisition rate is faster than expected frequencies of the nuisance signals (heart beat, respiration, Mayer waves [10]). The limited signal-to-noise ratio usually necessitates a certain number of trials to robustly identify an evoked response and the detection and estimation of the response evoked by a stimulus have been solved by two main techniques (see [11] for a review).

V. Van Toi et al. (Eds.): 4th International Conference on Biomedical Engineering in Vietnam, IFMBE Proceedings 40, pp. 250–253, 2013.
www.springerlink.com

The first is block averaging whereby low and high frequency noise components are band-pass filtered out of the data while saving the frequency support of the evoked response. The second uses the convolution of the experimental paradigm with a prototype hemodynamic response. It can be applied to both blocked and event-related designs and relies on the general linear model (GLM, see below for more details on this approach) where drifts are introduced in the model as regressors.

B. *Statistical Parametric Mapping in NIRS*

A central difficulty in applying a univariate GLM approach to DOI (i.e. a separate GLM for each measurement type on each channel) is the sparseness of measurements: accurate accounting for false positives due to multiple testing is not straightforward. This has been addressed recently by Ye et al. [12]. With dense measurements and pre-processing for spatio-temporal whitening of the noise, one method correcting for false positives due to multiple measurements in the GLM was addressed for t-statistics by considering the excursion probability for an *inhomogeneous* random field, similar to that found in voxel-based morphometry in fMRI, and using a technical result referred to as Sun's tube formula [13].

C. *Modeling the Hemodynamic Response*

The biomedical relevance of NIRS and other hemodynamic imaging modalities lies in the physiological interpretation of the data. It has been known for over a century [14] that neural activity evokes changes in cerebral blood flow and volume. Functional increases in oxygen metabolic demand, associated with disproportionately large [15] evoked increases in blood flow and volume, result in a local increase (decrease) in HbO_2 (HbR) tissue concentration.

These changes in concentration of natural contrast agents HbO_2 and HbR evoked by stimuli have been used for functional mapping of the brain by DOI and fMRI [16] in the last decades. However, the interpretation of these signals is hindered by incomplete characterization of the mechanisms underlying the hemodynamic response, a topic which still evokes controversy [17], [18] or [19], [20].

Because they share a common physiological origin, multimodal hemodynamic measurements can be combined in the framework of a biophysical model which describes the interaction between vascular and metabolic dynamics and their effect on measurable quantities. The physical relationship between pressure, flow and volume in fluids can be applied to hemodynamics in the brain [21–23]. Models of gas diffusion and cellular metabolism can describe oxygen transport and consumption by neuron. The so-called balloon models have been developed to explain the blood oxygen level dependent (BOLD) fMRI signal [24]. They were later expanded by Mandeville et al. [25] who explained the underlying physiology by explicit modeling of time-varying resistance and compliance of the so-called windkessel (elastic vascular compartment). In preliminary studies, these were used to study the mechanisms underlying the hemodynamic response from optical data [26], [27].

III. Application to Aging

A. *Optical Modeling in Aging Animals*

As a case study and to establish the applicability of optical methods to the aging brain, we performed a multimodal study in rats of different ages.

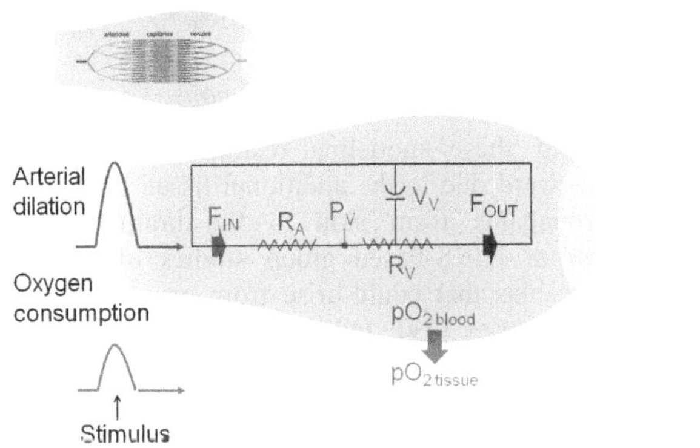

Fig. 1 Schematic representation of the biophysical model of brain hemodynamic response.

Briefly, young (Y, 4-5 months), old (O, 24-32 months) and very-old (VO, 40- 41 months) male and female LOU/c rats were studied. Functional images were recorded with multiple wavelength illumination (525, 590, 637 nm). For laser speckle imaging, a near infrared 780 nm laser diode was used. The data set was used to derive measured changes in HbO, HbR, HbT and blood flow following stimulations. In data presented below, a single short stimulation train (1 sec at 3 Hz) was used for parameter estimations leading to a single input, multiple output model (SIMO). A first order differential form of the biophysical model shown in Fig. 1 was used to describe measurements. Bayesian inversion of the models was done following previous work [28]. Given measures with Gaussian additive noise, we estimated the parameters, using expectation maximization (EM) algorithm under the assumption of Gaussian posterior density for the parameters described by their conditional mean and covariance.

Two main conclusions could be gathered from these studies. First, using extensive simulations, data confirmed that biophysical model parameters could be estimated using the data sets measured with optical imaging. Second, we confirmed that some of the parameters estimated change with age. In particular model estimations show that vessel compliance was reduced with age in the brain and that the baseline metabolic consumption was also reduced (Table 1) in agreement with the literature.

Table 1 Parameters changing with age.

Parameter	Young	Old	Very Old
β, "Inverse compliance"	2.26±0.75	3.29±1.32	3.39±0.69
M_0, CMRO2	3.6±0.2	1.9±1.1	1.7±1.0

These results show the potential of optical imaging towards the assessment of brain health. However, translation of these invasive animal measures to humans is difficult due to additional methodological difficulties.

B. Optical Modeling in Aged Humans

Applying these modeling results to humans is not straightforward due to the additional tissue layers in which light propagates from skin to the brain. A potential limitation of NIRS-based group studies of aging is the systematic bias that could arise from group differences in optical properties [29]. Indeed, hemoglobin concentration changes calculated from the modified Beer-Lambert law

$$\Phi_\Delta(t,\lambda) = -\ln\left(\frac{I(t,\lambda)}{I_0(\lambda)}\right) = -\ln\left(\frac{\Phi(t,\lambda)}{\Phi_0(\lambda)}\right) = d\cdot DPF(\lambda)\cdot\Delta\mu_a \quad (1)$$

require estimations of the optical path length at each wavelength [30], [31]. Therefore an age-related difference in DPF [29] or partial volume factor, if not quantified in the application of Eq. (1), could be the cause of a difference in estimated HbO_2/HbR responses. A robust quantification of age-specific effects on path length and partial volume is thus needed.

To address this issue, both young and old subjects were scanned in MRI and time-domain diffuse optical imaging. The time-domain technique, by assuming homogeneous medium, can be used to estimate the factor: DPF(λ). Separately, Monte-Carlo simulations of light propagation in a realistic anatomical geometry derived from the MRI scans can also be used to assess age-related changes in DPF(λ). In Fig. 2 we present the results of both assessments. We observe a small change of DPF with age (much smaller than previous results from Duncan et al.).

Accounting for the partial volume effect, changes during cognitive tasks can be integrated into a biophysical model with the goal of estimating physiologically relevant parameters. The application of our modeling work in animal models to humans will form the basis of our future work on the subject.

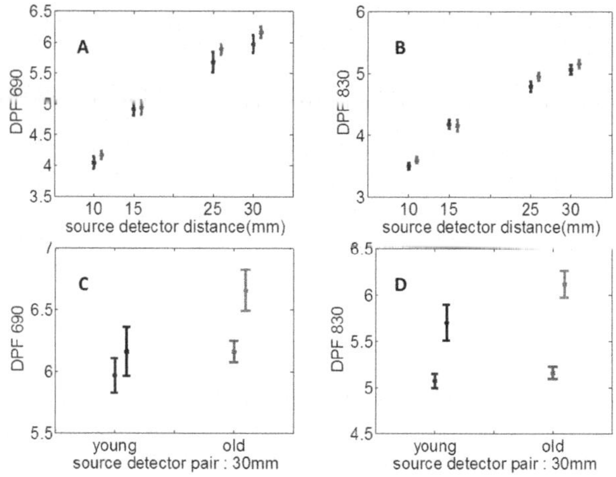

Fig. 2 Time domain measure of differential path length factor at different source-detector distances for 690nm (A) and 830nm (B) for young (blue) and older (red) subjects. Comparison of these values with Monte-Carlo derived DPFs for the 30 mm distance (C,D) (black and cyan).

IV. Conclusions

In order to improve the interpretation of hemodynamic signals especially those measured in atypical and heterogeneous populations, multimodality and modeling of the hemodynamic response are promising approaches. In light of recent evidence that vascular regulation could be altered with healthy aging as well as disease, medical research and potentially even diagnosis could benefit from the use of hemodynamic imaging with an adequate model of the underlying physiology. Our results show that optical imaging is a technique that may yield information useful for diagnosis but much remains to be done to establish a methodology for robust measures of hemodynamic changes with age.

Acknowledgment

P. Pouliot and F. Lesage by the Emerging Team Grant Program of the Canadian Institutes of Health Research, and F. Lesage also by the MDEIE and the Discovery Grant Program of the Natural Sciences and Engineering Research Council of Canada.

References

1. F. Jöbsis, "Noninvasive, infrared monitoring of cerebral and myocardial oxygen sufficiency and circulatory parameters," Science, vol. 198, no. 4323, pp. 1264-1267, 1977.
2. A. Yodh and B. Chance, "Spectroscopy and imaging with diffusing light," Physics Today, vol. 48, no. 3, pp. 40, 34, 1995.

3. D. A. Boas, K. Chen, D. Grebert, and M. A. Franceschini, "Improving the diffuse optical imaging spatial resolution of the cerebral hemodynamic response to brain activation in humans," Optics Letters, vol. 29, no. 13, pp. 1506-1508, Jul. 2004.
4. J. P. Culver, A. M. Siegel, J. J. Stott, and D. A. Boas, "Volumetric diffuse optical tomography of brain activity," Optics Letters, vol. 28, no. 21, pp. 2061-2063, Nov. 2003.
5. M. A. Franceschini and D. A. Boas, "Noninvasive measurement of neuronal activity with near-infrared optical imaging," NeuroImage, vol. 21, no. 1, pp. 372-386, Jan. 2004.
6. D. T. Delpy and M. Cope, "Quantification in tissue near–infrared spectroscopy," Philosophical Transactions of the Royal Society of London. Series B: Biological Sciences, vol. 352, no. 1354, pp. 649-659, 1997.
7. C. Kurth, J. Steven, D. Benaron, and B. Chance, "Near-infrared monitoring of the cerebral circulation," Journal of Clinical Monitoring and Computing, vol. 9, no. 3, pp. 163-170, Jul. 1993.
8. J. S. Wyatt, M. Cope, D. T. Delpy, and et al., "Quantitation of cerebral blood volume in human infants by near-infrared spectroscopy," Journal of Applied Physiology, vol. 68, no. 3, pp. 1086-1091, Mar. 1990.
9. F. Rosenow and H. Luders, "Presurgical evaluation of epilepsy," Brain, vol. 124, no. 9, pp. 1683-1700, Sep. 2001
10. 10 H. Nilsson and C. Aalkjaer, "Vasomotion: mechanisms and physiological importance," Molecular Interventions, vol. 3, no. 2, pp. 79-89, 51, Mar. 2003.
11. T. J. Huppert, S. G. Diamond, M. A. Franceschini, and D. A. Boas, "HomER: a review of time-series analysis methods for near-infrared spectroscopy of the brain," Applied Optics, vol. 48, no. 10, pp. D280-298, Apr. 2009.
12. J. Ye, S. Tak, K. Jang, J. Jung, and J. Jang, "NIRS-SPM: Statistical parametric mapping for near-infrared spectroscopy," NeuroImage, vol. 44, no. 2, pp. 428-447, Jan. 2009.
13. J. Sun, "Tail Probabilities of the Maxima of Gaussian Random Fields," The Annals of Probability, vol. 21, no. 1, pp. 34-71, Jan. 1993.
14. C. S. Roy and C. S. Sherrington, "On the regulation of the blood-supply of the brain," vol. 11, no. 1-2, pp. 85-158, 1890.
15. P. T. Fox and M. E. Raichle, "Focal physiological uncoupling of cerebral blood flow and oxidative metabolism during somatosensory stimulation in human subjects," Proceedings of the National Academy of Sciences of the United States of America, vol. 83, no. 4, pp. 1140-1144, 1986.
16. S. Ogawa, D. W. Tank, R. Menon, and et al., "Intrinsic signal changes accompanying sensory stimulation: functional brain mapping with magnetic resonance imaging," Proceedings of the National Academy of Sciences of the United States of America, vol. 89, no. 13, pp. 5951-5955, 1992.
17. A. Devor, E. M. C. Hillman, P. Tian, and et al., "Stimulus-induced changes in blood flow and 2-Deoxyglucose uptake dissociate in ipsilateral somatosensory cortex," J. Neurosci., vol. 28, no. 53, pp. 14347-14357, Dec. 2008.
18. J. Rossier, "Wiring and plumbing in the brain," Frontiers in Human Neuroscience, vol. 3, p. 2, 2009.
19. M. L. Schroeter, T. Kupka, T. Mildner, K. Uludağ, and D. Y. von Cramon, "Investigating the post-stimulus undershoot of the BOLD signal--a simultaneous fMRI and fNIRS study," NeuroImage, vol. 30, no. 2, pp. 349-358, Apr. 2006.
20. J. J. Chen and G. B. Pike, "Origins of the BOLD post-stimulus undershoot," NeuroImage, vol. 46, no. 3, pp. 559-568, Jul. 2009.
21. E. W. Washburn, "The dynamics of capillary flow," Physical Review, vol. 17, no. 3, p. 273, Mar. 1921.
22. T. J. Huppert, M. S. Allen, H. Benav, P. B. Jones, and D. A. Boas, "A multicompartment vascular model for inferring baseline and functional changes in cerebral oxygen metabolism and arterial dilation," Journal of Cerebral Blood Flow and Metabolism, vol. 27, no. 6, pp. 1262-1279, Jun. 2007.
23. D. A. Boas, S. R. Jones, A. Devor, T. J. Huppert, and A. M. Dale, "A vascular anatomical network model of the spatio-temporal response to brain activation," NeuroImage, vol. 40, no. 3, pp. 1116-1129, Apr. 2008.
24. R. B. Buxton, E. C. Wong, and L. R. Frank, "Dynamics of blood flow and oxygenation changes during brain activation: the balloon model," Magnetic Resonance in Medicine, vol. 39, no. 6, pp. 855-864, Jun. 1998.
25. J. B. Mandeville, J. J. Marota, C. Ayata, and et al., "Evidence of a cerebrovascular postarteriole windkessel with delayed compliance," Journal of Cerebral Blood Flow and Metabolism, vol. 19, no. 6, pp. 679-689, 1999.
26. D. A. Boas, G. Strangman, J. P. Culver, and et al., "Can the cerebral metabolic rate of oxygen be estimated with near-infrared spectroscopy?," Physics in Medicine and Biology, vol. 48, no. 15, pp. 2405-2418, Aug. 2003.
27. U. E. Emir, C. Ozturk, and A. Akin, "Multimodal investigation of fMRI and fNIRS derived breath hold BOLD signals with an expanded balloon model," Physiological Measurement, vol. 29, no. 1, pp. 49-63, Jan. 2008.
28. K. J. Friston, A. P. Holmes, J. B. Poline, and et al., "Analysis of fMRI time-series revisited," NeuroImage, vol. 2, no. 1, pp. 45-53, Mar. 1995.
29. A. Duncan, J. H. Meek, M. Clemence, and et al., "Measurement of cranial optical path length as a function of age using phase resolved near infrared spectroscopy," Pediatric Research, vol. 39, no. 5, pp. 889-894, May 1996.
30. G. Strangman, M. A. Franceschini, and D. A. Boas, "Factors affecting the accuracy of near-infrared spectroscopy concentration calculations for focal changes in oxygenation parameters," NeuroImage, vol. 18, no. 4, pp. 865-879, Apr. 2003.
31. D. A. Boas, T. Gaudette, G. Strangman, X. Cheng, J. J. Marota, and J. B. Mandeville, "The accuracy of near infrared spectroscopy and imaging during focal changes in cerebral hemodynamics," NeuroImage, vol. 13, no. 1, pp. 76-90, Jan. 2001.
32. D. Mehagnoul-Schipper, B. van der Kallen, W. Colier, and et al., "Simultaneous measurements of cerebral oxygenation changes during brain activation by near-infrared spectroscopy and functional magnetic resonance imaging in healthy young and elderly subjects," Human Brain Mapping, vol. 16, no. 1, pp. 14-23, May 2002.
33. C. Hock, K. Villringer, F. Müller-Spahn, and et al., "Decrease in parietal cerebral hemoglobin oxygenation during performance of a verbal fluency task in patients with Alzheimer's disease monitored by means of near-infrared spectroscopy (NIRS)--correlation with simultaneous rCBF-PET measurements," Brain Research, vol. 755, no. 2, pp. 293-303, May 1997.
34. B. L. Edlow et al., "The effects of healthy aging on cerebral hemodynamic responses to posture change," Physiological Measurement, vol. 31, no. 4, pp. 477-495, Apr. 2010.
35. B. M. Ances, C. L. Liang, O. Leontiev, and et al., "Effects of aging on cerebral blood flow, oxygen metabolism, and blood oxygenation level dependent responses to visual stimulation," Human Brain Mapping, vol. 30, no. 4, pp. 1120-1132, 2009.
36. D. J. Mehagnoul-Schipper, L. C. Vloet, W. N. Colier, W. H. Hoefnagels, and R. W. Jansen, "Cerebral oxygenation declines in healthy elderly subjects in response to assuming the upright position," Stroke; a Journal of Cerebral Circulation, vol. 31, no. 7, pp. 1615-1620, Jul. 2000.
37. M. C. Henry-Feugeas, F. Onen, and E. S. Claeys, "Classifying late-onset dementia with MRI: is arteriosclerotic brain degeneration the most common cause of Alzheimer's syndrome?," Clinical Interventions in Aging, vol. 3, no. 1, pp. 187-199, 2008.

Visualization of Survival and Migration of Transplanted Stem Cells by MRI

Tetsuji Yamaoka[1,2,*], Carlos A. Agudelo[1], and Yoichi Tachibana[1]

[1] Department of Biomedical Engineering, National Cerebral and Cardiovascular Renter, Research Institute, 5-7-1 Fujishiro-dai, Suita, Osaka 565-8565, Japan
[2] JST, CREST, 5 Sanbancho, Chiyoda-ku, Tokyo 102-0075, Japan

Abstract — **Stem cell transplantation therapy has been becoming one of the most attractive medical treatments for repairing ischemic tissues in limbs, myocardial infarction and the brain. To understand the mechanism implicated during the healing, a cell tracking system under MRI would be very useful in the field of fundamental research and clinical applications.**

An MRI contrast agent composed of bioinert polymeric carriers such as poly(vinyl alcohol) or dextran and gadolinium chelate was synthesized by activating their hydroxyl groups with 1,1'-carbonylbis-1H-imidazole (polymeric-Gd). Three kinds of cells, NIH-3T3, rat bone marrow-derived mesenchymal stem cells (BMSCs) and rat endothelial progenitor cells (EPCs) were labeled with the polymeric-Gd by the electroporation under a very mild condition. The amount of the delivered contrast agent, cytotoxicity, and the intracellular stability of the polymeric Gd was evaluated in vitro. These labeled cells were injected hindlimb ischemia rat models and the MR contrast at a given amount of cells injected was evaluated.

Our results revealed at the first time that the stem cell migration in the ischemic or nonischemic cells were very different and even the cell survival period of time after transplantation was greatly affected by the ischemic situation of the rats.

Keywords — **MRI, cell tracking, stem cells, survival, migration.**

I. Introduction

The recent article survey shows a increasing interest in the stem cell transplantation therapy (1-3) for various diseases such as ischemic limb (4), infarcted myocardium (5,6), and diabetic retinopathy (7). Especially, transplant-tation of autologous cell transplantation using bone marrow- or adipose tissue-derived mesenchymal stem cells is immunologically safe and promising clinically. However, the detailed behavior of the transplanted stem cells is still unknown. Even the engraftment efficiency and survival period of the stem cells was not fully clarified. One of the usual methods in analyzing them is immunostaining, but it cannot be used for autologous system because there is no phenotypic difference between transplanted cells and host cells. In recent years, then, noninvasive tracking system for cell transplantation is attracting great attention (8). Fluorescence- or bioluminescence-based optical imaging is very popular in recent years since for example green fluorescent protein (GFP)-transgenic animal and GFP-positive cells are commercially available, and they were easily detected under the in vivo optical imaging instruments for small animals.

In contrast, magnetic resonance imaging (MRI) is more promising system because of the high resolution, no limitation in the animal size but the cells should be labeled with some contrast agents to be observed. In this decade, superparamagnetic iron oxide particles (SPIO) have been widely used for cell labeling due to its high sensitivity (9, 10). Rice, H. E. et al. reported the homing of adipose derived stem cells in cerebral infarction (11). Targeted cells were usually labeled with SPIO only by co-culturing with SPIO. However, the taken up particles are known to be exocytized as time, and the SPIO may recruit macrophages. Since the free SPIO would continue to show the MR contrast, leading to a misunderstanding (12). For example, Amsalem, Y. et al. reported that the observed MRI signals after transplantation of SPIO-labeled MSCs were not attributed to transplanted cells but to cardiac macrophages which took up the released SPIO from transplanted cells (13).

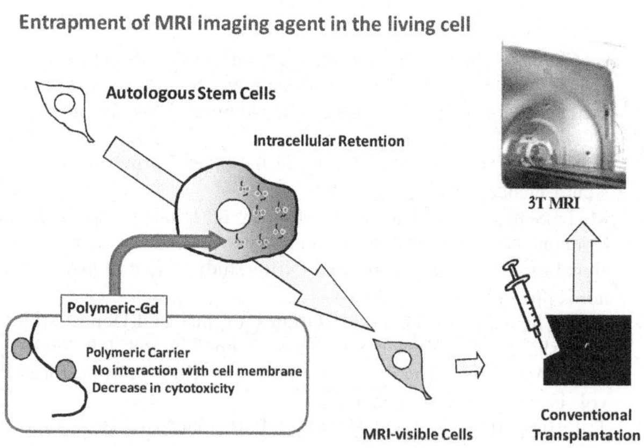

Fig. 1 Our strategy in MRI cell tracking.

* Corresponding author.

V. Van Toi et al. (Eds.): 4th International Conference on Biomedical Engineering in Vietnam, IFMBE Proceedings 40, pp. 254–257, 2013.
www.springerlink.com

In the present study, a novel water soluble contrast agent has been designed in order to track only the living cells in vio. Gd-DOTA (1,4,7,10-tetraazacyclo- dodecane-N,N',N",N"'-tetraacetic acid) was conjugated to bioinert and highly water soluble polymeric carriers such as poly(vinyl alcohol) (PVA) or dextran (14, 15). The intracellular delivery system of the conjugates was established using electroporation, and the cytotoxicity, intracellular stability, body distribution, and MR-imaging ability of the contrast agent were studied in vitro and in vivo. The conjugates are expected to be eliminated from the tissues without interacting with macrophages when they come out of the cells upon the cell death (Scheme 1).

II. Experimantal

A. Synthesis of Polymeric Gd

Synthesys of the polymeric contrast agents with different introduction ratio of Gd in the case of PVA carrier are shown in Scheme 1. A mixture of PVA and a given concentration of carbonyl diimidazole (CDI) were stirred in anhydrous DMSO at room temperature under a nitrogen atmosphere for 4 h. Then, 1,3-propanediamine was added to the mixture and further stirred at room temperature for 1 day and dialyzed with Spectra/Pore membrane in distilled water three times. The remaining solution was lyophilized to give **2**. PVA-diamine was reacted with DOTA-NHS-ester in 80 mL of anhydrous DMSO at room temperature for 1 day under a nitrogen atmosphere. The reaction mixture was dialyzed in distilled water three times, and lyophilized to give PVA-diamine-DOTA (**3**). The solution of **3** was then treated with 1.5 mole eq. of gadolinium chloride to the DOTA by adding dropwise while stirred and dialyzed in distilled water three times, and lyophilized to give Gd-PVA (**4a-d**).

For labeling Gd-PVA with FITC, PVA-diamine was mixed with small amount of FITC-NHS-ester and stirred in DMSO for 1 day under a nitrogen atmosphere. The reaction mixture was dialyzed, lyophilized to give FITC-PVA-diamine, and subjected to the following DOTA reaction as is shown in Scheme 1.

B. Three Different Types of Cells

NIH-3T3, rat bone marrow-derived mesenchymal stem cells (BMSCs) and rat endothelial progenitor cells (EPCs) were used for evaluating the cytotoxicity, cell labeling potential, and imaging efficiency of the polymeric-PVA. NIH3T3 cells were grown in Dulbecco's modified Eagle's medium with low glucose (DMEM) supplemented with 10 % bovine calf serum, 100 units/ml of penicillin, and 100 units/ml of streptomycin at 37 °C, 10 % CO2 atmosphere.

BMSCs were isolated from rat marrow aspirates and cultured in DMEM supplemented with 10% bovine calf serum, 5% horse serum, 100 units/ml penicillin, and 100 units/ml streptomycin at 37°C under 10% CO_2 atmosphere. All experiments were performed using fourth-passage rMSCs.

EPCs were also isolated from bone marrow. Bone marrow was flushed from femurs and tibias of F344 rats (4 weeks old, male) after previous cytokine inducted mobilization of bone marrow-derived endothelial progenitor cells by using granulocyte colony-stimulating factor (G-CSF, 200 μg/kg/day, 5 days subcutaneously injected). CD34 and FLK-1 positive bone marrow cells were isolated by means of magnetic beads coated with antibody CD34 and FLK-1 by using Biotin Labeling kit-NH2. Cells were placed in fibronectin coated dishes and cultured with an endothelial cell basal medium (EBM-2) supplemented with EGM-2 Single Quots.

C. Cell Labeling by Electroporation

The cells were cultured in a petri dish at for 1 day. An arbitrary amount of polymeric Gd is added to the culture medium, and electrical pulses were applied to cells using an electroporator (CUY-21; NEPPA GENE, Tokyo, Japan). Rectangular electrical pulses (field strength 300 V/cm, number of pulses 10, pulse duration 5 ms) were applied to cells using two parallel electrodes with a 5 mm gap. Cells were incubated for 1 h and washed with PBS twice.

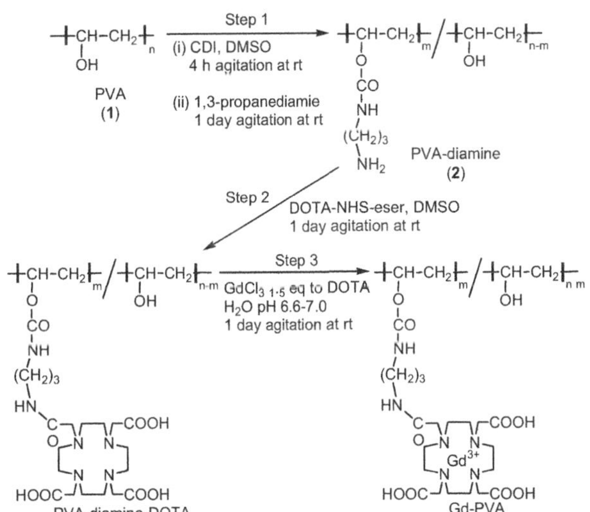

Scheme 1

To check whether the polymeric Gd stays in cells for a long period of time, the cells labeled with FITC-polymeric-Gd were seeded in a petri dish and cultured during 10 days without subculture. Time course of fluorescent intensity for the lysate of cultured cells was measured and the amount of polymeric-Gd in the cells was evaluated (excitation 430 nm, emission 540 nm). Time course of fluorescent intensity represented the stability of polymeric-Gd in the cells.

D. *MR Imaging of Labeled rMSCs in Vivo*

Unilateral hind limb ischemia was induced in 8-week-old male F344 rats by ligating and resecting the left femoral artery, vein, and their branches under isoflurane anesthesia (1.5% in air). Rats were then transplanted with the labeled stem cells and MRI measurements were performed at 0, 1, 2, 5, 7, and 14 days using a 1.5 T-MR scanner. Rats were again anesthetized with isoflorane and imaged in the dorsal position. Images were acquired in a coronal plane using the spin echo sequence. The segmented image stacks were used to generate a three dimensional (3D) model of the transplanted cells, gelatin sponge, and abdominal body of the rats using 3D-Doctor.

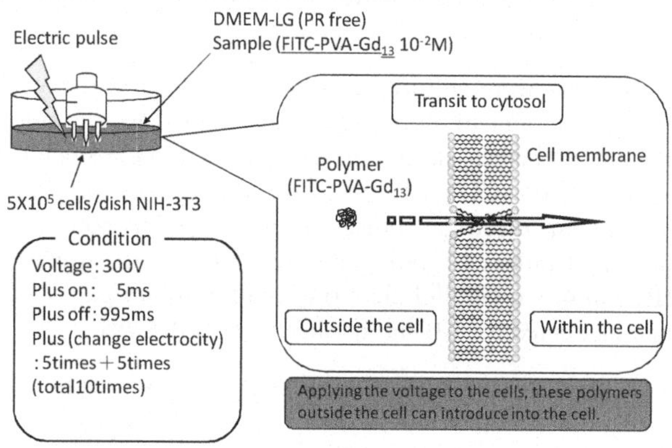

Fig. 2 Delivery of polymeric Gd contrast agent to NIH3T3, BMSCs and EPCs.

III. RESULTS AND DISCUSSION

A. *Intracellular Stability of Polymeric Gd*

To check the intracellular stability of the polymeric-Gd, for a long period of time, the time course of fluorescent intensity for the cultured cells was evaluated. Almost all cells were labeled byt the electroporation and the interestingly the delivered FITC-labeled polymeric Gd were detected only in the sytosolic complartment but not in the nucleic compartment even after the cell mytosis accompanied by the disappearance of the nucleic membrane.

The intracellular stability of polymeric Gd in the cells was excellent, and only about 5 % of the fluorescence intensity was reduced even after 1 month. These high intracellular stability was confirmed also by measuring the Gd molecules directly wity the ICP-MS.

B. *In Vivo Fate of Free SPIO and Free Gd-PVA*

The body distribution of the free polymeric Gd was compare with that of free SPIO. In order to track the living cells, the "free" contrast agents should be eliminated from the site of localization since the contrast agent comes out of cells upon the cell death. Solutions of the polymeric Gd and SPIO were directly injected into rat femoral muscles and on day 0, 3 and 6, MR image was analyzed. Bright signal attributed to polymeric Gd was rapidly decreased and slightly observed on day 3. In contrast, the dark signal due to SPIO remained at the same area and was clearly observed even 10 days after injection.

The particular substrates are hard to diffuse in the tissue and in addition the long lasting contrast may be attributed to the phagocytes engulfing the injected SPIO as is previously reported (16). In contrast, a very rapid clearance of polymeric Gd was observed. The result indicates that the cell tracking using SPIO-labeled cells may lead to misunderstanding of the in vivo fate of transplanted cells while newly developed water soluble MRI contrast agents would overcome this problem and are useful for in vivo living cell tracking.

C. *In Vivo Cell Tracking under 1.5 T MRI*

In figure 3, a typical MR image of the transplanted cells was shown. The cells were injected to the femoral muscle of the left hindlimb ischemia rat and observed after 1 day. We successfully constructed a 3D model from a series of two dimensional slices, as shown in Figure 3.

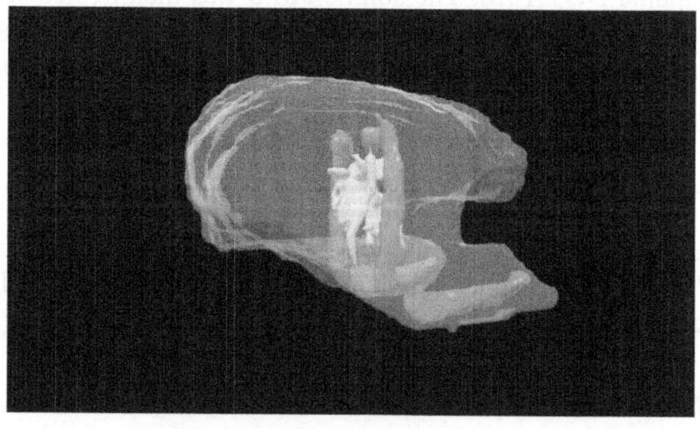

Fig. 3 MR image of the transplanted cells.

The signal attributed to the transplanted cells was weakened in about two weeks (data not shown). In the past, several groups have demonstrated MSC tracking using in vivo optical imaging method. Farrell et al reported that MSCs transplanted subcutaneously could be tracked in vivo for 4 weeks after transplantation (17). Furthermore,

SPIO-labeled MSCs transplanted into the infarcted heart of rats could be tracked for at least 16 weeks (18). However, our data was the first quantitative result of the cell survival in vivo under the MRI. It is of prime importance to know whether this change was resulted from (1) the cell death or (2) the contrast dilution due to the cell growth and cell migration.

D. Survial of Transplanted Cells

We succeeded to measure the amount of Gd in the urine with the ICP-MS. Because or materials composed of highly bioinert polymers, even when the polymeric-Gd leaked out of the cells upon cell death, the accumulation in the tissue and uptake by the macrophage was expected to be effectively suppressed. By measuring the amount of the excreted amout of Gd, the above mentioned weakening of the contrast was found to result from the cell death in vivo mainly. To our knowledge, it is the first report to monitor the stem cell death after transplantation using the MRI system. We believed that our new cell tracking system will be powerful tool to study the stem cell transplantation therapy in model animal and even clinically in the future.

IV. Conclusions

MRI as a method for tracking transplanted stem cells by using our novel water-soluble contrast agent brought us very powerful information which was not able to be observed with the conventional particle-type contrast agent. Now we are trying to find out the relationship between the cell injection method, cell migration and distribution thereafter, and blood flow recovery in the ischemic tissue.

Acknowledgment

This work was supported by grants-in-aid from the Ministry of Health, Labour, and Welfare of Japan (Health and Labour Sciences Research Grants, Research on Nanotechnical Medicine) and the Research Grant for Cadiovascular Diseases (18A-2) from the Japan Association for the Advancement of Medical Equipment. The authors thank Jun-ichiro Enmi, Takayuki Ose, Hajime Fukuda, and Akihide Yamamoto for their cooperation.

References

1. Perin EC, Dohmann HF, Borojevic R, Silva SA, Sousa AL, Mesquita CT, Rossi MI, Carvalho AC, Dutra HS, Dohmann HJ, Silva GV, Belem L, Vivacqua R, Rangel FO, Esporcatte R, Geng YJ, Vaughn WK, Assad JA, Mesquita ET, Willerson JT., Circulation 2003;107(18):2294-2302.
2. Schmid C, Schleuning M, Schwerdtfeger R, Hertenstein B, Mischak-Weissinger E, Bunjes D, Harsdorf SV, Scheid C, Holtick U, Greinix H, Keil F, Schneider B, Sandherr M, Bug G, Tischer J, Ledderose G, Hallek M, Hiddemann W, Kolb HJ., Blood 2006;108(3):1092-1099.
3. Slavin S, Nagler A, Naparstek E, Kapelushnik Y, Aker M, Cividalli G, Varadi G, Kirschbaum M, Ackerstein A, Samuel S, Amar A, Brautbar C, Ben-Tal O, Eldor A, Or R., Blood 1998;91(3):756-763.
4. Kim SW, Han H, Chae GT, Lee SH, Bo S, Yoon JH, Lee YS, Lee KS, Park HK, Kang KS., Stem Cells 2006;24(6):1620-1626.
5. 5. Stamm C, Westphal B, Kleine HD, Petzsch M, Kittner C, Klinge H, Schumichen C, Nienaber CA, Freund M, Steinhoff G. Autologous bone-marrow stem-cell transplantation for myocardial regeneration. Lancet 2003;361(9351):45-46.
6. Strauer BE, Brehm M, Zeus T, Kostering M, Hernandez A, Sorg RV, Kogler G, Wernet P., Circulation 2002;106(15):1913-1918.
7. Grant MB, May WS, Caballero S, Brown GA, Guthrie SM, Mames RN, Byrne BJ, Vaught T, Spoerri PE, Peck AB, Scott EW., Nat Med 2002;8(6):607-612.
8. 9. Beeres SL, Bengel FM, Bartunek J, Atsma DE, Hill JM, Vanderheyden M, Penicka M, Schalij MJ, Wijns W, Bax JJ. Role of imaging in cardiac stem cell therapy. J Am Coll Cardiol 2007;49(11):1137-1148.
9. Kraitchman DL, Tatsumi M, Gilson WD, Ishimori T, Kedziorek D, Walczak P, Segars WP, Chen HH, Fritzges D, Izbudak I, Young RG, Marcelino M, Pittenger MF, Solaiyappan M, Boston RC, Tsui BM, Wahl RL, Bulte JW., Circulation 2005;112(10):1451-1461.
10. Hill JM, Dick AJ, Raman VK, Thompson RB, Yu ZX, Hinds KA, Pessanha BS, Guttman MA, Varney TR, Martin BJ, Dunbar CE, McVeigh ER, Lederman RJ., Circulation 2003;108(8):1009-1014.
11. Rice HE, Hsu EW, Sheng H, Evenson DA, Freemerman AJ, Safford KM, Provenzale JM, Warner DS, Johnson GA., Am J Roentgenol 2007;188(4):1101-1108.
12. Hoshino K, Ly HQ, Frangioni JV, Hajjar RJ., Prog Cardiovasc Dis 2007;49(6):414-420.
13. Amsalem Y, Mardor Y, Feinberg MS, Landa N, Miller L, Daniels D, Ocherashvilli A, Holbova R, Yosef O, Barbash IM, Leor J., Circulation 2007;116(11 Suppl):I38-45.
14. Tachibana Y, Enmi J, Mahara M, Iida H, Yamaoka T., Contrast Media and Molecular Imaging, 5, 309-317(2010)
15. Agudelo CA, Tachibana Y, Teramoto N, Iida H, Yamaoka T., Tissue Eng Part A. 17(15/16), 2079-2089 (2011)
16. Amsalem Y, Mardor Y, Feinberg MS, Landa N, Miller L, Daniels D, Ocherashvilli A, Holbova R, Yosef O, Barbash IM, Leor J. , Circulation 2007;116(11 Suppl):I38-45.
17. Farrell E, Wielopolski P, Pavljasevic P, van Tiel S, Jahr H, Verhaar J, et al. Effects of iron oxide incorporation for long term cell tracking on MSC differentiation in vitro and in vivo. Biochem Biophys Res Commun 2008;369:1076-1081.
18. Stuckey DJ, Carr CA, Martin-Rendon E, Tyler DJ, Willmott C, Cassidy PJ, et al. Iron particles for noninvasive monitoring of bone marrow stromal cell engraftment into, and isolation of viable engrafted donor cells from, the heart. Stem Cells 2006;24:1968-1975.

Author: Tetsuji Yamaoka

Institute: National Cerebral and Cardiovascular Center Research Institute
Street: 5-7-1, Fujishiro-dai
City: Suita 565-8565
Country: Japan
Email: yamtet@ri.ncvc.go.jp

Drug Delivery System Using Polymers (Lactide–Co-Glycolide)

Gilson Khang

Department of Polymer Science and Technology, Chonbuk National University, 567 Baekje-daero, Deokjin-gu, Jeonju, 561-756, Korea

Abstract — **For last decade, we are developing the novel local drug delivery devices using biodegradable polymers, especially polylactide (PLA) and poly(D,L-lactide-co-glycolide) (PLGA) have been developed in our labaortory due to its relatively good biocompatibility, asily controlled biodegradability, good processability and only FDA approved synthetic degradable polymers. The relationship between various kinds of drug and different types of geometrical devices and pharmacological activity are proposed and discussed for the application of pharmaceutics formulations. Also, local drug delivery devices proposed in these works are introduced in view of preparation method, drug release behavior, biocompatibility, pharmacological effect, and animal studies. In conclusion, we can control the drug release profiles varying with the preparation, formulation and geometrical parameters. It is very important to design a suitable formulation for the wanted period of bioactive molecules loaded in biodegradable polymers for the local delivery of drug. The drug release is affected by many factors such as hydrophilicity of drug, electric charge of drug, drug loading amount, polymer molecular weight, the monomer composition, the size of implants, the applied fabrication techniques, and so on.**

Keywords — **local drug delivery system, delivery devices, biodegradable polymers, pharmaceutics.**

I. INTRODUCTION

It has been recognized that biodegradable polymers have become increasingly important in the development of drug delivery system (DDS) during the past two decades. Massive researches have been done to designed appropriate devices like microspheres (MSs) and nanospheres, injectable form, wafer, tablet, and so on using synthetic and natural biodegradable polymers.[1] In the 1970s, simply and mainly long term controlled release devices of contraceptive steroids, local anesthetic, narcotic antagonist, anticancer and antimalarials drugs were developed to test the possibility of the pharmacological activities and elimination of the inconvenience of the repeated injection. Beginning of1980s, rapid and innovative progress in the area of biotechonolgy, especially cell and cloning technology, made possible the successive and massive production of therapeutic peptides and proteins like cytokine, monoclonal antibody, hormone and growth factors. So, DDS using biodegradable polymers were extensively studied and commercialized for these peptides and proteins to achieve efficiency and to increase patient compliance. It can be avoided the difficulties associated with parental and oral delivery and patient compliance problems. Above DDS method would deliver the drug at a continuous rate, and reduce the dose-dependent toxicity by minimizing the fluctuation in plasma concentration.

Biodegradable polymers for local delivery system for DDS have been applied. Among of these biodegradable polymers, one of the most significant candidates for the development of the biodegradable polymeric controlled release system is the poly(a-hydroxy acid)s family such as poly(glycolide) (PGA), poly(L-lactide) (PLA) and its copolymers as poly(L-lactide-co-glycolide) (PLGA; chemical structure) which is only approved by the Food and Drug Administration (FDA) due to its controllable biodegradability and relatively good biocompatibility.[2] It provides many advantages such as regulating varying degradation period according to the molecular weight and mole fraction of lactide and glycolide in the copolymer, especially PLGA, producing toxicologically safe by-products that are further eliminated by the normal metabolic pathways.

Typical and successful commercial product is Lupron Depot® (leuprorelin acetate) and Decapeptil® (tryptorelin) composed of LH-RH agonist in biodegradable PLGA for the treatment of advanced prostatic cancer and endometriosis. Another successful commercial product is Gliadel® wafer which composed of polyanhyride polymer and BCNU (1,3-bis(2-chloroethyl)-1-nitrosourea) for the treatment of malignant brain tumor. This localized and controlled delivery device of anticancer agent using biodegradable polymeric implant provides to solve the problem of low penetration of blood brain barrier (BBB).[3]

This presentation will be discussed for recent our works on the development of various drug delivery applications with like MSs, microcapsule, nanoparticle, wafers, pellet, beads, multiple-layered beads, implants, fiber, scaffolds, and films applied using PLGA, PLA and PHVB biodegradable polymers during last 5 years.[4] Water soluble small molecule drugs [gentamicin sulfate (GS), fentanyl citrate (FC), BCNU, pamidronate (ADP), 5-fluorouracil (5-FU), azidothymidine (AZT), 1,25$(OH)_2$ vitamin D_3], water insoluble small molecule drugs [fent-anyl, ipriflavone (IP)

V. Van Toi et al. (Eds.): 4th International Conference on Biomedical Engineering in Vietnam, IFMBE Proceedings 40, pp. 258–261, 2013.
www.springerlink.com

and nifedipine], and water soluble large molecule drug[nerve growth factor (NGF),[5] vascular endothelial growth factor,[6] brain derived neurotrophic factor and Japanese encephalitis virus (JEV)] were applied for the development of local DDS devices.

Table 1 Typical Biodegradable Polymers Used In DDS.

Natural Biodegradable Polymers	Synthetic BiodegradablePolymers
- Polypeptides and proteins: Albumin, fibrinogen, gelatin, Collagen, etc - Polysaccharides: Hyaluronic acid, starch, chitosan - Virus and living cells: Erythrocytes, fibroblast, Myoblasts, etc.	- Aliphatic polyesters of hydroxy acids:PLA, PGA, PLGA,poly(hydroxybutyric acid),poly(e-caprolactone) - Polyorthoester - Poly(alkylcarbonate) - Polyaminoacids - Polyanhydrides - Polyacrylamides - Poly(alkyl-a-cyanoacrylate)s - Etc.

II. Applications

A. JEV Vaccine Loaded PLGA MSs for the Oral Immunization

The extensive studies for biodegradable MSs with incorporated antigens as oral vaccine are of particular interests in making currently available vaccines more effective and in developing new vaccines against infections for which no currently exist.

The control of some fabrication parameters for the preparation for JEV vaccine as a model vaccine loaded PLGA MSs have been investigated such as the types of emulsifier (polyvinylalcohol (PVA) and sodium dodecyl sulfate (SDS)), the concentrations of emulsifier, agitation speed, and the concentration of PLGA in W/O/W method. Also, the effect of surface morphology on biodegradation of PLGA has been observed. For the oral vaccination of JEV/PLGA MSs via the across the gastrointestinal tract as Peyers patches, the size of MSs must be below around 15.0 mm. For the satisfaction of this size, the rate of agitation of 2000 rpm could be minimum stirring speed in this study with 5.0 w/v% of PLGA concentration, and 100 mL of 1.0 wt% PVA. To investigate the effect of emulsifier types on the surface morphology of JEV vaccine loaded PLGA MSs, two kinds of emulsifiers as PVA and SDS were applied. In conclusion, the formulation and the process variables play important roles in morphology of PLGA MSs, biodegradation of PLGA MSs and resulting the release pattern of drug.[7]

B. GS Loaded PLGA MSs

Controlled GS releasing MSs manufactured from biodegradable PLGA (75:25 by mole ratio) were prepared with an oil/oil solvent evaporation method. The MSs of different size (30~350 mm) were obtained with varying the experiment conditions, and the shape of MSs was smooth and spherical. The efficiency of encapsulation was over 81%. The effects of the preparation conditions on the size of MSs have been investigated. In vitro release studies showed that different release patterns and release rates could be achieved by simply modifying factors in the preparation conditions such as polymer concentration, surfactant concentration, molecular weight of PLGA and initial amount of drug. PLGA MSs with 20% of initial drug loading, 0.2% (w/w) of surfactant concentration and 50% (w/v) of PLGA concentration, were free from initial burst effect and a near-zero order sustained release was observed for over 60 days. The release of GS loaded MSs with larger size like over 300 mm was more prolonged over 2 month, whereas small size batch like below 100 mm was observed burst effect. This study demonstrated that the release pattern of drug from MSs could be improved by optimizing the preparation conditions of MSs.[8]

C. Glycolide Monomer Containing GA-Loaded PLGA Microparticles by Melt Extrusion

For the achievement of more precise release pattern of GS, we developed GS-loaded PLGA microparticle (GSMP) containing glycolide monomer (GM) prepared by melt-extrusion method. After the preparation of polymer blend by melt extrusion, the powders of different sizes (90~1000 mm) were obtained by freezer mill as shown in Figure 8. In vitro drug delivery release study was performed in pH 7.4 phosphate buffered saline (PBS). GSMP released from 2 days to 7 days in case of the highest loading amount of GM (10%) and showed a near-zero-order with initial burst. GM affected to increase of GS release during the in vitro release test, which was the positive result of what would be expected based upon decrease of pH of medium. The morphological evaluations of samples were characterized by SEM. GM did not distinctive influence to change of morphology by the analysis of GPC and DSC, respectively. Bacterial inhibition zone test was established to identify antibiotic activity of GS.

From these results, we expected that containing GM would be a good dosage form with the sustained release pattern to deliver the antibiotic for the prevention of infections after surgery for 1 week or more. This locally sustained delivery form of scattering powders at infection site will be may be decreasing the side effects comparison to oral dosage forms with high dosage and frequent administration. Moreover, this delivery system has an advantage which does not need to remove the any materials after surgery due to its spontaneous degradation property by human body fluid.[9]

D. *Preparation of Fentanyl-Loaded PLGA MSs*

Fentanyl loaded biodegradable PLGA MSs were prepared to study the possibility for long acting local anesthesia. We developed the fentanyl base (FB, slightly water soluble)-loaded PLGA MSs by means of conventional O/W solvent evaporation method. The size of MSs was in the range of 10~150 mm. The lowest porous cross-sectional morphology and the highest encapsulation efficiency were obtained by using gelatin as an emulsifier. The influences of several preparation parameters, such as solvent types (MC and ethyl acetate), emulsifier types (gelatin and PVA), molecular weights and the concentrations of PLGA, and initial drug loading amount, etc, have been observed in the release pattern of fentanyl. The release of fentanyl in in vitro was more prolonged over 25 days, with close to zero order patterns by controlling the preparation parameters. We also investigated the physicochemical properties of fentanyl-loaded PLGA MSs by XRD and DSC.

From the results of preparation conditions, the encapsulation efficiencies of fentanyl were between 61.5 and 99.8%, depending on the particular formulation. The total MSs recovered amount varied between 52.8 and 87.6%. It was observed that the encapsulation efficiency and the yield decreased as a function of the increase in initial drug loading amount. As increasing of PLGA concentration from 3 to 20%, encapsulation efficiency was increased from 62.7 to 99.8%. Similar results were also obtained when decreasing the solvent volume. We assumed that the dominating loss of fentanyl must be due to transport of droplets of the O phase to the W phase. The increased viscosity in the O phase caused by the increased PLGA concentration or decreased solvent volume will decrease the loss transport of fentanyl and contribute to the enhanced entrapment efficiencies. Moreover, the O droplets containing fentanyl formed from the W phase were very small and the diffusion amount of fentanyl to the external phase during the solvent evaporation was relatively small, explaining the high encapsulation efficiency obtained. The pattern of drug release depends on various factors, such as initial drug loading ratio, polymer concentration, and solvent volume in W phase. Generally, small size and fast drug release are attributed to more water uptake, swelling ratio, and polymer degradation. In contrast, in our case, fentanyl release rate from the all batches decreased with decreasing MSs size. It might be suggested that the drug release profiles would be affected by the morphology of the MSs increased concentration of O phase made porous stable MSs and resulted in denser MSs. In conclusion, this sustained, besides, constant localized release system can potentially provide anesthesia for a longer period than injection or topical administration.[10]

E. *Characteristic of Nifedipine Loaded PLGA Wafer*

Nifedipine also was chosen as the model drug for the PLGA wafer for the local delivery device due to practically insoluble drug in water with solubility less than 10mg/ml, a well known and most widely used coronary vasodilator from the group of dehydropyrine derivatives. Biodegradable wafers were prepared with PLGA oligomer (50: 50 mole ratio, molecular weight 5000 g/mole) by direct compression method for the sustained release of nifedipine to investigate the possibility of the treatment of hypertension. PLGA wafers were prepared by altering initial drug/polymer loading ratio, wafer thickness, and HPMC content, and their morphology and release pattern have been investigated. These wafers showed steady static release pattern for 11 days, and various biphasic release patterns could be obtained by altering the composition of wafers such as addition of matrix binder as HPMC to the PLGA wafer to reduce release rate of initial phase. The onset of polymer mass only occurred after 4 days and about 40% of mass loss was observed after 11 days nifedipine release. This system had advantage in terms of simplicity in design and controling of drug release rate and may be useful as an implantable dosage form.[11,12]

F. *Treatment of Aural Cholestoma Using ADP-Loaded PLGA Wafer as Local Delivery System.*

Implantable biodegradable wafers were prepared with ADP-loaded PLGA (75:25 mole ration by lactide to glycolide, molecular weight: 20000 g/mole) by direct compression method for the sustained release of ADP to investigate the possibility for the treatment of bone resorption.[92-94] The release pattern of ADP/PLGA wafers were observed by HPLC and the wafers were implanted the Mongolian gerbils mastoid. The release rate of APD increased with increase of its initial loading amount as shown in Figure 19. We also measured the osteoclast index (Figure 20), i.e, number of osteoclast cell per total bone length and the surface area of osteoclast cell per total bone surface in experimental cholestoma, in which APD/PLGA wafers were implanted. The result indicated these wafers could reduce the osteoclast activities in experimental aural cholestoma. It may suggest the possibility for the treatment of bone resorption by implantable dosage form.[13]

III. CONCLUSIONS

Until recently, many biodegradable polymers come from natural and synthetic have been tested and applied for the development for drug carrier of DDS based on the following three mechanisms: mucosal absorption, controlled release and targeting. Among of these polymers, PLGA and PLA seems to be most desirable for the drug delivery device as many different types. Also, we can relatively and easily control the drug release profiles varying with the preparation, formulation and geometrical parameters. Moreover, any types of drug such as water-soluble, water-insoluble, small, large molecule, negatively or positively charged, and so on were successfully applicable to achieve linear sustained release from short period (1~3 days) to long period (over 2 months), in other words, it is very important to design a suitable formulation for the predetermined releasing period of bioactive molecules loaded biodegradable devices. The form of device can be MSs, microcapsule, nanoparticle, wafers, pellet, beads, multiple-layered beads, implants, fiber, scaffolds, and films for the purpose of local delivery in the research area of drug delivery and tissue engineering. It is also considered the drug release is affected by many factors such as hydrophilicity of drug, electric charge of drug, drug loading amount, polymer molecular weight, the monomer composition, the size of implants, the applied fabrication techniques, and so on.

It is well known that the development of new drug needs lots of money (average: over 10 million US dollar per one drug) and long time (average: above 9 years) whereas the development of DDS for potent generic drug might be need relatively small investment and short time. Also, one core technology can be applicable to many drugs to meet the market within time frame. From these reasons, the research on DDS for potent generic drug has less risk and high return than new drug development. For the next DDS generation, DDS will provide more complex release control such as a stimulus control, sensor control, and targeting using "intelligent" PLGA or PLA biodegradable materials responsive to external stimuli and more complex devices mimicking viruses or living cells such as drug self-producing system without need for a periodic drug supply must be necessary.

REFERENCES

1. Eom S, Yoo S.C, Song B.J et al. (2010) Effect of preparation conditions on release behavior of raloxifen HCl-loaded PLGA microsheres. Int J Tissue Reg 1:28-36
2. Kim S. J, Hong H. H, Kim S. H et al. (2010) Reduction of inflammatory reaction of PLGA using fibrin; in vivo study. Polymer(Korea) 34:63-68
3. Okada U, Yamamoto M, Heya T et al. (1994) Drug delivery using biodegradable microspheres. J. Control. Rel 28:121-129
4. Kim M. H, Hong H. N, Hong J. P et al. (2010) The effect of VEGF on the myogenic differentiation of adipose tissue derived stem cells within thermosensitive hydrogel matrices. Biomaterials 31:1213-1218
5. Jeon E. K, Shim J. Y, Whang H. J et al. (2001) Preparation and characterization of different types of NGF-loaded delivery carriers for nerve regeneration. Biomater. Res 5:23-34
6. Khang G, Park C. S, Rhee John. M et al. (2001) Preparation and characterization of dimineralized bone particle impregnated PLA scaffold. Korea Polymer J 9:267-276
7. Park J. H, Eom S, Ahn S. I et al. (2009) Dextran/PLGA double-layered microspheres for sustained release of doxorubicin. Tissue Eng. Regen. Med 6:588-594
8. Eom S, Yoo S.C, Kim Y.K et al. (2010) Effect of Cosurfactants on the Release Behavior of Zaltoprofen-loaded PLGA Microspheres in In Vitro : Preparation and Characterization . Polymer(Korea) 34:333-340
9. Cho W. H, Lee Y. M, Lim D. K et al. (2011) Characterization and Sustained-release Formulation of Valsartan Tablet using Hydroxypropylmethyl- cellulose with various Viscosity. Int J Tissue Reg 2:30-37
10. . Choi H. S, Shin H, Khang G et al. (2001) Quantitative analysis of fentanyl in rat plasma by gas chromatography with nitrogen phosporus detection. J. Chromatography B 765:63-69
11. Seo S, Choi H. S, Lee D. H et al. (2001) Characteristics of nifedipine loaded PLGA wafer, Polymer(Korea) 25:884-892
12. Youn J. Y, Ku J, Kim B. S et al. (2008) Controlled release of nifedipine from osmotic pellet based on porous membrane. Polymer(Korea) 32:328-330
13. Yoo J. Y, Kim S. W, Khang G et al. (2002) Effect of pamidronate/PLGA wafer on cholesteatoma in in vivo. Biomater. Res 6:115-123

Enhanced and Specific Internalization of Polymeric Nanoparticles to Cells

Kazuhiko Ishihara[1,2,*], Yuriko Tsukamoto[1], Yusuke Goto[1], and Yuuki Inoue[1]

[1] Department of Materials Engineering, The University of Tokyo, Tokyo 113-8656, Japan
[2] Department of Bioengineering, The University of Tokyo, Tokyo 113-8656, Japan

Abstract — **We designed new polymer nanoparticles embedding quantum dots with artificial cell membrane-surface as a highly sensitive bioimaging probe. These nanoparticles were prepared by assembling phospholipid polymer as a platform and oligopeptide as a bioaffinity moiety on the surface of the nanoparticles. They showed high resistance to non-specific cellular uptake from HeLa cells due to the nature of phospholipid polymer with phosphorylcholine groups. On the other hand, when arginine octapeptide was immobilized on the surface, they could permeate the membrane of HeLa cells effectively and good fluorescence based on quantum dots could be observed. Cytotoxicity was not produced by these nanoparticles even after immobilization of octapeptide. Thus we obtained stable fluorescent polymer nanoparticles covered with artificial cell membrane, which are useful as an excellent bioimaging probe evaluation for biomolecular function in the target cells.**

Keywords — **Nanoparticles, cellular internalization, phospholipid polymers, oligopeptide, quantum dots.**

I. Introduction

Understanding cellular response from view points of chemical reactions between biomolecules inside cells will provide useful information in the fields of medicine, pharmacology and biology. In general, the role of a biomolecule is evaluated by immobilizing it on the supporting materials for analysis. Fluorescence measurements with fluorescent probes are useful for detecting the reaction. However, the specific interaction between biomolecule and cells has not been precisely assessed because the supporting materials themselves also interact with the cells and cause the unwished phenomenon: for example, considerable reactions at interfaces and nonselective cellular uptake of the nanoparticles. In addition, the long-term monitoring is essential for kinetically analyzing biomolecules. Thus, there is a critical need to fabricate the cytocompatible materials having the abilities to suppress the non-specific interactions between the materials themselves and cell and to be constantly traced by non-invasive fluorescence microscopy.

* Corresponding author.

Quantum dots (QDs) have gained much interest as a promising alternative to organic dyes for biological imaging. QDs ranging in size between 2 nm and 6 nm have unique optical properties: material- and size-dependent emission spectra, a wide absorption spectrum, high quantum yields, simultaneous multi color emissions, and especially excellent resistance to photobleaching. This photostability is a critical feature in most fluorescence applications, particularly for long-term monitoring of labeled substances, and is an area in which QDs have a singular advantage over organic dyes. As QDs themselves are hydrophobic, the key to developing QDs as a tool in biological systems is to achieve good dispersion ability in an aqueous medium, and compatibility with biological components including cells. In spite of numerous efforts, the problems of cytotoxicity and the nonselective cellular uptake of QDs remain.

We have been researching for preparing new polymer materials for medical and pharmaceutical application based on the molecular design with bioinspired concept [1, 2]. 2-Methacryloyloxyethyl phosphorylcholine (MPC) polymers, which have representative phospholipid polar groups in the side chains, have been synthesized by free radical copolymerization with hydrophobic monomers as bio-compatible materials. The MPC polymers have found a number of biomedical applications [3] because they can provide an artificial cell membrane through coating, blending with other polymers, and grafting to the polymer substrates. By controlling the composition of the MPC units in the polymer and the molecular weight of the polymer, water-soluble and amphiphilic polymers such as poly (MPC-co-*n*-butyl methacrylate (BMA)-co-ω-methacryloyloxy poly (ethylene glycol) oxycarbonyl 4-nitrophenoyl (MEONP)) (PMBN) could be synthesized [4]. We have reported the preparation of the PMBN-covered nanoparticles using amphiphilic PMBN and water-insoluble polymers such as poly(L-lactic acid) (PLA) or polystyrene as a core material, and their applications in various biological fields [5]. These nanoparticles are easy to produce, and show the bioinert abilities to suppress nonspecific protein adsorption, to avoid phagocytosis from macrophage-like cells, and to sustain the activity of antibodies immobilized onto their surfaces [6]. That is, the PMBN coverings suppress non-specific interactions between bulk materials and cell, and create a specific affinity by ligand molecules immobilized on the

V. Van Toi et al. (Eds.): 4th International Conference on Biomedical Engineering in Vietnam, IFMBE Proceedings 40, pp. 262–265, 2013.
www.springerlink.com

surface. Additionally, it remains possible that various organic-inorganic hybrid nanocomposites could be fabricated by incorporating inorganic substances inside the PLA matrix. Thus, PMBN, PLA and QDs are the best candidate-materials for preparing tools to overcome the above-mentioned problems. We also reported that our polymer nanoparticles containing QDs with an artificial cell membrane-surface (PMBN/PLA/QD: Fig. 1) had the abilities to highly resist to nonselective cellular uptake from HeLa cells and to permeate the membrane of HeLa cells effectively when arginine octapeptide (R8) was immobilized on the surface of the nanoparticles [7]. In this study, we continued these studies to obtain a better understanding of the nature of PMBN/PLA/QD. We clarified effects of chemical structure of oligopeptide on internalization to cells.

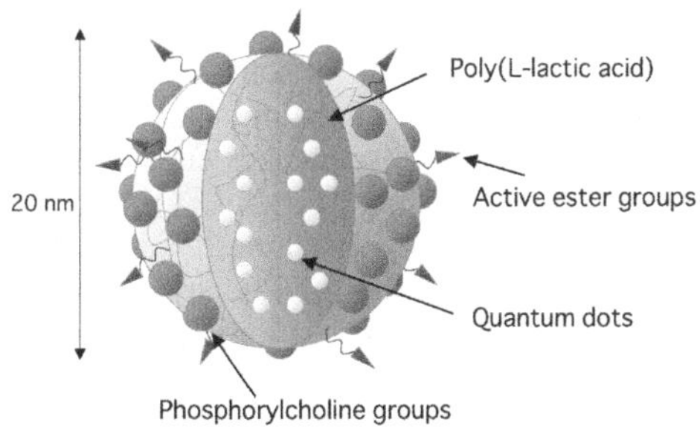

Fig. 1 Schematic concept of PMBN/PLA/QD.

II. EXPERIMANTAL

A. *Preparation on Polymer Nanoparticles Embedding QDs*

The PMBN/PLA/QD were prepared as follows. The PMBN (Fig. 2) was synthesized by a conventional radical copolymerization of corresponding monomers using a previously reported method [4-7]. An aqueous solution (20 mL) containing 200 mg of the PMBN (10 mg/mL) was placed in a glass bottle, and the solution was stirred at 400 rpm while being cooled in an ice bath. One hundred microliters of trioctylphohine oxide (TOPO)-coated ZnS-shell CdSe-core QDs (the original QDs) suspended in toluene were dried under a vacuum. The remaining solidified QDs were suspended again in 1 mL dichloromethane with 0.1

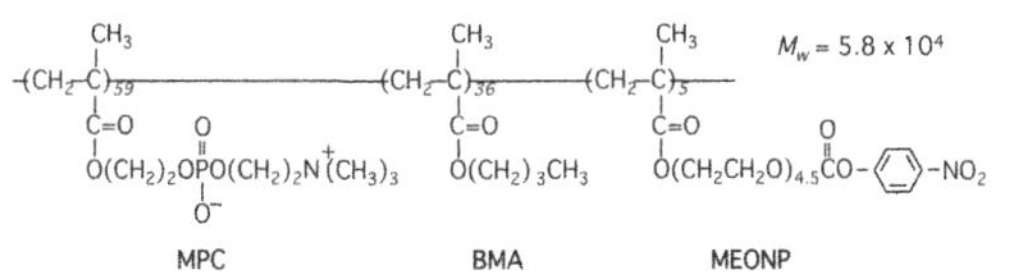

Fig. 2 Chemical structure of PMBN.

mg/mL PLA. The suspension was then added to the aqueous PMBN solution. This mixture was sonicated using a probe-type sonicator for 30 min under cool conditions, and then it was maintained under reduced pressure for 20 min to evaporate the dichloromethane. Since the solution contained both the formed PMBN/PLA/QD and the remaining PMBN after evaporation, the nanoparticles in aqueous medium were ultracentifuged at 50,000 rpm for 2 h. The supernatant was discarded and the PMBN/PLA/QDs as precipitate were resuspended in water.

B. *Immobilization of R8 Peptide to PMBN/PLA/QDs for Assessing Cell Penetration*

The R8 was dissolved in PBS to prepare a 1 μM solution. After mixing 1 mL of the PMBN/PLA/QD solution (2 μM) and 1 mL of the R8 solution, the resulting solution was stored for 24 h at 4° C. As a control, the active ester groups on PMBN/PLA/QD for immobilization were blocked with a glycine (3 mg/mL) aqueous solution to avoid reaction with the proteins in the cellular environment. Before dosing with these QDs, the QDs were put through a filter with a 0.45μm pore diameter. Other octapeptides were also immobilized on the PMBN/PLAQD as the same procedure of R8 case.

C. *Cell Culture, QD Treatment, Cellular Imaging, and Cytotoxicity Test*

HeLa cells were grown at 37° C and 5% CO_2 in Dulbecco's Modified Eagle Medium (D-MEM; Gibco), supplemented with 10% fetal bovine serum (Gibco) and were passaged twice before the start of experiments. For confocal microscopy, HeLa cells were seeded into 3 cm glass bottom culture dishes at 6.0 x 10^4 cells/mL in 1 mL of culture medium and were grown to 40-60% confluency before QD treatment. The medium in the 3 cm glass bottom culture dishes was replaced with 1 mL of D-MEM containing 100 nM R8-immobilized-PMBN/PLA/QD or glycine(G)-immobilized-PMBN/PLA/QD and incubated for 24 hours. After 24 hours of dosing with QDs, the 3 cm culture dishes were rinsed three times with 1 mL of PBS and the medium was replaced with 1 mL of pre-warmed PBS. QDs were imaged in HeLa cells treated by formalin using a laser scanning confocal microscope (LSM-510; Carl Zeiss) with a 60x objective. QDs were excited with a 488 nm Ar laser line with filter-based emission channels of 520-550 nm. Confocal-DIC images of HeLa cells were obtained by excitation with a 633 nm HeNe laser line. Images were captured using Carl Zeiss software. The full cellular thickness (approximately 25 mm) was scanned in 0.3 μm steps. Uptake of QDs by HeLa cells was analyzed using a fluorescence microscope (IX71; Olympus, Tokyo, Japan) equipped with a 20x objective lens using Hg lamp as a light source. Fluorescence was collected by the same objective, passed through a filter ranging from 510 nm to 550 nm.

D. The Assessment of the Abilities of Various Octapeptides as a Cell Penetrating Peptide

To assess the main factor determining the abilities of oligopeptides as a cell penetrating peptide, we selected various octapeptides that consist of one kind of amino acid (octatyrosine (Y8), octaasparagine (N8), octaglutamic acid (E8), octahistidine (H8), octalysine (K8), or octaarginine (R8)). These peptides were immobilized with PMBN/PLA/QD, respectively. HeLa cell (seeded into 12-well plates at 1.0 x 10^4 cell/well in 1 mL of culture medium) was incubated with each octapeptide-immobilized PMBN/PLA/QD for 5 h. The quantification of the uptake of PMBN/PLA/QD was performed with the fluorescence microscope.

To observe the effect of sequence of octapeptide on cellular internalization, we used various octapeptide composed of glycine and alginate. That is, G8, G6R2, G4R4, G2R6, and R8 were examined.

III. Results and Discussion

A. Characterization of PMBN/PLA/QD

The PMBN/PLA/QD obtained has good dispersion ability in water and phosphate buffered saline (PBS, pH 7.4), and has maintained high fluorescence. The absorption and emission spectra were very similar to those of uncoated commercial QDs in toluene. We confirmed that the encapsulation of QDs provided extended stability, and we observed no change in solubility or fluorescence over one year (stored at 4 °C). These indicate that our PMBN/PLA/QD does not degrade at least one year. We also tested whether these encapsulations are stable at various pHs that may be found in the different cellular environments required in bioconjugation chemistries. It was found that the PMBN/PLA/QD is stable between pH 4.0 and pH 9.0. The PMBN/PLA/QD was much more resistant to photo bleaching compared with the typical organic dye, fluoresciene isothiocyanate. Overall, the encapsulation of QDs using PMBN and PLA did not have any influence on the optical properties of QDs.

B. Cell Penetrating Ability of R8-Immobilized PMBN/PLA/QD

Subsequently, octaarginine (R8) was immobilized to the PMBN/PLA/QD (R8-PMBN/PLA/QD) for evaluation of its cell penetrating ability. R8 is well known as cell membrane penetration enhancer. It has the ability to translocate through cell membranes in a manner that does not involve the typical endocytic pathways of internalization. A fluorescence microscopy image of HeLa cells that had been incubated with R8-PMBN/PLA/QD. The R8-PMBN/PLA/QD clearly was associated with the HeLa cells and was found internalized at a perinuclear location.

C. Effect of Chemical Structure of Octapeptide on Cellular Internalization

To confirm the selectivity of uptake to cells, effect of chemical structure of octapeptide was examined. Many research groups reported that arginine-rich peptide is useful as a cell penetrating peptide [8]. However, the abilities of other oligopeptides to penetrate the cell membrane have not been well characterized. Thus, we evaluated the function of various octapeptides as a cell penetrating peptide by using our PMBN/PLA/QD as an analyzing tool. We selected the octapeptides having simple sequences constructed from just one kind of amino acid, tyrosin (Y, hydrophobic), asparagine (N, hydrophilic), glutamic acid (E, hydrophilic and anionic), histidine (H, hydrophilic and weakly cationic), lysine (K, hydrophilic and cationic), or arginine (R, hydrophobic and cationic). As shown in Fig. 3, only K8 and R8 immobilized PMBN/PLA/QD could internalize in HeLa cells. It is well known that R8 and K8 work as a cell penetrating peptide. This result indicated that the hydrophilic and cationic nature of oligopeptides play a key role in the cell membrane permeation. In order to understand the mechanism of cell penetration induced by these hydrophilic and cationic oligopeptides, it will be required to investigate the relationship between surface cationic density on the nanoparticles and cell membrane permeation.

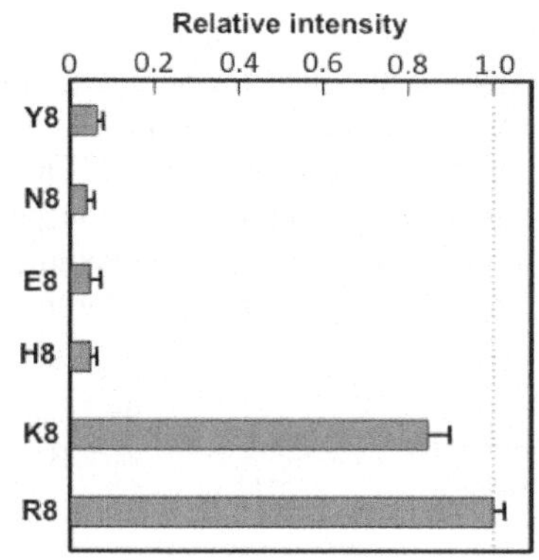

Fig. 3 Cellular uptake of various octapeptides-immobilized PMBN/PLA/QDs.

The PMBN/PLA/QD have no cytotoxicity for 3 days even after internalization in HeLa cells. However, there are some possibilities of the induction of inflammation reaction even when no cytotoxicity appears. Thus, more studies are required with respect to the inflammatory response. When the relative expression of TNF-α mRNA to GAPDH mRNA was examined in RAW264.7 cells incubated with R8 or G-PMBN/PLA/QD for a day, no significant difference between negative control and PMBN/PLA/QDs in the expression of TNF-α mRNA was observed. This indicated that phosphorylcholine groups on the surface of PMBN/PLA/QD suppress the inflammatory reaction in RAW264.7 cells even when internalizing into cells due to the cell penetrating peptides.

D. Effect of sequence and density of R8 on cellular internalization

It is important to understand the effect of sequence of octapeptide composed of inert and active amino residue for internalization to cells. We prepared a series of octapeptide by glycine(G) and arginine(R) as G8, GGGGGGRR(G6R2), GGGGRRRR(G4R4), GGRRRRRR(G2R6), and R8. Although the surface ζ-potential of original PNBN/PLA/QD was slightly negative, it altered by immobilization of octapeptide on the surface. In Fig. 4, image of cells examined with fluorescence microscopy is shown. After applying the various nanoparticles, they started to internalize into the cells. However, we observed that only R8-immobilized nanoparticles could internalize into cells. The R8 has a lot of cationic groups, however, the surface ζ-potential of the nanoparticles was almost zero. Fig. 5 shows quantitative analysis of PMBN/PLA/QD nanoparticles with octapeptide in the cells. It is clearly shown that the number of arginine residue is important and effective for internalization.

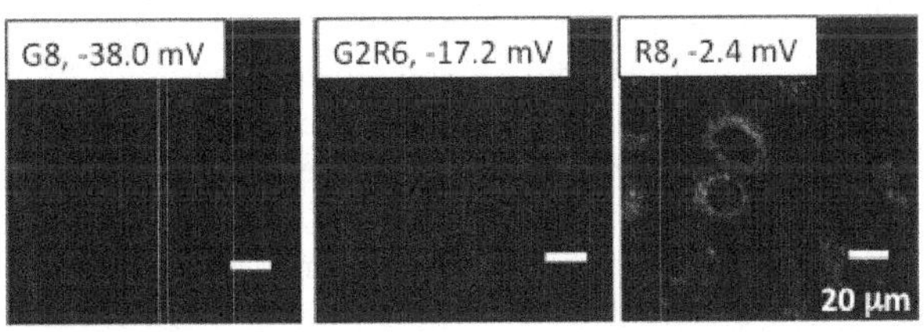

Fig. 4 Effect of octapeptide sequence of glycine (G) and arginine (R) on internalization into cells. The value of ζ-potential is indicated.

IV. Conclusions

The PMBN/PLA/QD is good probe for evaluation of biomolecules to understand their performance. And R8-peptide is effective to induce cell internalization. Moreover, R8-PMBN/PLA/QD is a most suitable material for conducting the kinetic analysis of cell membrane permeation.

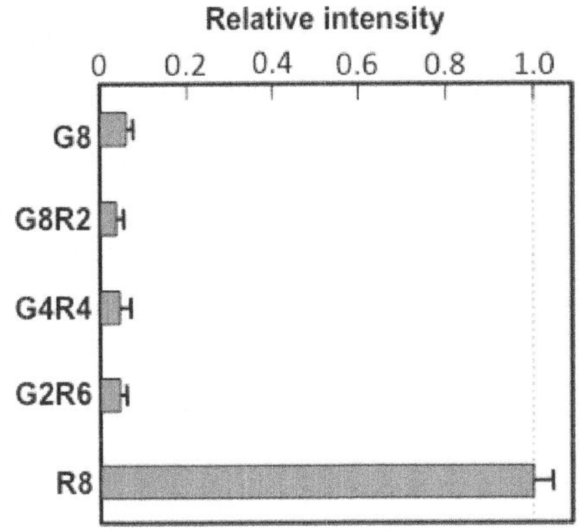

Fig. 5 Cellular uptake of various octapeptide-PMBN/PLA/ QDs.

Acknowledgment

A part of this research was supported by a Grant-in-Aid for Scientific Research on Innovative Areas ''Nanomedicine Molecular Science'' (No. 2306) from Ministry of Education, Culture, Sports, Science, and Technology of Japan.

References

1. Ishihara K, Ueda T, Nakabayashi N (1990) Preparation of phospholipid polymers and their properties as polymeric hydrogel membrane. Polym J 22: 355–360
2. Ueda T, Oshida H, Kurita K et al (1992) Preparation of 2-methacryloyloxyethyl phosphorylcholine copolymers with alkyl methacrylates and their blood compatibility. Polym J 24: 1259–1269
3. Iwasaki Y, Ishihara K (2005) Phosphorylcholine-containing polymers for biomedical applications. Anal Bioanal Chem 381. 534-546
4. Konno T, Watanabe J, Ishihara K (2004) Conjugation of enzymes on polymer nanoparticles covered with phosphorylcholine groups. Biomacromolecules 5: 342–347
5. Matsuno R, Ishihara K (2011) Integrated functional nanocolloids covered with artificial cell membranes for biomedical applications. Nano Today 6: 61-74
6. Goto Y, Matsuno R, Konno T et al (2008) Polymer nanoparticles covered with phosphorylcholine groups and immobilized with antibody. Biomacromolecules 9: 828-833
7. Goto Y, Matsuno R, Konno T et al (2008) Artificial cell membrane-covered nanoparticles embedding quantum dots as bioimaging probe. Biomacromolecules 9: 3252-3257
8. El-Sayed A, Khalil I A, Kogure K et al (2008) Octaarginine- and octalysine-modified nanoparticles have different modes of endosomal escape. J Biol Chem 283: 23450-23461

Author: Kazuhiko Ishihara

Institute: The University of Tokyo
Street: 7-3-1, Hongo, Bunkyo-ku
City: Tokyo 113-8656
Country: Japan
Email: ishihara@mpc.t.u-tokyo.ac.jp

Wireless Fire Security System for Hospitals

Kamran Hameed[1], Muhammad Muzammil Khan[1], I. Shahrukh Ateeq[1], Arsalan Khan[1], and Muhammad Ali[2]

[1] Sir Syed University of Engineering and Technology, Department of Biomedical Engineering, Karachi, Pakistan
[2] Karachi Institute of Radiotherapy and Nuclear Medicine, Department of Biomedical Engineering, Karachi, Pakistan

Abstract — **Fire security system always provides high security against fire. Conventional fire security systems are wired and they usually contain smoke detectors and control panel. The control panel is sometimes hard to operate. Not everyone can understand the use of the control panel. In hospitals and other health care institutions very expensive equipments are normally installed. So each and every organization needs proper insurance and safety for their equipments and patients. Normally fire is detected using smoke detectors, but after detecting the fire the operator or the user should know the exact place of fire in the organization. We have introduced a system that will detect and show the exact place of wire (for example in floor or any room). It is basically two channel device that can detect fire or on rooms or floors. This system is very easy to operate and it also operates on RF signals. Everyone can easily operate its portable receiver. Wireless fire security system for hospitals is also very low cost device. It is the best way to insure the proper safety against fire. The major problem of third world countries is electricity or energy, this fire security system has the solution of this problem. This security system can be operated on both AC as well as on rechargeable DC batteries. The DC batteries can operate all the seven smoke detectors attached through each transmitter. The battery is also very cheap and easy to access in local places. All the components of this fire security system are easily available. It is the best fire security device at low cost.**

Keywords — **Smoke detectors, control panel, safety, portable receiver and wireless security.**

I. INTRODUCTION

Fire alarm system always promotes efficient security with perfect protection from fire hazards by using ring led indicator fire alarm and smoke detectors. Conventional fire control panels helps to find out any symptoms of fire occurrence. Automatic fire alarm system informs you the chances of fire and the heat detectors catch the immediate.

Getting a fire alarm system is not only required for the safety or security purpose but also to get insured! Sounds strange but true! Recently, several commercial insurance agencies have come up with a condition that they will approve insurance for your company / industry only if you have got reliable and genuine fire safety equipment. This way, getting installed a fire security system or a smoke detector has become mandatory at several business places.

Almost all the business or industries are open for public and comprise of several employees. In that case, there should be some kind of monitoring service for the employees if a fire occurs. To serve the purpose, the manual pulling stations and manual fire safety measures seem to be outdated. Here comes the advanced fire alarm system into play. This device not only monitors the spot areas where fire occurs but also informs the public / people / employees about the situation and helps in controlling the same with inbuilt or auto-fitted sprinklers. This way, a 24x7 monitoring is very crucial for your business even if you are not present at the workplace. [1].

Automatic fire alarm system and hybrid fire detection system tem with the smoke detectors provide you safety with modern security and safety system with emergency fire security sound systems. Multi stored buildings, flats and shopping complexes are always at a risk of hazards and extreme heat from various processes working out there.

There are many recent developments in the security detection and locating. With recent advances, fire alarms improved its way of diagnostics with the reliability of the alarm system. In conventional fire alarm system, the cable line is out of the control panel. These alarms can help people from comfortable escape from a burning building as well as saving your business. The fire may damage you property as well as your personal belongings such as files, important documents of an office etc. Thus it can quickly burn away everything you have worked to build, whether in home or a business.

So due to the need of fire security systems we have designed a system that can operate without wired connections and AC power too. Wireless Fire security system totally works on DC batteries as well as on AC supply. The data transmits in the form of radio frequency signal without wires. [2].

To provide the proper safety against fire; the detection of fire is more important. Normally fire is detected using smoke detectors, but after detecting the fire the operator or the user should know the exact place of fire in the organization. We have introduced a system that will detect and show the exact place of wire (for example in floor or

V. Van Toi et al. (Eds.): 4th International Conference on Biomedical Engineering in Vietnam, IFMBE Proceedings 40, pp. 266–269, 2013.
www.springerlink.com

any room). It is basically two channel device that can detect fire or on rooms or floors [3].

Anyone can use it either for 2 floors or 2 rooms. The output will be displays on LCD. The whole system operates on the wireless transmission. RF signals are used in this system in order to show the output. We can attach seven smoke detectors in one floor or room (as per requirements), so total fourteen smoke detectors are used in this system to provide proper safety. We can also connect the temperature sensor or pressure sensor with the transmitters of the system.

The whole system works on DC as well as on AC. Rechargeable batteries are installed in the system. We can easily charge the system using 9 volt battery charger. The overall system is very easy to install and use anywhere [4].

II. Design Methodology

Basic working of the system

Wire fire security system works on three basic phases: *Transmitter:* Smoke detector first detects the smoke or fire in the place which is attached to the transmitter through wire to supply the voltages to the smoke detector. When the smoke detector detects the smoke the transmitter indicates the LED for fire emergency and it can also operate the external circuit with the help of built in relay in the transmitter. Transmitter then transmits the signals through radio frequency to the transceiver.

Transceiver: Transceiver receives the signals from transmitter and then sends them to the portable detector. The basic function of transceiver is to extend the range of the fire security system. It also operates through DC rechargeable battery.

Portable receiver: Portable receiver can be carried an where within the range (70-75 meters), It is easy to use and any one can operate it. It receives the signals from the transceiver and shows the information or emergency signals on the LCD screen. It also operates through the DC rechargeable battery as well as with the AC power [5].

Basic designing of the system

Detection/Transmitter phase: In detection phase first of all smoke detectors are required in order to detect smoke. A smoke detector is a device that detects smoke, typically as an indicator of fire. Commercial, industrial, and mass residential devices issue a signal to a fire alarm system, while household detectors, known as smoke alarms, generally issue a local audible and/or visual alarm from the detector itself. Smoke detectors are typically housed in disk-shaped plastic enclosure about 150 millimeters (6 in) in diameter and 25 millimeters (1 in) thick, but the shape can vary by manufacturer or product line. Then the smoke detector is attached with the transmitter circuit. Transmitter consists of LEDs and relays, capacitors are used to make continues pulse of 1 minute in to 3 seconds in order to transmit the signals. RF transmitter is also used with the rechargeable battery inside the transmitter's box [6].

Transceiver phase: Transceiver circuit is a small circuit that is used to increase the range of the system. It consists of receiver that contains RF module and remote / transmitter that will transmit signals to the portable receiver. It also contains relay and capacitors for delay and switching. This whole box operates on rechargeable DC battery.

Portable receiver phase: Finally, we have worked on the portable receiver. It contains LCD and RF module to receive the RF signals from transceiver. It also consists of controller IC so we have programmed the controller IC in order to display the required result on the LCD screen. Programming has been done in C-language using its suitable compiler. Portable receiver also contains relay and buzzer for emergency sound. The whole portable receiver is operated through the DC rechargeable batteries.

Programming stages of PIC microcontroller

The programming of the fire security system has been done on C language using its specific suitable compler. Thebasic stages of the programming are shown in fig 1.

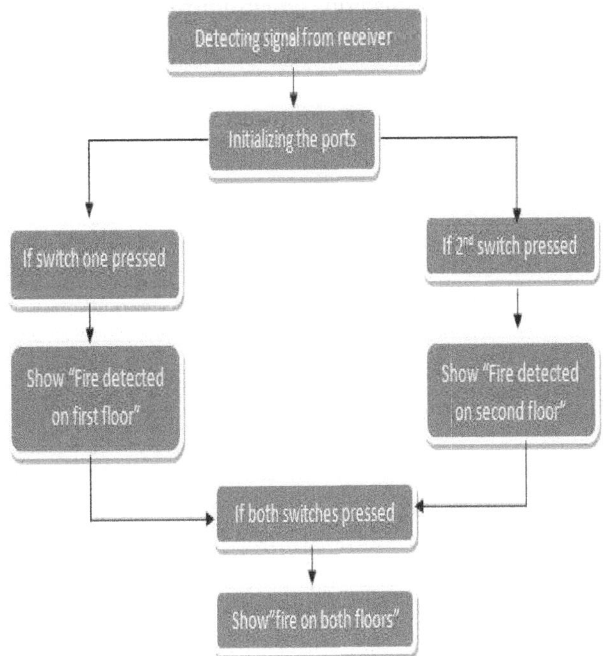

Fig. 1 Programming stages of microcontroller.

Representation of the fire security system

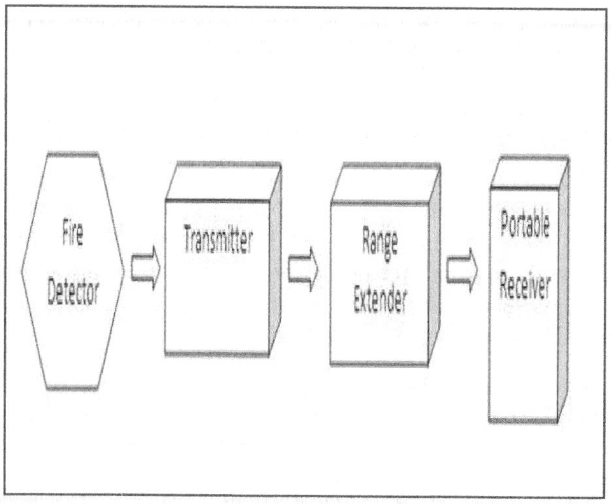

Fig. 2 Block diagram of fire security system for hospitals.

Fig. 2 shows the basic flow for fire security system, Designed system may use sixteen smoke detectors for covering four floor of four room at each floor, Whenever smoke detector sense the fire the corresponding smoke detector generates the interrupt signal which is picked by transmitter while at receiving hand portable receiver receive the transmitted signal which indicates the floor of the room in which fire has been sensed by the detector. The functionality of the security system for room one has been discussed in fig. 3.

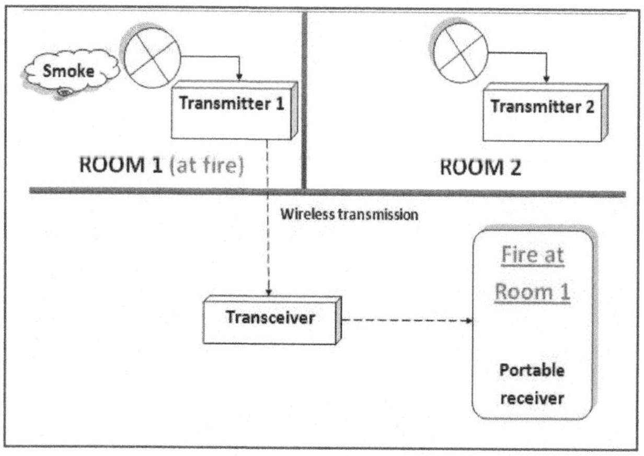

Fig. 3 Functional diagram of fire security system for hospitals.

Completion phases of the fire security system

Stages	*Design methodology*
Transmitter	Interfacing smoke detectors with the RF transmitter. Connecting transmitter with the relays and adjusting the voltage requirements of the circuit. Operating circuit through 9 volt re-chargeable battery.
Range extender	Interfacing the receiver of one RF circuit with another RF circuit. Operating both circuits with 9 volt re-chargeable battery. Designing circuit of charger to charge the device. Connecting relays.
Portable receiver	Interfacing receiver with the micro-controller and LCD. Programming of micro-controller. Operating whole circuit with 9 volt re-chargeable battery. Designing the circuit.
Finalization of the project	Casing and final testing of device in the hospital.

F. Basic technology used for RF signals

We have used car door lock system in this project to transfer the signals through radio frequency. As we know cardoor lock system is easily available in Pakistan and it is also very cheap. So in order to reduce the cost we have not usedconventional RF modules of 100 meters or more range. We have combined 2 car security systems to increase the range of the fire system. We have interfaced the circuit of car door lock system with the PIC microcontroller.

III. Conclusion

Wireless hospital's fire security system provides full security against fire because it can detect the fire at any specific area and sends the signal immediately to the portable receiver. The receiver can be carried anywhere within the range of the system that is nearly 75 meters. RF signals transmit the signals very accurately without any false triggering. External circuits can also be installed with this system because an additional option is also given for connecting the external circuit [5].

Fig. 4 shows the complete designed project i.e., Wireless fire security system for hospitals, this system can not only be installed in hospitals may also be installed in home, in industries like Chemical, Automobile, Food Manufacturing etc.

Fig. 4 Wireless fire security system for hospitals.

IV. Discussion

Transmitter Phase: Transmitter phase contains the battery that is a rechargeable and basic circuit component. Then the smoke detector is attached with the transmitter circuit. Transmitter consists of LEDs and relays, capacitors are used to make continues pulse of 1 minute in to 3 seconds in order to transmit the signals. RF transmitter is also used with the rechargeable battery inside the transmitter's box [7].

Transceiver Phase: Transceiver circuit is a small circuit that is used to increase the range of the system. It consists of receiver that contains RF module and remote / transmittethat will transmit signals to the portable receiver. It also contains relay and capacitors for delay and switching. Thiswhole box operates on rechargeable DC battery.

Receiver Phase: It contains LCD and RF module to receive the RF signals from transceiver. It also contains PIC IC so we have programmed the PIC IC in order to display the required result on the LCD screen. Programming has been done in C-language using its suitable compiler. Portable receiver also contains relay and buzzer for emergency sound. The whole portable receiver is operated through the DC rechargeable batteries [8].

V. Improvements for future

There are certain sets of possibilities to improve this fire security system. Some of these possibilities are described , *Make an external Circuit:* In order to provide effective fire extinguishing technique we can make a fire extinguisher circuit at the out put of the project.

Extend the Range: Current range of the project is about 75 meters that can be extended if you receiver of RF moulator is available of the certain frequency.

Sensitive smoke detectors: If we attach the sensitive smoke detectors with the transmitter than it will detect the smoke much efficiently and quickly.

Interfacing with computer: We can transmit signals to the computer also. For this a special circuitry is required that will interface the receiver with the computer.

Installing vibrator: A vibrator can also be installed in order to get the output on the portable receiver.

Acknowledgment

We are very indebted to Dr. M. A. Haleem Chairman of Biomedical Engineering department who gave us his complete support and resources utilization. His esteemed assistance and untiring efforts were commendable.

References

1. "Smoke alarm safety tips" safety information, National fire protection association, year 2009
2. Joel williams, "wireless smoke detectors on ezine" on march 1, 2010.
3. "Fire technology", volume 20 #1"(2010) springer 1984 02.
4. "Scientific and technical information", 2008, nasa center for Aero space information (casi).
5. Residential smoke alarm performance, thomas cleary, Building and fire research laboratory, national institute of Standards and technology, ul smoke and fire dynamics Seminar, november, 2007.
6. "Smoke detectors and americium-241 fact sheet" (2009) Canadian nuclear society.
7. New york city fire department (2008) "carbon monoxide Poisoning".
8. "smoke detector information", (2008) barre city fire Department.
9. "Smoke alarm safety tips" safety information (2009) National fire protection association.

Wavelet Coefficient Average Value for Prediction of Motor Control Area of Human Brain Using fNIRS

T.N. Nguyen[1], T.H. Nguyen[2], and T.T. Nguyen[2]

[1] Industrial Electronics Department, Cao Thang Technical College, Ministry of Industry and Trade, Vietnam
[2] Biomedical Engineering, International University - Vietnam National University, HCM, Vietnam

Abstract — **This paper proposed a average threshold algorithm for predicting motor control area of human brain using fNIRS technique. Firstly, data obtained from human brain were pre-processed to reduce noise and artifacts. Thus, the data were analyzed using a decomposition wavelet algorithm with a Bior 5.5 function to produce wavelet coefficients. From these coefficients, one can create a coefficient threshold to determine motor control area as well as to distinguish two brain hemispheres based on oxy hemoglobin (oxy-Hb) concentration. Experimental results for predicting motor control area confirmed the feasibility of the proposed method.**

Keywords — **Wavelet Coefficient, average value, fNIRS signal and motor control area.**

I. INTRODUCTION

Human brain has a complex structure with many nerve cells (around 100 billion neurons), so it is a challenging problem in biological computing [1]. There have been non-invasive technologies such as function Near-Infrared Spectroscopy (fNIRS), function Magnetic Resonance Imaging (fMRI) and ElectroEncephaloGraphy (EEG), which have been applied for patient diagnose and disabled people rehabilitation in recent years. An example of rehabilitation is that information obtained from human brain using one of the EEG or fNIRS technique could be used to perform shared control of motion wheelchairs [2-4].

Uncertain characters (noises, artifacts, interferences) of EEG signals or fMRI images often appear, thus the mechanism of motor processing function remains unclear and developed predictive models showed recognition inaccuracy and error. Furthermore, fMRI technique acquired only control data in supine position due to nature of fMRI machine leading to inappropriate conditions for data acquisition in real life [5, 6].

In recent years, fNIRS technology has been appeared as an alternative brain-based experimental technique [7] to measuring human thoughts and activities. Moreover, there is no position obligation (e.g. supine position) during experiment. Therefore, the subject feels free for performing his or her brain activities. In literature, this technique has been successfully used to study human brain functions such as assessment of motor task from everyday living, athletic performance, and recovery from neurological illness [8] or assessment of verbal fluency [9] or quantification of brain function during finger tapping [10], etc. However, to the best of our knowledge, there is no or few application of fNIRS technique to quantify the motor control signals leading to autonomous wheelchair control or brain simulator.

Recently, wavelet transform has been utilized to process many problems related to brain analysis [11]. In practice, there are a lot of research about brain signals based on fNIRS technique, but they are really difficult to recognize their features due to noises and artifacts [12, 13]. Moreover, a wavelet approach was proposed to remove motion artifacts from fNIRS signals [14].

In fact, detection of brain motor control arae has been a big challenge to better understanding of brain motor functions. In this paper, an average threshold algorithm of wavelet coefficients was investigated to predict brain motor control area and to distinguish the activity of the left and right brain using a Bior 5.5 wavelet function corresponding to oxy-Hb concentration of human brain.

II. MATERIALS AND METHODS

A. Data acquisition

Our method consists of four main procedures: (1) fNIRS data acquisition; (2) pre-processing; (3) decomposition wavelet and (4) average value determination. First, brain oxygenation signals during motor control task were acquired from a FOIRE 3000 fNIRS machine (Shimadzu Corporation, Japan). Second, raw signals were processed using wavelet algorithms and filters. After processing, data were analyzed and calculated to determine threshold value based on wavelet coefficients.

Three subjects (males, 20 average years old, 65kg average weight) were participated into this study. The

V. Van Toi et al. (Eds.): 4th International Conference on Biomedical Engineering in Vietnam, IFMBE Proceedings 40, pp. 270–274, 2013.
www.springerlink.com

subjects informed consent agreement after reading and understanding of the experiment protocol and the fNIRS technique as shown in Fig. 1. The subjects were demanded to perform a 5kg weight lifting motion as motor activity.

Rest	Task	Rest
10sec	20sec	10sec

Fig. 1 Protocol of each measure time

Two 2x4 matrices set up at the left and right brain sides to produce channels, as shown in Fig. 2a and Fig. 2b, for observing oxy hemoglobin (Hb) concentration.

Fig. 2a Ten channels (from 1 to 10) obtained at the left brain.

Fig. 2b Ten channels (from 11 to 20) obtained at the right brain.

The transmitter and receiver probes with a set of holder were mounted on the left and right hemispheres of a subject and signals of oxy Hb, deoxy Hb and total Hb obtained as shown in Fig. 3a and Fig. 3b.

Fig. 3a Subject installed with probes.

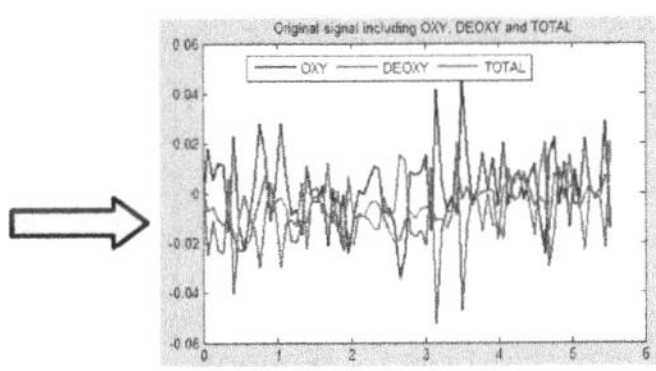

Fig. 3b Signals of oxy Hb, deoxy Hb and total Hb.

B. Signal Pre-processing

To reduce noise and artifacts (measure, environment and machine effect) and unknown frequencies problem of brain signals, a Savitzky-Golay method was applied. The fNIRS point output signal of the smoother $x[k]$ is as follows:

$$x[k] = \frac{\sum_{i=-n}^{n} A_i x[k+i]}{\sum_{i=-n}^{n} A_i} \quad (1)$$

in which A_i is a set of integers and $n,k = 0, 1, 2,\ldots$. The smoothed signals were connected to wavelet inputs to produce wavelet coefficients for brain area prediction.

C. Decomposition Wavelet Algorithm

Discrete wavelet transformation W employed to calculate its coefficients is presented as follows:

$$y[n] = \sum_{k=0}^{N-1} x[n].w[k-n] \quad (2)$$

in which, w is the filter and $n,k = 0, 1, 2,\ldots$.

In the wavelet decomposition (WD) algorithm, one can decompose the signal into a coarse approximation and detail information. In particular, the discrete signal $x[n]$ is passed through both a half band low-pass filter $h[n]$ and a half band high-pass filter $g[n]$. Therefore, the signal can be sampled by 2, it means that it is discarded every each sample. The formulas can mathematically be expressed as follows:

$$d_i[k] = \sum_{k=0}^{N-1} x[n].g[2k-n] \quad (3)$$

$$a_i[k] = \sum_{k=0}^{N-1} x[n].h[2k-n] \quad (4)$$

in which, d_i and a_i are also called the approximate and detail coefficients of wavelet decomposition.

The method of decomposition filters by down-sampling of 2 is shown in Fig. 4. After decomposition, each of down-sampling will produce a half band filter.

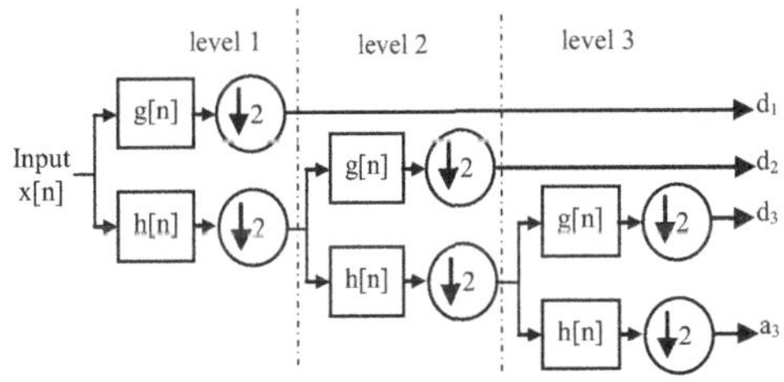

Fig. 4 Wavelet decomposition algorithm with band filters divided by 2.

D. Average Value Determination

Channel data obtained from two brain sides were analyzed using the wavelet composition algorithm to produce the approximate and detail coefficients. Therefore an average value of each detail coefficient will be calculated for prediction of motor control area. The equation of the average of a detail coefficient can be presented as follows:

$$m_j = \frac{\sum_{i=1}^{N} d_i[k]}{N} \quad (5)$$

in which $m_j, j=1,2,\ldots$ is the average value.

In this paper, the wavelet decomposition algorithm was utilized to create approximation and detail coefficients. Based on the detail coefficients, the average of each detail coefficient of each channel was employed for detecting channel corresponding to motor control area and distinguishing the difference of two brain hemispheres corresponding to weight lifting activities.

III. RESULTS AND DISCUSSION

A. Experiments-1: fNIRS signals obtained from subjects were passed through a smoother.

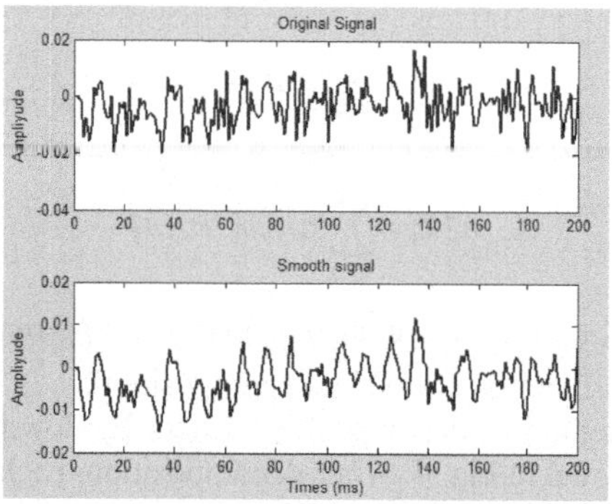

Fig. 5 Original signal and the smoothed signal.

In this case, $n = 3$ was chosen to smooth fNIRS signals using Equation (1). An original signal and the smoothed signal were shown in Fig. 5.

B. Experiments-2: Determining brain motor control area.

In this paper, a wavelet decomposition transform with the bior 5.5 function as a mother wavelet was utilized. The wavelet signals were decomposed into 10 levels to determine coefficients from d_1 to d_{10} and a_{10}.

Moreover, a matrix of the transmitter and receiver probes was designed to measure at the left brain side and its channels were generated as shown in Fig. 6.

With the arrangement of eight pairs of the transceiver and receiver probes on two left and right hemispheres as shown in Fig. 3, all 20 channels were collected. In particular, there are 10 channels (from 1 to 10) of the left and 10 channels (from 11 to 20) of the right. In order to find motor area of human brain, all channels were analyzed using the wavelet decomposition algorithm in Equations (3) and (4). After analyzing 10 channels of signals, the details coefficients (from d_1 to d_{10}) were obtained. Next the coefficients were averaged to produce a threshold.

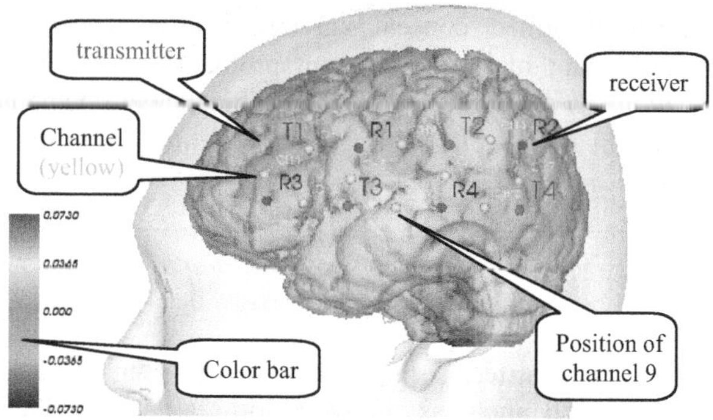

Fig. 6 Schematic of measured matrix including red transmitter, blue receiver and yellow channels at the left brain.

The measurements were repeated five times and then five different values m_j were calculated Table 1 showed the average values of 10 d_{10} channels and the average values Ave(M) were also calculated in 5 times of subject-1, in which the amplitudes of channels 6 and 9 are greater than the remaining channels and m_{10} /Ci is the average value of a channel.

Table 1 The average value m_{10} of the detail d_{10} - subject 1.

	Time-1	Time-2	Time-3	Time-4	Time-5	Ave(M)
m_{10} /C1	0,04926	0,15372	0,28106	0,27304	0,13580	0,17858
m_{10}/C2	-0,04366	0,18498	0,14113	0,27350	0,10948	0,13309
m_{10}/C3	0,10586	0,12191	0,12415	0,13766	0,12200	0,12232
m_{10}/C4	0,16408	0,20663	0,26338	0,20213	0,18965	0,20517
m_{10}/C5	0,19564	0,16110	0,06296	0,30295	0,11133	0,16680
m_{10}/C6	0,28340	0,39505	0,20791	0,34281	0,24151	**0,29414**
m_{10}/C7	0,20246	0,10988	0,06581	-0,02493	0,14229	0,09910
m_{10}/C8	0,31063	0,29007	0,30557	0,46017	0,15746	0,30478
m_{10}/C9	0,36518	0,46838	0,21668	0,49929	0,19673	**0,34925**
m_{10}/C10	0,32204	0,31245	0,18932	0,28298	0,18912	0,25918

In similarity, fNIRS data of the subject-2 were calculated and produced the average values of the detail-10 as shown in Table 2. However, channel 9 has the high amplitude and channel 6 is a litle bit lower corresponding to the high oxy Hb concentration.

Table 2 The average value m_{10} of the detail d_{10} - subject 2.

	Time-1	Time-2	Time-3	Time-4	Time-5	Ave(M)
m_{10}/C1	0,30862	0,79849	0,66522	0,74806	0,20530	0,54514
m_{10}/C2	0,18257	0,53088	0,48813	0,38760	0,05163	0,32816
m_{10}/C3	0,33732	0,52464	0,54771	0,68281	0,02196	0,42289
m_{10}/C4	0,23012	0,65847	0,53123	0,52142	0,13079	0,41440
m_{10}/C5	0,37774	0,51301	0,34862	0,35021	0,20432	0,35878
m_{10}/C6	0,36691	0,59628	0,40955	0,48889	0,23310	**0,41895**
m_{10}/C7	0,34614	0,69340	0,40308	0,67014	0,06843	0,43624
m_{10}/C8	0,34969	0,56098	0,38070	0,34396	0,23535	0,37414
m_{10}/C9	0,68848	0,73320	0,40516	0,40895	0,28514	**0,50419**
m_{10}/C10	0,62040	0,79541	0,32756	0,55140	0,16799	0,49255

The average values of the detail-10 of the subject-3 were analyzed as shown in Table 3, in which the amplitudes of channels 6 and 9 are a bit higher.

Table 3 The average value m_{10} of the detail d_{10} - subject 3.

	Time-1	Time-2	Time-3	Time-4	Time-5	Ave(M)
m_{10}/C1	0,00435	0,12627	0,07431	0,10176	0,03908	0,06915
m_{10}/C2	0,05855	0,16065	0,07233	0,22296	0,05316	0,11353
m_{10}/C3	0,05317	0,10234	0,05311	0,33319	-0,06150	0,09606
m_{10}/C4	-0,02763	0,09071	0,05240	0,09078	-0,01543	0,03817
m_{10}/C5	0,11635	0,10425	0,11881	0,15828	0,15735	0,13101
m_{10}/C6	0,12656	0,10474	0,08076	0,13986	-0,00542	**0,08930**
m_{10}/C7	0,11961	0,18434	0,16082	0,15240	0,02069	0,12757
m_{10}/C8	0,03312	0,13925	0,08486	0,21831	0,02045	0,09920
m_{10}/C9	0,07440	0,26008	0,05410	0,27998	0,04365	**0,14244**
m_{10}/C10	0,12909	0,18447	0,03187	0,20720	-0,02930	0,10467

Moreover, the signal shapes of channels 6 and 9 corresponding to their detail coefficients basically showed the same phases as shown in Fig.7 and Fig. 8. This means that the channels 6 and 9 are reliable to say that they are measured at motor control position.

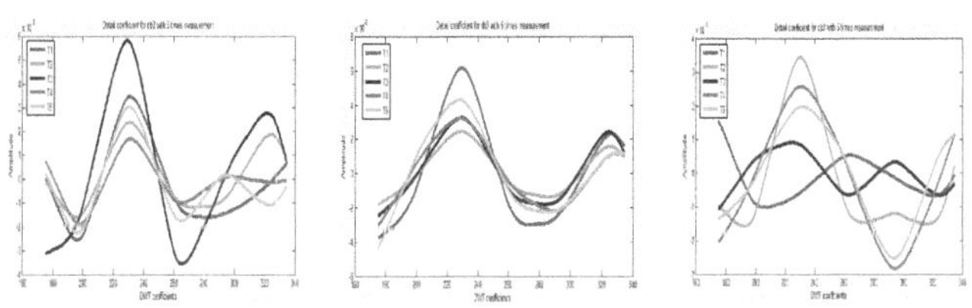

Fig. 7 Signals with the same phases of channel 6 of 3 subjects.

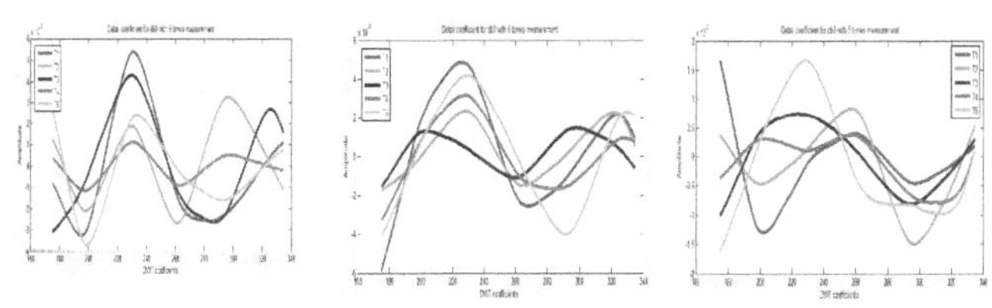

Fig. 8 Signals with the same phases of channel 9 of 3 subjects.

C. Experiments-3: Distinguishing the activity of the left and right brain sides.

From the analysed results as in *Experiments-2*, we can choose the channel 9 which represents the left brain. Similarly, one can choose the channel 19 corresponding to the right brain side. In addition, the Ave(M) values of two channels 9 and 19 have the high and stble amplitudes on three subjects as shown in Table 4.

Table 4 The Ave(M) values of details at channel 9 channel 19.

Subjects	Weight lifting	Channel 9	Channel 19
1	5 kg	**0,34925**	**0,1442**
2	5 kg	**0,50419**	**0,3791**
3	5 kg	**0,14244**	**0,1051**

From the values in Table 4, a bar graph, as shown in Fig. 9, say that the activity of the left brain is always more active than that of the right brain side.

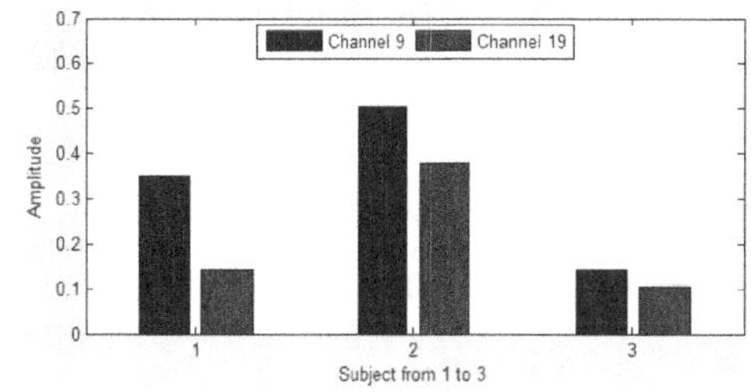

Fig. 9 The comparison of the amplitudes of channel 9 and channel 19 preformed on 3 subjects.

In addition to the results in Fig. 9, the average of detail coefficients from d_1 to d_{10} was calculate and their waveforms are similar, in which the amplitudes of channel 9 are greater than that of channel 19 as shown in Fig. 10. This is one more information to conclude that the left brain has higher oxy Hb concentration than the right brain.

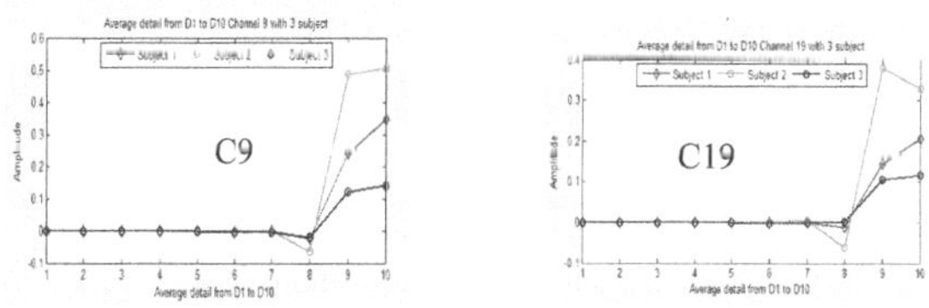

Fig. 10 The waveforms of coefficients from d_1 to d_{10} from 3 subjects.

IV. CONCLUSION

An average algorithm was proposed to calculate average values of detail coefficients for motor control area prediction and the distinction of two brain hemispheres. Firstly fNIRS data were smoothed to reduce noise. Moreover, a wavelet decomposition algorithm was employed to calculate wavelet coefficients. From these detail coefficients, the average values of channels 9 and 19 allow us recognize the motor control area as well as the activity of the left and right brain sides. Experimental results showed to illustrate the effectiveness of the proposed average method.

ACKNOWLEDGMENT

The authors would like to acknowledge the support of the Shimadzu through the Shimadzu-IU Research Collaborative Program. We would also like to thank Thien and Binh of Biomedical Engineering Department at IU for their support during data collection.

REFERENCES

1. J. Wolpaw, N. Birbaumer, D. McFarland, G. Pfurtschellere, and T. Vaughan, "Brain-computer Interfaces for Communication and Control," Clinical Neurophysiology, pp. 767-791, 2002.
2. J. Wolpaw, N. Birbaumer, D. McFarland, G. Pfurtschellere, and T. Vaughan, "Brain-computer interfaces for communication and control," in Clinical Neurophysiology, 2002, pp. 767-791.
3. N. Ince, F. Goksu, A. Tewfik, and S. Arica, "Adapting subject specific motor imagery EEG patterns in space-time-frequency for a brain computer interface," in Biomedical Signal Processing and Control, 2009, pp. 236-246.
4. N. Weiskopf, F. Scharnowski, R. Veit, R. Goebel, N. Birbaumer, and K. Mathiak, "Self-regulation of local brain activity using real-time functional magnetic resonance imaging (fMRI)," in Journal of Physiology, 2004, pp. 357-373.
5. N. Ince, F. Goksu, A. Tewfik, and S. Arica, "Adapting subject specific motor imagery EEG patterns in space–time–frequency for a brain computer interface," Biomedical Signal Processing and Contro, pp. 236–246, 2009.
6. N. Weiskopf, F. Scharnowski, R. Veit, R. Goebel, N. Birbaumer, and K. Mathiak, "Self-regulation of local brain activity using real-time functional magnetic resonance imaging (fMRI)," Journal of Physiology, pp. 357-373, 2004.
7. S. Lloyd-Fox, A. Blasi, and ElwellCE, "Illuminating the developing brain: The past, present and future of functional near infrared spectroscopy," Neuroscience and Biobehavioral, pp. 269–284, 2010.
8. D. Leff, F. Orihuela-Espina, C. Elwell, T. Athanasiou, D. Delpy, A. Darzi, and G. Yang, "Assessment of the cerebral cortex during motor task behaviours in adults: A systematic review of functional near infrared spectroscopy studies," NeuroImage, pp. 2922–2936, 2011.
9. M. Schecklmann, A. Ehlis, M. Plichta, and A. Fallgatter, " Influence of muscle activity on brain oxygenation during verbal fluency assessed with functional near-infrared spec-troscopy.," Neuroscience, pp. 434–442, 2010.
10. T. Sato, M. Ito, T. Suto, M. Kameyama, M. Suda, Y. Yamagi-sh, A. Ohshima, T. Uehara, M. Fukuda, and M. M. A. "Time courses of brain activation and their implications for function: A multichannel near-infrared spectroscopy study during finger tapping," Neuroscience Research, pp. 297-304 2007.
11. X.-G. Hu, L.-T. Liu, J. Hinderer, and H.-P. Sun, "Wavelet filter analysis of local atmospheric pressure effects on gravity variations," Springer-Verlag, pp. 447-459, 2005.
12. L. Chunguang, L. Tao, I. Yoshio, and S. Kyoko, "Evaluation of a Bimanual-Coordinated Upper-Limbs Training System Based on the Near Infrared Spectroscopic Signals on Brain," in 32nd Annual International Conference of the IEEE EMBS, 2010, pp. 6625-6628.
13. H. Ayaz, M. Izzetoglu, P. A. Shewokis, and B. Onaral, "Sliding-window Motion Artifact Rejection for Functional Near-Infrared Spectroscopy," in 32nd Annual International Conference of the IEEE EMBS, 2010, pp. 6567-6570.
14. B. Molavi and G. A. Dumont, "Wavelet Based Motion Artifact Removal for Functional Near Infrared Spectroscopy," in 32nd Annual International Conference of the IEEE EMBS, 2010, pp. 5-8.

Author: Nguyen Thanh Nghia
Institute: Cao Thang college.
Street: 65 Huynh Thyc Khang
City: Ho Chi Minh
Country: Viet Nam
Email: thanhnghiakdd@yahoo.com

A Mean Threshold Algorithm for Human Eye Blinking Detection Using EEG

T. Nguyen[1], T.H. Nguyen[2], K.Q.D. Truong[2], and Toi Van Vo[2]

[1] Industrial Electronics Department, Cao Thang Technical College, Ministry of Industry and Trade, Vietnam
[2] Biomedical Engineering Department, International University, Vietnam National University, Vietnam

Abstract — **A mean threshold algorithm was proposed to detect eye blinking using ElectroEncephaloGraphy (EEG) in this paper. Firstly, the activity of eye blinking related to the delta area of human brain was investigated using EEG technology. Before analyzing the EEG data, original data were smoothed to reduce noise or artifacts by a band-pass filter. Secondly, the proposed threshold method was applied to determine a threshold value which can be used to distinguish cases of eye blinking and opened eye. Moreover, the phenomenon of eye blinking can be useful for diagnosing eye diseases such as dry eye or congenital glaucoma. Experimental results showed that the proposed eye blinking threshold approach is the effectiveness.**

Keywords — **EEG signal, eye blinking, threshold value.**

I. INTRODUCTION

Human brain plays an important role in controlling all body activities. Moreover, it is a complex structure in which different nerve cells (around 100 billion neurons) communicate from one to another with or without external excitations to make control decisions (cognition, motion, pattern recognition, etc.) [1]. For these reasons, non-invasive technologies such as EEG, functional Magnetic Resonance Imaging (fMRI) and functional Near-infrared Spectroscopy (fNIRS) have been investigated to quantify motor processing function of human brain [2-5]. Thus the exploration of these technologies can allow us to perform autonomous wheelchair control or brain simulator leading to improve or recover the motor/cognitive functions of tetraplegic patients with spinal cord injuries and degenerative nerve diseases [3, 6, 7].

In recent years, EEG technique for exploring human brain has been quickly developed and also the complexity of human brain has attracted many researchers to discover it by non-invasive technologies. In fact, many Brain-computer Interface (BCI) applications as well as Brain-based diagnoses have been successfully applied. In particular, a BCI system can allow people communicate and control external devices without using traditional motor output pathways [8-10]. It means that one can translate brain activities into messages or commands to control devices [11-14].

An EEG system has been used to collect eye blinking and opened eye data of a subject at delta position of human brain. After that, a threshold algorithm has been employed to determine the eye blinking problem. In addition, the mean threshold allows us to be able to count eye blinking times and to control a typical device or to diagnose eye problems.

II. MATERIALS AND METHODS

A. Data Acquisition

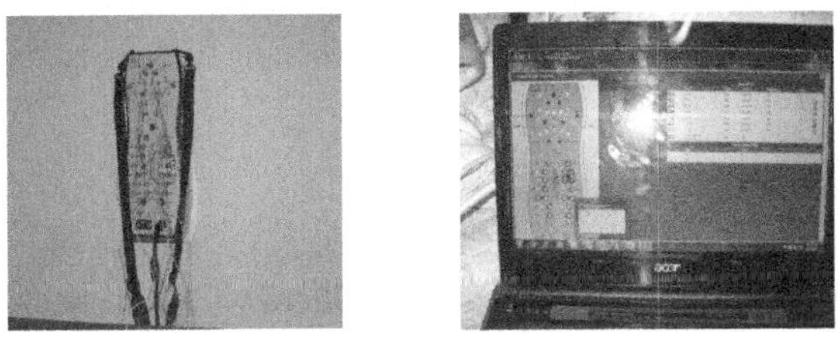

Fig. 1 The Alice 5 system connected to a laptop.

Data at delta area of human brain were obtained using EEG technology from an Alice 5 system as shown in Fig. 1. This system, which was located at Community Health Care Center (CHAC), was invested by Biomedical Engineering Department, International University (IU), VNU.

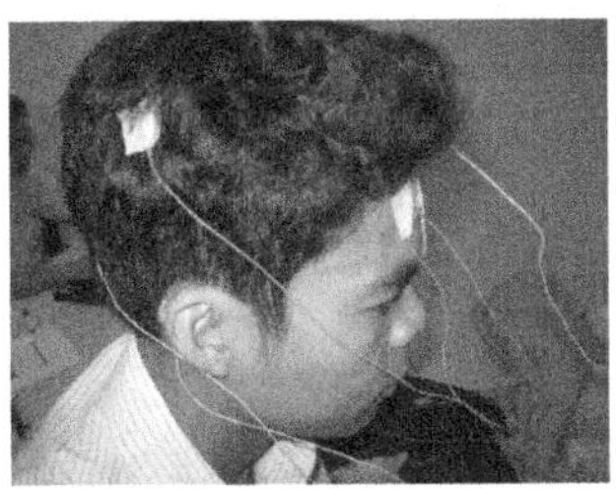

Fig. 2a A subject was installed electrodes for obtaining data.

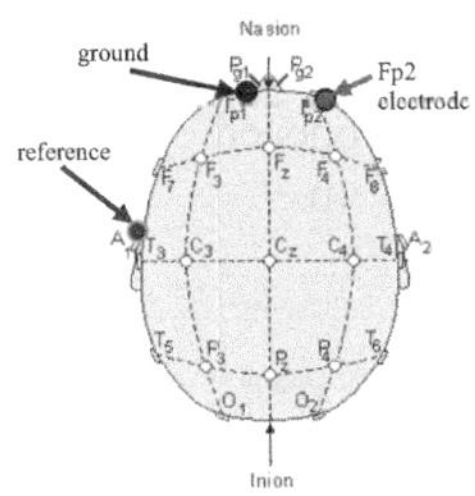

Fig. 2a Three electrodes were installed at 3 positions.

One subject (male, 25 years old, 58 kg weights) was participated into this study. The subject informed consents agreement after reading and understanding of the

V. Van Toi et al. (Eds.): 4th International Conference on Biomedical Engineering in Vietnam, IFMBE Proceedings 40, pp. 275–279, 2013.
www.springerlink.com

experiment protocol and the EEG technique as shown in Fig. 2. Offline data were obtained from the Alice 5 system when the subject performed his eyes blinking times (open and close his eyes). The first task of the subject is about 15 seconds for opening his eyes according to the protocol designed in Fig. 3. The second is that the subject carried out his task about 15 seconds each time for eye blinking case and the subject did with the same protocol in Fig. 3, but weeks and days of the second task are different from the first task.

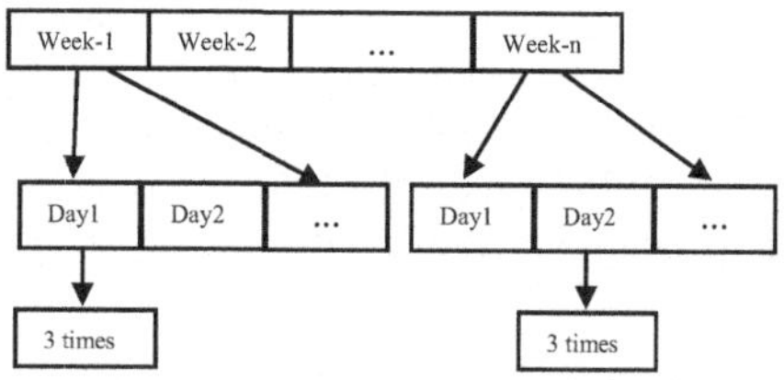

Fig. 3 Protocol of the subject for opened eye and eye blinking tasks.

The subject was installed with electrodes at some positions such as at Fp2, Ground and Reference (A_1) on his head as shown in Fig. 2a and Fig. 2b and he did two cases of eye blinking and opened eye to obtain EEG signals for analyzing to find threshold value.

B. Signal Pre-processing

In the EEG signal processing, original signal will be passed through the band-pass filter with an impulse response. In order to obtain the output of the filter, $y[n]$ the convolution operation of EEG signal and the impulse response are employed using the following formula:

$$y[n] = x[n] * h[n] = \sum_{n=-\infty}^{\infty} x[k].h[n-k] \qquad (1)$$

where $x[n]$ is EEG signal and $h[n]$ is the impulse response, $n,k = 1,2,\ldots N$.

In this paper, the number of EEG signal samples is N=2500) and the EEG signals and the filtered signals in cases of opened eye and eye blinking were shown in Fig. 4a and Fig. 4b.

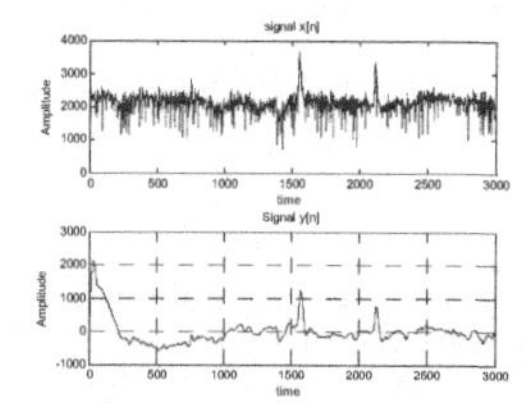

Fig. 4a Filtered signal in case of opened eye.

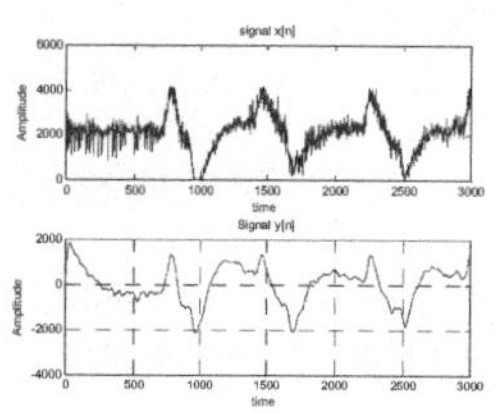

Fig. 4b Filtered signal in case of eye blinking.

After filtering noise by the band-pass filter, EEG signals will be analysed and calculated to determine a threshold for eye blinking recognition.

C. Threshold Determination

In this project, a threshold algorithm will be applied to determine cases of opened eye and eye blinking. In addition, we can count times of eye blinking of a subject. In order to determine a threshold for counting eye blinking times, firstly, we can calculate the average value M of opened eye signals using the following equation:

$$M = \frac{\sum_{n=1}^{N} y(n)}{N} \qquad (2)$$

where y(n) is sets of EEG signals and N denote the number of samples.

The second is that the standard deviation SD in case of opened eye signal can be calculated as follows:

$$SD = \frac{\sqrt{\sum_{n=1}^{N} (y(n) - M)}}{N} \qquad (3)$$

A threshold algorithm using equations (2) and (3) is built to determine cases of opened eye and eye blinking as well as the amount of eye blinking times as follows:

$$THR = M - a * SD \qquad (4)$$

where a is the coefficient of the standard deviation.

This paper showed the detection of states of eye blinking based on the change of amplitude of delta signal with its frequency range (0.5 to 3.5 Hz) at Fp2 position. Therefore the threshold determined based on EEG signals in the opened eye case plays an important role in detecting eye blinking case.

III. RESULTS AND DISCUSSION

A. Experiment 1: The subject opened his eye about 15 seconds and did this two weeks (3 times a week).

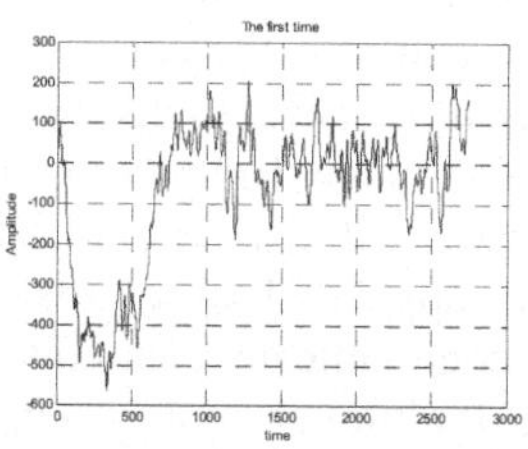

Fig. 5a Opened eye experiment in the first time of week-1.

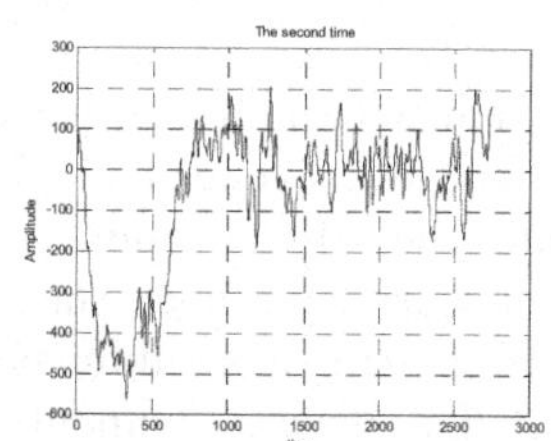

Fig. 5b Opened eye experiment in the second time of week-1.

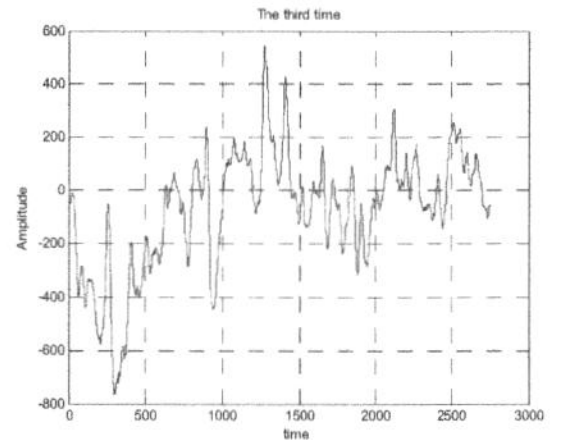

Fig. 5c Opened eye experiment in the third time of week-1.

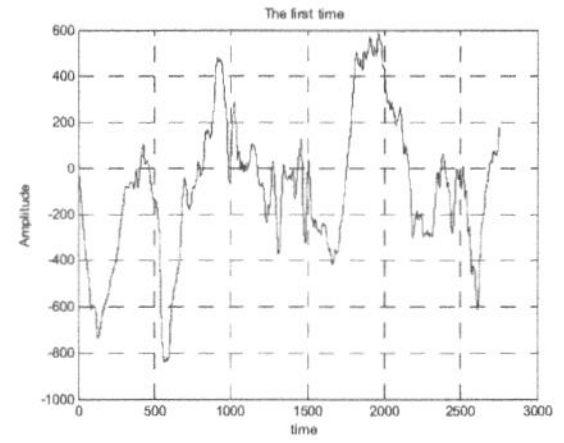

Fig. 6a Opened eye experiment in the first time of week-2.

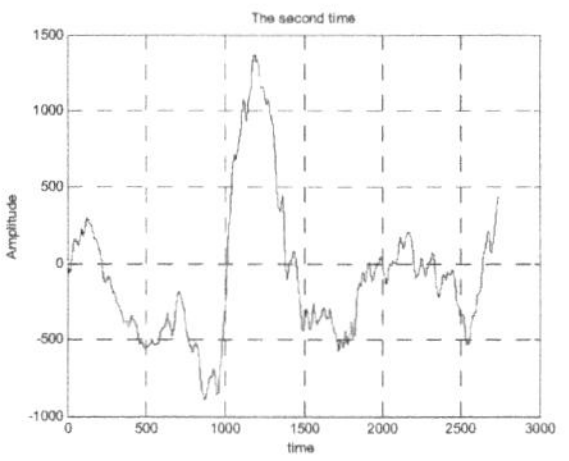

Fig. 6b Opened eye experiment in the second time of week-2.

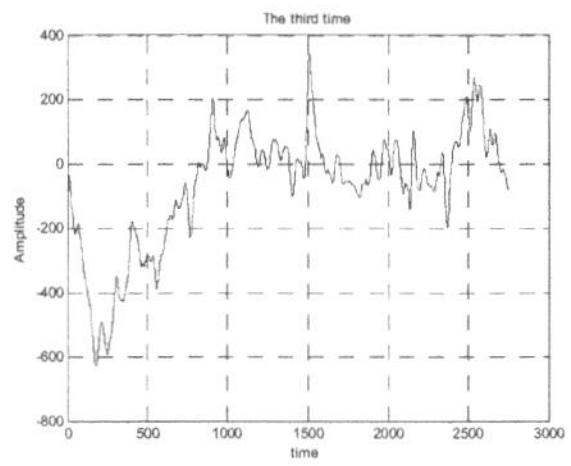

Fig. 6c Opened eye experiment in the third time of week-2.

The subject attended to obtain EEG signals, in which he was taken data 3 times as shown in Fig. 5a, Fig. 5b and Fig. 5c in the first week and the second week is that he still did the experiment 3 times similar to the first week and results showed in Fig. 6a, Fig. 6b and Fig. 6c.

B. Experiment 2: The subject blinked hid eyes about 15 seconds and the experiment was performed in two weeks (3 times a week).

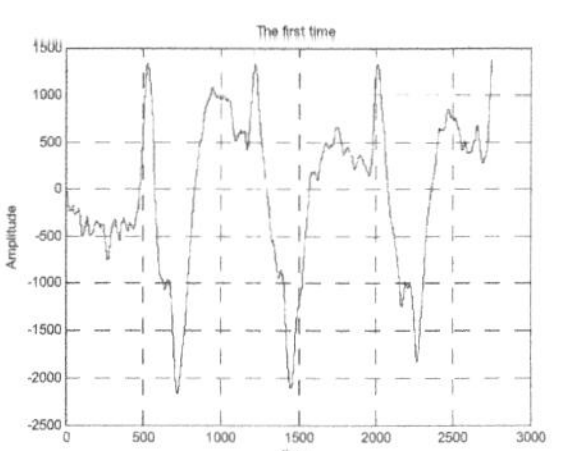

Fig. 7a Eye blinking experiment in the first time of week-1.

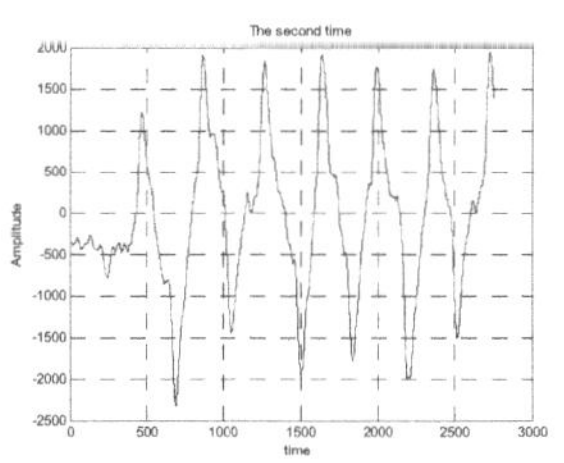

Fig. 7b Eye blinking experiment in the second time of week-1.

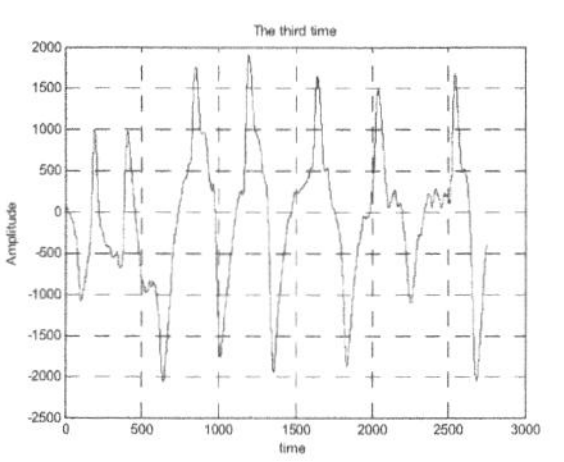

Fig. 7c Eye blinking experiment in the third time of week-1.

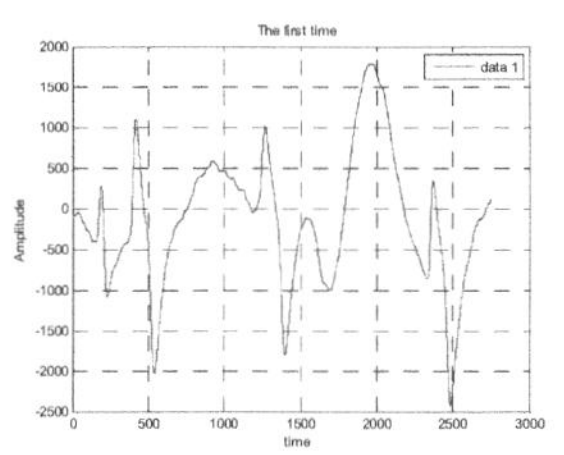

Fig. 8a Eye blinking experiment in the first time of week-2.

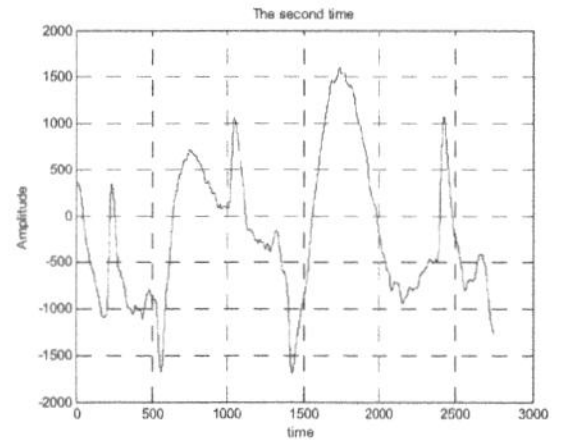

Fig. 8b Eye blinking experiment in the second time of week-2.

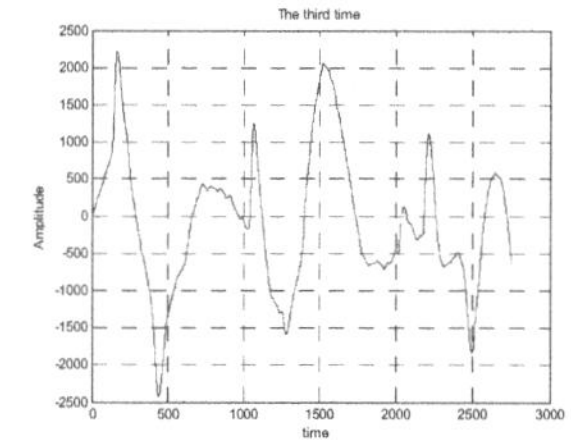

Fig. 8c Eye blinking experiment in the third time of week-2.

Similarly, the subject performed eye blinking experiments in two weeks and then the obtained results were shown in Figs. 7a, 7b, 7c, 8a, 8b and 8c.

Based on these obtained data, one can calculate threshold values. In particular, from the equation (2), (3) and (4), the average value M_B of the first week was determined in the case of eye blinking and the threshold value *THR* was calculated in the case of opened eye as shown in Table 1. In order to reduce the error of eye blinking recognition, the threshold was proposed to be as follows:

$$\text{Min}_B < THR < \text{Min}_O \quad (5)$$

where Min_B is the minimum amplitude of the eye blinking signal and Min_O denotes the minimum amplitude of the openeded eye case.

Table 1 Experimental results of the first week.

Eye Blinking			**Opened Eye**			
Times	Min_B	M_B	Min_O	M_O	*SD*	*THR*
1	-2116	-48,79	-562,7	-81,97	+231,4	-1014
2	-2332	-64,04	-567,7	-69,74	177,9	-780,8
3	-2066	-95,65	-765	-73,21	222,5	-963,3
Average	-2171	**-69,49**	-631,7	-74,93	210,6	**-917,3**

Substituting the equation (4) into the equation (5), one calculated M_O and *SD* using the equations (2) and (3) and then the coefficient was determined, $2,6 < a < 10$. After that we can choose the suitable coefficient, a so that the eye blinking threshold *THR* can be determined when $a = 4$ is chosen as shown in Table 1.

Table 2 Experimental results of the second week.

Eye Blinking			**Opened Eye**			
Times	Min	M_B	Min	M_O	*SD*	*THR*
1	-2422	-127,1	- 844	-88,58	300,5	-1290
2	-1688	-159,6	-890,2	-74,43	463,7	-1794
3	-2438	-75,09	- 628	-71,44	185,4	-813,6
Average	-2182	**-120,6**	-787,3	-78,15	316,5	**-1342**

In similarity, the coefficient a still chosen is 4 in the case of the second week, the threshold value was obtained as shown in Table 2.

The threshold created to recognize eye blinking was calculated by taking the average of THR in two weeks and one obtained THR = -1130. When this threshold is applied into eye blinking tasks, we can recognize eye blinking times as shown in Fig. 9.

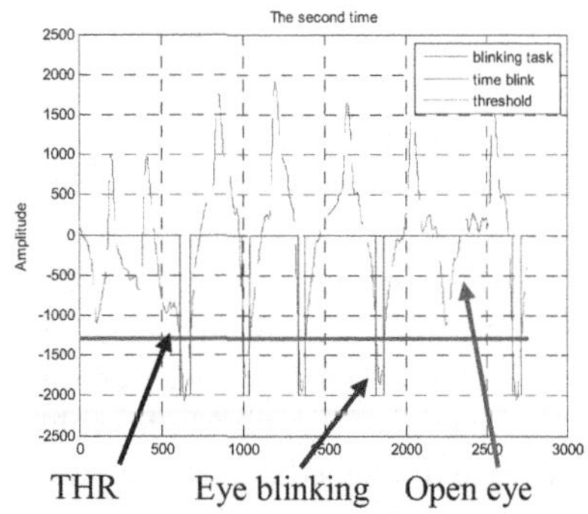

Fig. 9 Presenting the threshold THR, eye blinking and open eye.

After determining, one can count the number of eye blinking times of the same subject who carried out experiments in two weeks as shown in Figs. 10a, 10b, 10c, 11a, 11b and 11c.

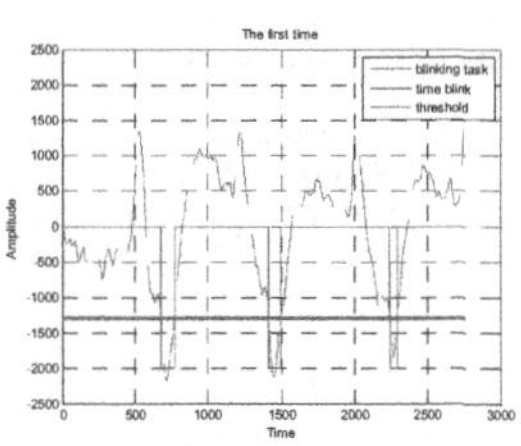

Fig. 10a Eye blinking experiment in the first time of week-1.

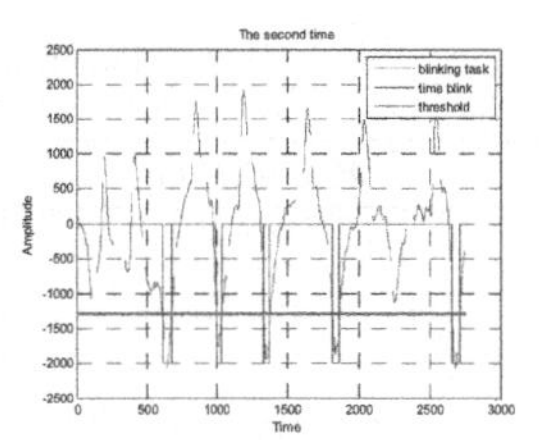

Fig. 10b Eye blinking experiment in the second time of week-1.

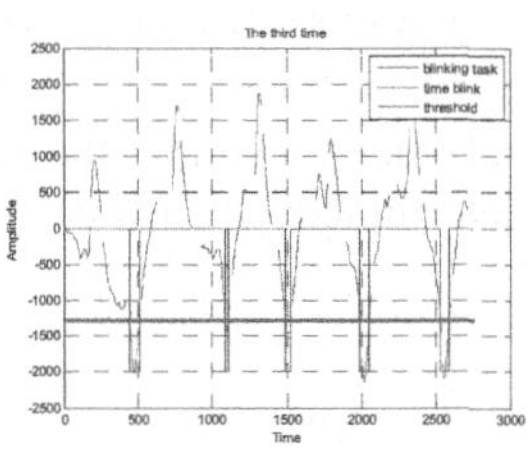

Fig. 10c. Eye blinking experiment in the third time of week-1.

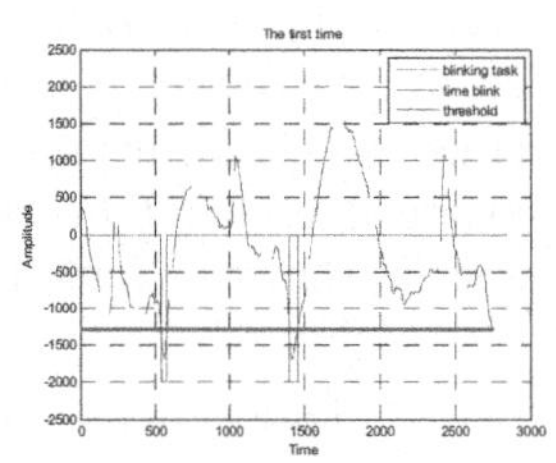

Fig. 11a Eye blinking experiment in the first time of week-2.

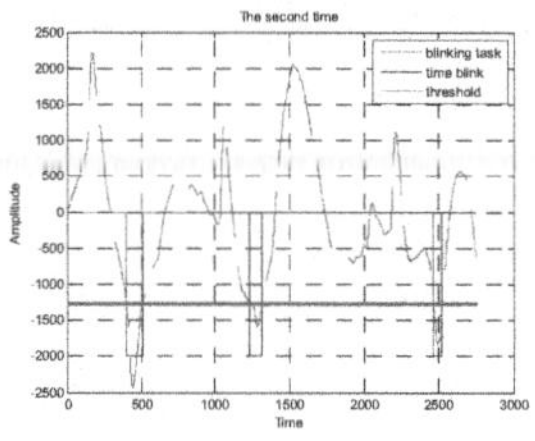

Fig. 11b Eye blinking experiment in the second time of week-2.

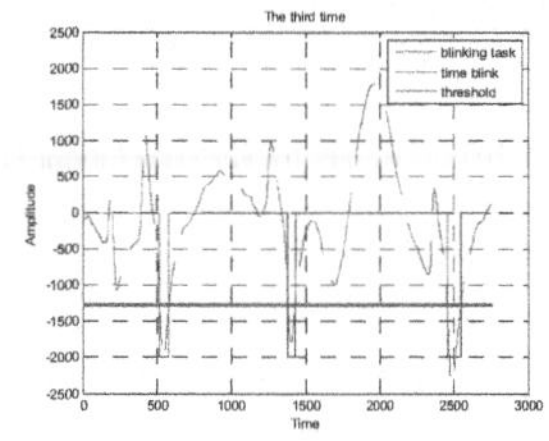

Fig. 11c Eye blinking experiment in the third time of week-2.

From the mean eye blinking threshold, one can know when human eye is open or blinking. Moreover, we can count eye blinking times for diagnosis of eye problems or eye blinking times can be used to control a device such as switch a light on/off based on the number of eye blinking times installed initially.

IV. Conclusion

In this paper, a threshold algorithm was utilized to determine eye blinking of human. Offline EEG data collected from the subject were filtered using a pass-band filter. After that the filtered data were calculated and analysed to eye blinking threshold. In particular, the proposed threshold algorithm was employed for both opened eye and eye blinking and then found a threshold value for eye blinking recognition based on the threshold value of opened eye case. Experimental results showed to illustrate the effectiveness of the proposed threshold method.

Acknowledgment

The authors would like to acknowledge the support of the CHAC-IU Research Collaborative Program. We would also like to thank Duy and Khai students for their support during data collection.

References

1. J. Wolpaw, N. Birbaumer, D. McFarland, G. Pfurtschellere, and T. Vaughan, "Brain-computer Interfaces for Communication and Control," Clinical Neurophysiology, pp. 767-791, 2002.
2. N. Ince, F. Goksu, A. Tewfik, and S. Arica, "Adapting Subject Specific Motor Imagery EEG Patterns in Space-time-frequency for a Brain Computer Interface," Biomedical Signal Processing and Control, pp. 236-246, 2009.
3. N. Weiskopf, F. Scharnowski, R. Veit, R. Goebel, and N. Birbaumer, "Self-regulation of Local Brain Activity Using Real-time Functional Magnetic Resonance Imaging (fMRI)," Journal of Physiology, pp. 357-373, 2004.

4. N. Ince, F. Goksu, A. Tewfik, and S. Arica, "Adapting subject specific motor imagery EEG patterns in space–time–frequency for a brain computer interface," Biomedical Signal Processing and Contro, pp. 236–246, 2009.
5. N. Weiskopf, F. Scharnowski, R. Veit, R. Goebel, N. Birbaumer, and K. Mathiak, "Self-regulation of local brain activity using real-time functional magnetic resonance imaging (fMRI)," Journal of Physiology, pp. 357-373, 2004.
6. S. Lloyd-Fox, A. Blasi, and ElwellCE, "Illuminating the developing brain: The past, present and future of functional near infrared spectroscopy," in Neuroscience and Biobehavioral, 2010, pp. 269-284.
7. D. Leff, F. Orihuela-Espina, C. Elwell, T. Athanasiou, D. Delpy, A. Darzi, and G. Yang, "Assessment of the cerebral cortex during motor task behaviours in adults: A systematic review of functional near infrared spectroscopy (fNIRS) studies," in NeuroImage, 2011, pp. 2922-2936.
8. G. E. Fabiani, D. J. McFarland, J. R. Wolpaw, and G. Pfurtscheller, "Conversion of EEG Activity into Cursor Movement by a Brain-computer Interface," Transactions on Neural Systems and Rehabilitation Engineering, vol. 12, pp. 331-338, 2004.
9. C. Guger, W. Harkam, C. Hertnaes, and G. Pfurtscheller, "Prosthetic Control by an EEG-based Brain-Computer Interface (BCI)," IEEE Transactions on Robotics, vol. 21, 2005.
10. D. J. Krusienski and J. J. Shih, "A Case Study on the Relation Between Electroencephalographic and Electrocorticographic Event-Related Potentials," in the 32nd Annual International Conference of the IEEE EMBS, 2010.
11. D. J. McFarland and J. R. Wolpaw, "EEG-Based Communication and Control:Speed-Accuracy Relationships," Applied Psychophysiology and Biofeedback, vol. 28, pp. 217-231, 2003.
12. B. Blankertz, G. Dornhege, M. Krauledat, K.-R. Müller, V. Kunzmann, F. Losch, and G. Curio, "The Berlin Brain-Computer Interface:EEG-Based Communication Without Subject Training," IEEE Transactions on Neural Systems And Rehabilitation Engineering, vol. 14, pp. 147-152, 2006.
13. X. Gao, D. Xu, M. Cheng, and S. Gao, "A BCI-based Environmental Controller For The Motion-Disabled," IEEE Transactions On Neural Systems And Rehabilitation Engineering, vol. 11, pp. 137-140, 2003.
14. D. JR, J. TP, K. WJ, Y. S, M. S, H. JC, and S. TJ, "Measuring the Variability of Event-related BOLD Signals," in Third International Workshop on Independent Component Analysis and Signal Separation, 2001.

Use macro [author address] to enter the address of the corresponding author:

Author: Nguyen Trung
Institute: Cao Thang Technical College
Street: 65 Huynh Thuc Khang
City: HoChiMinh city
Country: Vietnam
Email: nguyenvantrung201087@yahoo.com

Linear Regression Algorithm for Hand Tapping Recognition Using Functional Near Infrared Spectroscopy

C.Q. Ngo[1], T.H. Nguyen[2], and Toi Van Vo[2]

[1] Faculty of Electrical and Electronics Engineering, University of Technical Education HCMC, Vietnam
[2] Biomedical Engineering Department, International University, Vietnam National University, Vietnam

Abstract — **This paper proposed a linear regression (LR) algorithm for hand tapping recognition using functional Near Infrared Spectroscopy (fNIRS). Brain data with noise and artifacts were re-processed to obtain data smoothy using a Savitzky-Golay filter. The smoothy data were calculated using the proposed LR algorithm in order to produce the angular coefficients of the straight lines which correspond to oxygen-Hemoglobin (Oxy-Hb) concentration. Therefore, one can distinguish the right and left hand tapping tasks based on the different angular coefficients of the lines corresponding to the difference of the right and left brain Oxy-Hb. In addition, the difference of the left and right brain activities were determined based on comparing the angular coefficients. Experimental results showed to illustrate the effectiveness of the proposed method.**

Keywords — **Hand tapping, linear regression algorithm and left and right hand brain.**

I. Introduction

Cognitive neuroscience of human brain has played an important role in recent decades. Especially, non-invasive technologies, such as Near-Infrared Spectroscopy (NIRS), Magnetic Resonance Imaging (MRI) and ElectroEncephalo Graphy (EEG), for patient diagnose and disabled people rehabilitation, have applied. These technologies have also attracted researchers with many significant approaches to explore many problems related to human brain. An example of rehabilitation is that information obtained from human brain using one of the EEG or fNIRS technique could be used to perform shared control of motion wheelchairs [1-3]. For patients, a brain simulator can lead to improve or to recover the motor/cognitive functions of tetraplegic patients with spinal cord injuries and degenerative nerve diseases.

Recently, fNIRS technology has been appeared as an alternative brain-based experimental technique [4] to measure human thoughts or motor control activities. This non-invasive technique projects near infrared light into the brain to measure cerebral hemodynamics as well as to detect localized blood volume and oxygenation changes. In literature, this technique has been successfully used to study human brain functions such as assessment of motor task from everyday living, athletic performance, and recovery from neurological illness [5] or assessment of verbal fluency [6] or quantification of brain function during finger tapping [7], etc. However, to the best of our knowledge, there is no or few application of fNIRS technique to quantify the motor control signals leading to autonomous wheelchair control or brain simulator.

Savitzky-Golay filtes have been used much in smoothing noisy signals and images. In this Savitzky-Golay filte, the coefficients of the local least-square polynomial fit are pre-computed [8-10]. In this paper, we use this Savavitzky-Golay filter to reduce noise as well as artifacts for recognizing left or right hand tapping states based on oxygen hemoglobin concentration.

In fact, the detection of hand tapping features has been a big challenge for researchers to understand more about brain motor functions. In this project, a linear regression algorithm was investigated to determine these hand tapping features and the difference of the left and right brain oxy Hb concentration through the angular coefficients of the straight lines corresponding to hand tapping times of the left and right hand sides. Experimental results were demonstrated based on the proposed linear regression method.

II. Materials and Methods

A. Data Acquisition

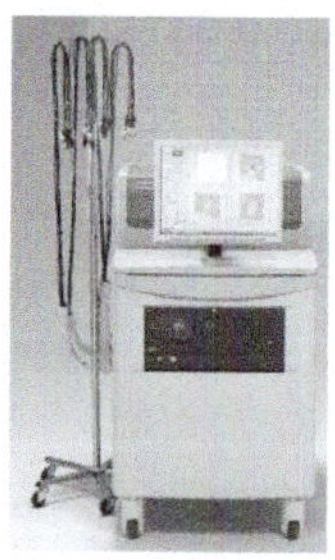

Fig. 1 FOIRE – 3000 system using fNIRS technology.

Fig. 2 A holder with probes was installed with human head.

V. Van Toi et al. (Eds.): 4th International Conference on Biomedical Engineering in Vietnam, IFMBE Proceedings 40, pp. 280–285, 2013.
www.springerlink.com

Oxy-Hb changes in motor control area of human brain were captured from a set of the holder with 20 channels for two brain halves using the fNIRS technique as shown in Fig. 2. A machine of FOIRE – 3000 series of Shimadzu Corporation in Japan located at Room 104 of Biomedical Engineering Department, International University, Vietnam, as shown in Fig.1. This system operates at three different wavelengths of 780 nm, 805 nm and 830 nm. Moreover, the distance between pairs of emitter and detector probes was set at 3 cm and all probes were attached with holders arranged on different sides of human hemispheres dependent on user.

Rest-20s	Task-20s	Rest-20s

Fig. 3 Setting up Rest and Task time of measurements.

One subject (male, 25 years old, 60 kg weights, right-handed) was participated into this study. The subject informed consents agreement after reading and understanding of the experiment protocol and the fNIRS technique as shown in Fig. 3. The subject was demanded to perform hand tapping motions, both left and right sides as motor activities. In these hand tapping motions, a protocol includes 20s (Rest) – 20s (Task) – 20s (Rest), it means that the subject relaxed in 20s, tapped his hand up/down about 10 times in 20s, and then rested 20s, as shown in Fig. 3. He did with this protocol 5 times.

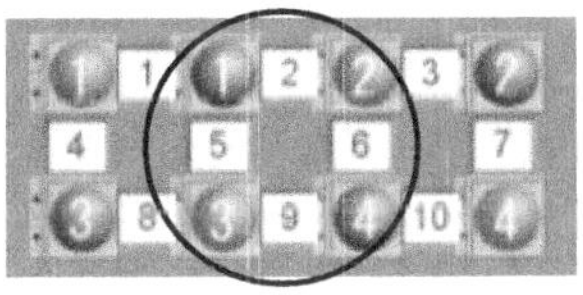

Fig. 4a Channels (2,5,6 and 9) predicted are installed the motor control field at the left brain side.

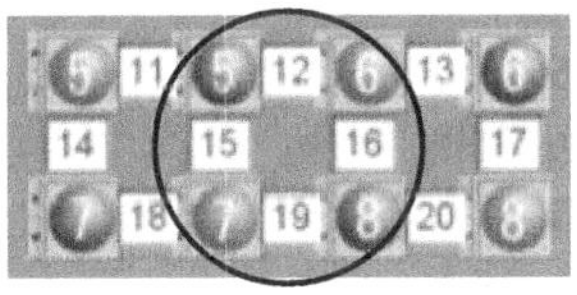

Fig. 4b Channels (12, 15, 16 and 19) predicted are installed the motor control field at the right side.

Measured data on 10 channels of the left brain side and that of the opposite side will be obtained for hand tapping recognition. However, we just choose 4 channels for each side which has reliable information to analyse and to estimate their hand tapping features. In particular, the left brain channels are 2, 5, 6, 9 and the 12, 15, 16, 19 channels are of the right brain side as shown in Fig. 4a and Fig. 4b. In addition, the linear regression algorithm was employed for each fNIRS data corresponding to one channel to obtain the slope of straight line. Therefore, eight slopes were obtained from eight channels (2, 5, 6, 9 and 12, 15, 16, 19). After that one biggest slope of four slopes at the left brain was chosen and similarly the biggest one at the left brain was considered.

B. Data Pre – processing

Brain data of a subject acquired from the channels have noise and artifacts as shown in Fig. 5a and Fig. 5b. In particular, the noise and artifacts can be generated from brain blood flow or the effect of measure environment. In addition, the fNIRS technology is also less sensitive to motion artifacts which are caused by head and body movement or around environment. In order to obtain brain data with less noise and artifacts, a filter as the Savitzky – Golay filter may be applied.

A Savitzky – Golay filter [11] was employed to make brain signal more smoothy in the time domain. This filter used a sliding window based on the samples of the brain signal. In this paper, the simplest digital filter replaced each data value d_i, in which d_i is the value of fNIRS data. The method is that the linear combination of a few nearby points was carried out to obtain point values g_i at the filter output using the following equation:

$$g_i = \sum_{n=-n_L}^{n_R} s_n d_{i+n} \quad (1)$$

where n_L and n_R are the number of brain data points of the left and the right, respectively and s_n denotes the coefficient of the filter, $n=-n_L,\ldots,n_R$.

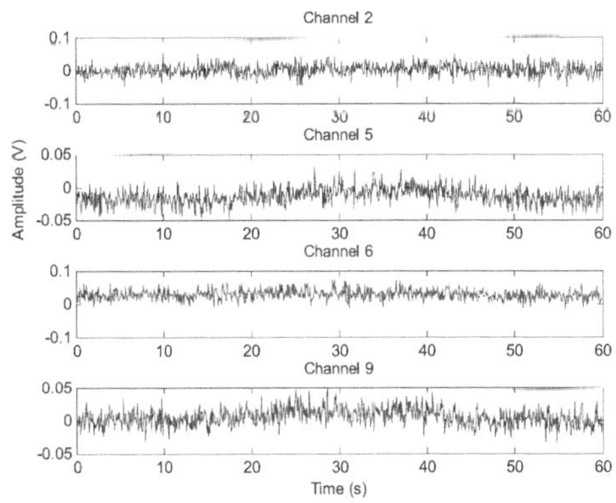

Fig. 5a Original signals of channels 2, 5, 6, 9 (left hemisphere) which were chosen to analyse for hand tapping recognition.

When $n_L = n_R$, the moving average filter is the following equation, $s_n = 1/(n_L + n_R + 1)$. In the Savitzky – Golay filter, for each point d_i, a polynomial will be fit to all (n_L + n_R +1) points, named frame size, in the moving window using the least-square method and g_i is the value of that

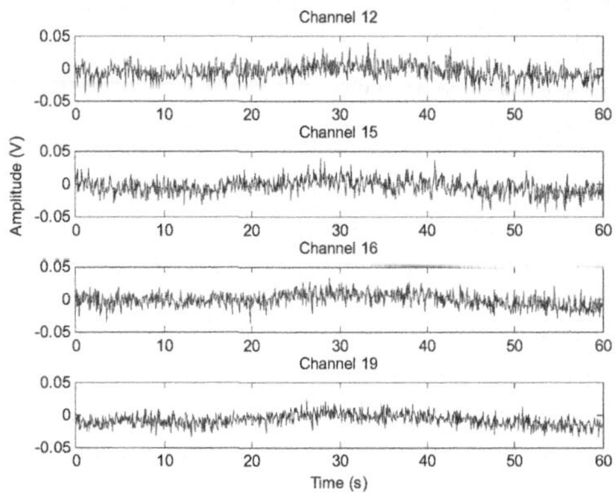

Fig. 5b Original signals of channels 12, 15, 16, 19 (right hemisphere) which were chosen to analyse for hand tapping recognition.

polynomial at position i. Therefore, the polynomial of order M is converted from ($a_0 + a_1 i + \cdots + a_M j^M$) into ($d_{-n_L}, \cdots, d_{n_R}$). In particular, the design matrix for this problem is performed as follows:

$$A_{ij} = i^j \tag{2}$$

in which $i = -n_L, \ldots, n_R$ and $j = 0,1,2,\ldots,M$

In order to determine the polynomial coefficients, $\boldsymbol{a} = a_0, a_1, \cdots, a_M$, one can present the following normal equation:

$$(\boldsymbol{A}^T \boldsymbol{A})\boldsymbol{a} = \boldsymbol{A}^T \boldsymbol{d} \tag{3}$$

where $\boldsymbol{d} = d_{-n_L}, \cdots, d_{n_R}$ are data points

thus

$$\boldsymbol{a} = (\boldsymbol{A}^T \boldsymbol{A})^{-1}(\boldsymbol{A}^T \boldsymbol{d}) \tag{4}$$

We also present the equation $(\mathbf{A}^T\mathbf{A})$ using equation (2) as follows:

$$\{\boldsymbol{A}^T \boldsymbol{A}\}_{ij} = \sum_{k=-n_L}^{n_R} A_{ki} A_{kj} = \sum_{k=-n_L}^{n_R} k^{i+j} \tag{5}$$

where $k = -n_L, \ldots, n_R$.

The normal equation for point data $\boldsymbol{d}$ can be presented as follow:

$$\{\boldsymbol{A}^T \boldsymbol{d}\}_j = \sum_{k=-n_L}^{n_R} A_{ij} d_k = \sum_{k=-n_L}^{n_R} k^j d_k \tag{6}$$

Because s_n is the component a_0 when $\boldsymbol{d}$ is replaced by the unit vector $\boldsymbol{e}_n$, we have s_n by calculating the following equation:

$$s_n = \{(\boldsymbol{A}^T \boldsymbol{A})^{-1}(\boldsymbol{A}^T \boldsymbol{e}_n)\}_0 = \sum_{m=0}^{M} \{(\boldsymbol{A}^T \boldsymbol{A})^{-1}\}_{0m} n^m \tag{7}$$

In this project, the Savitzky – Golay filter was applied to fNIRS data to reduce noise and artifacts which usually exist during measurements. Brain data after filtering was calculated using a linear regression algorithm for hand tapping recognition as well as the different determination of the left and right brain activities.

C. Linear Regression Algorithm

Linear regression algorithm presents the relationship between amplitude and time of a signal. In this paper, the linear regression algorithm was applied to allow us to analyse brain data corresponding to hand tapping times. From the analysed data, one can distinguish the difference of the left and right hand tapping times as well as the difference of the oxy Hb concentration in the left and right hemispheres.

Assume that we have two-dimensional data, $(x_1, y_1), \ldots, (x_n, y_n)$, where each of x and y has no information about the other. Our problem will be to fit a straight line generated by a typical data and then find out the relationship between x and y as show in Fig. 6. Based on the angular coefficients of the straight lines analysed, one can estimate its operation such as hand tapping task. In particular, the equation of a straight line is represented as follows:

$$y = ax + b \tag{8}$$

where a and b are the coefficients of the straight line.

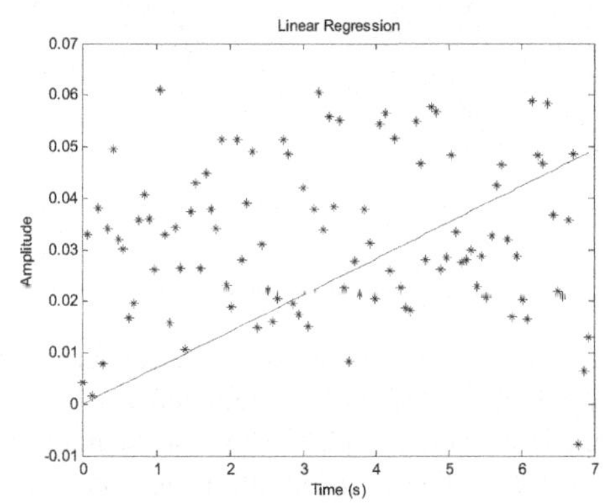

Fig. 6 Blue points is original data which will be calculated to become the red straight line using a linear regression method.

For the objective of the linear regression, assume that there are the ith data points which will not be on the straight line, so they can be considered as error points, e_i. Therefore, a measure of the discrepancy can be determined by the Sum of the Squared Errors (SSE) and calculated as follows:

$$\sum_{i=1}^{n} e_i^2 \tag{9}$$

Assume that a and b are chosen to minimize the SSE and its equation, s is written as follows:

$$s = \sum_{i=1}^{n} e_i^2 = \sum_{i=1}^{n} (y_i - a - bx_i)^2 \tag{10}$$

The differential of the first order of the SSE to the parameters a, b is worked out and we obtain:

$$\frac{\partial s}{\partial a} = -2\sum_{i=1}^{n} (y_i - a - bx_i) \tag{11a}$$

$$\frac{\partial s}{\partial b} = -2\sum_{i=1}^{n} x_i(y_i - a - bx_i) \tag{11b}$$

Both the equations (11a) and (11b) are set to zero, and then divided by n, in which $\bar{x} = \frac{1}{n}\sum_{i=1}^{n} x_i$, we have:

$$a + b\bar{x} = \bar{y} \tag{12a}$$

$$a\bar{x} + b\overline{x^2} = \overline{xy} \tag{12b}$$

Finally, the coefficients of the straight line are determined as follows:

$$b = \frac{\overline{xy} - \overline{x}\,\overline{y}}{\overline{x^2} - (\overline{x})^2} \tag{13}$$

and

$$a = \bar{y} - b\bar{x} \tag{14}$$

From these coefficients, their slopes will be calculated to determine problems of left hand (LH) tapping or right hand (RH) tapping as well as the different operation of two hemispheres given brain data measured on the same subject using fNIRS technology.

III. RESULTS AND DISCUSSION

In this paper, there are two main problems which need to be analysed for hand tapping recognition related to human brain activities at motor control area. Before analyzing these problems, brain data were pre-processed using a filter to eliminate noise and artifacts. In this case, the Savitzky-Golay filter is chosen with the polynomial order of M=3 and the frame size of 5 (n_L=2, n_R=2) to produce brain data smoother as shown in Fig. 7.

For determining the left or right hand tapping and the difference of the left or right activities, the installed protocol to measure oxy-Hb changes on the subject is 20s (Rest), 20s (Task) and 20s (Rest). In fact, a subject invited to carry out experiments about hand tapping 5 times. In the time data, in order to investigate hand tapping problems related to human brain activity, we considered the time data from 0sec to 7sec which relates to the up/down hand motion. To do this, a linear regression algorithm was applied to calculate the angular coefficient of a straight line on a set of 100-points data. In practice, we obtained a lot of data sets in many times, both the Left Hand (LH) tapping and the right one.

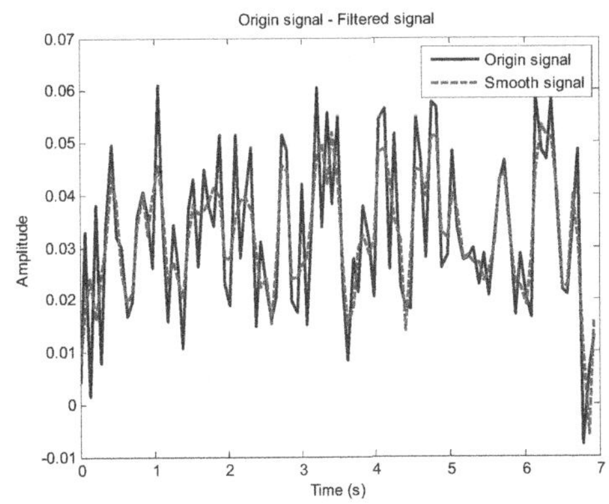

Fig. 7 The original signal is blue and the filtered signal is red.

For the determination of hand tapping problems, experiments were worked out as follows:

- Tapping the LH and measuring both the left and right brain sides 5 times.
- Tapping the Right Hand (RH) and measuring both the left and right brain sides 5 times.

In order to recognize hand tapping tasks, the equation (8) using the linear regression method was employed, in which the coefficients of the straight lines were calculated. In this case, the paper showed experiments of hand tapping, particularly the subject tapped his LH and then the left brain side was investigated. As the result, the coefficients are a_{LL} = 0.007 and b_{LL} = 0 (LL means the LH– Left brain side). Therefore the equation of the straight line was presented as $y_{LL} = 0.007x_{LL}$. In similarity, the equation of straight line for determining the LH tapping, at the right brain side was calculated as $y_{LR} = 0.0011x_{LR}$ (LR means the LH– Right brain side). Another application based on the same principle is that the equations used to distinguish the left and right brain activities when doing RH tapping were calculated as $y_{RL} = 0.0071x_{RL}$ and $y_{RR} = 0.0048x_{RR}$.

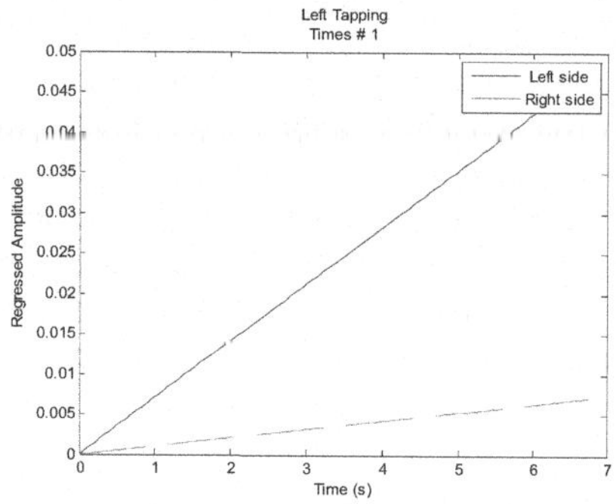

Fig. 8a Left tapping to investigate oxy-Hb changes of the left and right brain sides at the first times.

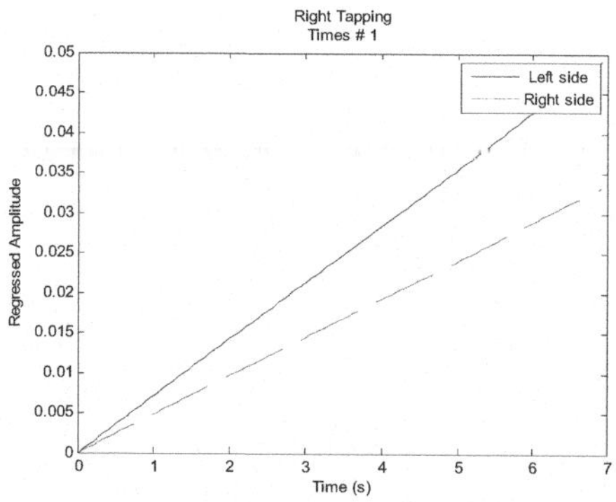

Fig. 8b Right tapping to investigate oxy-Hb changes of the left and right brain sides at the first times.

In this paper, there are two problems of hand tapping recognition based on the oxy-Hb concentration: firstly the left brain is always more active than the right brain and secondly recognition of the LH tapping or the RH tapping. In order to recognize these problems, a comparison of the activity of the left and right brain sides corresponding to the left and right hand tapping times was performed. In particular, experimental results analysed to recognize whether the LH tapping or the RH tapping as well as to determine the difference of the left and right brains as shown in Fig. 8a and Fig. 8b.

In this Fig. 8a, when the LH tapping was performed up/down, the slope of the blue line corresponding to the left brain oxy-Hb concentration is greater than that of the red line based on the oxy-Hb concentration of the right brain side. In similarity, Fig. 8b is the RH tapping plot which shows the blue and red straight lines, in which the slope of the blue line is still greater than that of the red. In addition with Fig. 8a and Fig. 8b, the angular coefficient values of tapping two hands showed in Table 1 and Table 2 which gave us a conclusion that the left brain activity is always greater than the right brain one.

Table 1 Case of the LH tapping, in which α_L (a_{LL}) and α_R (α_{LR}) are the angular coefficients of the straight lines.

Times	α_L (Corresponding to measured data of the left brain)	α_R (Corresponding to measured data of the right brain)
1	0.0070	0.0011
2	0.0081	0.0016
3	0.0072	0.0011
4	0.0071	0.0017
5	0.0051	0.0006

Table 2 Case of the RH tapping, in which β_L (β_{RL}) and β_R (β_{RR}) are the angular coefficients of the straight lines.

Times	β_L (Corresponding to measured data of the left brain)	β_R (Corresponding to measured data of the right brain)
1	0.0071	0.0048
2	0.0084	0.0071
3	0.0070	0.0048
4	0.0067	0.0045
5	0.0077	0.0057

In case of determining the LH or RH tapping, when the subject is performing the LH tapping and we measure the oxy-Hb concentration of the right brain side, the slope of the red line in Fig. 8a is always smaller than that of the red line in Fig. 8b. In this case, the red straight lines were built based on oxy-Hb data measured on the right brain of the subject.

Table 3 The comparison of the angular coefficients α_R and β_R corresponding to the oxy-Hb concentration of the two brain sides.

Times	α_R (corresponding to data of the right brain when doing LH tapping)	β_R (corresponding to data of the right brain when doing RH tapping)
1	0.0011	0.0048
2	0.0016	0.0071
3	0.0011	0.0048
4	0.0017	0.0045
5	0.0006	0.0057
Average	0.00122	0.00538

Table 3 also shows the angular coefficient values of the oxy-Hb concentration in cases of the left and right tapping times observed on the right brain. The average of slopes corresponding to the RH tapping is about 4 times greater than that of slopes corresponding to the LH tapping. This means that the slope of the line (the LH tapping) is always less than that of the line (the RH tapping) corresponding to measuring at the same right brain. In order to give us more information, Fig. 9 shows the angular coefficients of 5 measured times on the same subject. All the coefficients, β_R

are greater than α_R. This leads to a conclusion that the difference of the left and right tapping tasks can be recognized based on the difference of the oxy-Hb concentration of the same right brain.

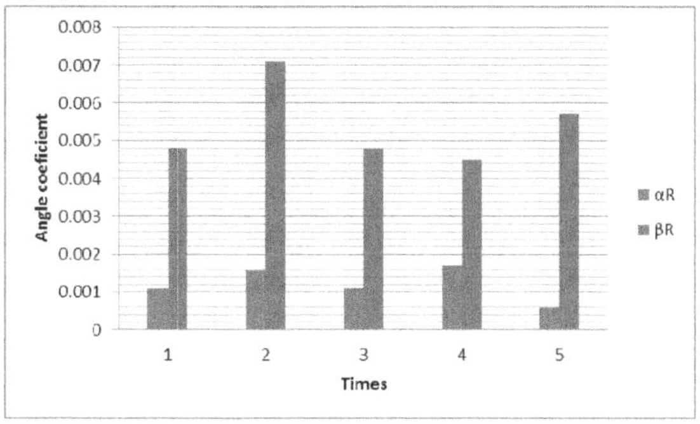

Fig. 9 The blue bars are of the oxy-Hb concentration at the right brain when doing the LH tapping and the red ones are of the oxy-Hb concentration at the right brain when doing the RH tapping.

IV. Conclusion

In this paper, original brain signals of hand tapping times were filtered by the Savitzky – Golay filter to produce the smoothy signals. Moreover, the smoothy signals of the left and right hand tapping times were analysed using the linear regression algorithm to distinguish when the LH or the RH moves up/down based on oxy-Hb changes in human brain. Experimental results with 5 hand tapping times showed that one could distinguish when the subject was doing the left or right hand tapping tasks based on different slope coefficients of the straight lines. In addition, we could estimate that the left brain activity is more active than the right one.

Acknowledgment

The authors would like to acknowledge the support of the Shimadzu through the Shimadzu-IU Research Collaborative Program. We would also like to thank students of Biomedical Engineering Department at IU for their support during data collection.

References

1. J. Wolpaw, N. Birbaumer, D. McFarland, G. Pfurtschellere, and T. Vaughan, "Brain-computer interfaces for communication and control," in Clinical Neurophysiology, 2002, pp. 767-791.
2. N. Ince, F. Goksu, A. Tewfik, and S. Arica, "Adapting subject specific motor imagery EEG patterns in space-time-frequency for a brain computer interface," in Biomedical Signal Processing and Control, 2009, pp. 236-246.
3. N. Weiskopf, F. Scharnowski, R. Veit, R. Goebel, N. Birbaumer, and K. Mathiak, "Self-regulation of local brain activity using real-time functional magnetic resonance imaging (fMRI)," in Journal of Physiology, 2004, pp. 357-373.
4. S. Lloyd-Fox, A. Blasi, and ElwellCE, "Illuminating the developing brain: The past, present and future of functional near infrared spectroscopy," in Neuroscience and Biobe., 2010, pp. 269-284.
5. D. Leff, F. Orihuela-Espina, C. Elwell, T. Athanasiou, D. Delpy, A. Darzi, and G. Yang, "Assessment of the cerebral cortex during motor task behaviours in adults: A systematic review of functional near infrared spectroscopy (fNIRS) studies," NeuroImage, pp. 2922–2936, 2011.
6. M. Schecklmann, A. Ehlis, M. Plichta, and A. Fallgatter, " Influence of muscle activity on brain oxygenation during verbal fluency assessed with functional near-infrared spec-troscopy.," Neuroscience, pp. 434–442, 2010.
7. T. Sato, M. Ito, T. Suto, M. Kameyama, M. Suda, Y. Yamagi-sh, A. Ohshima, T. Uehara, M. Fukuda, and M. M. c. o. b. a. a. t. i. f. f. A. m. n.-i. s. s. d. f. tapping, "Time courses of brain activation and their implications for function: A multichannel near-infrared spectroscopy study during finger tapping," Neuroscience Research, pp. 297-304 2007.
8. A. Savitzky and S. a. D. o. D. b. S. L. S. Procedures, "Smoothing and Differentiation of Data by Simplified Least Squares Procedures," Analy. Chemistry, vol. 36, pp. 1627-1639, 1964.
9. J. Steinier, Y. Termonia, and J. Deltour, "Comments on Smoothing and Differentiation of Data by Simplified Least Square Procedure," Analytical Chemistry, vol. 44, pp. 1906-1909, 1972.
10. P. A. Gorry, "General Least-Squares Smoothing and Differentiation by the Convolution (Savitzky-Golay) Method," Analytical Chemistry, vol. 62, pp. 570-573, 1990.
11. W. H.Press, S. A. Teukoisky, W. T. Vetterling, and B. P. Flannery, Numerical recipes The art of scientific computing: Cambridge university press, 2007.

Use macro [author address] to enter the address of the corresponding author:

Author: Ngo Quoc Cuong
Institute: University of Technical Education HCMC
Street: 1 Vo Van Ngan, Thu Duc
City: HoChiMinh city
Country: Vietnam
Email: ngoquoccuong175@gmail.com

Design a System of Home Health Care Telemedicine for Blood Pressure Measurement

N.P. Nam, N.H.M. Tam, L.G. Loc, and Vo Van Toi

International University, Biomedical Engineering Department, HoChiMinh, VietNam

Abstract — **We design and implement a system of home health care Telemedicine for Blood Pressure Measurement including Tele-Blood Pressures having a capable of transmitting data over the Internet via an ADSL or 3G and server applications to store and process data. The standard of** Health Level Seven **is applied to exchange information.**

Keywords — **Telemedicine, Home Care, Blood Pressure, Health Level Seven.**

I. INTRODUCTION

Today, telemedicine, the electronic transmission of health information from distance, has been increasingly used to provide efficiencies in the delivery of **home health care**. Home tele-healthcare is health services delivered to patients' homes in part by telecommunication devices, e.g., telephone, and telecommunication-ready healthcare monitors, e.g., blood pressure cuff [1]. According to the World Telemedicine Association, Hypertension is one of the most common worldwide diseases afflicting humans and it should be monitored every days. Since high blood pressure can damage many parts of the human body such as heart and it usually has no symptoms, controlling patients' blood pressure is a key to reduce these risks. In Vietnam, according to recent estimates, nearly 80.9% of 12 million people with hypertension disease don't control their blood pressure [2].

This paper introduces a design for Telemedicine Health Home Care for Blood Pressure Measurement including hardware and software of a monitor blood pressure having a capable of transmitting data over the Internet. This device allows a patient to measure his blood pressure on his arm at home by himself. The data will then be sent automatically via an ADSL or 3G connection to a server. At a clinic office, a healthcare provider can follow in real time the measurement displayed on a computer monitor as if he was next to the patient. The healthcare provider can also retrieve the data later to process them, e.g. to graph them. The data can also be stored in a memory card inside the device and can be sent later to the healthcare provider. Several patients (for example, the members of a family) can use the same device. The ID of each patient can be registered into the device to allow the healthcare provider to identify which data belong to whom. In case of serious illness which needs further test, the doctor will ask patients to hospital. In addition, measurement blood pressure at home by this device also has a number of potential advantages such as avoiding of the "white-coat effect", graphing and reading multiple blood pressures over a long time at different times of the day.

II. METHODOLOGY

Home telehealthcare services for Blood Pressure Measurement are based in Client-Server architecture [3]. It has server applications which receive incoming blood pressure results from the clients and then storages into database. The device client is responsible for measuring blood pressure from patient and transmitting them through Internet. Fig. 1 shows the system schematic.

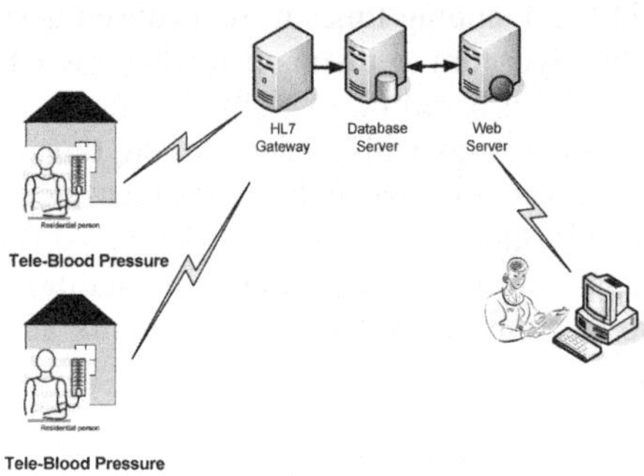

Fig. 1 System schematic.

A. Client Side

The client side includes hardware devices of blood pressure measurement and a module communication via Internet called **Tele-Blood Pressure**. Firmware application on this hardware platform will measure the patient's blood pressure using the oscillometric method [4]. The measurement is performed automatically after the patient wraps the cuff on the upper arm and presses a button to start implementing the measure. Users can change the peak pressure pump to suit the individual blood pressure conditions so can get a most accurate measurement results. After inflating/deflating to measure the vibration signal of the upper arm artery, the firmware will calculate the Systolic/

V. Van Toi et al. (Eds.): 4th International Conference on Biomedical Engineering in Vietnam, IFMBE Proceedings 40, pp. 286–289, 2013.
www.springerlink.com

Diastolic / Heart Rate results and show them on the screen. These results are stored on a historical EEPROM memory. The microprocessor will also construct an HL7 message for the results. Next, the system creates a TCP connection to the server and then sends the message using TCP / IP protocol. The information transmission has been completed when the client side receives an ACK message from the server.

There are some utility functions under the selection menu displayed on the screen of the medical device. The Settings menu allows the user to select types of Internet connection, the User ID on device, configure parameters. The History menu allows users to review previous measurements of each patient.

B. Server Side

Software on the server will acquire data sent from the Tele-Blood Pressure and store the data in the database of patients. Each doctor has an account and password to access this database, view the graph of the results measured over time. However, the system has the access permissions. The doctor can only see the results of the patients in their monitoring. The doctor's view of data will be made through the Web. And from there, they can give instructions to the patient treatment. Then through the Web, patients sign in to their account in the system, will get this treatment guidelines.

III. DESIGN

In order to develop the Telemedicine system, we propose a design solution including hardware and software. Hardware design will be implemented for measuring blood pressure and send the results to the Internet server. The software on the server will perform receive blood pressure data, management access database of patients.

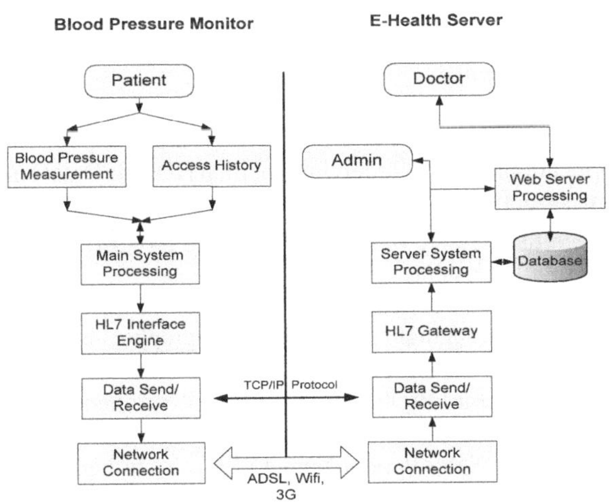

Fig. 2 Information Flow in system.

Fig.2 describes the the flow of blood pressure results from the patient is created and transferred to the server via the network connection.

A. Client Side

The **Tele-Blood Pressure** is designed on the principle of common blood pressure measurement in the electronic medical device. It is based on the principle of oscillometric pressure non-invasive method in the upper arm. It will determine the systolic blood pressure value, diastolic blood pressure value and pulse of the heart.

Design includes several blocks, together to operate as pump, valve, cuff, pressure sensor, amplifier, filter circuit, A/D converter (ADC), LCD, USB interface and network communication module. The component blocks are controlled by an ATmega128 microcontroller called the central processing unit (CPU). Fig.3 describes the connection between the component blocks in Tele-Blood Pressure.

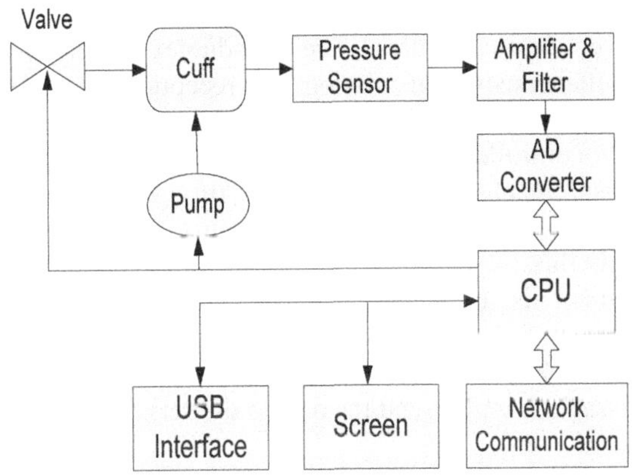

Fig. 3 Block Diagram of Tele-Blood Pressure.

The pump is controlled by the ATMega128, which will inflate air pressure to compress the cuff on patient's arm wrapped. Then the pressure sensor, which is built under Stress Strain Gauge technology, will produce output voltage varies with the pressure in the cuff. The ouput voltage from the sensor will be taken to the circuit of amplifier and filter. Next, the analog signal will be converted to digital data by the ADC for processing of CPU. The pressure in the cuff is pumped to stop blood flow in the hand arteries. After that, CPU control valve for pressure in the cuff diminishes. The appearing blood flow causes the generation of sounds or blood volume oscillations in the tested arteries, according to the heart rate. The blood volume oscillations cause air pressure oscillations in the cuff tightly occluding the arm. The pressure signal from the sensor is amplified up and passes through two filters. The low-pass filter exposes the slowly changing pressure in the cuff and this signal (P) is sent to the ADC. The high pass filter catches the pressure oscillations appearing in the cuff according to the flowing blood, amplified and next sent the oscillation signal (Po) to the ADC. Both signals are sampled at a frequency 100 Hz and with a resolution of 12 bits. The microprocessor collects and stores the cuff's pressure samples $P(ti)$ and amplitude values of the pressure oscillations $Po(ti)$ in time ti. Next, it finds the maximum amplitude of the oscillations $Pomax = max(Po(ti))$, and calculates the moments tm, tn with $Po(tm) = As\ Pomax$, $Po(tn) = Bd\ Pomax$, where As, Bd are respective coefficients necessary for the systolic and the diastolic pressure calculations. After all, it will calculate the results are: $Psystolic = P(tm)$ and $Pdiastolic = P(tn)$.

The results will be stored in EEPROM memory and displayed on LCD screen. The most important point is that the results will be sent to the server via the network communication module. In order to transmit and receive data, we used the RTL8019AS, a highly integrated Ethernet Controller which offers a solution to implement a Plug and Play NE2000 compatible adapter with full-duplex and power down features. The full-duplex function enables simultaneously transmission and reception on the twisted-pair link to a full-duplex Ethernet switching hub. The microcontroller will make an HL7 message for the measurement results. Base on this Ethernet hardware platform, it opens a TCP connection to the server and then sends this message with TCP / IP protocol. The TCP / IP protocol is used here to ensure that the sending of information from the Tele-Blood Pressure to the server have high reliability. If in the case the server does not get the message, it will resend the data to the server again.

B. Server Software

Server software is built on .NET framework 4.0 with Windows 2008 .Net Server operating system. Programming language used to realize the software is Visual C# .NET. The database management of the program is based on SQL Server 2008 which supports a variety of connection technologies such as ODBC, ADO/OLEDB and ADO.NET. Furthermore, SQL Server 2008 platform provides comprehensive Data warehouse and can be expanded. For security reasons, the database is fully protected against unauthorized access and is password protected and encrypted, whereas the whole application is password protected with several access levels depending on user groups.

We use Mirth 3rd Party software for HL7 interface gateway that allows for message filtering, transforming, and routing. Creating HL7 interfaces for our systems becomes easy using the rich client interface and channel creation wizard which associates applications with Mirth engine components. All configurations are done through a Java client that can connect remotely, and channels are easily configured, imported and exported using this interface. Once channels are properly configured, they can then be deployed, which means that they are actively running. They can be viewed on the status page of the application, and they can be started, stopped, or paused from this location. Various statistics, processed messages, and system events can also be viewed via the status page.

C. HL7 Message Standard

HL7 (Health Level Seven) is a standard for exchanging information between medical applications. This standard defines a format for the transmission of health-related information. HL7 messages are in ASCII format. Each HL7 message consists of one or more segments. A carriage return character (**\r**, which is 0D in hexadecimal) separates one segment from another. Each segment is displayed on a different line of text. Each HL7 segment consists of one or more fields. A pipe (**|**) character is used to separate one fields from another. If a field contains other fields, these sub-fields are normally separated by ^ characters.

In our system, HL7 message structure is used with a format consists of four segments as shown below.

MSH - Message Header
PID- Patient Identification Segment
PV1 - Patient Visit Segment
OBR - Observation Request Result Segment
OBX - Observation/Result segment

The **MSH** segment contains information about the message itself. This information includes the sender and receiver of the message, the type of message, and the date and time it was sent. Every HL7 message specifies MSH as its first segment.

MSH segment : MSH|^~\&|<Sending Application (HD)>|<Sending facility(HD)>|<Receiving Application(HD)>|<Receiving Facility(HD)> |<Date/Time of message(TS) | <Security(ST)> | <Message Type (MSG)> |<Message Control ID(ST)>|<Processing ID(PT)> | <Version ID(VID)>

The **PID** segment contains demographic information about the patient, such as name, patient ID and address.

PID segment : PID|<Set ID - PID(SI)>|<Patient ID(CX)>|<Patient Identifier List(CX)>|<Alternate patient ID - 00106(CX)>|<Patient Name(XPN)>|<Mother's Maiden Name(XPN)>|<Date/Time of Birth(TS)>|<Administrative Sex(IS)>

The **OBR** segment transmits information about an exam, diagnostic study/observation. An OBR segment should always be included in orders where a request is made for a given set of observations.

OBR segment : OBR|<Set ID - OBR(SI)>|<Placer Order Number(EI)>|<Filler Order Number +(EI)>|<Universal Service ID(CE)>|<Priority - OBR(ID)>|<Requested Date/Time(TS)>|<Observation Date/Time(TS)> | <Observation End Date/Time(TS)>|<Collection Volume *(CQ)> |<Collector Identifier *(XCN)>|<Specimen Action Cod*(ID)>|<Danger Code(CE)> |<Relevant Clinical Info(ST)>|<Specimen Received Date/Time *(TS)> |<Specimen Source *(CM)> |<Ordering Provider(XCN)> |<Order Callback Phone Number(XTN)>|<Placer Field 1(ST)> |<Placer Field 2(ST)>|<Filler Field 1 + (ST)> |<Filler Field 2 + (ST)>|<Results Rpt/Status Chng-Date /Time+(TS)>|<Charge to Practice + (CM)>|<Diagnostic Serv Sect ID(ID)>|<Result Status +(ID)>

The **OBX** segment is used to report information within report messages, which must be transmitted to the server system.

OBX segment: OBX|<Set ID - OBX (SI)>|<Value Type(ID)>|<Observation Identifier(CE)> | <Observation Sub-ID(ST)>|<Observation Value(Varies)>|<Units(CE)> |<References Range(ST)> |<Abnormal Flags(IS)> |<Probability(NM)>|<Nature of Abnormal Test(ID)> |<Observation Result Status(ID)>

A sample HL7 message from our system is shown below:

```
MSH|^~@&|EZLBPM|BME|eHealth|BME|2011052710370
0+0700^S||ORU^R01^ORU_R01|TBP0001-
20110527103700-01|P^1|2.5|1||SU||VN|Unicode
PID|1||PA-
0001^^^^PI^BME||NGUYEN^A^VAN^^^BS^L^A||194801
04^D|M
OBR|1||TBP0001-1|MBP^Measure Blood Pressure^MDC
|||20110527103700+0700^S| ||||||||||||||BG|F
OBX|1|NM|SYST||120|mmHg|||||F
OBX|2|NM|DIAS||90|mmHg|||||F
OBX|3|NM|PULS||70|bpm|||||F
```

IV. Conclusions

The proposed Telemedicine Health Home care system consists of computer hardware Tele-Blood Pressure and data management software on the server. This system has been tested and evaluated in a long time for the accuracy of medical devices and the stability of the telecommunication means. The results from the system use are very promising thus encouraging us to continue the development and improvement of the system in order to be able to cover additional future needs.

We are continuing to improve the interface of the user dashboards via web. The warning functions for the doctor will be added to website. When the blood pressure result of the patient's over a certain threshold, then the website will display the color warnings to their doctor. Then the doctor can give an advice to the patient. And when this feature is completed, we hope to put it to practical. This system will be very helpful for the treatment of hypertension in Vietnam, thereby improving Vietnamese's quality of life.

References

1. Current Principles and Practices of Telemedicine and e-Health, Volume 131 Studies in Health Technology and Informatics, Editor: R. Latifi, February 2008, 296 pp., hardcover ISBN: 978-1-58603-806-9
2. Heath statistics in 2010 of Vietnamese ministry of heath, http://www.moh.gov.vn.
3. Design and Implementation of Embedded Medical Data Transmission Platform based on HL7/XML System. Ning LIU, Jianjian LI.

The source of errors in the oscillometric blood pressure measurements, Metrology and Measurement Systems Vol.XV, Number 4 (2008), Barbara Juroszek.

Wireless Nurse Calling System

N.P. Nam, N.H.M. Tam, L.G. Loc, and Vo Van Toi

International University, Biomedical Engineering Department, HoChiMinh, VietNam

Abstract — **In this article we describe the design a wireless nurse call system for visually and audibly alerting nursing staff for use in hospitals, clinics, assisted living and senior living. If action from nurses is too slow, this system can make an alert or send an immediate Short Message Services (SMS) message to nurses and administrators advising them of the tardy response. The logging program has been designed to record the time and location of all calls and be very easy to operate to monitor, record and print reports.**

Keywords — **Nurse calling, wireless, management.**

I. INTRODUCTION

In recent years, the communication and information management in hospitals has attracted lots of interests. Until now in Vietnam, the nurse call solutions are essentially independent systems without network connection as well as management. So, they cannot alert the nurses effective, reliable, and any mistake may cause dangers for patients. Today, with the development of embedded technology and the popularity of Local Area Network (LAN) in hospitals, these information systems should be integrated with the LAN to enhance the functions of communication and management [1]. This paper introduces an innovative solution to integrate wireless connectivity Ethernet and Short Message Services (SMS) into the nurse call system. The hospital server software remotely manages and monitor the controller and displays of nurse call system at the departments.

This improvement system has not only the basic functions of a nurse call system, but also new functions, enhanced reliability of the system and strict management. This system includes three main parts:

1. **Call Display Controller:** The controller is placed in nurse room of each treatment zone in the hospital. It receives all signals from the nurse call buttons in system. When receive signals, the controller will display room number and bed number needs help with alarm until the nurse pushes the received button. This controller has an Ethernet port assigned an IP address so that it can connect to the LAN of the hospital for transmitting data to the management server.

2. **Nurse Call Button:** These buttons wirelessly connect with the Call Display Controller in nurse room. So the button can be attached anywhere near the bed or in the toilet of each room. Each button is assigned a private address inside. When patients need to call the nurse, they can press the button, a call signal will be sent to the controller via radio, and the controller will recognize the address and display a message for nurses to support patient.

3. **Management Server:** The server is connected to all display controllers in the treatment zones of hospital through the LAN. The server software collects data of the nurse calls and their response times. Then, the hospital can manage and evaluate the work of the nurses [2]. In addition, the software also has SMS feature, calls for more support from other nurses in case of many calls of many patients at a time in a treatment zone.

II. ARCHITECTURE OVERVIEW

The designed nurse call system is a multiple connection system which includes three types of digital communication: radio link, Ethernet and SMS. Figure 1 describes our system diagram. Communication between the Nurse Call Buttons and the Call Display Controller is performed on the radio link. The data is sent by radio link on frequency 433MHz ISM with FSK modulation techniques. Communication between these two components is two-way half-duplex.

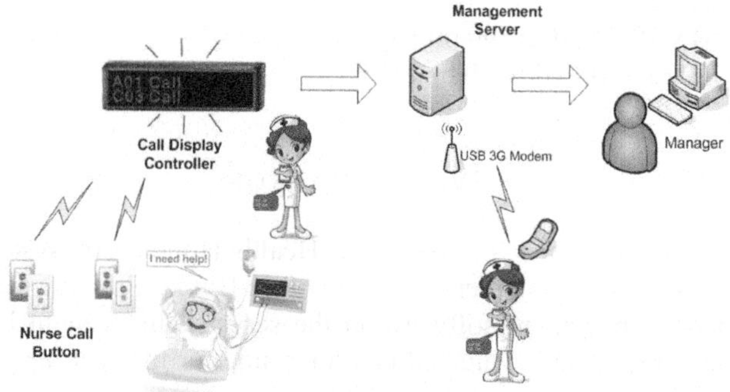

Fig. 1 System diagram.

V. Van Toi et al. (Eds.): 4th International Conference on Biomedical Engineering in Vietnam, IFMBE Proceedings 40, pp. 290–293, 2013.
www.springerlink.com

Data transmission on this connection using two mechanisms: Multiple Access Collision Avoidance Mechanism and Acknowledging Mechanism. When a nurse call button is pressed, the microcontroller insides the button will save the state of requirement and create a data frame which contains the node address and the state control for transmitting to the controller and display. The button only transmits data when the controller and display is willing to receive and does not transmit or receive any other data in that time. This is called principle of LBT (Listening Before Transfer). Before the packet is transmitted, the button will check if other buttons are transmitting or not. If there are signal on the channel, it will wait until the ongoing transmission is complete. To check for the other buttons finished transmission or not, it will check the input signal after a random period. When the channel is free, this node will send the data frame.

Acknowledging (ACK) Mechanism is a mechanism to notify the data transmission results. When the display controller receives data of a node, it will send an ACK message to the sender. When the sender does not receive the ACK, it will consider the receiver did not receive the message and it will resend the message after a timeout. This mechanism reduces the risk of data loss during transmission between two points.

Communication from the call display controller to the server is performed on Ethernet. Ethernet is generally used in IEEE 802.3. The network uses a bus or star topology and supports data transfer rates of 10 Mbps. It is one of the most widely implemented LAN standards. Data is transmitted over the network in discrete frames of between 64 and 1518 bytes in length. Ethernet supports a wide protocol types, including IP, TCP, UDP ... To prevent data loss, when two or more controllers attempt to send packets at the same time, Ethernet detects collisions. All controllers immediately stop transmitting and wait a randomly determined period of time before they attempt to transmit again.

The Management Server software running on a computer, monitor the status of nurse call systems at departments in the hospital. The structure of the server has six components: Signaling Gateway, HL7 Gateway, Processing Unit, Database, SMS Paging Bridge, and Report Server. Figure 2 describes the software block diagram.

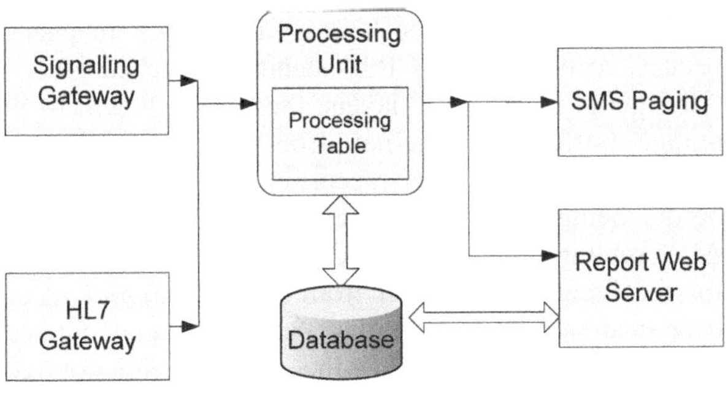

Fig. 2 Blocks server diagram.

The function of each block:

- **Signaling Gateway:** It receives all of the call events, the reminder events and the response events from the clients i.e. the call display controller. To ensure the reliability of the packets sent from client to server with no loss, the protocol is implemented in TCP/IP. After receiving a TCP packet, this block will transmit information to the Processing Unit and store data into the database.
- **HL7 Gateway:** This block receives information at the bed of patients, such as name, age, sex, ADT status (Admit, Discharge, and Transfer). This information is sent to the server according to the HL7 standard (**Health Level Seven**) from the patient management software in the hospital [3]. When receiving an HL7 packet, HL7 gateway separates the information of the beds and updates on the database.
- **Processing Unit:** This is the central processing unit used to manage the nurse calls in the departments of hospital. In this block, **Processing Table** records the nurse call of the bed at the present time. If multiple calls are sent to the Call Display Controller simultaneously and these calls are not responded within a specified period, based on this Processing Table, the server program will analyze the information and send a support request via SMS Paging Bridge.
- **SMS Paging Bridge** will send SMS for notifying support request to nurse's mobile phone. Sending SMS is implemented through GSM Modem. The GSM modem is a specialized type of modem which accepts a SIM card. When a GSM modem is connected to the server computer, this allows the computer to use the GSM modem to communicate over the mobile network. It used for sending and receiving SMS and MMS messages. The server program control the GSM Modem via AT Commands.
- **Reporting Server:** Manager can manage via web browser and see the periodic reports about the response of the nurse to calls. When this block is requested, it will query the database, generate reports and send that report to web browser at PC of the hospital manager through HTTP protocol.

III. Design

A. Design Hardware

The radio link in the nurse call system is based on RF transceiver module HM-TR. This module supports radio data communications with half duplex and FSK modulation method. The frequency domain is in 315/433/868/915MHz

ISM band. The radius of transmit and receive area is over 300m. This module includes an ATmega48 microcontroller to perform data communication and control components transmitted and received FSK radio signal through the SPI port.

In the design of the nurse call button, RF HM-TR module will act as the main controller. ATmega48's EEPROM memory will configure a unique 24-bit address for buttons. When button is pressed, the program in microcontroller will detect RF signal, if there is no RF signal from the other nodes, it will start to send control commands to the RF blocks through SPI port. Then, the RF block will modulate data and transmit to the controller and display.

The design of the call display controller has many more components. Overall block diagram of this controller as shown in Fig.3 below.

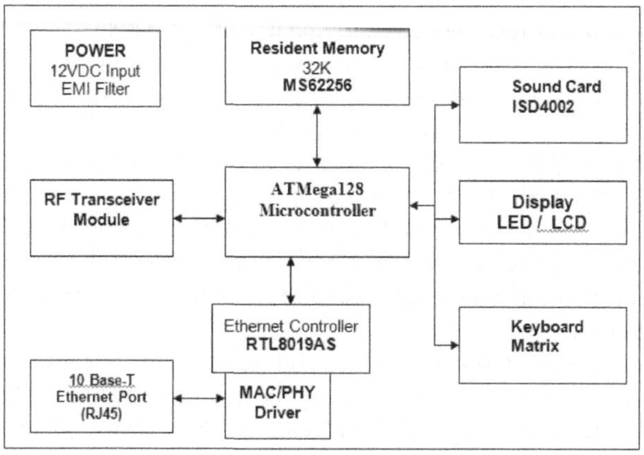

Fig. 3 Block Diagram of Call Display Controller.

In this diagram, ATMega128 Microcontroller acts as central processing. It connects to the RF transceiver Module for transmitting and receiving data, controls Ethernet Controller RTL8019AS, and show information to LED/LCD display with sound card. RF Transceiver Module, which is the RF HM-TR module, will receive commands and data via Tx/Rx of the microcontroller's UART.

Ethernet Interface block is designed based on Ethernet controller RTL8019AS, which compliant with IEEE802.3, 10BaseT and support Full-Duplex Ethernet. Controller firmware will perform a TCP/IP stack as show in Fig.4.

The protocol stack contain: Signaling protocol for transmission of nurse call status data from the controller to the server computer; TCP/UDP protocol in transport layer responses for reliable end to end message transmission; IP protocol in Internet layer responses for IP packet transmission; ARP protocol in Internet layer responses for the address resolution from IP to MAC and the Ethernet

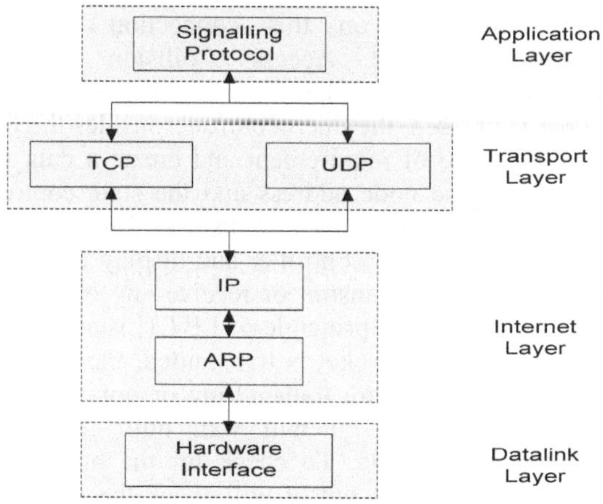

Fig. 4 Protocol stack for network communication.

driver in the data link layer responses for controlling the operation of RTL8019AS.

The main control loop of firmware program on the ATMega128 performs many main repetitive tasks: check if the RF HM-TR Module is receiving data from the nurse call button or not, and if so, then read the data. The microcontroller will send back ACK command to module to confirm for required button. When data is received from RF block, the TCP/IP control program will act as Client in TCP Server/Client model to request opening a TCP connection to the server. Basic data unit of TCP is called segment. This segment has a pair of parameters which are port number of source and port number of destination. Each process applied in a client will access TCP service via one port. This port with an IP address will create a unique socket in network. TCP service is provided by logical link between a pair of socket, each socket can link with many different far sockets. Before of the data transmission, they set up the link and when no longer data transmission needs, the link will be released. Process in opening connection has three-way handshake, according to TCP standard. Then program will put this data into frame which is designed consistent with application, and then push in TCP/IP stack to pack into packet, according to TCP/IP standard attached SYN flag. After getting through IP layers, these packets will be put in output buffer of Ethernet Controller RTL8019AS by microcontroller to pack according to Ethernet standard before being ejected Ethernet network to send to server. After sending data packet, TCP/IP stack program will wait for receiving ACK packet from server to report receiving data successfully. In case that the sender does not receive ACK, it will consider that the recipient did not receive massage and it will resend that data. This mechanism is used to reduce the risk of data loss during transmission between two points. Finally, client program will send an FIN packet to request closing TCP connection to server.

On the controller, the information of the bed number and the room number of patients called the nurse will be shown the LED screen. The LED screen has three rows and show number of rooms and beds which calls nurse to the display. At the same time, the IC sound ISD4002 announces this information through a loudspeaker. The ISD4002-120 circuit is a recorder, which can store 120 second of sound information. The shortest message can be 20ms. We can predefine where in the memory we want to store segments of rooms and beds information. The control of export sound will assemble and announce sound segments of rooms and beds that are recorded on the memory of ISD4002-120.

B. Design Software

As given above, the software has two modules that get the input data. This is Signaling Gateway and HL7 Gateway. Based on the Winsock in Windows OS, the Signal Gateway program will implement with the following procedures: Call Socket() function to create a socket, calls Bind() function to bind socket with a port. Next, it calls Listen() function to wait until a client connects to port. When a client is connected, listen function releases the control, and the gateway application calls Accept() function to confirm client connection. Its Receiving() function receives information of Nurse Call with client. After nurse call data receiving process is completed, Signaling Gateway server application calls Closesocket() function to close created socket.

HL7 Gateway is designed based on HL7/XML Interface Engine. The HL7 system is an electronic data transfer agreement of different application in the medical field, and an industry ANSI standard formulated and approved by the implementation of HL7 organization. The HL7/XML Interface Engine is a group of communication function call or control the process that support HL7. It supports TCP/IP communications protocol. HL7 Gateway will base on API functions of HL7/XML Interface Engine to receive HL7 Messages and analyze to get patient information at beds like Name, Age, Sex, and ADT Status.

The database management program is designed based on SQL Server 2008. SQL Server 2008 supports to build application using connectivity technologies available such as ODBC, ADO/OLEDB and ADO .NET. SQL Server 2008 provides comprehensive data platform and can be expanded. It allows to update database faster and can be upgraded when the amount of data and users are increasing, while supply analysis information to everyone.

With multi-threading programming method, the Processing Unit will handle parallel processing of the receiving information, updating database and Processing Table, and controlling data sent via SMS Paging Bridge. So the processing speed of the program will be guaranteed to meet real-time events of the nurse call.

In Reporting Server block, when receiving the request, it will access directly to the database through SQL query function to retrieve data for generating report and sending to the manager in the hospital. With the mechanism to access databases directly, it will reduce the load of Processing Unit, contribute to optimal performance of the entire server software.

IV. Conclusions

Nurse call system in this paper is designed with the new architecture and features compared to the current nurse call systems on the market. It does not only provide convenient, reliable and efficient management for direct user but also creates benefits for the deployment and installation as well as for hospitals.

The system provides flexibility. It can provide customer specific requirements, the scale of each unit, from small-scale private clinics to large hospitals. When the scale and scope of the system increases, the possible installation is guaranteed to be easy, and the structure and operation of the system remains consistent. The system has integrated hardware and software, so it is easy to comply with the requirements of hospital. Hospital can reset the region, department, or rearrange the nurse rooms for relevant.

The simplicity of system deployment is enhanced. The installation, setup, connection of wireless buttons in a network is easy, simple and can avoid the cables through walls and ceilings. It will save considerable cost, time of installation and maintenance. The system also is scalable in the future of the hospital.

References

1. An Extension to Existing Nurse Call Systems: Event Manager Server. Mehmet S. Unluturk, and Kaan Kurtel. A.
2. Composition of New Features into a Wireless Nurse Call System. Antoni Morey I Pasqual
3. HL7 Standards, Health Level Seven International, at http://www.hl7.org/implement/standards/index.cfm

Survey of Personnel Who Are Operating, Repairing and Maintaining Medical Equipment in Some Hospitals in Vietnam

Nguyen H.M. Tam[1], Nguyen T.M. Linh[2], Pham L.T. Vy[3], and Vo Van Toi[1]

[1] Biomedical Engineering Department
[2] School of Business of International University of Vietnam National Universities – Ho Chi Minh City, Vietnam
[3] Pennsylvania State University, USA

Abstract — **Our objective is to investigate and analyze the appropriateness of personnel who operate, maintain and repair medical equipment in some hospitals in Vietnam. Recommendations are proposed for actions regarding clinical engineering, and the link between research, education and entrepreneurship.**

Keywords — **Medical Devices, Clinical Engineering, Biomedical Engineering, Hospital Personnel.**

I. Introduction

The hospital regulations issued by the Vietnam Ministry of Health [1] require that all hospitals at the provincial level and above must have a department in charge of medical equipment and all medical centers at the district level must have one staff in charge of medical equipment in the Human Resource Office. Under those regulations, the medical equipment department plays a very important role in counseling the Board of Directors about investment, purchasing plan, management and utilization of medical equipment and supplies in the hospital. According to some incomplete statistics in 2007 [2] regarding medical equipment staff, in the provincial and district hospitals, the proportion of medical equipment staff who have a good technical background is very low: only 6% are engineers, 35% are technicians or technical workers, 59% are doctors, pharmacists or nurses. However, those statistics have not investigated deeply into the current working status of those staff. This paper provides results of a survey of this issue and also gives recommendations to improve medical equipment staff in hospitals in Vietnam.

II. Methodology

We executed the following steps:

1. Review published information related to the subject,
2. Design the questionnaire and form the survey team,
3. Work with a focus group which consisted of doctors, engineers, technicians, hospital directors to evaluate and re-edit the questionnaire accordingly,
4. Carry out the survey on-site (Note: for each surveyed unit, the survey team worked with at least two staff: one physician and one engineer or technician. If they provide controversial answers for the same question, the team discussed with each participant to reach a consensus),
5. Analyze survey results,
6. Discuss thorough those results with experts who are senior doctors in hospitals, lecturers at medical universities and chief engineers in a medical company.

III. Results and discussion

We surveyed 21 hospitals including: Bach Mai Hospital, Hospital for Traumatology & Orthopaedics, HCMC, Choray Hospital, Saigon General Hospital, General Hospital –Lam Dong Province, Hung Vuong Hospital, Nguyen Tri Phuong Hospital, MEDIC Medical Center, Nhi Dong I Hospital, Hospital of Post and Telecoms I, Tu Du Hospital, Viet Duc Hospital, General Hospital –Khanh Hoa Province, General Hospital –Khanh Hoa Province, Gia Dinh Hospital, General Hospital –Binh Duong Province, Eastern Military-Civilian Hospital, (General Hospital of Post and Telecoms), Binh An General Hospital, Odonto-Stomatology Hospital, Nhi Dong II Hospital, District 9 Hospital.

In those hospitals, 6 are premium hospitals (class I and special class), 12 are class II hospitals and 3 are class III hospitals. The geographic distribution of surveyed hospitals is: Ho Chi Minh City (14), Hanoi (3) and elsewhere (4).

In total, we contacted and discussed with 35 medical doctors, 15 pharmacists, 20 engineers and 30 technicians.

Question: Which of the followings are the duties of the staff of your medical center?

Table 1 below shows the results. The horizontal header indicates the position of the answerers and the left column indices their duties. The obtained results were expressed in percentage, which is the ratio of the number of hospitals a

V. Van Toi et al. (Eds.): 4th International Conference on Biomedical Engineering in Vietnam, IFMBE Proceedings 40, pp. 294–297, 2013.
www.springerlink.com

duty was checked on the total number of hospitals surveyed. For instance, the results show that 76% of surveyed hospitals assigned engineers to do the planning whereas all hospitals (100%) assigned this task to doctors or pharmacists.

Table 1. The duties of the staff in the surveyed medical establishments.

(multiple answers are allowed)	Engineer	Technician	Doctor / Pharmacist	Hired from outside
Do planning	76%	0%	100%	0%
Operate medical equipment (ME)	33%	71%	100%	0%
Repair ME	76%	90%	0%	48%
Test quality of ME	71%	71%	10%	43%
Standardize ME	52%	29%	14%	67%
Train staff	48%	14%	24%	67%
Do research	14%	14%	62%	5%
Do inventory of ME	62%	81%	5%	5%
Others:...	5%	5%	0%	0%

Fig.1 below is the bar chart of the results of the table above.

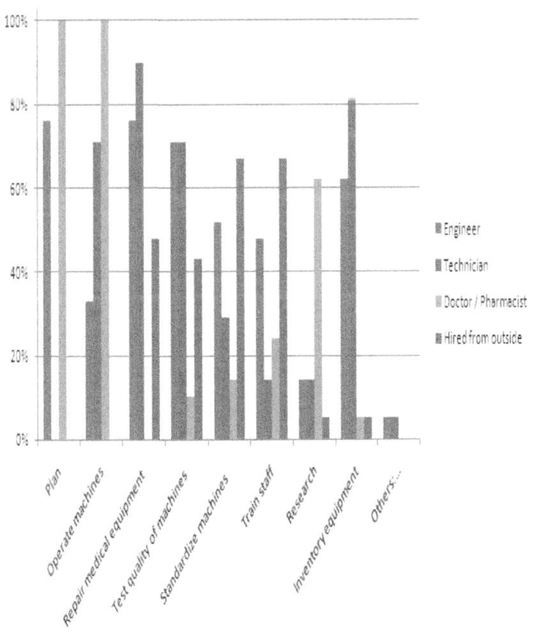

Fig. 1 Comparison of staff's duties in hospitals.

Analysis of the duties of engineers:

- Based on those results, we found that the engineers who are in charge of medical equipment in the hospitals work mainly in the following groups of duties: <u>Group I</u> (in more than 70% of surveyed hospitals) that includes making purchase plans, repairing and maintaining medical equipment, and controlling the quality of medical equipment; and <u>Group II</u> (in more than 48% of surveyed hospitals) that consists of standardizing machines, inventory and training staff. The discussions with the engineers in the medical equipment departments revealed that to handle this kind of job well, engineers must have the following skills and knowledge in: soft-skills including planning and presentation skills (so that they can perform such tasks as making purchase plans, repairing, maintaining medical equipment and training staff); electronic engineering, mechanical engineering and information technology, principles of medical equipment's operation (to carry out such tasks as repairing, testing and standardizing machines); and foreign language.

- 76% of hospitals employ engineers in planning and contributing ideas in the selection of medical equipment but the medical doctors are the ones who make the final recommendations what to be purchased. Interviews with engineers in hospitals revealed that even an engineer who has good education in modern medical technology is not allowed to do the recommendations. Interviews with medical equipment companies revealed that they approach primarily the doctors to introduce any product with new features, or new equipment either by personal contacts or in seminars.

- Only 33% of hospitals employed engineers in the operation of medical equipment. This is a very low figure although most hospitals we surveyed are the main hospitals located in HCMC and Hanoi. Therefore advanced medical equipment such as CT-Scanner, fMRI, or digital X-ray etc ... are not always operated by engineers who have both technical and medical knowledge.

- Currently, just 14% of hospitals have engineers who get involved in research. Even though our survey does not determine what kind of research, this is such a modest number in a very favorable environment for doing scientific and clinical research.

Analysis of the duties of technicians:

- Survey results show that the main tasks of the technicians working in the medical equipment-supplies department are to repair (90% of hospitals), inventory (81% of hospitals) and test machines' quality (71% of hospitals). Through discussions with technicians, we found that since the level of technicians is limited, they only have the ability to repair minor damages related to the power supply or mechanical faults but are not prepared to deal with the issues related to the system and software. In addition, to new equipment, most of the technicians are not able to understand the principles of their operations. Other reasons include: limitations in a foreign language (most documents are in English) and in basic knowledge related to medical equipment such as optics, medical image processing and information technology.

- The survey results also showed that technicians operate medical equipment in 71% of hospitals. In fact, this number is not really meaningful because we made a mistake in designing the survey. We realized only later that in the hospitals in Vietnam, there are two types of technicians. The first type is the staff working in specialized units and having an associate degree or lower. The second type is the experts working in the departments of the hospital such as department of medical equipment administration. In the questions we designed, we did not make clear what kind of technicians we want to survey. According to the hospital management regulations issued by Vietnam Ministry of Health, technicians in faculties are in charge of operating medical equipment. As a result, 100% hospitals employ technicians to operate medical equipment in reality. Particularly, the technicians will operate X-ray Scanners, electrocardiographs, electroencephalographs, and equipment for diagnostics. Several hospitals having experienced technicians will have them operate CT-Scanners.

Analysis of the duties of Doctors / Pharmacists

- In all hospitals (100%), doctors are involved in operating medical equipment: from Ultrasound, CT-Scanners, fMRI, endoscopy system to X-ray Scanners. This has several advantages because the doctors have medical knowledge to understand the results from such medical equipment, from which they can make decision to use a function of medical equipment to clarify a particular doubt. Nevertheless, letting doctors operate medical equipment has some drawbacks. Specifically, it wastes the time of the doctors who are rare in Vietnam. In addition, the operation of some medical equipment requires elaborated techniques to get an accurate diagnosis result, for example, in operating fMRI, PET-CET, then knowledge of general doctors is not appropriate. In hospitals, making purchase plan of medical equipment is of medical doctors. Because the heads of departments of medical equipment administration in some hospitals are doctors or pharmacists, the selection of medical equipment's configurations to be purchased is jointly carried out by doctors and pharmacists. Even though those doctors or pharmacists can understand the basic requirements of new medical equipment's technical configurations; they often do not have a general and systematic knowledge of medical equipment. Therefore, new equipment bought may not match or be compatible with existing equipment, or equipment will be purchased in the future. Sometimes, new equipment may not run due to the lack of facility or other indispensable accessories.

Analysis of the duties of people hired from outside

- Survey results also show that over 50% of hospitals hire outside people to repair medical equipment, test and standardize machines as well as to train staff. This is such a potential market when hospitals are expanding to meet demand. The Ministry of Health also plans to set up 3 maintenance, repairing and staff training centers in North, Central and South of Vietnam [3].

Question: what is the average number of people who is responsible for medical equipment in your hospital?

Surveyed results in 21 hospitals show that:

Position	Percentage (%)	Number of people
Engineer	29	78
Technician	50	135
Administrative Staff	9	25
Doctor	4	10
Pharmacist	3	7
Technician with high-school degree	6	17

Notes: The percentage is calculated as follows: in 21 surveyed hospitals, the total number of full-time staff in charge of medical equipment is 272. In these hospitals there are 78 engineers who account for 29% of the total. Fig. 2 below shows the pie-chart of the above results.

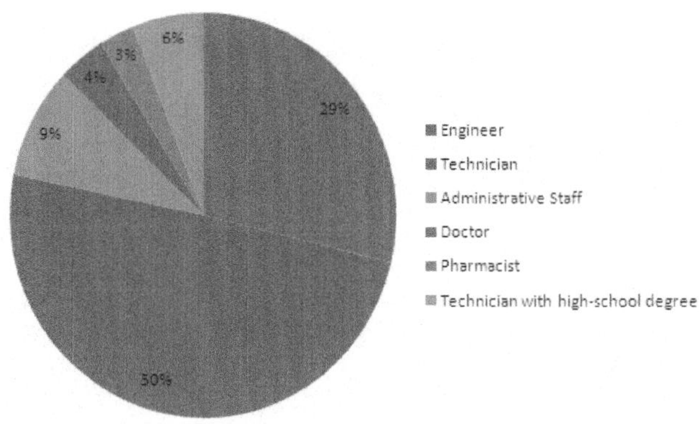

Fig. 2 Percentage of personnel's qualifications in the department of medical equipment administration.

According to this result, the current rate of engineers and staff working in the department of medical equipment administration is 29%. This figure is much higher than the rate of only 6% engineers in the survey of the Ministry of Health in the end of 2007 [2]. One of the reasons is that we focused only in large hospitals, while the results of the Ministry of Health may include small ones.

In 21 surveyed hospitals, the total number of engineers working in departments of medical equipment administration is 78 people, so there are 3.71 engineers per hospital, on the average. In our survey, the percentage of hospitals with at least one engineer working in the department of medical equipment administration is only 75%. In a private foreign-owned hospital in Ho Chi Minh City which has 100 beds, its department of medical equipment administration has 4 engineers (1 electrical engineer, 2 electronic engineers and 1 biomedical engineer). Another hospital has 2 staff in its department of medical equipment administration. From these results we and the consulting team believe that the percentage of engineers with knowledge of medical equipment working at the department of medical equipment administration in hospitals should be 80%. As a result, with over a thousand large and small hospitals in Vietnam, we need many more biomedical engineers now and in the future.

IV. Conclusions

We found that while the governmental policy regarding medical equipment for the period of 2002-2010, amended by the Ministry of Health and extended until 2020 is very valuable, the current situation is mainly due to the lack of appropriate actions. One example is the unattractive salaries and limited chance for an engineer working in the hospitals to advance in his/her future career. On the other hand, often they are not well prepared before entering in such a field.

One action needs to be taken immediately is the education program for Clinical Engineers. Other advanced countries and organizations including World Health Organization (WHO) put great efforts to promote this profession. Clinical engineers are the biomedical engineers equipped with management and leadership skills. They are capable to advise medical doctors in medical acts and clinical research. They are mainly in charge of the quality control of medical equipment, the personnel management and the welfare of patients in a hospital.

Furthermore, a field in applied science, engineering and technology in a country can be advanced sustainably if the country has a related industry go along. Often great medical inventions are the results of academic research. Hence, an invention is valuable only if it becomes a commercially viable product. An action that encourages not only the transfer of technology but also the creation of small or startup companies are vital. To that end, a strong link among research, education and entrepreneurship i.e., among universities, hospitals and companies must be implemented. This link is sustainable only if the frontier between public and private is clearly defined.

Acknowledgment

We would like to express our sincere thanks to: International University of Vietnam National University (VNU-HCMC) and the Fulbright program; Associate Professor, Doctor Tran Diep Tuan, Vice Rector of the University Medicine and Pharmacy, HCMC; Doctor Phan Thanh Hai, Director of MEDIC Medical Center (Hoa Hao Hospital); and Associate Professor, Doctor Le Thi Tuyet Lan, Chairwoman of Ho Chi Minh City Respiratory Society for their valuable.

References

1. Hospital Regulations, issued together with Decision No. 1895/1997/QD-BYT dated 19/09/1997 of the Minister of Health.
2. Report of the Ministry of Health in "A Symposium on training and providing solutions to develop personnel specializing in medical equipment field "on March 10, 2009, source: website of the Ministry of Health.
3. Notice No. 130/TB-VPCP, dated 17/04/2009 of the Prime Minister.

Author: Vo Van Toi
Institute: Biomedical Engineering Department, International University of Vietnam National Universities, Ho Chi Minh City, Vietnam
Street: Linh Trung Ward, Thu Duc Dist.
City: Ho Chi Minh
Country: Vietnam
Email: vvtoi@hcmiu.edu.vn

Graphene Alloy with Nano Carbon: Novel Nano Material for Biosensor?

Nguyen Chanh Khe[1], Le Van Giat[1], Mai Ngoc Tuan Anh[1], Bui Thi Thu Thao[2], Bui Quang Thuan[2], Hoang Duc Truong[2], Bui Van Nam[2], and Hieu Dinh[3]

[1] Research Laboratories of Saigon Hi-Tech Park (SHTP Labs), Saigon Hi Tech Park, D. 9, Ho Chi Minh City, Vietnam
[2] International Hi-Tech Business Inc., Dao Duy Anh, Hanoi, Vietnam
[3] K Tube Technologies, 2150 Monterey Hwy, San Jose, CA95025

Abstract — **We have discovered an unusual type of nano carbon which exhibits super high electrical conductivity over well known SWNT (single walled carbon nano tube) and graphene nano platelets. The new nano carbon material is suspected to be a graphene hybrid composite with some portions of tubular nano carbon including carbon nano tube, carbon nano rod and carbon nano wire. As the tubular shape nano carbon is always observed at low catalyst portion (MS/CS ratio, MS= Metal Source , CS= Carbon Source) while graphene related structure is always observed at much higher concentration of the catalyst, we assume that the extra amount of catalyst could work as a divider to separate single sheet of carbon into graphene products . The nano carbon materials of the present study are synthesized by the unoxidizing pyrrollysis of salt-milled solid carbon sources using specific catalyst...**

Keywords — **graphene, graphene alloy, tubular nano carbon, carbon nano wire, carbon nano rod.**

I. Introduction

According to Wikepedia, graphene is an allotrope of carbon having a structure of approximately one-atom-thick planar sheets of sp2-bonded carbon atoms in a honeycomb crystal lattice. Graphene can be made by CVD process pf methane gas CH4 onto a thin film of copper [1]. It is actually a catalytic support growth process [2] well known in the preparation of carbon nano tube. However, the catalytic support growth process may have the difficulty in control the amount of catalyst needed for gas phase carbon source. We had reported the process of making carbon nano tube from solid state carbon sources in previous studies [3], [4]. This process obviously can offer the multiple advantages such as greater feeding of raw material, better control of the amount of the catalyst which is usually metal element as well as the positioning of catalyst molecules onto the surface of carbon source (CS) component prior to the baking process in an unoxidizing environment.

In the previous studies, the solid phase synthetic process shows the capability of producing all kinds of tubular nano carbon products such as hollow tube, solid tube including carbon nano rod, carbon nano wire, cloud shape carbon nano fiber, the short tube in the range of few hundred nano meter.

In the present study, we report a new way to make graphene and or graphene related materials. Just using specific catalyst from iron relatives, we are able to convert a solid state carbon source into graphene product in a reaction chamber filled up with unoxidizing gas when the amount of catalyst is high enough. This kind of catalyst only gives one graphene structure certified by FE-SEM, TEM and XRD. The specific pyrollysis products also successfully demonstrated the catalytic effect in producing electron from hydrogen gas. These facts will be described in details as following.

II. Experimental results and discussion

Preparation of pyrollysis products

As previously reported [3], the solid carbon source are selected from a wide variety of natural resources including wood, paddy husk, coffee bean, The carbon source is roughly ground into wood chip then into fine particle. The particles are further milled with table salt as milling media then washed and dried, re-ground into final raw materials which were soaked into the metal salt catalyst solution and dried out into fine precursor before under going a bake process in N2 at 1000C for 1 hr. The impure products were purified with 3M HCl at room temperature to remove metal element. The pure products were exposed to several measurements such as electrical resistivity, FE-SEM, TEM, TGA, and Raman.

Measurement of bulk resistivity 0.5 g of the pyrollysis product was pored and strongly pressed into the cylinder having diameter of 30mm, forming a solid block following the procedure described somewhere[6]. The electrical resistivity was calculated based on measured cylinder diameter, packing length and resistor R. Always, commercially available SWNTs, MWNTs, grapheme nano platelets from Cheaptubes LLC (USA) were used for measurement reference.

V. Van Toi et al. (Eds.): 4th International Conference on Biomedical Engineering in Vietnam, IFMBE Proceedings 40, pp. 298–300, 2013.
www.springerlink.com

Effect of catalyst on the electrical resistivity

Fig.1 displays the effect of MS/CS ratio on the electrical resistivity (ohm-cm) of the pyrollysis products prepared from GTM wood using two different kinds of metal catalyst: Ni and Fe

Electrical resistivity compared to that of Ni catalyst. Both catalysts show minimal resistivity in the vicinity of MS/CS = 0.10 -0.20.

Fig.2 (a) and (b) shows the FE-SEM images of pyrollysis products made out of light catalyst portion (MS/CS= 0.06) and heavy catalyst portion (MS/CS=2.5) respectively for iron catalyst. It is observed from these two Fig. 2(a) and 2(b) that the light portion of Fe catalyst gives rise to tubular shape nano carbon while heavily doped catalyst gives more "flake" shape or graphene look.

Fig.3 (a) shows TEM image of pyrrolysis product from heavy portion of Fe catalyst (MS.CS=0.48).

Fig.3 (b) shows TEM image of pyrrolysis product from middle heavy portion of Fe catalyst (MS.CS=0.24). Both of them show graphene like structure. Fig. 3(c) shows TEM image of pyrrolysis product from light portion of Ni catalyst (MS.CS=0.05). It is clearly in this case that the tubular shape nano carbon products were formed. In the tubular shape and one can see two distinguished tubes; the solid tube might be carbon nano rod and carbon nano wire, but the hollow tube must be carbon nano tube.

It should be concluded right here that the catalyst plays a key role in the physical structure and electrical properties of the nano carbon products coming out from the pyrollysis reaction of the solid state carbon sources. Especially, for specific catalyst such as Fe, the graphene is more easily to form at heavy concentration of catalyst. Such kind of specific catalyst somehow creates a physical divider to separate out the layer of nano carbon to form graphene products.

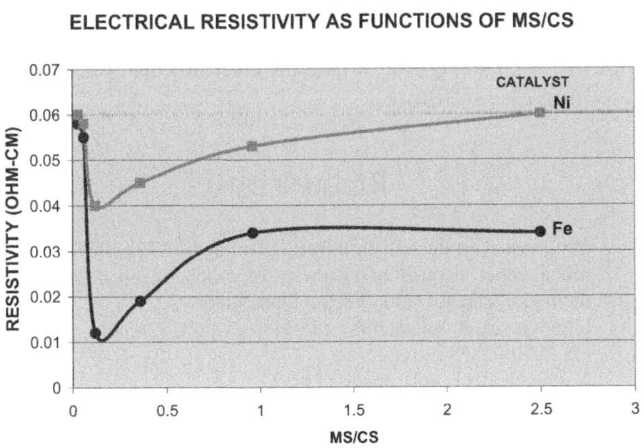

Fig. 1 Effect of MS/CS ratio on electrical resistivity for two different catalysts.

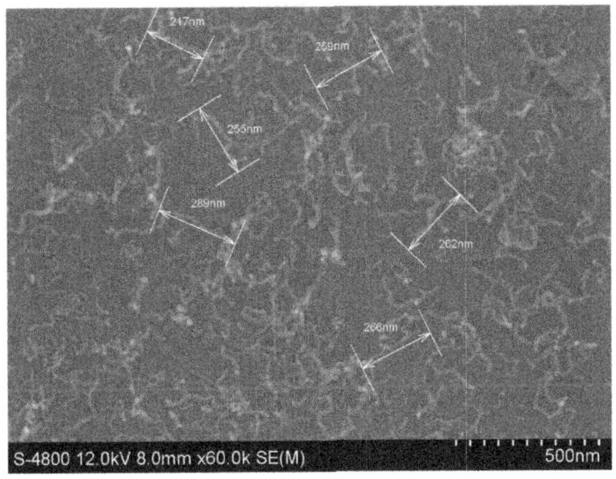

Fig. 2 (a) FE-SEM of pyrollysis product made out of Fe catalyst with the MS/CS ratio=0.06 (light portion).

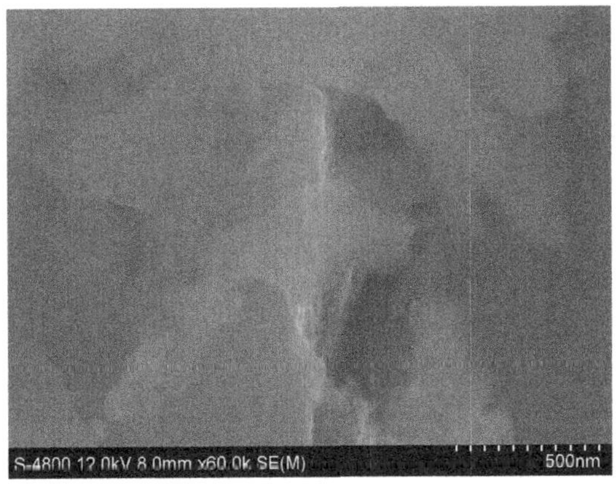

Fig. 2 (b) FE-SEM of pyrollysis product made out of Fe catalyst with the MS/CS ratio=2.5 (heavy portion).

Fig. 3 (a) TEM image of product from heavy doped Fe catalyst (MS/CS=0.24).

Generally speaking, PEMFC (Proton Exchange Membrane Fuel cell) using the Pt catalyst to ionize H2 gas into proton H+ and electron which contributes to the electricity source. For practical application purpose, the Pt powder has to be formed into nano scale and adsorb onto a surface of a conductive support, usually, carbon black Vulcan XR72 or XR72C (Cabot) in order to successfully deliver electrochemically generated electron to the outside circuit system. In the present study, we replaced the Vulcan XR72 or Vulcan XR72C with the pyrollysis nano carbon material made out of Fe catalyst in a heavy dose, for example MS/CS>0.12 as above described to see and found out that the new graphene alloy successfully demonstrated a significant improvement on electrocatalyst performance. This result will be studied in details in another future report. However, these findings suggest a new direction of design biosensor such as glucometer and alcohol detector or any kinds of chemicals containing –OH group. This hydroxyl group –OH will react with nano Pt adsorbed on the surface of graphene alloy to give rise better signals due to the high electroconductivity of graphene alloy.

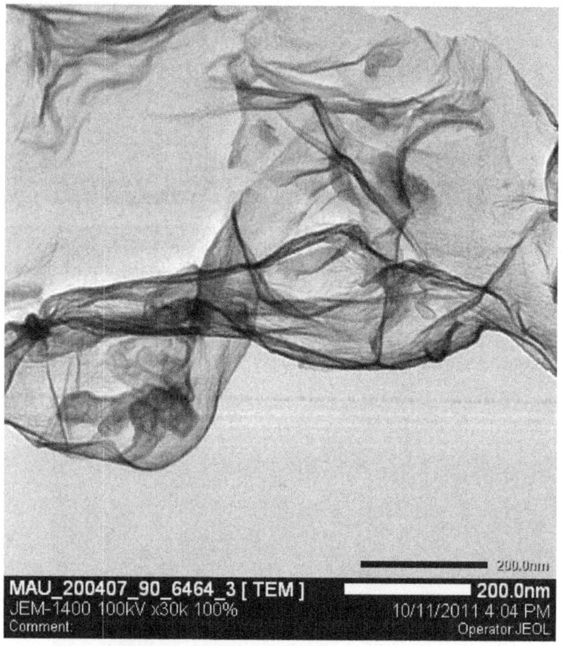

Fig. 3 (b) TEM image of product from heavy doped Fe catalyst (MS/CS=0.48).

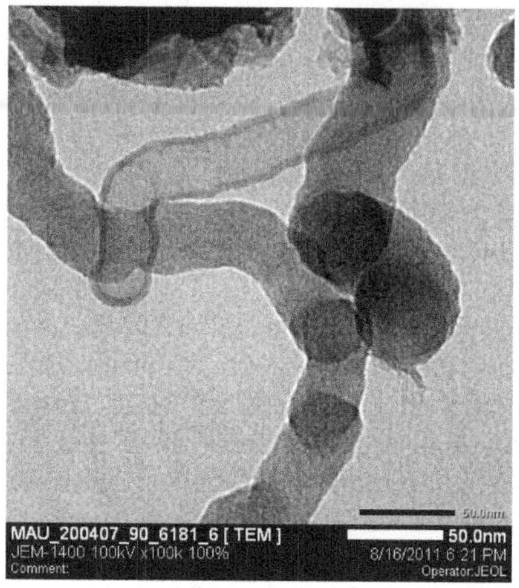

Fig. 3(c) TEM image of paralysis product made out of low portion of Ni catalyst (MS/CS=0.05).

III. Conclusion

It is concluded that certain catalyst such iron related can convert the solid state carbon source into a new type of nano carbon during pyrollysis under unoxidizing environment. These nano carbon materials are graphene related, showing very high electro conductivity, and improved compatibility to form certain kind of nano composite which can be successfully used as electro catalyst for PEM FC. The layer structure of these graphene related materials need further investigation, however the graphene like structure might be formed due to the excess amount of catalyst which could act as dividers to separate the carbon sheet into more single layer that graphite oxide researchers targeted.

The superior electroconductivity of the Graphene alloy in the present study shows very strong electrocatalyst sensitivity regarding to Pt, promising as biosensor for any substance containinghydroxyl –OH group, suggesting the capabilities of designing a glucometer, alcohol detector and something else in future.

References

1. http://uqam.academia.edu/FilipBogdanPopescu/Papers/928005/Chemical_vapor_deposition_synthesis_of_graphene_on_copper_with_methanol_ethanol_and_propanol_precursors
2. https://research.sabanciuniv.edu/51/1/stvkaf01726.pdf
3. US 2010/0247419 A1
4. Nanotech 2011, vol.1, page 171-173

Protein-Fatty Acid Conjugate for Self-assembled Nanoparticles in Drug Delivery

Beom-Jin Lee[1] and Phuong Ha-Lien Tran[1,2]

[1] College of Pharmacy, Kangwon National University, Chuncheon, Korea
[2] Department of Biomedical Engineering, International University - Vietnam National University, Ho Chi Minh City, Vietnam

Abstract — **Novel biocompatible protein-oleic acid conjugate was prepared to form self-assembled nanoparticles (NP) when exposed to an aqueous medium. The conjugation of protein-fatty acid was confirmed using FTIR and NMR. The physicochemical properties of NP were characterized by the experimental conditions such as pH, temperature and the amount of protein/fatty acid. The self-assembled NP had a uniformly spherical shape with particle size (150-200 nm) and showed good physical stability. This NP had no any potential cellular toxicity. Paclitaxel was chosen as model drug for anti-cancer effect using self-assembled NP. Drug release was controlled over an extended period of time. Furthermore, surface-modified NP using folic acid showed a higher cancer-cell killing effect as compared with Taxol®. The *in vivo* pharmacokinetic parameters and biodistribution of the NP containing paclitaxel also indicated longer blood circulation.**

Keywords — **conjugation; biocompatible nanoparticles; self-assembly; controlled drug release; actively surface targeting; anti-tumor effect.**

I. Introduction

NPs can be defined as solid colloidal particles, produced by mechanical or chemical means from synthetic or natural polymers, which are typically in the nanometric size range (1 to 1000 nm). NPs are currently developed for many pharmaceutical reasons to minimize drug degradation upon administration, prevent undesirable side effects and increase drug bioavailability and target drug to pathological area [1].

Various nanoparticle (NP) drug delivery systems (DDS) enter into blood circulation for systemically and interact with a receptor at the target site and organ. Nanoparticles can deliver numerous medicines for local or systemic

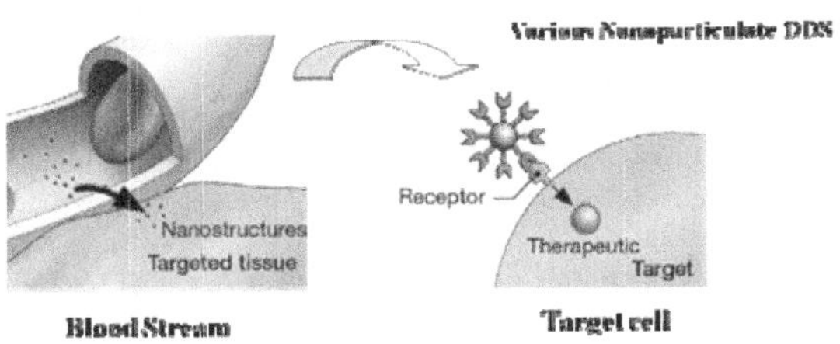

Fig. 1 Nanoparticles functionalized nanoparticles interacting with target cell via blood circulation [2].

effects [2]. Furthermore, the functionalized NP using various targeting molecules gives better accumulations in target tissues via the bloodstream.

Recently, various nanosystems such as nanocrystal (oral) and mainly parentally liposomes, micelles, quantum dots, polymeric and metal NPs, PEGylated asparaginase, albumin-paclitaxel bound nanoparticles, superparamagnetic iron oxide nanoparticles as a MRI contrasting agent, dendrimers, microcapsules, cells, cell ghosts, lipoproteins, and many different nanoassemblies are available. Some lists of US FDA-approved commercial NP DDS are given in Table 1.

Table 1. US FDA-approved NP DDS.

Drug or therapeutic agent (trade name)	Indication
Liposomal amphotericin B (Ambisome®, Albecet® and Amphotec®)	Fungal infections, leishmaniasis
PEG-adenosine deaminase (Pegademase®)	Severe combined immunodeficiency disease
PEG-stabilized liposomal doxorubicin (Doxil® and Evacet™)	Kaposi's sarcoma, refractory ovarian cancer
Liposomal cytosine arabinoside (DepoCyt®)	Lymphomatous meningitis, neoplastic meningitis
Interleukin 2-diptheria toxin fusion protein (Denileikin diffitox/Ontak®)	Cutaneous T cell lymphoma
Lipiosomal vertporfin (Visudyne®)	Wet macular degeneration
PEG-interferon α-2b (Pegasys®)	Hepatitis C
PEG-granulocyte colony stimulating factor (Neulasta®)	Chemotherapy associated neutropenia
Protein bound paclitaxel (Abraxane®)	Metastatic breast cancer
PEG L-asparaginase (Oncaspar®)	Acute lymphocytic leukemia
PEG aptanib (Macugen®)	Wet macular degeneration
Pemetrexed (Alimta®)	Malignant pleural mesothelioma
Fenofibrate (TriCor®)	Mixed lipid disorders
Ferumoxytol (Feraheme®)	Iron deficiency anemia in chronic kidney disease patients

V. Van Toi et al. (Eds.): 4th International Conference on Biomedical Engineering in Vietnam, IFMBE Proceedings 40, pp. 301–303, 2013.
www.springerlink.com

In the preparation of NP, numerous materials and excipients have been widely investigated in pharmaceutical industry. The polymeric materials such as poly (DL-lactide-co-glycolide) (PLGA), poly (ε-caprolactone) (PCL), poly (alkylcyanoacrylates), poly (styrene-co-maleic anhydride), poly (divinylether-co-maleic anhydride), poly (vinyl alcohol), poly (ethylene glycol) are some examples of synthetic and non-immunogenic polymers extensively used for NP preparation. Similarly, poly (amino acids), albumin, gelatin, hyaluronic acid, dextran, starch, and chitosan are some of the natural biodegradable polymers. The synthesis and pharmaceutical applications of new biocompatible biomaterials are recently of hot issues for delivery of various medicines including anticancer agents for optimal therapy. However, the safety and biocompatibility are important criteria for the selection of carriers [2].

There are several methods on the preparation of NPs to load drug. Many NPs are spontaneously constructed via self-assembly of many amphiphilic materials [3]. The self-assembly is a thermodynamic and, consequently, a reversible process that is entropically driven by the release of ordered water from hydrophobic blocks. A variety of drugs can be physically incorporated into the core by engineering the structure of the core-forming segment. Furthermore, NPs generally and self-assembled NPs particularly can be actively targeted to tumors for receptor-mediated uptake by specifically recognizing and binding target tissues or cells via the surface-attached specific ligand like a "vector" molecule. Thus, the formation of NP via self-assembly using amphiphilic materials has versatile advantages in pharmaceutical industry.

For this goal, gelatin as a model protein was modified by conjugating liphophilic molecules like fatty acid to form self-assembled NP when exposed to aqueous media. Gelatin is a protein material obtained by hydrolytic degradation of naturally occurring collagen and is well known for its non-toxic, non-irritant and biodegradable properties, good body compatibility and unique gelling properties [4].

II. Experimental

The gelatin-fatty acid conjugates were then synthesized for the preparation of self-assembled NPs. Anticancer agent (paclitaxel) was chosen as a model drug to encapsulate NPs for tumor treatment. The physicochemical properties, cellular toxicity and pharmacokinetics of NPs were evaluated.

The protein-fatty acid conjugates were prepared and then successfully identified using ^{1}H NMR spectroscopy and Fourier transform infrared spectroscopy (FT-IR). When this conjugate was exposed to an aqueous media, uniform and spherical NP was formed via self-assembly. The drug-loaded NP was also loaded with high efficiency. The NP was also spherical with negative zeta potential. The drug loaded in NP was amorphous when characterized using powder X-ray diffraction (PXRD). Drug release was controlled over an extended period of times with enhanced dissolution because of solubilization of drug inside micellar layers as compared drug alone and Taxol®. Folic acid has a very high affinity for its cell surface receptor and it moves into the cell cytoplasm, which is an advantage for more efficient intracellular delivery of anticancer agents than using a cell membrane marker that is not cell internalized [5]. The drug-loaded NPs with folic acid specifically located on the surface had better cancer cell killing effect by circulating in blood over longer times. The pharma cokinetics and biodistribution of such drug-loaded NPs are mainly determined by the surface properties, size, and stability of the self-assembled NPs, and are less affected by the properties of the loaded drug.

III. Conclusions

The protein-fatty acid NPs were successfully formed via self-assembly of hydrophilic-hydrophobic moieties. The drug-loaded NP showed numerous pharmaceutical advantages. Drug release rate was controlled over longer period of time. Drug solubility was also highly increased. Most of all, the NP and functionalized NP had enhanced cancer-cell killing effect even at a low drug concentration. The current self-assembled protein-oleic acid NP can be used to deliver for numerous model drugs for pharma ceutical applications.

Acknowledgment

This work was supported by the Korean Health Technology R&D Project, Ministry for Health and Welfare, Korea (A092019).

References

1. Maury, P.A, Reinhoudt, D.N, Huskens, J, 2007. Assembly of nanoparticles on patterned surfaces by noncovalent interaction. Curr Opin Colloid Interface Sci. 13: 74-80.
2. Kateb, B, Chiu , K, Black , K et al. et al. (2011) Nanoplatforms for constructing new approaches to cancer treatment, imaging, and drug delivery: What should be the policy?. NeuroImage 54 (2011) S106–S124.
3. Gao, F.P, Zhang, H.Z, Liu, L.R, Wang, Y.S, Jiang, Q, Yang, X.D, Zhang, Q.Q, 2008. Preparation and physicochemical characteristics of self-assembled nanoparticles of deoxycholic acid modified-carboxymethyl curdlan conjugates. Carbohyd. Polym. 71: 606-613.

4. Curcio, M, Spizzirri, U.G, Iemma, F. Puoci, F. Cirillo G. Parisi, O.I. Picci, N. 2010, Grafted thermo-responsive gelatin microspheres as delivery systems in triggered drug release, European Journal of Pharmaceutics and Biopharmaceutics 76 : 48–55.
5. Stella, B, Arpicco, S, Peracchia, M.T, Desmaele, D, Hoebeke, J, Renoir, M, D'angelo, J, Cattel, L, Couvreur, P, 2000. Design of folic acid-conjugated nanoparticles for drug-targeting. J. Pharm. Sci. 89: 1452-1464.

Author: Beom-Jin Lee
Institute: Kangwon National University
Street: 1 Kangwon University Road
City: Chuncheon
Country: Korea
Email: beomjinlee@gmail.com

pH-Sensitive Polymeric Systems for Controlling Drug Release in Nocturnal Asthma Treatment

Phuong Ha-Lien Tran[1,2,*], Thao Truong-Dinh Tran[2], Van Toi Vo[1], and Beom-Jin Lee[2]

[1] Department of Biomedical Engineering, International University - Vietnam National University, Ho Chi Minh City, Vietnam
[2] College of Pharmacy, Kangwon National University, Chuncheon 200-701, Korea

Abstract — **The aim of our study was to design pH-sensitive pulsatile drug delivery system which exposes a pulsed release of drug. Polyacrylic acid (PAA) is a pH-responsive weak polyelectrolyte polymer that exhibits a highly coiled conformation at low pH but a highly extended state at high pH. Binary blends of PAA and ethylcellulose (EC) with different ratios were used for coatings as pH sensitive polymeric systems efficiently controlling the release of drug from the coated film through a design of pulsed release with some pause intervals. Imaging analyses using scanning electron microscopy, near-infrared imaging, confocal laser scanning microscopy, and Fourier transform infrared spectroscopy confirmed the release behavior of the drug and fumaric acid from coated tablets.**

Keywords — **Pulsatile drug delivery systems, pH-responsive, weak polyelectrolyte polymer, structural transformation.**

I. Introduction

In recent decades, chronotherapy, which is another term for controlled drug delivery systems, has been investigated to allow the pulsed release of the drug and the synchronization of drug release with circadian rhythms of the body to achieve maximum health benefits and minimum harmful effects [1]. For example, moderate to severe asthma, particularly where severe nocturnal symptoms can occur during the night, may not be adequately controlled even with high doses (over 1,000 μg per day) of inhaled steroids [2]. In addition, conventional β-agonist aerosol medications possess a rather short duration of action. Thus, an advanced drug delivery system that synchronizes asthma treatment with circadian biochemical changes and has a 24-h duration of activity would be effective.

Bambuterol (BAM) was selected as a model drug. BAM belongs to a group of long-acting β2-agonists and enables asthma treatment by once-daily dosing via the oral route. Because asthma worsens or asthma attacks frequently occur during nighttime in most patients, evening dosing using a pulsed release of BAM can be effective for nocturnal asthma treatment.Currently, numerous immediate released BAM tablets are commercially available. So far, there has been no pulsed delivery system that synchronizes the circadian release of BAM in a controlled manner after a lag time. Polymeric coating techniques have been widely used in pharmaceuticals for many purposes including most applications for controlled release of drugs in the preparation of various dosage forms [3,4].

In this study, we designed a novel pH-sensitive oral delivery system using nonenteric polymeric coatings of a core drug loaded tablet; these coatings consisted of a binary blend of ethyl cellulose (EC) and poly(acrylic acid) (PAA) and provided a pulsed release of BAM with some pause intervals. PAA is a pH-responsive weak polyelectrolyte polymer that exhibits a highly coiled conformation at low pH but a highly extended state at high pH [5,6]. In other words, PAA is uncharged at low pH but becomes highly anionic as the pH is increased. Furthermore, the PAA was combined with EC to efficiently control the release of BAM from the coated film. This binary blend of coated film was expected to render the drug sensitive to pH conditions, creating pause intervals when the PAA was tightly coiled. Furthermore, an internal acidic agent such as fumaric acid can be added to core tablets to create an acidic microenvironmental pH (pH), which controls this special type of pulsed drug release even at the higher pH of the gastrointestinal tract.

The cross-sections of the coated films at different intervals of the dissolution test were characterized by scanning electron microscopy (SEM). Near-infrared imaging (NIR imaging) and confocal laser scanning microscopy (CLSM) were also utilized to investigate the pulsed behaviors of the tablets' coated films upon dissolution. The molecular interaction of the coated films was also investigated using Fourier transform infrared (FTIR) spectroscopy.

II. Experimental

A. Materials

BAM powder was received fromKorea United Pharm. Inc.(Seoul, Korea). EC (45 cps), PAA and fumaric acid were obtained from Sigma (St. Louis, MO, USA). Hydroxypropyl

* Corresponding author.

V. Van Toi et al. (Eds.): 4th International Conference on Biomedical Engineering in Vietnam, IFMBE Proceedings 40, pp. 304–308, 2013.
www.springerlink.com

methylcellulose (HPMC) was supplied by Richwood Inc. (Seoul, Korea). The solvents used were high-performance liquid chromatography (HPLC) grade. All other chemicals were of analytical grade and used without further purification.

B. Preparation of Core Tablets

Core tablets (160 mg) were prepared by the direct compressing method. All of drugs and excipients weredried under vacuum at room temperature before use. BAM, HPMC, fumaric acid, and other agents were thoroughly mixed and passed through a sieve. Thereafter, the homogeneous mixtures were directly compressed intotablets using a rotary tablet machine (Korea Machine, Anyang, Korea). The theoretical drug contentper core tablet was 20 mg.

C. Preparation of Coating Solutions

PAA and plasticizer were initially dissolved in ethanol; EC was added to this mixture; and the solution was stirred completely until homogeneous mixture was obtained. Plasticizer was selected as a plasticizer at 5% or 10% (w/w compared with the total weight of polymers in the formulations). For the evaluation of tablet-coated films using CLSM, fluorescein (1% w/w compared with the polymer weight) was added into the coating solution.

D. In vitro dissolution

Dissolution studies of coated tablet were conductedusing the USP I apparatus (100 rpm, 37°C, and a 900-mldissolution medium) with a dissolution tester (DCM-D12,DCM Korea, Siheung, Korea). Tablets wereexposed to the simulated gastric fluid (pH 1.2) for the first2 h and then switched to simulated intestinal fluid (pH 6.8).Samples were withdrawn at predetermined intervals andreplaced with an equivalent amount of fresh medium tomaintain a constant dissolution volume. The samples werefiltered through a 0.45-μm membrane filter. The concentration of drug was determined by HPLC.

E. Morphology of Coated Films

Cross sectional views of the tablet coatings before and during thedissolution test at different times were observed with a scanning electron microscopy (SEM). Thedried tablets were coated with gold under an argon atmosphere using a Jeol JFC-1,100 sputter coater (Jeol, Japan) for about2 min. Micrographs were taken with a Cambridge Stereo Scan200 (London, UK) at an accelerating voltage of 15 kV.

F. Near-Infrared Imaging (NIR) Imaging of Coated Films

Tablets were removedfrom the dissolution vesselsat different timescorresponding to predetermined intervals given in SEM. After removing the excessivewater from the samples using tissue paper, the tablets weredried at room temperature in the dark overnight. The coatedfilms of the dried samples were then observed using a MatrixNIR™ chemical imaging microscope and Isys™ 3.1.1chemical imaging software (Spectral Dimensions) to analyzethe spatial distribution of PAA in the coatings.

G. Confocal Laser Scanning Microscopy (CLSM) of Coated Films

CLSM (FLUOVIEW-FV300, Olympus, Japan) wasutilized for taking images. Coated tablets containing fluoresceinwere exposed to the dissolution test andthen collected at different timescorresponding to predetermined intervals given in SEM. The excessive water of the samples was gentlyremoved using tissue paper. The wetted tablet was then driedand placed onto the glass slide to observe the CLSM imagesof coated films at 488 nm using an argon ion laser.

H. Fourier Transform Infrared (FTIR) Spectroscopy

A FTIR spectrophotometer (Excalibur Series UMA-500, Bio-Rad, USA) was used to investigate the molecular interaction of the coated films at different times correspondingto predetermined intervalsgiven in SEM. Thewavelength was scanned from 500 to 4,000 cm-1with aresolution of 2 cm−1. KBr pellets were prepared by gentlymixing 1 mg of the sample with 200 mg of KBr.

III. RESULTS AND DISCUSSION

A. Release Profiles of Drug

The extent of drug release rate from coated tablets after the lag time depended on the levels of coating layer and plasticizer as well as different ratios of PAA/EC. Figure 1 shows the drug release profiles after 8 h of the formulations of C1, C2, C3 and C4. As the amount of PAA relative to EC in the coating formulation with the same amount of plasticizer increased, the drug release rates highly increased (C1, C2, and C3). This is actually an obvious result because the increased amount of PAA simultaneously reduced the amount of EC in the coating formulation, hence leading to decrease the hydrophobic nature of this polymer, a factor obstructing tablets to imbibe water. Furthermore, for the same ratio between EC and PAA, the higher amount of plasticizer in the formulation contributed to the slower release of drug (C3 compared to C4). This dissolution result indicated the PAA transformation property upon the change of pH environment because increasing plasticizer amount facilitated a higher penetration of pH 1.2 fluid into coated tablets due to hydrophilic property of plasticizer, leading to a higher extent of highly coiled state of PAA, and therefore, subsequently required a longer period for extended transformation of PAA which induced the drug release when the coated tablets were switched to intestinal fluid. Based on the release profiles of the drug, the C4 formulation showing a 135-min lag time followed by a complete drug release could deliver the drug in a pulsatile manner and was chosen for the investigation of the pharmaceutical mechanism of coated films using instrumental analyses.

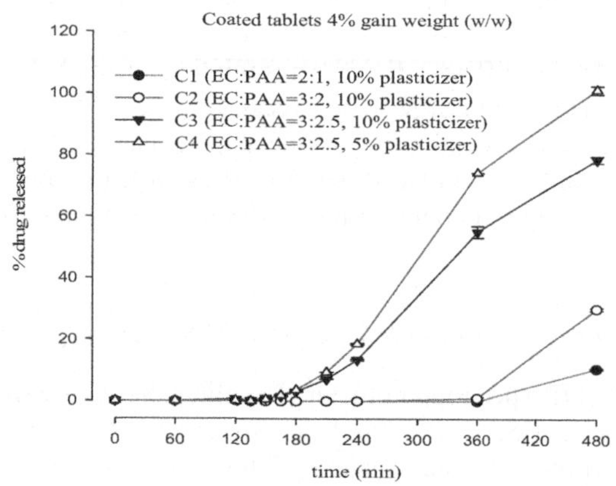

Fig. 1 Release profiles of the drug from coated tablets.

Interestingly, the pulsed release profiles of the drug from coated tablets showed some pause intervals based on the structural transformation of PAA during dissolution (Fig. 2). When the fumaric acid was retained in the core tablet, it gave a more acidic microenvironmental pH (pHM) so that the PAA was kept in a highly coiled state in the coated film, which hindered drug release. Drug release was then increased in a pulsed manner when the dissolution fluid was switched to intestinal fluid. The pH modulation between core (pH=3 by fumaric acid) and outer intestinal media (pH=6.8) could induce the structural transformation of PAA in a coated film for the pulsed release with some pause intervals. A complete drug release after a lag time ensured that this pH sensitive delivery system was suitable for the asthmatic patient during night sleep against nocturnal asthma.

The pH modulation between core (pH=3 by fumaric acid) and outer intestinal media (pH=6.8) could induce the structural transformation of PAA in a coated film for the pulsed release with some pause intervals. A complete drug

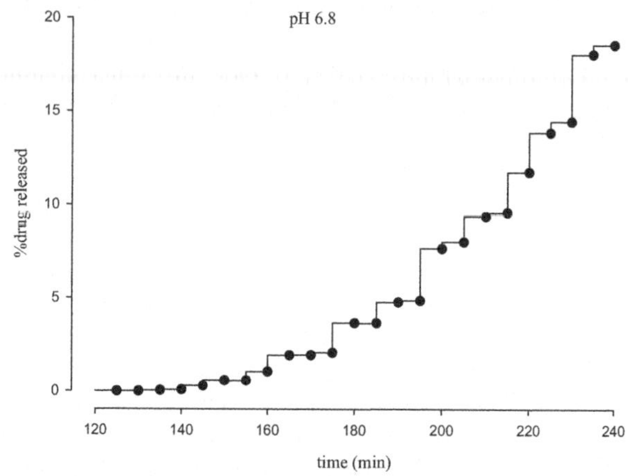

Fig. 2 Release profiles of the drug from coated tablets (5% plasticizer), showing the pause intervals.

release after a lag time ensured that this pH sensitive delivery system was suitable for the asthmatic patient during night sleep against nocturnal asthma.

B. Mechanism of the Coated Film through Instrumental Analyses

The cross-sections of the coated tablets removed from the dissolution test at different interval times were compared with those of the coated tablet before the test (Fig. 3). The photos show two differentiate layers indicating coating film and core tablet. The coating films before the test was homogenous and without any crystals observed. However, the coating films of the tablets taken out at 145, 180, and 235 min showed the numerous microcrystals on this layer; whereas much fewer crystals appeared on the coating film of the samples taken out at 155 and 230 min. The appearance or disappearance of the microcrystals indicated the pause intervals occurred in the drug release profiles.

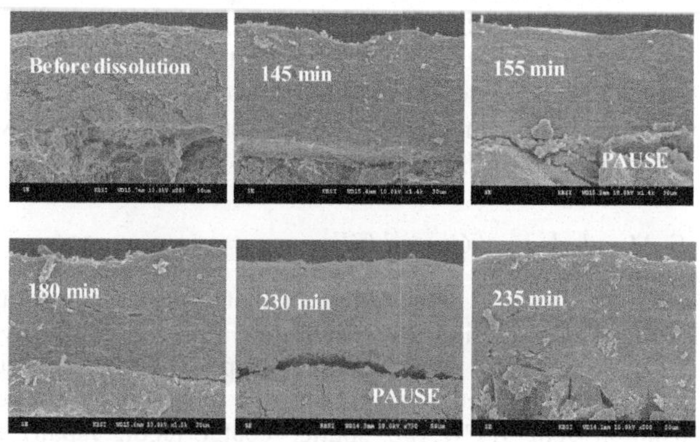

Fig. 3 Cross sections of coated films (5% plasticizers) by SEM at different times.

Moreover, both CLSM and NIR images were clearly matched with the pulsed release behaviors of the drug (Fig.4. and Fig.5., respectively). For CLSM, the fluorescence of samples was reduced or ordinary at some positions corresponding to the pHM at more acidic value transforming PAA to highly coiled state hindering drug release or higher pH of intestinal fluid changing PAA to extended state, respectively. Similarly, for NIR imaging, the high absorbance of the PAA was fixed throughout all of the samples; the color red in the NIR images represents the concentration of the PAA. The intensity of the red on the samples decreased at 145 min, 180 min and 235 min and increased at 155 min and 230 min, indicating the transformation of PAA into highly coiled or highly extended state for the pulse release of drug with pause intervals, respectively.

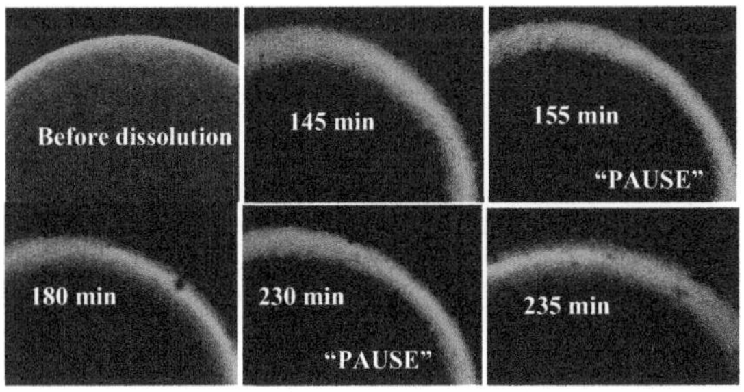

Fig. 4 CLSM images of coated films (5% plasticizers) at different times.

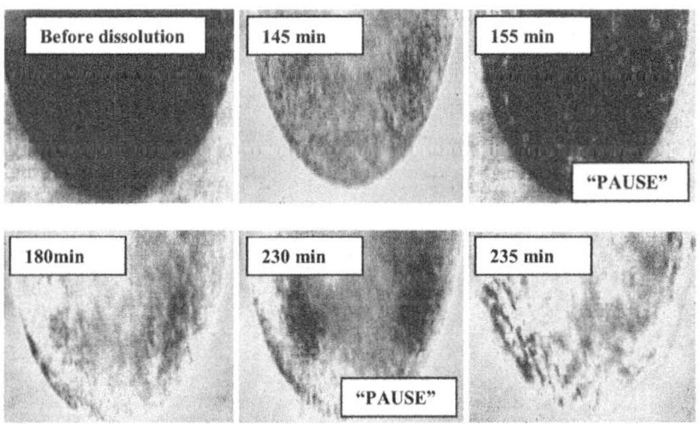

Fig. 5 Spatial distributions of PAA in coated films (5% plasticizers) at different times using NIR spectroscopy.

The FTIR spectra of the coated films also figured out the structural transformation of PAA upon pH changes at different release times. The FTIR spectra in this study (Fig.6.) showed that the intensity of the COOH peak in coated films during the dissolution test was higher in the cases of samples at 155 and 230 min, corresponding to the lower intensity of COO− band of these samples. On the other hand, the COOH peak intensity was lower and the COO− band intensity was much higher for the samples at 145 and 180 min.

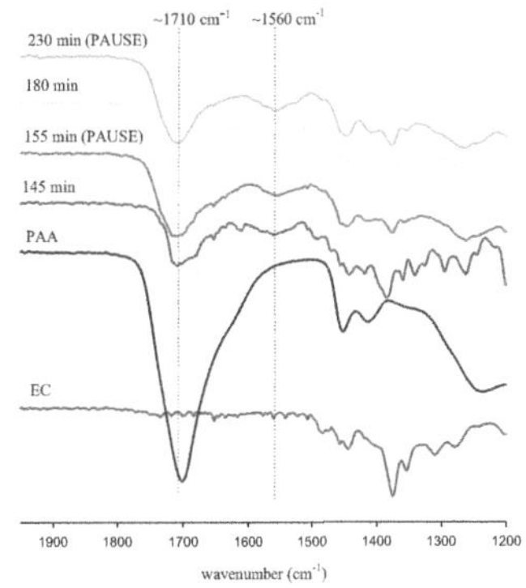

Fig. 6 FTIR spectra of coated films (5% plasticizer) at different times.

IV. Conclusions

The current pH-sensitive pulsatile system could be used to deliver bambuterol after a determined lag time to treat asthmatic patient during night sleep.

Acknowledgment

The authors would like to thank the 2009 Small Business Administration, BK21 program and a grant from the Korean Health Technology R&D Project, Ministry of Health and Welfare, Republic of Korea for financial support; the Central Research Laboratory (Kangwon National University) for the use of the FTIR and SEM and the Research Institute of Pharmaceutical Sciences, Kangwon National University, for the use of their HPLC systems; KBSI (Chuncheon, Korea) for the CLSM and KAIST (Daejeon, Korea) for the NIR imaging.

References

1. Roy P, Shahiwala A. Multiparticulate formulation approach to pulsatiledrug delivery:current perspectives. J Contr Rel.2009; 134:74–80.
2. Wallaert B, Brun P, Ostinelli J, Murciano D, Champel F, Blaive B, et al. The French bambuterol study group. A comparison of two long acting β-agonists, oral bambuterol and inhaled salmeterol, in the treatment of moderate to severe asthmatic patients with nocturnal symptoms. Respir Med.1999; 93:33–38.
3. Lee B-J, Min G-H. Oral controlled release of melatonin using polymer-reinforced and coated alginate beads. Int J Pharm.1996; 144:37–46.
4. Cao Q-R, Choi H-G, Kim D-C, Lee B-J. Release behavior and photo-image of nifedipine tablet coated with high viscosity grade hydroxypropylmethylcellulose: effect of coating conditions. Int J Pharm. 2004; 274:107–117.

5. Suh J, Paik H-J, Hwang BK. Ionization of poly(ethylenimine) and poly(allylamine) at various pH's. Bioorg Chem.1994; 22:318–327.
6. Grunlan JC, Liu L, Regev O. Weak polyelectrolyte control of carbon nanotube dispersion in water.J Collo Int Sci.2008;317:346–349.

Author: Phuong Ha Lien Tran

Institute: Department of Biomedical Engineering, International University - Vietnam National University, Ho Chi Minh City, Vietnam
Email: thlphuong@hcmiu.edu.vn

Application of l_1 Regularization for High-Quality Reconstruction of Ultrasound Tomography

Tan Tran-Duc[1], Nguyen Linh-Trung[1], Michael L. Oelze[2], and Minh N. Do[2]

[1] Fac. Electronics & Telecommunications, VNU University of Engineering & Technology, Hanoi, Vietnam
[2] Dept. Electrical & Computer Engineering, University of Illinois at Urbana-Champaign, Urbana-Champaign, USA

Abstract — **Ultrasound tomography based on inverse scattering has the capability to resolve structures which are smaller than the wavelength of the incident wave, as opposed to conventional ultrasound imaging using echo method. Some material properties such as sound contrast are very useful to detect small objects. Born Iterative Method (BIM) based on first-order Born approximation has been introduced as an efficient diffraction tomography approach. However, this method has a high complexity because it has to solve large iterative forward and inverse problems. In this paper, we propose to replace Tikhonov regularization by l_1-regularized least squares problem (LSP) in solving the inverse problem in BIM. As a result, the quality of reconstruction is improved and the complexity is reduced.**

Keywords — **Ultrasound, tomography, inverse scattering, Born Iterative Method, Tikhonov regularization.**

I. Introduction and state-of-the-arts

Ultrasound imaging and tomography play important roles in clinical detection. Ultrasound image acquisition is mostly based on a pulse echo method that uses the time of light energy reflected by the boundaries of the object under imaging [2]. By extending the number of angles around the object, inverse scattering based techniques offer better quality of image reconstruction under strong scattering [2].

Works in ultrasound tomography often focus on calculating the size of tissue (scattering area) and the speed of sound crossing the object being imaged. At present, there are only a few commercialized tomography devices. The reason is that state-of-the-art inverse scattering techniques have high computational complexity as well as limited efficiency. Born Iterative Method (BIM) and Distorted Born Iterative Method (DBIM) are well-known for diffraction tomography [1].

DBIM is more sensitive to noise, though it offers faster convergence as compared to that of BIM. In addition, the computational complexity of these methods is high due to their use of iterative forward and inverse processes. In [2], edge detection during the iterative process was introduced in order to speed up the convergence and to enhance the quality of reconstruction, but the complexity and noise sensitivity issues remain. In [3], the multi-level fast multi-pole algorithm (MLFMA) was applied to the forward solver for further speed up the reconstruction process. However, MLFMA requires high set-up costs that make it difficult to implement in practice.

In conventional BIM, Tikhonov regularization (a.k.a., l_2 regularization) [4] was employed in solving the inverse problem. This method does not de-noise well, as we will later show in this paper. Based on this observation, we realize that if we solve the inverse problem well in terms of the speed and the quality of reconstruction, then we can reduce the number of the forward problem that follows in the iterative process. Thus, in this paper, we replace Tikhonov regularization by the l_1-regularized least squares problem (LSP) [5]. We also show that this modification helps improve the quality of reconstruction and reduce the computational complexity.

II. Materials and methods

A. Born Iterative Method (BIM)

Fig. 1 schematically shows geometrical and acoustical configuration of the ultrasound tomography system.

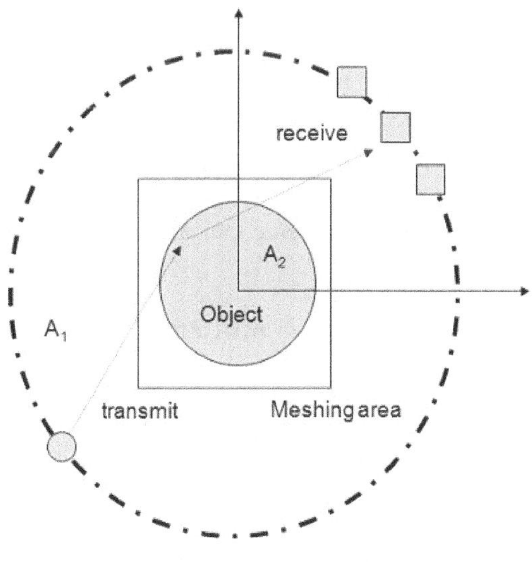

Fig. 1 Geometrical and acoustical configuration.

V. Van Toi et al. (Eds.): 4th International Conference on Biomedical Engineering in Vietnam, IFMBE Proceedings 40, pp. 309–312, 2013.
www.springerlink.com

The region of interest (ROI) consists of the reconstructing object, centered at the origin of a 2-D space and discretized into $N \times N$ square pixels of side . The number of transmitters and receivers are N_t and N_r, respectively. With the circular scatter area as shown in Fig. 1, the object function can be computed by

$$O(r) = \begin{cases} \omega^2\left(\frac{1}{c_1^2} - \frac{1}{c_0^2}\right), |r| \le R \\ 0, \quad |r| > R \end{cases} \quad (1)$$

Where c_1 and c_2 are the speed of sound in the object and in the water, respectively, f is the ultrasound frequency, ω is the angular frequency (i.e., $\omega = 2\pi f$), and R is the radius of the object.

We set up a measurement configuration for transmitters and receivers in order to obtain the scattered data. At an instance, only one transmitter and one receiver are active to for a corresponding measured data value. Assumed that the density variation is negligible, the inhomogeneous wave equation is given by

$$\left(\nabla^2 + k_0^2(r)\right)p(r) = -O(r)p(r), \quad (2)$$

where $k_0 = \omega/c_0$ is the wavenumber in the reference medium (i.e., water), and $p(r)$ is the total pressure field.

By solving (2), the scattering pressure can be obtained in intergral form, in term of the Green's function, as

$$p^{sc}(r) = p(r) - p^{inc}(r) = \iint O(r)\, p(r) G(|r - r'|) \quad (3)$$

where $p^{inc}(r)$ in the incident pressure, and is the free space Green's function. Equation (3) can be solved by method of moment utilizing the sinc basis and delta functions [6]. The pressure ingrid points can be presented in an $N^2 \times 1$ vector

$$\boldsymbol{p} = \boldsymbol{p}^{inc} + \boldsymbol{C}\boldsymbol{D}(\overline{\boldsymbol{O}})\overline{\boldsymbol{p}}, \quad (4)$$

and the scattered pressure can also be calculated in a scalar value

$$p^{sc} = \overline{\boldsymbol{B}}_i\boldsymbol{D}(\overline{\boldsymbol{O}})\overline{\boldsymbol{p}} \quad (5)$$

where $\overline{\boldsymbol{B}}_i$ is a $1 \times N^2$ vector converted from a matrix formed by Green's coefficient $G_0(r, r')$ from each pixel to the ith receiver, $\boldsymbol{C}$ is an $N^2 \times N^2$ matrix formed by Green's coefficient among all pixels in the meshing area, and $\boldsymbol{D}(\cdot)$ is an operator converting a vector into a diagonal matrix. Detailed calculations of $\overline{\boldsymbol{B}}_i$ and $\boldsymbol{C}$ can be found in [6].

If N_t transmitters and N_r receivers are used, then the scattered pressure signal vector of size $N_tN_r \times 1$ can be obtained from (5) as

$$p^{sc} = \overline{\boldsymbol{B}}_i\boldsymbol{D}(\overline{\boldsymbol{O}})\overline{\boldsymbol{p}} = \boldsymbol{M}\overline{\boldsymbol{O}} \quad (6)$$

where $\boldsymbol{M} = \boldsymbol{B}\boldsymbol{D}(\overline{\boldsymbol{p}})$ is the matrix whose size is $N_tN_r \times N^2$.

These data are processed using BIM to reconstruct the speed of sound contrast. By this way, we can detect if any tissue exists in the medium. BIM uses Born approximation to compute iterative solutions of a nonlinear inverse scattering problem. If we use multiple transmitters and receivers, the object function $\overline{\boldsymbol{O}}$ can be estimated using Tikhonov regularization

$$\overline{\boldsymbol{O}} = arg\min_{O}\|\overline{\boldsymbol{p}} - \overline{\boldsymbol{M}}\overline{\boldsymbol{O}}\|_2^2 + \gamma\|\overline{\boldsymbol{O}}\|_2^2 \quad (7)$$

where γ is the regularization parameter that needs to be chosen carefully because it contributes mostly to the stability of the system [7]. High values of γ make the reconstructed image rough. However, small values of γ lead to high computational complexity.

The solution of Tikhonov regularization problem can be calculated directly, or indirectly using an iterative method. The iterative method is more efficient than the direct one, especially when $\overline{\boldsymbol{M}}$ is sparse or has a special form (e.g., partial Fourier or wavelet matrices).

B. Proposed Method

Recently, l_1-regularization based methods have been applied widely for sparse signal reconstruction or feature selection. In [5], authors extended to l_1-regularized least squares programs (LSP) in order to reformulate them as convex quadratic programs, and then solved them by a specialized interior-point method. The advantage of this approach is that it can solve both sparse and dense problems with high accuracy at some small additional computational costs.

The object function $O(r)$ contains the information of the sound contrast of the target image. In Section II.A, $O(r)$ was updated at each iteration by Tikhonov regularization. Now, we reformulate the problem as the following LSP in order to improve the quality of solution $O(r)$:

$$\overline{\boldsymbol{O}} = arg\min_{O}\|\overline{\boldsymbol{p}} - \overline{\boldsymbol{M}}\overline{\boldsymbol{O}}\|_2^2 + \gamma\|\overline{\boldsymbol{O}}\|_1 \quad (8)$$

where $\overline{\boldsymbol{M}}$, $\overline{\boldsymbol{p}}$, and γ are respectively the measurement matrix, the scattered field vector, and the regularization parameter. Our method can be summarized in Algorithm 1. In this algorithm, the relative residual error is defined by

Algorithm 1. l_1-Regularization for high-quality reconstruction.

Choose initial values: $\overline{\boldsymbol{O}}_0 = \overline{\boldsymbol{0}}$ and $\overline{\boldsymbol{p}}_0 = \overline{\boldsymbol{p}}^{inc}$ as shown in (10)
Calculate $\overline{\boldsymbol{B}}$ and $\overline{\boldsymbol{C}}$ [6]
while $n < N_{\max}$ or RRE $< \epsilon$
1. Calculate $\overline{\boldsymbol{p}}$ corresponds to $\overline{\boldsymbol{O}}_n$ using (4)
2. Calculate RRE corresponds to $\overline{\boldsymbol{O}}_n$
3. Calculate a new value of $\overline{\boldsymbol{O}}_{n+1}$ by solving (8)
4. $n = n + 1$
End

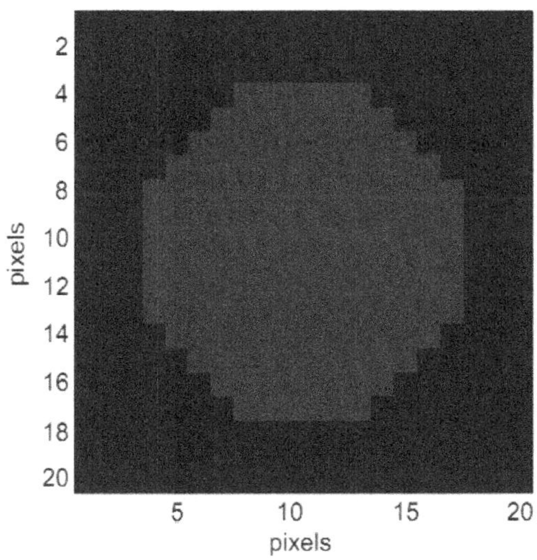

Fig. 2 Ideal object function.

$$RRE = \frac{\|\overline{\boldsymbol{M}}\overline{\boldsymbol{O}} - \overline{\boldsymbol{p}}^{sc}_{measured}\|}{\|\overline{\boldsymbol{p}}^{sc}_{measured}\|} \quad (9)$$

RRE is computed in each iterative step.

The loop will be terminated when RRE is smaller than a desired tolerance or when the number of iterative steps reaches some preset N_{max}.

III. RESULTS AND DISCUSSIONS

Simulated data were created for infinite circular cylinderusing 20×20 pixels. The number of transmitters and receivers are 40 and 20, respectively. The scattering field is affected by 5% Gaussian noise. The radius of the cylinder is 5λ, the frequency of ultrasound signal is 1 MHz, and the sound contrast is 2%.

The incident pressure for a Bessel beam of zero order in two dimensional case is

$$\overline{\boldsymbol{p}}^{inc} = J_0(k_0|\boldsymbol{r} - \boldsymbol{r}_k|) \quad (10)$$

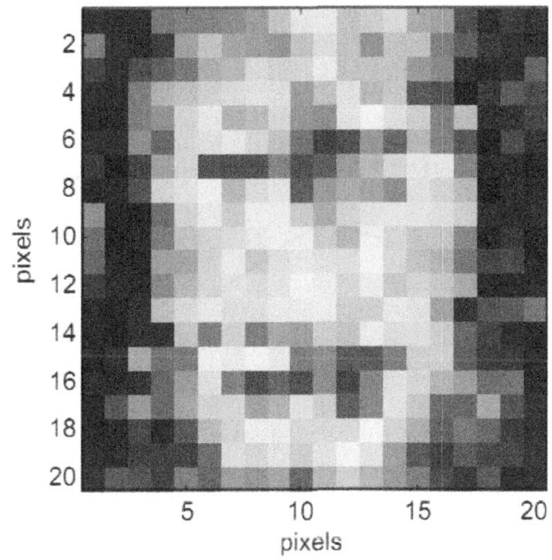

Fig. 3 Reconstructed object function after 5 iterations, using conventional method.

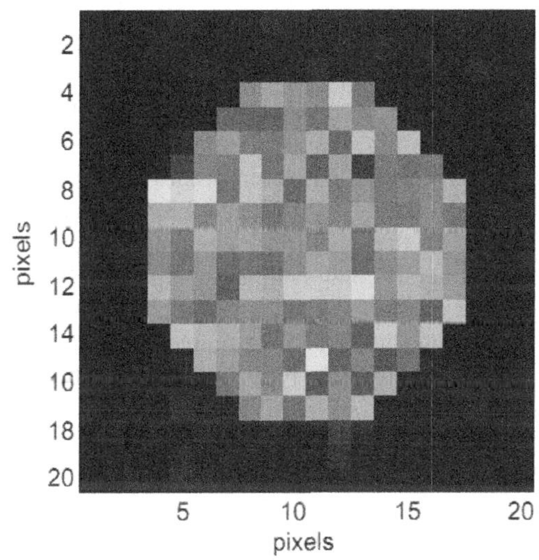

Fig. 4 Reconstructed object function after 5 iterations, using proposed method.

where J_0 is the 0th order Bessel function and $|\boldsymbol{r} - \boldsymbol{r}_k|$ is the distance between the transmitter and the kth point in the ROI.

Fig. 2 is the ideal object function corresponding to (1). Figs. 3 and 4 are reconstructed object functions after 5 iterations, using the conventional and proposed methods, respectively. They indicate that using l_1-regularization can greatly enhance the quality of reconstruction as compared to using the l_2-regularization.

We can also compare the performance the conventional and proposed systems using the performance error

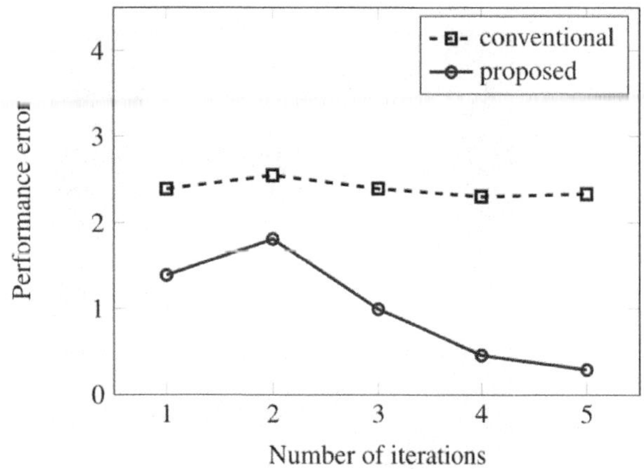

Fig. 5 Performance comparison between the conventional and proposed methods.

$$e = \sum_{i=1}^{N}\sum_{j=1}^{N}\frac{\left|C_{ij} - \hat{C}_{ij}\right|}{C_{ij}} \tag{11}$$

where C_{ij} and $\hat{C}_{ij}$ are the ideal and estimated ultrasound velocities at pixel (i,j).

Fig. 5 presents the error performance of these two methods. It shows that, by applying the LSP, the reconstruction quality has been improved. Although the computational complexity for each LSP reconstruction is 1.2 times larger than that in the conventional one, we have reduced the overall computational complexity for the same reconstructed quality. In the conventional method, after the 4th BIM iteration, the error performance is saturated due to the noise background. It means that even if we further increase the number of BIM iterations, we could hardly improve the reconstruct quality.

IV. Conclusions

This paper has successfully applied LSP to BIM in order to reduce the computational complexity and to improve the quality of the construction of the sound contrast. The BIM is chosen here to solve the inverse scattering problem due to its strength against noise. A simulation scenario of sound contrast reconstruction has been performed to show the outperformance of this method. This work will be further developed using experimental data.

Acknowledgment

This work was supported by Project QG.10.40 granted by Vietnam National University Hanoi.

References

1. Devaney AJ (1982) Inversion formula for inverse scattering within the Born approximation. Optics Letters 7:111-112.
2. Haddadin OS, Ebbini ES (1995) Solution to the inverse scattering problem using a modified distorted Born iterative algorithm. Proceedings of IEEE Ultrasonics Symposium, 1411-1414.
3. Hesford AJ, Chew WC (2010) Fast inverse scattering solutions using the distorted Born iterative method and the multilevel fast multipole algorithm. J Acous Soc America, 128:679-690.
4. Golub GH, Hansen PC, O'Leary DP (1999) Tikhonov regularization and total least squares. J Acous Soc America, 21:185-194.
5. Kim SJ, Koh K, Lustig M, Boyd S, Gorinevsky D (2007) A method for large-scale l1- regularized least squares. IEEE Journal on Selected Topics in Signal Processing, 1:606-617.
6. Tracy ML, Johson SA (1983) Inverse scattering solutions by a sinc basis, multiple source, moment method–Part II: Numerical evaluations. Ultrasonic Imaging 5:376-392.
7. Lavarello R, Oelze M (2008) A study on the reconstruction of moderate contrast targets using the distorted Born iterative method. IEEE Transaction of Ultrasonic, Ferroelectric, and Frequency Control 55:112-124.

Author: Tran Duc-Tan
Institute: VNU University of Engineering and Technology
Vietnam National University Hanoi (VNU)
Street: E3, 144 Xuan Thuy, Cau Giay
City: Hanoi
Country: Vietnam
Email: tantd@vnu.edu.vn

Complex Shear Modulus Estimation Using Maximum Likelihood Ensemble Filters

Tan Tran-Duc[1], Yue Wang[3], Nguyen Linh-Trung[1], Minh N. Do[2], and Michael F. Insana[3]

[1] Fac. Electronics & Telecommunications, VNU University of Engineering & Technology, Hanoi, Vietnam
[2] Dept. Electrical & Computer Engineering, University of Illinois at Urbana-Champaign, Urbana-Champaign, USA
[3] Dept. Bioengineering, University of Illinois at Urbana-Champaign, Urbana-Champaign, USA

Abstract — **Ultrasound shear wave imaging can be used to estimate quantitative information about soft tissue mechanical properties, specifically the complex shear modulus (CSM). It is promising because it has the potential to bridge molecular, cellular, and tissue biology and to support for medical diagnoses. However, there is a lack of efficient methods for CSM estimation due to high complexity and low reconstruction's quality. To overcome these disadvantages, in this paper, we have applied the Bayesian-approach-based Maximum Likelihood Ensemble Filter (MLEF) for CSM estimation. The error performance is below 10%. Both simulated and experimental data were tested in order to assess the proposed method.**

Keywords — **Ultrasound shear wave imaging, maximum likelihood ensemble filter, complex shear modulus, vibrating needle.**

I. INTRODUCTION

Elasticity imaging combines the application of a mechanical force stimulus and the use of a medical imaging device to track the resulting tissue deformation. Among various elasticity imaging techniques, shear wave imaging allows to measure mechanical information of tissues such as viscoelasticity, anisotropy, and nonlinearity. Of particular interest in medical diagnosis, viscoelasticity is related to pathological change in soft tissues. Quantitative information about the complex shear modulus (CSM), which yields both the shear wave number and the attenuation simultaneously, can be estimated by using ultrasound shear wave imaging [1]. It is a promising imaging technique because it has the potential to bridge molecular, cellular, and tissue biology and to support for medical diagnoses.

Recent CSM techniques apply acoustic radiation, surface vibration, or deep vibration as the mechanical force stimulus to generate the shear wave [2-5]. In [5], the authors use a vibrating needle for shear wave stimulus and estimate the CSM by fitting shear wave velocities of varying frequencies to the dispersion equation. The CSM can also be estimated by direct algebraic Helmholtz inversion [6]. However, these state-of-the-art methods for CSM estimation are limited to long processing time and low signal-to-noise ratio (SNR). These limitations can be overcome by using a Bayesian approach [7] that provides a better estimate of the velocity in low SNR and, hence, a better estimate of the CSM. However, this latter method currently deals with homogeneous media only and would miss the existence of tumors.

In this paper, we extend the method in [7] to heretogeneous media. Specifically, we propose to use the maximum likelihood ensemble filter (MLEF) for more accurate estimates of the CSM while reducing the computational complexity. MLEF is an efficient Bayesian method developed by Zupanski [8].

II. MATERIALS AND METHODS

A. Shear Wave Propagation

The stimulating needle vibrates along the vertical (z) axis. The vibration frequency f_0 is chosen in the range from 50 to 150 Hz. This vibration creates a cylindrical shear wave that propagates for several millimeters. The velocity of the shear wave can be tracked using an array transducer system with a high sampling frequency (e.g., 20 Mega-samples/s).

For a totally elastic medium, the shear wave velocity is given by

$$c_s = \sqrt{\mu/\rho} \tag{1}$$

where μ is the elastic shear constant and ρ is the density of the medium.

For a viscoelastic medium, we have to apply Kelvin-Voigt rheological model [5] in which the CSM is defined as

$$\mu = \mu_1 - i\omega_0\eta, \tag{2}$$

where ω_0 is the vibration angular frequency in rads/s ($\omega_0 = 2\pi f_0$), μ_1 is the elastic shear constant and η is the dynamic viscous constant defined the Kelvin-Voigt model.

The complex wavenumber for viscoelastic media is given by

$$k_s' = \sqrt{\rho\omega^2/\mu} \tag{3}$$

V. Van Toi et al. (Eds.): 4th International Conference on Biomedical Engineering in Vietnam, IFMBE Proceedings 40, pp. 313–316, 2013.
www.springerlink.com

Since k_s' is complex, it can be written as

$$k_s' = k_s - i\alpha, \quad (4)$$

where α is the shear wave attenuation coefficient. From (3) and (4), by estimating k_s and α, we can obtain the CSM.

Under a cylindrical assumption and with certain vibration frequency f_0, the particle velocity $v(r,t)$ is spatiotemporal function of the tracking location r and time t, obtained by

$$v(r,t) = \frac{1}{\sqrt{r-r_0}} A e^{-\alpha(r-r_0)} \cos[\omega t - k_s(r-r_0) - \phi], \quad (5)$$

where r_0 is the needle position, A is the amplitude of source excitation, and φ is the initial temporal phase.

B. Proposed Estimation of CSM Using MLEF

Now, we will apply the MLEF to estimate α and k_s; hence, the CSM. First, a discrete formulation of the particle velocity in (5) is given by

$$v_n(r) = v_{n-1}\cos(\omega\Delta t) - \frac{1}{\sqrt{r-r_0}} A e^{-\alpha(r-r_0)} \times \sin[\omega(n-1)\Delta t - k_s(r-r_0) - \phi]\sin(\omega\Delta t), \quad (6)$$

where n is the discrete time index and Δt is sampling cycle.

Then, by tracking the velocity at each spatial location, the state equations can be constructed from Eq. (6), as shown below:

$$x = \begin{bmatrix} \boldsymbol{v}_n \\ \boldsymbol{\theta}_n \end{bmatrix} = \begin{bmatrix} \mathcal{F}(\boldsymbol{v}_{n-1}, \boldsymbol{\theta}_{n-1}) \\ \boldsymbol{\theta}_{n-1} \end{bmatrix} \quad (7)$$

where parameter $\boldsymbol{\theta}_n$, to be estimated, is defined as, $\boldsymbol{\theta}_n = [\boldsymbol{\alpha}^T, \mathbf{k}_s^T, \phi, r, A]^T$, $\mathcal{F}$ is a nonlinear function modeling the spatial shear wave dynamics. The length of vectors $\mathbf{v}_n$, $\boldsymbol{\alpha}$ and $\mathbf{k}_s$ equals to the number of spatial locations. We can assume that $\boldsymbol{\theta}_n$ would not be changed during the time of the experiment; hence, $\boldsymbol{\theta}_n = \boldsymbol{\theta}_{n-1}$ as shown in (7).

During the experiment, we can only measure the velocities at every spatial location. Under ideal Doppler imaging, the measurements are given by

$$\boldsymbol{y}_n = \begin{bmatrix} \hat{\boldsymbol{v}}_n \\ \mathbf{0} \end{bmatrix} = [\boldsymbol{I} \quad \boldsymbol{O}] \begin{bmatrix} \boldsymbol{v}_n \\ \boldsymbol{\theta}_n \end{bmatrix} + \begin{bmatrix} \boldsymbol{\varepsilon} \\ 0 \end{bmatrix} \quad (8)$$

where $\boldsymbol{\varepsilon}$ is the measurement noise is zero-mean Gaussian with variance R.

Algorithm 1. MLEF-based CSM estimation.

Initialization
Initialize state x_0 and error square-root covariance matrix $\boldsymbol{P}_0^{1/2}$
$n = 0$
While $n < N_{\max}$ **do**
Forecast
1. Estimate state vector $\boldsymbol{x}$ using (7)
2. Calculute error square-root covaricance matrix $\boldsymbol{P}^{1/2}$ using (9)
Analysis
3. Calculate observed perturbation matrix $\boldsymbol{Z}$ using (10)
4. Update state vector $\boldsymbol{x}_+$ using (11)
5. Update error square-root covariance matrix $\boldsymbol{P}_+^{1/2}$ (12)
% prepare for the next loop
$n = n + 1$
$\mathbf{x} = \mathbf{x}_+$ and $\boldsymbol{P}^{1/2} = \boldsymbol{P}_+^{1/2}$
End while

The MLEF consists of two main steps: forecast and analysis, corresponding to prediction and correction in Kalman filtering. The forecast step estimates the current state vector $\mathbf{x}$ according to (7), and the error square-root covariance matrix $\boldsymbol{P}^{1/2}$ of size $N \times s$, according to

$$\boldsymbol{P}^{1/2} = \begin{bmatrix} \mathcal{F}(\boldsymbol{x}+\boldsymbol{p}_1) - \mathcal{F}(\boldsymbol{x}) \\ \mathcal{F}(\boldsymbol{x}+\boldsymbol{p}_2) - \mathcal{F}(\boldsymbol{x}) \\ \cdots \\ \mathcal{F}(\boldsymbol{x}+\boldsymbol{p}_s) - \mathcal{F}(\boldsymbol{x}) \end{bmatrix}^T \quad (9)$$

where $\mathbf{p}_1,\ldots,\mathbf{p}_s$ are the columns of the previous error square-root covariance matrix, N is the length state vector $\mathbf{x}$, s is the number of ensembles (generally, $s = N$).

The analysis step updates the state vector and the error square-root covariance matrix. To do so, we first calculate the observed perturbation matrix $\mathbf{Z}$ by

$$\boldsymbol{Z} = \boldsymbol{R}^{1/2} \begin{bmatrix} \mathcal{H}(\boldsymbol{x}+\boldsymbol{p}_1) - \mathcal{H}(\boldsymbol{x}) \\ \mathcal{H}(\boldsymbol{x}+\boldsymbol{p}_2) - \mathcal{H}(\boldsymbol{x}) \\ \cdots \\ \mathcal{H}(\boldsymbol{x}+\boldsymbol{p}_s) - \mathcal{H}(\boldsymbol{x}) \end{bmatrix}^T \quad (10)$$

where functions as an ideal Doppler imaging system, and thus $\mathcal{H} = [\boldsymbol{I} \quad \mathbf{0}]$. Then, the state vector is updated by

$$\boldsymbol{x}_+ = \boldsymbol{x} + \boldsymbol{P}^{1/2}[\boldsymbol{I} + \boldsymbol{Z}^T\boldsymbol{Z}]^{-1}\boldsymbol{Z}^T[\boldsymbol{R}^{-1/2}(\boldsymbol{y} - \mathcal{H}(\boldsymbol{x}))] \quad (11)$$

and the error square-root covariance matrix is updated by

$$\boldsymbol{P}_+^{1/2} = \boldsymbol{P}^{1/2}[\boldsymbol{I} + \boldsymbol{Z}^T\boldsymbol{Z}]^{-1/2} \quad (12)$$

Since the number of ensembles is far fewer than the length of the state vector ($s = N$), the computational complexity is reduced as compared to Kalman filtering. Our scheme can be summarized in Algorithm 1. In this algorithm, $N_{\max}$ is the length of the sampled velocities at a spatial location.

III. RESULTS AND DISCUSSIONS

A. Homogeneous Medium - Experimental Data

We first test the proposed algorithm using experimental data from homogeneous tissue-like gelatin. A needle with diameter of 1.5 mm was set to vibrate along the vertical axis, by using a mechanical actuator with excitation frequency $f_0 = 100$ Hz. The sampling frequency is 10 kHz. After sampling, the data were smoothed using a 5th order moving-average filter. The data were collected over 43 spatial locations and tracked by an array transducer with 0.46 mm/pitch. Thus, the total length of the radial line is 19.32 mm. The first spatial location, r_0, was assumed to be close to the needle; $r_0 = 0.4$ mm. The values of the amplitude, A, and the phase, ϕ, were estimated using the first cycle of the particle velocity at the first spatial location.

These parameter were then used for initializing the initial state vector of the MLEF. The initial error square-root covariance matrix was generated randomly. After only tens of iterations, the velocity was denoised and the estimates of the attenuation and shear wave number quickly converged toward the control values $\alpha = 37.5\text{Npm}^{-1}$ and $k_s = 776\text{m}^{-1}$, as shown in Figs. 1 and 2. These figures show that the CSM is approximately constant with spatial locations. It agrees with the experimental condition. Consequently, we can estimate the elastic modulus, $\mu_1 = 650$ Pa, and the dynamic viscosity, $\eta = 0.1$ Pa/s.

Fig. 3 shows the Dopler estimates of particle velocities in galentin as compared to MLEF's ones. It is obvious that MLEF filtered the noise out of the desired signal. Similarly, the shear wave was also de-noised, as shown in Fig. 4.

B. Heterogeneous Medium - Simulated Data

For heterogeneous materials, a simulated scenario was created wherein we inserted an inclusion into the object then tested the ability of inclusion detection. The inclusion was set to be from spatial location 15 to 23 along the radial line. The needle with diameter of 1.5 mm was set to vibrate with frequency $f_0 = 100$ Hz. The data were also collected over 43 spatial locations, the total length of the radial line is 19.32 mm, and the first spatial location r_0 is 0.4 mm. For the surrounding environment, $\mu_1 = 650$ Pa and $\eta = 0.1$ Pa/s.

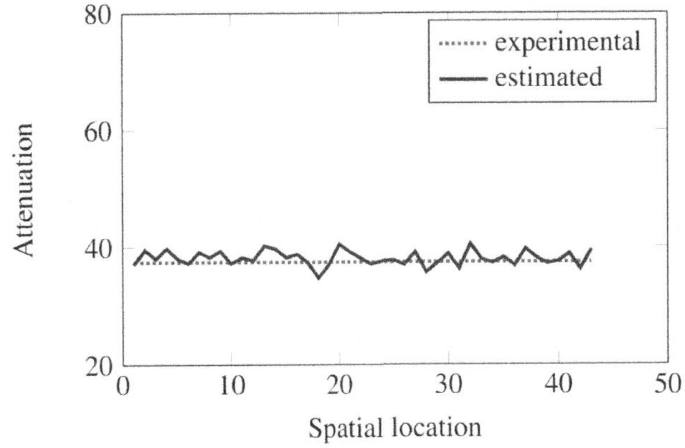

Fig. 1 Shear attenuation vs. spatial location.

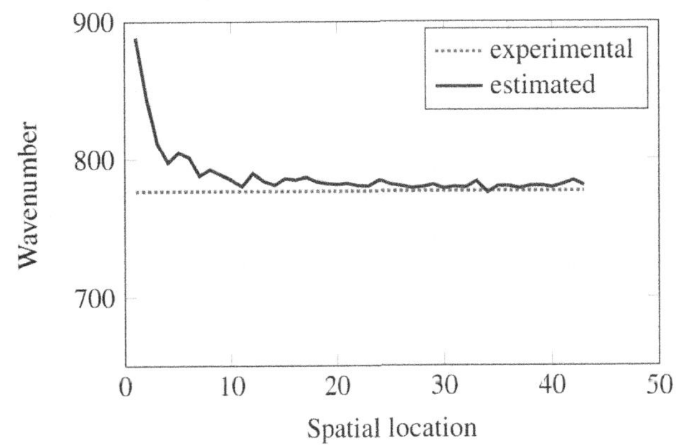

Fig. 2 Shear wavenumber vs. spatial location.

For the inclusion, $\mu_1 = 800$ Pa and $\eta = 0.2$ Pa/s. These values formed the window functions of the attenuation and the wavenumber as shown by the dotted lines in Figs. 5 and 6. The simulated results in these figures indicate that the inclusion was successfully detected.

IV. CONCLUSIONS

We have shown that using the MLEF can improve the ability of ultrasonic shear wave imaging in assessing mechanical properties of soft tissues. Our method only requires a single vibration frequency to accurately estimate the CSM. In addition, it extends the algorithm proposed in [3] by taking into account the location-varying characteristics of the attenuation and shear wave number. It performs well in both homogeneous tissue-like materials and materials with inclusions. This will facilitate clinical diagnoses such as cancer detection.

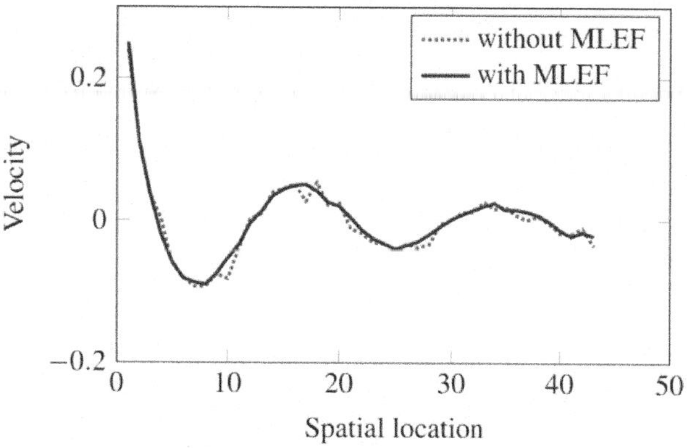

Fig. 3 Particle velocity vs. spatial location.

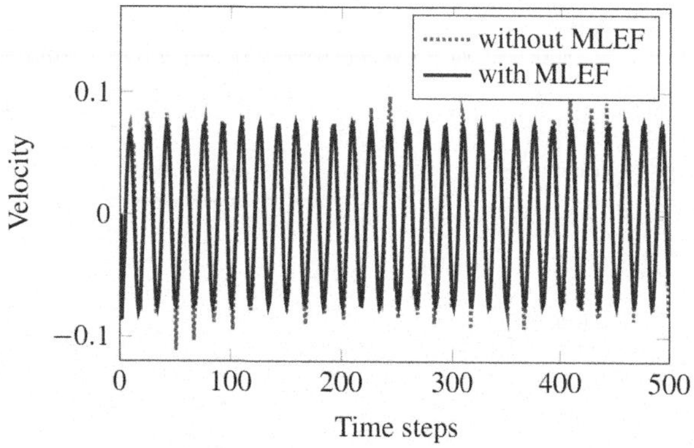

Fig. 4 Particle velocities at the 10th location.

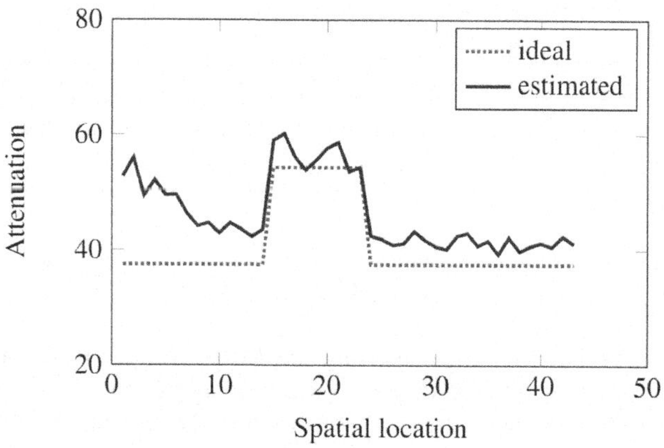

Fig. 5 Shear wave attenuation vs. spatial location.

Acknowledgment

This work was supported by Project QG.10.40 granted by Vietnam National University Hanoi.

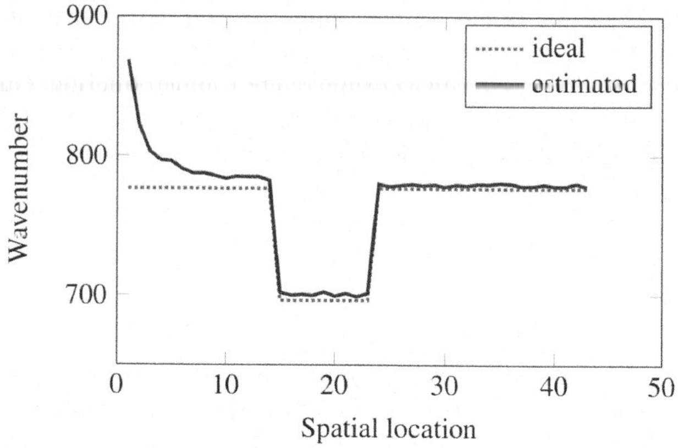

Fig. 6 Shear wavenumber vs. spatial location.

References

1. Chen S, Fatemi M, Greenleaf JF (2004). Quantifying elasticity and viscosity from measurement of shear wave speed dispersion. Journal of Acoustic Soc Am, 115:2781-2785.
2. Chen S, Fatemi M, Greenleaf JF (2007). Real time shear velocity imaging using sonoelastograpgic techniques. Ultrasound Medical Biology, 115:2781-2785.
3. Nightingale K, McAleavey S, Trahey G (2003). Shear-wave generation using acoustic radiation force: in vivo and ex vivo results. Ultrasound in Medicine and Biology, 29:1715-1738.
4. Kanai MH (2009). Propagation of vibration caused by electrical excitation in the normal human heart. Ultrasound Medical Biology, 35:936-948.
5. Orescanin M, Insana MF (2011). 3-D FDTD simulation of shear waves for evaluation of complex modulus imaging. IEEE Trans. Ultrasonics, Ferroelectrics, and Frequency Control, 58:389-398.
6. Orescanin M, Insana MF (2010). Model-based complex shear modulus reconstruction: A Bayesian approach. IEEE Int'l Ultrasonics Symposium, 61-64.
7. Orescanin M, Insana MF (2010). Shear Modulus Estimation With Vibrating With Needle Stimulation. IEEE Trans. Ultrasonics, Ferroelectrics, and Frequency Control, 57:1358-1367.
8. Zupanski, M (2005). Maximum Likelihood Ensemble Filter: Theoretical Aspects. Monthly Weather Review, 133:1710-1726

Author: Tran Duc-Tan
Institute: VNU University of Engineering and Technology
Vietnam National University Hanoi (VNU)
Street: E3, 144 Xuan Thuy, Cau Giay
City: Hanoi
Country: Vietnam
Email: tantd@vnu.edu.vn

Separation of Nonstationary EEG Epileptic Seizures Using Time-Frequency-Based Blind Signal Processing Techniques

Nguyen Thi Thuy-Duong[1], Nguyen Linh-Trung[1], Tan Tran-Duc[1], and Boualem Boashash[2,3]

[1] Fac. Electronics & Telecommunications, VNU University of Engineering & Technology, Vietnam
[2] Perinatal Research Centre, University of Queensland, Australia
[3] College of Engineering, Qatar University, Qatar

Abstract — **Epilepsy is a neural disorder in which the electrical discharge in the brain is abnormal, synchronized and excessive. Scalp Electroencephalogram (EEG) is often used in the diagnosis and treatment of epilepsy by examining the epileptic seizures and epileptic spikes. By modeling the signal acquired at each electrode of the EEG measurement system as a linear combination of source signals generated in the brain, we can apply Blind Source Separation (BSS) techniques to separate the seizures from other signals. Alternating Columns - Diagonal Centers (AC-DC) and Second-Order-Blind Identification (SOBI) are well-known BSS algorithms and have been previously applied to the separation of seizures. However, the seizure signals in new-born babies exhibit nonstationary second order statistics. In this paper, we concentrate on applying two time-frequency (TF) based algorithms: TF-SOBI and TF-UBSS to seizure separation. These algorithms are more appropriate for analyzing nonstationary signals and have not been previously applied to studies of EEG-based seizures.**

Keywords — **epileptic seizures, EEG, nonstationary sources, time-frequency representations, under-determined blind separation.**

I. Introduction

Epilepsy is a neural disorder characterized by an enduring predisposition to generate epileptic seizures and its neurobiologic, cognitive, psychological, and social consequences. An epileptic seizure is the abnormal, synchronized and excessive electrical discharge in the brain [1]. Scalp electroencephalogram (EEG) is the recording of the temporal electrical brain activity through a set of scalp electrodes, thus it is useful to localize epileptogenic zones.

However, the EEG-based epileptic diagnosis faces difficulty because EEG signals are often disturbed by artifacts, such as: eye movements, muscle activity and heart activity. For new-born babies, it is even more difficult because the baby skull is too small for normal EEG measurement set-ups, hence fewer number of electrodes are used.

Each recorded EEG signal can be modeled as a linear combination of source signals generated in the brain (seizures, background neural activity, artifacts, etc.) [2]. Hence, it is appropriate to we can apply blind source separation (BSS) techniques to separate the seizure from other signals in the EEG data. BSS aims at recovering several source signals from several observed only linear mixtures of the source signals while the information about the mixing system is unavailable. Assumptions for BSS to work depend on the mixing model and statistical properties of the source signals. Well-known algorithms such as INFOMAX and FastICA are based on the assumption that the sources are independent and there is at most one Gaussian source.

The assumption of independence among source signals can be relaxed to uncorrelation while using additional information about the source autocorrelation. Two well-known BSS algorithms using these assumptions are Alternating Columns - Diagonal Centers (AC-DC) [3] and Second-Order-Blind Identification (SOBI) [4]. Thanks to using only second order statistic (SOS) information, the algorithm complexity and signal length can be reduced. These algorithms have been previously applied to EEG seizure separation.

The seizure signals in new born babies, however, exhibit a nonstationary SOS, unlike those in adults. It means that the power spectral densities (PSDs) of the source signals vary in time. In such cases, AC-DC or SOBI are not appropriate for seizure separation because they only exploit the time average of PSD. To overcome this problem of nonstationarity, in this paper, we apply two other time-frequency (TF) based BSS algorithms: TF-SOBI [5] and TF-UBSS [6]. Time-frequency signal processing provide tools to analyze nonstationary signals. The latter algorithm was designed for the "underdetermined" scenarios wherein the number of source signals is higher than the number of electrodes.

II. Materials and Methods

A. EEG data

Within the scope of this research, we study two epileptic EEG datasets (simulated and real) disturbed by an eye-blink artifact, which is the most popular factor of distortion in the EEG recording process. The first one contains four

V. Van Toi et al. (Eds.): 4th International Conference on Biomedical Engineering in Vietnam, IFMBE Proceedings 40, pp. 317–323, 2013.
www.springerlink.com

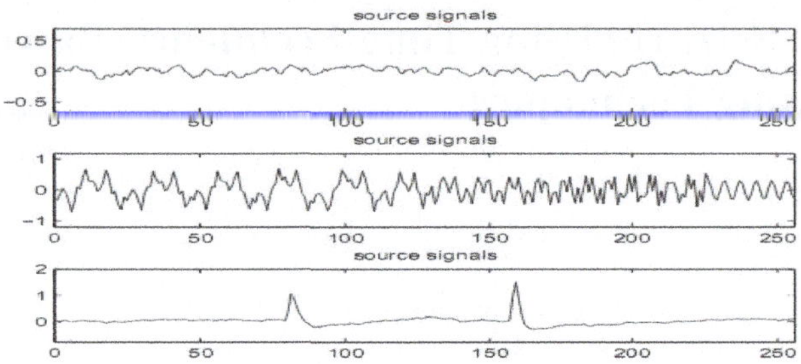

Fig. 1 The first data set: Simulated source signals in the time domain. Top: EEG background. Middle: seizure. Bottom: eye-blink artefact.

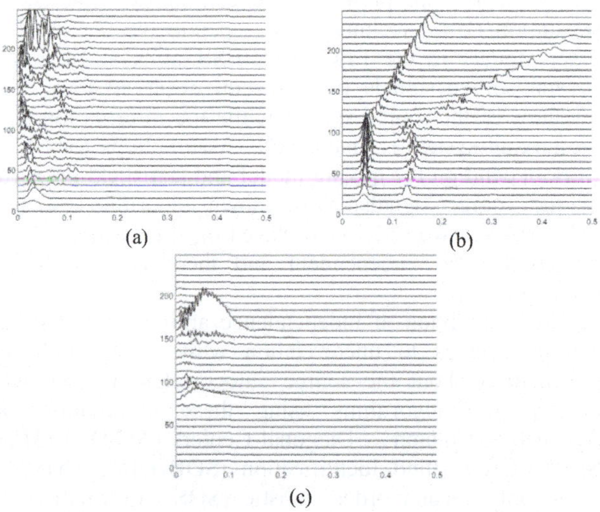

Fig. 2 The first data set: Simulated source signals in the TF domain. Top-left: EEG background. Top-right: seizure. Bottom: eye-blink artefact.

simulated EEG signals each of which is a mixture of three sources: EEG background, eye-blink artifact and seizure.

The EEG background and the seizure are generated using MATLAB following [7]. Figs. 1 and 2 show these signals in the time domain and TF domain, respectively. We can clearly see how the frequency of the seizure varies in time.

The second dataset includes 13 channel outputs of EEG recorded on a four-day-old new-born baby having epilepsy. Fig. 3 shows these signals in time, of duration 500 samples. The seizure occurs evidently in channels 2 and 11 and can also be seen in some other channels, e.g., 3 and 9). The eye artifact appears near sample 250 of channels 4, 5 and 13.

B. BSS algorithms base on TF methods

a) BSS methods and TF distribution

Suppose that we have n source signals and m received signals. The source and received signal vectors are denoted by $\mathbf{s}(t) = [s_1(t), \ldots, s_n(t)]^T$ and $\mathbf{x}(t) = [x_1(t), \ldots, x_m(t)]^T$ where $(\cdot)^T$ denotes the transpose

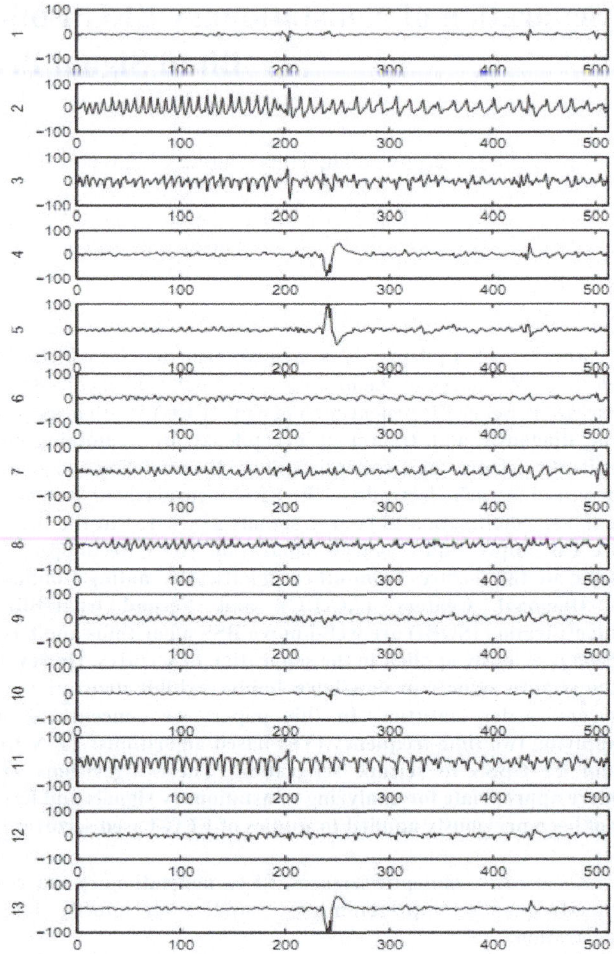

Fig. 3 The real EEG dataset.

operator. Under the model of instantaneous linear mixture, the received signals are often expressed by:

$$\mathbf{x}(t) = \mathbf{A}\mathbf{s}(t) + \boldsymbol{\eta}(t) \quad (1)$$

where $\mathbf{A} = [\mathbf{a}_1, \mathbf{a}_2, \ldots, \mathbf{a}_n]$ is the mixing matrix and $\boldsymbol{\eta}(t)$ is additive noise vector. The goal of BSS is to estimate $\mathbf{s}(t)$ from the given $\mathbf{x}(t)$.

As the above introduction, our algorithms are based on the SOS information, thus, we need three assumptions to make it possible: (i) the source signals are mutually uncorrelated or their cross-correlations are equal to zero, (ii) the additive noise $\boldsymbol{\eta}(t)$ is stationary, white, and zero mean, and (iii) the $m \times n$ mixing matrix $\mathbf{A}$ is assumed to have full column rank but is otherwise unknown.

To deal with nonstationary EEG signals, the signal model in (1) can be extended by incorporating the TF distribution (TFD) in order to exploit information of a signal in the time domain and the frequency domain simultaneously. Further, to deal with vectors of signals, such as $\mathbf{x}(t)$, meaning to combine with the spatial information, it is necessary to extend the TFD to spatial TFD (STFD), which is defined as below:

$$\mathbf{D}_{\mathbf{xx}}(t,f) \triangleq \begin{bmatrix} D_{x_1x_1}(t,f) & \cdots & D_{x_1x_m}(t,f) \\ \vdots & \ddots & \vdots \\ D_{x_mx_1}(t,f) & \cdots & D_{x_mx_m}(t,f) \end{bmatrix} \quad (2)$$

where $D_{x_ix_j}(t,f)\ (i,j = 1,\ldots,m)$ are cross-TFD of $x_i(t)$ and $x_j(t)$. When $i = j$, it becomes auto-TFD of one signal.

By applying STFD to both sides of (1) with assumption of a noise-free environment, we obtain the following equation:

$$\mathbf{D}_{\mathbf{xx}}(t,f) = \mathbf{A}\mathbf{D}_{\mathbf{ss}}(t,f)\mathbf{A}^H \quad (3)$$

where $\mathbf{D}_{\mathbf{ss}}(t,f)$ and $\mathbf{D}_{\mathbf{xx}}(t,f)$ are the source and signal STFD matrices, respectively. Note that the size of $\mathbf{D}_{\mathbf{xx}}(t,f)$ is $m \times m$, and of $\mathbf{D}_{\mathbf{ss}}(t,f)$ is $n \times n$.

In the TF domain, cross-terms occur when calculating the auto-terms by using TFD bilinear operation. It does not represent the "true" energy concentration of the signal. Thus, the efficiency of TF based algorithms will depend on the kind of TFDs used. A common TFD is the Short Time Fourier Transform (STFT), defined for signal $s(t)$as below

$$\rho_{\text{spec}}(t,f) = \left|\int_{\infty}^{\infty} s(\tau)\, h(\tau - t) e^{-j2\pi f\tau} d\tau\right|^2 \quad (4)$$

where $h(\tau - t)$ is the window function. The advantage of the STFT is no cross-terms, but its drawback is the low energy resolution. Another common kind of TFD is the Wigner Ville Distribution (WVD)

$$W(t,f) = \int_{-\infty}^{\infty} s\left(t + \frac{\tau}{2}\right) s^*(t - \frac{\tau}{2}) e^{-j2\pi f\tau} d\tau \quad (5)$$

In contrast to STFT, WVD has a high energy resolution but the cross-term will appear. To utilize the both advantages of STFT and WVD, one can use another TFD called the Masked Wigner Ville Distribution (MWVD) defined as below

$$\rho_{\text{mwvd}}(t,f) = \rho_{\text{spec}}(t,f).W(t,f) \quad (6)$$

In the case of the multi-signal, we use two concepts: auto-source and cross-source points [6]. An auto-source point is the point at which only source presents, and a cross-source point is at which several source signals overlap.

b) TF-based BSS algorithms

This section is focus on two TF-based algorithms, TF-SOBI [5] and TF-UBSS [6]. First, we describe SOBI as the fundamental algorithm to develop TF-SOBI.

After whitening the signals by matrix $\widehat{\mathbf{W}}$, SOBI computes the joint diagonalizer $\widehat{\mathbf{U}}$ of a set of covariance matrices $\{\underline{\widehat{\mathbf{R}}}(\tau_j)|j = 1,\ldots,K\}$. Due to the whitening process, $\widehat{\mathbf{U}}$ is unitary; then the mixing matrix can be calculated by the multiplication of pseudo-inverse of the whitening matrix with the diagonalizing matrix $\widehat{\mathbf{A}} = \widehat{\mathbf{W}}^{\#}\widehat{\mathbf{U}}$. Finally, the source signals are estimated as $\hat{\mathbf{s}}(t) = \widehat{\mathbf{A}}^{\#}\mathbf{x}(t)$.

Algorithm 1: TF-SOBI algorithm

1. Estimate the covariance matrix $\widehat{\mathbf{R}}_{\mathbf{x}}$
2. Estimate the whitening matrix $\mathbf{W}$
3. Whiten the observed signals $\mathbf{z}(t) = \mathbf{W}\mathbf{x}(t)$
4. Calculate the STDF of $\mathbf{z}(t)$
5. Choose K auto-source points
6. Calculate the joint diagonalizer $\mathbf{U}$ of K STFD matrices
7. Calculate the mixing matrix $\mathbf{A} = \mathbf{W}^{\#}\mathbf{U}$
8. Estimate the source signals $\mathbf{s}(t) = \mathbf{U}^{\#}\mathbf{W}\mathbf{x}(t)$

Algorithm 2: TF-UBSS algorithm

1. Compute STFD matrices and reduce the noise using appropriate threshold
2. Select auto-source points
3. Cluster auto-source points to obtain TF source signatures
4. Estimate source TFDs based on source signatures
5. Synthesize the source waveforms based on their TFDs

The principle of TF-SOBI algorithm is the same as SOBI. However, instead of diagonalizing a set of covariance matrices, TF-SOBI jointly diagonalizes the set of STFD matrices at the auto-source points. The main steps of TF-SOBI are summarized in Algorithm 1.

TF-UBSS has some attractive features and could be applied in the under-determined case, when there are more source signals than observed signals. After separating the auto-source TF points from cross-source TF points, for each obtained auto-source point, we compute the main eigen-vectors of the auto-source STFD matrix, then clustering these vectors into different classes; each class contains the auto-source points of only a single source signal. Next, by using the eigenvalues of their STFD matrices, we estimate the TFD of each source signal. Finally, the source signal is then recovered from its TFD estimate using TF synthesis. The main steps of TF-UBSS are described in Algorithm 2. Depend on the kind of STFD used in the algorithm, the details of each process can be changed appropriately.

C. Application of TF-SOBI and TF-UBSS for seizure separation

a) TF-SOBI for seizure separation

From [7], newborn EEG seizures are generally characterized as nonstationary, repetitive, rhythmical spike activity which is often multifocal with slow migration across the brain. Besides, there is a significant overlap of EEG signals in the TF domain. Hence, we need some adjustments to overcome these specific limitations when applying TF-SOBI and TF-UBSS algorithms to the separation process.

The TF-SOBI algorithm proposed in [5] does not mention the kind of TFD used in STFD calculation and neither how to choose the auto-source points. In our simulations, TF-SOBI computes STFD using MWVD due to its advantages in both resolution and cross-term elimination. After the whitening step, the STFD can be represented as:

$$\mathbf{D}_{\mathbf{zz}}^{\mathrm{mwvd}}(t,f) = \mathbf{U}\mathbf{D}_{\mathbf{ss}}^{\mathrm{mwvd}}(t,f)\mathbf{U}^H \tag{7}$$

To choose auto-source points for each time-line (t_j, f) in the TF domain, we retain the points satisfying these following conditions:

$$\frac{\max_f \|\mathbf{D}_{\mathbf{zz}}^{\mathrm{mwvd}}(t_j,f)\|}{\max\|\mathbf{D}_{\mathbf{zz}}^{\mathrm{mwvd}}\|} > \epsilon_1 \tag{8}$$

$$\frac{\|\mathbf{D}_{\mathbf{zz}}^{\mathrm{mwvd}}(t_j,f_k)\|}{\max_f \|\mathbf{D}_{\mathbf{zz}}^{\mathrm{mwvd}}(t_j,f)\|} > \epsilon_2 \tag{9}$$

Condition (8) eliminates the time-line which does not contains any auto-source point. Condition (9) is used to pick up the auto-source points in each remaining time-line. In the simulations, two above thresholds are set as $\epsilon_1 = 0.1$ and $\epsilon_2 = 0.6$.

b) TF-UBSS for seizure separation

In this research, we use STFT to compute the STFD matrix because of its low computational cost. Equation (3) can be rewritten as:

$$\mathbf{D}_{\mathbf{x}}(t,f) = \mathbf{A}\mathbf{D}_{\mathbf{s}}(t,f) \tag{10}$$

where $\mathbf{D}_{\mathbf{x}}(t,f)$ and $\mathbf{D}_{\mathbf{s}}(t,f)$ are the STFT vectors of source and observed signals, respectively.

Because the STFD matrix of an auto-source point is diagonal, it can be represented by the principal eigenvector of the mixing matrix and the source signal auto-TFD. Thus, its STFT vector can be reduced as

$$\mathbf{D}_{\mathbf{x}}(t,f) = \mathbf{a}_i \mathbf{S}_{\mathbf{a}_i}(t,f), \quad \forall (t_a,f_a) \in \Omega_i \tag{11}$$

where $\mathbf{a}_i$ is the ith column of the mixing matrix $\mathbf{A}$, and Ω_i is the ith class of the corresponding auto-source point $(i = 1, \dots, n)$. Equation (11) shows that point $(t_a, f_a) \in \Omega_i$ has

$$\frac{\mathbf{D}_{\mathbf{x}}(t_a,f_a)}{\|\mathbf{D}_{\mathbf{x}}(t_a,f_a)\|} = \frac{\mathbf{a}_i}{\|\mathbf{a}_i\|} \tag{12}$$

Let $\mathbf{a}(t_a, f_a) = \frac{\mathbf{D}_{\mathbf{x}}(t_a,f_a)}{\|\mathbf{D}_{\mathbf{x}}(t_a,f_a)\|}$ is a vector which characterizes the source holding (t_a, f_a). Next, we cluster these vectors into n classes using the common clustering algorithm K-means [8]. After that, the mixing matrix $\mathbf{A}$ can be estimated [6]. Each column of $\mathbf{A}$ is the centroid of the corresponding class. Then, the source signals can be calculated using this mixing matrix.

To select the auto-source point, we use the statistical method introduced in [9]. In the TF domain, we consider a T length window in which the signals is assumed to be stationary, thus, the equation (1) becomes

$$\mathbf{X}(t,f) = \mathbf{A}(f)\mathbf{S}(t,f) + \boldsymbol{\eta}(t,f) \tag{13}$$

The covariance matrix of each point in the TF domain is computed as:

$$\mathbf{R}_x(t,f) = \mathbf{A}(f)\mathbf{R}_s(t,f)\mathbf{A}^H(f) + \mathbf{R}_\eta(t,f) \tag{14}$$

If there is a single m^{th} source in this window, Equation (14) can be rewritten as

$$\mathbf{R}_{xm}(f) = \boldsymbol{a}_m(f)\boldsymbol{a}_m^H(f)\boldsymbol{\sigma}_{sm}^2(f) + \sigma_\eta^2 \mathbf{I}_N \tag{15}$$

where $\boldsymbol{a}_m(f)$ is the mth column of matrix $\mathbf{A}(f)$ and $\boldsymbol{\sigma}_{sm}^2(f)$ is the energy spectrum distribution of the mth source. Hence, $\mathbf{R}_{xm}(f)$ has a rank of 1 and $\boldsymbol{a}_m(f)$ is proportional to the eigenvector of $\mathbf{R}_{xm}$ corresponding to its maximum eigenvalue $\beta_{1,m}$:

$$\beta_{1,m}(f) = \boldsymbol{\sigma}_{sm}^2(f) \parallel \boldsymbol{a}_m^H(f) \parallel^2 + \sigma_\eta^2 \tag{16}$$

This property is used to verify if a point is auto-source or not based on the following hypotheses

$$\begin{aligned} &H_0: \mathbf{X}(t,f) \sim N(0, \sigma_\eta^2 \mathbf{I}_N) \\ &H_1: \mathbf{X}(t,f) \sim N(0, u^H(f)\sigma_s^2(f)\sigma_\eta^2 \mathbf{I}_N) \\ &H_2: \mathbf{X}(t,f) \sim N(0, \mathbf{R}_x(f)) \end{aligned} \tag{17}$$

where H_0, H_1, and H_2 are noise, auto-source and cross-source, respectively. Two thresholds are selected to classify the TF points as following

$$\mathbf{T}_1 = \left(\frac{\beta_1}{\sigma_\eta^2} - \log\frac{\beta_1}{\sigma_\eta^2}\right) \underset{H_0}{\overset{H_1}{\gtrless}} \gamma_1 \tag{18}$$

$$\mathbf{T}_2 = \sum_{i=2}^{N}\left(\frac{\beta_1}{\sigma_\eta^2} - \log\frac{\beta_1}{\sigma_\eta^2}\right) \underset{H_2}{\overset{H_1}{\gtrless}} \gamma_2 \tag{19}$$

Equation (18) is used to eliminate noise and retain the points having high concentrated energy. Equation (19) is used to choose the auto-source from the remaining points of (18).

III. RESULTS AND DISCUSSIONS

In this section, we compare the results of three algorithms: SOBI, TF-SOBI, and TF-UBSS when applying to two data sets and obtain some attracting features.

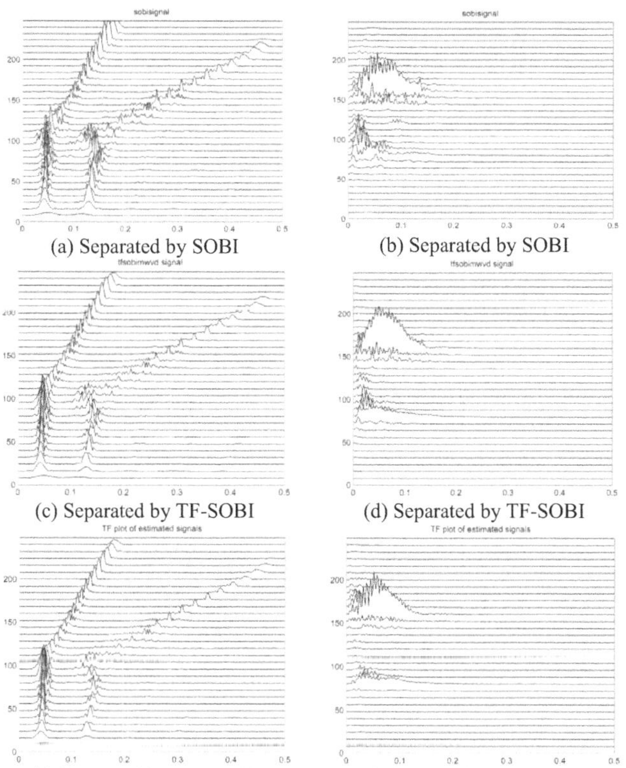

(a) Separated by SOBI (b) Separated by SOBI

(c) Separated by TF-SOBI (d) Separated by TF-SOBI

(e) Separated by TF-UBSS (f) Separated by TF-UBSS

Fig. 4 The results of TF-BSS algorithms in comparation with the original simulated EEG signals.

A. Simulated EEG data

Simulated EEG dataset consists of four EEG signals mixed from three sources. Our target is estimating these sources by SOBI, TF-SOBI, and TF-UBSS algorithms. Fig. 4 shows their results in sequences. It is difficult to distinguish between the original and estimated signals in the time domain. Thus, we use the TF representation instead of time domain.

From this data set, SOBI algorithm cannot estimates the source signals as good as the others. In the interval from about sample 50 to 200 and the normalized frequency range between 0 and 0.15, there is a large overlapping spectrum of the seizure and the eye artifact. In this region, the separated signals using SOBI shows attenuation in amplitude, while TF-SOBI and TF-UBSS can offer good recovery. This figure demonstrates that the SOBI method does not yield the good results in the case of nonstationary.

In spite of the success in separating the mixed signals, the TF algorithms exist some drawbacks left to further study:

1. The threshold of the TF algorithms needs to be chosen manually. Up to now, there has been not any optimal technique to compute the threshold.

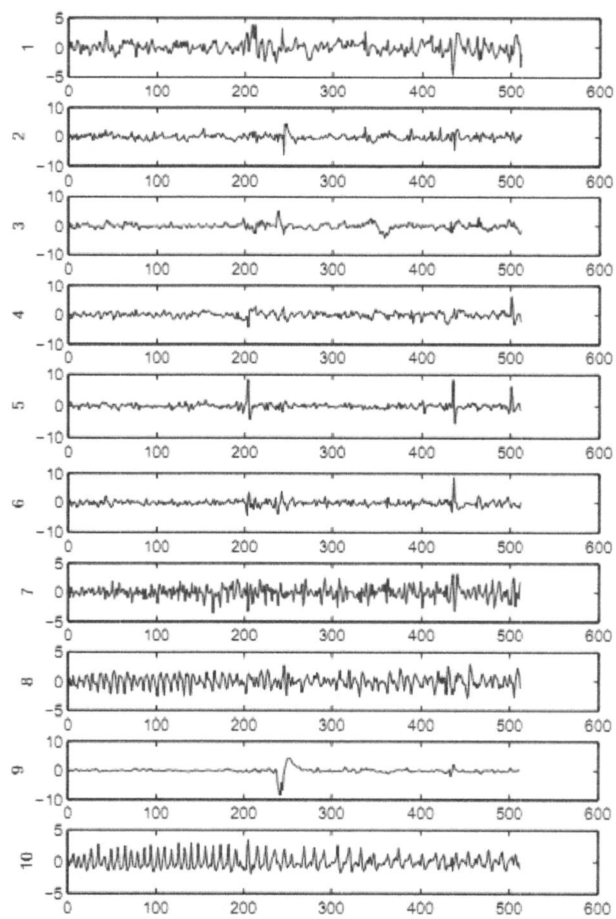

Fig. 5 The results of SOBI algorithm with the real EEG data set.

2. TF-UBSS require a precise estimation of the number of the source signals. In our work, TF-UBSS used K-means for clustering and source estimation. However, K-means is not stable, and then makes TF-UBSS low in efficiency.

B. Real EEG data

The results when applying SOBI, TF-SOBI, TF-UBSS to the real EEG data set are shown in Figs. 5, 6, and 7, respectively. In Fig. 5, we can see that SOBI algorithm separates successfully the eye artifact (channel 9). However, the seizure is not recovered completely (the seizure occurs in both channels 8 and 10). It means that SOBI does not work well with these seizure signals.

In Fig. 6, TF-SOBI shows a better estimated seizure in comparison with SOBI (only in channel 10), but the eye artifact is not (channels 5 and 9).

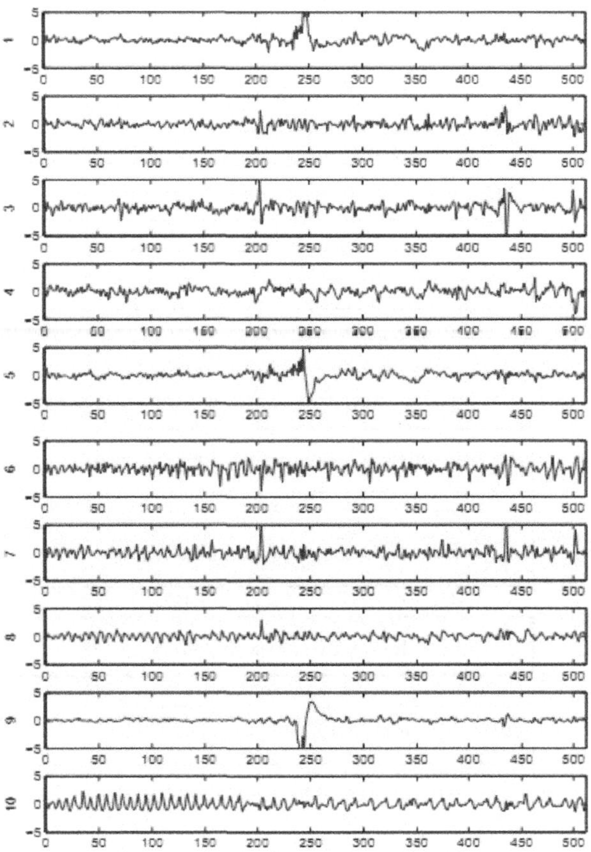

Fig. 6 The results of TF-SOBI algorithm with the real EEG data set.

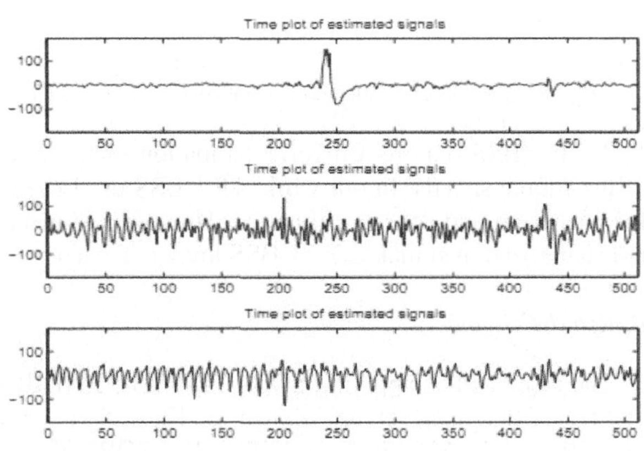

Fig. 7 The results of TF-UBSS algorithm with the real EEG data set.

By choosing the number of sources to be 3, Fig. 7 shows the good estimations of the both seizure and eye artifact in channels 1 and 3, respectively.

IV. Conclusions

This paper presents a new approach for seizure separation using TF, TF-SOBI and TF-UBSS algorithms. These methods, in theoretically, have some considerable features that can be utilized to obtain good separation results.

The simulation results demonstrated that in the case of nonstationary seizure, TF-SOBI and TF-UBSS algorithms show the better performance comparison with SOBI. If the source number can be estimated exactly, TF-UBSS algorithm can outperform TF-SOBI.

However, two proposed methods have some limitations need to be further improved. In the future works, we will integrate the EMD method [10] with these algorithms to overcome their drawbacks.

Acknowledgments

This work was supported by Project QG-10.40 granted by Vietnam National University Hanoi (VNU). The authors greatly thank Paul Colditz and Mostefa Mesbah from the Perinatal Research Centre for the kind provision of the new-born EEG data and other invaluable advice and support. The authors are also graceful Nguyen Van Du and Le Huy Phuong from the VNU University of Engineering and Technology in performing the simulations during their undergraduate studies.

References

1. Fisher RS et al. (2005) Epileptic seizures and epilepsy: definitions proposed by the international league against epilepsy (ILAE) and the international bureau for epilepsy (IBE). Epilepsia, 46:470–472.
2. Sanei S, Chambers J (2007), EEG Signal Processing. John Wiley & Sons.
3. Hesse CW, James CJ (2006) On semi-blind source separation using spatial constraints with applications in EEG analysis. IEEE Trans Biomed. Eng., 53(12):2525–2534.
4. Tang AC, Sutherland MT, McKinney CJ (2004) Validation of SOBI components from high-density EEG. NeuroImage, 25:539-553.
5. Belouchrani A, Amin MG (1998) Blind source separation based on time-frequency signal representations. IEEE Trans Sig Proc, 46(11): 2888–2897.
6. Linh-Trung N, Abed-Meraim K, Belouchrani A, Boashash B (2005) Separating more sources than sensors using time-frequency distributions," EURASIP J Appl Sig Proc, 2005:2828–2847.
7. Rankine L (2005) Newborn EEG Seizure Detection Using Adaptive Time-Frequency Signal Processing. PhD thesis, Queensland U Tech.
8. Gersho A, Gray RM (1992) Vector quantization and signal compression, Springer, Netherlands.

9. Shlomo D, Joseph T, Miki AT (2006). Speech source separation in convolutive environments using space-time-frequency analysis. EURASIP J Appl Sig Proc, 2006(1):1–11.
10. Huang NE et al. (1998) The empirical mode decomposition and the Hilbert spectrum for nonlinear and non-stationary time series analysis. Proc Royal Soc A: Math, Phy & Eng Sci, 454(1971):903– 905.

Author: Nguyen Linh-Trung
Institute: VNU University of Engineering and Technology
Vietnam National University Hanoi (VNU)
Street: E3, 144 Xuan Thuy, Cau Giay
City: Hanoi
Country: Vietnam
Email: nltrung@vnu.edu.vn

Meta-reg: A Computational Metaheuristic Framework to Improve SVM-Based Prediction of Regulatory Activity

Dong Do Duc[1,*], Huan Hoang Xuan[2], and Huy Q. Dinh[3,4,*]

[1] Institute of Information Technology, Vietnam National University, Hanoi, Vietnam
[2] College of Technology, Vietnam National University of Hanoi, Vietnam
[3] Center for Integrative Bioinformatics, Max F. Perutz Laboratories, Vienna
[4] Gregor Mendel Institute of Molecular Plant Biology, Vienna, Austria
dongdoduc@vnu.edu.vn, huy.dinh@univie.ac.at

Abstract — **Gene regulatory activity prediction is an important step to understand which Transcription Factors (TFs) are important for regulation of gene expression in cells. The development of recent high throughput technologies and machine learning approaches allow us to archive this task more efficiently. Support Vector Machine (SVM) has been successfully applied for the case of predicting gene regulatory activity in *Drosophila* embryonic development. Here, we introduce meta-heuristic approaches to select the best parameters for regulatory prediction from transcription factor binding profiles. Experimental results show that our approach outperforms existing methods and the potentials for further analysis beyond the prediction.**

Keywords — **Gene regulation, machine learning, support vector machine, ant colony optimization.**

I. Introduction

Understanding how a gene is regulated is one of the central tasks of molecular biology. Not long time after the discovery of DNA structure, the regulatory information that regulates gene expression patterns was evidenced to appear in a DNA fragment called enhancers or cis-regulatory modules (CRMs) [1]. CRMs contain binding sites to which specific transcriptional factors (TFs) can bind to carry out their regulatory functions. As a result, understanding the CRMs activity is crucial to gain insight into the gene regulatory mechanism.

ChIP (Chromatin Immunoprecipitation) is a popular-used technique for profiling the transcriptional factor bindings in eukaryotes. First, the cross-linking of transcriptional factors and binding DNA are sheared into number of fragments. Then, a specific antibody is added and immunoprecipitated to the protein-DNA complex. It is then to be purified that allowed the separation between DNA and proteins. Last, the ChIP-DNA fragments are used for further quantification steps. Boosted by recent high-throughput data technologies like microarray or next-generation sequencing, it becomes more powerful especially in terms of genomic resolution. Recent work [2] has introduced for the first time a comprehensive experiment of transcriptional factor bindings at time-specific manner in the model organism, *Drosophila melanogaster*. Using tiling microarray incorporated with ChIP, called ChIP-on-chip, they successfully built up high-resolution atlas of mesodermal cis-regulatory modules. Furthermore, they found correlation between expression-specific information of a CRM and its transcriptional factor binding profiles represented by ChIP hybridization signals by computational approach. It is not one of the first evidences but the first at genome-wide scale. Thus, the regulatory code in cells that regulates gene expression patterns can be modeled using computational methods. It has been become a hot branch in computational biology and bioinformatics [1,3].

Machine learning is a widely used method for data prediction based on the computational model learning from the known training data [3]. Among many of available methods, Support Vector Machine (SVM, [4]) is considered the best one in terms of its performance as well as its popularity in applications, especially in biology and medicine [3]. The method uses the kernel function to find the best separation (i.e. hyper-plane) between normally two opposite categories by solving an optimization problem in computer science. Here, SVM was used in [2] SVM to learn the prediction model from the training data, and then apply for inferring the expression category of unseen CRMs.

In SVM application, parameters are very sensitive with the prediction performance. Hence, choosing the best suitable parameters plays an important role in producing a good predictor. Recently, metaheuristic is being used as an efficient method to explore the parameter space in computational learning. The traditional grid search was used in [2] for optimizing parameters to obtain up to 80% accuracy under the leave-one-out cross validation (LOOCV) framework. By searching over a sparse parameter space, it can find good parameters for a specific instance in a quick

* Corresponding authors.

V. Van Toi et al. (Eds.): 4th International Conference on Biomedical Engineering in Vietnam, IFMBE Proceedings 40, pp. 324–327, 2013.
www.springerlink.com

running time. However, the tradeoff is that the accuracy is not very good for some cases especially with those where the expression pattern need to be precisely determined. Hence, the more thoroughly searching is in need to enhance the prediction performance especially with those cases mentioned above.

A genetic algorithm (GA, [5]) is a traditional approach inspired by biological evolution model as an iterative search solution for optimization problem. There is always one population of candidate solutions at the current iteration, and that population is evolved according to the so-called genetic operators like mutation and crossover. The evolution process tends to keep the good solutions with higher probability than the worse ones according to the so-called fitness function. Recently, Ant Colony Optimization (ACO, [6]) is a meta-heuristic technique based on simulating the behaviors of the ant colony via the pheromone trails on the path from the food source to the nest and vice versa. With the incorporation of the reinforcement learning via the pheromones and the random search based on the heuristic information, the ACO algorithms guarantees to find the optimal solution in the definite running time. Hence, GA and ACO are promising for search the parameter space intensively in order to find the best prediction performance.

This paper investigates metaheuristic approaches based on ACO and GA in order to select well-suited parameters that can lead to a better prediction performance. Our approach gained outperformed results (5-10%) in terms of prediction accuracy for well-known dataset. Thus, our results hint that searching the right parameters is necessary in biological prediction and our proposed approach offers an efficient way for this issue.

II. Prediction problem

A. Biological Data

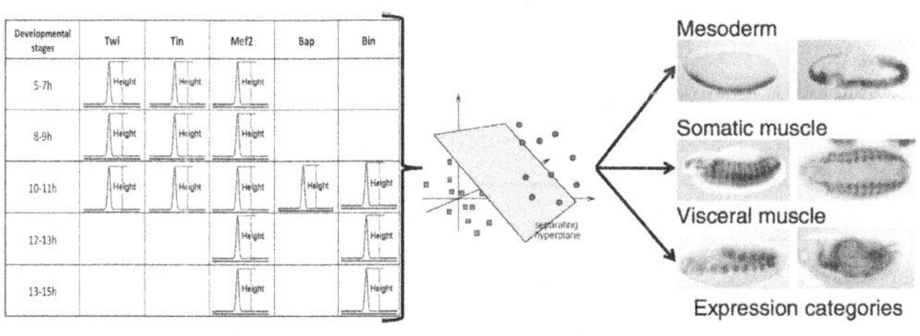

Fig. 1 Using transcriptional factor binding ChIP profiles for predicting the regulatory activity of CRMs. The ChIP peak heights were measured at early time points (5-7,8-9,10-11h) for Twist and Tinman, late time points (10-11,12-13,13-15h) for Biniou, all those times points for Mef2 and only 10-11h for Bagpipe. The classified regulatory activities are corresponding to three embryonic development stages (mesoderm, somatic muscle visceral muscle or ambiguity between two of them).

A part of the figure is from [1] and http://www.dtreg.com/svm.htm

By tiling microarray technology followed by ChIP, [2] profiled the binding landscapes for five important transcription factors in *Drosophila* (Twist, Tinman, Mef2, Bagpipe and Biniou) at 5 critical time-points during embryogenesis development. The microarray chip hybridization signals for each genomic region is detected over the overlapping tiles by Hidden Markov Model method (TileMap, [7]), called peaks. Each peak height is normalized and represented by a real value at log2 scale that indicates the activeness of CRMs in terms of gene regulation.

Hence, in order to understand the influence of the transcriptional factor binding on the gene expression, [2] used 310 CRMs with the known expression patterns (mesoderm, somatic muscle, visceral muscle, called meso, sm and vm in the paper) obtained from the popular database (REDFly, [8]) including the mixed patterns like mesoderm and somatic muscle (called meso_sm) or somatic muscle and visceral muscle (called sm_vm).

B. Regulatory Prediction Problem

The well-known SVM framework was used in [2] under radial kernel function to evaluate the prediction performance of transcriptional regulatory activity for each CRM based on its binding profiles under cross-validation (LOOCV) context. They found out a good correlation between the profiles of the 5 TFs at those mentioned time points and the corresponding expression categories (meso, vm, sm, meso_sm, vm_sm) of the CRMs. To simplify the unnecessary complication from multiple-class classification, only binary SVM was used. Thus, one classifier learns the rule on the training set of CRMs excluding the testing CRM given the binary expression label (e.g meso or non-meso); and then tries to predict the label of the testing CRM. From the predicted labels for each CRM, the summarized accuracy for the prediction method is computed.

In order to boost the prediction performance, the traditional grid search was used to find the best suitable parameters (cost C and gamma γ) for the radial kernel SVM. The former is for the trade-off between fitting error minimization and classification margin maximization whereas the later is for the kernel function especially for high-dimensional data The parameter space is searched starting with default parameters to find the better solution as the prediction accuracy. However, the grid search has some limitations including not searching intensively enough the parameter space nor trying to learn the good parameters for the next search.

Here, we introduce the meta-heuristic searching methods based on the GA and ACO frameworks that can circumvent those disadvantages and gain a better prediction performance by deriving the optimal parameters.

III. META-REG METHOD

A. Parameter Representation

The SVM parameters are represented as N-bit binary vectors consisting of first bits for cost C and others for the γ (here we chose N=51 including 24 bits for C and 27 for γ). That allows us to apply the computational exploration method to build each element of the vector and then convert it to the real-valued parameters. For each parameter pair (C, γ), the prediction accuracy is computed accordingly. Thus, the pair whose highest accuracy is considered as the optimal solution.

B. ACO-REG: ACO for Regulatory Prediction

Each artificial ant starts from the start node and travel through 2N binary nodes (i, v_i), i = 1..N, v_i = (0,1) to build up a solution as a N-bit binary vector $v_1v_2..v_N$ (Fig 2). At each node, the pheromone trail $\tau_{i,v}$ is deposited during ant colony's travels and is lost if no ant visited. The pheromone trails represent how likely the nodes are chosen for a good parameters and is updated during the travel of ant colony.

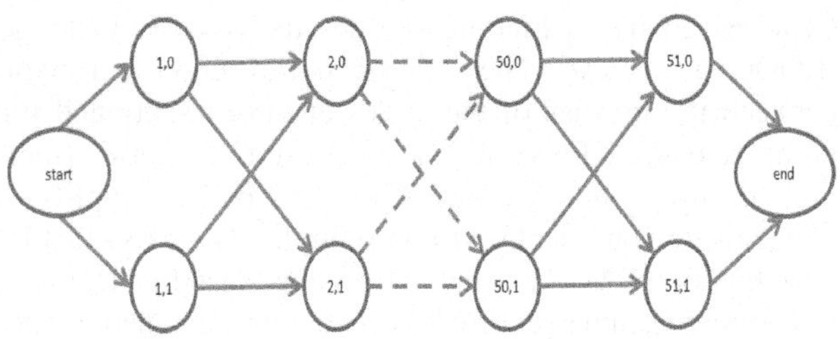

Fig. 2 The ACO graph consisting of 2 start and end nodes and 2N (N=51) nodes: the upper nodes (i,0) is for 0-bit , he lower (i,1) for 1-bit with i = 1..N. Each path from start to end including N binary nodes will build up a N-bit vector for the parameters.

The node (i, v) is chosen for each path according to the probabilistic rule. The ant k choose either (i,0) or (i,1) for the ith node based on the probability

$$p_{i,v}^{k} = \frac{\tau_{i,v}}{\tau_{i,0} + \tau_{i,1}} \quad (1)$$

After one iteration is finished, the pheromone trails are updated based on the best solution at the current iteration according to the state-of-the-art Smoothed Max-Min Ant System introduced by [9] :

$$\tau_{i,v} = (1-\rho)\tau_{i,v} + \begin{cases} \rho\tau_{\max} & if\ node(i,v) \in s^{i-best} \\ \rho\tau_{\min} & otherwise \end{cases} \quad (2)$$

where $\tau_{\max}$, $\tau_{\min}$ are the input parameters, $s^{i\text{-}best}$ is the best solution at the current iteration. In experiment, we set $\tau_{\max}$ =1.0 and $\tau_{\min} = \frac{\tau_{\max}}{1000}$.

In a nutshell, the overall ACO-REG is as followings:

i. Initialization

ii. Repeating until the termination conditions is met

a) Each ant will build a binary-node path. They will be converted into a real-valued parameters.

b) Updating pheromones according to SMMAS rule

c) Updating the best parameters corresponding to the best prediction accuracy.

C. GA-REG: GA for Regulatory Prediction

Besides ACO, we also try a well-established Genetic Algorithm (GA) approach [10] for searching the best well-suited SVM parameters. The fitness function used is the balanced prediction accuracy that is average of the sensitivity and the specificity (explaining on the later section).

The population of solution is the set of N-bit vectors (N=51, the same setup in ACO). The implementation of GA algorithm is derived from R package *genalg*[1].

IV. EXPERIMENTAL RESULTS

A. Evaluation

We used a known dataset that consists of 310 CRMs with known regulatory activity determined by the CRM activity database with the expression driven by the published REDFly database [8]. For each data, the accuracy of the prediction is measured based on whether the specific expression category is correctly predicted or not.

Table 1. Confusion matrix of the prediction results.

Predicted/ Known fact	Positive	Negative
Positive	TP	FP
Negative	FN	TN

To evaluate such type of data, we use the so-called Balanced Accuracy (BACC) as the average of Sensitivity ($TP / (TP+FN)$) and Specificity($TN / (FP+TN)$) of the prediction results. It will give us the good evaluation of the

[1] http://cran.r-project.org/web/packages/genalg/index.html

SVM prediction for the imbalanced data where the number of positive and negative CRMs is not similar. The SVM computation is implemented based on the R package *e1071*[2].

B. Method Comparative Results

Both of our metaheuristic approaches (GA and ACO) outperform the results of grid search method in previous study in terms of balanced accuracy (Fig 3). Most of the cases we gained from 5-10% better accuracy except the Meso_SM where we are only better than 1%. Interestingly, both GA and ACO gained very similar results in 3 over 5 datasets where the expression category is unique (Meso, SM or VM). In two other cases of ambiguous expression, ACO is slightly better than GA. It showed that in case of precisely expression category, the computational prediction is more difficult due to necessary information lacking.

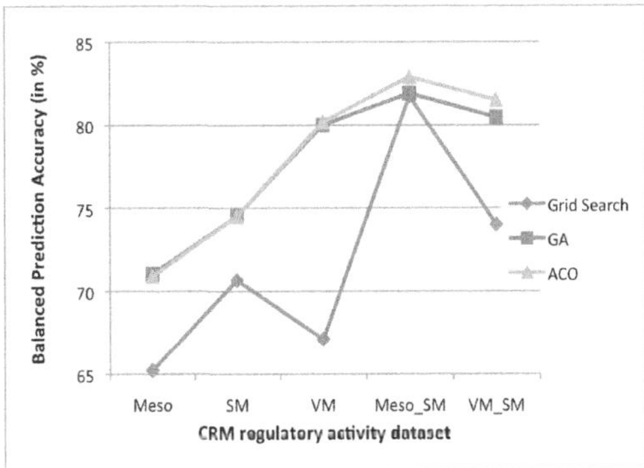

Fig. 3 Prediction accuracy comparison between three SVM-based prediction parameter optimization methods.

All together indicates the necessary of intensive search in order to find the optimal parameters for SVM prediction in case of gene regulatory data.

V. Discussions

In summary, we introduce better methods to improve the gene regulatory activity prediction of cis-regulatory modules based on time-course transcriptional factor binding. Our approaches gained the improvements up to 10% prediction accuracy compared to the results reported by the original work [2]. We then show that the optimization of prediction parameters needs to be considered in most of the expression categories particularly in case of unambiguous gene expression.

Boosted by high-throughput technologies especially next-generation sequencing (ChIP-Seq, [11]), more and more transcriptional factor data will be reported in a very rapid pace. Hence, it raises the need of computational optimization like in SVM prediction. Our future work is to challenge those available datasets and to make an automatic parameter tuning method for biological prediction. It can also help for further analysis like determining which factor/features is important for a given gene expression category or development stage.

Acknowledgment

This work is partially supported by Vietnam National Foundation for Science and Technology Development (NAFOSTED). We thank Thanh Hai Dang (University of Antwerpen, Belgium) for useful comments and proof reading of the manuscript.

References

1. A. Stark (2009), "Learning the transcriptional regulatory code," Mol. Syst. Biol., vol. 5, p. 329.
2. R. P. Zinzen, C. Girardot, J. Gagneur, M. Braun, and E. E. Furlong (2009), "Combinatorial binding predicts spatio-temporal cis-regulatory activity," Nature, vol. 462, pp. 65–70.
3. A. Ben-Hur, C. S. Ong, S. Sonnenburg, B. Scholkopf, and G. Ratsch (2008), "Support vector machines and kernels for computational biology," PLoS Comput. Biol., vol. 4, p. e1000173.
4. C. Cortes and V. Vapnik (1995), "Support-vector networks," Machine Learning, vol. 20, pp. 273–297, 10.1007/BF00994018. [Online]. Available: http://dx.doi.org/10.1007/BF00994018
5. Reeves, C (1995). Genetic Algorithms and Combinatorial Optimisation: Applications of Modern Heuristic Techniques, In V.J. Rayward Smith (Eds), Alfred Waller Ltd, Henley-on-Thames, UK.
6. M. Dorigo and T. Stutzle (2004). Ant Colony Optimization. The MIT Press, Cambridge, Masachusetts.
7. H. Ji and W. H. Wong, "TileMap: create chromosomal map of tiling array hybridizations," Bioinformatics, vol. 21, pp. 3629–3636, Sep 2005.
8. M. S. Halfon, S. M. Gallo, and C. M. Bergman, "REDfly 2.0: an integrated database of cis-regulatory modules and transcription factor binding sites in Drosophila," Nucleic Acids Res., vol. 36, pp. D594–598, Jan 2008.
9. D. Do Duc, H. Q. Dinh, and H. Hoang Xuan (2008). On the pheromone update rules of ant colony optimization approaches for the job shop scheduling problem. In Proceedings of the 11th Pacific Rim International Conference on Multi-Agents: Intelligent Agents and Multi-Agent Systems, PRIMA '08, pages 153–160, Berlin, Heidelberg, 2008. Springer-Verlag.
10. C.-L. Huang and C.-J. Wang (2006), "A ga-based feature selection and parameters optimization for support vector machines," Expert Systems with Applications, vol. 31, no. 2, pp. 231 – 240.
11. P. J. Park, "ChIP-seq: advantages and challenges of a maturing technology," Nat. Rev. Genet., vol. 10, pp. 669–680, Oct 2009.

Corresponding Author: Dong Do Duc
Institute: Institute of Information Technology, Vietnam National Univeristy, Hanoi
Street: 144 Xuan Thuy City: Hanoi
Country: Vietnam Email: dongdoduc@vnu.edu.vn

[2] http://cran.r-project.org/web/packages/e1071/index.html

Advances in Computational Identification and Modeling of DNA Regulatory Elements in the Human Genome

Dongwon Lee and Michael A. Beer

Johns Hopkins University, Department of Biomedical Engineering and McKusick-Nathans Institute of Genetic Medicine, Baltimore, MD, USA

Abstract — **Identification of DNA regulatory elements in the human genome remains a significant challenge. Variation in these regulatory elements can contribute to disease in many ways by altering protein levels. Enhancers constitute an important class of these DNA regulatory elements, and a major component of current research is focused on a more complete understanding of enhancer function and improved techniques for enhancer detection. We recently developed a computational approach to identify enhancers from primary DNA sequence using a support vector machine (kmer-SVM) framework. Here we show that the kmer-SVM model can accurately predict tissue specific enhancer activity without any prior knowledge about TF binding sites. We adapt this approach to predict genomic TF binding data generated by the ENCODE project, showing that genomic MYC binding can be accurately predicted from local DNA sequence with the kmer-SVM. We find similar accuracy with an SVM using PWMs representing known TF binding specificities. By integrating Chip-seq and expression data, we show that while much of MYC binding is shared between ENCODE cell types and is promoter proximal, cell-type specific MYC binding is distal and is correlated with enhanced cell-specific expression of nearby (~50kb) genes. The distinction between shared and cell-specific MYC binding is determined by DNA sequence variation around the canonical MYC binding site, which by itself cannot distinguish cell-specific binding events. These results suggest that tissue specific enhancer activity is specified by primary DNA sequence, that local sequence context controls tissue specific activity through cooperative TF interactions, and that local context sequence features can be identified from genomic binding data.**

Keywords — **computational biology, genomics, transcriptional regulation, enhancers.**

I. Introduction

DNA regulatory elements are widely suspected to play a significant role in human health and disease, but detecting and understanding the function of DNA regulatory elements in the human genome remains a significant challenge. These elements include enhancers, repressors, and insulators which regulate the expression of associated genes. Recent genome wide studies have found that 80% of common human variants with significant disease associations (p-value<10^{-8}) lie in intergenic or intronic regions and are thus predicted to be regulatory[1]. As a result, a major component of future NIH research development [2] is focused on a more complete understanding of the biology of genomes. Specifically, the ENCODE project [3] is moving toward producing a comprehensive genome wide map of protein binding (including histone modifications, transcription factor binding, and chromatin accessibility) in a wide range of cell types and tissues. A central limitation to a comprehensive understanding of this growing database is our general inability to identify the short regions within a broad (~1kbp) regulatory locus that specify its function through direct molecular interactions. A predictive model of ENCODE data at the primary sequence level would enable a significant enhancement of our understanding of the impact of human sequence variation on phenotypic diversity and disease risk and mechanisms.

Enhancers are gene regulatory sequences that can control transcriptional activities at a distance, independent of their position and orientation with respect to affected genes [4]. Enhancer activity is modulated by interactions between sequence specific DNA binding proteins and sequence elements in the enhancer. The leading model of this process is that regulatory regions are bound by clusters of sequence specific transcription factors (TFs) which cooperatively bind and stabilize a complex of co-activator proteins that then modulate the activity of RNA Polymerase II at the gene promoter through direct contacts mediated by DNA looping. Since individual transcription factor binding sites (TFBSs) can be relatively short and degenerate, TFBSs tend to be clustered to achieve precise temporal and developmental specificity [5]. Factors bound to enhancer sequences often interact with common coactivators, which in turn recruit the basal transcription machinery [6,7]. Models of TF cooperativity are limited by our incomplete understanding of the binding specificity of most TFs, often represented by a position weight matrix (PWM). *De novo* identification of putative PWMs from expression or binding data has met with some success, particularly in yeast [8],

V. Van Toi et al. (Eds.): 4th International Conference on Biomedical Engineering in Vietnam, IFMBE Proceedings 40, pp. 328–331, 2013.
www.springerlink.com

but extended sequence and complexity has limited applications to mammalian enhancers.

Recently we have developed an alternative approach to predict enhancers using a complete set of oligomers or "kmers" as features in a support vector machine (SVM) [9]. We show that the kmer-SVM can accurately predict regulatory sequences without any prior knowledge about TF binding sites. The biological relevance of the predicted enhancers is supported by the fact that most of the predictive sequence features identified by our methods are binding sites of TFs known to play a role in the relevant context, the enriched predictive sequence features are much more evolutionarily conserved within the enhancers than less predictive sequence features, and the sequence features are significantly more spatially clustered in the enhancers than would be expected by chance, also a well-known characteristic of functional binding sites.

Here, we adapt this approach to predict tissue specific enhancer activity from genomic TF binding data generated by the ENCODE project, specifically binding data for the MYC TF, which plays a major role in the regulation of cellular proliferation. By integrating Chip-seq and expression data, we show that while much of MYC binding is shared between distinct cell types and is promoter proximal, cell-type specific MYC binding is distal and is correlated with enhanced cell-specific expression of nearby (~50kb) genes. The distinction between shared and cell-specific MYC binding is determined by DNA sequence variation around the canonical MYC binding site, which by itself cannot distinguish cell-specific binding events. We show that genomic MYC binding can be accurately predicted from local DNA sequence with the kmer-SVM. We find similar accuracy with an SVM using PWMs representing known TF binding specificities. These results suggest that tissue specific enhancer activity is specified by primary DNA sequence, that local sequence context controls tissue specific activity through cooperative TF interactions, and that local context sequence features can be identified from genomic binding data.

II. DEFINING A SET OF CELL-SPECIFIC ENHANCERS

One of the most challenging questions in enhancer biology is understanding the mechanisms of differential TF binding in different cell types. To address this question we used genomic binding data from the ENCODE project in five different cell types. We chose to study the TF MYC because of its role in cancer and because its DNA binding specificity is well studied. Using Chip-seq data[10-12] for MYC in the five cell lines (Gm12878, Helas3, Hepg2, Huvec, and K562), we defined a set of well defined MYC bound peaks in each tissue using the MACS algorithm[13]. Many of these peaks are shared between all cell types and are promoter proximal. To define a set of cell-type specific MYC binding regions, we selected the top 500 bound regions in each cell-type that were unbound in all other tissues. After the common MYC bound regions, this is the second largest pattern of genomic MYC binding. While the common MYC binding is usually near promoters, these cell-specific binding regions are on average ~50kb from the nearest gene TSS. To determine whether these cell-specific MYC bound regions were acting as distal enhancers, we compared the expression of the nearest gene in the MYC bound cell-type to the expression of that same gene in all of the unbound cell-types. As shown in Fig. 1, the expression in the MYC bound cell-type is significantly higher, indicating that these MYC bound regions are acting as cell-specific enhancers, and further, that MYC is playing an activating role in these enhancers.

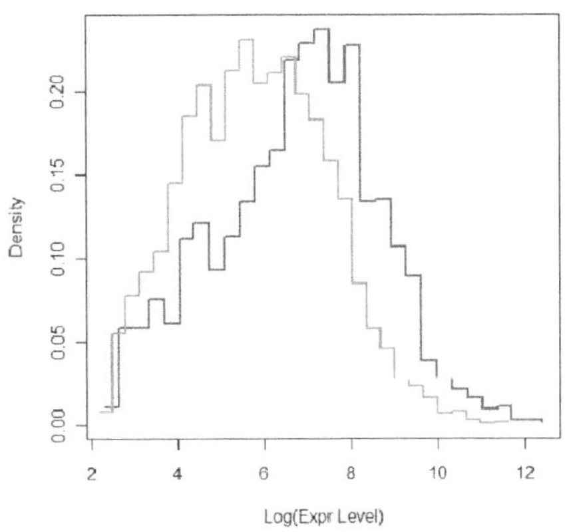

Fig. 1 Expression of genes nearest a cell-specific MYC bound element is enhanced in the MYC bound cell-type (red) relative to the expression in all other cell-types (green), demonstrating that differentially MYC bound regions are acting as distal enhancers.

For comparative analysis we also defined a set of the top 10k peaks in Gm12878 (mostly promoter proximal peaks), the set of the top 5k CTCF bound peaks in Gm12878, and the union of all of the cell-specific MYC bound peaks.

III. KMER-SVM METHOD FOR PREDICTING ENHANCERS

We have previously developed a method for predicting enhancers bound by the transcriptional co-activator P300 in mouse dissected tissue using a kmer-SVM model, for details see [9]. Briefly, an SVM finds a decision boundary that maximally distinguishes two sets of data, here a positive (enhancer or TF bound) and negative (random genomic) sequence set. Weights, w_i, determine the contribution of each feature to this boundary, according to:

$$S(\mathbf{x}) = \mathbf{w} \cdot \mathbf{x} + b = \sum_{i=1}^{N} w_i x_i + b,$$

where $S(\mathbf{x})$ is the SVM score whose sign determines the predicted class of sequence $\mathbf{x}$.

Once the set of sequence features, x_i, is specified, the weights are optimized to maximize the separation between the two classes. We use as sequence features the full set of *k*-mers of varying length (6-mers and 7-mers perform very similarly, here we will concentrate on 6-mers for ease of interpretation). We can also use as a negative set a set of enhancers bound in a different cell-type.

To evaluate classification performance, we use a 5-fold cross validation method. Initially, the data set to be classified is randomly partitioned into five subsets. One subset is then reserved as a test data set, and the SVM weights are trained on sequences in the remaining four subsets. The SVM is then used to predict the reserved test data set to assess its accuracy. This process is repeated five times so that every sequence element is classified in one test set. Because there is a trade-off between specificity (the accuracy of positively classified enhancers) and sensitivity (the fraction of positive enhancers detected), we measure the quality of the classifier by calculating the area under the ROC curve (auROC), which plots the True Positive Rate for prediction vs. the False Positive Rate as the SVM score threshold varies, shown for several cases in Figure 2.

IV. PREDICTING CELL-SPECIFIC ENHANCERS FROM SEQUENCE

Our main result is that the kmer-SVM is capable of predicting the cell-specific binding of MYC at enhancers with high accuracy. In Figure 2A we show first the classification of all MYC bound regions in Gm cells using (GC content and length matched) random genomic sequence as a negative set. Most of these peaks are MYC bound in most cell-types. This MYC bound set is classified accurately (AUC=.894), and the MYC binding site CACGTG is the highest weight feature. Notably, however, a single CACGTG kmer or even the score for the MYC PWM classifies very poorly (black curve in Fig. 2A with AUC=.580, random guessing gives AUC=0.5), indicating that even the strongest common (active in many cell-types) MYC enhancers are specified by combinatorial interactions between MYC and diverse TFs which recognize other kmers. In contrast, Fig 2B shows classification results for CTCF, which is very well classified by a single PWM representing the CTCF binding site. In all these ROC plots we are comparing the performance of the kmer-SVM and and SVM using PWM scores for a set of 811 known TF binding sites as features (PWM-SVM) [9], using PWMs from databases of known TFs. While some situations might be well modeled by a single PWM, our method does not rely on knowing TF binding specificities, and classifies comparably or better in all cases. We also compare classification of the union of all the MYC cell-specific enhancers vs. random genomic sequence. The most predictive kmers (highest positive and negative weights) learned by the SVM for this case (union) are shown in Table 1. As expected, the strongest binding site for MYC (CACGTG, red) ranks highest. This is a significant point: the feature which most accurately predicts MYC binding in any tissue is CACGTG, however, this feature will have little power to predict tissue specificity. Similar E-box (CANNTG, red) elements also score highly. Next, we show the classification of the cell-specific MYC peaks in each of the five tissues, using the union of the other four sets of cell-specific MYC peaks as the negative set, in Fig 2D-H. Again the kmer-SVM and PWM-SVM perform similarly, and both models are able to accurately predict the cell-specific binding of MYC.

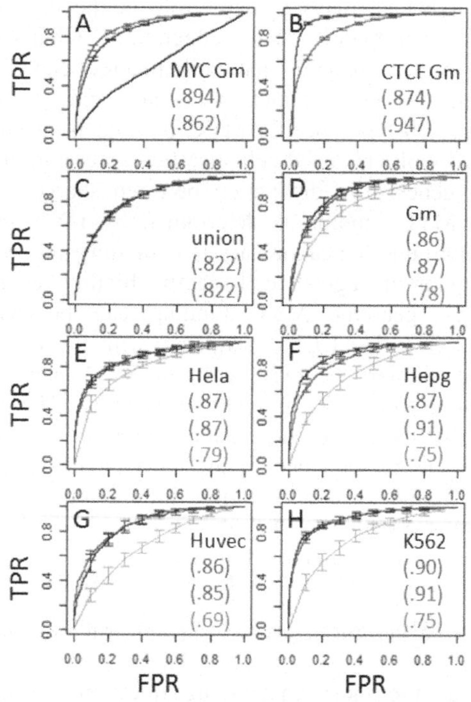

Fig. 2 Classification results for cell-specific MYC bound regions. Each plot shows the ROC curve for kmer-SVM (red), PWM-SVM (blue), and randomPWM-SVM (green), with AUC in parentheses for each method. Both kmer-SVM and PWM-SVM generally classify with high accuracy. (A) Top 10k MYC peaks in Gm cells. Also shown is performance of a single MYC PWM (black), which predicts only slightly better than random. (B) Top 5k MYC CTCF peaks in Gm. (C) Union of all cell-specific MYC peaks. (D-H) Cell-specific peaks in Gm, Hela, Hepg, Huvec, and K562 cells are predicted accurately, where the SVM learns different combinations of significant cofactor binding sites in each cell-type.

To test whether the database of known TF PWMs is contributing to the good performance of the PWM-SVM, we also repeated the PWM-SVM classification using a set of 811 PWMs generated by randomizing the columns of the known PWM matrices. As shown in Figure 2D-H, the random PWM-SVM performs significantly less well than PWM-SVM, indicating that the binding sites recognized by the co-factors which are responsible for modulating MYC binding in the different tissues are largely present in the 811 PWM database. The kmer-SVM represents these binding sites in an unbiased way by completely covering the kmer feature space.

Table 1. Sequence features which determine cell-specific activity.

Most predictive kmers in tissue specific enhancers and in union enhancers

union		Gm12878		Helas3		Hepg		Huvec		K562	
CACGTG	4.6	CACCTG	2.1	ACGTCA	2.3	CAAACA	2.5	AGGAAA	3.7	AGATAA	5.2
CTTATC	4.5	ACACCT	2	TGCCAA	1.8	ACTTTG	2.4	CAGGAA	3.2	CTTATC	5
AGATAA	4.1	GACACC	1.5	GAGTCA	1.7	AGCAAA	1.9	ATTTCC	2.9	TGATAA	3
ACGTGG	3.9	AAACCA	1.3	CTGGCA	1.6	AACAAA	1.8	AGGAAG	2.7	GAGATA	2.8
CACATG	3.5	AAACCC	1.1	GACTCA	1.4	CTTTGA	1.7	AGGATG	2.1	ATAAGG	2.5
GAGTCA	2.9	ACCACA	1.1	ATAAAT	1.3	AAAGTC	1.7	GAGGAA	2	GATAAA	2.5
ACGTGC	2.9	CCCATC	1.1	AATTCC	1.2	GCAAAC	1.6	ACAGGA	1.9	CTGATA	2.4
GACTCA	2.6	GGGGAC	1	ATGCCA	1.2	GATCAA	1.5	CATTTC	1.8	CAGATA	2
TGATAA	2.4	AGTTTC	1	ATATTA	1.2	AAGGCC	1.5	GCAGGA	1.6	AGATAG	1.6
AGCACG	2.4	CTGAAA	1	AATAAA	1.1	AGTCCA	1.5	GAGTCA	1.4	AAGATA	1.4
GGGATA	-1.3	GAGATA	-1.2	AGAGCA	-1.1	CTTATC	-1.1	ACACCT	-0.9	GAATCA	-1.1
GACCCC	-1.3	AGGAAT	-1.2	CAGGAA	-1.1	GAGTCA	-1.2	CACCTG	-1	CATTCC	-1.1
AGCATC	-1.5	TGATAA	-1.3	ATTTCC	1.2	AGGAAG	1.2	CTTATC	1.1	AGGAAA	-1.1
AGACCC	-1.6	AGGTCA	-1.4	CACGTG	-1.2	GACTCA	-1.2	CACATG	-1.2	AAACAA	-1.2
CAGGTA	-1.8	CTTATC	-1.8	AGGATG	-1.4	CACACC	-1.3	ACACCC	-1.2	GAGTCA	-1.3
AATTCA	-1.8	AGATAA	-2.2	AGGAAG	-1.6	AGATAA	-1.3	AGATAA	-1.4	ATTCCA	-1.5

Table 1 shows the kmer weights for the most predictive sequence features which determine tissue specificity, and implicates the combinatorial interaction of a set of known TFs in determining cell-specific enhancer activity (MYC, GATA, AP1, and Ets). The MYC binding site (CACGTG, red) and the GATA factor binding site (GATAA, blue) are predictive of MYC binding overall (union). But specific MYC binding sites have negative weights in Helas3 and Huvec cells, and instead these cell types have active enhancers defined by AP1/Fos/Jun binding sites (GAGCTA, green) and Ets family member binding sites (AGGAA, orange). The GATA factor sites also positively specify a set of K562-specific enhancers, and negatively specify Gm enhancers.

V. Conclusions

By analyzing genomic binding data for the transcription factor MYC, we have shown that differential genomic occupancy of MYC defines a set of cell-specific enhancers, and that our sequence based kmer-SVM model can predict human cell-specific enhancer activity with high accuracy. Cell-specificity of MYC activity is regulated by the presence of distinct sequence features bound by cofactors which are present (or absent) in the cell-specific enhancers. This analysis makes specific hypotheses about cooperative cell-specific interactions between MYC, GATA, AP1, and Ets family members in determining MYC enhancer activation.

Acknowledgment

M. Beer was supported by the Searle Scholars Program and in part by NS062972 (NIH).

References

1. Manolio T A (20 (2010) Genome-wide association studies and assessment of the risk of disease. N. Eng. J. Med. 363:166-76.
2. Green ED, Guyer MS (2011) Charting a course for genomic medicine from base pairs to bedside. Nature 470:204-13.
3. Myers RM, Stamatoyannopoulos J, Snyder M, et al, (2011) A user's guide to the encyclopedia of DNA elements (ENCODE). PLOS Biology 9:e1001046.
4. Banerji J (1981) Expression of a beta-globin gene is enhanced by remote SV40 DNA sequences. Cell 27: 299-308.
5. Kadonaga, JT (2004). Regulation of RNA Polymerase II Transcription by Sequence-Specific DNA Binding Factors. Cell 116: 247-257.
6. Blackwood EM, and Kadonaga JT (1998) Going the Distance: A Current View of Enhancer Action. Science 281: 60-63.
7. Carter D, Chakalova L, Osborne CS, Dai Y, and Fraser P (2002) Long-range chromatin regulatory interactions in vivo. Nat Genet 32: 623-626.
8. Beer MA and Tavazoie S (2004) Predicting gene expression from sequence. Cell 117:185-98. Lee D, Karchin R, and Beer MA. (2011) Discriminative prediction of mammalian enhancers from DNA sequence. Genome Res, advance online publication August 29.
9. Lee D, Karchin R, and Beer MA. (2011) Discriminative prediction of mammalian enhancers from DNA sequence. Genome Res, advance online publication August 29.
10. Song L, Zhang Z, Grasfeder LL, Boyle AP, et al. (2011) Open chromatin defined by DNaseI and FAIRE identifies regulatory elements that shape cell type identity. Genome Res 21:1757-67.
11. Boyle AP, Song L, Lee BK, et al. (2011) High-resolution genome-wide in vivo footprinting of diverse transcription factors in human cells. Genome Res 21, 456-464.
12. McDaniell R, Lee BK, Song L, Liu Z, Boyle AP, et al. (2010) Heritable individual-specific and allele-specific chromatin signatures in humans. Science 328, 235-239.
13. Zhang Y, Liu T, Meyer CA, Eeckhoute J, Johnson DS, et al. (2008) Model-based Analysis of ChIP-Seq (MACS). Genome Biol 9: R137

Author: Michael A. Beer
Institute: Johns Hopkins University
Street: 733 N. Broadway, Suite 573
City: Baltimore, MD 21212
Country: USA
Email: mbeer@jhu.edu

Data Mining in Uniform Hospital Discharge Data Set Using Rough Set Model

M. Park

College of Nursing/Chungnam National University, Daejeon, Republic of Korea

Abstract — **Purpose: The purpose of this study were to apply rough set model to nursing knowledge discovery process. Method: Data mining based on rough set model was conducted on a large clinical data set containing Nursing Minimum Data Set elements. Randomized patient data were selected from Uniform Hospital Discharge Data which had the frequently used nursing diagnoses. Patient and care characteristics including nursing diagnoses, interventions and outcomes were analyzed to derive the decision rules. Results: Number of comorbidity, marital status, nursing diagnosis related to risk for infection and nursing intervention related to infection protection, and discharge status were the predictors to determine the length of stay. Age, impaired skin integrity, pain, and discharge status were identified as valuable predictors for nursing outcome, relived pain. Age, pain, potential for infection, marital status, and primary disease were identified as important predictors for mortality. Conclusion: This study demonstrated the utilization of Rough Set Model through a large data set with standardized language format to identify the contribution of specific care to patient's health.**

Keywords — **Data, Data mining, Nursing.**

I. Introduction

There is wide spread concern about the quality and effectiveness of health care, the relationship of health care workers' knowledge and quality outcome, the high cost of health care, and the access to health care. Without collecting information systematically, having standardized classification system, and analyzing collected clinical data, healthcare providers cannot demonstrate their contribution to patient health. One of the efforts to determine contribution to patient outcome is the development of the Minimum Data Set (MDS).

Rough set model which was introduced in the early 1980's is one of the data mining methods. Rough set is a set theory that classifies objects into sets based on attributes of the objects. The classification can then be used to access sets of objects where objects are roughly equal or overlap. Rough set is a non-statistical methodology for data analysis and concerns the classificatory analysis of imprecise, uncertain, or incomplete information acquired from experience. Basically, rough set theory deals with the approximation of sets that are difficult to describe with the available information. Rough set evaluates the importance of conditional attributes, reduces redundant attributes, determines minimum subsets of attributes ensuring satisfactory classification of objects, creates models for objects, and results in the form close to natural language (decision rules) allowing researchers understand and interpret the outcome more easily. One of the main advantages of rough set theory is that it does not need any preliminary or additional information or statistical assumptions about data compared with other statistical methods [1].

Therefore, rough set theory can be used easily for healthcare data which have a variety of attributes, missing data, and incomplete data [2]. The purposes of this study were to apply Rough Set Model to knowledge discovery process using large clinical data and to identify the utilization of data mining skill for clinical decision making.

II. Materials and Method

Data mining based on rough set model was conducted on a large clinical data set, Uniform Hospital Discharge Data Set containing Nursing Minimum Data Set elements. Randomized patient data were selected from year 2000 database which had at least one of the five most frequently used nursing diagnoses [3]. Sixteen most frequently used nursing interventions [4] were identified for specific five nursing diagnoses. Seventeen most frequently used nursing outcomes [5] were identified for specific five nursing diagnoses. Some redundancy data variables were excluded. Real values such as ICD 10 were categorized into groups. All missing values were designed for specific values to fill up the field. Patient characteristics and care service characteristics including nursing diagnoses, interventions and outcomes. The rules were evaluated with the testing set data for their predictability. ROC (Receiver Operating Characteristics) was derived to see the discriminatory performance of rules.

V. Van Toi et al. (Eds.): 4th International Conference on Biomedical Engineering in Vietnam, IFMBE Proceedings 40, pp. 332–334, 2013.
www.springerlink.com

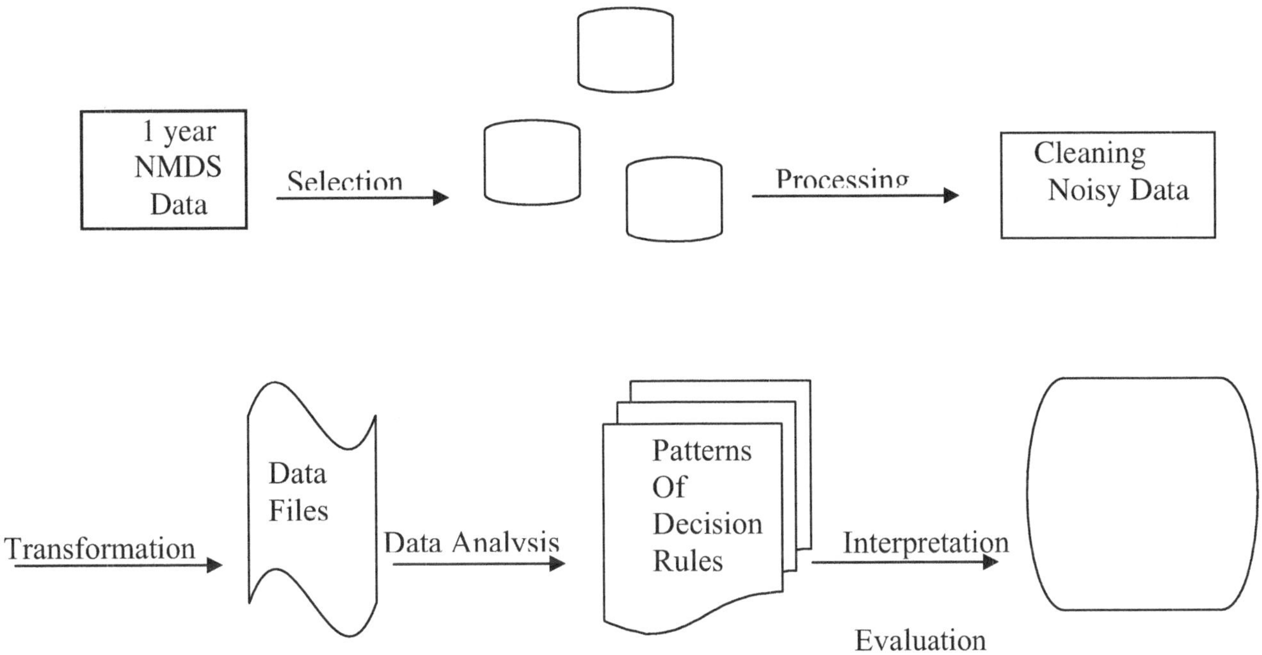

Fig. 1 Steps of the Rough Set process.

III. Results and discussions

The five most common nursing diagnosis was potential for infection, followed by *pain*, *potential for injury*, *impaired skin integrity*, and *immobility*. The intervention most frequently used was *pain management*. Other frequently used interventions were *infection protection*, *analgesic administration*, and *skin surveillance*.

Number of co-morbidity, marital status, nursing diagnosis related to risk for infection and nursing intervention related to infection protection, and discharge status were the predictors that could determine the length of stay. Age, impaired skin integrity, pain, and discharge status were identified as valuable predictors for nursing outcome, *relived pain*. Age, pain, potential for infection, marital status, and primary disease were identified as important predictors for mortality. An example of the decision rule indicates that if patients had less than 2 comorbidity, were married, had a risk for infection as nursing diagnosis and received infection protection as nursing intervention, and discharged to home, patients had less than 3 days of length of stay. It describes exactly 23.4 % of patients with LOS $\leq$ 3. Data mining found 7 rules for nursing outcome, *Relieved Pain*. Age, impaired skin integrity, pain, and discharge status were identified as valuable predictors. Applying the rules to the test set resulted in the area of under ROC curve of 90.1% with standard error of 2.3% showing the high predictability. Data mining found 15 rules for being expired at hospital. Age, pain, potential for infection, marital status, and primary disease were identified as valuable predictors for mortality. Applying the rules to the test set resulted in the area of under ROC curve of 91.1% with standard error of 1.8% showing the high predictability.

This study explored the potential of utilization of data mining tool to discover the information and knowledge in a nursing minimum data set. One NMDS including discharge summary data was mined with data mining program based on Rough Sets theory. The strength of this method are follow as. First, the patterns of decision rules are readily interpretable and can give possible new insight how various attributes interact and serve as baseline data for decision making. Second, a decision is made with high accuracy.

(Comorbidity $\leq$ 2) & (Marital Status = M) & (INFEC =1) & (NIC4=1) & (DISSAT=1) => (LOS$\leq$3)

⇩

If Comorbidity $\leq$ 2 and Marital Status = M and INFEC =1 and NIC4=1 and DISSAT=1
Then LOS $\leq$ 3
With Coverage 23.4% Accuracy 87.2%

Fig. 2 An example of decision Rule.

Third, no decision is made for a small number of patients due to insufficient information.

The limitations of this method also need to be noted for further development. First, this program is under development and need further refinement. Second, the time for data analysis increases considerably with the number of decision rule and data. The support of hardware is needed to cut down the time for efficiency of data analysis. The most important prerequisites of effective data mining are the selection of right data set and variables to analyze. The clear and in-depth understanding of selected data sets should be followed. The cleaning and preparing according to the purpose of analysis is one of important process for data miming.

IV. Conclusions

This study demonstrated the utilization of data mining method through a large clinical data set with standardized language format. The value of large clinical database is directly related to the contribution of clinical practice to knowledge development. Data from the clinical settings can produce meaningful information and knowledge that can improve the quality of care. Such information and knowledge can empower clinicians in their clinical decision-making and problem solving.

Acknowledgment

This research was supported by Basic Science Research Program through the National Research Foundation of Korea(NRF) funded by the Ministry of Education, Science and Technology (2010-0024922).

References

1. Tam, S., Cheing, G. L. Y., & Hui-Chan, C. W. Y. (2004). Predicting osteoarthritic knee rehabilitation outcome by using a prediction model developed by data mining techniques. Int J Rehabil Re 27: 65-69
2. Kusiak, A. (2000). Computational Intelligence in Design and Manufacturing. John Wiley, New York
3. North American Nursing Diagnosis Association International. (2009). Nursing Diagnoses: Definitions and classification 2009-2011. Wiley-Blackwell, Philadelphia
4. McCloskey, J. C., & Bulechek, G. M. (2008). Nursing Interventions Classification (NIC) (5th ed). Mosby, St. Louis
5. Moorhead S, Johnson M, Maas M, Swanson E. (2008). Nursing Outcome Classification (4th ed). Mosby, St. Louis

Author: Myonghwa Park
Institute: College of Nursing/Chungnam National University
Street: 55, Munhwa-ro, Jung-gu
City: Daejeon
Country: Republic of Korea
Email: mhpark@cnu.ac.kr

Medical Image Classification and Symptoms Detection Using Fuzzy NARX Technique

Ho Pham Huy Anh[1,*] and Le Tan Loi[2]

[1] Ho Chi Minh City University of Technology, Ho Chi Minh City, Viet Nam
hphanh@hcmut.edu.vn
[2] Nguyen Trai Hospital, Ho Chi Minh City, Viet Nam

Abstract — **In this paper the novel neural NARX technique was applied for classification and detection purposes. Decision making was performed in two stages: feature extraction using the principal component analysis (PCA) and the neural NARX model trained with the backpropagation method. The performance of the neural NARX-based classifier was evaluated in terms of training and classification accuracies. The results confirmed that the proposed neural NARX-based classifier has potential in detecting the tumors.**

Keywords — **medical image classification, neural NARX technique, principal component analysis (PCA), back propagation (BP) method, sympton tetection.**

I. Introduction

The conventional method in medicine for brain MR images classification and tumor detection up to now based on human inspection. However operator-assisted classification methods are impractical for a large amounts of data and also non-reproducible. MR images also always contain a noise caused by operator performance which can cause seriously inaccurate classification. The use of artificial intelligent techniques, for instance, neural networks, have shown great potential in this field. Hence, in this paper the novel neural NARX technique was applied for classification and detection purposes. Decision making was performed in two stages: feature extraction using the principal component analysis (PCA) and the neural NARX model trained with the backpropagation method. The performance of the neural NARX-based classifier was evaluated in terms of training and classification accuracies. The results confirmed that the proposed neural NARX-based classifier has potential in detecting the tumors.

Nowadays automated classification and detection of tumors in various medical images demands high accuracy since it deals with human life. Also, computer assistance is highly sought in medical institutions due to it could improve the results of humans in such a domain where the false negative cases must be at a very low rate. It has been proven that double reading of medical images could lead to better tumor detection. But the cost incurred in double reading is very high, therefore good software to assist humans in medical institutions is of great interest.

This paper proposed the automated classification of brain magnetic resonance images (MRI) based on prior knowledge of pixel intensity and anatomical features. Since up to now there were no widely accepted methods, hence automatic and reliable methods for tumor detection are of great demand. The application of neural fuzzy systems in the classification and detection of data for MRI problems are not fully utilized yet. These include the clustering and classification techniques especially for MR images problems with great scale of data which consumes time and energy if completed manually. Thus, deeply understanding the detection, classification and clustering techniques is quite necessary to the developments of neural fuzzy applications particularly in medical-related problems. In addition, fuzzy set theory plays an important role in dealing with uncertainty when making decisions in medical applications [1]. A specific approach in neuro fuzzy development is the adaptive neuro fuzzy inference system (ANFIS), which has shown significant results in modelling nonlinear functions. The ANFIS learns features in the data set and adjusts the system parameters adapt to a given error criterion [2]. Successful implementations of ANFIS in biomedical engineering have been reported, for classification [3] and data analysis [4][8].

Rami J.Oweis et al., (2005) [5] introduced the pixel classification of biomedical image using neuro fuzzy approach. The method based on the spatial properties of the image features and makes use of multi-scaled representations of the image. The classes are represented as fuzzy sets with degrees of memberships. Each pixel is assigned a degree of membership for each of the three fuzzy subsets. Classified pixels are finally shown as three separate images each representing a set. Proposed method showed rather complex but ontained high quality classification for MR images.

* Corresponding author.

V. Van Toi et al. (Eds.): 4th International Conference on Biomedical Engineering in Vietnam, IFMBE Proceedings 40, pp. 335–342, 2013.
www.springerlink.com

Benamrane et al., (2006) [6] proposed an approach for detection and classification of anomalies in medical images. The idea is to combine three tools: Neural Networks, Fuzzy Logic and Genetic Algorithms in a hybrid system. The Neural Networks and Fuzzy Logic are coupled in one system called Fuzzy Neural Networks. The Genetic Algorithm adds the property of total research like an initialization of the Fuzzy Neural training algorithm. After applying the growing region algorithm to extract regions, the Fuzzy Neural Network detects the suspect regions, which are interpreted by the Fuzzy Neural Network of specification.

Castellanos et al., (2000) [7] have evaluated a segmentation technique for magnetic resonance (MR) images of the brain based on the adaptive fuzzy leader clustering (AFLC) algorithm. This approach performs vector quantization by updating the winning prototype of a competitive network through an unsupervised learning process. Segmentation of MR images is formulated as an unsupervised vector quantization process, where the value of a vigilance parameter restricts the number of prototypes representing the feature vectors.

Since all these techniques have their advantages and drawbacks, it needs more research in this domain and maybe some combination of the already existing methods or some new image processing techniques will lead to better results. Hence, in this paper the novel fuzzy NARX technique [10][11] was firstly applied for classification and detection purposes. Decision making was performed in two stages: feature extraction using the principal component analysis (PCA) and the fuzzy NARX model trained with the backpropagation method. The performance of the fuzzy NARX-based classifier was evaluated in terms of training and classification accuracies. The results confirmed that the proposed fuzzy NARX-based classifier has potential in detecting the tumors.

This paper is organized as follows. In Section 1, some overview including the background, problem statement, and some literature reviews, related works and also theories regarding this paper are included. Magnetic resonance imaging is introduced in Section 2. Section 3 contains a research methodology of the proposed method, the extraction of a region of interest, transformation of image into a form of data and also data fuzzy NARX modelling. The fuzzy NARX architecture and software development and implementation for classification are discussed in Section 4. Section 5 presents the result from the simulation and discusses the performance analysis. Finally, the conclusion and some recommendations for future work are expressed in the Section 6.

II. MAGNETIC RESONANCE IMAGING

Magnetic resonance imaging (MRI) is a procedure used to produce high resolution images of the inside of the body, including the brain. MRI is a type of *nuclear magnetic resonance* (NMR), which is used by chemists to study the properties of molecules. The technique is called magnetic resonance imaging rather than nuclear magnetic resonance imaging because of the public's negative associations with the word *nuclear* in the late 1970's. MRI, like other imaging techniques such as CAT and PET, is primarily a tomography imaging technique ("tomo" means slice). This means that it produces an image in a thin slice through the body. This form of imaging is, in some respects, very much like a real anatomical slice through the body. In fact, to the untrained eye, an MRI image may be indistinguishable from a slice taken from a post-mortem body. However, unlike a post-mortem slice, the MRI is formed by an MRI camera or scanner and the slices are composed of groups of picture elements or *pixels* displayed on a computer monitor. In MRI the subject or patient is placed inside the scanner which is made up of a moveable bed-like structure and a large hollow tube. The tube contains a thick coil of wire that generates a very intense magnetic field (*between 1.5 and 4 Tesla*) which is strong enough to accelerate a paper clip to near lethal velocities across the distance of a small room. In order to generate a magnetic field of this strength the coil is cooled to near absolute zero with liquid helium. This very strong magnetic field is then used to align the hydrogen nuclei of the tissue to be imaged. Also placed near the tissue to be imaged is another coil of wire called an RF coil (RF because it generates/reads radio frequency fields). The RF coil is used to both change the energy state of the hydrogen nuclei and to record the RF output of these perturbations. The latter are the raw data of an MRI which are recorded by the computer, transformed and then displayed [9].

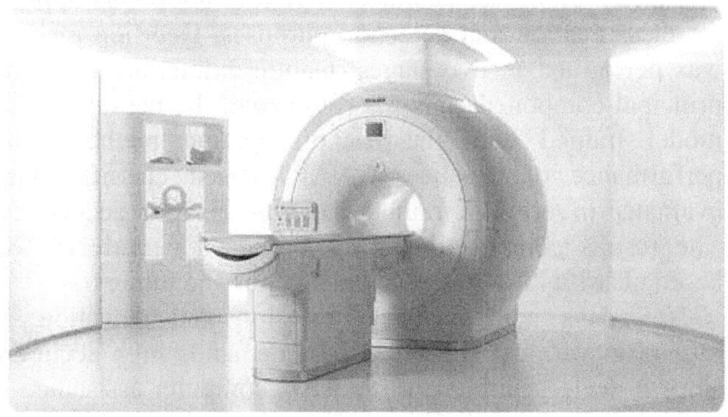

Fig. 1 Structure of MRI machine

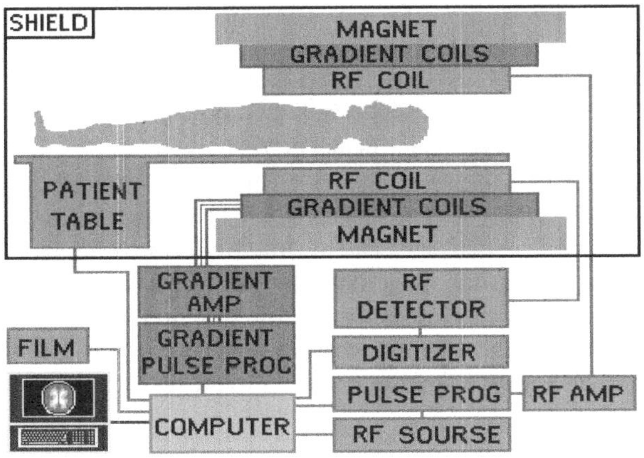

Fig. 2 Schematic diagram of MRI machine

Fig. 1 shows a picture of an MRI and Fig. 2 shows the schematic of an MRI machine. In MRI, water molecules give off radio signals which are converted into high resolution images that look like a picture shown in fig. 3.

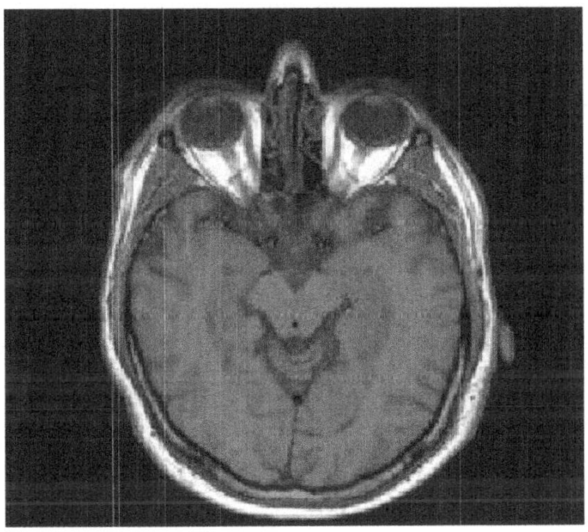

Fig. 3 Brain MR Image

The conventional method in medicine for brain MR images classification and tumor detection is human inspection. Operator-assisted classification methods are impractical for large amounts of data and are also non-reproducible. MR images also always contain a noise caused by operator performance which can lead to serious inaccuracies classification. The MR images data is by nature, a huge, complex and cognitive process. Accurate diagnosis of MR images data is not an easy task and is always time consuming. In some extreme scenario, diagnosis with wrong result and delay in delivery of a correct diagnosis decision could occur due to the complexity and cognitive process of which it is involved.

The use of artificial intelligent techniques for instance, neural networks, fuzzy logic, neuro fuzzy has shown great potential in this field. With the involvement of soft computing, the pattern matching, classification and detection of algorithms which have direct applications in many medical problems have become much easier to be implemented and diagnosed. Hence, this paper tries to find out how the neuro fuzzy system or the hybrid combination of neural network and fuzzy logic techniques can be applied in the classification and detection purposes, particularly on the medical field.

III. RESEARCH METHODOLOGY

Generally, two main phases are employed in this paper which is theoretical and practical phase. Theoretical phase involves the process of reading and understanding, reviewing of theories, studying and surveying the proposed fuzzy NARX model, classification and detection techniques on medical image fields. Forwardly, the practical phase involves medical image processing, fuzzy NARX model coding, simulation, assumptions, adaptation and training of the neuron and etc by using MATLAB.

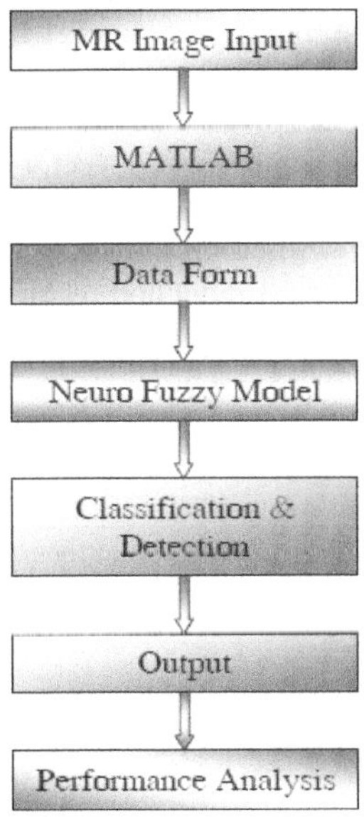

Fig. 4 Fuzzy NARX-based MRI classification process

Fig.4 shown that there are 7 stages involved in the proposed model which starts from the data input to output. The first stage is the image processing system. Basically in image processing system, image acquisition and enhancement are steps that need to be done. In this paper, these two steps are skipped and all the images are collected from available resources. The proposed model requires converting the image into a format capable of being manipulated by the computer. The MR images are converted into matrices form by using MATLAB. Then, the fuzzy NARX model is developed using MATLAB programming. After the fuzzy NARX model is successfully developed, the classification of the MR images starts. Symptoms detection phase will follow once the output from the classification technique is done and lastly, performance based on the result will be analyzed at the end of the development phase. The proposed brain MR images classification and detection method is shown in Figure 4.

In this paper, the principal component analysis (PCA) is used as a feature extraction algorithm. The PCA is one of the most successful techniques that have been used in image recognition and compression. The purpose of PCA is to reduce the large dimensionality of the data. MR image recognition systems find the identity of a given test image according to their memory. The memory of an MR image detector is simulated by a training set. In this paper, the training database consists of a set of MR images. Hence, the task of the MR image detector is to find the most similar feature vector among the training set to the given test image's feature vector.

In the training phase, feature vectors are extracted for each image in the training set. Let Ω1 be a training image of image 1 which got a pixel resolution of *M* x *N* (*M* rows, *N* columns). As to extract PCA features of Ω1, initially convert the image into a pixel vector Φ1 by compressing each of the *M* rows into a single vector. The dimensionality of the vector Φ1 will be *M* x*N*. Thus, the PCA algorithm is used as a dimensional reduction technique transformed the

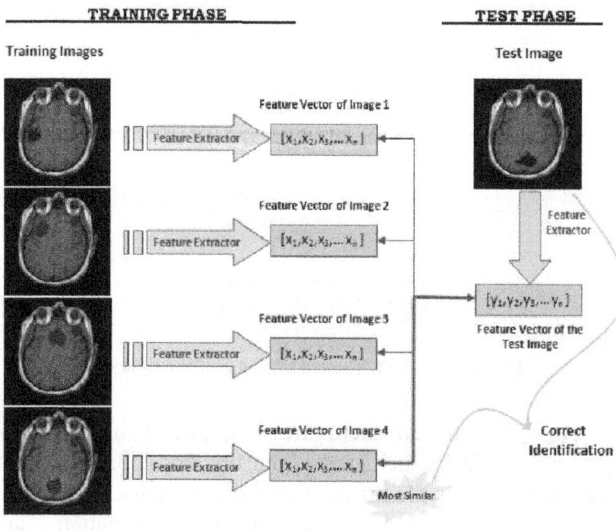

Fig. 5 Diagram of a MR image detection process

vector Φ1 to a vector ω1 which has a dimensionality *d* where *d* << *M* x *N*. For each training image Ωi, these feature vectors ωi are calculated and stored.

In the evaluating phase, the feature vector ωj of the test image Ωj is computed using PCA. In order to identify the test image Ωj, the similarities between ωj and all of the feature vectors ωi's in the training set are computed based on Euclidean distance. The identity of the most similar ωi is the output of the image detector. If $i = j$, it means that the MR image j has correctly identified, otherwise if $i \neq j$, it means that the MR image j has misclassified. Diagram of the MR image recognition system implemented is shown in Figure 5.

IV. Fuzzy NARX Model Implementation

A. *Proposed Inverse Fuzzy NARX Classifier of the MR Images*

The proposed IFN model of the magnetic resonance image (MRI) presented in this paper is improved by combining the approximating capability of the fuzzy system with the powerful predictive and adaptive potentiality of the nonlinear NARX structure. The resulting model establishes a nonlinear relationship between the past inputs and outputs and the predicted output, while the system prediction output is a combination of the system output produced by the real inputs and the historical behaviors of the system. This can be expressed as:

$$\hat{y}(k) = f\left(y(k-1),...,y(k-n_a),u(k-n_d),...,u(k-n_b-n_d)\right) \quad (1)$$

Here, n_a and n_b are the maximum lag considered for the output and input terms, respectively, n_d is the discrete dead time, and *f* represents the mapping of the fuzzy model.

The structure of the proposed DC motor IFN model interpolates between the local linear, time-invariant (LTI) ARX models as:

Rule j: if $z_1(k)$ *is* $A_{1,j}$ *and ... and* $z_n(k)$ *is* $A_{n,j}$ then

$$\hat{y}_j(k) = \sum_{i=1}^{n_a} a_i^j y_j(k-i) + \sum_{i=1}^{n_b} b_i^j u(k-i-n_d) + c^j \quad (2)$$

where $z_i(k)$, $i=1...n$ is the element of the $Z(k)$ "*scheduling vector*" which is usually a subset of the $X(k)$ regressor that contains the variables relevant to the nonlinear behaviors of the system. In this paper, $X(k)$ regressor contains all of the inputs of the inverse fuzzy NARX model

$$Z(k) \in X(k) = \{y(k-1),...,y(k-n_a),u(k-n_d),...,u(k-n_b-n_d)\} \quad (3)$$

The $f_j(q(k))$ consequent function contains all the regressors $q(k)=[X(k)\ 1]$,

$$f_j(q(k)) = \sum_{i=1}^{n_a} a_i^j y(k-i) + \sum_{i=1}^{n_b} b_i^j u(k-i-n_d) + c^j \quad (4)$$

In the simplest case, the NARX type zero-order fuzzy model (singleton or Sugeno fuzzy model which isn't applied in this paper) is formulated by the simple rule consequents:

Rule j : if $z_1(k)$ *is* $A_{1,j}$ *and...and* $z_n(k)$ *is* $A_{n,j}$ then

$$\hat{y}(k) = c^j \quad (5)$$

with $z_i(k)$, $i=1...n$ is the element of the $Z(k)$ regressor containing all of the inputs of the DC motor IFN model:

$$Z(k) = X(k) = \{y(k-1),...,y(k-n_a), u(k-n_d),...u(k-n_b-n_d)\} \quad (6)$$

Thus the difference between the fuzzy NARX model and the classic TS Fuzzy model method is that the output from the TS fuzzy model is linear and constant, and the output from the NARX fuzzy model is the NARX function. However, both of these methods have the same fuzzy inference structure (FIS).

B. *MGA-Based MRI IFN Model Identification*

The problem of modeling the nonlinear and dynamic system always attracts the attention of researcher. Some research has been published using a fuzzy model based on expert knowledge. Unfortunately the resulting fuzzy model was often too complex to be applied in practice and thus only simulation was carried out. Fig. 6 initially presents the block scheme for the modeling and identification of a MGA-based inverse fuzzy NARX model using experimental input-output training data. MGA stands for Modified Genetic Algorithm and the MGA will be described later in the section 4.3.

This proposed approach can help to simplify the modeling procedure for nonlinear systems. Genetic algorithm (GA) is applied to optimize the FIS structure and other parameters of proposed fuzzy model. However the poor experimental result proves that the GA-based TS fuzzy model is incapable of modeling all nonlinear, dynamic features of the dynamic system. Recently the fuzzy/neural NARX model has been successfully applied to identify nonlinear, dynamic system [10-11].

The block diagram presented in Fig.6 illustrate the MGA-based MRI IFN model identification. The error *e(k)=U(k)-Uh(k)* is used by the MGA algorithm to calculate the Fitness value (see Equation (7)) in order to identify and optimize parameters of the proposed MRI IFN model.

$$F_j = 10^4.(\frac{1}{M}\sum_{k=1}^{M}(y(k) - \hat{y}_j(k))^2)^{-1} \quad (7)$$

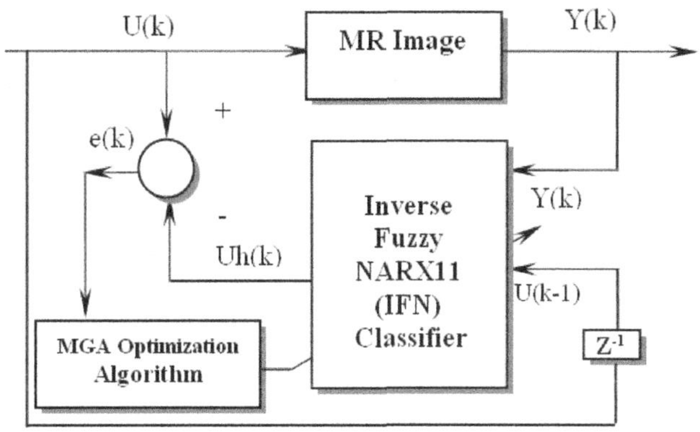

Fig. 6 Block diagram of the MGA-based MR Image inverse fuzzy NARX classifier identification.

C. *MGA-Based Inverse Fuzzy NARX Classifier Identification*

The classic GA involves three basic operations: reproduction, crossover and mutation. As to derive a solution to a near optimal problem, GA creates a sequence of populations which corresponds to numerical values of a particular variable. Each population represents a potential solution of the problem in question. Selection is the process by which chromosomes in population containing better fitness value having greater probability of reproducing. In this paper, the roulette-wheel selection scheme is used. Through selection, chromosomes encoded with better fitness are chosen for recombination to yield off-springs for successive generations. Then natural evolution (including Crossover and Mutation) of the population will be continued until a desired termination or error criterion achieved. Resulting in a final generation contained of highly fitted chromosomes represent the optimal solution to the searching problems.

In recent years, the GA has continued to be improved upon and has been applied successfully to identify and optimize different nonlinear, dynamic systems [12-13]. However the inappropriate choice of operators and parameters used in GA process makes the GA susceptible to premature convergence. The focus of this paper is to attempts to simultaneously apply two improved strategies as a means to overcome these problems.

(1) *Extinction strategy.*
(2) *Elitist strategy.*

Fig.7 shows a procedure for the MGA-based identification process. The proposed Modified Genetic algorithm (MGA) adopts all of the advanced strategies that were used to modify the classic GA. The elitist strategy ensures a steady increase in the maximum fitness value. The extinction strategy prevents the searching process from becoming trapped in local optima. Consequently, the overall efficiency and the optimum solution are greatly improved when these modifications are used.

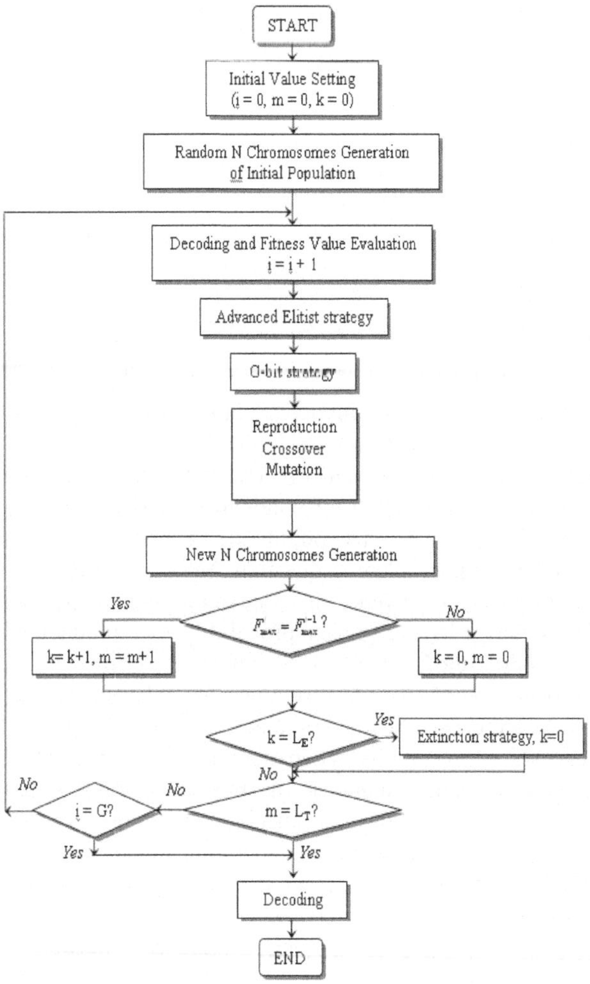

Fig. 7 Flow chart of the MGA-based Identification Process.

To run a MGA, suitable encoding needs to be carefully completed for each of the parameters and bounds. For this task the parameters given in Table 1 are used with the ranges and precision parameters that are shown. Binary encoding is used because it allows the MGA more flexibility in searching the solution space thoroughly. The number of membership functions is limited to odd integers, which are inclusive between (3–9) when using the MGA-based MRI inverse fuzzy NARX11 model and between (3–5) when the MGA-based MRI inverse fuzzy NARX model identification is used. Experimentally, this was considered to be a reasonable constraint to apply. The advantage of doing this is that this parameter can be captured in just one to two bits per variable.

Table 1 MGA-based MRI IFN Classifier Parameters used for Encoding.

Parameter	Range	Precision	No. of Bits
Number of Membership Functions	3-9	2	2
Membership Function Scaling	0.1 – 1.0	0.01	7
Membership Function Spacing	-1 - 1	2	1
Rule-Base Scaling	0.1 – 1.0	0.01	7
Rule-Base Spacing	-1 - 1	2	1
Input Scaling	0 - 100	0.1	10
Output Scaling	0 - 1000	0.1	14
Rule-Base Angle	0 - 2π	π/512	11

V. EXPERIMENTAL RESULTS

Various experiments were performed and the sizes of the training and testing sets were determined by taking into consideration the classification accuracies. The data set was divided into two separate data sets – the training data set (20 subjects) and the testing data set (15 subjects). The training data set was used to train the fuzzy NARX model, whereas the testing data set was used to verify the accuracy and the effectiveness of the trained fuzzy NARX model for the detection of brain tumors.

The classifier is able to differentiate between normal and abnormal brain. There were a total of 27 fuzzy rules in the architecture of the fuzzy NARX model using a 3-type (generalized bell, triangular and pi) shaped membership function. The fuzzy NARX model was implemented by using MATLAB software package. The fuzzy NARX model used 20 training data in 40 training periods and the step size for parameter adaptation had an initial value of 0.01. At the end of 40 training periods, the final error fitness convergence value is 2.5x10-3. The overall result regarding to the fuzzy NARX model fitness values can be seen in Table 2.

Table 2 The fuzzy NARX fitness values.

Fitness Values	Average
Training Fitness Values	0.00238
Evaluating Fitness Values	0.00542

Concerning performance analysis, we see that in a real world domain, just like the one used in the present study, all of the features used in the descriptions of instances may have different levels of relevancy. Therefore, in the present study changes of the final (after training) membership functions with respect to the initial (before training) membership functions of the input parameters were examined. The final (after training) membership functions of fuzzy NARX model are shown in Figure 8. The proposed method gives a framework to detect more quickly and comfortably brain tumors using fuzzy NARX model method for the classification.

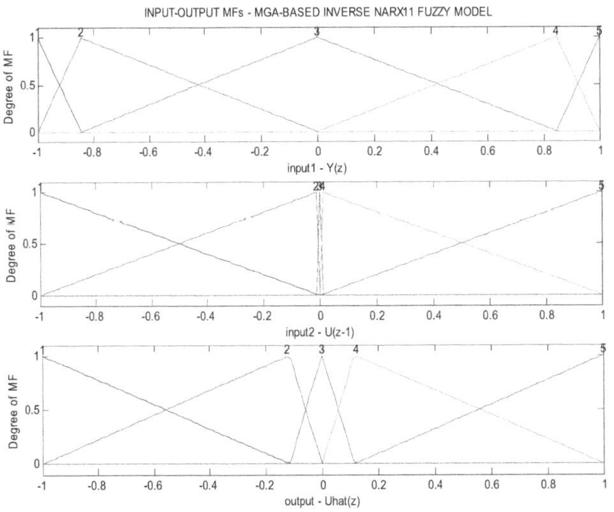

Fig. 8 Membership function after training

After training, 15 testing data was used to validate the accuracy of the fuzzy NARX classifier for the detection of brain tumors. The test results of the fuzzy NARX model are presented in Table 3. According to Table 3, 1 subject was classified incorrectly by the fuzzy NARX model. The test performance of the fuzzy NARX model was determined by the computation of the statistical parameters such as specificity, sensitivity and accuracy. The values of these statistical parameters are given in Table 3. As seen from that table, the fuzzy NARX model classified MRI subjects with the accuracy of 93.33%.

Table 3 The fuzzy NARX classifier validation results

Type of Data	Number of Sublects	Number of Correctly Classified Subjects
MR Images	15	14

In summary the resulting procedure of this research is as follows:

i. Involve the fuzzy NARX model coding, simulation, adaptation and training using the MATLAB.

ii. Start with converting the brain medical images into a form of 1D data using MATLAB.

iii. The brain medical images also have to be pre-processed using PCA to reduce the dimensionality of the data.

iv. Classify brain medical images based on the region of abnormal tissues using fuzzy NARX model.

v. The proposed technique will be coded in MATLAB and subsequently to be trained and tested with the brain medical images data.

In future we intend to design a combining multiple classifiers that use different features such as texture features, tumor shape and size features, etc in order to get an accurate classification of the normal and abnormal brain image. A combined feature is more meaningful than the one feature representation alone and can significantly improve the performance..

VI. Conclusions

Medical image classification is an important thing in medicine. It allows for biological structures to be isolated non-invasively. Whether it is for diagnostic purposes or practically applied in image guided surgeries, image classification has many forms and uses. Unfortunately, currently there is no classification strategy that can accommodate all its applications. The system that has been designed is used for classifying and detecting tumors in brain MR images. The system involves two major modules, one that performs classification and the other one that detects the tumors from the images. Proposed fuzzy NARX model is suitable and feasible to be used as supportive tools for medical problems. The combination of both, fuzzy and ARX model or known as fuzzy NARX model has been tested and evaluated for classification and detection of MR images.

Based on the result in Section 5, it can be concluded that this project has achieved the objectives and problem statements as set earlier in Section 2. There is a need to apply the classification and detection techniques in many fields especially in the medical field whereby the data is large and time consuming to sort or classify manually. Furthermore, it is very crucial to determine the accuracy and precision of the output of the analysis, especially in the medical problems or field such as lung cancer, breast cancer and etc. Experimental result indicates that the novel fuzzy NARX technique is workable with an accuracy greater than 90%. This technique is fast in execution, efficient in classification and easy in implementation.

Acknowledgment

This paper was partially supported by the NAFOSTED, Viet Nam and the DCSELAB, VNU-HCM, Viet Nam.

References

1. L.I. Kuncheva, F. Steimann, "Fuzzy diagnosis," Artificial Intelligence in Medicine, vol. 16, pp. 121-128, 1999.
2. J.-S.R. Jang, "ANFIS: Adaptive-network-based fuzzy inference system," IEEE Transactions on Systems, Man, and Cybernetics, vol. 23(3), pp. 665-685, 1993.
3. S.Y. Belal, A.F.G. Taktak, A.J. Nevill, S.A. Spencer, D. Roden, S. Bevan, "Automatic detection of distorted plethysmogram pulses in neonates and paediatric patients using an adaptive-network-based fuzzy inference system," Artificial Intelligence in Medicine, vol. 24, pp. 149-165, 2002.
4. I. Virant-Klun, J. Virant, "Fuzzy logic alternative for analysis in the biomedical sciences," Computers and Biomedical Research, vol. 32, pp. 305-321, 1999.
5. Rami J. Oweis and Muna J.Sunna', "A Combined Neuro Fuzzy Approach for Classifying Image Pixels in Medical Applications". Journal of Electrical Engineering, Vol.56, No.5-6, 2005.
6. N.Benamrane, A. Aribi, L.Kraoula, "Fuzzy Neural Networks and Genetic Algorithms for Medical Images Interpretation". IEEE Proceedings of the Geometric Modeling and Imaging, 2006.
7. Ramiro Castellanos and Sunanda Mitra, "Segmentation of magnetic resonance images using a neuro-fuzzy algorithm". IEEE Symposium on Computer-Based Medical Systems, 2000.
8. Panella, M.; Rizzi, A.; Mascioli, F.M.F.; Martinelli, G., ANFIS synthesis by hyperplane clustering, IFSA World Congress and 20th NAFIPS International Conference, vol.1: 340 -345, 2001.
9. Joseph P. Hornak, The Basics of MRI, 1996-2008.
10. Anh, H.P.H. and Ahn, K.K., 2009, "Identification of the pneumatic artificial muscle manipulators by MGA-based nonlinear NARX fuzzy model," IFAC Journal of MECHATRONICS, Volume 19, Issue 1, pp. 106-133 (February 2009).
11. Anh, H.P.H. and Ahn, K.K., 2010, "Inverse Double NARX Fuzzy Modeling for System Identification," IEEE/ASME Journal of MECHATRONICS, Volume 15, Issue 1, pp. 136-148 (February 2010).
12. S.K. Oh, W. Pedrycz, H.S. Park, "A New Approach to the Development of Genetically Optimized Multilayer Fuzzy Polynomial Neural Networks," IEEE Trans. on Industrial Electronics, vol. 53, no. 4, pp. 1309- 1321, Aug 2006.
13. S.-B. Roh, W. Pedrycz, S.K. Oh, "Genetic Optimization of Fuzzy Polynomial Neural Networks," IEEE Trans. on Industrial Electronics, vol. 54, no. 4, pp. 2209-2218 , Aug 2007.

Author: Ho Pham Huy Anh, Le Tan Loi
Institute: HCM City University of Technology, VNU-HCM
Street: 268 Ly Thuong Kiet street, 10Dist., HCM City
City: Ho Chi Minh City
Country: Viet Nam
Email: hophamhuyanh@hotmail.com

Performance Study of GPU- and FPGA-Based 3-D Monte Carlo Computation Used in Dynamic Radiation Topotherapy

T.T. Le[1] and J.H. Levan[2]

[1] Electrical Engineering Department, San José State University, San Jose, California 95192-0084 U.S.A.
[2] Midwest Nuclear Proton TopoRx Research Group, Chicago, Illinois 60077, U.S.A.

Abstract — **The development of cluster and parallel computers has increased the popularity of implementing the Monte Carlo computational method in medical and scientific applications. However, having access to these high-performance systems is not easy and a reasonable size system is not yet able to solve many popular Monte Carlo problems near real-time. The Field Programmable Gate Arrays (FPGAs) have been proved to be useful in speeding up the Monte Carlo algorithm since they are able to execute the problem with a high degree of parallelism. In recent years, the Graphic Processing Units (GPUs) have been proved to be useful for many general-purpose calculations due to their parallel and pipelined architecture, which is able to accelerate various algorithms. Since Monte Carlo simulations require intensive computations, GPU architecture can also deliver great performance for Monte Carlo simulations similarly to the FPGA. This paper explores performance characteristics of the FPGA and GPU in computational intensive 3-D Monte Carlo applications. The study is part of our current research on the study of dynamic radiation topotherapy with stationary sweeping fan beam and translational moving microbeam on stationary treatment couch. We performed our initial study with photon interactions on 3-D water phantoms and the initial results (without intensive performance tuning) showed that performance of the 16-core NVidia GeForce 8400 GS GPU was about 2 times better than single-core 3-GHz Intel Xeon and about 4 times slower than the Xilinx Virtex-5 FPGA. We learned that out-of-core memory contention is an important performance drawback. Although we performed our study on photon transport, we expected the results to be similar for charged-particle transport (our interest in dynamic topotherapy) since only core Monte Carlo calculations were implemented on the GPU and the FPGA.**

Keywords — **3-D Monte Carlo, Field Programmable Gate Array, Graphic Processing Unit, Dynamic Radiation Topotherapy.**

I. INTRODUCTION

Monte Carlo (MC) is the most powerful computational method used in solving nuclear particle transport problems [1-2]. There is substantial evidence that Monte Carlo method is also a powerful computational scheme for solving medical physics problems such as dose calculations in radiotherapy treatment and planning [3-6]. Monte Carlo method can simulate individual particles in 3-dimentional with complex geometries. The method uses microscopic physical data and statistical algorithms for dose calculations that can apply on a wide range of clinically important conditions including complex beam geometries and tissue heterogeneities. The method can be used with charged particles (proton and electron therapy) and non-charged particles (photon and neutron therapy) simulationsand the accuracy of the calculation can be predefined by setting the target errors andincreasing the number of histories for better averages.

More than 85% of radiation treatments are photon therapies and more than 10% are electron therapies. The proton and neutron therapies are rarely used due to their new developments and so the cost. Photon beams with high energies have moderate to long ranges (tens of centimeters) and so are mostly used for treatments of internal tumors, while electron beams are useful for treatments of shallow cancers because of the limitation in their penetration power in the energy ranges used. Since high-energy proton beams can deposit most of their energy at a range defined by the beam's energy, a treatment planner can tailor a proton beam to deliver mostof its radiation dose into the tumor while avoiding healthysurrounding tissue. For radiation-resistant tumors, the neutron radiation treatment is more effective than the photon one, but the method is relatively new [7].

The success of a radiation therapy depends on the accurate computation of dose distribution. Monte Carlo simulations can provide an accurate solution but the method has high computation time that prevents them from being used in treatment planning. This highlights the need for accelerating the computation to enable the use of Monte Carlo models in radiotherapy treatment and planning.

This paper describes our initial study on the performance of Graphic Processing Unit (GPU)[8] and Field Programmable Gate Array (FPGA)[9, 10] in 3-D Monte Carlo computations, which can be used in any type of radiation treatment as well as other applications in radiation physics. The paper however is limited to experiments that were performed for thevalidation of our dynamic radiation

V. Van Toi et al. (Eds.): 4th International Conference on Biomedical Engineering in Vietnam, IFMBE Proceedings 40, pp. 343–346, 2013.
www.springerlink.com

topotherapy study [11, 12]. Our work on FPGA- and GPU-based 3-D Monte Carlo is still in progress and final results are subject to future publications.For this initial study, we developed our own simple Monte Carlo program for a simple 3-D cubic geometry. The program was run on a single-core 3-GHz Xeon CPU [13] and the results were compared with physical measurements. The major computation operations of the program were implemented on the Xilinx Virtex-5 FPGA [14] and on the 16-core NVidia GeForce 8400 GS GPU cards [15], but the overall execution wasstill under the control of the host CPU. The simulation elapsed-time of the 3-GHz Xeon CPU, the Xilinx Virtex-5, and the GeForce 8400 GS GPU were then measured and compared.

II. Concepts of Dynamic Radiation Topotherapy

Topotherapy is one of the new procedures for helical tomotherapy [16]. The helical tomotherapy involves the delivery of radiation by a continuous fan beam with the treatment couch set in motion. Topotherapy however uses a fixed-angle modulated beam to accommodate the continuous and translational motion of a couch through the radiation beam.

Our topotherapy study explores the applications of topology and topography concepts in developing a new topo-radiotherapy treatment procedure. This approach can be associated with either a stationary sweeping fan beam or a translational moving microbeam while the treatment couch is in a fixed position [11, 12]. The treated tumor is looked at from the mapping topographic "eye" of the radiation beam involved as the surface of the earth viewed from an orbiting satellite passing above. The ultrathin scanning radiation beam "sees" the tumor topographic configuration at different source-object distances, and the radiation beam would be programmed to "stop" over on a nanospot of the object according to principles of energy-time-dose relationships in radiation physics.

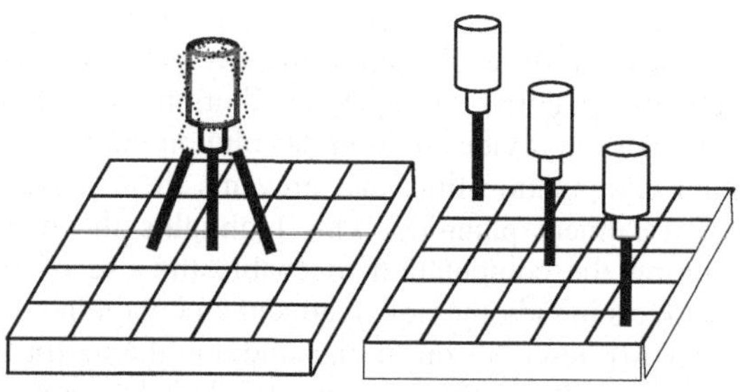

Fig. 1a Sweeping Fan Beam Fig. 1b Translational Scanning

Although the topotherapy procedures are more practical with electron than with photon, we performed our experiments with photon due to the unavailability of electron radiotherapy machines. Our measurements were being carried out with a Cyberknife equipped with a collimating system mounted on a robotic arm, adaptable to a set of exchangeable circular cones. Figures 1a and 1b show our dosimetric measurement experiments using 5mm collimator. The surface of the tumor was arranged as 3-by-3, 5-by-5, and 7-by-7 (odd-numbered) pixel matrix located at 80 cm source-axis distances (SAD). In the sweeping fan beam configuration, the radiation source was stationary in space, but spread out from one tumor pixel border to the other and repeated this sweeping motion in the X and Y directions. In the translational scanning configuration, the radiation source moved from one tumor pixel border to the other by scanning all pixels involved in either X or Y direction.The dose-volume optimization was provided by a treatment planning system, the peripheral pixel doses were measured with Kodak EDR-2 film, and the central axis pixel doses were measured with Exradin ion chambers and Kodak EDR-2 film. The experiment and simulation results are summarized in Section VI of the paper.

III. Monte Carlo Implementation

A Monte Carlo program was developed for modeling the proposed topotherapy procedure. The Monte Carlo computations are based on the radiative transport equation shown in the equation below [17]:

$$\frac{1}{v}\frac{\partial}{\partial t}\phi(r,\Omega,t)=-\Omega\cdot\nabla\phi(r,\Omega,t)-\left[\mu_a(r)+\mu_s(r)\right]\phi(r,\Omega,t)$$
$$+\int_{4\pi}\phi(r,\Omega,t)\mu_s(r,\Omega'\rightarrow\Omega)\,d\Omega'+S(r,\Omega,t)$$

where v is the speed of light, ϕ is the photon directional fluent, $\mu_a$is absorption coefficient, μ_s is scattering coefficient, S is photon source, r represents the photon position, t represents time, and Ω is the photon propagation direction.

Thesample tumor volume was divided into a large number of small voxels in three-dimensional geometric configuration. The voxel volume (or resolution) was defined based on the photon mean free path.In general, the sampling and computation procedure in Monte Carlo can be classified into three categories interacting to each other. These 3 categories are the samplings for the physics of the particle, the calculations of geometry-related particle status, and the calculations of particle contribution to the interested results. The samplings for particle physics include the samplings for collision location, collision type, collision energy and direction, secondary particle generations, status of incident and target particles, etc... The calculations of geometry-related particle status include the search for particle location, medium, and problem boundaries, etc...

The overall procedure of Monte Carlo simulation includes 3main loops. The outermost loop iterates over a number of time steps for time-dependent tissue optical properties. The middle loop iterates over a number of independent batches per time step, and the innermost loop iterates over a number of histories (or particles) per batch.For each history, a "while loop" is used to sample and calculate particle distance to next collision, particle location, boundaries, transmission or reflection, scattering or absorption, secondary particle generation, particle survival roulette, etc...By recording the dose deposited by each of the millions of histories tracked, the method reports a detailed map of the dose distribution over the target geometry. The larger the number of histories, the smaller the sampling error and the closer the approximation is to the continuous solution.

The main disadvantage of Monte Carlo methods is the lengthy calculation times required to obtain meaningful results. A high-performance personal computer would take several days to complete the calculation for a particular case.For Monte Carlo methods to be useful, the computationtime must meet the planning time, which is less than an hour or so for clinics to manage real-world caseloads. Dedicated cluster system can provide the required performance but it would be too high in power consumptionand equipment cost. With the availability of real-time treatment planning, the treatment would be based on the most recent images of the treated volume. This will take into account the changes in tissue properties as the treatment is in progress.

The Monte Carlo program that we implemented has cubicgeometry and the problem physics properties include absorption coefficient, scattering coefficient, refractive index, and anisotropy factor. Since absorption was the only interested event in this study, the absorption was saved in a 3D absorption array and other results were eliminated.For sweeping fan beam problem, the radiation source was fixed above the phantom center pixel and initial photon incident angles to all pixel surface centers were pre-calculated and stored. For translational scanning beam problem, the initial photon beam was perpendicular all surface centers of the pixels. Although several calculations can be eliminated due to the homogeneousproperties of our experiment geometry, we still implemented these calculations in the code at this moment to reserve for future experiments. In our program, the two highest computational costs involved inthe tracking of each photon particle were (1) the calculations of travel distance that involve logarithm, division,and multiplication operations, and (2) the calculations of direction cosine that involve trigonometry, square, square root, division, and multiplication operations.

IV. Implementing on the FPGA

Implementing a Monte Carlo program on an FPGA requires architecture exploration in order to match required datapaths to available resources on the FPGA. In order to take advantage of the number of Look Up Tables (LUTs)and memories available on the FPGA, trigonometry and logarithmoperations were converted into lookup tables with linear interpolation. The operation procedure of the Monte Carlo program was not directly mapped into the hardware, but computational entities from the method were viewed as computational primitives controlled by a control unit. Timing properties of computational primitives were measured and the overall MC procedure was programmed based on the dataflow model. In order to reduce the required memory, only final calculated doses were recorded.

To maximize the performance, the program should be implemented with both pipelining and parallel operations. We did not utilize the pipelining technique in this experiment, but only parallelized the computationsteps by scheduling the computation primitives. This is one of the main reasons that we did not get high performance as expected and is subject to further improvements.

For the current experiment, all computations related to the tracking of photon position were implemented inthree independent primitives connected in locked-step series. With this implementation, each primitive can operate on different photon at the same time. The computations related to direction and absorption updates wereimplemented in other three primitives connected in parallel. Each photon particle would go through four of the above six major computational primitives. Timing measurements were performed on all primitives and the information was used by the control circuit in steering the data through the computation stages.

Although Virtex-5 can have up to 16.4 Mbits of integrated block memory, the lack of on-chip memory was a problem in our current FPGA implementation since we used 3-D absorption array. This is another reason that we did not get high performance as expected and is also subject to further improvement.

V. Programming the Graphic Processing Unit

Graphic Processing Unit (GPU) is specially designed for processing of high-definition high frame rate and so has good performance on computationally intensive parallel applications [18]. GPU is a highly parallel, highly multithreaded multiprocessor architecture that can provide real-time visual interaction with computed objects via graphics images and video. A GPU consists of a number of streaming multiprocessors (SMs) and eachmultiprocessor has a number of streaming processors (SPs).

The NVIDIA GeForce 8400 GS GPU used in our study is the 80nm technology processor with PCI Express x16 interface [15]. The GPU has 2 multiprocessors and each multiprocessor has 8 streaming processors with 450 MHz clock speed. The GPU has 256 MB Video Memory with DDR2 SDRAM technology. By using single-instruction multiple-thread (SIMT) scheme, all threads of a GPU perform the same sequence of Monte Carlo instructions for a number of photon particles.

In order to optimize the performance, program data was mapped to different memory hierarchy based on its nature. Data that is updated frequently and is private to each particle (such as particle position, direction cosines, etc…)was stored in the processor registers and shared memory. The unchanged data (such as material properties, problem geometry, etc…) was stored in the GPU constant memory, and data that is public to all photon particles (such as absorption dose) was stored in the global/device memory.

The code was compiled and installed on the GPU based on CUDA programming model [19]. Although multiple thread blocks should be launched on each multiprocessor for performance optimization, we launched only one thread block for each multiprocessor due to the memory limitation. Each thread block contains 64 threads and so each streaming processor executes 8 particles per launch.

VI. Results and conclusions

For the purpose of validating the experiments on dynamic radiation topotherapy, problems with3-by-3, 5-by-5, and 7-by-7 pixels were simulated and the ratios of Dose/Dose$_{max}$at pre-defined z-layers were compared with the measurements. The calculated doses are in fact the relative dose averages over the x-y planes. We executed the same problems on a single-core 3-GHz Intel Xeon system, a host system witha Virtex-5 FPGA card, and the same host system with a GeForce 8400 GS Graphics Card. The results showed that performance of the GeForce 8400 was about 2 times better than the 3-GHz Xeon but was about 4 times slower than the Virtex-5 FPGA. We also learnedfrom the experiments that out-of-core memory contention is an important performance drawback on the FPGA. Performance of the FPGAcan also be further improved by applying pipelining techniques to computational primitives. Although we performed our study on photon transport, we expect the results to be similar for charged-particle transport since the Bragg Peak effects and multiple scattering interactions of protons in a medium would not change the relative performance of the CPU, the FPGA, and the GPU.

Acknowledgment

Special thank goes to Robert P. Laureckas and Mylinh T. Dang at Rosalind Franklin University of Medicine and Science, for their excellent work in getting measurement data at the Christ Hospital using CyberKnife Robotic Radiosurgery System.

References

1. N. Metropolis and S. Ulam (1949). The Monte Carlo Method. Journal of the American Statistical Association, pp. 335–341, 1949.
2. J. Briesmeister (1993). MCNP TM–A General Monte Carlo N-Particle Transport Code. Los Alamos National Laboratory Manual LA-12625-M, Los Alamos, New Mexico, USA , 1993.
3. E. Heath, J. Seuntjens, and D. Sheikh-Bagheri (2004). Dosimetric Evaluation of the Clinical Implementation of the first Commercial IMRT Monte Carlo Treatment Planning System at 6 MV. Medical Physics 31, p. 2771, 2004.
4. C. Ma, E. Mok, A. Kapur, T. Pawlicki, D. Findley, S. Brain, K. Forster, and A. Boyer (1999). Clinical Implementation of a Monte Carlo Treatment Planning System. Medical Physics 26, p. 2133, 1999.
5. S. Coyle, K. Thomas, T. Naughton, M. Charles, and W. Toms (2006). Distributed Monte Carlo Simulation of Light Transportation in Tissue. Proceedings of 20th IEEE International Symposium onParallel andDistributed Processing, p. 4, 2006.
6. Ilic, RD., Lalic, D., and Stankovic, SJ. (2002). SRNA-Monte Carlo Codes for Proton Transport in Combined and Voxelized Geometries. Nuclear Technology & Radiation Protocol, 2002, 1-2, pp. 27-36.
7. Mayles, P; Rosenwald, JC; Nahum, A (2007). Handbook of Radiation therapy Physics. Taylor & Francis. ISBN 9-780-750-308-601.
8. David Luebke and Greg Humphreys (2007). How GPUs Work. IEEE Computer, 2007.
9. Jonathan Rose, Abbas El Gamal, Alberto Sangiovanni-vincentelli (1993). Architecture of Field-Programmable Gate Arrays. Proceedings of the IEEE. Vol 81. No 7, July 1993.
10. HartmutF. W. Sadrozinski, Jinyuan Wu (2010). Applications of Field Programmable Gate Arrays in Scientific Research. Taylor & Francis. ISBN 978-143-984-133-4.
11. J. H. LeVan, T. T. Le, R. P. Laureckas, and M. T. Dang (2007). A New Dynamic Topotherapy Using Stationary Sweeping Radiation Fan Beam and Moving Scanning Microbeam. International Journal of Radiation Oncology Biology Physics, Volume 69, Issue 3, November 2007.
12. Hoa Van Le and Thuy T. Le (2009). Cyberknife Stereotactic Radiosurgery System as a Simulation Device for Proton Scanning Beam Dynamic Topotherapy. Proceedings of the 8th National Conference on Nuclear Science and Technology, 2009.
13. Dual-Core Intel Xeon Processor 5160 Data Sheet. Intel Corporation. http://ark.intel.com/cpu.aspx?groupId=27219.
14. Andrew Percey (2007). Advantages of the Virtex-5 FPGA 6-Input LUT Architecture. Xilinx White Paper, December 19, 2007.
15. NVIDIA GeForce 8400M GT. www.notebookcheck.net/NVidia-GeForce-8400M-GT.3708.0.html.
16. Gonzalez, VJ., Bucholz, DJ., Langen, KM., et al. (2006). Evaluation of Two Tomotherapy-Based Techniques for the Delivery of Whole-Breast Intensity-Modulated Radiation Therapy. International Journal of Radiation Oncology Biology Physics, 2006, 65 (1), pp. 284-290.
17. L.-H. Wang, S. L. Jacques, and L.-Q. Zheng (1995). Monte Carlo Modeling of Photon Transport in Multi-Layered Tissues. Computer Methods and Programs in Biomedicine 47, 131–146 (1995).
18. E. Alerstam, T. Svensson & S. Andersson-Engels (2008). Parallel Computing with Graphics Processing Units for High-Speed Monte Carlo Simulation of Photon Migration, J. Biomed. Opt. 13, 060504 (2008).UDA Programming Guide 2.3. NVIDIA Corporation, 2009

A Useful Application of Self-recorded Photos by Mobile Phones in Maxillofacial Surgery: Case Report

Fereydoun Pourdanesh[1], Ashraf Sayyedi[2], and Masoud Yaghmaei[1]

[1] Shahid Beheshti University, Department of Oral and Maxillofacial Surgery, School of Dentistry, Tehran, Iran
[2] Shahid Beheshti University, Department of Prosthodontics, School of Dentistry, Tehran, Iran

Abstract — **In recent years, we have seen a some sort of medical literature about using of cell phones in different fields of medical sciences. The aim of this article to present the utility of patients coming to their clinicians will cell phone photos that serve to some extent as a substitute for missing medical records. The serial self –recorded photos by two of patients in two different fields of maxillofacial surgery were selected. Our cases have not any documented records from local hospitals at the time of admission to our clinic.**

The comprehensive study of the photos taken by patients showed it serve to some extent as an alternate for missing medical records. In cases where records do not exist or a new clinician has not access to the former clinician and the patients cannot explain what treatment they received, use of photos taken by the patient may help doctors to make better decisions.

Keywords — **cellular phone, medical record, decision making.**

I. Introduction

In recent years, we have seen a some sort of medical literature about using of cell phones in different fields of medical sciences. Although there are many reports of the use of mobile phone as a telehealth to send and receive short text message (SMS) in medicine[1-6] but there are a few reports of mobile photos being used for access to patients information[7-9].

Moumouldis et al, in 2007, described a novel use of photo messaging in the assessment of nasal fractures[7] . Schreier et [al in 2008, reported a mobile phone based teledermatology system to support self management of the patient suffering from psoriasis[11]. Donald et al in 2009 reported the use of mobile phones in dietary assessment in young people with type 1 diabetes (T1DM) [9], Walters et al in 2010 reported a mobile phone –based care model for outpatient cardiac rehabilitation[10]. and Hayn et al reported the results of the preliminary study about mobile phone –based teledermatologic compliance management in 20 patients suffering psoriasis[11]. Recently, using mobile phone as a part of telecare system or telemedicine for the self management of health condition is going to be increased. Meanwhile, in all above mentioned studies the physicians defined the design of study and they asked of their patients to take photos, but in this present work, self – recorded photos by patients showed it serve to some extent as an alternate for missing medical records. The aim of this article is to report our experience on two patients, who presented some part of their history of present illness by self –recorded photos.

II. Material and Methods

The serial self –recorded photos by two of patients in two different fields of maxillofacial surgery were selected.

A. Case1

A 17- year – old boy was presented to our department for treatment of erythematous swelling on his left cheek. He had a visible scar in left alar grove of the nose, when was asked about the history of his present illness, he showed the several photos in his cell phone which clearly showed the previous procedures he had undergone and the progress of his disease in the last few months (fig 1). Mean while these photos gave us more details of missing medical records compare than what patient had described before that. Self – recorded photos showed that we are confronting with a complicated chronic infection, and helped us to make a more proper decision. Mucormycosis was the final diagnosis based upon biopsy and tissue culture.

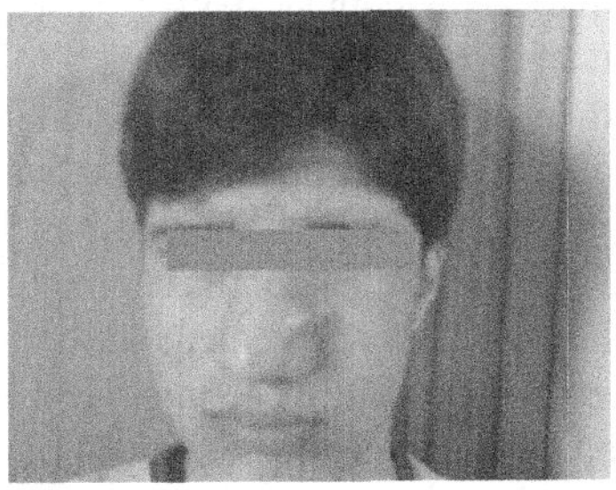

Fig. 1 A sample of pictures which was taken by patient in his cell phone before admission to our department.

V. Van Toi et al. (Eds.): 4th International Conference on Biomedical Engineering in Vietnam, IFMBE Proceedings 40, pp. 347–349, 2013.
www.springerlink.com

B. Case2

A 20-year- old man was referred with chief complaint of unaesthetic scar on his right frontal area due to previous car accident (12 months ago). Again when asked him about the history of present illness, he could hardly explain the procedures he had undergone. This case had also no medical report, when we request him for any photographs of his face before this event, he said, "I myself took many photos from time of accident till now, and all are in my cell phone" (fig. 2).these photos gave us more detailed information about the type of operation and also healing process phenomena in this patient and of course helping us to make a better decisions on choosing among the treatment options (scar revision, w plasty ,micro dermabrasion, laser therapy, skin graft, tissue expander, and so on). Finally after complete evaluation of patient's photos and his expectation, we decided to choose self-inflated tissue expander installation in the first phase.

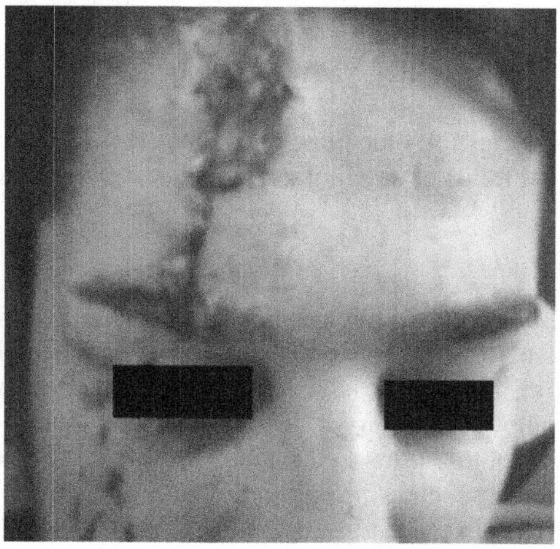

Fig. 2 A sample of pictures which was taken by patient in his cell phone before admission to our department.

III. Discussion

Using cell phones as popular devices is well documented to help doctors and patients in medicine [1-14]. Mao et al in 2008 reported using of Mobile phone text messaging for pharmaceutical care in a hospital in China was effective[1]. Moumouldis et al in 2007, described a novel use of photo messaging in the assessment of nasal fractures, based on their study, There was little agreement between photographic and clinical assessment and the clinical examination being the 'gold standard' method of assessment.[7] Shreier et al presented a mobile phone based telemedical compliance management system for psoriasis patients using special soft ware, patient can acquire health parameters such as their body weight ,take photos of their psoriasis lesions, and report side effect. The data are automatically sent to a monitoring center. for continuous clinical monitoring with periodic assessment of the state of the disease on psoriasis patient, they are provided to the patient physician via the use of Web –browser[8], and Donald et al reported their study on 15 adolescent patient with type one diabetic mellitus. They were asked to complete food, diary for 4 days and to photograph all food and drink consumed during that period. They concluded photographs taken using mobile phone may be a useful tool in the diabetic management of T1DM in adolescents [9]. Hayn et al concluded that mobile photos simplify the patient – physician communication and shorten the mean duration of a regular outpatient follow up. And also, it facilitates individual therapy adjustment based on the availability of photos from lesions in regular intervals [11]. But in one study the results showed there was no evidence to support the use of photographs sent by email or text message for the follow-up of facial lacerations sutured in the emergency department and despite the request of doctors, no patients sent a photograph[14].In our present work, without our request, the patients took a several photos in different times and saved in their mobile phones to show doctors and this is valuable to check the progress of the diseases .

Today's, with advances in technology, nearly all cell phone provides the possibility of taking photos by patient, so he or she can be able to record any visible change in his or her maxillofacial region and also the progress and outcome of the disease respectively. so it seems, in some cases serial photos might be very useful to show the progress of the disease, onset and extent of the infection, post surgical course, and outcome of interventions in traumatic patients. Therefore, photographs taken using mobile phones may be a useful tool in describing the history of present illness and helping to make better decision.

IV. Conclusions

Regarding to this limited study, the comprehensive study of the photos taken by patients showed it serve to some extent as an alternate for missing medical records.

In cases where records do not exist or a new clinician has not access to the former clinician and the patients cannot explain what treatment they received, use of photos taken by the patient may help doctors to make better decisions and self –recorded photos might be an alternate to some medical record missing.

References

1. Mao Y, Zhang Y , Zhai S[2008]. Mobile phone text messaging for pharmaceutical care in a hospital in China. J Telemed Telecare; 14:410-414
2. Ferrer-Roca O, Cárdenas A, Diaz-Cardama A[2004], Pulido P. Mobile phone text messaging in the management of diabetes. J Telemed Telecare;10:282–5
3. Joo NS, Kim BT[2007]. Mobile phone short message service messaging for behavior modification in a community-based weight control programme in Korea. J Telemed Telecare;13:416–20
4. Dunbar PJ, Madigan D, Grohskopf LA, et al.[2003] A two-way messaging system to enhance antiretroviral adherence. J Am Med Inform Assoc;
5. Scherr D, Zweiker R, Kollmann A, Kastner P, Schreier G, Fruhwald FM2006][. Mobile phone-based surveillance of cardiac patients at home. J Telemed Telecare;12:255–61
6. Ryan D, Cobern W, Wheeler J, Price D, Tarassenko L. [2005] Mobile phone technology in the management of asthma. J Telemed Telecare;11 (Suppl. 1):43–6
7. Moumoulidis I, Mani N, Patel H, Leong P. A novel use of photo messaging in the assessment of nasal fractures. J Telemed Telecare. 2007;13(8):387-90 PubMed PMID: 18078548
8. Schreier, Gunter, Hayn, Dieter, Kastner,et al [2008].A mobile phone based teledermatology system to support self management patient suffering psoriasis, engineering in medicine and biology scoity,EMBS, 30th annual international of the IEEE
9. Donald H, Franklin V, Greene S[2009]. the use of mobile phones in dietary assessment in young people with type I. journal of human nutrition and diabetics , , 22:3,256-257
10. Darren L Walters D L, Sarela A , Fairfull A[2010] ,Et al, BMC cardiovascular disorders, http/www.biomedcentral.com (1471-2261/10/5)
11. Hayn D,Koller S,et al,[2009p[. Mobile phone- based teledermatologic - Compliance management – preliminary results of the TELECOMP study, medical informatics in a united and healthy Europe,doi;10.3233/978-1-60750-044-5-468
12. Bhachu A S. ,Hine N A., etal2008 , Tele care system for the self management of health condition , (ASSETS), http://doi.acm.org/10.1145/1414471.1414484
13. Schermer M[2009],Telecare and self-management: opportunity to change the paradigm ?, jMed Ethics 2009,35;688-691

Corresponding author

Fereydoun Pourdanesh,DDS,MS,department of oral and

Maxillofacial surgery,school of dentistry,Tehran ,Iran

E-mail:drpourdanesh@gmail.com

Cell:+98 912 254 5297

Computer-Aid Cobb Measurement of Scoliosis Using Deformable Model with Fuzzy Spatial Relations

N.B. Toan, T.Q.D. Khoa, and Vo Van Toi

Biomedical Engineering Department, International University of Vietnam National Universities, Ho Chi Minh City, Vietnam

Abstract — **Scoliosis is a disorder that causes an abnormal curve of the spine, or backbone. Treatment decisions for scoliosis are based on consideration of the patient's physiologic maturity, curve severity, curve location, cosmetic concerns, and the estimated potential for progression. In order to reduce the observer variability in radiographic scoliosis assessment, we developed a semi-automatic method for measuring the Cobb angle using deformable model with fuzzy spatial relations integrated and our algorithm applying on the final contour of the model. In experiment, the method proves it is faster and requires less manipulation than the manual method. Also, this method does not require any special qualifications, skills or trainings to use it. Therefore it can be developed to make portable software which could be widely used by the healthcare providers.**

Keywords — **Scoliosis, Cobb angle, Deformable model, Fuzzy spatial relations.**

I. INTRODUCTION

Scoliosis is a disorder that causes an abnormal curve of the spine, or backbone. Treatment decisions for scoliosis are based on consideration of the patient's physiologic maturity, curve severity, curve location, cosmetic concerns, and the estimated potential for progression. The Cobb method is the accepted standard technique world-wide to measure the curvature of the spine [2]. Lines are drawn parallel to the upper and lower vertebral endplates of those uppermost and lowermost vertebrae which tilt towards the curve concavity. The angle formed by these two lines is known as the Cobb angle and is in degree.

The Cobb angle is important to monitor the scoliosis. If it is 15° or less, the patients only need regular check-up. Between 20° and 40°, the patient will probably need a back brace. Over 40°, the surgeon will likely recommend surgical intervention [3, 4].

To reduce variability of Cobb angle measurement by reducing user intervention and bias, many research have developed a computer-aided system for scoliosis assessment. Zhang et al. [3, 4] developed a system in which the regions of interest (ROI) that fit to the two end-vertebrae need to be assigned manually to estimate the Cobb angle. Chockalingam et al.[7] also proposed a computer method, in which at least 16 points must be assigned manually to define the vertebral edges. Allen et al. [6] utilized active shape models in a learning system to auto-detect the spine.

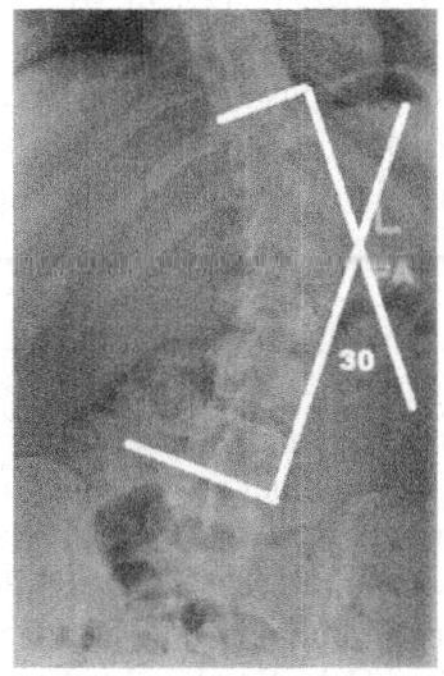

Fig. 1 Cobb angle measurement.

Still the accuracy of Zhang and Chockalingam' methods depended on how well users select points or edges which is often not familiar to operators and takes time. In other hand, Allen's method requires training and large data to deploy the active shape models. Therefore, it needs specialists to maintain and its accuracy depends on training set.

The method used in this study requires little user intervention and no training. User skill has no effect on final result. So it would take no time to be executed and can be utilized by anyone.

II. MATERIALS AND METHOD

A. Materials

The radiographs used in this study were acquired at the MEDIC Medical Center and the Hospital for Trauma-Orthopedics, both are in Ho Chi Minh City, Vietnam. The radiographs are standard, digital and anonymous.

In order to evaluate the software that we developed, we used a total of 20 radiographs of 4 patients (3 females and 1 male, age between 13-17 years old) admitted in the scoliosis hospitals between 2010 and 2011. All of the radiographs were taken using Canon DR–CXDI computed radiography machine.

V. Van Toi et al. (Eds.): 4th International Conference on Biomedical Engineering in Vietnam, IFMBE Proceedings 40, pp. 350–353, 2013.
www.springerlink.com

B. Method

We developed a semi-automatic method for measuring the Cobb angle using the deformable model with fuzzy spatial relations integrated to define the outline of the spine and then applying our algorithm on the final contour of the model to calculate the Cobb angle.

a) Deformable models integrated with fuzzy spatial relations:

Deformable models [11], also called active contours or snakes, are curves moving towards the object boundaries or image features which we want to extract. The evolution of a deformable model can be written as the minimization of the following energy:

$$E(X) = E_{int}(X) + E_{ext}(X), \quad (1)$$

where X is the deformable contour, E_{int} is the internal energy that specifies the smoothness constrain of the contour and E_{ext} is the external energy that drives the contour towards image edges.

Olivier Colliot et al. proposed a methodology to introduce prior constraints based on fuzzy spatial relations in a deformable model [10]. Spatial relations are represented as fuzzy subsets of the image space and are integrated in the deformable model as a new external force.

The external force F_{ext} in Eq. (1) is replaced with a force describing both edge information and structural constraints:

$$F_{ext} = \lambda F_C + \nu F_R,$$

where F_C is a classical data term that will drive the model towards the edges, F_R is a force associated to spatial relations and λ and ν are weighting coefficients. The spatial relation we used in this study is "around the spine with distance Δ". Assuming that there is a line A going through the middle of the spine, let P be a point of the space, $d_A(P)$ be the distant from A to P, we need to make a fuzzy set so that $d_A(P)$ will have high membership if its value goes near Δ.

Let R be a fuzzy set representing this spatial relation and μR its membership function, the fuzzy subset is defined as:

$$\mu_R = \begin{cases} \dfrac{dA(P)}{(\Delta-\varepsilon)} & \text{if } d_A(P) \le \Delta - \varepsilon \\ 1 & \text{if } \Delta - \varepsilon < d_A(P) < \Delta + \varepsilon \\ \dfrac{dA(P) - (\Delta+\varepsilon)}{(\Delta-\varepsilon)} & \text{if } d_A(P) \ge \Delta + \varepsilon \end{cases}$$

(ε is a small value which creates a padding space for the snake)

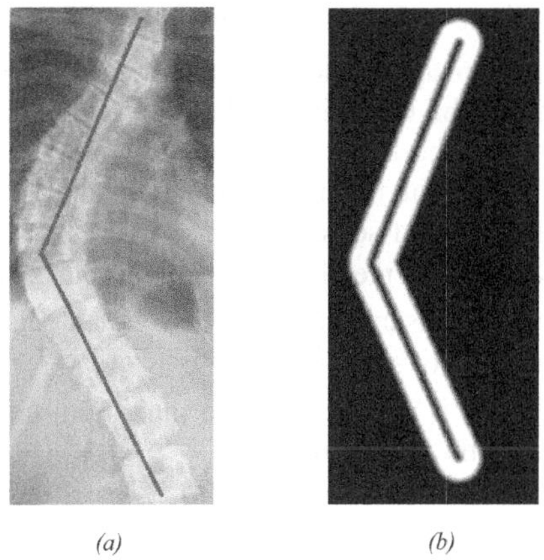

Fig. 2 Spatial relationship "around the spine".
(a) Line A going through the middle of the spine
(b) Fuzzy set representing the spatial relation "has distance Δ from line A".
Its thickness is defined by ε.

The force FR should drive the deformable model towards regions where the relation is fulfilled and thus be directed towards high values of μR. Using the fuzzy set as potential, we have

$$\mathbf{F}_R(P) = -(1 - \mu_R(P))\frac{\nabla P_R(P)}{\|\nabla P_R(P)\|}$$

where $P_R = 1 - \mu_R$

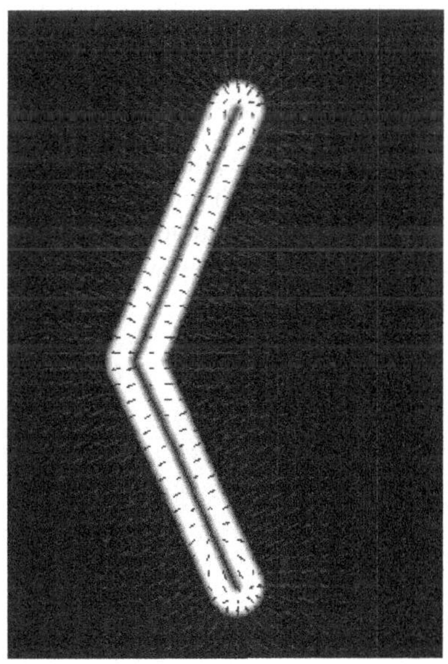

Fig. 3 F_R force field.

b) Algorithm to calculate the Cobb angle:

From the result contour of the deformable model, we have 2 edges of the spine outbound – left edge and right edge (top and the bottom part of the contour are omitted). We choose one of them (in this study, the longer one for more accurate) and select n points Ai ($0 < i < n$ with i, n $\in$ N) of it to divide the edge into n short straight lines $\overline{A_iA_{i+1}}$ with the same length. Then we calculate the angle between each 2 of straight lines $(\overline{A_iA_{i+1}, A_jA_{j+1}})$ ($0 < i, j < n$ with $i \neq j$ and i, j, n $\in$ N). The maximum angle between these lines is the Cobb angle:

$$\text{Cobb angle} = \max_{0 \le i \le n, 0 \le j \le n} \theta(\overline{A_iA_{i+1}, A_jA_{j+1}})$$

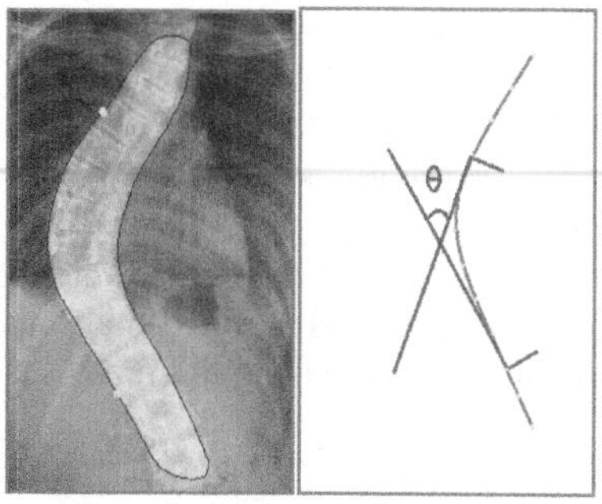

(a) (b)

Fig. 4 Find the Cobb angle

(a) Final contour of the deformable model - (b) Find the maximum angles between n short straight lines

c) Implement the method:

We had object A – a line going through the middle of the spine - defined by letting the user choose 6 points on the images (top left, top right, middle left, middle right, bottom left, bottom right). The 2 lines were made by connecting each of 2 midpoints form A.

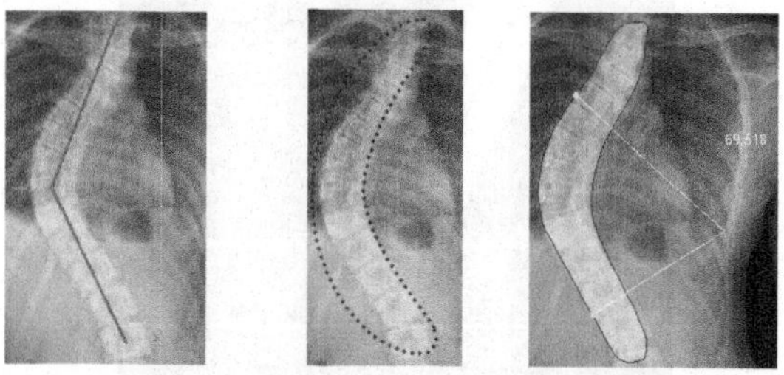

(a) (b) (c)

Fig. 5 Run the method.
(a) 6 points which user chooses on the image - (b) Initial contour of the snake – (c) Result of the snake and Cobb angle's value

The initial contour of the snake is also created based on these 6 points. After running the snake integrated with spatial relation, we got the result contour which formed by 100 points. Each of 2 points connected together to make 100 short lines. We calculated the Cobb angle using our algorithm on those lines.

III. RESULTS AND DISCUSSIONS

To determine the accuracy and reliability of the developed software we conducted the following comparative study. We performed the measurements of the Cobb angle using (1) the method presented here, (2) ruler and protractor (by hand). Each measurement was repeated twice. Twenty radiographs obtained as aforementioned were used in this study.

A. Result

The obtained results performed on 20 radiographs showed that the Cobb angle of 2 cases was smaller than 20°, the angles of 10 cases were between 20° and 40°, the angles of 8 cases were more than 40°. The averages and standard deviations for all cases are shown in Table 1.

Table 1 Averages and standard deviations (in degrees) of Cobb angles of 20 cases obtained by 2 different methods.

Radiograph no.	Our method	Manual method
1	68.92±0.31	64.58±1.14
2	48.9±0.12	46.21±1.05
3	58.47±1.27	54.39±1.28
4	39.44±1.61	38.51±2.79
5	20.35±0.03	20.5±1.22
6	31.6±1.7	30±1.79
7	43±0.4	36.09±1.12
8	40.99±0.28	38.1±1.13
9	10.25±0.05	10.42±2.11
10	36.15±0.25	29.24±0.35
11	4.9±0.7	3.81±0.91
12	62.76±1.24	60.9±0.18

Table 1 (*continued*)

13	44.72±2.46	49.49±0.48
14	46.8±3.8	45.46±0.84
15	31.4±0.62	30.12±0.38
16	6.38±3.48	4.57±1.63
17	25.59±0.43	24.89±1.04
18	27.2±0.34	29.93±0.77
19	45.46±0.54	45.81±6
20	46.48±1.87	42.76±0.73

B. *Discussion*

In analyzing those results we found that:

- The results obtained with our method yielded smaller standard deviation than those obtained by hand.

- With some exceptions, the differences of the results obtained with 2 methods were less than 5 o which is acceptable for scoliosis assessment.

- In some cases, there were tissues over the spine which made the snake harder to converge at the edge of the spine. We have to choose the initial 6 points more carefully and closer to the edge.

- The average time to perform each measurement was about 5 minutes with the manual method, and only a few seconds for our method. Our method is faster and required less manipulation.

- Our method is more accurate than the manual method because when the operator used the manual method, they had to choose the right vertebrae up and down to start measuring. In some hard cases, they have to guess and choose the most suitable vertebrae in their opinion. This problem does not occur in our method.

IV. Conclusion

In order to assist the surgeons in the management of the scoliosis we developed a semi-automatic method for determining the Cobb angle. The performance tests using our method showed that our method yielded more reliable results than manual method and allowed the operators to obtain faster results.

Moreover, it does not require the operators to use their judgment to estimate the edges of the vertebrae. Therefore it reduces the observer variability in radiographic scoliosis assessment.

References

1. Lonstein JE (1994) Adolescent idiopathic scoliosis. Lancet 344:1407–1412
2. Cobb JR (1948) Outline for the study of scoliosis. American Academy of Orthopaedic Surgeons. Instructional course lectures 5:261-275
3. Junhua Zhang, Edmond Lou, Lawrence H.Le et al (2008) Automatic Cobb Measurement of Scoliosis Based on Fuzzy Hough Transform with Vertebral Shape Prior. Journal of Digital Imaging 22:463-472
4. Junhua Zhang, Edmond Lou, Douglas L.Hill et al (2009) Computer-aided assessment of scoliosis on posteroanterior radiographs. Med BiolEngComput 48(2):185-95
5. PruijsJEH, Hageman MAPE, Keessen W et al (1994) Variation in Cobb angle measurements in scoliosis. Skeletal Radiol 23:517–520
6. Allen S, Parent E, et al (2008) Validity and reliability of active shape models for the estimation of Cobb angle in patients with adolescent idiopathic scoliosis. J Digit Imaging 21:208–218
7. Chockalingam N, Dangerfield PH, Giakas G et al (2002) Computer-assisted Cobb measurement of scoliosis. Eur Spine J 11:353–357
8. Isabelle Bloch, Olivier Colliot , Oscar Camara , Thierry Geraud (2005) Fusion of Spatial Relationships for Guiding Recognition, Example of Brain Structure Recognition in 3D MRI
9. Isabelle Bloch (1999) Fuzzy relative position between objects in image processing a morphological approach
10. Olivier Colliot, Oscar Camara, Isabelle Bloch (2006) Integration of Fuzzy spatial relations in deformable model. Pattern Recognition 39 1401 – 1414
11. M. Kass, A. Witkin, D. Terzopoulos (1987) Snakes: active contour models. Int. J. Comput. Vision 1 (4) 321–331.

Author: Nguyen Bao Toan
Institute: Biomedical Engineering Department, International University of Vietnam National Universities
Street: Quarter 6, Linh Trung, Thu Duc District
City: Ho Chi Minh
Country: Vietnam
Email: nbtoan@hcmiu.edu.vn

Structural Bioinformatics Approaches to Chemical Senses

Alejandro Giorgetti[1,2]

[1] Department of Biotechnology, University of Verona, strada Le Grazie 15, 37134, Verona, Italy
[2] German Research School for Simulation Science, FZ-Jülich and RWTH Aachen, Jülich, Germany

Abstract — **A deep characterization of the molecular mechanisms underlying the functioning of the receptors involved in chemical sensing, is fundamental for gaining insights into the way our cells communicate with the external world. Unfortunately, the paucity of experimental structural information hampers such characterization. Protein structural bioinformatics, combined with computational biology and experimental techniques is the key to face these challenges. Here I will show our recent works regarding the bitter-taste signaling pathway. Moreover, I will also present some elements of the state of art of membrane protein structural bioinformatics field.**

Keywords — **chemical senses, structural bioinformatics, docking, bitter taste receptors.**

I. Introduction

Life would not be possible without the ability of living organisms to communicate with their surroundings and respond to changes within them. Cells are able to capture and decode environmental stimuli into biologically signals. Consequently, communicating mechanisms, able to detect stimuli coming from the outside world are of fundamental importance for the survival of living beings. A deep understanding of the molecular mechanisms underlying signal transduction is thus needed for a complete characterization of the way our cells communicate with the rest of the world. Three different sensorial systems associated with the nose, mouth and the trigeminal chemosensory system, are dedicated to the detection of chemicals in the environment. The olfactory system detects airborne molecules called odours., providing information about food, self, other people, animals, plants, and many other aspects of the environment.. The taste system detects ingested, primarily water-soluble molecules called tastants. Tastants provide information about the quality, quantity, pleasantness, and safety of ingested food. The trigeminal chemosensory system provides information about irritating or noxious molecules that come into contact with skin or mucous membranes of the eyes, nose and mouth. All three of these chemosensory systems rely on receptors that interact with the relevant molecules and fire the signalling event, thus transmitting the effects of chemical stimuli to appropriate regions of the central nervous system. In the last few years we have concentrated our attention particularly in the olfactory and bitter taste perception mechanisms. Indeed these proteins belong to the G protein-coupled receptors (GPCRs) family, which is the largest membrane-bound receptor family expressed by mammalians (encompassing more than 1% of the genome). Moreover, more than half of the GPCRs codified in mammalian genomes are olfactory receptors (ORs) [1][2][3].

The second largest sensorial GPCR subfamily is the bitter-taste receptor family, formed by about 30 members in the human genome. The latter were shown to be fundamental during evolution for preventing the ingestion of toxic substances. Nowadays, these receptors are important not only in food industry, but also because bitter taste receptors have been recently found to express along the entire digestive pathway with astonishing protective functions [4]. This wealth of sequences, ligands, and mutations are in net contrast with the small amount of structural information available. Indeed, till a few years ago, rhodopsin was the only structurally characterized GPCR and is still considered a prototypical member of the superfamily. The first X-ray structure of rhodopsin reflected the dark adapted ground-state of the bovine receptor, captured in four different crystals of 2.8 Å [5], 2.65 Å [6] and 2.2 Å [7], 3.4 Å [8], 4.15 Å [9]. Very recently the structure of a GPCR in its empty state, Opsin, has been crystallized [10], opening in this way an exciting new era in the studies of GPCRs. This new structure of rhodopsin followed another very recent key event, the crystallization of a second GPCR, the beta2-adrenergic receptor [11][12][13][14].

This extremely low number of membrane proteins with known structure is due to the fact that membrane proteins are very difficult to study because they are inserted into lipid bilayers surrounding the cell and its sub-compartments. In these conditions they expose to the polar outer and inner environments portions of different sizes, completely changing the biophysics with respect to soluble proteins. Thus, when isolated from membranes, membrane proteins are generally less stable than globular ones.

In the last few years, several efforts were carried out for the determination of new crystal structures of membrane

V. Van Toi et al. (Eds.): 4th International Conference on Biomedical Engineering in Vietnam, IFMBE Proceedings 40, pp. 354–358, 2013.
www.springerlink.com

proteins. Although the improvements in the technologies allowed the determination of several structures, the gap between the known sequences and the solved structures is still enormous. Furthermore, albeit their extreme importance, crystal structures only provide a static image of a protein under conditions that sometimes are far from being physiological.

Thus, the combination of existing crystal structures, computational biology techniques and molecular biology validating experiments, may be the key to face the challenges of bridging the gap between the characterized membrane proteins with and without solved structure.

Here I will review some of the efforts of the last years aimed at the characterization at the structural level of a bitter taste receptor for which the crystal structure is not known.

I will also introduce the concept of the membrane protein structural bioinformatics, that is, modelling techniques that use the most recent membrane proteins crystal structures as input for the modelling of membrane proteins in different activation states. For the studied case I will review also the experiments carried out in order to validate the proposed hypothesis.

II. Protein structural bioinformatics

Proteins are in general defined as 'the major macromolecular constituent of cells. A linear polymer of amino acids linked together by peptide bonds in a specific sequence' [15]. The latter definition of a protein, albeit technically correct, lacks of the most important part of the definition, that is, proteins are the product of evolution. In fact, proteins are indeed polypeptides, but formed following a very precise relation sequence-structure-function. The structure/fold of a protein is then determined by its amino acid sequence and, therefore, upon mutation of one amino acid with another may affect the structure in different ways. The extremely weak folding energy landscape of a protein might be easily altered by mutations, giving place to unfolded proteins. However, for functional proteins, the structure-function relationship has been preserved during evolution thus, the proteins can only contain mutations that did not alter too much the global folding energy landscape with respect to their ancestor sequence. From those considerations it was concluded that evolutionarily related proteins, that diverged from a common ancestor via the accumulation of small changes, cannot but have similar structure, where mutations have been accommodated only causing small local rearrangements. Therefore if we can ensure that two proteins are homologous, that is evolutionary related, the structure of one can be used as a template for the building up of the structure of the other. This forms the basis of the technique known as comparative or homology modeling [16].

25 years ago, Chothia and Lesk, in a seminal article [17] have aligned the sequences and structures of all the proteins with known structure, finding a correlation between the evolutionary distance and structural divergence between evolutionary related proteins. This work settled up the basis of the comparative (or homology) modeling technique, a method that allows the prediction of protein structures using as a template a member of the family for which the three-dimensional structure (3D-structure) is known.

All these ideas were shown to be valid through the years for soluble proteins. Indeed, the comparative modeling technique has not always been considered valid when applied to membrane proteins [18]. The main criticism regarded the low amount of membrane proteins with known three-dimensional structure and the enormous evolutionary gap that must be filled in order to produce models of the most studied membrane proteins. However, in the last few years X-ray crystallography reached very high levels of applicability in the field of membrane proteins and new, more refined structures, have been solved by several groups. Indeed, this great advancement in the crystallographic techniques produced about 298 structures of unique membrane proteins for a total of 842 structures. At this point, we have checked for the existence of correlation between evolutionary relationship and structure similarity. Indeed, we have recently (Suku et al. in preparation) followed the same protocol as Chothia and Lesk [17] but considering just membrane proteins. Using the LGA server (http://proteinmodel.org/) we have aligned (considering structures and sequences) the residues present in the core of all the membrane proteins with known three-dimensional structure and produced the graph of Figure 1. The structural divergence between two evolutionary correlated proteins is measured as their Root Mean Square Deviation (RMSD).

Figure 1 shows the RMSD of the backbone of the core of pairs of evolutionarily related proteins as a function of the percent of identity between their amino acid sequences.. For practical purposes we considered as the core of the proteins, all the amino acids present in secondary structure elements and those regions not diverging for more than 3 Å, as Chothia and Lesk did [17].

The important thing is that, based on Figure 1 the homology modeling procedure is valid also for membrane proteins. In this regard, the astonishing improvements in membrane proteins crystallography together with comparative modeling techniques will allow the characterization of an enormous amount of membrane protein in the near future.

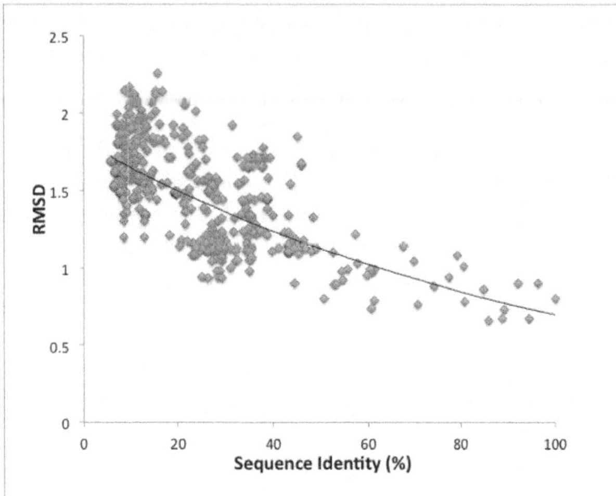

Fig. 1 Root Mean Square Deviations (RMSD) vs. sequence identity of the core of membrane proteins with known three-dimensional structure (see main text for details).

III. Applications: bitter taste receptors

Mammals during evolution have been prevented from ingesting toxic compounds because of their strong bitter taste [19][20][21][22]. This protection mechanism has been carried out for millions of years by a family of about 30 bitter taste receptors (TAS2Rs) expressed in taste receptor cells [23][24][25]. TAS2Rs was shown to belong to the super family of GPCR receptors, albeit their low sequence identity with rhodopsin, for example [23][24][25]. The binding of a bitter compound to its cognate target TAS2R, is able to fire a downstream cascade of events inside the cell, typical of GPCRs signaling pathways [26], leading to the production of an electrical signal, i.e. bitter taste perception [19]. Thus, albeit natural selection decreased its constraints, bitter taste is becoming more and more important regarding food taste, cuisine and pharmaceutical matters. One of the most interesting bitter taste receptors, also from the evolutionary point of view, is the human receptor for phenylthiocarbamide (PTC) and propylthiouracil (PROP) molecules, i.e. TAS2R38 receptor. Indeed, among several polymorphisms annotated for this receptor, the most pronounced ones affect its perception of the phenylthiocarbamide (PTC) compound. In fact, differences in the perception of PTC has divided the human population into tasters and non-tasters. Albeit these results were reported in the early nineties, polymorphisms in the hTAS2R38 gene underlying the observed phenotype were identified only recently by Kim and coworkers [27]. Indeed, the taster/non-taster quality originate in three hTAS2R38 non-synonymous polymorphisms, i.e. the haplotypes code for either the amino acids PAV (P^{49}, A^{262},I^{296}) constituting the taster variant of hTAS2R38 or AVI in the corresponding positions for the non-taster variant [28][27]. Up to now, the molecular/structural basis of bitter taste sensing was analyzed by very few studies, i.e. computational studies [18][29]. First principle [18] and homology modeling approaches based on bovine rhodopsin [29] have been used to predict the structure of the widely studied olfactory receptors [28][30][27][31].

Very recently [32] we have used a combined experimental/computational iterative approach with the aim at identifying hTAS2R38 residues involved in binding to one of its main agonists, i.e. PTC, as well as in receptor activation. We used state-of-the-art bioinformatics approaches based on multiple sequence alignment across the whole family of GPCRs combined with structural bionformatics tools; we also used the homology modeling techniques because sequences at their own were not likely to be sufficient to identify residues in the binding site, as ligands pockets vary largely in position and orientation across this family [33]. Furthermore extensive virtual docking experiments were carried out to predict the putative binding cavities for PTC. In fact, homology modeling and molecular docking has been shown to guide satisfactorily the design of site-directed mutagenesis experiments, in spite of the little power of the structural predictions [32]. Indeed, the proposed receptor positions were then studied and validated/rejected by site-directed mutagenesis experiments and measurements of receptor activation by recording intracellular calcium levels following agonist administration.

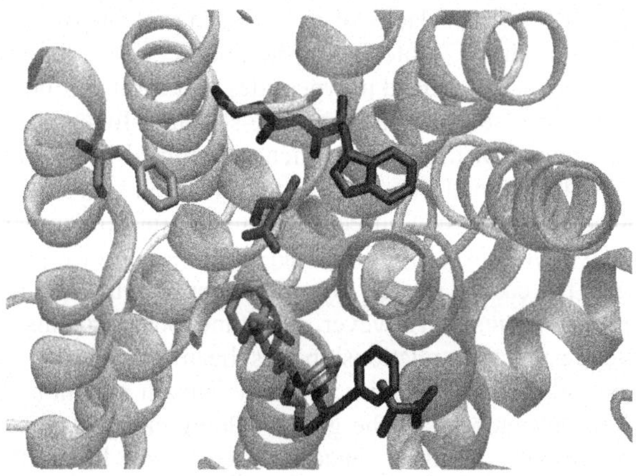

Fig. 2 Model of the hTAS2R38 bitter taste receptor with the ligand (PTC) docked in the binding cavity.

We thus have proposed that hTASR38 activation upon PTC binding is reminiscent of the transition of the G-protein/opsin complex to free rhodopsin [34]. Indeed, we were able to identify some of the residues directly involved in the interaction with the ligand, those that define the shape of the binding cavity and, more important, we were able to identify the residues participating in receptor activation. In our model, TMs 5, 6 and 7 change conformation upon ligand binding, in particular TM6 tilts around the helical bundle upon G-protein binding. Similar sequences of events also have been suggested to play a role for activation of all GPCRs [35].

IV. Perspectives: olfactory receptors

Olfactory receptors (OR) possess high affinity for thousands of volatile molecules associated with odour. With such a large number of different ORs, the olfactory system is capable to discriminate between 10 000 different odours: one odorant can activate numerous types of ORs, while a single OR can be activated by several different odorants. ORs in the olfactory sensory neurons of the nasal epithelium translate odorants into neural signals. Each OR is thought to be specialized to recognize physicochemical features, such as functional groups or molecular size, of odorant molecules; these features are then translated into a neural signal that in turn leads to an olfactory perception. The physicochemical features of an odorant, therefore, are a key determinant of the olfactory percept. The rules governing the translation of molecule into percept, however, remain largely unknown. To develop a theory that predicts olfactory percept from molecular structure, we must first be able to predict OR activation from molecular structure. The overall sequences of mammalian ORs are diverse, with amino acid sequence similarity between different ORs ranging from less than 40% to more than 90%.

Deciphering olfactory encoding requires not only a thorough description of the ligands that activate each odorant receptor (OR) but also a description of the putative binding cavities of the latter. In mammalian systems, however, ligands are known for fewer than 90 of over 1400 human and mouse ORs, greatly limiting our understanding of olfactory coding. Recently we have developed a database, i.e. OlfactionDB (http://molsim.sci.univr.it/OlfactionDB), that includes all the known receptor/ligand pairs found in literature. The next phase of our work will include the use of chemoinformatic techniques combined with homology modelling, molecular dynamics simulations and multiscale approaches, i.e. we are developing and *ad hoc* hybrid coarse grained/ molecular mechanics simulation approach (Nguyen in preparation), for the development of a predictive model, relating physicochemical odorant properties, OR sequences and/or structures, and their interactions. The obtained results may provide the basis for translating odorants into receptor neuron responses and unravelling mammalian odor coding.

V. Conclusion

A deep insight into the molecular mechanisms underlying signal transduction is needed for a complete characterization of the way we communicate with the rest of the world. A complete characterization implies the passage through the structural features of the membrane proteins. Unfortunately, as of today, the structural biology scientific community still finds several inconvenient to systematically solve the structure of membrane proteins, albeit giant steps forward were carried out in the last few years. Moreover, we have to consider that the main challenges for the near future will include the development and application of methods that permit the full description at the molecular/structural level of large protein complexes, most of all including membrane proteins with unknown structure.

Here I have described an example for which an experimental/computational multidisciplinary approach was shown to be the key for the gaining of insights into a very complex system. In this case not only 'static' structural elements were identified, but also putative dynamical mechanisms that may imply large conformational changes. Indeed, advancements in experimental structural biology, extensively combined with state-of-art computational biology tools and model-guided molecular biology experiments may become extremely effective for the characterization of complex molecular mechanisms including membrane proteins. These observations make us confident that more difficult challenges of structural/ functional characterization can be undertaken in short time and that these approaches may provide a great improvement to our understanding of cell and molecular biology events.

References

1. Kroeze, W.K., Sheffler, D.J. & Roth, B.L. 2003, G-protein-coupled receptors at a glance. *J. Cell. Sci* **116**, 4867-4869
2. Buck, L. & Axel, R. ,1991,A novel multigene family may encode odorant receptors: a molecular basis for odor recognition. *Cell* **65**, 175-187.
3. Takeda,S. et al. (2002) Identification of G protein-coupled receptor genes from the human genome sequence. *FEBS Lett*, 520, 97-101.
4. Chen MC, Wu SV, Reeve JR Jr., Rozengurt E (2006) Bitter stimuli induce Ca2+ signaling and CCK release in enteroendocrine STC-1 cells: Role of L-type voltage-sensitive Ca2+ channels. Am J Physiol Cell Physiol 291:C726–C739.
5. Palczewski,K et al. (2000) Crystal structure of rhodopsin: A G protein-coupled receptor. *Science*, 289, 739-745.
6. Li,J. et al. (2004) Structure of bovine rhodopsin in a trigonal crystal form. *J. Mol. Biol*, 343, 1409-1438.

7. Okada,T. et al. (2004) The retinal conformation and its environment in rhodopsin in light of a new 2.2 A crystal structure. *J. Mol. Biol*, 342, 571-583.
8. Standfuss,J. et al. (2007) Crystal structure of a thermally stable rhodopsin mutant. *J. Mol. Biol*, 372, 1179-1188.
9. Salom,D. et al. (2006) Crystal structure of a photoactivated deprotonated intermediate of rhodopsin. *Proc. Natl. Acad. Sci. U.S.A*, 103, 16123-16128.
10. Park,J.H. et al. (2008) Crystal structure of the ligand-free G-protein-coupled receptor opsin. *Nature*, 454, 183-187.
11. Bokoch,M.P. et al. (2010) Ligand-specific regulation of the extracellular surface of a G-protein-coupled receptor. *Nature*, 463, 108-112.
12. Cherezov,V. et al. (2007) High-resolution crystal structure of an engineered human beta2-adrenergic G protein-coupled receptor. *Science*, 318, 1258-1265.
13. Hanson,M.A. et al. (2008) A specific cholesterol binding site is established by the 2.8 A structure of the human beta2-adrenergic receptor. *Structure*, 16, 897-905.
14. Rasmussen,S.G.F. et al. (2007) Crystal structure of the human beta2 adrenergic G-protein-coupled receptor. *Nature*, 450, 383-387.
15. Alberts B., Johnson A., Lewis J., Raff M., Roberts K., and Walter P. (2002) *Molecular Biology of the Cell (4th edition)*, Garland Science, ISBN-10: 0-8153-3218-1, New York.
16. Tramontano,A. (2006) The role of molecular modelling in biomedical research. *FEBS Letters*, 580, 2928-2934.
17. Chothia,C. and Lesk,A.M. (1986) The relation between the divergence of sequence and structure in proteins. *EMBO J*, 5, 823-826.
18. Floriano,W.B. et al. (2006) Modeling the human PTC bitter-taste receptor interactions with bitter tastants. *J Mol Model*, 12, 931-941.
19. Behrens, M. and Meyerhof,W. (2009) Mammalian bitter taste perception. *Results Probl Cell Differ*, 47, 203-220.
20. Meyerhof,W. (2005) Elucidation of mammalian bitter taste. *Rev. Physiol. Biochem. Pharmacol*, 154, 37-72
21. Mueller,K. L. et al. (2005) The receptors and coding logic for bitter taste. *Nature*, 434, 225-229
22. Soranzo,N. et al. (2005) Positive selection on a high-sensitivity allele of the human bitter-taste receptor TAS2R16. *Curr. Biol*, 15, 1257-1265.
23. Adler,E. et al. (2000) A novel family of mammalian taste receptors. *Cell*, 100, 693-702.
24. Chandrashekar,J et al. (2000) T2Rs function as bitter taste receptors. *Cell*, 100, 703-711.
25. Matsunami,H. et al. (2000) A family of candidate taste receptors in human and mouse. *Nature*, 404, 601-604.
26. Chandrashekar,J et al. (2006) The receptors and cells for mammalian taste. *Nature*, 444, 288-294.
27. Kim,U. K. et al. (2003) Positional cloning of the human quantitative trait locus underlying taste sensitivity to phenylthiocarbamide. *Science*, 299, 1221-1225.
28. Bufe,B. et al. (2005) The molecular basis of individual differences in phenylthiocarbamide and propylthiouracil bitterness perception. *Curr. Biol*, 15, 322-327.
29. Miguet,L. et al. (2006) Computational studies of ligand-receptor interactions in bitter taste receptors. *J. Recept. Signal Transduct. Res*, 26, 611-630.
30. Khafizov,K. et al. (2007) Ligand specificity of odorant receptors. *J Mol Model*, 13, 401-409.
31. Kleinau,G. et al. (2007) Implications for molecular mechanisms of glycoprotein hormone receptors using a new sequence-structure-function analysis resource. *Mol. Endocrinol*, 21, 574-580.
32. Biarnés,X. et al. (2010) Insights into the binding of Phenyltiocarbamide (PTC) agonist to its target human TAS2R38 bitter receptor. *PLoS ONE*, 5, e12394.
33. Jaakola,V. P. et al. (2008) The 2.6 angstrom crystal structure of a human A2A adenosine receptor bound to an antagonist. *Science*, 322, 1211-1217.
34. Scheerer,P. et al. (2008) Crystal structure of opsin in its G protein-interacting conformation. *Nature*, 455, 497-502.
35. Altenbach,C. et al. (2008) High-resolution distance mapping in rhodopsin reveals the pattern of helix movement due to activation. *Proc. Natl. Acad. Sci. U.S.A*, 105, 7439-7444.

Author: Alejandro Giorgetti
Institute: Dept of Biotechnology, University of Verona
Street: Ca' Vignal 1, strada 'Le Grazie' 15
City: Verona
Country: Italy
Email: alejandro.giorgetti@univr.it

Uncertainty and Symmetries in DNA Sequences

C. Cattani

Department of Mathematics, University of Salerno, Fisciano (SA) Italy

Abstract — **In this paper, similarities and distinctions between random sequences and DNA sequences are analized. It will be shown that by using the complex unit roots representation, and some methods of signal analysis, DNA sequences look as random sequences with some fractal shape (in their DNA walks). A short review on the main results about the long range statistical dependence and the main parameters of complessity will be also given. However, by using wavelet transform it is possible to give some more information about a possible hidden rule concerning the nucleotide distribution. In fact, clusters of wavelet coefficients clearly show the differences between random sequences and DNA sequences.**

Keywords — **Wavelet, fractal dimension, complexity, random walk, Ulam spiral.**

I. Introduction

In recent years, there has been a rising interest for the statistical analysis of DNA sequences by using methods from signal analysis and frequency analysis [1-44]. Almost all these papers are aiming to visualize the existence of regular patterns in nucleotide distributions [1,4,6,7,9,12,13,16-23,25,29-32,35-37,41,42,45-49] and the existence of long range correlation [2,3,5,8,10,11,14,15,24-28,33,34,38-41,43,44]. At the moment, we can assume that nucleotides look randomly distributed, their random walks feature some fractal shape, and the power spectrum could be either $1/f$ or $1/f^2$ but the form of the correlation function is still debated. However there are still many doubts and open debates about the interpretation and validity of these results. In particular, it seems clear that most results depend on the numerical representation (with or without redundancy) of the symbolic sequence of nucleotides, on the length of the sequence from a few base pairs (bp) to billions bp, and on the intrinsic nature of DNA concerning the distinction between coding and non-coding regions. In particular, when we take into account also non-coding regions, all methods give rise to different conclusions. On the other hand, we can't ignore that, in the more evolved organisms, DNA chains have a non-coding region which is more than $90-95\%$ of the sequence length, thus, by ignoring this part, we would get some doubtful conclusions. Some interesting attempts to add more information concerning the nucleotide distribution and the existence of patterns were given by the wavelet analysis of DNA (see e.g. [1,3,4,12,13,17,46-48,50-53]). In particular in [12,13, 46,47,50,51] it has been shown that although the DNA sequence looks very similar to a random sequence the wavelet coefficients of the DNA walks are clearly distributed according to some rule while the random walks are not. This suggest us that close nucleotides have to fulfill some mathematical constraints. Wavelet theory [54-57], due to the localization properties of wavelets, is a powerful method for the local analysis of signals thus being an expedient tool for singling out the dependence of each nucleotide with the closest. Also multifractal nature of the time series [58-60] can be easily detected by wavelet analysis [57]. In particular, by focussing our wavelet analysis on archae and bacteria, whose DNA is 90% made by coding regions, we will show that first order wavelet coefficients of DNA walks are distributed along some featuring clusters, thus emphasizing the distinction of DNA from the random sequences.

In details, materials and methods are given in section 2. Section 3 deals with frequency analysis and some elementary plots of correlation based on the indicator map. Some of the most popular parameters of complexity, as randomness, complexity, fractal dimension, information entropy, are discussed in section 4. Some remarks on Fourier analysis and power spectrum are given in section 5. Wavelet analysis and the short wavelet transform are described in section 6. In this section plots of wavelet clustering are also given to show the existence of simmetries.

II. Materials and methods

A. Materials

In the following, we analize sequences of DNA, concerning the following bacteria

b1: Mycoplasma putrefaciens KS1 chromosome, complete genome. DNA, circular, length 832603 bp, [61], accession NC 015946.

b2: Mortierella, verticillata, mitochondrion, complete genome. dsDNA, circular, length 58745 bp, [61], accession NC 006838.

V. Van Toi et al. (Eds.): 4th International Conference on Biomedical Engineering in Vietnam, IFMBE Proceedings 40, pp. 359–369, 2013.
www.springerlink.com

b3: Blattabacterium sp. (Periplaneta Americana) str. BPLAN, complete genome. DNA, circular, length 636994 nt, [61], accession NC 013418.

and archaea:

h1: Aeropyrum pernix K1, complete genome. DNA, circular, 1669696 bp, [61], accession BA000002.3.

h2: Acidianus hospitalis W1, complete genome. DNA, circular, 2137654 bp, [61], accession CP002535.

h3: Acidilobus saccharovorans 345-15. complete genome. DNA, circular, 1496453 bp, [61], accession CP001742.1.

Archaea are a group of elementary single-cell microorganisms, having no cell nucleus or any other membrane-bound organelles within their cells. They are similar to bacteria, since they have the same size and shape (apart few exceptions) and the generally similar cell structure. The most peculiar feature of archaea is that they can live also in some environments with extreme life conditions (thus being considered as extremophiles [62,63]). It has been also recognized that the archaea may be the most ancient organisms on the Earth, so that archaea, and eukaryotes probably diverged early from an ancestral colony of organisms.

B. Methods

Our analysis of DNA is based on both the 4-alphabet of nucleotides

$$A_1 \stackrel{def}{=} \{A\ ,C\ ,G\ ,T\} \tag{1}$$

being the nucleotides (nucleic acids): adenine (A), cytosine (C), guanine (G), thymine (T), the 20-alphabet of amino acids (see e.g. [46,47])

$$A_3 = \{M\ ,E\ ,Q\ ,D\ ,R\ ,T\ ,N\ ,H\ ,V\ ,G\ ,L\ ,\ S\ ,P\ ,F\ ,I\ ,C\ ,A\ ,K\ ,Y\ ,W\} \tag{2}$$

and their corresponding complex numerical representation [12,13,46,47].

In general, the ℓ -length alphabet A_ℓ is defined as follows: let the ℓ -length DNA word defined by the ℓ -combination with repetitions of the 4 nucleotides (1). For each fixed length ℓ there are 4^ℓ words, however not all of them can be considered, from biological point of view, as independent instances, for this we define the ℓ -length alphabet as the set of ℓ -length independent words w_j:

$$A_\ell \stackrel{def}{=} \{w_1\ ,w_2\ ,\dots\ ,w_{M_\ell}\},\qquad M_\ell \stackrel{def}{=} |A_\ell| \le 4^\ell$$

with $|\dots|$ cardinality of the set and

$$\ell \stackrel{def}{=} length(w_j)\ \ ,\ \ (j = 1,\dots,M_\ell)\ .$$

For instance with $\ell = 1$, the alphabet is given by Eq. (1), with $\ell = 3$ the number of 3-combinations with repetitions is $4^3 = 64$ words (named *codons*), however the only significant words are the 20 *amino acids* of (2).

In our analysis we will review the following statistical-mathematical methods:

1) Frequency analysis (Sec. 3), it is used to easily visualize some correlations. By using th frequency we can define, for large sequences, same complexity parameters like randomness, fractal dimension, complexity, entropy. These parameters enable us to classify sequences when we compare one to another, however, when taken singularly, they can be hardly used to understand about the intrinsic complexity of a sequence. Frequency analysis shows that, in DNA sequences, nucleotides are randomly distributed with parameters of complexity (entropy, fractal dimension,..) that, more or less, give the same information
2) Fourier analysis (Sec. 5) it is used to understand about the long range correlation in DNA. Some more information can be obtained by the power spectrum analysis of the complex representation of DNA.
3) Wavelet analysis (Sec. 6), due to its localization property, will be used to show that some intrinsic properties of DNA can be singled out by analyzing the short wavelet transform coefficients. It is shown that short wavelet transform gives the better approach for the analysis of short range correlation on nucleotides. Although they seem apparently randomly distributed, it is seen that on a short range, there is a strong dependence. For this reason DNA sequences can be considered as Markov chains.

III. Frequency Analysis

In this section we will consider some elementary plots from where it is possible to visualize auto-correlation, distribution law of nucleotides and to measure some fundamental parameters by using frequency count.

Let S_N be a N -length ordered linear sequence of nucleotides (1) and A_ℓ the chosen alphabet, a DNA sequence of ℓ-length words in the ℓ-alphabet is the map

$$D_\ell(S_N) = N \times A_\ell$$

so that

$$D_\ell(S_N) \stackrel{def}{=} \{x_h\}_{h=1,\dots,N} \;, \qquad (x_h \in A_\ell;\; N < \infty)$$

being

$$x_h \stackrel{def}{=} (h, w_j) \;,\; (h = 1,2,\dots,N;\; w_j \in A_\ell) \qquad (3)$$

the ℓ-length word w_j at the position h (for simplicity we dropped the second index j in the definition of x_h).

When $\ell = 1$, trivially is $D_1(S_N) = N \times A_1 = S_N$, and $|D_1(S_N)| = |S_N| = N$ while, in general, for $\ell > 1$, it is $|D_\ell(S_N)| < |S_N|$.

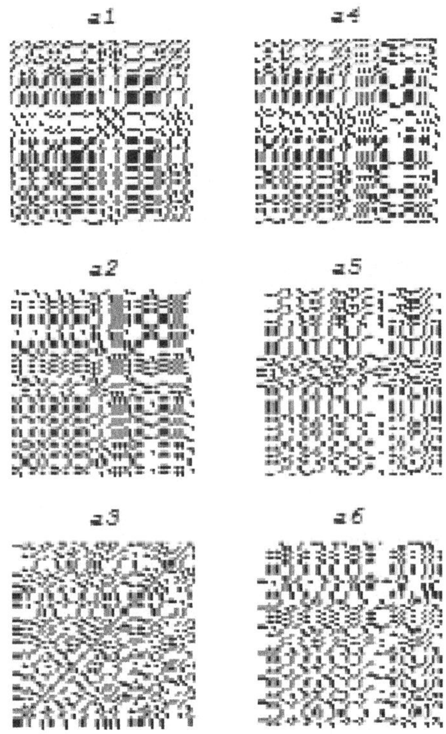

Fig. 1 Indicator matrix for the map (4) on the first 70 bp of: a1) b1 × b1 a2) h2 × h2 a3) random sequence rr × rr a4) b1 × h2 a5) b1 × rr a6) h2 × rr .

Given two DNA sequences $D_\ell(S_N)$ and $D_\ell(S'_N)$, the indicator function (for $S_N \equiv S'_N$ see e.g. [12,13,46,47,64,65]), based on the definition given by Voss [39,40], is the map

$$u: D_\ell(S_N) \times D_\ell(S'_N) \to \{0\;,1\} \qquad (4)$$

where, according to (3), it is

$$x_h = x_k \Leftrightarrow (w_j, h) = (w_i, k) \Leftrightarrow w_j = w_i \;, \quad (w_j, w_i \in A_\ell)$$

such that

$$u(x_h,\, x_k) \stackrel{def}{=} \begin{cases} 1 & if \quad x_h = x_k \\ 0 & if \quad x_h \neq x_k \end{cases} , \qquad (5)$$
$$(x_h \in D_\ell(S_N) \;, \quad x_k \in D_\ell(S'_N)) \;,$$

The indicator matrix ([12,13,46,47,64,65]), defined as

$$u_{hk} \stackrel{def}{=} u(x_h, x_k) \;,$$
$$(x_h \in D_\ell(S_N) \;, x_k \in D_\ell(S'_N); h, k = 1, \dots, N) \;,$$

can be used to obtain the two dimensional dot-plot (Figs. 1-2), which is a special case of the so-called recurrence plot [66].

Fig. 2 Indicator matrix for the map (4) on the first 300 amino acids of: a1) b1 × b1 a2) h2 × h2 a3) random sequence rr × rr a4) b1 × h2 a5) b1 × rr a6) h2 × rr .

A particular indicator map is when the first sequence is the ℓ-alphabet

$$u^*: A_\ell \times D_\ell(S'_N) \to \{0\;,1\} \qquad (6)$$

so that

$$u^*_{hk} \stackrel{def}{=} u^*(w_h,\, x_k) = \begin{cases} 1 & if \quad w_h = x_k \\ 0 & if \quad w_h \neq x_k \end{cases} , \qquad (7)$$
$$(w_h \in A_\ell \;, \quad x_k \in D_\ell(S'_N)) \;,$$

as a special case we re-obtain the Voss indicator:

$$u^*: A_1 \times D_1(S_N) \to \{0\ ,1\}\ .$$

Thanks to (7) , we can define the frequency of the ℓ -word w_h at the position k in the sequence $D_\ell(S_N)$ as the marginal frequency

$$\nu_h(k) \stackrel{def}{=} \frac{1}{k}\sum_{j=1}^{k} u^*_{hj}\ , \qquad (8)$$

so that

$$\sum_{h=1}^{M_\ell} \nu_h(k) = 1.$$

As approximated value of the probability to find the ℓ -word w_h at the position k in the sequence $D_\ell(S_N)$, we can take

$$p_h(k) \cong \nu_h(k)\ . \qquad (9)$$

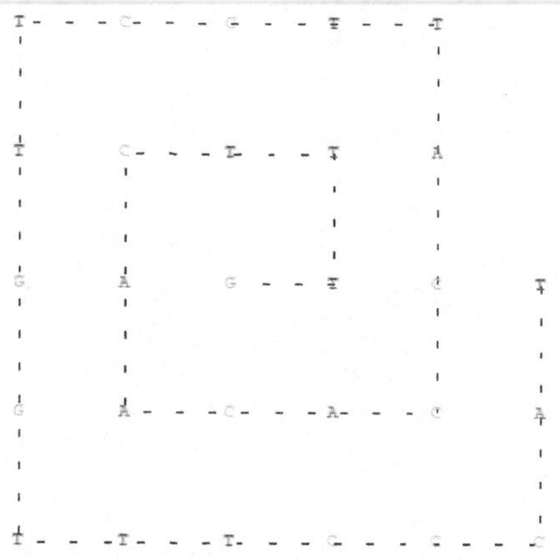

Fig. 3 Distribution of nucleotides on a rectangular spiral.

If we plot the indicator matrix of some bacteria and compare it with a pseudo-random and periodic sequence, we can see that (Fig. 1-2)

1. the main diagonal is a symmetry axis for the plot
2. there are some motifs which are repeated at different scales like in a fractal;
3. periodicity is detected by parallel lines to the main diagonal;
4. empty spaces are more distributed than filled spaces, in the sense that the indicator matrix is a sparse matrix (having more 0's than 1's);
5. square-like islands where black spots are more concentrated show the persistence of a nucleotide.

A. Spiral Plot

We can visualize the distribution of words, by mapping each word of $D_\ell(S_N)$ into the integer points of a spiral (Fig. 3,4) in R^2, as the Ulam spiral [67]:

$$D_\ell(S_N) \mapsto (x, y)\ , (x, y \in Z)$$

These plots can immediately visualize the different distribution of nucleotides, for instance in b1(Fig. 5) there is a major distribution of A, T while on the contrary in h3, the higher frequency belongs to C, G (Fig. 6).

Fig. 4 Spiral plot for a 10000 length random DNA sequence

Fig. 5 Spiral plot for the first 10000 nucleotides of b1

Fig. 6 Spiral plot for the first 10000 nuceotides of h3.

IV. PARAMETERS OF COMPLEXITY

In this section we define some parameters, based on frequency distribution, which can measure the complexity of a DNA by computing the complexity of its representation in the complex plane (for a more detailed analysis see [46,47] and references therein). In particular, for the n -length sequence, we define (see also [12,13,46,47]):

1) *Randomness*:

$$R \stackrel{def}{=} 1-\sigma(\nu_1(n),\nu_2(n),\ldots,\nu_{M_\ell}(n))$$

being σ the variance, so that $R=1$ for random sequences and $R=0$ for a non-random sequences

2) *Complexity*:

$$K=\frac{1}{n}\log\frac{n!}{\nu_1(n)!\nu_2(n)!\ldots\nu_{M_\ell}(n)!}$$

3) *Fractal dimension*: is computed on the dot-plot, by the box counting algorithm [8,9], as the average of the number $p(n)$ of $1's$ in the randomly taken $n\times n$ minors of the $N\times N$ indicator matrix u_{hk}

$$D=\frac{1}{2N}\sum_{n=2}^{N}\frac{\log p(n)}{\log n}\ .$$

4) *Normalized Shannon entropy* [68,69]: is defined, over the alphabet A_ℓ as

$$H(n)=-\frac{1}{\log\ell}\sum_{w\in A_\ell}p_w(n)\log p_w(n)\quad \text{if } p_w(n)$$

$$H(n)=0 \text{ if } p_w(n)=0 \qquad (10)$$

where $p_w(n)$ should be computed for large sequences. According to Eqs. (8), (9), we will approximate its value with

$$p_w(n)\cong\frac{1}{n}\sum_{i=1}^{n}u_{wi},\quad (w\in A_\ell, 1\le n\le N)$$

Table 1 Randomness, complexity, fractal dimension, entropy in A_1 (in bold the low-highest values).

	R	K	D	H
Mycoplasma p.	**0.696**	**1.151**	**1.283**	**0.877**
Mortierella v.	0.779	1.285	1.296	0.976
Blattabact.	0.743	1.197	1.287	0.911
Aeropyrum p.	0.982	1.212	1.288	0.922
Acidianus h.	0.828	1.231	1.290	0.937
Acidilobus s.	0.934	**1.296**	1.297	**0.984**
random	**0.999**	1.295	**1.298**	**0.984**

All these global parameters give, more or less, the same information concerning the complexity of the sequence (see Table 1) however, if we compute the randomness index over the frequencies of amino acids in the A_3 alphabet then we can observe a different distribution of values. Over the first 30000 nucleotides corresponding to 10000 amino acids, we have the randomness value of Table 2

Table 2 Randomness of amino acids distribution.

Mycoplasma p.	0.946
Mortierella v.	0.938
Blattabacterium	0.953
Aeropyrum pernix	0.962
Acidianus hospitalis	**0.916**
Acidilobus s.	0.950
pseudo − random	**0.963**

V. FOURIER ANALYSIS

In order to improve our analysis of correlation in DNA, we have to introduce a digitalization of the symbolic alphabet and apply the usual methods of signal analysis. For the numerical representation, and signal analysis methods, there have been many different approaches [2-5,7,11-15,20,22,24-28,38-41,43,46,47] some of them also by trying to avoid redundancy (see e.g. [42]). In this section, the complex roots representation is proposed .

A. Complex Root Representation

The complex (digital) representation of a DNA sequence of words is the map of the symbolic sequence of words into a set of complex numbers and is defined as

$$D_\ell(S_N) \overset{\rho}{\rightarrow} C$$

such that for each $x_h \in D_\ell(S_N)$ it is $\rho(x_h) \in C$. The complex root representation of $D_\ell(S_N)$ is the sequence of complex numbers $\mathbf{Y}_{M_\ell} = \{y_h\}_{h=1,\dots,M_\ell}$ defined as

$$y_h = \rho(x_h) \overset{def}{=} e^{\frac{2\pi i(j-1)}{|A_\ell|}} \quad ,$$

$$(j = 1, \dots, |A_\ell|, \; h = 1, \dots, M_\ell) \tag{11}$$

with $i = \sqrt{-1}$ the imaginary unit and $M_\ell = |A_\ell|$. There follows that, independently on the alphabet, it is

$$|y_h| = |e^{2\pi i(j-1)/|A_\ell|}| = 1 \quad , \quad (\forall \ell; \; h = 1, \dots, M_\ell)$$

being all complex roots, of the unit, located on the unit circle of the complex plane C^1. Therefore the complex representation of a DNA sequence is a sequence of complex numbers

$$y_h = \xi_h + \eta_h i \quad , \quad \xi_h = \Re(y_h) \; , \qquad \eta_h = \Im(y_h)$$

with y_h given by (11).

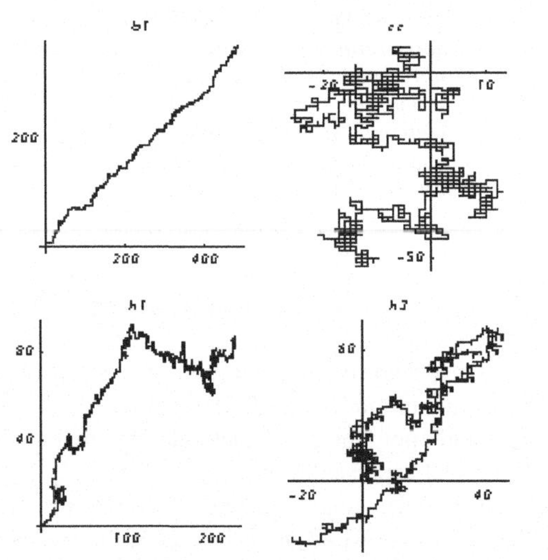

Fig. 7 Walks on the first 1700 nucleotides: b1) Mycoplasma putrefaciens , rr) random sequence, h1) Aeropyrum pernix, h3) Acidilobus saccharovorans.

B. Random Walks

Random walk on the complex sequence $\boldsymbol{Y}_N$ is defined as the series $\boldsymbol{Z}_N = \{z_n\}_{n=1,\dots,N}$

$$z_n \overset{def}{=} \sum_{k=1,\dots,n} y_k \quad , \quad n = 1, \dots, N \tag{12}$$

which is the cumulative sum

$$\{y_1 \; , y_1 + y_2 \; , \dots , \sum_{s=1}^{n} y_s \dots , \sum_{s=1}^{N} y_s\} \; .$$

When $y_k = \rho(x_k)$ with $\boldsymbol{Y}_k \in D(S_N)$ we will properly call these walks as DNA walk. When the $\boldsymbol{Y}_k$ are randomly generated we will call them random walks.

It has been observed that DNA walks have the typical shape of fractals (see Figs. 7,8), and this fractal property can be seen in all alphabets [46,47].

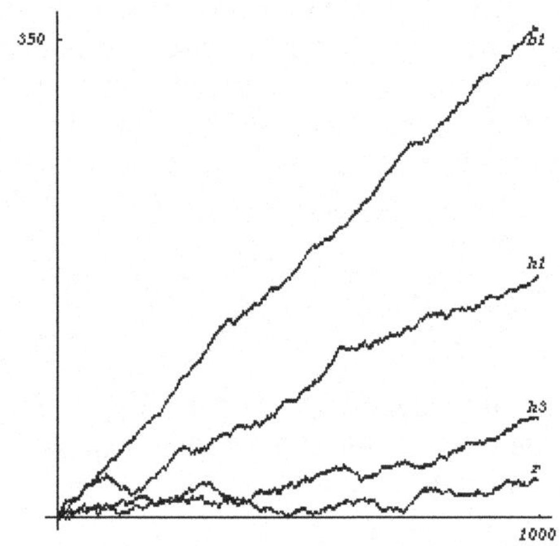

Fig. 8 Absolute value of walks on the first 1000 nucleotides: b1) Mycoplasma putrefaciens , r) random sequence, h1) Aeropyrum pernix, h3) Acidilobus saccharovorans.

C. Power Spectrum

Power spectrum method is one of the most popular method to investigate autocorrelation in time series (signal) analysis [28, 39]. Some other popular methods are: the direct computation of the correlation function, the analysis of variance [33], later improved by the detrended fluctuation analysis [34],the mutual information function [24], the wavelet method [3-5, 57].

For a given numerical sequence $\{y_0, \, y_1, \, \dots \, , \, y_{N-1}\}$, with variance σ^2 the autocorrelation at the distance k, $(k = 0, \dots, N-1)$ is (see e.g. [8])

$$c_k \overset{def}{=} \frac{1}{\sigma^2}\left(\frac{1}{N-k}\sum_{i=0}^{N-k-1} y_i y_{i+k} - \frac{1}{(N-k)^2}\sum_{i=0}^{N-k-1} y_i \sum_{i=0}^{N-k-1} y_{i+k}\right) \quad (13)$$

with $k = 0, \dots, N-1$. This function can be approximated [29] by

$$c_k^* \overset{def}{=} \sum_{i=F}^{N-1-k} \frac{1}{N-F-k} u(x_i, \quad x_{i+1+k})$$

with the indicator as given in the previous section.

If $c_k = 0$ there is no linear correlation, $c_k > 0$ means that there is a strong (linear) correlation, anticorrelation when $c_k < 0$, while $c_0 = 1$ doesn't give any information about correlations. A true random process has a vanishing correlation $c_h = \delta_{0h}$ and its power spectrum S_h is constant. Its integral gives the Brownian motion (random walk) whose power spectrum is proportional to $1/k^2$.

In [3,4, 43] it has been shown that correlations in DNA are linear. However, the main problem of this measure is that it strongly depends on the representation, on the length of the sequence and, for non binary representation, it is affected by spurious results [8]. Moreover, the definition (13) holds only for real values of the representation. It has been also noticed [8, 41] that to have accuracy and to avoid statistical fluctuations in the computation of the autocorrelation function a long sequence is needed. The statistical fluctuation is $\varepsilon = \frac{1}{\sqrt{N}}$ so that the autocorrelation is measured by $c_k \pm \frac{1}{\sqrt{N}}$. Therefore the shorter is the sequence and the larger is its fluctuation.

The power spectrum of the numerical sequence $\{y_k\}_{k=0,\dots,N-1}$ is defined as the Fourier transform of c_k, as given by (13)

$$S_k \overset{def}{=} \hat{c}_k = \sum_{n=0}^{N-1} c_n e^{-2\pi i n k/N} .$$

However, a simplified definition [16] is based on the direct discrete Fourier transform of the sequence $\{y_k\}_{k=0,\dots,N-1}$ so that

$$S_k \overset{def}{=} \textstyle\sum_{s=0}^{k-1} |\hat{y}_s|^2 \quad (14)$$

with

$$\hat{y}_s = \frac{1}{N}\sum_{n=0}^{N-1} y_n e^{-2\pi i n s/N} \quad , \quad s = 0, \dots, N-1 .$$

The power spectrum of a stationary sequence, gives an indirect measure of the autocorrelation. A correlation, can be detected if the fluctuations can be described by a power law so that, S_k is locally approximated by the straight line

$$\alpha \frac{k}{\max_{1\le k\le k_{max}} [\alpha k]} \quad , \quad 1 \le k \le k_{\max} .$$

The fluctuation exponent α, with its values, characterizes a sequence as

1. anti-correlated: $\alpha < 1/2$
2. uncorrelated (white noise): $\alpha \cong 1/2$
3. correlated (long range correlated): $\alpha > 1/2$
4. $1/f$ noise: $\alpha \cong 1$
5. non-stationary, random-walk like: $\alpha > 1$
6. Brownian noise: $\alpha \cong 3/2$.

For the human DNA it was observed [7] a long range correlation, only for coding regions, with $\alpha = 0.61$. However, this value can be also seen for dog's and candida DNA [12, 13, 47] for the complete sequence (coding and noncoding regions), even if, by including the noncoding regions, this value is a little bit higher being $\alpha \cong 0.65$ for the dog's, and $\alpha \cong 0.62$ for the candida's DNA respectively.

The power spectrum $1/k$ has been observed in DNA sequences [26, 28] however it is not yet clear how should be the correlation function (step-function, power law decay, white-noise), and in particular if there exists a single length scale or a multi-length scale [27].

The biological explanation of long-range correlations can be explained by the existence of heterogeneity in DNA (i.e. different density distribution of bases). The main still unclear questioning is about the power law spectrum: $1/k$ [38] or $1/k^2$ [27] and the form of the autocorrelation function.

VI. Wavelet Analysis

Wavelet analysis is based on the discrete wavelet transform of a signal and the following interpretation of the fluctuation of wavelet coefficients. These coefficients give a description of local abrupt changes and variance. Wavelet analysis has been already extensively applied to the analysis of biological signals [3,4,27,44] focussing on the complexity and heterogeneity.

We will consider in the following the Haar wavelet basis (see e.g. [12,13,46,47]) made by scaling functions:

$$\begin{cases} \varphi_k^n(x) \overset{def}{=} 2^{n/2}\varphi(2^n x - k) \;, (0 \le k \le 2^n - 1) \;, \\ \psi(2^n x - k) = \begin{cases} 1 & , x \in \Omega_k^n \;, \quad \Omega_k^n \overset{def}{=} \left[\frac{k}{2^n}, \frac{k+1}{2^n}\right) \;, \\ 0 & , x \notin \Omega_k^n \;, \end{cases} \end{cases} \tag{15}$$

and the *Haar wavelets*:

$$\begin{cases} \psi_k^n(x) \overset{def}{=} 2^{n/2}\psi(2^n x - k) \;, \; ||\psi_k^n(x)||_{L^2} = 1 \;, \\ \psi(2^n x - k) \overset{def}{=} \begin{cases} -1 \;, & x \in \left[\frac{k}{2^n}, \frac{k+1/2}{2^n}\right) \;, \\ 1 \;, & x \in \left[\frac{k+1/2}{2^n}, \frac{k+1}{2^n}\right) \;, \\ 0 \;, & elsewhere \;. \end{cases} \end{cases} \tag{16}$$

The *discrete Haar wavelet transform* is the $N \times N$ matrix $W^N : K^N \subset \ell^2 \to K^N \subset \ell^2$ which maps the time series $\boldsymbol{Y}$ into the vector of *wavelet coefficients* $\boldsymbol{\beta} = \{\alpha\,,\, \beta_k^n\}$:

$$\begin{cases} W_N \boldsymbol{Y} & = \boldsymbol{\beta}_N \\ \boldsymbol{\beta}_N & \overset{def}{=} \{\alpha, \beta_0^0, \dots, \beta_{2^{M-1}-1}^{M-1}\} \;, \\ \boldsymbol{Y} & \overset{def}{=} \{Y_0,\, Y_1,\, \dots\,,\, Y_{N-1}\} \;, \quad (2^M = N) \;. \end{cases} \tag{17}$$

The matrix W_N can be easily computed by some recursive product [12,13,46,47] so that with $N = 4, M = 2$, we have

$$W_4 = \begin{pmatrix} \frac{1}{2} & \frac{1}{2} & \frac{1}{2} & \frac{1}{2} \\ -\frac{1}{2} & -\frac{1}{2} & \frac{1}{2} & \frac{1}{2} \\ -\frac{1}{\sqrt{2}} & \frac{1}{\sqrt{2}} & 0 & 0 \\ 0 & 0 & -\frac{1}{\sqrt{2}} & \frac{1}{\sqrt{2}} \end{pmatrix}. \tag{18}$$

From Eq. (17) with $M = 2, N = 4$, by explicit computation, we have

$$\alpha = \frac{1}{4}(Y_0 + Y_1 + Y_2 + Y_3)$$

and [46,47]

$$\begin{cases} \beta_0^0 & = \frac{1}{2}(Y_2 - Y_0 + Y_3 - Y_1) \\ \beta_0^1 & = \frac{1}{\sqrt{2}}(Y_0 - Y_1) \,, \\ \beta_1^1 & = \frac{1}{\sqrt{2}}(Y_3 - Y_2) \,. \end{cases}$$

So that the first wavelet coefficient α represents the average value of the sequence, while β's , also called detail coefficients, are strictly connected with the first order properties (jumps and variance) of the discrete time-series.

In the following we will consider the p-parameter short wavelet transform which consists in the subdivision of the DNA sequence into p-length segments and apply the wavelet transform to each segment [12,13,46,47] With $p = 4$, from the $N = 2^M$-length complex vector $\boldsymbol{Y}$, which is subdivided into 2^{M-2} segments, the 4-parameter short Haar wavelet transform, locally performed by the matrix (18) , gives the complex vector

$$W_4(\boldsymbol{Y}) = \boldsymbol{\beta} + i\,\boldsymbol{\beta}^*$$

with

$$\boldsymbol{\beta} = W_4\Re(\boldsymbol{Y}) = (\alpha \;, \beta_0^0 \;, \beta_0^1 \;, \beta_1^1) \;,$$
$$\boldsymbol{\beta}^* = W_4\Im(\boldsymbol{Y}) = \left(\alpha^* \;, {\beta^*}_0^0 \;, {\beta^*}_0^1 \;, {\beta^*}_1^1\right) .$$

From there we obtain the cluster of points

$$(W_4\Re(\boldsymbol{Y}^s), W_4\Im(\boldsymbol{Y}^s)), \qquad s = 0, \dots, 2^M - 2$$

in the 4-dimensional space

$$(\alpha \;, \alpha^*) \times \left(\beta_0^0 \;, {\beta^*}_0^0\right) \times \left(\beta_0^1 \;, {\beta^*}_0^1\right) \times \left(\beta_1^1 \;, {\beta^*}_1^1\right) .$$

This algorithm enables us to construct clusters of wavelet coefficients and to study the correlation between the real

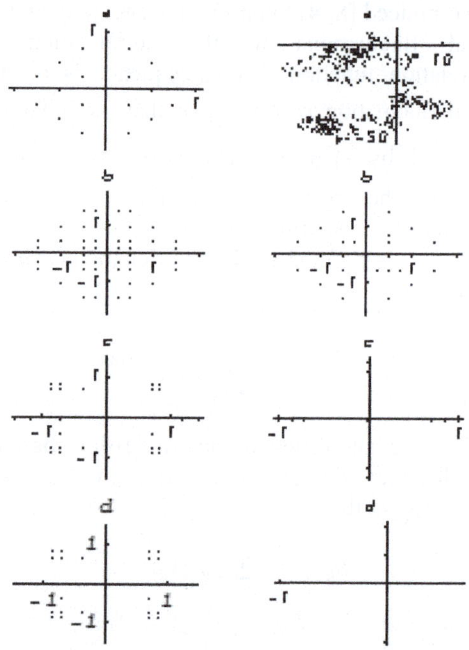

Fig. 9 Cluster analysis of the 4-th short Haar wavelet transform of a 4000-length random sequence (left) and its 2000-length random walk (right) : a) (α, α^*); b) $\left(\beta_0^0, {\beta^*}_0^0\right)$; c) $\left(\beta_0^1, {\beta^*}_0^1\right)$; d) $\left(\beta_1^1, {\beta^*}_1^1\right)$.

and imaginary coefficients of the DNA representation and DNA walk. It has been observed [12,13,46,47] that some more interesting symmetries arise from the plots of wavelet coefficients of DNA walks.

As can be seen from Fig. 9, random walks and random sequences have a very special patterns.

The cluster algorithm applied to the DNA walks, shows that the values of the wavelet coefficients belong to some discrete finite sets (Fig. 10).

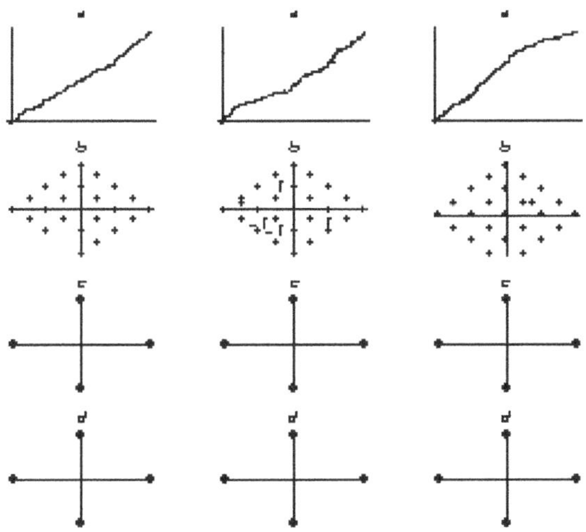

Fig. 10 Cluster analysis of the 4 th short Haar wavelet transform of the complex representation for a DNA walk on the first 2000 nucleotides of: b1(left column), b2 (middle column), b3 (right column) in the planes: a) (α, α^{*}) ; b) $(\beta^{0}_{0}, \beta^{*0}_{0})$; c) $(\beta^{1}_{0}, \beta^{*1}_{0})$; d) $(\beta^{1}_{1}, \beta^{*1}_{1})$.

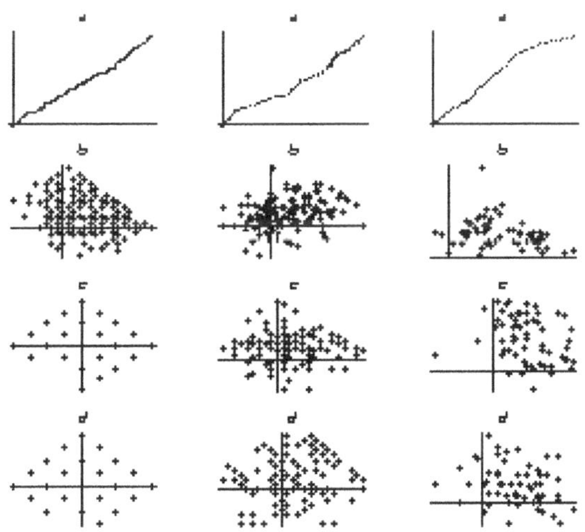

Fig. 11 Cluster analysis of the 8-th (left), 16-th (middle colum), 32-th (right) short Haar wavelet transform of the DNA walk on the first 2000 nucleotides of b1 in the planes: : a) (α, α^{*}) ; b) $(\beta^{0}_{0}, \beta^{*0}_{0})$; c) $(\beta^{1}_{0}, \beta^{*1}_{0})$; d) $(\beta^{1}_{1}, \beta^{*1}_{1})$.

However, it should be noticed that this symmetry on detail coefficients is lost for wavelet transform on longer segments (Fig. 11) at least for the first order coefficients while it still holds true at the highest order (Fig. 12).

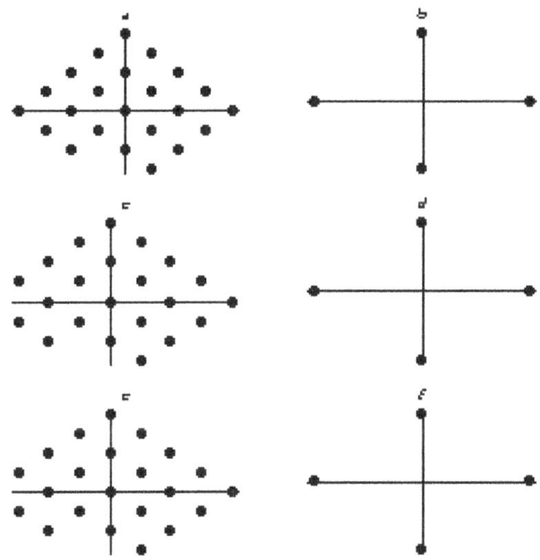

Fig. 12 Cluster analysis of the 8-th (first raw), 16-th (middle raw), 32-th (below) short Haar wavelet transform of the DNA walk on the first 2000 nucleotides of b1 in the planes: : a) $(\beta^{4}_{0}, \beta^{*4}_{0})$; b) $(\beta^{5}_{0}, \beta^{*5}_{0})$; c) $(\beta^{8}_{0}, \beta^{*8}_{0})$; d) $(\beta^{9}_{0}, \beta^{*9}_{0})$; e) $(\beta^{16}_{0}, \beta^{*16}_{0})$; f) $(\beta^{17}_{0}, \beta^{*17}_{0})$.

There follows that DNA sequences have to be considered as Markov chain with short range dependence, in other words any acid nucleic is attached to the chain on the base of a correlation of the previous acid nucleic. In other words, if we look for a dependence rule on the DNA nucleotides this dependence might be summarized by a function as

$$x_{n+1} = f(x_n) \quad , \quad (n = 1, \dots, N) \; .$$

VII. Conclusions

In this paper some of the most popular statistical methods of signal analysis for DNA sequences have been shortly analyzed. We have noticed that DNA sequences are very similar to random sequences so that they have the same nearly the same complexity values and power spectrum. Some minor differences can be seen in the evolution so that the more evolved DNA is little bit far from randomness. However there are no possibilities to single out intrinsic differences of DNA from random sequences. Only by using wavelet clustering we have observed that DNA coefficients tend to assume some special values comparing with the corresponding values of random sequences. These special clusters imply the existence of some local dependence of nucleotides.

References

1. Altaiski M., Mornev O., Polozov R. (1996) Wavelet analysis of DNA sequence, Genetic Analysis, 12 : 165-168.
2. Anastassiou D. (2000) Frequency-domain analysis of biomolecular sequence, Bioinformatics, 16, 12: 1073-1081.
3. Arneado A, Bacry E, Graves P.V, Muzy J.F (1995) Characterizing long-range correlations in DNA sequences from wavelet analysis, Phys. Rev. Lett., 74 : 3293-3296.
4. Arneado A, D'Aubenton-Carafa Y, Audit B, Bacry E., Muzy J.F, Thermes C (1998) What can we learn with wavelets about DNA sequences?, Physica A, 249 : 439-448.
5. Audit B, Vaillant C, Arneodo A, d'Aubenton-Carafa Y, Thermes C (2002) Long range Correlations between DNA Bending Sites: Relation to the Structure and Dynamics of Nucleosomes, J. Mol. Biol. 316 : 903-918.
6. Bai-Lin Hao (2000) Fractals from genomes-exact solutions of a biology inspired problem, Physica A, 282 : 225-246.
7. Berger J.A, Mitra S.K, Carli M, Neri A (2004) Visualization and analysis of DNA sequences using DNA walks, Journal of The Franklin Institutes, 341 : 37-53.
8. Bernaola-Galván P, Román-Roldán R, Oliver J L (1996) Compositional segmentation and long-range fractal correlations in DNA sequences, Phys. Rev. E 55, 5 : 5181-5189.
9. Berthelsen C. L, Glazier J. A, Skolnick M.H (1992) Global fractal dimension of human DNA sequences treated as pseudorandom walks, Phys. Rev. A 45, 12 : 8902-8913.
10. Borstnik B, Pumpernik D, Lukman D (1993) Analysis of apparent 1/ spectrum in DNA sequences, Europhys Lett. 23: 389-394.
11. Buldyrev S.V, Goldberger A. L, Havlin A.L, Peng C-K, Simons M, Sciortino F, Stanley H.E (1995) Long-range fractal correlations in DNA, Phys. Rev. E 51: 5084-5091.
12. Cattani C (2010) Wavelet Algorithms for DNA Analysis, Chapter 35, in Algorithms in Computational Molecular Biology: Techniques, Approaches and Applications, (Wiley Series in Bioinformatics) by M. Elloumi and A.Y. Zomaya, John Wiley & Sons: 799-842.
13. Cattani C (2010) Fractals and Hidden Symmetries in DNA, Mathematical Problems in Engineering, vol. 2010: 1-31. doi:10.1155/2010/507056
14. Cheever E.A, Searls D.B, Karanaratne W, Overton G.C (1989) Using signal processing techniques for DNA sequence comparison, Proc. 15th Annu. Northeast Bioeng. Conf.: 173-174.
15. Cristea P.D (2003) Large scale features in DNA genomic signals, Signal Processing, 83: 871-888.
16. Coward E (1997) Equivalence of two Fourier methods for biological sequences, Journal of Mathematical Biology, 36: 64-70.
17. Dodin G, Vandergheynst P, Levoir P, Cordier C, Marcourt L (2000) Fourier and Wavelet Transform Analysis, a Tool for Visualizing Regular Patterns in DNA Sequences, J. Theor. Biol., 206: 323-326.
18. Ferrer-i-Cancho R, Forns N (2010) The self-organization of genomes, Complexity, 15: 34-36.
19. Fitch J.P, Sokhansanj B (2000) Genomic engineering: moving beyond DNA sequence to function, Proc. IEEE 88, (12): 1949-1971.
20. Gates M.A (1985) Simpler DNA sequence representations, Nature 316, 219. doi:10.1038/316219a0
21. Gates M.A (1986) A simple way to look at DNA, J. Theor. Biol., 119: 319-328.
22. Jiang X.-Y, Lavenier D., Yau S.S.-T (2008) Coding Region Prediction Based on a Universal DNA Sequence Representation Method, Journal of Computational Biology 15, 10: 1237-1256.
23. Hamori E, Ruskin J (1983) H Curves, A Novel Method of Representation of Nucleotide Series Especially Suited for Long DNA Sequences, The Journal of Biological Chemistry 258, 2: 1318-1327.
24. Herzel H, Trifonov E.N, Weiss O, Grosse I. (1998) Interpreting correlations in biosequences, Physica A 249: 449-459.
25. Karlin S, Brendel V (1993) Patchiness and correlations in DNA sequence, Science 259: 677-680.
26. Li W (1997) The complexity of DNA: the measure of compositional heterogenity in DNA sequence and measures of complexity, Complexity 3: 33-37.
27. Li W (1997) The study of correlation structures of DNA sequences: a critical review, Computer Chem. 21 , 4: 257-271.
28. Li W, Kaneko K (1992) Long-range correlations and partial 1/ spectrum in a noncoding DNA sequence, Europhys. Lett. 17: 655-660.
29. McNally J.G, Mazza D (2010) Fractal geometry in the nucleus', EMBO J., 29: 2-3.
30. Metze K, Lorand-Metze I, Leite N.J, Adam R.L (2009) Goodness-of-fit of the fractal dimension as a prognostic factor, Cell Oncol., 31 : 503-504.
31. Metze K (2010) Fractal dimension of chromatin and cancer prognosis, Epigenomics, 2 (5): 601-604.
32. Misteli T (2009) Self-organization in the genome, Proc. Natl. Acad. Sciences USA, 106: 6885-6886.
33. Peng C-K, Buldryev S.V, Goldberg A.L, Havlin S, Sciortino F, Simons M, Stanley H.E (1992) Long-range correlatins in nucleotide sequences, Nature 356: 168-170.
34. Peng C-K, Buldryev S.V, Havlin S, Simons M, Stanley H.E, Goldberg A.L (1994) Mosaic organization of DNA nucleotides, Phys. Rev. E 49: 1685-1689.
35. Solé R.V (2010) Genome size, self-organization and DNA's dark matter, Complexity, 16: 20-23,
36. Szczepanski J, Michalek T (2003) Random Fields Approach to the Study of DNA Chains, Journal of Biological Physics 29: 39-54.
37. Takahashi M (1989) A fractal model of chromosomes and chromosomal DNA replication, J. Theor. Biol., 141: 117-136.
38. Vaidyanathan P.P, Yoon B-J (2004) The role of signal-processing concepts in genomics and proteomics, Journal of The Franklin Institute, 341: 111-135.
39. Voss R.F (1992) Evolution of Long-Range Fractal Correlations and 1 Noise in DNA Base Sequences, Physical Review Letters, 68, 25: 3805-3808.
40. Voss R.F (1992) Long-Range Fractal Correlations in DNA introns and exons, Fractals, 2 : 1-6.
41. Weiss O, Herzel H (1998) Correlations in protein sequences and property codes, J. Theor. Biol. 190: 341-353.

42. Yau S.S-T, Wang J, Niknejad A, Lu C, Jin N, Ho Y-K, (2003) DNA sequence representation without degeneracy, Nucleid Acids Res., 31: 3078-3080.
43. Yu Z.G, Anh W.W, Wang B (2001) Correlation property of length sequences based on global structure of the complex genome, Phys. Rev. E, 63: 011903.
44. Zhang M (1995) Exploratory analysis of long genomic DNA sequences using the wavelet transform: examples using polyomavirus genomes, Genome Sequencing and Analysis Conference VI: 72-85.
45. Aldrich P.R., Horsley R. K., Turcic S. M. (2011) Symmetry in the Language of Gene Expression: A Survey of Gene Promoter Networks in Multiple Bacterial Species and Non- Regulons, Symmetry, 3: 1-20.
46. Cattani C (2011) On the Existence of Wavelet Symmetries in Archaea DNA, Computational and Mathematical Methods in Medicine, vol. 2011: 1-37. doi:10.1155/2011/673934
47. Cattani C (2012) Complexity and Simmetries in DNA sequences, Chapter 22, in Handbook of biological discovery, (Wiley Series in Bioinformatics) by M. Elloumi and A.Y. Zomaya, John Wiley & Sons: 700-742.
48. Arneado A, D'Aubenton-Carafa Y, Bacry E, Graves P.V, Muzy J.F, Thermes C. (1996) Wavelet based fractal analysis of DNA sequences?, Physica D, 96: 291-320.
49. Gee H (2001) A journey into the genome: what's there, Nature, http://www.nature.com/nsu/010215/010215-3.html.
50. Cattani C (2008) Complex representation of DNA sequences, Communications in Computer and Information Science, Proceedings of the Bioinformatics Research and Development Second International Conference, BIRD 2008 Vienna, Austria, July 7-9, 2008, M. Elloumi et al. (Eds), Springer-Verlag Berlin Heidelberg, CCIS 13: 528-537.
51. Cattani C (2010) Harmonic Wavelet Approximation of Random, Fractal and High Frequency Signals, Telecommunication Systems 43, 3-4: 207-217.
52. Murray K.B, Gorse D, Thornton J.M (2002) Wavelet Transform for the characterization and detection of repeating motifs, JMB, J. Mol. Biol. 316: 341-363.
53. Tsonis A.A, Kumar P, Elsner J.B, Tsonis P.A (1996) Wavelet Analysis of DNA sequences, Physical Review E, 53: 1828-1834.
54. Cattani C, Rushchitsky J.J. (2007) Wavelet and Wave Analysis as applied to Materials with Micro or Nanostructure, Series on Advances in Mathematics for Applied Sciences, World Scientific, Singapore, 74.
55. Daubechies I (1992) Ten Lectures on wavelets SIAM, Philadelphia, PA.
56. Percival D.B, Walden A.T (2000) Wavelet Methods for Time Series Analysis, Cambridge University Press.
57. Wornell G (1996) Signal Processing with Fractals: A Wavelet-Based Approach, Prentice Hall.
58. Li M (2010) Fractal Time Series-A Tutorial Review, Mathematical Problems in Engineering, vol. 2010: 1--26. doi:10.1155/2010/157264
59. Li M, Li J.-Y (2010) On the predictability of long-range dependent series, Mathematical Problems in Engineering, vol. 2010: 1--9. doi:10.1155/2010/397454
60. Li M, Lim S.C (2010) Power spectrum of generalized Cauchy Process, Telecommunication Systems, 43 (3-4): 219-222.
61. National Center for Biotechnology Information, http://www.ncbi.nml.nih.gov/GenBank; Genome Browser, http://genome.ucsc.edu; European Informatics Institute, http://www.ebi.ac.uk; Ensembl, http://www.ensembl.org.
62. Howland J.L (2000) The Surprising Archaea, Oxford University Press, New York and Oxford.
63. Madigan M.T, Marrs B.L (1997) Extremophiles', Scientific American, 4: 82--87.
64. Cattani C, Pierro G (2011) Complexity on Acute Myeloid Leukemia mRNA Transcript Variant, Mathematical Problems in Engineering, vol. 2011: 1-16. doi:10.1155/2011/379873
65. Cattani C, Pierro G, Altieri G (2011) Entropy and multi-fractality for the myeloma multiple TET 2 gene, Mathematical Problems in Engineering, MPE vol. 2011: 1-17. doi:10.1155/2011/193761
66. Eckmann J.P, Kamphorst S.O, Ruelle D (1987) Recurrence Plots of Dynamical Systems, Europhysics Letters, 5: 973-977.
67. Stein M, Ulam S.M (1967) An Observation on the Distribution of Primes, American Mathematical Monthly (Mathematical Association of America) 74, (1): 43–44.
68. Shannon C.E (1948) A Mathematical Theory of Communication, The Bell System Technical Journal, 27: 379-423, 623-656.
69. Yulmetyev R.M, Emelyanova N.A, Gafarov F.M (2004), Dynamical Shannon entropy and information Tsallis entropy in complex systems, Physica A, 341: 649-676.

Carlo Cattani
Department of Mathematics, University of Salerno,
Via Ponte Don Melillo,
84084 Fisciano (SA)
Italy
ccattani@unisa.it

A Numerical Model of Metastatic Colon Cancer

Alan T. Lefor

Department of Surgery, Jichi Medical University, Tochigi Japan

Abstract — **Mathematical models of tumor growth have become an important approach to understanding tumor biology. To date, such models have been applied to only a few tumor types. The goal of this study is to apply a tumor model to liver metastases, to understand if the model is useful for this very common, clinically important type of tumor. CT scan and MRI scan tumor growth data will be extracted from routine imaging studies, examining tumor size in patients who received no treatment as well as those who were treated with a variety of modalities. Calculations of tumor growth velocity, diffusion and proliferation will be calculated. These parameters will then be correlated with survival data.**

Keywords — **Mathematical model, liver metastases, proliferation, survival.**

I. INTRODUCTION

Tumor growth is a complex phenomenon that is the end result of interactions among a wide range of factors including cell growth, cell dispersion, vascular supply, and host immune factors. Using mathematical models to understand the behavior of cancer has generally proceeded in two directions, including descriptive models and mechanistic models (1). Descriptive models may seem most intuitive to clinical oncologists, and examine the gross characteristics of a tumor such as its size, growth pattern, and overall dynamics. Mechanistic models focus on the various processes that lead to tumor growth in an effort to understand the relative contributions of the various components to overall tumor behavior. Both of these approaches are important for further understanding of tumor biology.

Descriptive models of tumors are especially useful when the model is refined by patient data, enabling the development of patient-specific therapy based on their own tumor growth pattern. In a study of 32 patients with gliomas, investigators found that dynamic insight from pre-treatment imaging may be quantitatively useful in predicting the survival of individual patients (41). A mathematical model was first developed, using a reaction-diffusion partial differential equation that equated the rate of change of glioma cell density to the net dispersal of glioma cells plus the proliferation. This model allowed quantification of the extent of the tumor as well as predicting radial expansion on routine imaging. The net result of combining this model with patient imaging data is an understanding of the gliomaphenotype for each patient yielding patient-specific prognostic information. Using two sets of imaging data allowed measurement of the velocity of radial expansion of the tumor, and then using the model, net rates of dispersal and proliferation were calculated.

Further studies of this approach are important to test the model with other tumors and then, if it is validated, to use this model to benefit other patient populations.

II. MATERIALS AND METHODS

A. Patient Data

B. Mathematical Model

The mathematical model developed by Swanson et al (2,3,4) has been used to study the growth of gliomas based on imaging data and may be useful for the understanding of other tumor types. The fundamental model is as follows:

Rate of change of glioma cell concentration =

Net dispersal of glioma cells + Net Proliferation of glioma cells

This model is then described using the following equation:

$$\frac{\partial C}{\partial t} = \nabla \cdot D(x)\nabla C + \rho C(1 - (\frac{C}{K}))$$

In this model, D is the spatial invasion rate, determined from the imaging data. The proliferation rate is given by ρ. Using Fishers approximation, the velocity of growth is calculated from

$$v = 2\sqrt{D\rho}$$

Then, using measured velocity from imaging data, and a measurement of D/ρ from the comparison of the T1 and T2 weighted images, we have two equations in two unknowns, allowing the calculation of D and ρindividually.

V. Van Toi et al. (Eds.): 4th International Conference on Biomedical Engineering in Vietnam, IFMBE Proceedings 40, pp. 370–371, 2013.
www.springerlink.com

III. Results

A. Measured Quantities

The essential feature of this model is the use of T1 Gd enhanced images and T2 images to estimate the amount of diffusion beyond the main tumor mass. Radius of the lesions is measured on both images and recorded. These measured quantities are shown in Table 1 as a preliminary result for only 1 patient.

Table 1 Measured Quantities-Patient Data.

Patient	Day	T1-Radius (mm)	T2-Radius (mm)	Size (mm)
1	0	20	22	
	92			30

The linear velocity is calculated from the rate of growth at the two time points used. The "index of invisibility" is then calculated from the difference in the size of the lesions on the T1 and T2 images, using an estimate of 16% for diffusion of cells beyond the size on the T1 image. Using the values calculated for velocity and the ratio of D/ρ, it is then possible to calculate D and ρ individually.

Table 2 Derived Quantities.

Patient	D/ρ	D (mm2/yr)	ρ	Velocity
1	0.1	6.4	64.2	40.6

IV. Conclusions

This model was originally developed for the study of gliomas, in a carefully conducted series of investigations by Swanson et al (2-4). The purpose of the present study is to evaluate this diffusion-proliferation model in the study of liver metastases. While there may be some similarities between the growth of metastases in solid organs such as the brain and liver, there also may be considerable differences. The aim of this study is to test this model and then modify it further to try and develop a useful mathematical model for the growth of metastases in the liver. More data is necessary to evaluate and modify this model.

The application of techniques from the physical sciences to problems in oncology comes at a time of significant advances in the application of computing techniques to problems in physical sciences. There is an extraordinary parallel between what is going on in the field of numerical relativity and Computational Oncology. In numerical relativity, computers are being used to solve complex differential equations in an attempt to further understand the theory of Relativity. This has only become recently possible with the development of supercomputers and new computational algorithms. The situation in the development of complex tumor models using systems of partial differential equations is remarkably similar, and we can only hope that the future of this exciting new field will ultimately provide a benefit to our patients from the synergistic activity between investigators in the physical sciences and those in oncology.

References

1. Lefor AT. Computational Oncology. Jpn J ClinOncol (2011) 41:937-47.
2. Swanson KR, Harpold HLP, Peacock DL et al. Velocity of radial expansion of contrast-enhancing gliomas and the effectiveness of radiotherapy in individual patients: A proof of principle. ClinOncol (2008) 20:301-8.
3. Wang CH, Rockhill JK, Mrugala M et al. Prognostic significance of growth kinetics in newly diagnosed glioblastomas revealed by combining serial imaging with a novel biomethematical model. Cancer Res (2009) 69: 9133-40.
4. Swanson KR, Rostomily RC and Alvord EC. A mathematical modeling tool for predicting survival of individual patients following resection of glioblastoma: A proof of principle. Br J Cancer (2008) 98:113

Symmetry and Folded Structures of Biomolecules

T.X. Hoang[1], N.T.T. Nhung[1], J.R. Banavar[2], and A. Maritan[3]

[1] Institute of Physics, Vietnam Academy of Science and Technology, Hanoi, Vietnam
[2] College of Computer, Mathematical and Natural Sciences, University of Maryland, College Park, MD
[3] Dipartimento di Fisica "G. Galilei", Universita di Padova, Padova, Italia

Abstract — **Our work suggests that protein native state structures occupy a novel phase of matter corresponding to the marginally compact conformations of a flexible tube. This phase arises from common attributes of proteins and is independent of amino acid sequences. Our approach provides a simple and unified framework to understand protein folding and amyloid formation. With a constraint on the local radius of curvature, the tube model is also shown to have ground state conformations similar to that of DNA toroids.**

Keywords — **protein folding, amyloid formation, tube model, buried area, marginally compact phase, DNA toroid.**

I. INTRODUCTION

The phases of inanimate matter are governed by the laws of physics that can be essentially understood in terms of the principles of geometry and symmetry [1]. The fixed and finite number of crystal structures are determined by the requirements that the constituent atoms are periodic and space-filling. Living matter are also governed by physical law. DNA and proteins are molecules providing key mechanisms for life: storing information, replication, natural selection and functionality. A related question is whether the considerations of symmetry and geometry can provide a better understanding of the structures and behaviors of these molecules.

Proteins, the workhorse molecules of life, play a central role in all living organisms. For example, enzymes catalyze biochemical reactions, proteins such as hemoglobin transport small molecules such as oxygen, hormonal proteins transfer signals and gene regulatory proteins bind to the DNA and switch the genes on or off. Under physiological conditions, proteins fold rapidly and reproducibly into their native state structures [2]. The folding is predominantly driven by the hydrophobic effect -- the aversion to water of some amino acid side chains leading to the creation of a hydrophobic core in the compact folded state. In addition, other interactions such as hydrogen bonds, Van der Waals interactions and electrostatic effects play a role in the folding process.

There are millions of proteins differed by their amino acid sequences. Yet, the number of protein folds (or native structures) is limited and is estimated to be about a few thousand [3]. Protein structures are stable against mutations in the amino acid sequence. The native states of proteins have a large "basin of attraction" that facilitates fast folding. Furthermore, many proteins with uncorrelated amino acid sequences are found to be able to aggregate and form amyloids [4]. There have been much progress in understanding each of these phenomena, but a simple explanation to all of these phenomena has been lacking.

II. TUBES SUBJECT TO COMPACTION

Nature provides us a strong hint: DNA and proteins are chain molecules and are characterized by the common occurrence of helical conformations – the double helix in DNA and α-helix in proteins. From everyday experience, the helix is a natural, compact conformation of a short, flexible tube. It has been shown that the pitch to radius ratio of a tube curled up into an optimal space-filling helix coincides remarkably with that of protein α-helices [5]. A similar packing of a double-helix formed by two tubes also match that of DNA [6]. This suggests that the phase behavior of a flexible tube subject to compaction might be related to protein and DNA behaviors.

Let's consider a discrete version of a tube of thickness Δ represented by a chain of coins whose planes coincide with equally spaced circular cross-sections of the tube. The self-avoidance of the tube is implemented by the thee-body description [6,7]. The hydrophobic effect is taken into account by introducing an attractive pairwise interaction between the coin centers with an interaction range R_0. In [8] it was shown that when Δ is approximately equal to R_0, a short tube had relatively few low energy conformations compared to the generic compact phase ($\Delta << R_0$) and the swollen phase ($\Delta >> R_0$). In this novel marginally compact phase, one find only two types of structures, helices and zig-zag strands. Being marginally compact, these structures possess a liquid crystal like sensitivity because of being poised in the vicinity of a transition to the swollen phase.

One can ask a question what happens in the continuum limit, in which the coin centers are placed infinitesimally

V. Van Toi et al. (Eds.): 4th International Conference on Biomedical Engineering in Vietnam, IFMBE Proceedings 40, pp. 372–375, 2013.
www.springerlink.com

close while the tube radius Δ is kept constant. As a direct consequence, in the continuum limit, the two body interaction between pairs of coins becomes singular because there is an infinite number of pairs, close by along the chain, that are within the interaction range [7]. A singularity-free formulation for the attractive interaction can be derived from the following physical situation. Let us suppose that the tube is immersed in a poor solvent whose molecules are approximated by spherical balls of radius R. For any given tube configuration, there are regions of the tube surface that the solvent molecules can come in contact with and other regions that are unaccessible. Compaction can be provided by considering the buried surface area of the tube. The latter is well-defined in the continuum. Fig. 1 (a-f) depicts the conformations adopted by short tubes subject to compaction by maximizing the buried area [9]. Such an optimization requirement is generically encountered when a tube shows a higher affinity to itself than to a solvent, e.g. in poor solvent conditions. Strikingly, the conformations of choice are single, double and triple helices, all characterized by chirality and adopted by nature in the context of biomolecules such as proteins and DNA.

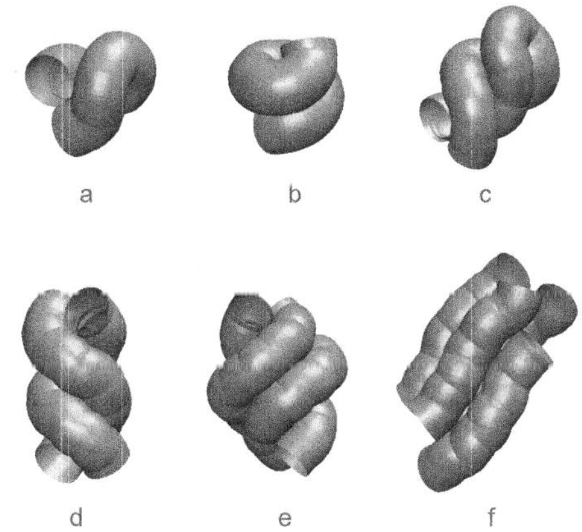

Fig. 1 Optimal conformations of solvophobic tubes [9]. The results are obtained by maximizing the buried area of the tubes, which are immersed in a poor solvent of spherical balls of radius R. Conformations shown in (a), (b), (c) are for single tubes. Conformations shown in (d), (e), (f) are for systems of two, three, and four tubes, respectively.

III. Proteins and Amyloid

Recent work [11] has probed the phase diagram of a short discrete tube mimicking a homopolymer made up of just one type of amino acid. Apart from the tube symmetry, this model of peptide is characterized by generic geometrical constraints arising from sterics and explicit backbone hydrogen bonding. The potential energy in the model is given by the sum of local bending energy penalty, hydrophobic energy and hydrogen bonding energy. The geometrical constraints on hydrogen bonds and sterics are derived from an analysis of protein PDB structures.

Fig. 2 shows the phase diagram of ground state conformations for a short homopolymer of length 24 in the peptide tube model [11]. Like for the simple tubes, in the vicinity of the swollen phase, one obtains a marginally compact phase on making small changes in the interaction parameters.

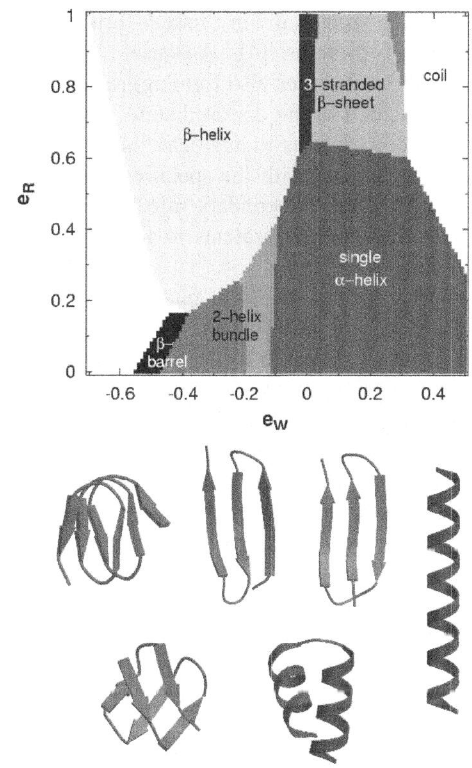

Fig. 2 (Top) Phase diagram of ground state conformations for a homopolymer chain of 24 amino acids in the tube model of protein [11]. The energy parameters in the model are local bending energy penalty, e_R, hydrophobic contact energy, e_W, and hydrogen bond energies. (Bottom) Ground state conformations obtained in the phase diagram.

The marginally compact phase has distinct structures including a single helix, a bundle of two helices, a helix formed by β-strands, a β-hairpin, three-stranded β-sheets with two distinct topologies and a β-barrel like conformation. These structures are the stable ground states in different parts of the phase diagram. Furthermore, conformations such as the β–α–β motif are found to be competitive local minima. The striking similarity between the observed structures and real protein structures suggests that our model captures the essential ingredients responsible for a limited menu of folds of protein native structures [3].

We hypothesize that the free energy landscape of proteins is pre-sculpted by geometry and symmetry. There are a few minima corresponding to structures in the menu of folds, each is associated with a large basin. The role of sequence is choose a structure from this menu. Indeed it is relatively easy to design sequences with just two types of amino acid that fold to target protein-like native structures [13].

A range of human diseases such as Alzheimer's, spongiform encephalopathies and light-chain amyloidosis lead to degenerative conditions and involve the deposition of plaque-like material in tissue arising from the aggregation of proteins [4]. A variety of proteins not involved in these diseases also form aggregates very similar to those implicated in the diseased state. This suggests that the tendency for proteins to aggregate is a generic property of polypeptide chains with the specific sequence of amino acids playing at best a secondary role. Can one understand this general tendency of proteins to form amyloids within our framework?

In order to assess the role played by the interaction between multiple short proteins, let us first consider our model homopolymer chain made up of 36 identical amino acids in the marginally compact phase of the refined tube model. The hydrophobic energy parameter and the local bending energy are chosen so that the ground state is a single long helix (Fig. 3(a)). On making two incisions in the chain to create three distinct chains each containing 12 amino acids, the ground state of the system appears to be a bundle of three helices. This helix structure however is stable only at very low temperatures. At intermediate temperatures, close to but lower than the temperature of the specific heat peak, it is destabilized in favor of aggregated β-helices (Figs. 3(b, b') and 3(c, c')) or sandwiches of β-sheets (Fig. 3(d,d')), due to entropic effects. Cutting a single chain into parts increases the entropy of the system. Unbonded chains are more flexible and this promotes the formation of interchain hydrogen bonds. The β-sheet structures also show an increased flexibility comparing to the helix bundle, and they have better kinetic accessibility from a disordered globule. While the appearance of β-sheet conformations in the case of three chains seems to have an entropic origin, it seems likely that the ground state of a system of multiple chains does in fact consist of aggregated β-sheets. Indeed simulations of 5 or 10 chains have shown that β-structures are the most likely choice [12].

That a β-sheet structure is a significant competitor with a large basin of attraction in a region where the stable phase is a helix reinforces the possibility that the interaction between several proteins could stabilize the formation of extended hydrogen bonded β-sheets via the aggregation of individual chains. These kinds of structures, which resemble the basic structures associated with amyloid fibrils, thus seem to belong to the general class of pre-determined folds, but this time for multiple proteins, and ought to be seen ubiquitously in generic proteins.

Our results show the generic tendency for multiple chains of amino acids to form aggregated amyloids rather than maintain their protein-like shape. This finding is consistent with experimental results on aggregation of proteins and peptides [4].

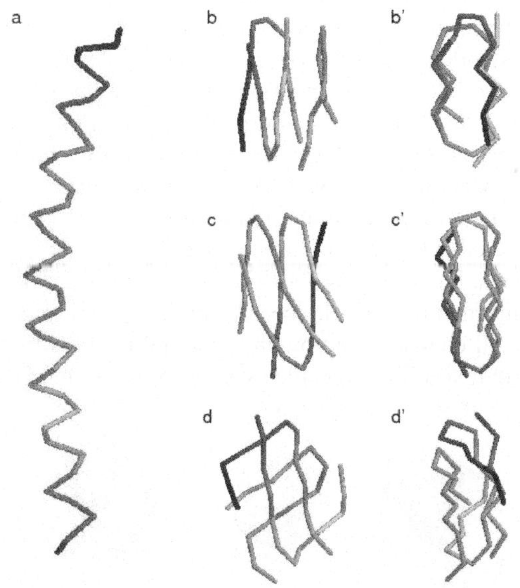

Fig. 3 The ground state of a model homopolymer chain of 36 amino acids is a single helix (a). After cutting the chain into three equal pieces each containing 12 amino acids, the ground state of the system is a bundle of three helices, but at intermediate temperatures, the chains are in favor of forming aggregated β-sheets (b,b',c,c',d,d'). β-sheet structures are found in amyloids.

IV. DNA TOROIDS

DNAs are long stiff polymer chains. Their persistence length is about 50 nm, while the diameter of the chain's cross section is about 2 nm. In aqueous solvent, DNA is negatively charged due to the phosphate groups that are soluble in water. It has been shown that multivalent cations of widely different types, such as Mn^{2+}, spermidine^{3+}, spermine^{4+} and $Co(NH_3)_6^{3+}$, induce a condensation of DNA into toroid-shaped structures [14]. These cations neutralize DNA negative charges and the collapse of the chain into toroidal condensates is due to hydrophobic effect. The outer diameter of these condensates can vary from 40 nm to 170 nm. A typical toroidal structure has a size of about 100 nm in outer diameter with a 30-nm hole. The size of the toroids is unaffected by the size of DNA molecules that form the condensates. Similar sized toroids are observed for DNA chains of few tens of kb to a genome size. Multiple DNA chains can condense on one toroid. The packing of DNA strands inside the toroids is highly ordered and has been shown to have primarily a hexagonal lattice arrangement [14]. It has been suggested that the organization of DNA in toroidal condensates follows a model of constant radius of curvature and the formation of the toroid is initiated by a spontaneous loop formation [15].

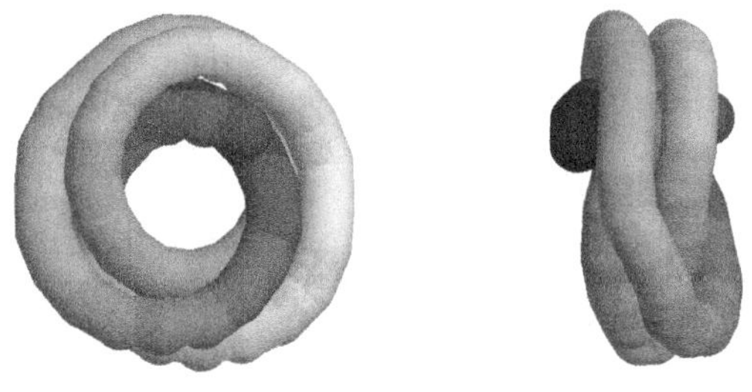

(a) (b)

Fig. 4 Two views of the lowest energy conformation obtained in the tube model of DNA. The chain shown is for N=60 beads, Δ=0.8, R_{int}=1.95 and R_{bend}=2.4.

We attempted to model a DNA as a chain of N beads representing a discrete tube of radius Δ. The bead spacing along the chain is 1. The hydrophobic effect is mimic by using a pairwise attractive potential of range R_{int} between non-consecutive beads. Stiffness is imposed by requiring that the local radius of curvature along the chain is never smaller than a value R_{bend}. We found that when R_{bend} is larger than 2Δ, the chain collapses in to a toroidal conformation at low temperatures (Fig. 4). Our result shows that like for DNA, the toroid phase is a natural choice for a self-attrating tube with significant stiffness.

V. Conclusions

In summary, within a simple, yet realistic, framework, we have shown that protein native state structures can arise from considerations of symmetry and geometry associated with the polypeptide chain. The limited number of folds arises from the tube symmetry of the chain molecule and geometrical constraints imposed by sterics and hydrogen bonds. Our studies suggest a stunning simplicity to the protein folding problem and also to amyloid structures. It is tempting to speculate that nature, through evolution and natural selection, has been able to exploit the marginally compact phase of flexible tubes under the constraints of quantum chemistry governing covalent and hydrogen bonds. Furthermore, the simple tube model can be extended to study DNA toroids.

Acknowledgment

This work has been supported by National Foundation for Science and Technology Development (NAFOSTED) grant 103.01-2010.11.

References

1. Chaikin PM, Lubensky TC (2000) Principles of condensed matter physics, Cambridge University Press, Cambridge
2. Anfinsen C (1973) Principles that govern the folding of protein chains. Science 181:223–230
3. Chothia C (1992) One thousand families for the molecular biologist. Nature (London) 357:543–544
4. Dobson CM (2002) Protein-misfolding diseases: Getting out of shape. Nature (London) 418:729-730
5. Maritan A, Micheletti C, Trovato A, and Banavar JR (2000) Optimal shapes of compact strings. Nature 406:287-290
6. Stasiak A, Maddocks JH (2000) Mathematics – Best packing in proteins and DNA. Nature (London) 406:251-253
7. Gonzalez O and Maddocks JH (1999) Global curvature, thickness and the ideal shapes of knots. Proc Natl Acad Sci USA 96:4769-4773
8. Banavar JR, Gonzalez O, Maddocks JH, Maritan A (2003) Self-Interactions of Strands and Sheets, J Stat Phys 110:35-50
9. Marenduzzo D, Flammini A, Trovato A, Banavar JR, Maritan A (2005) Physics of Thick Polymers. J Pol Sci 43:650-679
10. Banavar JR, Hoang TX, Maddocks JH, Maritan A, Poletto C, Stasiak A, Trovato A (2007) Structural motifs of biomolecules. PNAS 104:17283-17286
11. Hoang TX, Trovato A, Seno F, Banavar JR, and Maritan A (2004) Geometry and symmetry presculpt the free energy landscape of proteins Proc Natl Acad Sci USA 101:7960
12. Banavar JR, Hoang TX, Maritan A, Seno F, and Trovato A (2004) Unified perspective on proteins: A physics approach. Phys. Rev. E 70:041905
13. Hoang TX, Marsella L, Trovato A, Seno F, Banavar JR, Maritan A (2006), Proc. Natl. Acad. Sci. USA 103:6883-6888
14. Hud NV, Vilfan ID (2005) Toroidal DNA Condensates: Unraveling the fine structure and the role of nucleation in determining size. Annu Rev Biophys Biomol Struct 34:295-318.
15. Hud NV, Downing, KH, Balhorn R (1995), A constant radius of curvature model for the organization of DNA in toroidal condensates, Proc. Natl. Acad. Sci. USA 92:3581-3585.

Use macro [author address] to enter the address of the corresponding author:

Author: Trinh Xuan Hoang
Institute: Institute of Physics, Vietnam Academy of Science and Technology
Street: 10 Dao Tan, Ba Dinh
City: Hanoi
Country: Viet Nam
Email: hoang@iop.vast.ac.vn

Diffusion, Aggregation and Penetration: Modelling the Complex Cell Envelope of E.coli

Syma Khalid[1], Joseph Goose[2], Daniel A. Holdbrook[1], Thomas J. Piggot[1], and Mark S.P. Sansom[2]

[1] School of Chemistry, University of Southampton, Southampton, UK, SO17 1BJ
[2] Department of Biochemistry, University of Oxford, Oxford, UK, OX1 3DP

Abstract — ***E.coli* is a simple organism with a complex, dual-membrane cell envelope. Moreover the two membranes differ in their compositions and thus also their permeabilities. Both membranes contain multiple proteins that perform a variety of functions. While it is clear that the membrane is not simply and inert environment that simply serves to contain the proteins, the precise details of its role and the complex interplay between membrane and proteins are far from understood. Here we employ atomistic and coarse-grain simulations to focus on the interaction of melittin with model inner and outer membranes and aggregation and diffusion of proteins in the outer membrane. The complexity of the membrane composition in terms of lipids and proteins is addressed.**

Keywords — **bacteria, membrane protein, molecular dynamics, complexity, lipid.**

I. INTRODUCTION

Gram-negative bacteria such as *E.coli* are typically regarded as "simple" model organisms, yet their cell envelopes are surprisingly complex. The cells of these bacteria are protected by two membranes; the inner and outer membrane, which are separated by the periplasmic space. The two membranes differ in composition. The outer membrane is asymmetric in nature; it contains lipopolysaccharide (LPS) in the outer leaflet and a mixture of phospholipids in the inner leaflet. In contrast, the inner membrane contains phospholipids in both leaflets and is essentially symmetric. The periplasm is largely composed of peptidoglycan; a polymer of repeating sugar-peptide units. The precise composition of bacterial membranes in terms of the lipids and proteins differs from species to species and plays a key role in determining the relative permeabilities of the membranes.

One of the main roles of the cell envelope is to facilitate the correct functioning of the proteins embedded within it. Bacterial membrane proteins perform a wide range of functions including allowing the passage of ions across the membrane, enzymatic activity, host/pathogen interactions and surface recognition. Indeed, their role in host/pathogen interactions provides a possible route to vaccine development. Thus unraveling the complex behavior of these ubiquitous proteins is a key step in advancing our knowledge of many cellular processes. Furthermore, membrane proteins are of interest from a biotechnology perspective, for example, simple bacterial outer membrane proteins (OMPs) are being explored as components of biosensors, while more complex OMPs are potential antimicrobial targets. Recent experimental and theoretical studies have revealed that contrary to the traditional view of the cell membrane as a passive bystander in membrane protein function, it often plays a key role in protein folding, binding and aggregation. Thus, the details of the specific compositions of bacterial membranes and the interdependency of the lipid and protein components are key to the correct functioning of the cell envelope. However, experimental and simulation studies are often performed on simplified model systems such that the details of the complexity are lost.

Here we have focused our attention on two different aspects of the *E.coli* cell envelope: we describe (1) the interaction of melittin with models of the inner and outer membranes of *E.coli* and (2) the aggregation and diffusion behavior of outer membrane proteins in model membranes. We have retained much of the complexity of the membranes we have simulated, to study the details of membrane-protein interactions that are characteristic of *E.coli* cells. Below we discuss the two different sets of simulations.

II. ANTIMICROBIAL PEPTIDES

The interactions of antimicrobial peptides with eukaryotic cell membranes have been well-studied by experimental and computational methods (*1, 2*). However, the molecular-level interactions of these peptides with Gram-negative bacteria such as *E.coli* are more difficult to study, largely due to the complex nature of the asymmetric outer membrane. While cell lysis is thought to proceed via destruction of the inner membrane, it is the outer membrane that is the first barrier encountered by antimicrobial peptides as they attempt to reach their target site.

V. Van Toi et al. (Eds.): 4th International Conference on Biomedical Engineering in Vietnam, IFMBE Proceedings 40, pp. 376–379, 2013.
www.springerlink.com

In the present work, we employ atomistic molecular dynamics simulations to study the interaction of melittin, the major component of honey-bee toxin, with detailed models of the *E.coli* outer membrane (*3*). Experimental data (*4*), has indicated that upon interaction, melittin induces reorganisation of the LPS component of the bacterial outer membrane. We have simulated multiple copies of melittin in lipid bilayers composed of varying depths of LPS in the outer leaflet, and realistic mixtures of phospholipids in the inner leaflet. Our results reveal that melittin monomers are unable to penetrate into asymmetric LPS—phospholipid bilayers on the 200 ns timescale. The slowly diffusing LPS molecules in the outer leaflet are cross-linked by divalent cations and thus tightly packed together. They form a barrier that is not penetrable by melittin on the timescale of an atomistic simulation.

Thus, a second set of simulations was initiated in which a preformed pore formed composed of 4 melittin monomers was inserted into three different model membranes; (1) a phosphatidylethanolamine (PE) bilayer (2) a phosphate-dylcholine (PC) bilayer and (3) a LPS-PE bilayer. Three independent simulations, each of duration 100 ns were performed for each membrane system. The composition of the *E.coli* inner membrane is ~ 70-80% PE, while the inner leaflet of the outer membrane also has a similar composition, thus (1) and (3) are models of the inner and outer membranes respectively while (2) was chosen for comparison due to its larger headgroups and also as it is often used as a model membrane for experimental studies. Our results show that the small PE lipid headgroups are able to penetrate in between the melittin monomers such that they interact with the charged residues of the peptides. The loss of peptide-peptide interactions, due to the PE headgroup-peptide interactions, causes the melittin monomers to move apart. Such that after 100 ns, the inter-peptide distance is 5 nm in some cases, and thus the pore has essentially dissociated. In contrast, the larger headgroups of the PC lipids are unable to penetrate in between the melittin peptides, although some transient penetration of the lipid tails is observed. The peptides remain associated with each other in this environment. Furthermore, Cl^- ions from the bulk solution are observed to enter the pore such that they cross-link the basic residues of the melittin monomers leading to tighter association and reduced inter-peptide distance compared to the original pore.

Pores in the model outer membranes in which the outer leaflet was composed of LPS molecules and the inner leaflet of PE lipids exhibited interesting behavior. The small PE headgroups were able to penetrate in between the peptides from the inner leaflet side. However, the large headgroups of the lipid A moieties of the LPS molecules were unable to penetrate into the melittin pore. Lysine and arginine residues at the peptide C-terminal domain facing into the outer leaflet region formed long-lived (~100 ns) interactions with sugar moieties of the LPS molecules, such that one LPS molecule was often observed to interact with two peptide residues simultaneously. This strong interaction between the outer leaflet of the bilayer and the melittin monomers was sufficient to retain the structural integrity of the pore such that a water channel within the pore, extending from side of the bilayer to the other was present even after 100 ns.

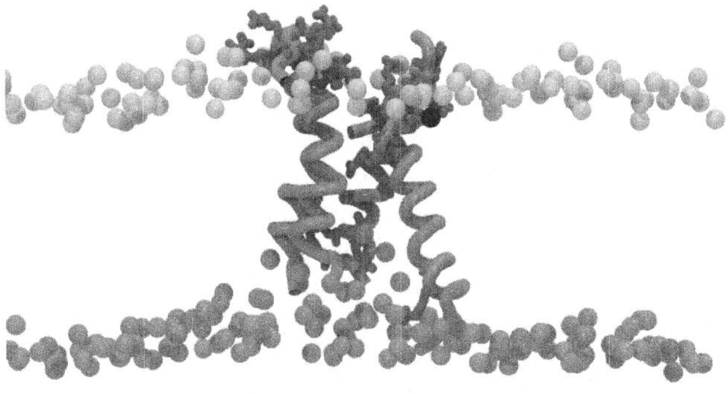

Fig. 1 Melittin tetrameric pore (pink) in a LPS (headgroups shown as yellow spheres) and POPE (headgroups shown as orange spheres) bilayer after 100 ns of simulation.

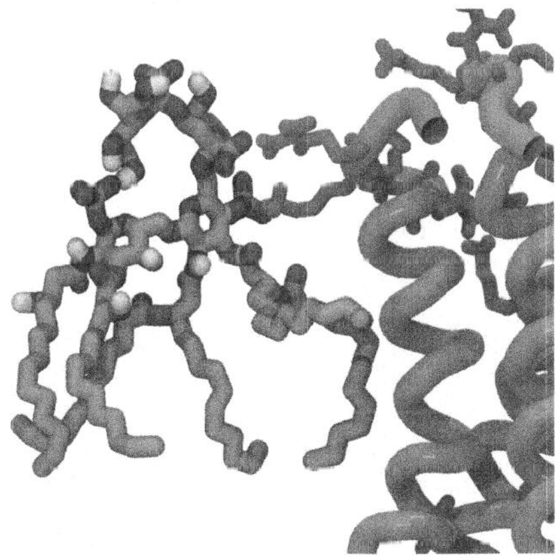

Fig. 2 An arginine and lysine residue (blue) from the same melittin peptide (pink) simultaneously interact with a single LPS molecule (red, white, blue, cyan and gold).

III. Aggregation and Diffusion of Outer Membrane Proteins

The *E.coli* outer membrane is a crowded environment containing over 30 different protein species. All known structures of outer membrane proteins (OMPs) reveal a β-barrel architecture. The functions of these proteins include passage of solutes, enzymes and adhesion. While molecular dynamics simulations of individual OMPs have revealed details of conformational dynamics and membrane interactions, the complex interplay between multiple proteins and the environment in which they reside, is almost totally uncharacterized at the molecular level of detail.

To address this, we have employed CG models to simulate vesicles composed of 32 outer membrane proteins (OMPs). 8 copies each of the proteins OmpA (*5*), EspP (*6*), FhuA (*7*) and OmpF (*8*) were embedded in vesicles composed of a mixture of lipids (PE, PG and cardiolipin), in the inner leaflet and DLPC lipids in the outer leaflet, as these provide a good approximation to the hydrophobic thickness of the lipid A moiety of LPS. The total system is composed of over 1 million particles.

The simulations reveal that the diffusion of OMPs and the surrounding lipids is closely related. Specifically we observe lipids that are closest to the protein (termed annular lipids) have the slowest diffusion rate, those in the next 'shell' are slightly faster and so on until the diffusion cannot be distinguished from bulk diffusion. Dimerisation of two OmpF molecules was also observed. Lipids at the interface of the two proteins were observed to 'flip' between the two proteins until they moved out of this region altogether. Removal of lipids from the protein interfacial region appeared to be the rate-determining step in dimer formation. We did not observe and changes in the bilayer thickness around the dimer compared to the individual OmpF monomers, but interestingly we did observe transient correlated motion of lipids near the OmpF proteins. This lipid motion seems to facilitate rotation of the proteins such that they are able to dimerise.

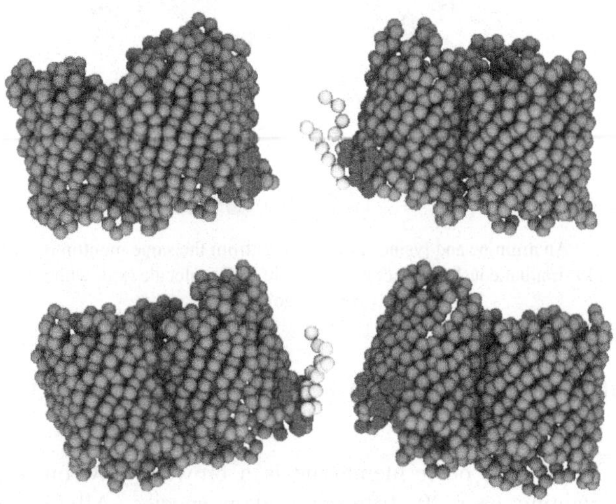

Fig. 3 OmpF proteins just prior to forming a dimer (blue and red). A lipid molecule (green) at the interfacial region 'flips' between the two proteins.

IV. Methods

Atomistic simulations

Atomistic simulations were performed using the GROMOS53A6 forcefield (*9*) within the GROMACS4 simulation package (*10*). Three independent simulations, each of at least 100 ns duration were performed for each membrane system to ensure reproducibility.

Coarse-grained simulations

Coarse-grained simulations were performed using the MARTINI forcefield (*11*) with the Bond & Sansom implementation for proteins (*12*). Each vesicle was simulated for at least 1μs.

V. Conclusions

Our results clearly show the marked influence of the composition of membranes on the dynamics of proteins that reside within them. The highly charged LPS molecule interacts with basic residues at the C-terninal domain of melittin to hold the peptide in place in a tetrameric pore. In the absence of LPS, the pore may dissociate depending upon the size of the headgroup of the phospholipids. This observation suggests that melittin may interact via different modes with the inner and outer membranes of *E.coli*.

Coarse-grain simulations of vesicles containing multiple copies of outer membrane proteins have shown that lipids nearer to the protein diffuse slower than 'bulk' membrane lipids. Furthermore the simulations show that correlated motion of lipids and proteins may facilitate protein aggregation. Both the melittin and outer membrane protein results reveal the need to incorporate the complexity of bacterial membranes for the study of membrane protein function.

Acknowledgment

This work was funded by the BBSRC OMSys project; grants BB/H000658/1 and BB/H000267/1.

References

1. Sengupta, D., Leontiadou, H., Mark, A. E., and Marrink, S. J. (2008) Toroidal pores formed by antimicrobial peptides show significant disorder, Biochim Biophys Acta 1778, 2308-2317.

2. Kuchinka, E., and Seelig, J. (1989) Interaction of melittin with phosphatidylcholine membranes. Binding isotherm and lipid head-group conformation, Biochemistry 28, 4216-4221.
3. Piggot, T. J., Holdbrook, D. A., and Khalid, S. (2011) Electroporation of the E. coli and S. Aureus Membranes: Molecular Dynamics Simulations of Complex Bacterial Membranes, The Journal of Physical Chemistry B 115, 13381-13388.
4. Hanulova, M., Andra, J., Garidel, P., Olak, C., Howe, J., Funari, S. S., Gutsmann, T., and Brandenburg, K. (2009) Interaction of Melittin with Phospholipid- and Lipopolysaccharide- Containing Model Membranes, Anti-Infective Agents in Medicinal Chemistry (Formerly 'Current Medicinal Chemistry - Anti-Infective Agents) 8, 17-27.
5. Pautsch, A., and Schulz, G. E. (1998) Structure of the outer membrane protein A transmembrane domain, Nature Structural Biology 5, 1013-1017.
6. Barnard, T. J., Dautin, N., Lukacik, P., Bernstein, H. D., and Buchanan, S. K. (2007) Autotransporter structure reveals intra-barrel cleavage followed by conformational changes, Nature Structural & Molecular Biology 14, 1214-1220.
7. Locher, K. P., Rees, B., Koebnik, R., Mitschler, A., Moulinier, L., Rosenbusch, J. P., and Moras, D. (1998) Transmembrane signaling across the ligand-gated FhuA receptor: crystal structures of free and ferrichrome-bound states reveal allosteric changes, Cell 95, 771-778.
8. Cowan, S. W., Schirmer, T., Rummel, G., Steiert, M., Ghosh, R., Pauptit, R. A., Jansonius, J. N., and Rosenbusch, J. P. (1992) Crystal structures explain functional properties of two E. coli porins, Nature 358, 727-733.
9. Oostenbrink, C., Villa, A., Mark, A. E., and Van Gunsteren, W. F. (2004) A biomolecular force field based on the free enthalpy of hydration and solvation: The GROMOS force-field parameter sets 53A5 and 53A6, Journal of Computational Chemistry 25, 1656-1676.
10. Hess, B., Kutzner, C., van der Spoel, D., and Lindahl, E. (2008) GROMACS 4: Algorithms for Highly Efficient, Load-Balanced, and Scalable Molecular Simulation, Journal of Chemical Theory and Computation 4, 435-447.
11. Marrink, S. J., de Vries, A. H., and Mark, A. E. (2004) Coarse Grained Model for Semiquantitative Lipid Simulations, The Journal of Physical Chemistry B 108, 750-760.
12. Bond, P. J., and Sansom, M. S. P. (2006) Insertion and Assembly of Membrane Proteins via Simulation, Journal of the American Chemical Society 128, 2697-2704.

Use macro [author address] to enter the address of the corresponding author:
Author: Syma Khalid
Institute: University of Southampton
Street: Highfield
City: Southampton
Country: United Kingdom
Email: S.Khalid@soton.ac.uk

Synthetic Infectious Prions

G. Legname and N. Le Tran Thanh

Department of Neurobiology, Scuola Internazionale Superiore di Studi Avanzati (SISSA), Trieste, Italy

Abstract — **Prion diseases are fatal and incurable neurodegenerative maladies of unique etiology. These disorders have attracted considerable attention for their unique infectivity pathogenic mechanism. Here we describe the milestone studies that have lead to the demonstration that prions are solely constituted of proteins. The formal demonstration has been achieved through several papers describing unconvertible evidence that recombinant prion protein refolded into amyloid fibrils can contain infectious species that lead to disease in animal models of prion coversion and replication. These molecules have been named as synthetic infectious prions.**

Keywords — **prion protein, recombinant proteins, prions, transgenic mice, synthetic prions.**

I. Introduction

Prion diseases, or transmissible spongiform encephalopathies (TSE), are neurodegenerative disorders, which affect human and several animals. The incubation time of the diseases is generally quite long, and once symptoms appear, the diseases progress rapidly, leading to brain damage and death. Till now, prion diseases are invariably fatal with no effective treatment. The causing agent of prion disease is known as prion (acronym for proteinaceous infectious particle), which cause disease within the central nervous system disrupting the normal tissue. Neuropathology of the disease is featured by neuronal loss, gliosis and spongiform changes. The prion protein (PrP) is highly conserved among species. The putative functions of normal cellular form, PrP^C and the mechanism of prion replication- a process in which PrP^C converts into the pathological form of the prion or PrP^{Sc}, have been intensively studied.

II. Results

In 2004 the production of synthetic prions via *in vitro* induction of misfolding and aggregation of bacterially expressed recombinant (rec) PrP was introduced. The material was formed from amyloidal mouse recPrP compose residues 89-230, demonstrating that conversion after polymerization of purified recPrP into amyloidal fibrils, that represent a subset of β sheet-rich structures, is sufficient for the generation of infectivity. The first mouse synthetic prion strain inoculated into transgenic (Tg) 9949 mouse line expressing N-terminal truncated mouse recPrP(Δ23-88) showed clinical disease. Furthermore, brain tissue from these sick mice contained prions that caused TSE disease when inoculated into wild-type mice and Tg mice overexpressing PrP. Neuropathological findings suggested that a novel prion strain was created (Figure 1). The incubation time, neuropathological lesion profiles and conformation stability (expressed as guanidinium hydrochloride half concentration values or Gdn1/2) indicated that synthetic prions differ from RML and many other prion strains such as scrapie or bovine spongiform encephalopathy.

These data accompanied with those available for naturally occurring strains indicated the existence of a reverse correlation between the stability and infectivity of prion isolated (Legname *et al.*, 2004; Legname *et al.*, 2005). Furthermore, after having produced PrP amyloid fibers under different experiments set-up, Colby and colleagues postulated that newly created synthetic prion strains would possess conformation stabilities and incubation period dictated by the properties of the amyloid preparation from which they were generated (Figure 2 and 3). The disease-causing PrP^{Sc} form is frequently distinguished from normal PrP^C by its partial resistance to digestion by proteases. Considerable evidence shows that protease-sensitive prions may occur naturally in humans and sheep. These results also show the importance of using alternate methods for detecting PrP^{Sc}, rather than employing only the presence of PK-resistant PrP (rPrP). The production of synthetic prions, which are sensitive to proteolysis but cause transmissible disease, is an important step toward understanding the role of protease-sensitive forms of PrP^{Sc} ($sPrP^{Sc}$) in the pathogenesis of prion disease. The studies mentioned here not only demonstrate the validity of the experimental systems reported earlier but they also extend the understanding of synthetic prions. Moreover, these findings establish that in the absence of detectable $rPrP^{Sc}$, $sPrP^{Sc}$ alone is sufficient to cause neurodegeneration. Exclusive reliance on the detection of $rPrP^{Sc}$ as a surrogate marker for prion infectivity may overlook the contribution of $sPrP^{Sc}$ to prion infectivity and the pathogenesis of prion disease.

V. Van Toi et al. (Eds.): 4th International Conference on Biomedical Engineering in Vietnam, IFMBE Proceedings 40, pp. 380–383, 2013.
www.springerlink.com

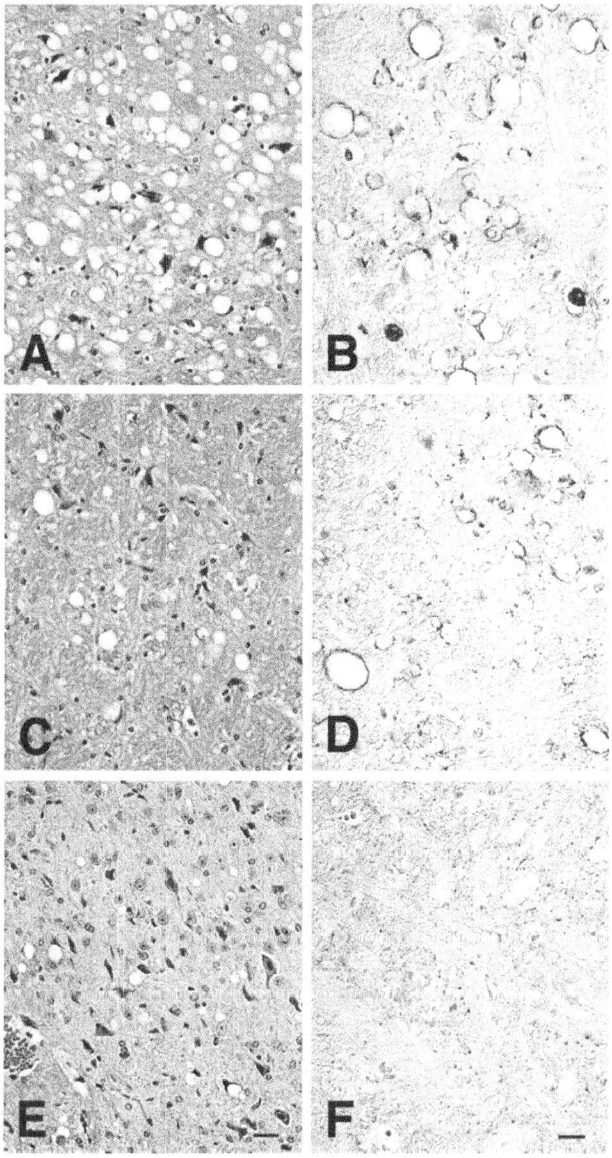

Fig. 1 The distinguishing neuropathological features of unseeded recPrP amyloid (A and B), seeded recPrP amyloid (C and D), and RML prions (E and F) in the pons of Tg9949 mice. Left column: hematoxylin and eosin (H&E) stain (scale bar, 50 μm); right column: immunohistochemistry of PrP^{Sc} by the hydrated autoclaving method using the PrP-specific HuM-R2 monoclonal antibody fragment (scale bar, 25 μm). (Legname et al., 2004).

In fact, Tg9949 mice inoculated with the unseeded amyloid fibers contain protease-sensitive prions that exhibit all the neuropathological features of prion disease (Figure 4, 5) (Colby *et al.*, 2010). Protease-sensitive prions could be detected using a so-called amyloid seeding assay (ASA) (Colby *et al.*, 2007).

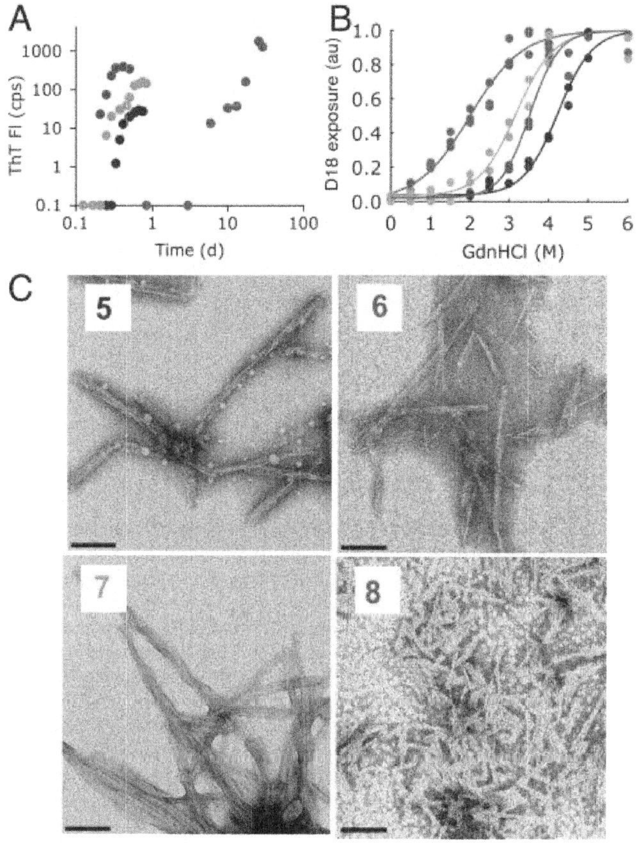

Fig. 2 Characterization of diverse PrP amyloid conformations. Formation of Amyloid 5 (blue), Amyloid 6 (green), Amyloid 7 (orange), and Amyloid 8 (red) was monitored by ThT fluorescence (A), followed by measurement of conformational stability by ELISA coupled with denaturation with increasing concentrations of GdnHCl (B). Electron micrographs (C) show morphology of Amyloids 5–8, as indicated. (Scale bars, 100 nm.). (Colby et al., 2009).

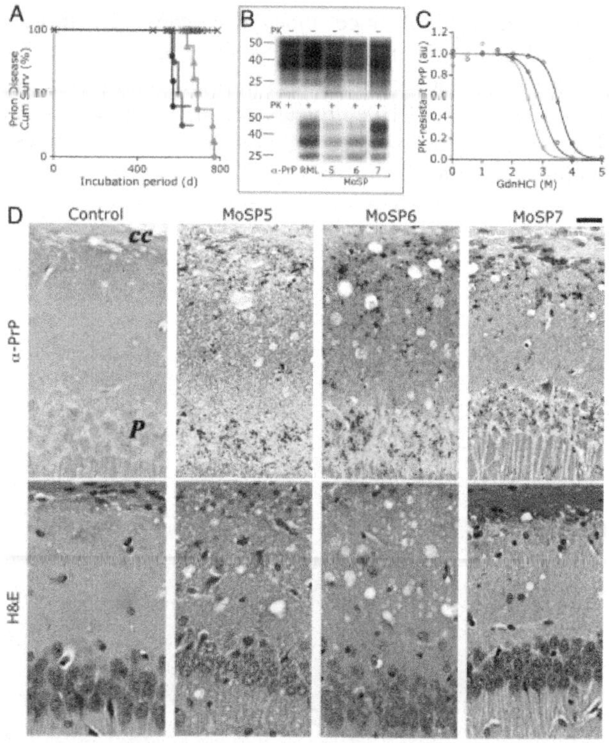

Fig. 3 Prion disease results from inoculation of Amyloids 5, 6, and 7 into Tg4053 mice. Amyloids 5 (blue), 6 (green), 7 (orange), and 8 (red) were inoculated into Tg4053 mice (n = 8–12 per inoculum). Monomeric, α-helical recPrP (purple) was inoculated as a control. Amyloids 5, 6, and 7 resulted in prion disease (A), which was confirmed by Western blotting for protease-resistant PrPSc (B) and neuropathology (D). The resulting isolates were designated "MoSP5," "MoSP6," and "MoSP7," respectively. The conformational stability of MoSP5 (blue), MoSP6 (green), and MoSP7 (orange) was measured by titration with GdnHCl followed by PK digestion (C). •, mouse with symptomatic prion disease; ▲, mouse with asymptomatic prion disease; ×, deceased mouse with no indication of prions. In B, a mouse infected with RML prions is shown as a control for PK-resistant PrP^{Sc}. Molecular masses based on the migration of protein standards are shown in kDa. Neuropathology included PrP deposits (dark brown, top row in D) and vacuolation (white holes, bottom row in D). cc, corpus callosum; P, pyramidal cell layer. (Scale bar, 30 μm.). (Colby et al., 2009).

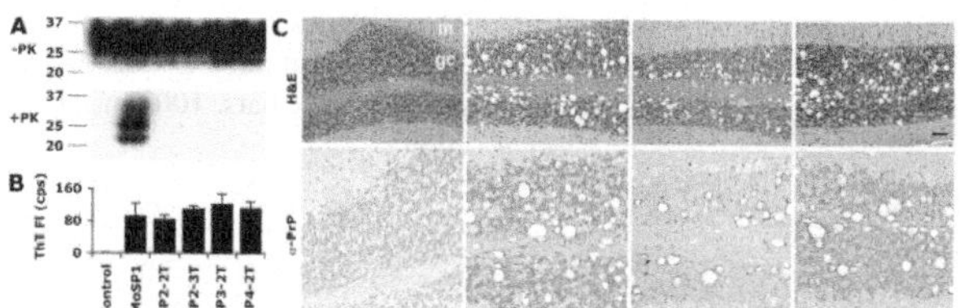

Fig. 4 Protease-sensitive synthetic prions are serially transmissible in Tg9949 mice. MoSP2-1T, MoSP3-1T and MoSP4-1T were serially passaged to new groups of Tg9949 mice for a second transmission (*2T*) by intracerebral inoculation of brain homogenate containing each isolate. MoSP2-2T was serially passaged an additional time for a third transmission (*3T*). The brains of Tg9949 mice containing MoSP2-2T, MoSP3-2T, or MoSP4-2T showed no protease-resistant PrP by Western blotting (*A*), but activity in the ASA (*B*) and neuropathology consistent with prion disease (*C*). Brain samples from mice inoculated either with MoSP1 (*Legname et al., 2004)* or with homogenates of uninfected Tg9949 mice are shown as controls. (*C*) Cerebellar sections were stained with H&E (*top row*) and α-PrP (*bottom row*). m, molecular layer; gc, granule cell layer. Each scale bar represents 100 μm and applies to the panels in the same row.

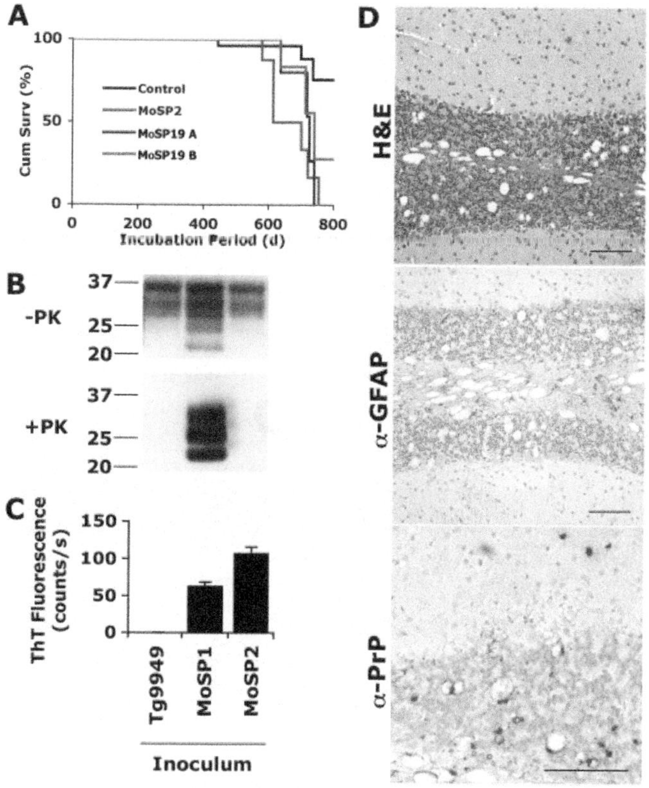

Fig. 5 Protease-sensitive synthetic prions are serially transmissible to Tg4053 mice. Tg4053 mice intracerebrally inoculated with MoSP2-1T (*red*), MoSP19A (*blue*) or MoSP19B (*purple*) developed signs of prion disease between 600–750 d (*A*). MoSP19A and MoSP19B are two Tg9949 brain isolates inoculated with Amyloid Prep 19. Tg4053 mice inoculated with protease-sensitive synthetic prions were significantly more likely to develop neurological dysfunction than mice inoculated with uninfected Tg9949 brain homogenate (*black, p<0.001*). PrP in the brains of these mice was sensitive to PK digestion (*B*) and active in the ASA (*C*). MoSP1 was used as a control. The brains of ill, MoSP2-inoculated Tg4053 mice showed neuropathology consistent with prion disease (*D*), including vacuolation (*top panel*), astrocytic gliosis (middle panel), and punctate PrP deposits (*bottom panel*). Scale bars represent 100 μm. (Colby *et al.*, 2010).

III. Conclusions

Overall, the impressive recent process in production of synthetic prions has open new venues of scientific research. Nevertheless, several key aspects of PrP^{Sc} infectivity remain unknown; particularly the detail in structure of infectious particles, and whether or not other elements such as environmental factors, DNA, or chaperones besides PrP is required for prion replication. Understanding the structure, functions of PrP, the molecular mechanism of PrP^{C}-PrP^{Sc} conversion may have important implications in understanding prion diseases, and thus leading to the identification of a treatment for these fatal disorders.

Acknowledgment

The research leading to presented results has received funding from the PRIORITY FP7 project (contract no.: 222887) to GL.

References

1. Legname G, Baskakov IV, Nguyen HO et al (2004) Synthetic mammalian prions. Science; 305:673–6.
2. Legname G, Nguyen HO, Baskakov IV et al (2005) Strain-specified characteristics of mouse synthetic prions. Proc Natl Acad Sci USA 102: 2168-73.
3. Colby DW, Giles K, Legname G et al (2009) Design and construction of diverse mammalian prion strains. Proc Natl Acad Sci USA 106: 20417-22.
4. Colby DW, Wain R, Baskakov IV, Legname G et al (2010) Protease-sensitive synthetic prions. PLoS Pathog 6:e1000736.
5. Colby DW, Zhang Q, Wang S, Groth D, Legname G, et al. (2007) Prion detection by an amyloid seeding assay. Proc Natl Acad Sci USA 104: 20914-9.

Author: Giuseppe Legname
Institute: SISSA
Street: via Bonomea 265
City: Trieste
Country: ITALY
Email: legname@sissa.it

Binding Drugs on Two Position of M2 Proton Channel and Its Mutants

T. Hieu Nguyen[1], Dzung Hoang[1], and Ly Le[2]

[1] Theoretical Physics Division, University of Science Ho Chi Minh City, Vietnam
[2] School of Biotechnology, Ho Chi Minh International University, Vietnam

Abstract— **M2 protein plays as an proton channel in influenza virus, and many candidates were proposed to inhibit its activity. Three new inhibitors were proposed recently by Du Q-S, et al. (2010), and their biding energy was higher than two commercial drug, Amantadine and Rimantadine. To investigate the potency of the new compound on drug-resistant variants, molecular docking was used to investigate how they bind in two positions of M2 proton channel and its mutants of H3N2 virus. The global docking on the M2 protein (residues 22-46) showed that the coexistence of two binding positions in NMR structure is possible, but the inhibitors favor the In pore position more than the outside of the pore. In the detail docking, mean binding energies of the new inhibitors were lower the commercial drugs, Rimantadine. The comparison between mean binding energy and intermolecular energy indicated that the torsional free energy has caused this decreasing energy. In another word, the adding new groups actually increase binding energy of inhibitors but it also makes them heavy and hard to rotate. In other hand, these new inhibitors were failed in binding on mutant D44A, and their binding energies were no different between wild type and mutants V27A, S31N and L26F. These mutants have suggested some hints to improve inhibitors' structure, and chosen a choice about how these mutants resist the inhibitors.**

Keywords— **M2 channel, mutants, binding energy, inhibitors, drug resistance.**

I. INTRODUCTION

M2 channel plays an important role in influenza viruses' life cycle. It is a homotetramer proton ion channel which allows virus to release its RNA genome into host cells [1]. Many researches have been conducted on the structure and mechanism of virus' M2 proton channel, which is an important target for antiviral drug design. The protein contains 97 amino acids, but only 24 of them (residue 22-46) are required for the proton transport activity [2,3] and the place in which inhibitors bind [4]. Two FDA approved M2 inhibitors are Rimantadine and Amantadine have been highly resisted by M2 current strains, but there is still no unanimity in binding sites of inhibitors. Therefore, designing new M2 channel inhibitors is an important task in the fight against current and future flu pandemics.

M2 has been found to have two binding sites which should be both considered in designing inhibitors. In prior studies, there are two structures solved by two different methods which are X-ray crystallography [7] and solution state NMR [5]. Both structures agree that the pH sensor His37 and Trp41 gate are key residues for opening or closing the M2 channel [5, 8-10], but the difference, which is still controversial, is the location of the inhibitors binding site. According to the two obtained structures, M2 channel has two positions that the adamantane inhibitors can probably bind. One is in pore of X-ray structure of M2 with amantadine drug [7], and the other is in lipid-facing pocket of NMR structure of M2 with rimantadine drug [5]. The surface plasmon resonance (SPR) experiments were conducted to measure the affinity of amantadine and rimantadine to M2 channel embedded in DMPC liposomes, and people found that the both proposed positions are feasible [11]. In which, the high-affinity corresponds to the binding in M2 pore, and the lower affinity binding site is the lipid-facing pocket. Furthermore, a solid state NMR studied by Cady et al. has shown that amantadine binding to both sites may exist [12]. Thus, study designing inhibitors for M2 should have been considered both two positions. On the other hand, the inhibitors, new and old, were proposed [6, 13], and their inhibition were only tested on out pore (new) [6] or in pore (old) [14] position. These studies provide important strategy for M2 channel inhibitors but the information is lacking several important factors that should be taken into account.

Drug resistance induced by mutations is also important issue in M2 channel inhibitor design. Because of the coexistence of two binding positions, docking inhibitors into M2 mutants will be separated into two case: (i) In lipid-facing pocket binding case, because the adamantane inhibitors interact with Asp44 [6] in this region, mutant D44A will be used for this studies; (ii) Inside the channel pore case, the mutants V27A, S31N, L26F will be tested with some new drugs. Inhibitors with two pharmacophore groups (Table.1-A3, A4 and A5) were proposed to increase binding ability [6, 13] in lipid-facing pocket, but they were only docked on wild type and near Asp44 residue. Though they are designed for this region, they still probably bind inside pore, and certainly interact with these mutations. The main goal of this study is docking adamantane-based inhibitors in both positions of wild type and mutants of M2 proton channel.

In prior study by Du Q-S, et al [6], five compounds (Table.01) have been docked at Asp44. These compounds

V. Van Toi et al. (Eds.): 4th International Conference on Biomedical Engineering in Vietnam, IFMBE Proceedings 40, pp. 384–390, 2013.
www.springerlink.com

Table 1 Structure and torsional free energy of adamantane-base drugs which were used for docking in this study. A1 and A2, amantadine and rimantadine, are commercial drugs. A3 was reported in ref [13]. A4 and A5 were proposed as the more effective drugs than A1 and A2 in ref [6].

A1	A2	A3	A4	A5
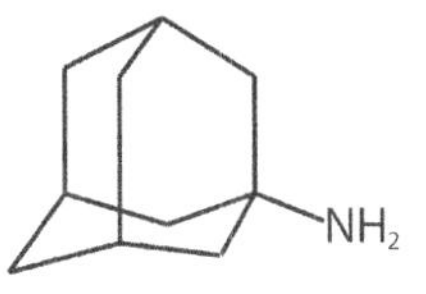	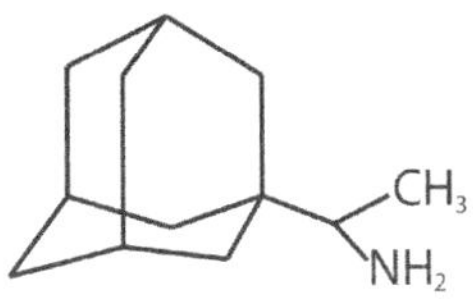	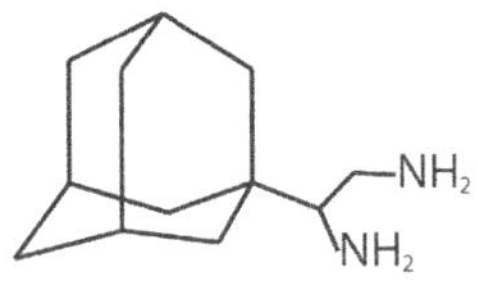	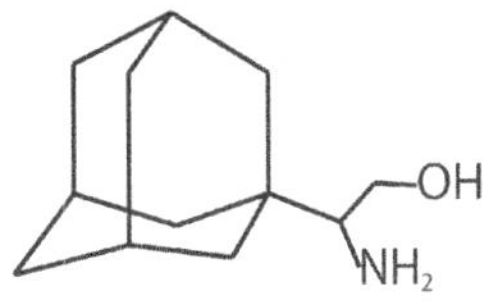	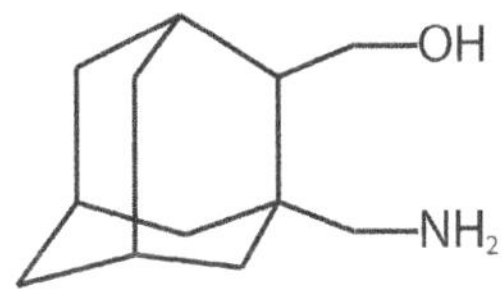
$E_{torsional}$= 0.30	0.60	1.19	1.19	1.19

reveal higher affinity to H1N1 M2 channel than even Amantadine and Rimantadine in both docking score for outside pore position and experimental EC_{50}. We choose these compounds for comparing binding affinity of ligands in two different binding sites and effects of the mutations on binding affinity of them.

II. MATERIALS AND METHODS

A. *Proteins and Inhibitors*

The NMR structures of M2 channel with rimantadine taken from Protein Data Bank, [PDB entry: 2RLF] was used for building wild type and mutants of channel as receptors, and adamantane inhibitors as ligands. The M2 protein and the inhibitors in 2RLF pdb file were separated by VMD 1.8.7 (Visual Molecular Dynamics), and these mutants were fixed from wild type in the same way. Five new inhibitors used for docking were constructed by ArgusLab 4.0 (Table 1).

B. *Docking Method*

Autodock 4.2 program with AMBER force field was used for all our studies, in which, rigid docking was main method. For generation, whole M2 protein would be docked as a receptor to find out which positions inhibitors could bind. Subsequently, a box 80-90-80 covered M2 protein from residue 23 to 46, and the starting point of inhibitor was at lipid-facing which is position of drug in 2RLF file (Fig.1). After that, we had two positions (Fig.2-A) for more detailed docking. One is lipid-facing position with a box having size 30-30-30, and the other is inside the pore with a box 50-50-50 (Fig.1). The difference size of these boxes is because of the difference of residues which the drugs bind in pore and the test which we want to conduct on the mutants spread out a larger area then that of lipid-facing case. The starting points of two docking case are results of whole protein docking.

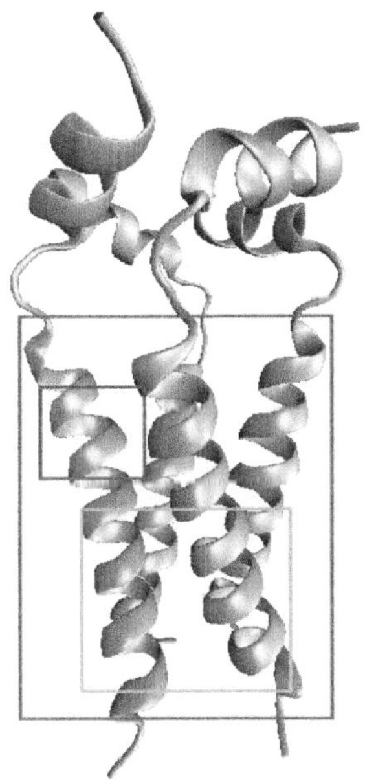

Fig. 1 Three cases of docking: The red correspond to the box 80-90-80, blue: 30-30-30 and yellow: 50-50-50

For analyzing results, two energy types were used. They are Binding Energy and Intermolecular Energy, which their relation is:

$$E_{binding} = E_{inter} + E_{torsional}$$

Which Torsional Energy is usually positive and constant in whole docking process, and Intermolecular is usually negative. The comparison between two energy types is useful for indirectly knowing Torsional Energy, how strong (E_{inter}) the affinities of these positions are, and estimating $(E_{binding})$ inhibitor constant K_i.

III. RESULTS

A. Inhibitors Bind at Both Position, but the in Pore Is the Main Location of Them

As expected, amantadine (A1) and rimantadine (A2) binded at both two positions (Fig.2-A), which were showed in SPR experiments [11], but the cluster percent showed that the in-pore is the main place which A1 and A2 bind, and A2 is more effective than A1 of course (Fig.2-B). The proposed drugs (A3, A4 and A5) were at the same situation. Furthermore, mean binding energy, compared between two cases, agreed with prior study [11] that the affinity of the in-pore is higher than the out pore except for A1. This occurred because the difference between mean binding energies of two situations is small, which can be seen more in detail at docking below.

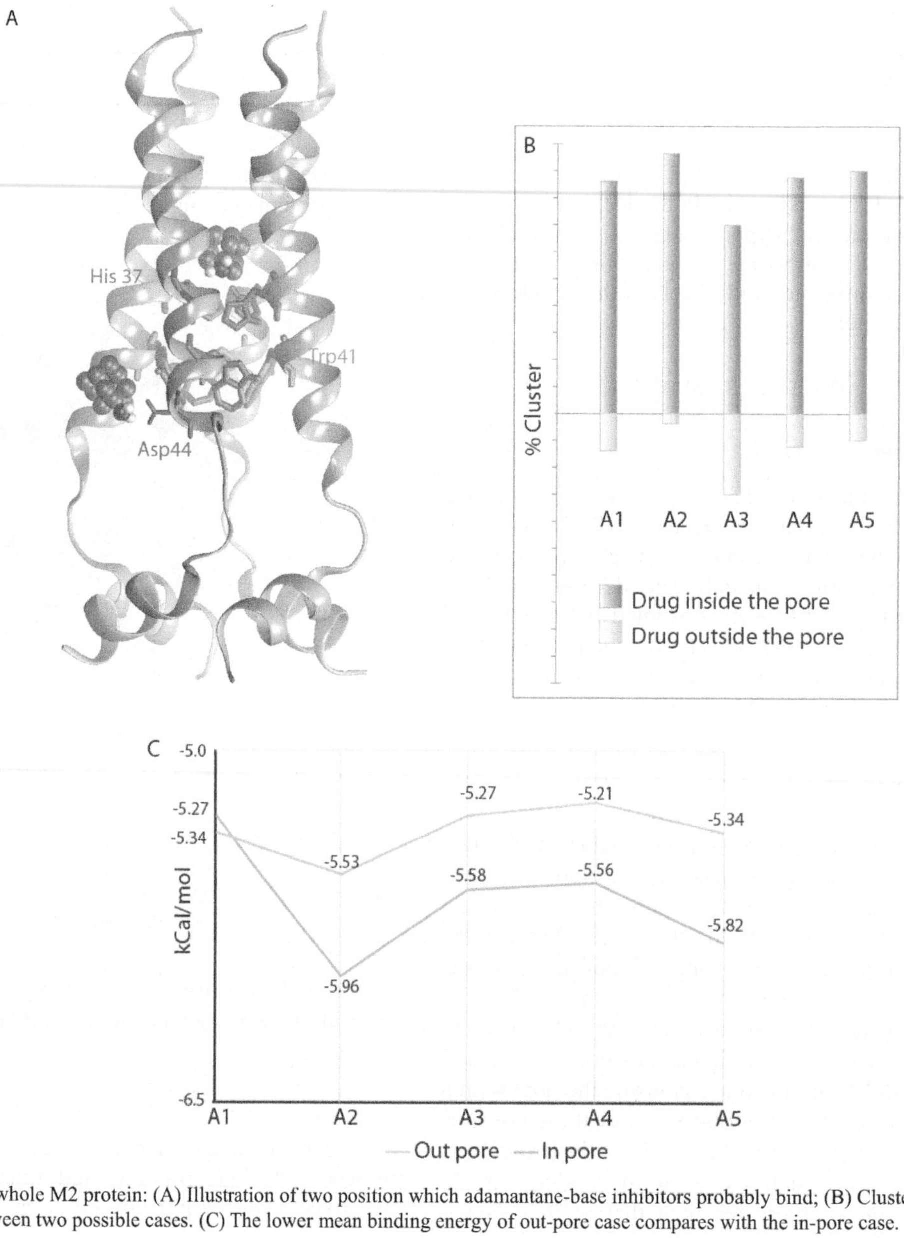

Fig. 2 Result of docking whole M2 protein: (A) Illustration of two position which adamantane-base inhibitors probably bind; (B) Cluster percent comparing between two possible cases. (C) The lower mean binding energy of out-pore case compares with the in-pore case.

B. The Decreasing of Binding Energy of A3-5

Mean binding energies in both case in and out pore have showed that Rimantadine (A2) is the best choice (Fig.2-C, 3 and 4). This is contrast with results in [6] which pointed out the higher binding energies than that of A1 and A2 of two pharmacophore group inhibitors (A3-A5). In our opinion, there are two reasons explaining for the decreasing of energies of A3, A4 and A5 compared with results in [6].

The first reason is the increasing torsion energy, caused by adding more phamarcophore group in adamantane cage. As mentioned before in materials and methods section, binding energies were separated into two part: Intermolecular energy and Torsion energy for analyzing purpose. In which, torsion energy rates with rotable bonds in inhibitors and is usually constant. Comparing intermolecular energies among inhibitors (Fig.3, 4) showed that A5 was the best choice, in another words, interaction between A5 and M2 protein was strongest, because of the more pharmacophore group in A5 than that in A1. In other hand, the rotable bonds of it increased because of this, and its torsion energy, therefore, increased too. Unfortunately, the increasing intermolecular energy of A5 (the same for A4 and A3) was not high enough to balance with the increasing of torsion energy; the binding energy, summed by them, therefore, was lower than those of A1 and A2. This is the main reason explaining for the decreasing of binding energies, and it noted that the adding pharmacophore group need to be considered more carefully.

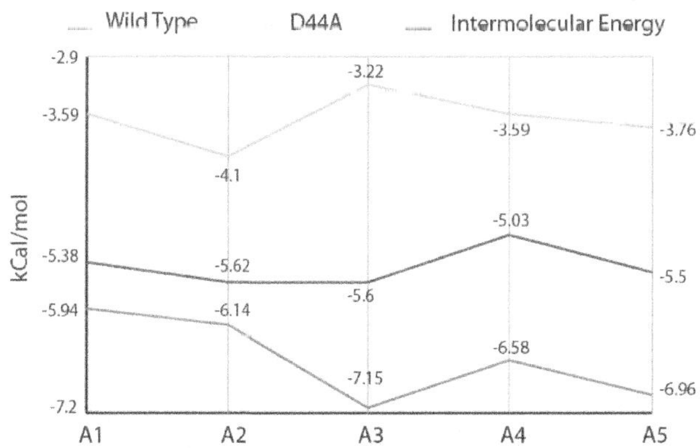

Fig. 3 Comparison mean binding energy of five drugs between the wild type M2 (red) and its mutant D44A (yellow), and between mean binding energy (red) and the lowest intermolecular energy (blue) of wild type.

The second is the decreasing of binding energy caused by meaning. As described in [6], two pharmacophore of A4 and A5 form the hydrogen bonds with two residues Asp44 and Arg45 on two adjacent helices of M2 protein (2-2) (Fig.6-A, C), which explains why their binding energies are so high. However, the out pore docking results showed that the almost conformations of inhibitors and protein were usually interaction between two pharmacophore groups and one residue (2-1) (Fig.6-B), or one pharmacophore group and two residues (1-2) (Fig.6-D), which had the lower binding energies than that in (2-2) case. Therefore, the binding energies is lowered when they were meant. This reason has just been right for out-pore case, so it is the sub-reason.

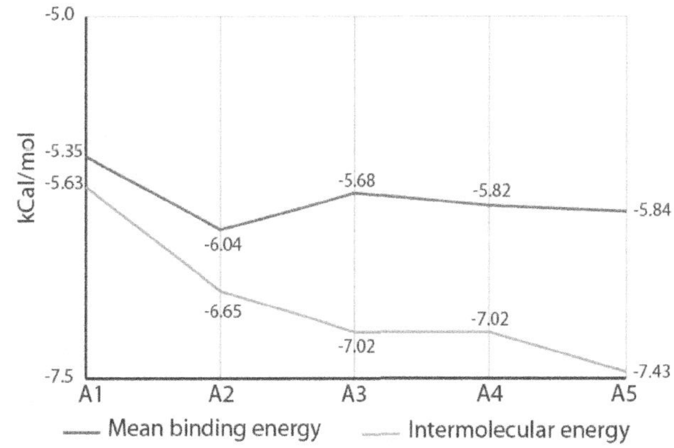

Fig. 4 Comparison between mean binding energy (red) and the lowest intermolecular energy (green) of adamantine-base inhibitors binding at in-pore position.

In other hand, it was not difficult to understand the low binding energy in out-pore docking results of A3. Although owning two amino groups, it hardly binds at two neighboring helices because of the positive polarization of both two amino groups of A3 and residue Arg45 (**Fig.5**).This result quite agree with [6].

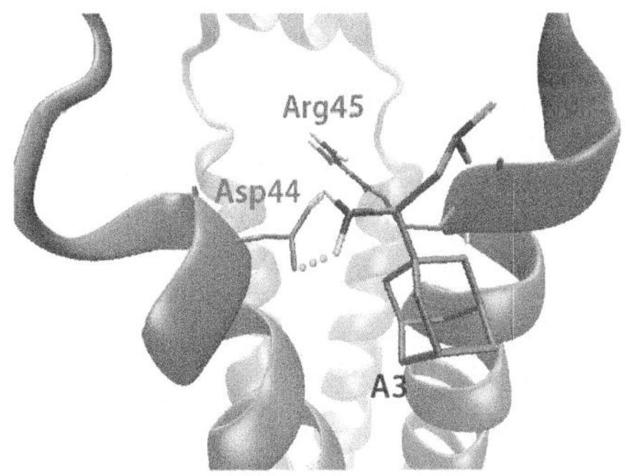

Fig. 5 Illustration of binding ability of A3 inhibitor. Although owning two amino groups, it hardly binds at two neighboring helices because of the positive polarization of both two amino groups of A3 and residue Arg45.

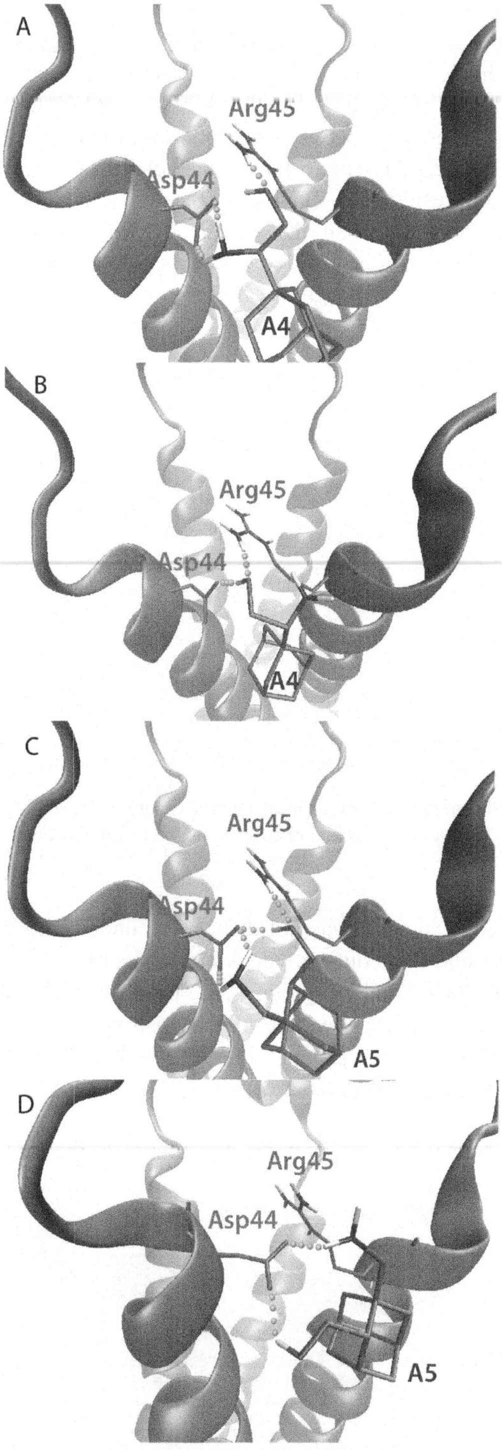

Fig. 6 Illustration of the reason of the decreasing mean binding energy: (A) Two pharmacophore groups of proposed drug A4 bind on two residues Asp44 and Arg45 through hydrogen bonds (green dotted lines); (B) One pharmacophore group of A4 bind on two residues Asp44 and Arg45; (C) Two pharmacophore groups of A5 bind on two residues Asp44 and Arg45; (D) Two pharmacophore groups of A5 bind on one residue Asp44.

C. Behavior of Inhibitors Inside the Pore of M2

Although A3-5 had lower binding energies than A2, they had some advantages in forming hydrogen bonds when they were docked inside the pore, the main binding position of inhibitors, of M2 channel. Frequency of appearance conformations,having hydrogen bonds, of inhibitors and residues inside the pore of M2 protein were showed on Figure 7. It proved that (i) adding pharmacophore group increased forming hydrogen bonds ability of inhibitors (comparing column A1-2 and A3-5); (ii) the hydroxyl group was more active than amino group because of higher electro negativity of oxygen than that of nitrogen (comparing column A3 and A4-5). These advantages of A3-5 helped them to 'anchoring' strongly inside the pore. Besides, which residue they bound inside the pore also pointed out the correlation between the 'branch's' shape of inhibitors and their in-pore positions. The inhibitors which have 'Y' shape branch (A2-4) usually bound at residue Ala30, and which have 'I' shape branch (A1 and A5) bound at Gly34 (Fig.8). And it looks like the 'Y' shape branch was 'stuck' with Ala30, and 'I' shape branch could easily penetrate the pore deeper and form H-bond with Gly34.

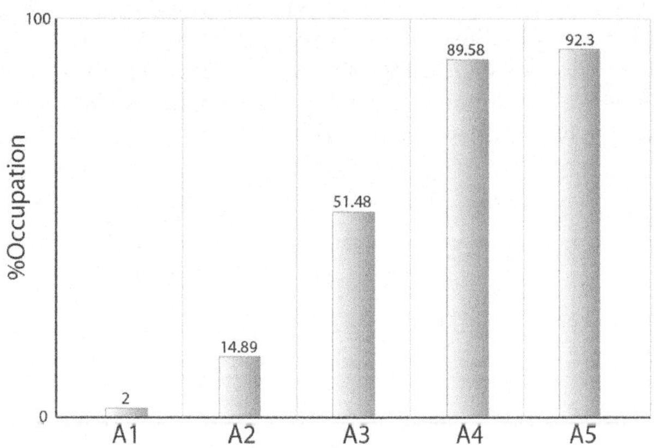

Fig. 7 Hydrogen bond occupation between residues inside the pore of M2 and inhibitors.

D. Losing Binding Ability of the New Proposed Inhibitors in Mutant D44A and the Unchanged Binding Energy in Mutants V27A, S31N and L26F

In the out-pore docking results for mutant D44A, the inhibitors totally lost their binding ability, even though with the new proposed drugs A4, A5 (Fig.3). This agree well with the fact that the main residue Asp44, Aspartic acid, for binding is replaced by a less polar residue Alanine. However, from the docking results of mutant D44A, we recognized the hydroxyl group is very active, because its polarization is stronger than amino group. Therefore, the hydroxyl groups of A4 and A5 still bind and form H-bonds with polar hydrogen of Arg45, while A1-3 are purely docked without any H-bond created (Fig.9).

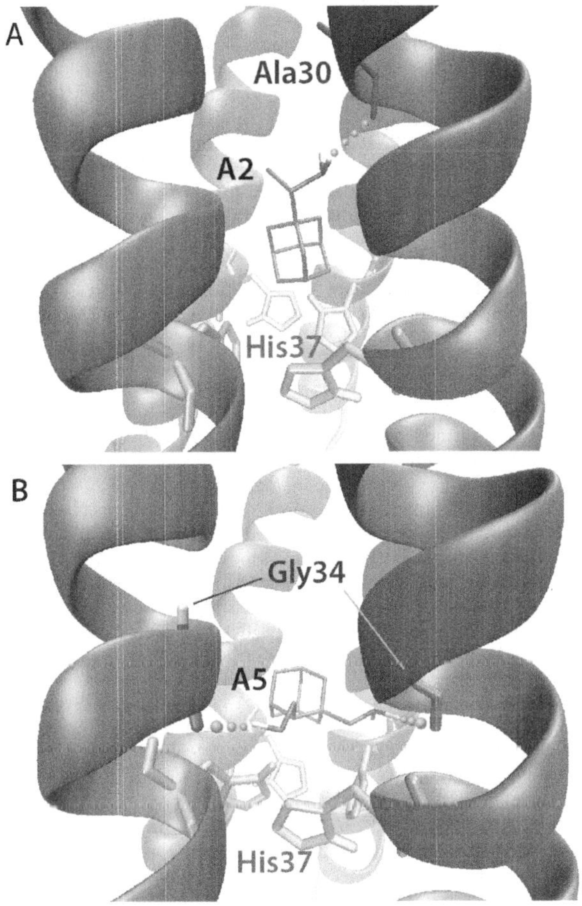

Fig. 8 Illustration of in-pore docking results: (A) A2, A3 and A4 tend to form H-bond with residue Ala30; (B) A1 and A5 tend to bind at residue Gly34.

Based on in-pore docking results, no significant difference between wild type and its mutants V27A, S31N and L26F has been revealed. This unchanged energy has suggested that these mutants may not influence to their binding energies, but decrease their function, which will be discussed more detail in conclusions section.

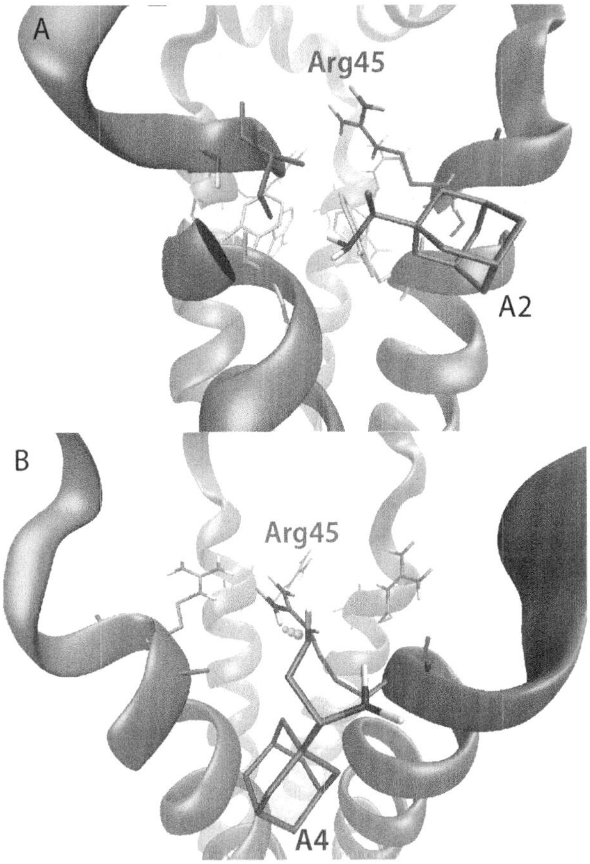

Fig. 9 Illustration of docking results of mutant D44A: (A) The relative position of A1-3 docking without the present of hydrogen bond; (B) The active hydroxyl group of A4, 5 helps them binding at Arg45, but they totally lost their function in this mutant. The green residues are Trp41, and the red ones are Ala44, the residue replacing for Asp44.

IV. Conclusion

As the global docking results, the inhibitors probably bound at two positions: inside and outside the pore of M2, in which in-pore was the main place. Therefore, the designing inhibitors needs to consider both sides binding ability of them. Example, 'Y' shape branch will help inhibitors binding strongly inside the pore (A2), but, for increasing out-pore binding energy of the inhibitors with M2, two pharmacophore group on two position of adamantane-cage are also needed (A5). So, a hint structure of inhibitors likes figure 10-A probably satisfying both requirements above. However, the *adding more pharmacophore group*, in another word, *more rotable bonds* could make them heavy and hard to rotate; therefore, changing pharmacophore group was a suitable choice in this situation. Hydroxyl group, being more active than amino group, is a good choice, so a inhibitor, having structure liking figure 10-B, may be a effecting inhibitor.

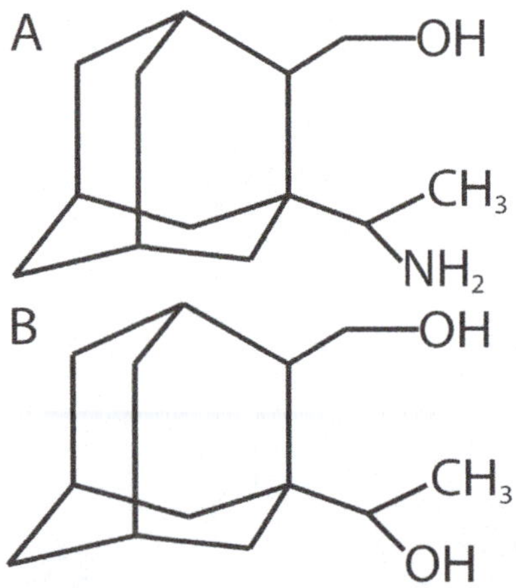

Fig. 10 Two suggested structure may improve binding ability of the inhibitirs.

There are two resistance mechanism of M2 dealing with adamantane inhibitors: (i) a more hydrophilic pore ,created by the mutations, will pushing the hydrophobic part of inhibitor out of the pore [17]; (ii) the mutation increasing diameter of the pore and allowing water going through it with or without the inhibitor [18]. According to the unchanging binding energy of in-pore mutant docking results, it suggested that the inhibitors still bind at pore but their locking pore ability are vanish.

V. Acknowledgments

This study was funded by the Vietnam's National Foundation for Science and Technology Development (NAFOSTED) under grant number 106.99-2010.03.

References

1. Helenius A (1992) Unpacking the incoming influenza virus. Cell 69:577–578.
2. Pinto LH, et al. (1997) A functionally defined model for the M2 proton channel of influenza A virus suggests a mechanism for its ion selectivity. Proc Natl Acad Sci USA 94:11301–11306.
3. Sakaguchi T, Tu Q, Pinto LH, Lamb RA (1997) The active oligomeric state of the minimalistic influenza virus M2 ion channel is a tetramer. Proc Natl Acad Sci USA 94:5000–5005.
4. Hay AJ, Wolstenholme AJ, Skehel JJ, Smith MH (1985) The molecular basis of the specific anti-influenza action of amantadine. EMBO J 4:3021–3024.
5. Schnell JR, Chou JJ (2008) Structure and mechanism of the M2 proton channel of influenza A virus. Nature 451: 591–595.
6. Du Q-S, Huang R-B, Wang Q-S, Chou K-C (2010) Designing Inhibitors of M2 Proton Channel against H1N1 Swine Influenza Virus. PloS ONE 5(2): e9388.
7. Stouffer AL, et al. (2008) Structural basis for the function and inhibition of an influenza virus proton channel. Nature 451:596–599.
8. Chen H, Wu Y, Voth GA (2007) Proton transport behavior through the influenza A M2 channel: Insights from molecular simulation. Biophys J 93:3470–3479.
9. Hu J, Fu R, Cross TA. (2007) The chemical and dynamical influence of the anti-viral drug amantadine on the M2 proton channel transmembrane domain. Biophys J 93:276–283.
10. Hu J, et al. (2006) Histidines, heart of the hydrogen ion channel from influenza A virus: Toward an understanding of conductance and proton selectivity. Proc Natl Acad Sci USA 103:6865–6870.
11. Matthew R. Rosenberg and Marco G. Casarotto (2010) Coexistence of two adamantane binding sites in the influenza A M2 ion channel. PNAS 107: 13866-13871.
12. Cady SD, et al. (2010) Structure of the amantadine binding site of influenza M2 proton channels in lipid bilayers. Nature 463:689–692.
13. Tataridis D, Fytas G, Kolocouris A, Fytas C, Kolocouris N, et al. (2007) Influence of an additional 2-amino substituent of the 1-aminoethyl pharmacophore group on the potency of rimantadine against influenza virus A. Bioorg Med Chem Lett 17: 692–696.
14. Pathumwadee Intharathep, et al. (2008) How amantadine and rimantadine inhibit proton transport in the M2 protein channel. Journal of Molecular Graphics and Modelling 27: 342-348.
15. Stamatiou G, Kolocouris A, Kolocouris N, Fytas G, Foscolos GB, et al. (2001) Novel 3-(2-Adamantyl)pyrrolidines with Potent Activity Against Influenza A Virus-Identification of Aminoadamantane Derivatives Bearing Two Pharmacophoric Amine Groups. Bioorg Med Chem Lett 11: 2137–2142.
16. Wei H, Wang CH, Du QS, Meng J, Chou KC (2009) Investigation into Adamantane-Based M2 Inhibitors with FB-QSAR. Medicinal Chemistry 5: 305–317.
17. Laohpongspaisan C, Rungrotmongkol T, Intharathep P, Malaisree M, Decha P, Aruksakunwong O, Sompornpisut P, Hannongbua S (2009) Why Amantadine Loses Its Function in Influenza M2 Mutants: MD Simulations. J Chem Inf Model 49: 847-52
18. Astrahan P, Kass I, Cooper MA, Arkin IT (2004) A novel method of resistance for influenza against a channel-blocking antiviral drug. Proteins 55: 251-7

Author: Hieu Nguyen T.
Institute: Institute for computational Science and Technology
Street: Ward 06, Linh Trung, Thu Duc District
City: Ho Chi Minh
Country: Vietnam
Email: nthieu.phys@gmail.com

Computational Study of Diseases Associated with Protein Aggregation

Mai Suan Li[1] and Ngo Son Tung[1,2]

[1] Institute of Physics, Polish Academy of Sciences, Al. Lotnikow 32/46, 02-668 Warsaw, Poland
[2] Institute for Computational Science and Technology, Quarter 6, Linh Trung Ward, Thu Duc District, Ho Chi Minh City, Vietnam

Abstract — **A number of diseases such as the Alzheimer's disease, Creutzfeldt-Jakob disease, type-2 diabetes are believed to be associated with the aggregation of proteins and amyloid peptides. In this mini-review we discuss general factors that govern the aggregation of polypeptide chains. It is shown that the extent of population of an ensemble of fibril-prone structures in the spectrum of conformations of an isolated protein, is the major determinant of fibril formation rates. Presently, available drugs help to mask the symptoms, but do not treat aggregation-associated diseases making it vital to develop drugs to cope them. Recent progress on design of top-leads for the Alzheimer's disease will be covered.**

Keywords — **Aggregation, Alzheimer's disease, drug design, polyphenols.**

I. Introduction

Proteins that are unrelated by sequence or structure aggregate to form amyloid-like fibrils with a characteristic cross b -structures, which are linked to a number of diseases such as Alzheimer's and prion-disorders [1]. The observation that almost any protein could form fibrils seemed to imply that fibril rates can be predicted solely based on sequence composition and the propensity to adopt global secondary structure. Such a conclusion has limited validity because it does not account for fluctuations that populate aggregation-prone structures. Despite the common structural characteristics of amyloid fibrils [1] the factors that determine the fibril formation tendencies are not understood. In this review we consider the main factors that control the aggregation process. We highlight the role of aggregation-prone ensemble of **N*** structures [2] in the folding landscape of the monomer in determining $\tau_{\rm fib}$ and the propensity of sequences to form fibrils.

The design of one new drug approved by FDA takes about 15 years and it costs about one billion USD. In order to shorten this process and make it less expensive one can use the computer simulations to identify leads and validate them as potential drugs. We will discuss recent progress on designing new leads for Alzheimer's disease (AD) with the main focus on natural products.

II. Factors governing fibrillogenesis of proteins

Protein folding and function take place in the environment crowded with biologicalmacromolecules. As a result proteins are exposed to intermolecular interactions that may lead to aggregation [3]. In all, about 20 proteins and polypeptides such as polylysine or polyglutamic acid peptides, myoglobin, SH3 e al are now implicated in amyloid formation in vivo [1]. In many cases protein aggregates take the form of amyloid fibrils, which appear as unbranched rod-like nanostructures with the diameter of an order of 10 nm and varying length [4]. A large body of evidence suggests that amyloid fibrils and associated oligomeric intermediates are related to a number of diseases, including Alzheimer's, Parkinson's, Huntington's, and prion diseases [1]. For example, in the case of the Alzheimer's disease the memory decline may result from the accumulation of the amyloid β (Aβ) peptide present in two forms - 40 (Aβ_{40}) and 42 (Aβ_{42}) amino acids of which are produced through endoproteolysis of the β-amyloid precursor transmembrane protein.

Although amyloid forming proteins and peptides exhibit no obvious sequence or structure homology, the common structural element shared by all amyloid fibrils is an extensive cross-b structure stabilized by backbone hydrogen bonds oriented parallel to the fibril axis (Fig. 1). Then an important question emerges is what are general principles that govern the fibril formation process? Experiments on fibril formation times ($\tau_{\rm fib}$) have been shown that global factors such as the hydrophobicity of side chains [5], net charge [6], patterns of polar and non-polar residues [7], frustration in secondary structure elements [8], and aromatic interactions [9] are very important. The initial hypothesis about the role of aromatic interactions in amyloid fibril formation, e.g., was based on the remarkable occurrence of aromatic residues in many amyloid-related proteins and short peptide fragments [9] as well as on the well-known role of aromatic stacking in processes of self-assembly in chemistry and biochemistry [10]. This hypothesis led to the suggestion that stacking of aromatic residues may play a role in acceleration of the assembly process in many cases of amyloid fibril formation. However, recent experiments

V. Van Toi et al. (Eds.): 4th International Conference on Biomedical Engineering in Vietnam, IFMBE Proceedings 40, pp. 391–395, 2013.
www.springerlink.com

suggest that aromatic pi-stacking interactions are not critical for Aβ aggregation or for the inhibition of Aβ aggregation [11].

Using the simple lattice model [12] we have shown that the population of the fibril-prone state **N*** (Fig. 2A), P_{N*}, is a key factor that governs aggregation of polypeptide chains [2] (the fibril-like structure of the six-chain system is shown in Fig. 2B). Namely, fibril formation time $\tau_{\rm fib}$ measured in Monte Carlo steps (MCS), depends on P_{N*} exponentially,

$$\tau_{\rm fib} = \tau^0_{\rm fib}\exp(-cP_{N*}) \quad (1)$$

For 8-bead sequences the prefactor $\tau^0_{\rm fib} \approx 1.014\times10^{10}$ MCS and 3.981×10^{11} MCS, and $c \approx 0.9$ and 1.0, for number of chains $N = 6$ and 10, respectively [2] (Fig. 2C). The important role of the population of **N*** is also revealed by all simulations by different force fields [13,14]. Enhancement of P_{N*} either by mutation or chemical cross linking should increase fibril formation rates. Indeed, a recent experiment [15] showed that the aggregation rate of Aβ_{1-40}-lactam[D23-K28], in which the residues D23 and K28 are chemically constrained by a lactam bridge, is nearly a 1000 times greater than in the wild-type. Since the salt bridge constraint increases the population of the **N*** conformation in the monomeric state [14], it follows from Eq. 1, $\tau_{\rm fib}$ should decrease. One can show that those molecules that have P_{N*} less than a few percents have low propensity to oligomerization [2,13]. From this prospect, our result (Eq. 1) is useful for elucidating the fibrillogenesis at the single-monomer level. This is of paramount importance because the fibril formation is an extremely slow process which is difficult for numerical study.

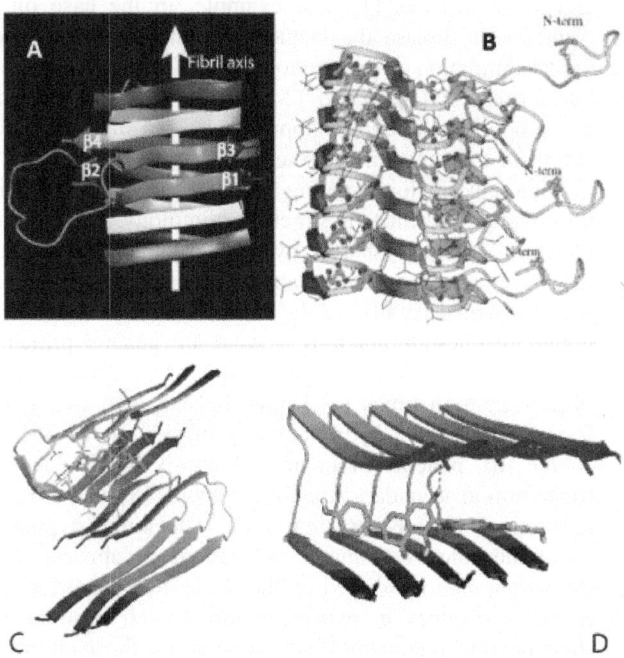

Fig. 1 (A) The fibril from the C-terminal domain 218-289 of the prion protein HET-s (PDB ID: 2RNM). (B) The same as in (A) but for amylin which is associated with the type-2 diabetes. (C) The same as in (A) but for 6 truncated peptides Aβ_{9-40}. The structure was kindly provided by R. Tycko. (D) The same as in (C) but for 5Aβ_{17-42} (PDB ID: 2BEG).

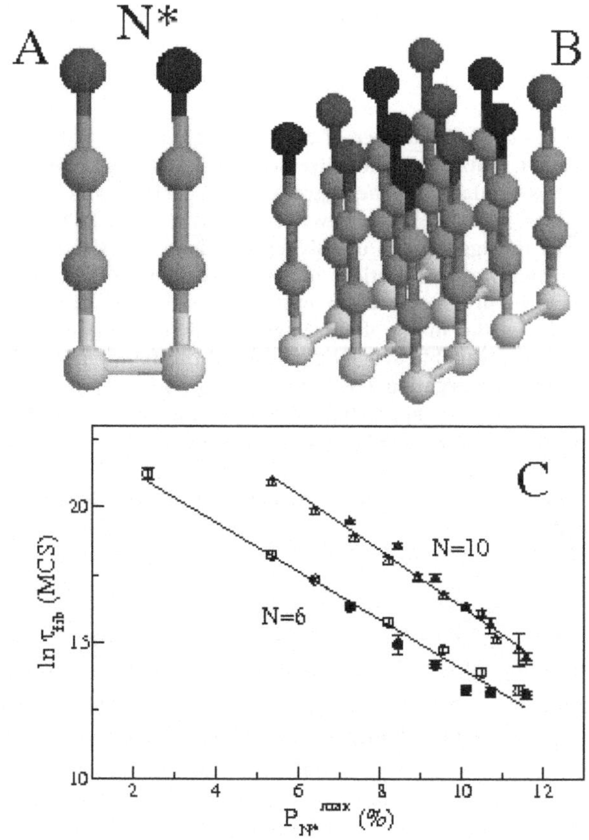

Fig. 2 (A) The 8-bead **N*** conformation in the lattice models.12 (B) The fibril-like structure for $N = 6$ monomers. (C) Dependence of $\tau_{\rm fib}$ on P_{N*} for N = 6 and 10. $\tau_{\rm fib}$ is measured in MCS and P_{N*} in %. The correlation coefficient for all fits $R \approx 0.98$. (C) is taken from Ref [2].

III. RECENT PROGRESS ON DESIGN OF NEW LEADS FOR AD

Two major hypotheses about the pathogenesis of AD have been proposed: Aβ peptide and tau protein. According to the Aβ hypothesis this peptide forms extra-cellular fibrils, which in turn aggregate to form senile plaques, one of the two major morphological hallmarks of AD. The tau protein hypothesis assumes that this protein, in its hyper-phosphorylated form, is the main component of the neurofibrillary tangles, the other hallmark lesion of AD [16].

The amyloid hypothesis states that the generation of Aβ is a key event of AD, and inhibiting this process may affect the disease progression. Aβ peptides are generated from the posttranslational processing of a large transmembrane protein, the amyloid precursor protein (APP), by two proteases, β- and γ-secretase, respectively [17]. From a structure-based perspective, β- rather than γ-secretase has appeared to be a suitable target for drug design purposes because the structure of β-secretase has been experimentally resolved, while γ-secretase has not been crystallized yet. The β-secretase APP-cleaving enzyme (BACE-1) catalyzes the rate-limiting step in the production of Aβ. BACE-1 inhibition represents a possible therapeutic strategy to drastically reduce Aβ levels [18]. Indeed, huge progress has been made in search for BACE-1 inhibitors as potential anti-AD drug candidates using structure-based approaches [19,20].

On the another hand, since AD is presumably associated with oligomerization of Aβ peptides [16], the second of strategy to cope with this disease is to develop compounds able to promote Aβ anti-aggregation and clearance. In this case Aβ peptides and their fibrils (Fig. 1C and D) serve as drug targets. Here we focus on inhibitors of this class of receptors.

Because Aβ is self-assembling, one can use short peptide fragments homologous to the fulllength wild-type protein [21–25] or N-Methylated peptides [26] as inhibitors. The binding affinity of beta-sheet breaker peptides KLVFF and LPFFD have been studied by experiment [21,22] and simulation [27] in detail. The binding free energy is presumably related to hydrophobicity of ligands in such a way that the higher is hydrophobicity the lower inhibitory capacity. The use of specific peptides to inhibit Aβ oligomerization and toxicity, although intriguing, has yet to progress beyond vivo models of amyloidosis.

Carbohydrate-containing compounds, polyamines, chaperones, metal chelators etc may be used to interfere Aβ fibrillogenesis [28,29]. One of the most important classes of potential leads is polyphenols [30] that represent a large group of natural and synthesized small molecules. They are composed of one ormore aromatic phenolic ringsmaking them susceptible to Aβ peptides. Natural polyphenols are phytochemicals found in high concentrations in wine, tea, nuts, and a wide variety of other plants. Some nutraceuticals, as shown by pre-clinical and certain clinical studies, may be of value as AD therapeutic [31,32]. Among them Curcumin (diferulomrthane) [33] (Fig. 3), ginkgo biloba [34] and (-)-Epigallocatechin-3-Gallate (EGCG) (green tea) [35] from the traditional Chinese and Indian medicines are reported to inhibit Aβ aggregation and to be capable against Aβ -inducedtoxicity. Ginkgo biloba tree extract includes many compounds including ginkgolides A, B, C and J (see Fig. 3 for chemical structure of ginkgolide A) which may protect against Aβ-induced synapse damage. Clinical trials are going in phase II and III for curcumin [36] and ginkgo biloba [37], respectively. Tannic acid (Fig. 3) is the most potent inhibitor of Aβ fibril formation [38] among polyphenols having the inhibition constant $IC_{50} \approx 0.01\mu m$, while Curcumin has $IC_{50} \approx 0.8\mu m$.

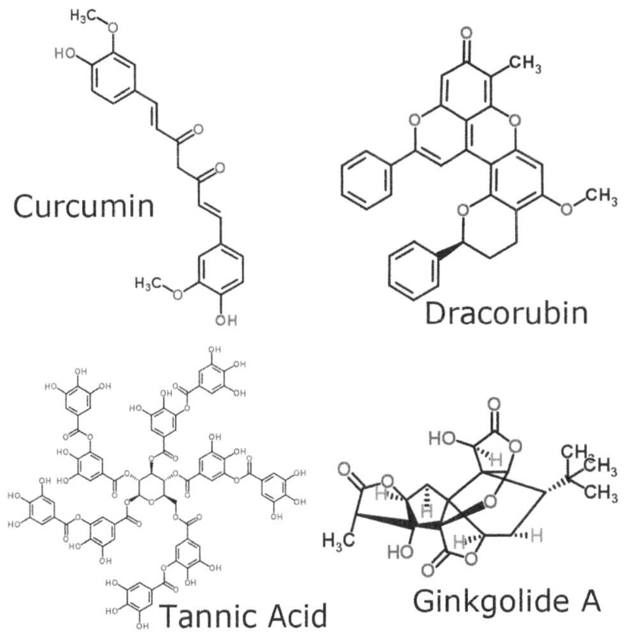

Fig. 3 Chemical structures of Curcumin, ginkgolide A, Tannic acid and Dracorubin (ID: 160270)

Recently we have conducted search for possible leads among compounds derived from Vietnamese herbs and plants for anti-aggregation of amyloid peptides [39]. This problem is of interest because although the traditional Vietnamese medicine shares common features with Chinese and Indian ones, it has many specific herbs due to difference in geography and soil quality [40]. The binding affinity of 342 compounds to full-length Aβ_{40} and Aβ_{42} peptides and their mature fibrils have been studied using Autodock Vina version 1.1. [41] Results obtained by the docking have been then refined by the more accurate Molecular Mechanic-Poisson Boltzmann Surface Area (MM-PBSA) method [42]. We have found that the champion lead is Dracorubin which has ID number 160270 in the Pubchem and Chemspider database (see http://pubchem.ncbi.nlm.nih.gov and http:// http://www.chemspider.com/) (Fig. 3). This compound shows even higher binding affinity to Aβ pepides than Curcumin having the binding free energy $\Delta G_{\rm bind} = -15.59$ kcal/mol [39]. Recall that Curcumin has the binding free energy $\Delta G_{\rm bind} \approx -13.3$ kcal/mol [43].

IV. Conclusions

We have reviewed the key factors such as charge, hydrophobicity, aromatic rings and population of the fibril-prone state in controlling aggregation of polypeptide chains. They are intrinsic and dependent on protein sequences. External conditions like temperature, pH, salt concentration, environment crowding etc also affect aggregation but they are not mentioned here. The recent *in-silico* and *in-vitro* development on search for potential drugs for AD has been briefly discussed. Further clinical tests for potent leads would be an important step in this direction.

V. Acknowledgment

We are very thankful to Nguyen Truong Co, Man Hoang Viet and D. Thirumalai for collaboration and discussion. This work was supported by the Department of Science and Technology of the Ho Chi Minh city and the Ministry of Science and Informatics in Poland (grant No 202-204-234).

References

1. Chiti, F.; Dobson, C. M. (2006) Protein misfolding, functional amyloid, and human disease. Annual Rev. Biochemistry 75:333–366.
2. Li, M. S.; Co, N. T.; Hu, C. K.; Straub, J. E.; Thirumalai, D. (2010) Determination of factors governing fibrillogenesis of polypeptide chains using lattice models. Phys. Rev. Lett. 105:218101.
3. Dobson, C. M. (2003) Protein folding and misfolding. Nature 426:884–890.
4. Tycko, R. (2004) Progress towards a molecular-level structural understanding of amyloid fibrils. Curr. Opin. Struct. Biol. 14:96–103.
5. Otzen, D. E.; Kristensen, O.; Oliveberg, M. (2000) Designed protein tetramer zipped together with a hydrophobic Alzheimer homology: A structural clue to amyloid assembly. Proc. Natl. Acad. Sci. (USA) 97:9907–9912.
6. Chiti, F.; Calamai, M.; Taddei, N.; Stefani, M.; Ramponi, G.; Dobson, C. M. (2002) Studies of the aggregation of mutant proteins in vitro provide insights into the genetics of amyloid diseases. Proc. Natl. Acad. Sci. (USA) 99:16419–16426.
7. West, M. W.; Wang, W. X.; Patterson, J.; Mancias, J. D.; Beasley, J. R.; Hecht, M. H. (1999) De novo amyloid proteins from designed combinatorial libraries. Proc. Natl. Acad. Sci. (USA) 96:11211–11216.
8. Kallberg, Y.; Gustafsson, M.; Persson, B.; Thyberg, J.; Johansson, J. (2001) Prediction of amyloid fibril-forming proteins. iJ. Chem. Biol. 276:12945–12950.
9. Gazit, E. (2002) A possible role for pi-stacking in the self-assembly of amyloid fibrils. FASEB 16:77–83 .
10. Tartaglia, G. G.; Cavalli, A.; Pellarin, R.; Caflisch, (2004) A. The role of aromaticity, exposed surface, and dipole moment in determining protein aggregation rates. Protein Sci 13:1939–1941.
11. Armstrong, A. H.; Chen, J.; McKoy, A. F.; Hecht, M. H. (2011) Mutations that replace aromatic side chains promote aggregation of the Alzheimer's A beta peptide. Biochemistry 50:4058–4067.
12. Li, M. S.; Klimov, D. K.; Straub, J. E.; Thirumalai, D. Probing the mechanisms of fibril formation using lattice models. J. Chem. Phys. 129:175101.
13. Nam, H. B.; Kouza, M.; Zung, H.; Li, M. S. (2010) Relationship between population of the fibrilprone conformation in the monomeric state and oligomer formation times of peptides: Insights from all-atom simulations. J. Chem. Phys. 132.165104.
14. Reddy, G.; Straub, J. E.; Thirumalai, D. (2009) Influence of preformed Asp23-Lys28 salt bridge on the conformational fluctuations of monomers and dimers of A beta peptides with implications for rates of fibril formation. J. Phys. Chem. B 113:1162–1172.
15. Sciarretta, K.; Gordon, D.; Petkova, A.; Tycko, R.; Meredith, S. (2005) A beta 40-Lactam(D23/K28) models a conformation highly favorable for nucleation of amyloid. Biochemistry 44:6003–6014.
16. Hardy, J.; Selkoe, D. J. (2002) Medicine - The amyloid hypothesis of Alzheimer's disease: Progress and problems on the road to therapeutics. Science 297:353–356.
17. W. P. Esler,; Wolfe, M. S. (2001) A portrait of Alzheimer secretase-new features and familiar faces. Science 293:1449–1454.
18. Ghosh, A. K.; Gemma, S.; Tang, J. (2008) Beta-Secretase as a therapeutic target for Alzheimer's disease. Neurotherapeutics 5:399–408.
19. Huang, W. H.; Sheng, R.; Hu, Y. Z. (2009) Progress in the development of nonpeptidomimetic BACE-1 inhibitors for Alzheimers's disease. Curr Med Chem 16:1806–1820.
20. Chiriano, G.; Sartini, A.; Mancini, F.; Andrisano, V.; Bolognesi, M. L.; Roberti, M.; Recanatini, M.; Carloni, P.; Cavalli, (2011) A. Sequential virtual screening approach to the identification of small organic molecules as potential BACE-1 inhibitors. Chem Biol Drug Des 77:268–271.
21. Tjerngber, L. O.; Naslund, J.; Lindqvist, F.; Jahannson, J.; Karlstrom, A. R.; Thyberg, J.; Terenius, L.; Nordstedt, C. (1996) Arrest of beta-amyloid beta-peptide fibril formation by a pentapeptide. J. Biol. Chem. 271:8545–8548.
22. Soto, C. M.; Kindy, S.; Baumann, M.; Frangione, B. (1996) Inhibition of Alzheimer's amyloidosis by peptides that prevent β-sheet conformation. Biochemical and Biophysical Research Communications 226:672–680.
23. Adessi, C.; Soto, C. (2002) Beta-sheet breaker strategy for the treatment of Alzheimer's disease. Drug Dev. Res. 56:184–193.
24. Li, H. Y.; Monien, M. B.; Lomakin, A.; Zemel, R.; Fradinger, E. A.; Tan, M. A.; Spring, S.M.; Urbanc, B.; Xie, C.W.; Benedek, G. B.; Bitan, G. (2010) Mechanistic investigation of the inhibition of A beta 42 assembly and neurotoxicity by A beta 42 C-terminal. Biochemistry 49:6358–6364.
25. Wu, C.; Murray, M. M.; Summer, S. L. B. L.; Condron, M. M.; Bitan, G.; Shea, J. E.; Bowers, M. T. (2009) The structure of A beta 42 C-terminal fragments probed by a combined experimental and theoretical study. J. Mol. Biol. 387:492–501.
26. Gordon, D. J.; Tappe, R.; Meredith, S. C. (2002) Design and characterization of a membrane permeable N-methyl amino acid-containing peptide that inhibits Aβ(1-40) fibrillogenesis. J. Pep. Res. 60:37–55.
27. Viet, M. H.; Ngo, S. T.; Lam, N. S.; Li, M. S. (2011) Inhibition of aggregation of amyloid peptides by beta-sheet breaker peptides and their binding affinity. J. Phys. Chem. B 115:7433–7446.
28. Hawkes, C. A.; Ng, V.; MacLaurin, J. (2009) Small molecue inhibitors of Aβ -aggregation and neurotoxity. Drug Dev. Res. 70:111–124.
29. Yamin, G.; Ono, K.; Inayathullah, M.; Teplow, D. B. (2008) Amyloid β-protein assembly as therapeutic target of Alzheimer's disease. Curr. Phar. Design 14:3231–3246.
30. Porat, Y.; Abramowitz, A.; Gazit, E. (2006) Inhibition of amyloid fibril formation by polyphenols: Structural similarity and aromatic interactions as a common inhibition mechanism. Chem. Biol. Drug Des 67:27–37.

31. Williams, P.; Sorribas, A.; Howes, M. J. R. (2011) Natural products as a sourse of Alzheimer's drug leads. Nat. Prod. Rep. 28:48–77.
32. Howes, M. J. R.; Houghton, P. J. (2003) Plants used in Chinese and Indian traditional medicine for improvement of memory and cognitive function. Pharmacology, Biochemistry and Behavior 75:513–527.
33. Yang, F. S.; Lim, G. P.; Begum, A. N.; Ubeda, O. J.; Simmons, M. R.; Ambegaokar, S. S.; Chen, P. P.; Kayed, R.; Glabe, C. G.; Frautschy, S. A.; Cole, G. M. (2005) Curcumin inhibits formation of amyloid beta oligomers and fibrils, binds plaques, and reduces amyloid in vivo. J. Biol. Chem. 280:5892–5901.
34. Luo, Y.; Smith, J. V.; Paramasivam, V.; Burdick, A.; Curry, K. J.; Buford, J. P.; Khan, L.; Netzer, W. J.; Xu, H. X.; Butko, P. (2002) Inhibition of amyloid-beta aggregation and caspase-3 activation by the Ginkgo biloba extract EGb761. Proc. Natl. Acad. Sci. 99:12197–12202.
35. Bastianetto, S.; Yao, Z. X.; Papadopoulos, V.; Quirion, R. (2006) Neuroprotective effects of green and black teas and their catechin gallate esters against beta-amyloid-induced toxicity. Eur J Neurosci 23:55–64.
36. Ringman, J. M.; Frautschy, S. A.; Cole, G. M.; Masterman, D. L.; Cummings, J. L. (2005) A potential role of the curry spice curcumin in Alzheimer's disease. Curr Alzheimer Res 2:131–136.
37. DeKosky, S. T. Ginkgo biloba prevention trial in order individuals; 2010.
38. Ono, K.; Hasegawa, K.; Naiki, H.; Yamada, M. (2004) Anti-amyloidogenic activity of tannic acid and its activity to destabilize Alzheimer's beta-amyloid fibrils in vitro. Biochim. Biophys. Acta 1690:193–202.
39. Ngo, S. T.; Li, M. S. (2011) Top-leads from natural products for treatment of Alzheimer's disease: docking and molecular dynamics study. Mol. Sim. (accepted).
40. Loi, D. T. (2004) Medical plants and medicines of Vietnam (in Vietnamese); Nha Xuat ban Y hoc, Hanoi.
41. Trott, O.; Olson, A. J. (2004) Improving the speed and accuracy of docking with a new scoring function, efficient optimization, and multithreading. Journal of Computational Chemistry 31:455–461.
42. Kollman, P.;Massova, I.; Reyes, C.; Kuhn, B.; Huo, S.; Chong, L.; Lee,M.; Lee, T.; Duan, Y.; Wang, W.; Donini, O.; Cieplak, P.; Srinivasan, J.; Case, D.; T.E. Cheatham, I. (200) Calculating structures and free energies of complex molecules: combining molecular mechanics and continuum models. Acc. Chem. Res. 33:889–897.
43. Ryu, E. K.; Choe, Y. S.; Lee, K. H.; Choi, Y.; Kim, B. T. (2006) Curcumin and dehydrozingerone derivatives: Synthesis, radiolabeling, and evaluation for beta-amyloid plaque imaging. J. Med. Chem. 49:6111–6119.

The address of the corresponding author:

Author: Mai Suan Li
Institute: Institute of Physics, Polish Academy of Sciences
Street: Al. Lotnikow 32/46, 02-668
City: Warsaw
Country: Poland
Email: masli@ifpan.edu.pl

FIDELITY:
Fuzzy Inferential Diagnostic Engine for on-LIne supporT to phYsicians

P. Sena[1], P.Attianese[1], M. Pappalardo[2], and F. Villecco[2]

[1] University of Salerno, Department of Pharmaceutical and Biomedical Sciences, Fisciano (SA), Italy
[2] University of Salerno, Department of Industrial Engineering, Fisciano (SA), Italy

Abstract — **A diagnostic engine for supporting physicians in analyzing symptoms and anamnesis of patients is proposed.**

The system uses fuzzy logic as the picklock to overcome difficulties typically encountered when dealing with a computer aided model for medical diagnosis, since the large number of parameters to be taken into account.

The model in its current formulation allows physicians to have immediate diagnostic hypotheses considering patient's anamnesis, symptoms, drug administered and previously formulated diagnoses, simultaneously accessing available databases and suggesting new diagnostic examinations.

An auto-adaptative fuzzy logic approach can be easily implemented, that could allow the model to progressively learn from each formulated diagnosis.

Keywords — **Medical diagnosis, fuzzy logic, modeling, computerized diagnosis.**

I. INTRODUCTION

In the web –based world, approach to life has changed in all its aspects. Accuracy of questions and rapidity of answers are two of the crucial items in a life more and more husty. No ability nor skillness seems to be sufficient, nobody has apparently time to wait for thoughtful answers or deeper analyses of questions arisen, not even caring of all risks associated with such an uncertainty.

Among the others, field of medicine is claimed to suffer from such an approach, as a wrong or improper answer may have a tremendous fall-out on the patients health. On the other hand, in situation such as those occurring in Emergency, seconds may be of a vital importance. The progress in medical sciences with the consequent increasing diagnosis accuracy from one side, and the continuously growing number of pharmaceutical products from the other side meet what new technologies appear to be able to offer in terms of databases availability, easiness of access to information, celerity of scientific international contacts [1-4].

It is a while that fuzzy logic [5] is successfully used in the process control design. Whatever the reason, either its ability to model imprecise and subjective notions or the easiness of building up controllers that do not need mathematical modeling of the process, fuzzy logic is nowadays applied in virtually all fields of science in the modern world [6-7].

The basis of a fuzzy logic controller is the representation of linguistic descriptions as membership functions indicating the degree a value belongs to the class wit the linguistic label. In fuzzy logic control algorithms, degrees of membership serve as inputs. Once the membership functions are defined, the actual input values are transformed to the degrees of membership (varying from 0 to 1) of the linguistic descriptors. This process is called fuzzyfication. The resulting fuzzyfied data are passed through an inference mechanism containing the rules for the output. Then, combined effect of all rules is evaluated by properly weighing each rule. Finally, the weighing will be used to finely tune the fuzzy controller for the defuzzyfication step [8-9].

Examples of applications are available in literature, ranging from aircraft engine to locomotive wheel sleep control, steam turbine start-up, SEP (symbol error probability) of HS-MRC (hybrid selection/maximal-ratio combining power supplier controller), up to domestic or industrial scale microwave ovens. Fuzzy logic is even used to the fed-batch cultures of microorganisms and for on-line control of feeding rate of substrates [10].

The number of parameters to account for when a medical diagnosis has to be done is definitely significant. Symptoms, ranging from fever to pain to not well defined uneasiness, drugs administered, patient anamnesis are to be considered altogether The analysis may also be complicated from previously formulated diagnoses that in turn could even be misleading. Moreover, all parameters are in a non numerical form, that would make almost impossible any formulation of a specific software.

Aim of this work is to show how a tool for helping in medical diagnosis can be implemented, basing on the powerfulness of the fuzzy logic able to handle a multiparametric system, as the analysis of the patient medical story and symptoms typically is.

V. Van Toi et al. (Eds.): 4th International Conference on Biomedical Engineering in Vietnam, IFMBE Proceedings 40, pp. 396–400, 2013.
www.springerlink.com

II. THE PROPOSAL

Complexity of the problem above pointed out may be overcome formulating a comprehensive model that puts together medical expertise and engineering approaches, allowing the implementation of a program that can, all in real time, handle logical variables, such as for instance symptoms, and address diagnoses even verifying drugs current availability, dosage and incompatibility for the patient under exam.

Schuh reviewed the use of fuzzy logic in medicine, specially in some electro-medical devices. Author also reported note of the first systems to support complex fuzzy diagnosis and successful attempts to diagnostics certain cancers and diabetes [11]. Croskerry outlined that a diagnosis support system must be able to reflect all of the performance of a doctor. Thus, the system should be able to do all the necessary considerations and to adapt to the various cases. Besides being undoubtedly of considerable complexity, the model should be based on two levels to be combined: i) "intuitive" (searching from database and selection of either possible combinations of diseases explaining all symptoms, or the possible rebuttal evidence): ii)"analytical" (everything which determines the weight of each hypothesis) [12]. Khanale and Ambilwade tested a fuzzy assisted system to support diagnoses in the area of the tiroidean dysfunctions [13]. Baig et al. uses fuzzy logic to describe the values of some tests that one would expect to find in specific diseases [14]. Wagholikar et al. demonstrated that fuzzy logic perfectly fits with the needs of medical decisions [15]. Raich et al, Gerstenkom T et al, Innocent P. R. et al., Kazem Sadegh-Zadeh and Rakus-Andersson, also report examples of application of fuzzy logic in medicine [16-20].

All that above strongly supports the idea at the base of this work of developing FIDELITY (Fuzzy Inferential Diagnostic Engine for on-LIne supporT to physicians), able to consider all parameters and intelligently guide the physician to the final diagnosis.

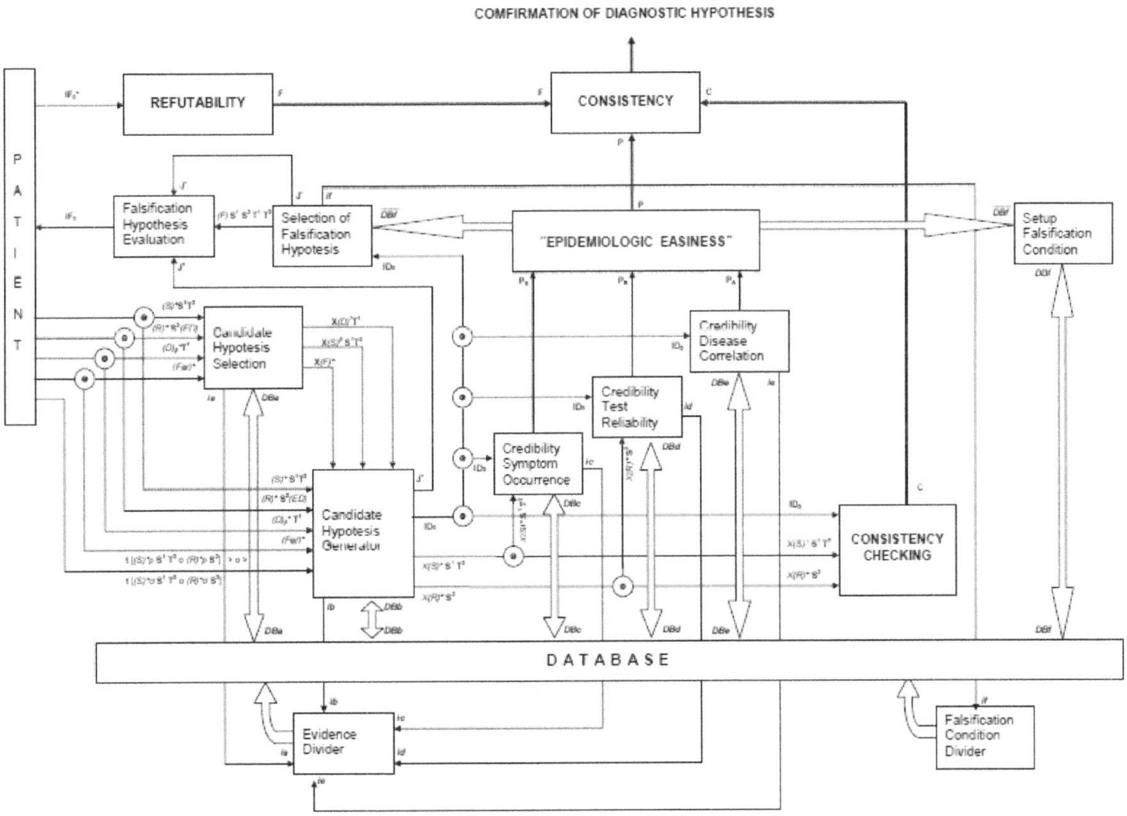

Fig. 1 The Model.

III. The Model

In its entirety, the diagnostic process is configured by FIDELITY as a succession of phases that, in fact, mimic and support the medical action in all its developments. According to this operating model, since from the preliminary stages FIDELITY suggests provisional diagnostics hypotheses, that then guide collection and organization of clinical data toward the final diagnosis. The model (Fig.1) has been structured in the following steps.

- STEP 1 - Data Collection

Information is collected for the patient's symptoms and anamnesis, i.e. presence of concurrent diseases, association with specific sub-populations of epidemiologic interest or use of drugs, as well as results of current exams. Such information includes intensity of symptoms reported and its relation, if any, with the phase of the day in which each symptom occurs. If information concerns the presence of a known pathology, the input will include its stage of evolution. In case of multiple symptoms reported by the patient, the input information may also consider the temporal evolution of symptoms intensity, and the order in which they appeared.

At the end of the patient interview, four inputs will be available:

- symptoms
- anamnesis
- diagnostic examinations
- administered drugs

- STEP 2 - Generation of Diagnostic Hypotheses

The generator selects diagnoses among all diseases present in the available databases and containing at least one of the inputs.

All diseases selected by the system are examined by FIDELITY, that evaluates each of them [C1, C2, ...] (according to the scale shown in Table 1), taking into account the correspondence between typical symptom profile and current input, intensity of inputs and their evolution with time. The disease with best rating [C*] is selected as the candidate hypothesis.

- STEP 3 - Epidemiological Evaluation of the Hypothesis

For the selected cases, it is evaluated on epidemiological base how much the disease is found in the selected population [P] (Fig 2), according to official WHO databases (or alternative, as an option).

Table 1 [C], [F]

n°	Scale
1	Very rarely
2	Rarely
3	Occasionally
4	Frequently
5	Very frequently

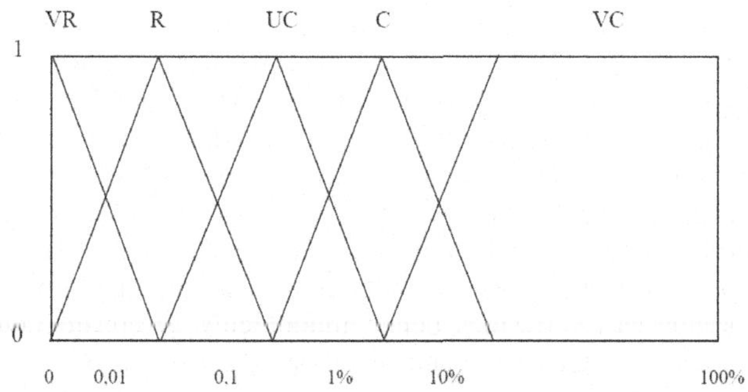

Fig. 2 [P].

Labels:

- very rare
- rare
- uncommon
- common
- very common

If a comorbidity conditions holds in the candidate hypothesis, or more hypotheses may be associable with any of the clinical conditions reported by the patient, it is necessary to consider a comorbidity index for the considered pathologies.

- STEP 4 - Refutability of Diagnostic Hypothesis

Given a diagnostic hypothesis candidate [C*], all symptoms and results of specific tests that should be present in clinical conditions under that hypothesis are examined. Among them, only those not reported by the patient in STEP 1 are considered, and only those whose presence or absence can significantly alter the value of the hypothesis [C*] are chosen.

The hypothesis of forgery is then checked in the patient's medical status. New symptoms are searched, new diagnostic tests are performed, specific conditions are looked for in the patient's anamnesis.

The search will give an index [F] (according to the scales shown in Table 1), that shall take into account how much the new symptoms occurred in the patient are in agreement or disagreement with the hypothesis [C*]

- STEP 5 - Validation of Diagnostic Hypothesis

In the last phase, FIDELITY critically reviews the parameters [C*], [P] and [F], through a FAM (Fuzzy Associative Matrix) and gives back. after defuzzification (Fig. 3), the reliability [X] of our diagnostic hypothesis [C*]

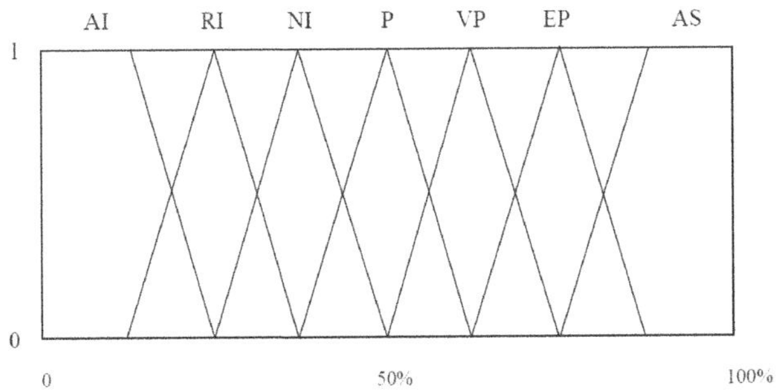

Fig. 3 Defuzzification [X].

Labels:

- almost impossible
- rather impossible
- not impossible
- possible
- very possible
- extremely possible
- almost sure

IV. Conclusions

FIDELITY makes available a large amount of medical information and, in turn, allows the patient to benefit of such an availability through the use of a computer system that assists the physician in his decisions.

FIDELITY provides medical assistance not only basing on the adequacy of available databases and the relevant management, but also on a functional interaction between the fuzzy system and the physician able to i) suggest the appropriate measures to collect data necessary to the diagnostic hypothesis definition, ii) confirm it by appropriate investigations and iii) prescribe, if necessary, the adequate therapy.

It is important to outline that a clinical diagnosis obtained with the help of a computer system must not be the result of a totally automatic process, but must always be validated by the final choice of a physician.

No system supporting medical diagnosis can replace the free and responsible decision of the physician: the role of FIDELITY is simply to drive data collection, making decisions, provide explanations procedures followed or conclusions reached. Since these latter depend on the way FIDELITY was built, they can not be considered exhaustive nor definitive.

Conclusions drawn by FIDELITY must thus be proposed rather than imposed, as a complete automation in the clinical decisions should be to avoided as not to introduce possible inappropriate or even dangerous suggestions.

References

1. Elstein A. S., Jason H. (1970) Study of the process of medical diagnosis: method and preliminary results, Educ Med Salud.; Oct-Dec 4 (4): 355-362
2. Elstein A. S. (2000) Clinical problem solving and decision psychology: comment on the epistemology of clinical reasoning, Acad Med.; Oct 75 (10 Suppl), 134–136
3. Green L. A., Phillips R. L., Fryer G. E. (2004) The nature of primary medical care, Oxford Textbook of Primary Medical Care; Oxford, Oxford University Press, Vol 1: 3–10
4. P. Sestini (2010) Epistemology and ethics of evidence-based medicine: putting goal-setting in the right place, Journal of Evaluation in Clinical Practice, Volume 16, issue 2, pp. 301-305, DOI: 10.1111/j.1365-2753.2009.01327.x
5. L.A. Zadeh (1965) Fuzzy sets, Information and Control, Volume 8, Issue 3, pp. 338-353, DOI: 10.1016/S0019-9958(65)90241-X
6. M. Pappalardo, A. Pellegrino, M. d'Amore, and F. Villecco, (2003) Design of a fuzzy control for a microwave monomode cavity to soot trap filter regeneration, in Intelligence in a Small World—Nanomaterials for the 21st Century, CRC- Press, Boca Raton, Fla, USA
7. Arcangelo Pellegrino, Francesco Villecco (2010), Design Optimization of a Natural Gas Substation with Intensification of the Energy Cycle, Mathematical Problems in Engineering, Hindawi, DOI:10.1155/2010/294102
8. Kosko B. (1996) Fuzzy Cognitive Maps, Int. Jou. Man-Machine Studies; 24, pp. 65–75
9. H. J. Zimmermann, (2009) *Fuzzy Set Theory and Its Applications*, Kluver Academic Publishers, Boston, Mass, USA.
10. P. P. Bonissone, V. Badami, K. H. Chiang, P. S. Khedkar, K. W. Marcelle, and M. J. Schutten, (1995) Industrial applications of fuzzy logic at general electric, Proceedings of the IEEE, vol. 83, no. 3, pp. 450–465
11. Ch. Schuh (2005) Fuzzy sets and their application in medicine, NAFIPS - Annual Meeting of the North American Fuzzy Information Processing Society ,pp.86-91, DOI 10.1109/NAFIPS.2005.1548513
12. Pat Croskerry (2009) A Universal Model of Diagnostic Reasoning, Academic Medicine, Volume 84, Issue 8, pp 1022-1028, DOI: 10.1097/ACM.0b013e3181ace703
13. P.B. Khanale and R.P. Ambilwade (2011) A Fuzzy Inference System for Diagnosis of Hypothyroidism, Journal of Artificial Intelligence, Vol 4, 1: 45-54
14. Faran Baig et al (2011) Design model of fuzzy logic medical diagnosis control system, IJCSE International Journal on Computer Science and Engineering, Vol. 3 No. 5, pp. 2093-2108
15. Kavishwar Wagholikar, Sanjeev Mangrulkar, Ashok Deshpande and Vijayraghavan Sundararajan. (2010) Evaluation of Fuzzy Relation Method for Medical Decision Support, Journal of Medical Systems, DOI 10.1007/s 10916-010-9472-5

16. Raich, V. V., Tripathi R. K., Bawa N. P. S., Dookhitram, K., Dalal S. K. (2011) Application of interval valued fuzzy matrices in medical diagnosis via a new approach, Multimedia Technology (ICMT), pp. 3440 – 3443, DOI: 10.1109/ICMT.2011.6002135
17. Gerstenkom T., Rakus E. (1994) An application of fuzzy set theory to differentiating the effectiveness of drugs in treatment of inflammation of genital organs, Fuzzy Sets and Systems, 68 (3) pp. 327-333, DOI: 10.1016/0165-0114(94)90187-2
18. Innocent P. R., John R. I. (2004) Computer aided fuzzy medical diagnosis, Information Science; 162: 81-104
19. Sadegh-Zadeh K. (2000) Fundamentals of clinical methodology. 4 – Diagnosis, Artificial Intelligence Med.; 20, pp. 227-241, DOI:10.1016/S0933-3657(00)00066-X
20. Rakus-Andersson E. (1999) A Fuzzy Decision Making Model Applied to the Choice of the Therapy in the Case of Symptoms not Disappearing after the Treatment, 4th Meeting Euro Working Group on Fuzzy Sets, 298-303

Author: Francesco Villecco
Institute: University of Salerno, Department of Industrial Engineering
Street: Via Ponte don Melillo
City: Fisciano (SA)
Country: Italy
Email: fvillecco@unisa.it

Retraction Note to: A Pipelined Analog – To – Digital Converter (ADC) Using Umc 0.25μm Technology for Pacemaker Application

Phan Vo Kim Anh and Hoang Trang

University of Technology, VietNam National University at HoChiMinh City, HoChiMinh City, VietNam
pvkanh@hcmut.edu.vn, hoangtrang@hcmut.edu.vn

V.V. Toi et al. (Eds.): 4th International Conference on Biomedical Engineering in Vietnam, IFMBE Proceedings 40, pp. 47–49, 2013.
www.springerlink.com

DOI 10.1007/ 978-3-642-32183-2_96

"This article "A Pipelined Analog – To – Digital Converter (ADC) Using Umc 0.25μm Technology for Pacemaker Application", written by Phan Vo Kim Anh and Hoang Trang, published in 4th International Conference on Biomedical Engineering in Vietnam IFMBE Proceedings Volume 40, 2013, pp 47-49, DOI (10.1007/978-3-642-32183-2_13) has been retracted at the request of the authors because it was submitted without the consent of one of the authors. In addition, the article contains data of unacknowledged scientists."

The original online version for this chapter can be found at
http://dx.doi.org/10.1007/978-3-642-32183-2_13

Author Index

L

M

N

O

P

R

S

T

Keyword Index

The manufacturer's authorised representative in the EU is Springer Nature Customer Service Centre GmbH, Europaplatz 3, 69115 Heidelberg, Germany. If you have any concerns regarding our products, please contact ProductSafety@springernature.com

Printed and bound by CPI Group (UK) Ltd, Croydon, CR0 4YY
25/09/2026
02222906-0004